Methods in Enzymology

Volume 232
HEMOGLOBINS
Part C
Biophysical Methods

METHODS IN ENZYMOLOGY

EDITORS-IN-CHIEF

John N. Abelson   Melvin I. Simon

DIVISION OF BIOLOGY
CALIFORNIA INSTITUTE OF TECHNOLOGY
PASADENA, CALIFORNIA

FOUNDING EDITORS

Sidney P. Colowick and Nathan O. Kaplan

Methods in Enzymology

Volume 232

Hemoglobins

Part C
Biophysical Methods

EDITED BY

Johannes Everse

DEPARTMENT OF BIOCHEMISTRY
AND MOLECULAR BIOLOGY
TEXAS TECH UNIVERSITY
HEALTH SCIENCES CENTER
LUBBOCK, TEXAS

Kim D. Vandegriff
Robert M. Winslow

DEPARTMENT OF MEDICINE
UNIVERSITY OF CALIFORNIA AT SAN DIEGO
VETERANS AFFAIRS MEDICAL CENTER
SAN DIEGO, CALIFORNIA

ACADEMIC PRESS
A Division of Harcourt Brace & Company
San Diego   New York   Boston   London   Sydney   Tokyo   Toronto

Academic Press, Inc.
525 B Street, Suite 1900, San Diego, California 92101-4495

United Kingdom Edition published by
Academic Press Limited
24–28 Oval Road, London NW1 7DX

International Standard Serial Number: 0076-6879

International Standard Book Number: 0-12-182133-1

PRINTED IN THE UNITED STATES OF AMERICA
94 95 96 97 98 99 MM 9 8 7 6 5 4 3 2 1

Table of Contents

Section I. Molecular Structure and Dynamics

Section II. Spectroscopy

Section III. Ligand Binding

Section IV. Mathematical Analysis and Modeling

Contributors to Volume 232

Article numbers are in parentheses following the names of contributors.
Affiliations listed are current.

GARY K. ACKERS (27), *Department of Biochemistry and Molecular Biophysics, Washington University School of Medicine, St. Louis, Missouri 63110*

BERNARD ALPERT (13, 14), *Laboratoire de Biologie Physico-Chimique, Université Paris VII, 75251 Paris, France*

ANJUM ANSARI (18), *Department of Physics, University of Illinois at Chicago, Chicago, Illinois 60680*

ROBERT H. AUSTIN (10), *Department of Physics, Princeton University, Princeton, New Jersey 08544*

ANDREA BELLELLI (5), *Centro di Biologia Molecolare del CNR, c/o Dipartimento di Scienze Biochemiche, Università "La Sapienza," Roma, Italy*

ROBERT L. BERGER (25), *Laboratory of Biophysical Chemistry, National Heart, Lung, and Blood Institute, National Institutes of Health, Bethesda, Maryland 20892*

ANTONIO BIANCONI (14), *Dipartimento di Medicina Sperimentale, Università del l'Aquila, 67100 L'Aquila, Italy*

MAURIZIO BRUNORI (5), *Dipartimento di Scienze Biochemiche, Università "La Sapienza," Roma, Italy*

WINSLOW S. CAUGHEY (9), *Department of Biochemistry and Molecular Biology, Colorado State University, Fort Collins, Colorado 80523*

HUI-LING CHUI (16), *Department of Chemistry, University of Rochester, Rochester, New York 14627*

AGOSTINA CONGIU-CASTELLANO (14), *Dipartimento di Fisica, Università degli Studi di Roma "La Sapienza," 00185 Roma, Italy*

NORMAN DAVIDS (25), *Department of Engineering Science and Mechanics, School of Engineering, Pennsylvania State University, University Park, Pennsylvania 16802*

JOHN DEAK (16), *Department of Chemistry, University of Rochester, Rochester, New York 14627*

EDWARD C. DELAND (30), *Department of Anesthesiology, University of California at Los Angeles, Los Angeles, California 90024*

STEFANO DELLA LONGA (14), *Dipartimento di Medicina Sperimentale, Università dell'Aquila, 67100 L'Aquila, Italy*

ROBERT M. DEUTSCH (18), *Laboratory of Chemical Physics, National Institute of Diabetes and Digestive and Kidney Diseases, National Institutes of Health, Bethesda, Maryland 20892*

ENRICO DI CERA (31), *Department of Biochemistry and Molecular Biophysics, Washington University School of Medicine, St. Louis, Missouri 63110*

AICHUN DONG (9), *Department of Biochemistry and Molecular Biology, Colorado State University, Fort Collins, Colorado 80523*

MICHAEL L. DOYLE (27), *Department of Macromolecular Sciences, SmithKline Beecham Pharmaceuticals, King of Prussia, Pennsylvania 19406*

JOAN J. ENGLANDER (3), *Johnson Research Foundation, Department of Biochemistry and Biophysics, University of Pennsylvania, Philadelphia, Pennsylvania 19104*

S. WALTER ENGLANDER (3), *Johnson Research Foundation, Department of Biochemistry and Biophysics, University of Pennsylvania, Philadelphia, Pennsylvania 19104*

JEHUDAH FEITELSON (7), *Department of Chemistry, The Hebrew University, Jerusalem, Israel*

FRANK A. FERRONE (15), *Department of Physics and Atmospheric Science, Drexel University, Philadelphia, Pennsylvania 19104*

ANTHONY L. FINK (1), *Department of Chemistry and Biochemistry, University of California at Santa Cruz, Santa Cruz, California 95064*

JOEL M. FRIEDMAN (11), *Department of Physiology and Biophysics, Albert Einstein College of Medicine, Yeshiva University, Bronx, New York 10461*

CARTER C. GIBSON (23), *National Institutes of Health, Bethesda, Maryland 20892*

YUJI GOTO (1), *Department of Biology, Faculty of Science, Osaka University, Toyonaka, Osaka 560, Japan*

ERIC R. HENRY (18), *Laboratory of Chemical Physics, National Institute of Diabetes and Digestive and Kidney Disease, National Institutes of Health, Bethesda, Maryland 20892*

RHODA ELISON HIRSCH (12), *Department of Medicine, Division of Hematology, and Department of Anatomy and Structural Biology, Albert Einstein College of Medicine, Bronx, New York 10461*

CHIEN HO (8), *Department of Biological Sciences, Carnegie Mellon University, Pittsburgh, Pennsylvania 15213*

JAMES HOFRICHTER (18), *Laboratory of Chemical Physics, National Institute of Diabetes and Digestive and Kidney Disease, National Institutes of Health, Bethesda, Maryland 20892*

KIYOHIRO IMAI (26), *Department of Physiology, Osaka University Medical School, Osaka 565, Japan*

MICHAEL L. JOHNSON (28), *Department of Pharmacology, University of Virginia, Charlottesville, Virginia 22908*

COLLEEN M. JONES (18), *Department of Chemistry, University of South Alabama, Mobile, Alabama 36688*

JEAN KISTER (6, 24), *INSERM, Hôpital de Bicêtre, 94275 Le Kremlin-Bicêtre, France*

TODD M. LARSEN (29), *Department of Biochemistry, University of Wisconsin—Madison, Madison, Wisconsin 53706*

HORNG-YUH LEE (29), *Department of Chemistry, University of Nebraska, Lincoln, Nebraska 68588*

ROBERT LIDDINGTON (2), *Laboratory of X-Ray Crystallography, Dana-Farber Cancer Institute, Boston, Massachusetts 02115*

MICHAEL C. MARDEN (6, 24), *INSERM, Hôpital de Bicêtre, 94275 Le Kremlin-Bicêtre, France*

JEAN-LOUIS MARTIN (19), *INSERM Unité 275, Laboratoire d'Optique Appliquée, Ecole Polytechnique, ENSTA, 91120 Palaiseau Cedex, France*

ANTONY J. MATHEWS (17), *Somatogen, Inc., Boulder, Colorado 80301*

GEORGE MCLENDON (7), *Department of Chemistry, University of Rochester, Rochester, New York 14627*

R. J. DWAYNE MILLER (16), *Department of Chemistry, University of Rochester, Rochester, New York 14627*

ALISON MURRAY (23), *Blood Research Division, Letterman Army Institute of Research, Presidio of San Francisco, California 94129*

DAVID W. MYERS (27), *Department of Biochemistry, Duke University Medical Center, Durham, North Carolina 27710*

JOHN S. OLSON (17), *Department of Biochemistry and Cell Biology, Rice University, Houston, Texas 77251*

LAWRENCE J. PARKHURST (29), *Department of Chemistry, University of Nebraska, Lincoln, Nebraska 68588*

MARCO PEREIRA (16), *Department of Chemistry, University of Rochester, Rochester, New York 14627*

MICHELE PERRELLA (21, 25), *Department of Science and Technology, School of Medicine, University of Milan, Milan 20133, Italy*

JANICE R. PERUSSI (8), *Departmento de Química, Universidade de São Paulo, IFQSC–VSP, CEP 13560-250-São Carlos-SP, Brazil*

SERGE PIN (4, 13, 14), *Laboratoire de Biologie Physico-Chimique, Université Paris VII, 75251 Paris, France*

CLAUDE POYART (6, 24), *INSERM, Hôpital de Bicêtre, 94275 Le Kremlin-Bicêtre, France*

LYNN RICHARD (16), *Department of Chemistry, University of Rochester, Rochester, New York 14627*

LUIGI ROSSI-BERNARDI (21), *Dipartimento di Scienze e Tecnologie Biomediche, Università di Milano, Milano 20133, Italy*

LEWIS J. ROTHBERG (10), *AT&T Bell Laboratories, Murray Hill, New Jersey 07974*

CATHERINE A. ROYER (4), *School of Pharmacy, University of Wisconsin—Madison, Madison, Wisconsin 53706*

VIJAY S. SHARMA (20), *Department of Medicine, University of California at San Diego, La Jolla, California 92093*

RICHARD I. SHRAGER (22, 27), *Physical Sciences Laboratory, Division of Computer Research and Technology, National Institutes of Health, Bethesda, Maryland 20892*

JOSEPH H. SOMMER (18), *Laboratory of Chemical Physics, National Institute of Diabetes and Digestive and Kidney Diseases, National Institutes of Health, Bethesda, Maryland 20892*

KIM D. VANDEGRIFF (22), *Department of Medicine, University of California at San Diego, Veterans Affairs Medical Center, San Diego, California 92161*

MARTEN H. VOS (19), *INSERM Unité 275, Laboratoire d'Optique Appliquée, Ecole Polytechnique, ENSTA, 91120 Palaiseau Cedex, France*

ROBERT M. WINSLOW (23), *Department of Medicine, University of California at San Diego, Veterans Affairs Medical Center, San Diego, California 92161*

CHRISTIAN ZENTZ (13), *Laboratoire de Biologie Physico-Chimique, Université Paris VII, 75251 Paris, France*

Preface

Much has happened since "Hemoglobins," Volume 76 of *Methods in Enzymology,* was published in 1981. Methods have been refined and new methods have been devised. At the time Volume 76 went to press, the general feeling in the hemoglobin "community" was that this venerable protein had contributed about all it could to our knowledge of fundamental protein chemistry, and scientists were turning their attention to other proteins.

Three forces have brought hemoglobin back to center stage. First, the expression of human globin genes in *Escherichia coli, Saccharomyces cerevisiae*, and transgenic animals has led to new approaches to hemoglobin chemistry. It is now possible to produce quickly and efficiently site-specific hemoglobin mutants that are being used to test hypotheses on structure–function relationships. The amount of information derived from such studies is staggering. Second, new techniques have been developed that have led to significant advances in the field since 1981. For example, laser photolysis has expanded from nanosecond to femtosecond time domains that are used to explore the very fast events associated with the hemoglobin molecule. Third, hemoglobin has been used in exciting new studies of cell-free oxygen carriers for eventual clinical use, currently being carried out in both academia and industry. Purification and characterization of hemoglobin and chemically or genetically modified hemoglobins are now of critical importance. This new application requires a detailed understanding of the oxygen transport function of hemoglobin and of its interactions with other biological systems.

Thus, what began as a modest effort to update methods described in Volume 76 quickly expanded into two new volumes, 231 and 232. The division of the chapters is somewhat arbitrary, but we settled on methods of biochemical and analytical focus for Volume 231 and those which deal with biophysical methods for Volume 232. As in Volume 76, authors were instructed to emphasize techniques, currently in use in their laboratories, in sufficient detail to allow the interested reader to implement those methods independently.

We wish to thank the authors for their contributions and cooperation and the staff of Academic Press for their assistance. Special thanks are due the Editors-in-Chief of *Methods in Enzymology* for encouraging the preparation of these volumes and to Shirley Light of Academic Press for her expert guidance and support during this work.

JOHANNES EVERSE
KIM D. VANDEGRIFF
ROBERT M. WINSLOW

METHODS IN ENZYMOLOGY

Section I

Molecular Structure and Dynamics

[1] Acid-Induced Folding of Heme Proteins

By YUJI GOTO and ANTHONY L. FINK

Introduction

One of the oldest known methods of denaturing proteins is by the addition of acids. However, the conformation of acid-denatured states varies from apparently fully unfolded to a significant amount of folded structure remaining. Although the acid denaturation of proteins has been studied extensively, the exact mechanism and the detailed conformation of acid-denatured proteins have not been determined.[1,2]

The conformation of the acid-denatured state of several proteins, including the heme proteins, myoglobin, apomyoglobin, and cytochrome c, depends on salt conditions.[3-6] Under conditions of low salt at pH 2, these proteins are substantially unfolded to a conformation (U_A) similar to that obtained with high concentrations of guanidine hydrochloride (Gdn-HCl). The addition of salt cooperatively stabilizes an intermediate conformational state (state A) having many properties of a molten globule. The molten globule state is a compact structure with a significant amount of secondary structure but with a largely disordered tertiary structure.[4,7-10] The molten globule has been proposed to be a significant intermediate state in protein folding,[7-10] and its participation in various *in vivo* processes has been suggested.[11,12] It has been pointed out that the current literature on the "molten globule" is rather confusing because it is used for various kinds of intermediate conformational states.[1,9] In this chapter, we consider the molten globule state to be an experimentally observed compact inter-

[1] P. S. Kim and R. L. Baldwin, *Annu. Rev. Biochem.* **59,** 631 (1990).
[2] K. A. Dill and D. Shortle, *Annu. Rev. Biochem.* **60,** 795 (1991).
[3] E. Stellwagen and J. Babul, *Biochemistry* **14,** 5135 (1975).
[4] M. Ohgushi and A. Wada, *FEBS Lett.* **124,** 21 (1983).
[5] Y. Goto and A. L. Fink, *Biochemistry* **28,** 945 (1989).
[6] Y. Goto, L. J. Calciano, and A. L. Fink, *Proc. Natl. Acad. Sci. U.S.A.* **87,** 573 (1990).
[7] D. A. Dolgikh, R. I. Gilmanshin, E. V. Brazhnikov, V. E. Bychkova, G. V. Semisotnov, S. Yu Venyaminov, and O. B. Ptitsyn, *FEBS Lett.* **136,** 311 (1981).
[8] K. Kuwajima, *Proteins* **6,** 87 (1989).
[9] H. Christensen and R. H. Pain, *Eur. Biophys. J.* **19,** 221 (1991).
[10] O. B. Ptitsyn and G. V. Semisotnov, *in* "Conformation and Forces in Protein Folding" (B. T. Nall and K. A. Dill, eds.), p. 155. Am. Assoc. Adv. Sci., Washington, DC, 1991.
[11] V. E. Bychkova, R. H. Pain, and O. B. Ptitsyn, *FEBS Lett.* **238,** 231 (1988).
[12] M.-J. Gething and J. Sambrook, *Nature (London)* **355,** 33 (1992).

mediate with substantial native-like secondary structure but with a largely disordered tertiary structure.

With the heme proteins, Goto and co-workers[6] found that the addition of strong acids to the solution of fully unfolded protein at pH 2 causes the "acid-induced refolding" transition. If one starts with a native protein at low ionic strength and lowers the pH by adding HCl, the protein initially changes to a relatively fully unfolded conformation, typically in the vicinity of pH 2 (acid-induced unfolding). Contrary to our expectation, as the pH decreases below 2 with increasing acid concentration, the protein refolds into the molten globule state (acid-induced refolding).

Goto et al.[13] showed that the effects of acids in stabilizing the molten globule state are similar to those of the corresponding salts at pH 2. Because the pK_a values of titratable groups are mostly above 3,[14] the similar effects of acid and corresponding salt indicated the importance of the anion in determining the conformation of acid-denatured proteins. Goto and co-workers[13,15-18] showed that the conformation of acid-denatured proteins is determined by a balance of electrostatic repulsions between positive residues, which favor the extended conformation, and the opposing forces, which stabilize the intermediate state. Anions, either from acid or salt, shield the former by direct binding to the positive charges, resulting in the manifestation of the latter. Thus, whereas the acid-induced unfolding largely depends on pH, the acid-induced refolding depends on the acid species and its concentration.

Because the acid-induced unfolding and refolding reactions are readily examined by using spectroscopic methods such as circular dichroism (CD) and fluorescence, it is straightforward to explore these intermediate states of proteins. In particular, by applying this method to heme proteins, it is possible to estimate the contribution of the heme group in stabilizing the intermediate states.

General Procedures

Choice of Acid

The acid-induced refolding reaction depends critically on the anion species and its concentration. The minimal anion concentration, $[A^-]$, at

[13] Y. Goto, N. Takahashi, and A. L. Fink, Biochemistry 29, 3480 (1990).
[14] C. Tanford, Adv. Protein Chem. 23, 121 (1968).
[15] Y. Goto and A. L. Fink, J. Mol. Biol. 214, 803 (1990).
[16] A. L. Fink, L. J. Calciano, Y. Goto, and D. R. Palleros, in "Current Research in Protein Chemistry" (J. Villafranca, ed.), p. 417. Academic Press, San Diego, CA, 1990.

a particular pH increases sharply below pH 2 by the relation: $[A^-] = 10^{-pH}$. Therefore, the choice of acid is important in these experiments. The results, which will be described below, indicated that a combination of HCl and perchloric acid ($HClO_4$) is the best to start the experiments.

The effects of various anions in stabilizing the molten globule state have been compared by using horse apomyoglobin and horse ferricytochrome c.[13] The proteins are initially unfolded by HCl alone at pH 2. Then, the effects of various anions are examined by following the acid- or salt-induced transition. Conformational change is monitored by using the ellipticity at 222 nm. In the case of cytochrome c, the Soret band absorption arising from the covalently bound heme group is also used. Values of the midpoint concentration (C_m) for the transition of cytochrome c induced by various acids and salts are shown in Table I. The values of C_m for the corresponding acid and salt are similar, showing the importance of anion in the acid- or salt-induced refolding transitions.

The effectiveness of various anions in stabilizing the molten globule state is consistent with the electroselectivity series, representing the order of affinity of anions to the positively charged anion-exchange resin.[19,20] The series obtained using apomyoglobin is consistent with that for cytochrome c, although the values of C_m differ slightly.

Thus, the effectiveness of various strong acids follows the series: H_2SO_4 > trichloroacetic acid > $HClO_4$ > HNO_3 > trifluoroacetic acid > HCl. However, the acids having strong absorption in the far- and near-ultraviolet (UV) regions (i.e., trichloroacetic acid, HNO_3) are inappropriate for the CD measurements. In the case of heme-containing protein, the Soret band absorption of heme group, reflecting the conformational state of the protein, can be monitored for such acids (Table I).

Because HCl has the poorest ability to stabilize the intermediate state, it is the acid of choice to check the maximal acid unfolding. If the stability of a protein against acids is low, the titration with HCl of the protein dialyzed against distilled water will show the initial acid-induced unfolding, the maximum unfolding being in the vicinity of pH 2, and the subsequent acid-induced refolding transition at lower pH. Other acids such as H_2SO_4 may show incomplete acid-induced unfolding because their anions tend to stabilize the intermediate state more strongly.

---

[17] A. L. Fink, L. J. Calciano, Y. Goto, and D. Palleros, in "Conformation and Forces in Protein Folding" (B. T. Nall and K. A. Dill, eds.), p. 169. Am. Assoc. Adv. Sci., Washington, DC, 1991.

[18] Y. Goto and S. Nishikiori, *J. Mol. Biol.* **222**, 679 (1991).

[19] H. P. Gregor, J. Belle, and R. A. Marcus, *J. Am. Chem. Soc.* **77**, 2713 (1955).

[20] D. T. Gjerde, G. Schmuchler, and J. S. Fritz, *J. Chromatogr.* **187**, 35 (1980).

TABLE I
ACID- OR SALT-INDUCED REFOLDING TRANSITION OF
CYTOCHROME c[a]

Acid	$C_m{}^b$ (mM)	Salt	$C_m{}^b$ (mM)
		$K_3Fe(CN)_6$	0.030 (0.028)
		$K_4Fe(CN)_6$	0.048 (0.041)
H_2SO_4	1.7 (1.6)	Na_2SO_4	1.2 (1.7)
TCAH	6.3 (6.6)	NaTCA	5.5 (4.9)
		NaSCN	nd (5.6)
$HClO_4$	8.1 (7.3)	$NaClO_4$	6.6 (6.5)
		NaI	nd (13)
HNO_3	nd (29)	$NaNO_3$	nd (22)
TFAH	84 (90)	NaTFA	22 (25)
		NaBr	nd (31)
HCl	114 (138)	NaCl	45 (45)
		KCl	47 (48)

[a] Transitions were measured in the presence of 18 mM HCl at 20° by using the change in ellipticity at 222 nm. The values in parentheses were obtained by using the change in Soret absorption at 394 nm. TCAH, Trichloroacetic acid; TFAH, trifluoroacetic acid; NaTCA, sodium trichloroacetate; NaTFA, sodium trifluoroacetate; nd, Not determined. [Adapted with permission from Y. Goto *et al., Biochemistry* **29**, 3480 (1990). Copyright 1990 American Chemical Society.]

[b] Midpoint concentration of transition.

If the acid-induced refolding is not observed with HCl, the effects of other acids should be examined. Perchloric acid is recommended because of its relatively high potential to induce the acid refolding and its extremely low absorption in the far-UV regions.

In the range of acid concentration where the cooperative refolding transition is almost over, proteins show a high tendency to form aggregates. The tendency depends on both protein and anion species.[13] The aggregation arises from the decrease in net charge of the protein due to anion binding, resulting in intermolecular hydrophobic interactions. Precipitation of proteins by trichloroacetic acid, which is often used for protein purification, may be explained by the mechanism described here.

Instruments

The acid-induced unfolding and refolding reactions can be followed by using any methods that are sensitive to the conformational state of proteins. However, it is important to characterize the conformational states by as many methods as possible in order to distinguish the different conformational states. Because of the propensity of the A states to aggregate it is desirable to use methods that are quite sensitive so that low protein concentrations may be used. At least two methods measuring the secondary and tertiary structures of protein should be used, because the intermediate states frequently have a secondary structure similar to that of native protein, but a largely disordered tertiary structure. Far-UV CD is the most convenient and useful method to follow the change in secondary structure. It requires a small volume of solution at a low protein concentration, that is, about 0.5 ml at 0.1 mg/ml for one spectral measurement. Near-UV CD, tryptophan fluorescence, absorbance by the heme group, and quenching of tryptophan fluorescence by the heme group can be used to follow the change in tertiary structure.

Preparation of Solutions

To carry out the HCl titration of a protein, the protein solution at neutral pH is first dialyzed against distilled or deionized water. Then the titration is carried out, usually with a constant volume of the protein solution mixed with a constant volume of acid at various concentrations. The conformation of the native protein should be checked by preparing a protein solution at neutral pH.

The conformational transitions below pH 2 are usually rapid (complete within a few minutes), and the measurements should be carried out soon after the preparation of the solution. Judging from the CD signal and the elution pattern of the denatured proteins by high-performance liquid chromatography,[13] 1 M HCl had no detectable effect on the covalent structure of the proteins for at least several hours. However, in the presence of a high concentration of acids, the time-dependent change of the signal should be checked carefully to confirm the stability of the marginal conformational states. Acid concentrations greater than 1 M will result in the acid-catalyzed hydrolysis of peptide bonds and must be carefully monitored. When the acid-induced unfolding transition above pH 2, in particular in the transition region, is measured, the time-dependent changes in the signal should be checked to confirm the equilibration. The pH value should be measured soon after the spectroscopic measurement.

HCl-Induced Refolding of Apomyoglobin

The HCl titration of apomyoglobin produces the stepwise unfolding and refolding transitions, providing a typical example of acid-induced refolding.[6,13,15]

Materials and Methods

Apomyoglobin is prepared from horse myoglobin (Sigma, St. Louis, MO) by 2-butanone extraction of the heme.[21] The content of the holoprotein remaining is <1%. Apomyoglobin, dialyzed extensively against distilled water, is filtered with a Millipore (Bedford, MA) filter (pore size of 0.22 μm) and used as a stock solution.

Conformational change is measured by far-UV CD and tryptophan fluorescence. All measurements are carried out at 20° in thermostatically controlled cell holders. Circular dichroism measurements are carried out with a Jasco spectropolarimeter (model J-500A) (Tokyo, Japan) or an Aviv spectropolarimeter (model 60DS) (Lakewood, NJ). The instruments are calibrated with ammonium (+)-10-camphor sulfonate.[22] The results are expressed as mean residue ellipticity $[\theta]$, which is defined as $[\theta] = 100\theta_{obs}/(lc)$, where θ_{obs} is the observed ellipticity in degrees, c is the concentration in residue moles per liter, and l is the length of the light path in centimeters. The CD spectra are measured with a 1-mm cell from 250 to 195 nm. Tryptophan fluorescence is measured with a Perkin-Elmer (Norwalk, CT) MPF4 instrument with excitation at 280 nm.

Typically, 0.1 ml of the protein solution, dissolved in distilled water, is mixed with 0.9 ml of HCl or buffer solution. Glycine (pH 2–3.5), sodium acetate (pH 3–6.5), and sodium phosphate (pH 6.5–7.5) buffers are also used to follow the initial unfolding transition above pH 2. The pH is measured with a Radiometer PHM83 (Copenhagen, Denmark) at 20°. Protein concentrations are determined spectrophotometrically. Extinction coefficients used to calculate the concentration of native metmyoglobin at 409 nm and native apomyoglobin at 280 nm are 1.6 × 10^5 and 1.43 × 10^4 M^{-1} cm^{-1}, respectively.[23]

HCl-Induced Unfolding and Refolding Transitions

Whereas the ellipticity at 222 nm indicates the amount of helical structure, the wavelength of maximal emission of tryptophan fluorescence indicates the solvent exposure of tryptophan residues. The CD spectrum of horse apomyoglobin at neutral pH shows minima at 222 and 208 nm, representative of a high α-helix content (about 60%). Horse myoglobin has two tryptophan residues at positions 7 and 14. Tryptophan fluorescence has a maximum at 335 nm under native conditions, indicating that

[21] K. D. Hapner, R. A. Bradshaw, C. R. Hartzell, and F. R. N. Gurd, *J. Biol. Chem.* **243**, 683 (1968).
[22] T. Takakuwa, T. Konno, and H. Meguro, *Anal. Sci.* **1**, 215 (1985).
[23] M. J. Crumpton and A. Polson, *J. Mol. Biol.* **11**, 722 (1965).

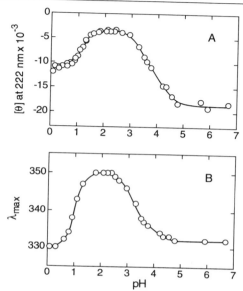

Fig. 1. Effects of increasing concentrations of HCl on the conformation of horse apomyo-globin. The HCl-induced unfolding and refolding transitions of apomyoglobin at 20° were monitored by the ellipticity at 222 nm (A) and the tryptophan fluorescence (B). (Taken and modified from Goto *et al.*[6])

the residues are buried in the interior of the protein molecule in the native state.

Figure 1 shows the acid-induced unfolding and refolding transitions of apomyoglobin induced by HCl alone, measured by the ellipticity at 222 nm (A) and the maximal wavelength of tryptophan fluorescence (B). Apomyoglobin started to unfold at pH 5, with loss of helical structure and exposure of the tryptophan residues. At pH 2, the protein was maximally unfolded (U_A) to an extent similar to that unfolded by 5 or 6 M Gdn-HCl. Further increase in HCl concentration resulted in the refolding to the A state, with the reformation of helical structure and the burial of tryptophan residues. Judging from the CD intensity at 222 nm, the helical content of the A state is about 30% in 1 M HCl.

Phase Diagram for Acidic Conformational States

To understand the mechanism of acid-induced unfolding and refolding transitions of apomyoglobin, the dependence on ionic strength was mea-

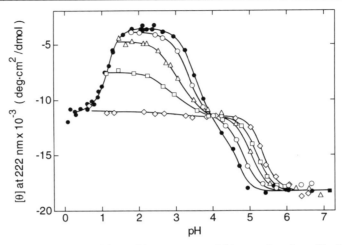

FIG. 2. Acid-induced transitions of horse apomyoglobin as a function of ionic strength (I), measured by the change in ellipticity [θ] at 222 nm at 20°. (●) Minimal ionic strength (HCl only); (○) $I = 0.02$; (△) $I = 0.05$; (□) $I = 0.1$; (◇) $I = 0.5$. The pH for conditions of minimal ionic strength was adjusted with HCl, and for the other conditions the pH was adjusted with 10 mM potassium acetate buffer above pH 3 and with HCl below pH 3. The ionic strength was controlled by KCl. A filled square at pH 7.2 shows the signal in 20 mM sodium phosphate buffer. [Taken from Y. Goto and A. L. Fink, *J. Mol. Biol.* **214**, 803 (1990), with permission.]

sured. Figure 2 shows the HCl-induced transition of apomyoglobin under different conditions of ionic strength, measured by the change in ellipticity at 222 nm. When the ionic strength of the solution was increased with KCl, the three-state nature of the unfolding transition became evident. Whereas the first transition [native (N) to A] occurred at higher pH as the ionic strength increased, the second unfolding transition (A to U_A) occurred at lower pH. The second transition disappeared at an ionic strength of 0.5. The acid-titration curves under conditions of different ionic strength showed a crossover point at pH 4, where the ellipticity is the same for the A state induced by HCl or for the A state induced by KCl at pH 2.

The data in Fig. 2 were most simply interpreted in terms of three conformational states, the N, A, and U_A states, with mean residue ellipticities at 222 nm of $-18,300$, $-11,500$, and $-3700°$ cm^2 dmol^{-1}, respectively. From Fig. 2, the midpoints of the transitions were calculated and then a phase diagram for the pH and ionic strength-dependent conformational states of apomyoglobin was constructed.

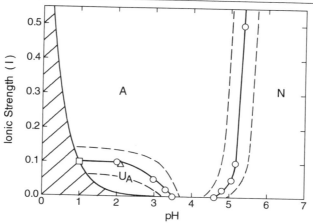

FIG. 3. Phase diagram for the native (N), U_A, and A states of horse apomyoglobin at 20°. The continuous lines show the boundaries between the conformational states, which were determined from the midpoints of the pH-induced transitions under the various salt conditions shown in Fig. 2 (○). The results of KCl-induced transition at pH 2.1 (△) and HCl-induced refolding transition (□) are also used.[6] Broken lines indicate the contour lines corresponding to 80 or 20% progress of the transition. The hatched area is prohibited due to the increase in the minimal ionic strength with decrease in pH. (Taken from Goto and Fink[15] with permission.)

Figure 3 shows the phase diagram of the three states. The hatched area is prohibited owing to the increase in the minimal ionic strength with decrease in pH ($I = [A^-] = 10^{-pH}$). The intermediate state (A) is stable at around pH 4 or below pH 3 under conditions of high salt. Figure 3 clearly shows that the acid-induced refolding comes from the increase in chloride concentration due to the increase in HCl concentration.

The A state of apomyoglobin observed here is the same state as the intermediate state observed by Hughson et al.[24,25] at pH 4 under conditions of low salt. The A state of apomyoglobin corresponds to the molten globule state observed for other proteins, although the hydrodynamic radius is increased by about 60% compared to that of the native protein, being higher than those of other proteins (10–20% for β-lactamase and cytochrome c).[6] Hughson et al.[24] studied the secondary structure present in the intermediate state by trapping slowly exchanging peptide NH protons and analyzing them by two-dimensional ^1H nuclear magnetic resonance (NMR). They indicated that a compact subdomain consisting of the A,

[24] F. M. Hughson, P. E. Wright, and R. L. Baldwin, Science 249, 1544 (1990).
[25] F. M. Hughson, D. Barrick, and R. L. Baldwin, Biochemistry 30, 4113 (1991).

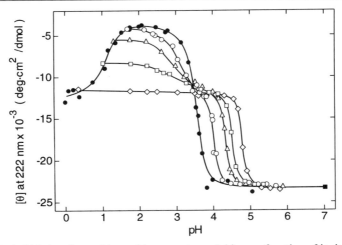

FIG. 4. Acid-induced transitions of horse metmyoglobin as a function of ionic strength, measured by the change in ellipticity [θ] at 222 nm at 20°. The symbols are the same as shown in Fig. 2.

G, and H helices is folded in the intermediate state, whereas the other subdomain comprising the B and E helices is unfolded.

HCl-Induced Refolding of Metmyoglobin

It is well known that the native structure of myoglobin is stabilized by the noncovalently bound heme group.[23,26] However, the role of the heme in the conformation and stability of the intermediate state is unknown. To elucidate the role of heme, the acid-induced transitions of myoglobin have been examined. Horse myoglobin obtained from Sigma (metmyoglobin form) is used. Measurements are carried out by using the ellipticity at 222 nm as in the case of apomyoglobin.

Phase Diagram of Myoglobin

Figure 4 shows the HCl-induced transition of myoglobin under different conditions of ionic strength, measured by the change in ellipticity at 222 nm. The ellipticity at 222 nm of the native state of metmyoglobin ($-23,500$) is larger than that of apomyoglobin ($-18,300$), indicating that metmyoglobin contains a higher amount of helical structure. In the absence of salt, unfolding and refolding transitions similar to those of apomyoglobin

[26] A. N. Schechter and C. J. Epstein, *J. Mol. Biol.* **35,** 567 (1968).

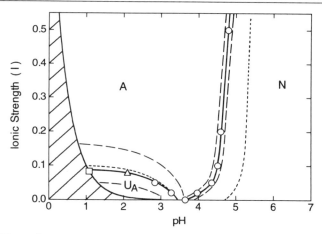

Fig. 5. Phase diagram for the native (N), U_A, and A states of horse metmyoglobin at 20°. The continuous lines show the boundaries between the conformational states, which were determined from the midpoints of the pH-induced transitions under the various salt conditions shown in Fig. 4 (○). The results of KCl-induced transition at pH 2.1 (△) and HCl-induced refolding transition (□) are also used. Broken lines indicate the contour lines corresponding to 80 or 20% progress of the transition. The hatched area is prohibited owing to the increase in the minimal ionic strength with decrease in pH. For comparison, the boundaries of apomyoglobin taken from Fig. 3 are indicated by dotted lines.

were observed. However, the cooperativity of the acid-unfolding transition was much higher, and the transition started at a pH value one unit lower than that of apomyoglobin. These results show that the heme increases the stability of the native state.

When the ionic strength of the solution was increased with KCl, the three-state nature of the unfolding transition became evident as in the case of apomyoglobin. Whereas the first unfolding transition (N to A) shifted to the higher pH regions, the second unfolding transition (A to U_A) shifted to lower pH. The dependence of the initial unfolding transition on salt was similar to that reported by Friend and Gurd.[27] At each ionic strength, the initial acid-induced unfolding transition starts at a pH value one unit lower than that of apomyoglobin. The cooperativity of the initial transition is higher than that of apomyoglobin. On the other hand, the second unfolding transition is similar to that of apomyoglobin.

Figure 5 shows the phase diagram for the conformational states of myoglobin. The phase diagram was constructed assuming the three states (i.e., N, A, and U_A) and applying the same procedure used for apomyo-

[27] S. H. Friend and F. R. N. Gurd, *Biochemistry* **18,** 4612 (1979).

globin. Mean residue ellipticities of N, A, and U_A states were $-23,500$, $-12,000$, and $-3700° cm^2 dmol^{-1}$, respectively. The ellipticities for the A and U_A states are essentially the same as those of apomyoglobin. The phase diagram shows that, whereas the native state of myoglobin is more stable than that of apomyoglobin against acid, the relative stability of the U and A states is independent of the heme group.

It has been reported that on the acid denaturation of myoglobin or hemoglobin, the heme group (hemin) dissociated from the protein and aggregated.[28–31] The yield of native metmyoglobin renatured from the acid denaturation decreased with length of time in the denaturing conditions, and the decrease in yield was explained by the formation of the hemin aggregate.[30] Although we cannot exclude the possibility that the aggregation of hemin prevents the specific interaction of monomeric hemin with the A state, the close similarity of the boundary between the U_A and A states of apomyoglobin with that of myoglobin suggests that the heme group does not interact significantly with the A state.

In the native structure of myoglobin, the heme group is accommodated in a hydrophobic box formed by helices.[32,33] As described above, Hughson et al.[24] indicated that a subdomain comprising B and E helices, which is in contact with the heme group, is unfolded in the A state. This may explain why heme does not interact with and stabilize the A state.

Conclusion

Acid-induced refolding in combination with acid-induced unfolding is a useful method to examine the behavior of proteins under acidic conditions. In particular with heme proteins, the comparison of apo- and holoproteins may elucidate the role of the heme group in the stability of intermediate states. Various acids differ in their potential to stabilize the intermediate state. Titration with HCl is a standard method to follow the maximum acid-induced unfolding and subsequent acid-induced refolding. In a case in which the acid-induced refolding is not observed with HCl, the effects of perchloric acid should be checked, because it is more efficient at inducing refolding, and the far-UV absorbance is extremely low. Phase diagrams can be prepared from the titrations under different conditions

[28] H. Polet and J. Steinhardt, *Biochemistry* **8**, 857 (1969).
[29] L. L. Shen and J. Hermans, Jr., *Biochemistry* **10**, 1836 (1972).
[30] L. L. Shen and J. Hermans, Jr., *Biochemistry* **10**, 1842 (1972).
[31] L. L. Shen and J. Hermans, Jr., *Biochemistry* **10**, 1845 (1972).
[32] J. C. Kendrew, R. E. Dickerson, B. E. Strandberg, R. G. Hart, D. R. Davis, D. C. Phillips, and V. C. Shore, *Nature* (*London*) **185**, 422 (1960).
[33] T. Takano, *J. Mol. Biol.* **110**, 537 (1977).

of ionic strength and are useful in considering the mechanism of the conformational transition. By adding urea or reducing the disulfide bonds, the method may be applicable even for proteins with high stability under acidic conditions.[16,17]

Acknowledgment

This work was supported by the Japan–U.S. Cooperative Science Program.

[2] X-Ray Crystallography of Partially Liganded Structures

By ROBERT LIDDINGTON

Introduction

The crystal lattice can be thought of as a strong allosteric effector that stabilizes one quaternary conformation. For example, Mozzarelli *et al.*[1] have measured the oxygen affinity of T-state crystals grown from polyethylene glycol (PEG) and found the affinity to be much lower than in solution: the $p50$ at 25° is 300 torr in crystals compared with 30–50 torr under typical aqueous conditions. Because of this low affinity it has been possible to produce crystals of molecules that are not the predominant species in aqueous solution, and from these studies better understand the chemical basis of low affinity in the T state and normal affinity in the R state.

The degree of stabilization by the crystal depends on the particular lattice and the stabilizing solution. Some conditions are sufficient to stabilize a fully liganded T state, but the addition of further effectors or drugs may be necessary to inhibit the quaternary structure switch. The manifestation of the quaternary structure switch is either cracking (Haurowitz's classic observation of crystals grown from high salt[2]) or dissolving [characteristic of low salt (PEG) crystals]. Both of these are of course disastrous for the crystallographer. The last weapons available to the crystallographer are glutaraldehyde and acrylamide, which can cross-link crystals such that they are immune to most outside influences, but will still bind ligand. However, loss of diffraction quality is likely, and doubts about the "biological relevance" will inevitably be raised.

[1] A. Mozzarelli, C. Rivetti, G. Rossi, E. Henry, and W. Eaton, *Nature (London)* **351,** 416 (1991).
[2] F. Haurowitz, *Hoppe-Seyler's Z. Physiol. Chem.* **254,** 266 (1938).

This chapter focuses on human hemoglobin (Hb), because this has been studied the most, and generally restricts the discussion to Hb A. Extrapolation to other species will not be exact, of course, but what follows may provide a reasonable and informed starting point for further studies. New results are expected shortly on the fish hemoglobins: it should be possible to produce fully liganded Root effect molecules in the T state. All of the intermediate-state crystals so far studied have had either the T or R quaternary organization, or slight deviations from it, with one exception: Silva et al.[3] have described a modified R state, which they call R2, that has a tertiary structure similar to the R state, but a different quaternary structure.

Crystallization and Preparation of Derivatives

Crystallization of proteins in general is well described in the monograph by MacPherson.[4] For hemoglobin, the first thing to consider is which quaternary state is desired. For the T state one may employ either high-salt or low-salt conditions (using PEG). The R state has normally been grown from high-salt conditions, but crystallization of the R2 state was from PEG.[3]

Crystallization from Polyethylene Glycol

For the T state, low salt has several advantages: (1) binding of inositol hexaphosphate (IHP) and 2,3-diphosphoglycerate (DPG) is inhibited by high salt; (2) anaerobic conditions are not required with PEG; and (3) oxidation is not a serious problem at pH values above 7.

Aliquots of a stock solution containing 50 g of PEG (M_r 8000) and 100 ml of H_2O are added to a buffered solution (10 mM potassium phosphate, pH 7.2) of oxyhemoglobin (~50 mg/ml), with a total volume of around 100–500 μl. The solutions should be immediately vortexed to prevent precipitation of the protein, and briefly spun in a benchtop centrifuge to remove any particles that might cause excessive nucleation. The solutions are then set down in sealed tubes in a quiet, draft-free environment. Storage in a Styrofoam box buffers the system against sudden temperature changes. Crystals grow in about 1 week at room temperature, but take longer at 4°. Typical conditions are 20–25% stock PEG (7–8%, w/v), but a range of conditions, perhaps 4–15% PEG (w/v) in 2% steps, should be tried initially. Once crystallization has begun, the process is rapid (~24

[3] M. Silva, P. Rogers, and A. Arnone, *J. Biol. Chem.* **267**, 17248 (1992).
[4] A. MacPherson, "Preparation and Analysis of Protein Crystals." Krieger, Malabar, FL 1989.

hr), and no subsequent growth occurs. The crystals belong to space group $P2_12_12$, with cell dimensions of $a = 95.8$, $b = 97.8$, $c = 65.5$ Å. Crystals may also be grown under anaerobic conditions,[5] or with the addition of dithionite, and the crystals produced are identical; these might be preferred with unstable hemoglobins, but it is not necessary for Hb A.

If supplies of protein are limited, then smaller scale trials can be simply achieved in a tissue culture plate that holds about 20–50 μl in each well, each individually sealed with a fragment of plastic coverslip and vacuum grease. The modern microscale vapor diffusion methods, which can deal with volumes as small as 1 μl, have also been used successfully, and may be useful for determining optimal conditions. The most convenient of these are the CrysChem "sitting drop" trays available from Charles Supper Company (Natick, MA). The tray is sealed with transparent tape to avoid time-consuming cleaning of coverslips and applications of vacuum grease. These have the further advantage that initial trials in the 1- to 2-μl range can be scaled up to 100 μl with the same apparatus. If the trials need to be done at 4°, it is a good idea to do all steps at this temperature to avoid condensation problems.

The addition of PEG to oxyhemoglobin solutions actually induces deoxygenation by an unknown mechanism. Korber[6] has followed this process spectroscopically and has shown that for a solution containing 10% PEG (w/v) the half-point of deoxygenation occurs after 40 hr at room temperature, but that the solution could be completely reoxygenated by shaking. The crystals probably grow in the deoxy form.

Oxygenation of Hemoglobin Crystals Grown from Polyethylene Glycol

DeoxyHb crystals grown from PEG will rapidly oxygenate and liquefy if they are removed from the crystallization liquor and mounted in capillaries in the presence of air (they can be safely mounted in capillaries under a nitrogen atmosphere). Transferring the crystals quickly into a large excess of 33% PEG (w/v) (buffered with phosphate) or higher prevents dissolution, and crystals are completely stable at room temperature when mounted in capillaries in the presence of air.[7] Mozzarelli et al.[1,8] have measured the oxygen affinity of stabilized crystals grown from PEG by spectroscopy: the $p50$ is temperature sensitive, varying between 70 torr at 4° and 300 torr at 25° (67% PEG), but independent of pH. Cooling would

[5] K. Ward, B. Wishner, E. Lattman, and W. Love, J. Mol. Biol. 98, 161 (1975).
[6] F. Korber, Ph.D Thesis, University of Leeds, England (1984).
[7] A. Brzozowski, Z. Derewenda, E. Dodson, G. Dodson, M. Grabowski, R. Liddington, T. Skarzynski, and D. Vallely, Nature (London) 307, 74 (1984).
[8] A. Mozzarelli, personal communication, 1992.

therefore seem to be an excellent method of producing any desired level of oxygenation. In preliminary experiments by the author, crystals stabilized in 33% PEG were mounted in the presence of air. Data sets were collected over several days to 2.5-Å resolution from a single crystal at 25 and 4°, and again at 25°C. Reversible oxygen binding was observed at the β-hemes, but the α-hemes were slowly oxidized, and by the end of the experiment fully so; dissolved crystals showed about 50% metHb content, consistent with complete oxidation at the α-hemes, but little oxidation at the β-hemes. Under conditions of half-oxygenation at room temperature, Brzozowski et al.[7] showed that oxygen bound predominantly to the α-hemes. This has since been repeated with PEG Hb cocrystals with IHP, DPG, and the drug bezafibrate.[9,10] In a mutant hemoglobin grown from PEG, oxidation also occurs predominantly at the α-hemes.[11]

Preparation of Other Polyethylene Glycol-Grown Hemoglobin Derivatives

Polyethylene glycol-grown Hb crystals, grown with IHP and stabilized in 33% stock PEG, are resistant to oxidation above neutral pH: an overnight soaking in 1 mM K_3Fe(CN)_6 at pH 7.5 does not produce full oxidation.[12] However, a shorter soak (~1 hr) at high concentration of oxidant (~100 mM) is effective. Large, but poorly formed, crystals can be employed to monitor the process; they can be removed, rinsed well in stock PEG, and dissolved in water. Visible absorption spectra will indicate the course of oxidation. Crystals for diffraction studies should be rinsed extensively in fresh stock PEG for several hours to remove excess oxidant. The diffracting power is unaffected by this treatment. Crystals grown in the absence of IHP are much less stable.

Hemoglobin crystals grown with IHP and stored for several weeks in stock PEG at pH ~5 oxidized spontaneously, but the exact time course is unknown.

Soaking metHb crystals in 1 mM NaN_3 induces severe cracks in the crystals, but they still diffract and show large changes in the diffraction pattern; data have not been collected on this species.

Polyethylene glycol crystals of deoxyHb treated with stock PEG rapidly dissolve when exposed to CO; treating the crystals first with glutaral-

[9] D. Waller and R. Liddington, Acta Crystallogr., Sect. B **B46**, 409 (1990).

[10] P. Elmsley and G. Dodson, personal communication, 1993.

[11] G. Fermi, M. Perutz, D. Williamson, P. Stein, and D. Shih, J. Mol. Biol. **226**, 883 (1992).

[12] R. Liddington, Z. Derewenda, R. Hubbard, E. Dodson, and G. Dodson, J. Mol. Biol. **228**, 551 (1992).

dehyde (1% w/w, 5 min) prevents this, but diffraction quality is reduced. The conditions have never been optimized nor diffraction data collected.

Because the T-state NO derivative can be made in solution in the presence of IHP, it should be possible to do this in the crystalline form. Rigorous exclusion of oxygen is required, and this has not been pursued.

High-Salt Hemoglobin Crystals and Derivatives

The methods described by Perutz[13] remain the preferred method of growing crystals from high salt. Anaerobic conditions are necessary for the growth of T-state deoxyHb; these have space group $P2_1$ with $a = 63.3$ Å, $b = 83.3$ Å, $c = 53.7$ Å, $\beta = 99.5°$, and one tetramer in the asymmetric unit. R-state oxyHb crystals have space group $P4_12_12$ with $a = 53.7$ Å, $c = 193.8$ Å, and half a tetramer in the asymmetric unit.

Anderson[14] showed that high-salt T-state deoxyHb crystals could be briefly exposed to oxygen without destroying the lattice. Anderson describes acrylamide treatment of crystals, but also states that this was not absolutely necessary for crystal integrity. Under these conditions Anderson reports that the hemes are most likely oxidized, not oxygenated. Abraham et al.[15] have studied high-salt crystals grown with the drug RSR-56 bound, and showed that these crystals can be fully oxygenated without loss of diffraction power; further, under reduced oxygen conditions, they have produced an α-oxygenated species, with the same character as that produced from PEG.

Cocrystallizations with Effectors and Drugs

Crystals can be grown under essentially identical conditions in the presence of IHP, DPG, or any of a large number of drugs that stabilize the T or R state. Low-salt conditions are required for DPG and IHP binding; simply adding a twofold molar excess of these to the hemoglobin solution is sufficient. The optimal conditions for crystal growth will change slightly. Drugs that are known to stabilize the T state can also be soaked into high-salt crystals at a concentration of 1–2 mM.[16]

Crystallization of Metal Hybrids

For metal hybrid hemoglobins, it is possible in principle either to crystallize the deoxy form and add ligand later or to crystallize the li-

[13] M. Perutz, J. Cryst. Growth 2, 54 (1968).
[14] L. Anderson, J. Mol. Biol. 79, 495 (1973).
[15] D. Abraham, R. Peascoe, R. Randad, and J. Panikker, J. Mol. Biol. 227, 480 (1992).
[16] M. Perutz, G. Fermi, D. Abraham, C. Poyart, and E. Bursaux, J. Am. Chem. Soc. 108, 1064 (1986).

ganded form directly. Both methods should be tried if possible. The first method can be simply achieved by mounting a deoxy crystal in a capillary with the large end open. Introduce a small amount of liquid at the large end to provide an air-tight seal. Fill a syringe with the desired gas, and introduce the gas slowly, so that it forces the liquid to the end of the capillary, then makes a bubble that bursts, expelling gas and immediately reestablishing an air-tight seal. Finally, some of the gas may be sucked back into the syringe, so that the liquid seal moves down the capillary, and the capillary sealed with wax or vacuum grease. This approach has been used to prepare crystals of α-FeCO,βCo[17] and α-Zn,β-FeO$_2$.[18]

Direct crystallization of the liganded form has been successful. If it is necessary to add IHP to drive the equilibrium to the T state, then crystallization from low salt is the method of choice. Crystallization of the liganded hybrid α-Ni,β-FeCOHb from PEG and IHP[19] produced a crystal form different from the normal deoxy form, although the quaternary structure was T: $P2_1$, $a = 63.18$ Å, $b = 82.26$ Å, $c = 55.06$ Å, $\beta = 98.42°$. The structural changes at the β subunits were larger than those found by Arnone's group,[20] who crystallized the deoxy form and introduced ligand later. It is not surprising that the crystal lattice inhibits tertiary changes to some extent, because we are using it explicitly to inhibit quaternary changes; in all the liganded T-state crystals so far studied, the T-state salt bridges are intact, in contrast to solution studies showing that Bohr protons are released at low pressures of oxygen in which the T state is the predominant species.[21]

Deoxy R State

High-salt R-state crystals of oxyHb or metHb crack when exposed to reducing agents. Certain mutations and chemical modifications can lock the molecule into the R state; some of these will withstand the removal of ligand, and some will crystallize in the absence of ligand. For example, in Hb Kempsey (Asp G1 $\beta \rightarrow$ Asn), the T state is not observed, presumably due to the loss of the H bond across the $\alpha_1\beta_2$ interface between Asp G1 β_2 and Tyr C7 α_1. Hemoglobins cleaved by carboxypeptidase to remove the C-terminal α-Arg or β-His cannot make the salt bridges that normally

[17] B. Luisi and N. Shibayama, *J. Mol. Biol.* **206**, 723 (1989).

[18] B. Luisi and R. Liddington, unpublished (1993).

[19] B. Luisi, R. Liddington, G. Fermi, and N. Shibayama, *J. Mol. Biol.* **214**, 7 (1990).

[20] A. Arnone, P. Rogers, B. Hoffman, J. Nocek, and D. Gingrich, *Symp Oxygen Binding Heme Proteins*, Pacific Grove, CA (Asilomar), PV1-2 (1988).

[21] M. Perutz, G. Fermi, B. Luisi, B. Shaanan, and R. Liddington, *Acc. Chem. Res.* **20**, 309 (1987).

stabilize the T state. Various modifications to Cys F9 β by maleimide derivatives reduce cooperativity. Luisi[22] has refined the earlier work on horse deoxy R-state hemoglobin stabilized by cross-linking with bis(N-maleimidomethyl) ether (BME). Careful purification of the derivative allowed growth of R-state metHb crystals that could be reduced without cracking. Crystals grow readily from Perutz's solutions A' and B, crystallizing in space group $C2$ with one dimer in the asymmetric unit. Crystals are frequently twinned and share the [001] crystal face, but suitable crystals for high-resolution study were obtained by carefully dissecting the twins with a fine-point scalpel. Exposure of these crystals to 2 mM dithionite in deoxygenated solution A' overnight reduces the crystals to the deoxy R state. These can be mounted in capillaries under nitrogen without loss of diffracting quality (d_{min} = 1.9 Å on a synchrotron source). In contrast to the earlier studies, Luisi found no evidence for derivitization of Cys F9 β, and the only site of attachment was Lys A14 β, a residue on the surface far from either heme. Furthermore, the molecule showed nearly normal cooperativity. Hence both α- and β-hemes can be analyzed with some confidence. Luisi was unable to crystallize human metHb derivatized in the same way. Coordinates of horse deoxy R-state crystals are available from B. Luisi (MRC Virology Unit, University of Glasgow, Scotland).

Data Collection and Refinement

For a detailed account of the theory and practice of modern protein crystallography, the reader is referred to two earlier volumes of *Methods in Enzymology* (Volumes 114 and 115).

Single crystals (several tenths of a millimeter in all dimensions) are generally required for high-resolution data collection. There are three resolution limits to consider. For detecting whether ligand has bound, a resolution of 5 Å is adequate. This can be achieved with a diffractometer. To define the structural effects of ligation it is possible to interpret difference Fouriers at about 3 Å to indicate some of the grosser movements in the globin, but for a full analysis it is essential to collect data to at least 2.5-Å resolution, in order to be able to apply reciprocal space least-squares refinement. A modern area detector and rotating anode source should be adequate for this. For the highest resolution, between 2 and 1.5 Å, depending on crystal size and quality, a synchrotron radiation source is necessary.

Molecular replacement is necessary if the hemoglobin crystallizes in a new form or if substantial cell dimension changes occur. An excellent

[22] B. Luisi, Ph.D. Thesis, University of Cambridge.

program suite, distributed by P. Fitzgerald (MERLOT),[23] is well documented and features hemoglobin examples. If the quaternary state is uncertain then the $\alpha_1\beta_1$ dimer should be used as the search model. This method can work even if the dimer represents one-quarter of the asymmetric unit. The highest resolution structure, ideally from the same conditions (high or low salt) and in the expected quaternary structure, should be chosen as the search model.

Most tertiary changes are going to be subtle; therefore careful analysis of the data is essential. Early studies used difference Fouriers to study structural changes, but it is now clear that small rotations and translations often occur within the crystal as a consequence of ligation. These movements produce confusing peaks in the difference Fourier, and high-resolution refinement followed by superposition of the atomic models is the best solution to this problem. For refinement, the methods of Konnert and Hendrickson,[24] and of Jack and Levitt,[25] have been used extensively. The tertiary fold of human Hb has now been so well defined that new structures should be well within the radius of convergence of these methods. If refinement does not proceed smoothly using conventional methods, the Derewenda method can be tried[26]: refit the original high-resolution model of T- or R-state Hb, using a least-squares fit on the central part of the latest refined model, and reinitiate refinement; iterate if necessary. This amounts to an accurate rigid body refinement, and has allowed refinements from R values of 0.47 down to 0.20 without manual intervention. After refitting the reference model, look carefully at the Fo–Fc map, as this is probably the most unbiased map and will be most informative about structural changes outside the radius of convergence. The modern refinements featuring molecular dynamics (e.g., XPLOR[27]) should be employed only by experienced crystallographers. If the conventional refinement does not proceed smoothly, then the problem more likely lies with the diffraction data, or the molecule has not been properly positioned in the cell. In the latter case rigid body refinement or the Derewenda method should be applied, followed by a new round of conventional refinement.

It is valuable, if possible, to have an isomorphous series of deoxy and liganded molecules in the same crystal form, in order to separate the effects of crystal contacts from changes induced by ligation.[12] Similarly,

[23] P. Fitzgerald, *J. Appl. Crystallogr.* **21,** 273 (1988).

[24] J. Konnert and W. Hendrickson, *Acta Crystallogr., Sect. A* **A36,** 344 (1980).

[25] A. Jack and M. Levitt, *Acta Crystallogr., Sect. A* **A34,** 931 (1978).

[26] Z. Derewenda, "Molecular Replacement." Daresbury Study Weekend, Daresbury Laboratory, Warrington, England 1985.

[27] A. Brunger, J. Kuriyan, and M. Karplus, *Science* **235,** 458 (1987).

comparisons between different species, even if only a few amino acids are different, are dangerous.

The questions of ligand occupancy and metHb content should be carefully addressed. Whenever possible there should be an independent (spectroscopic) estimate of ligand binding, by dissolving crystals after data collection and recording visible spectra. Ligand occupancy can be refined using the program PHARE in the Daresbury program suite,[28] treating the ligand as a heavy atom in a phased refinement. At the resolution limits achievable with protein crystals it is not possible to distinguish between B value and occupancy. The dependence of occupancy (occ) on the imposed B value (B_{imp}) is linear, and is given by

$$Occ(B_{imp}) = Occ(B = 10) + 0.3(B_{imp} - 10)$$

In this circumstance the best one can do is to impose a chemically sensible B value on the ligand, for example, equal to the B value of the iron atom. If ligand occupancy is significantly less than one, then the only effect on refinement is to underestimate the structural changes; a mixed structure does not give rise to resolved electron density peaks for the liganded and unliganded structures when the overall structural changes are in the subangstrom range.[29]

Heme conformation must be carefully considered too. High-resolution refinements demonstrate a clear influence of ligand binding on heme conformation. At resolutions of around 2.5 Å it is difficult to determine with confidence the degree of heme folding or doming.[30] At any resolution, the results of restrained-refinement must be carefully examined. The Konnert–Hendrickson and Jack–Levitt refinement programs restrain individual pyrrole rings to be planar, and allow flexibility about the bridging carbon atoms; this is the variability Hoard[31] found when studying a range of five-coordinate small molecule porphyrin structures. It is useful to examine omit maps in which the heme atoms are excluded from the phasing, and also to inspect the first Fo–Fc map derived from molecular replacement. At resolutions around 2 Å the iron–sixth ligand bond needs to be restrained. The Fe–N$^{\varepsilon}$-2 (His) bond need not be strongly restrained, as the rest of the histidine ring should define the position of the nitrogen atom rather well. In several intermediate structures the Fe–His bond seems to be stretched,[32] so it is worth investigating the effects of different restraints on this bond length.

[28] CCP4 Crystallographic Suite, Daresbury Laboratory, Warrington, England (1993).
[29] Z. Derewenda, unpublished calculations (1986).
[30] G. Fermi, *J. Mol. Biol.* **97**, 235 (1975).
[31] J. Hoard, *Science* **174**, 1295 (1971).
[32] R. Liddington, Z. Derewenda, G. Dodson, and D. Harris, *Nature (London)* **331**, 725 (1988).

Analysis of Results

Key Structural Parameters

Key structural parameters include (1) position of the iron atom with respect to both the plane through the pyrrole nitrogens, and through the heme "core" (pyrrole rings, bridging carbons, and first atom of side chains); (2) distortions of the heme, including the difference between planes of nitrogens and core, and symmetry of the pyrrole tilts; (3) length of the Fe–His bond, and orientation of the histidine with respect to the heme; (4) condition of the salt bridges; (5) condition of the $\alpha_1\beta_2$ interface; and (6) ligand occupancy and metHb content.

Structural Comparisons

Structures should be compared by overlapping deoxy and liganded structures in some suitable reference frame. Baldwin and Chothia[33] chose the $\alpha_1\beta_1$ interface; a related but improved reference frame, in light of the high-resolution structures, is to use parts of the B, G, and H helices. The main chain atoms of the α-chain residues 20–36, 98–112, and 118–134 overlap with a root mean square (RMS) difference of 0.29 Å between human deoxy T-state and oxy R-state Hb, well within the error expected from individual coordinate errors. For the β chain, residues 22–33, 104–116, and 125–139 overlap with an RMS difference of 0.32 Å. These comparisons can also give a sensible estimate of the likely error in atomic coordinates. Structural changes in the unoptimized helical parts of the molecule can then be considered significant if they exceed twice the RMS difference in the optimized helices; the expected error will be higher in the exposed corners. The heme is another obvious reference frame in which to consider structural changes, although expected errors increase rapidly with distance from the heme; it is usual to exclude the side-chain atoms beyond the first in this overlap. A further reference frame favored by the author is the F helix, because there is a distinct conformational change at the end of the F helix between the T and R states that allows for heme unfolding in the α subunit.[12,33] It is important to try several of these reference frames to try to understand structural changes. For structures refined at a resolution of around 2 Å, typical errors in coordinates are 0.1–0.2 Å for the well-defined parts of the molecule, and 0.05–0.1 Å for the iron atom.

The significance of concerted structural changes should also be considered. Although movements of individual atoms may not be significant,

[33] J. Baldwin and C. Chothia, J. Mol. Biol. **129**, 175 (1979).

the expected error in the movement of groups of N restrained atoms is reduced by a factor of $0.6N^{1/2}$, so that for a group of 25 atoms, the expected error is reduced by a factor of 3.[33]

A novel method of analyzing structural change has been described by Kundrot and Richards,[34] using a "distance matrix" approach. This method does not depend on a choice of reference frame, and should effectively detect concerted changes; it has not yet been used for hemoglobin.

There are several commercially available computer graphics programs with which to analyze the structural results [Quanta (MSI; Burlington, MA); Insight (Biosym, Inc.; Parsippany, NJ); Sybil (Tripos Assoc.; St. Louis, MO)], which will run on Silicon Graphics (Mountain View, CA) or Evans and Sutherland (Salt Lake City, UT) workstations. A computer graphics program that allows rapid alternation or "flashing" between deoxy and liganded structures is an effective method of detecting concerted movements.

Future Studies

Spectroscopy vs Crystallography

Mozzarelli et al.,[1] from their spectroscopic observations, detect little difference in affinity between α- and β-hemes in PEG-grown Hb crystals. In contrast, Brzozowski et al.[7] showed by X-ray diffraction that under conditions near half-saturation of oxygen, it is the α-hemes that are predominantly ligated. Crystallographic studies of PEG crystals grown with IHP, DPG, and the drug bezafibrate also show a similar behavior, and Abraham et al.[15] have shown the same with high-salt crystals. To resolve this issue, experiments are planned by the author and G. Dodson (University of York, England) in collaboration with A. Mozzarelli (University of Parma, Italy) and W. Eaton (NIH, Bethesda, MD).

Asymmetric Hybrids

At least one of the liganded asymmetric hybrids (those lacking symmetry about the molecular dyad) has unexpected cooperative free energy,[35] and its structure would be of great interest. The four subunits in most of the crystal forms of hemoglobin have unique intermolecular contacts. However, if the existence or absence of ligand does not change the outward appearance of the subunit, then a useless averaged structure would most

[34] C. E. Kundrot and F. M. Richards, J. Mol. Biol. **193,** 157 (1987).
[35] G. Ackers and F. Smith, Annu. Rev. Biophys. Chem. **16,** 583 (1987).

likely result. Luisi[36] has suggested a possible solution: engineer a different side chain onto, say, the liganded subunit, that is known to be involved in a particular crystal contact, and that would be unfavorable in the alternative contact. In principle, this would allow the asymmetrically liganded molecule to take up a unique orientation in the crystal lattice, but it would require that crystal growth be fast compared with dimer exchange (the time required for crystal growth can probably be reduced to a few hours).

Time-Resolved Studies

The field of time-resolved studies of protein crystals is in its infancy.[37] Using white radiation from synchrotron sources it should be possible to collect data sets in under 1 sec from existing sources and perhaps well into the microsecond range in the next few years. What sort of reactions could be studied on this time scale? Most events of ligand binding are still too fast, although it is conceivable to imagine following geminate recombination reactions at very low temperature (80 K) following laser flash photolysis. These experiments will be technically demanding. Quaternary changes exist on this time scale, but they are likely to destroy the crystal lattice.

[36] B. Luisi, personal communication, 1993.
[37] K. Moffat, *Annu. Rev. Biophys. Chem.* **18,** 309 (1989).

[3] Structure and Energy Change in Hemoglobin by Hydrogen Exchange Labeling

By S. Walter Englander and Joan J. Englander

Present thinking about allosteric phenomena rests on two fundamental concepts. One concerns the pervasive role of structure change in the control of function.[1-3] The second relates structure change to ligand binding in quantitative free-energy terms.[4,5]

How are these ideas properly applied to working molecular machines such as hemoglobin? A focus on energy relationships suggests the follow-

[1] J. Wyman and D. W. Allen, *J. Polym. Sci.* **7,** 499 (1951).
[2] J. Wyman, *Adv. Protein Chem.* **19,** 224 (1964).
[3] J. Wyman, *Q. Rev. Biophys.* **1,** 35 (1968).
[4] J. Monod, J. Wyman, and J. P. Changeux, *J. Mol. Biol.* **12,** 88 (1965).
[5] D. E. Koshland, G. Nemethy, and D. Filmer, *Biochemistry* **5,** 365 (1966).

ing point of view. Hemoglobin binds its initial oxygen ligands with reduced binding energy. This energy is not lost, but is used to bring about structure changes, that is, to raise parts of the protein to a higher energy level. The loss in binding energy must be transduced quantitatively into structure change energy. Internal hemoglobin mechanisms then come into play to move this energy through the protein in the form of structure changes. The structure changes both contain the energy and provide a pathway for transporting it to distant parts of the protein, to sites as yet unliganded. In this way, remote sites are prepared to bind subsequent ligands with higher energy. This final stage might be viewed as transducing the carried energy back into ligand-binding energy or as releasing constraining interactions so that the binding sites can exhibit their normal affinity.

To understand allosteric processes at the molecular level properly, it is certainly necessary to catalog the various structure changes involved. All possible methodologies have been and are being applied to this goal. It will also be necessary to localize and measure quantitatively the role of each structure change in terms of energy content and the carriage of energy from one local change to another. No general method for localizing change and for evaluating each change in terms of structural free energy has been available. It now appears that this kind of information is contained in the hydrogen exchange (HX) behavior of protein molecules.[6] The local unfolding model for protein hydrogen exchange[7] connects the exchange rate with local structural free energy. Thus the measurement of changes in HX rate may locate allosterically important changes and delineate the handling of allosteric energy in quantitative free-energy terms. This chapter deals with a methodology designed to extract that information.

Hydrogen Exchange Method

Hemoglobin carries hundreds of main-chain peptide NH groups that are in continual exchange with the hydrogens of solvent water. The different hydrogens exchange over a range of rates roughly 10 orders of magnitude wide due to differences in local structural stability and dynamics. These sites provide definable probe points at which nonperturbing measurements of HX rate can reveal details of structure, structure change, structural energy, and structural dynamics not available by other means.

---

[6] S. W. Englander, *Ann. N.Y. Acad. Sci.* **244**, 10 (1975).
[7] S. W. Englander, J. J. Englander, R. E. McKinnie, G. A. Ackers, G. J. Turner, J. A. Westrick, and S. J. Gill, *Science* **256**, 1684 (1992).

The HX behavior that encodes this kind of information can be inscribed on a protein by use of HX labeling and trapping methods.[7-9] Here a protein is exposed to hydrogen isotope exchange while it is engaged in whatever functional behavior one wishes to study. For example, the protein may be bound in an equilibrium complex with a functional partner[10,11] or the protein may be studied in some kinetic process such as folding from the denatured to the native state.[9,12] The H isotope-labeling pattern imposed during the interaction being studied can then be trapped and read out subsequently under other conditions chosen to optimize the analysis. This approach is referred to as HX labeling.

For proteins small enough to be accessible to modern nuclear magnetic resonance (NMR) analysis, HX patterns can in many cases be measured directly at high resolution, that is, at an amino acid-resolved level, by NMR spectroscopy of a protein that has been subjected to hydrogen–deuterium (H–D) exchange labeling. Larger proteins such as hemoglobin appear inaccessible to NMR analysis at this time. However, it has been possible to develop approaches that can deal with large proteins. Here functionally interesting sites are labeled by use of hydrogen–tritium (H–T) exchange, the labeling pattern is trapped, and the distribution of tritium label is then measured at medium resolution, that is, at a segment-resolved level, by fragmentation and separation methods.

Kinetic Labeling

To place H isotope label selectively on just those sites one chooses to target, HX labeling takes advantage of intrinsic differences in HX rate that correlate with the structural effects being studied. The selectively labeled sites can then be identified, their HX behavior can be measured, and their response to experimental challenges can be studied.

A so-called "kinetic labeling" method was initially used in studies with myoglobin to label and measure the HX behavior of fast-exchanging, non-hydrogen-bonded NHs.[13] To selectively label the fast-exchanging sites, myoglobin was exposed to labeling in tritiated water for a short time period, 25 min, equal to ~12 HX half-times for free peptides under the conditions used. The free tritiated water was removed by a short gel-filtration run. Measurement of the subsequent exchange-out of the bound

[8] S. W. Englander and C. Mauel, *J. Biol. Chem.* **247**, 2387 (1972).
[9] S. W. Englander and L. Mayne, *Annu. Rev. Biophys. Biomol. Struct.* **21**, 243 (1992).
[10] Y. Patterson, S. W. Englander, and H. Roder, *Science* **249**, 755 (1990).
[11] L. Mayne, Y. Paterson, D. Cerasoli, and S. W. Englander, *Biochemistry* **31**, 10678 (1992).
[12] R. L. Baldwin, *Curr. Opin. Struct. Biol.* **3**, 84 (1993).
[13] S. W. Englander and R. Staley, *J. Mol. Biol.* **45**, 277 (1969).

tritium of myoglobin then produced an HX curve that portrayed the number and exchange behavior of the free NHs. The normally occurring background of more slowly exchanging H-bonded NHs was not labeled and did not appear in the HX curve. Kinetic labeling was later extended to study the behavior of different kinetic subgroups among the H-bonded peptide NHs of myoglobin, especially their response to solution conditions.[14] An impressive example of kinetic labeling involves the use of these same H–T exchange methods to follow protein-folding processes on a time scale of minutes[15] and the more recent adaptation of these methods to H–D exchange labeling with NMR analysis to follow folding-dependent hydrogen bond formation on a millisecond time scale.[9,12]

Functional Labeling

By focusing on sites that *change* their exchange rates during any functional interaction, HX labeling can be used to distinguish and study the parts of a protein that are involved in the interaction. To identify and study the functionally interesting parts of a relatively small protein, one can use H–D exchange with NMR analysis to measure the relatively complete HX behavior of the protein in its different functional forms. We have used this approach to detect the parts of cytochrome *c* that are affected when its oxidation–reduction (redox) state is changed, and when it complexes with a protein partner.[10,11] Hydrogens that change their exchange rates and thus mark involved sites can then be directly identified. The HX behavior of larger proteins such as hemoglobin cannot yet be followed in such complete detail. Here a more complex "functional labeling" approach using tritium exchange has been devised in order to label selectively just those sites that change so that these interesting sites can, in this sense, be isolated for study.

Functional labeling uses a sequence of steps involving initial tritium labeling in the fast-exchanging form, then change in functional state, and subsequently an exchange-out chase in the slow-exchanging form. These operations selectively label just those sites that change their HX rate when the functional state of the protein is changed. Experiments so far reported continue to use tritium labeling, gel-filtration separations, and measurement by liquid scintillation counting. We describe the functional labeling approach after some discussion of the chemical and technological underpinnings of tritium-labeling experimentation.

[14] D. B. Calhoun and S. W. Englander, *Biochemistry* 24, 2095 (1985).
[15] P. S. Kim and R. L. Baldwin, *Biochemistry* 19, 6124 (1980).

Tritium Handling

An active laboratory project using tritium exchange can consume several curies of tritiated water in 1 year. Nevertheless, simple handling procedures can negate any biological risk. Tritium decay produces low-energy β rays, with a maximum path of only 6 μm through water, glass, or tissue and 6 mm through air. Thus direct radiation is not a danger. To avoid ingestion and skin contact, one should use vinyl gloves, laboratory coats, manual rather than oral pipetting, and proper storage and disposal practices. We obtain stock tritiated water packaged in 0.5-Ci amounts (1 Ci/ml) from commercial sources. Each 0.5-Ci package is opened as needed, transferred into a 1-ml free-standing, glass-stoppered flask, which is kept in a padded Mason jar, and stored frozen in the separated freezer compartment of a laboratory refrigerator. For use the stock tritiated water is placed in an efficient laboratory fume hood and allowed to thaw. A small working volume (microliter amounts; 1 μl $\approx$ 1 mCi) is taken by microsyringe into a short length of disposable polyethylene tubing (PE-90) press fitted onto a 20-gauge needle and then expelled and mixed into the experimental solution. Subsequent operations with the much lower level experimental solutions can be done on a laboratory bench with only normal precautions.

Gel Filtration

To rapidly remove free tritiated solvent from tritium-labeled hemoglobin samples after a timed exchange-in period or to move samples from one solvent to another, for example, in trapping the HX-labeling pattern (see Functional Labeling, below), one uses a short gel-filtration passage. For this purpose preswollen Sephadex G-25 gel, treated to remove fines (by settling in a large beaker) and trapped air (by vacuum), is packed in a column 1 cm in diameter by 3 to 8 cm in length. A 6-cm gel column can separate 0.3 ml of hemoglobin sample from tritiated water by a factor of ~10^6 in a 1-min passage. Commercially available glass columns (Fisher Scientific, Pittsburgh, PA) are inexpensive and convenient. Prepared columns can be used, washed, and reused almost indefinitely. Because the HX rate has a large temperature coefficient, temperature control of samples during timed exchange-in and exchange-out is important and can be simply achieved by incubating closed test tube samples in a water bath. We standardly use an experimental temperature of 0° at which HX is slow, perform experiments in a refrigerated cabinet, and maintain exchanging samples in an iced beaker. More discussion on column separation procedures and sample handling can be found in Englander and Englander.[16,17]

Hydrogen Exchange Chemistry

Peptide NH hydrogen exchange is catalyzed by H^+ ion below about pH 3 and by OH^- ion at higher pH. Thus in the normal physiological range of pH, the exchange rate can be expected to change approximately ~10-fold per pH unit, although hemoglobin, like most proteins, responds somewhat less,[18] presumably due to the counteracting change in protein charge.[19] A temperature change of 20° changes rates by another 10-fold. The manipulation of pH and temperature allows the HX rate to be set over a wide range. For example, the change from pH 5 and 0° to pH 9 and 40° changes the intrinsic chemical exchange rate by 1 millionfold. In laboratory experiments, exchange times between 0.5 min and several days can be easily handled, adding another factor of $~10^4$ to the measurable HX rate range.

These conditions are fortunately matched to cover the normally encountered range of $~10^{10}$ in protein HX rates. At pH 5 and 0°, peptide NHs exchange with a half-time of ~100 sec when freely exposed and are much slower when involved in hydrogen-bonded structure. Thus the exchange of structurally involved NHs with HX rates that span a wide dynamic range can be brought into a convenient laboratory time window by changing pH and temperature within limits that proteins tolerate easily.

The minimum HX rate of ~1 hr for freely exposed peptide NHs, reached at ~pH 3 and 0°, is well adapted for the trapping of protein HX patterns, and also for subsequent analysis by proteolysis and high-performance liquid chromatography (HPLC) procedures (see below). This same low-pH condition is widely used to optimize peptide HPLC separations, although the low temperature most suitable for HX-labeling analysis is unusual for peptide HPLC.

More detailed information on peptide NH hydrogen exchange behavior can be found in Englander et al.[16–20]

[16] S. W. Englander and J. J. Englander, this series, Vol. 26C, p. 406.
[17] S. W. Englander and J. J. Englander, this series, Vol. 49, p. 24.
[18] J. J. Englander and S. W. Englander, *Biochemistry* **26**, 1846 (1987).
[19] S. W. Englander and N. R. Kallenbach, *Q. Rev. Biophys.* **16**, 52 (1984).
[20] S. W. Englander, N. W. Downer, and H. Teitelbaum, *Annu. Rev. Biochem.* **41**, 903 (1972).

Functional Labeling

Figure 1 illustrates the functional labeling approach. In these experiments,[21-24] oxyhemoglobin is exposed to exchange-in labeling in tritiated water for 1 min (pH 7.4, 0°). Sites that exchange during this time period, both allosterically sensitive and insensitive sites, are tritiated. The protein is then switched to the deoxy form by removal of O_2 (minimal dithionite for a few seconds), and passed through a short, deoxygenated Sephadex column. The gel filtration removes the tritiated water (and dithionite). Bound tritium starts to exchange out. Tritium label on sites that are insensitive to the allosteric form exchange at the same rate in both oxy- and deoxyhemoglobin. Therefore, after an exchange-out time somewhat longer than 1 min, label on the insensitive sites is largely removed. However, tritium on sites that become much slower in deoxyHb is now locked in. Therefore, after some exchange-out time in the deoxy state longer than the exchange-in period, one has a sample with tritium label selectively placed on just those sites that are allosterically sensitive, that is, on a subpopulation of allosterically sensitive sites that are fast enough in oxyHb to become labeled in 1 min and that switch to slower exchange in deoxyHb.

The exchange behavior of these self-selectively labeled NHs can then be studied in deoxyHb by straightforward tritium exchange methods. Here samples are taken after increasing times of exchange-out in the deoxy form, free tritium exchanged into the solvent during the experimental exchange-out time is removed by a second gel-filtration run, the eluant sample is analyzed for protein concentration and carried tritium, and these data are computed in terms of number of peptide NHs per Hb molecule (or per protomeric $\alpha\beta$ dimer). A series of such points in time trace out the exchange curve of the selectively labeled allosterically sensitive sites in deoxyHb. The upper curve in Fig. 1A shows results obtained in this way.

Variously modified hemoglobins can be studied in the same way to determine, for example, the effect of chemical and mutational alterations at known positions on the exchange of these particular allosterically sensitive sites. Hydrogen exchange behavior of labeled sites can be studied as just described in deoxyHb. The same sites can be measured in liganded Hb by adding O_2 or CO back to the selectively labeled deoxyHb sample and following its sharply accelerated tritium exchange-out behavior.

[21] R. K. H. Liem, D. B. Calhoun, J. J. Englander, and S. W. Englander, *J. Biol. Chem.* **255**, 10687 (1980).

[22] G. Louie, T. Tran, J. J. Englander, and S. W. Englander, *J. Mol. Biol.* **201**, 755 (1988).

[23] G. Louie, J. J. Englander, and S. W. Englander, *J. Mol. Biol.* **201**, 765 (1988).

[24] J. J. Englander, J. R. Rogero, and S. W. Englander, *Anal. Biochem.* **147**, 234 (1985).

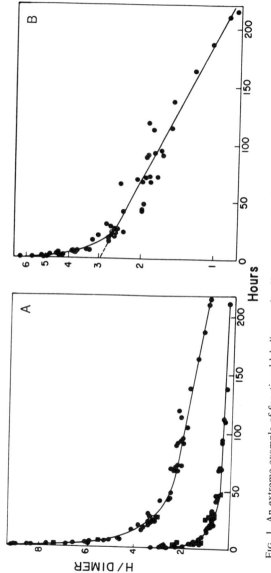

FIG. 1. An extreme example of functional labeling results. Hemoglobin in the fast-exchanging oxy form was labeled for 1 min in tritiated water, then deoxygenated and exchanged out. Sites that exchange in 1 min in oxyHb are seen to exchange out over hours and days in the deoxy form (A, upper curve). The lower curve in (A) is a background curve obtained by initially labeling in the deoxy form, then exchanging out as oxyHb. Subtraction of the background curve from the curve above it yields the curve for allosterically sensitive NHs (B). The faster phase was located at the β-chain COOH terminus, shown in Fig. 2. The very slow allosterically sensitive NHs have not yet been identified.

Experiments done in this way yield HX curves that are in some measure contaminated by allosterically insensitive NHs, which cannot be wholly removed by kinetic selection. For example, the upper curve in Fig. 1A is greatly enriched in allosterically sensitive NHs but also contains some insensitive NHs. The contribution of contaminating background NHs to the measured HX data can be defined by doing the selective labeling in the reverse direction. The initial exchange-in is performed in deoxyHb, in which the sensitive NHs are in their slow form and tend not to become labeled. However, the allosterically insensitive NHs, unaffected by the allosteric form, are labeled precisely as before. Oxygen or CO is then added to switch Hb to the liganded form, the sample is immediately passed through a gel-filtration column, and an exchange-out curve is measured. The data obtained in this way exhibit a background curve that contains just the same insensitive NHs as before, but in this case the sensitive NHs have been removed. The lower curve in Fig. 1A shows the background curve for the 1-min exchange-in experiment.

An improved approximation to the pure behavior of the sensitive NHs that were labeled can then be obtained by subtracting the bottom, background HX curve from the upper curve containing both sensitive and insensitive NHs (Fig. 1B). This is most useful for obtaining meaningful data during the early exchange-out period, for example in the first 5 or 10 min when exchange-in was 1 min.

The exchange-out curve improves as the difference between HX rate in the oxy and deoxy forms increases. Conversely, as the HX rates of a given set of NHs in oxy and deoxy forms approach each other, the ability to distinguish them kinetically decreases. Fortunately, according to the local unfolding model (see below), rates that change little represent segments of the protein that experience little change in structural free energy, and therefore appear to be relatively unimportant in the allosteric mechanism. The parts of the protein that experience large rate changes, that is, that participate most significantly in carrying allosteric energy, are the easiest to define.

Functional labeling methods in the low-resolution mode just described have been used to catalog the HX behavior of all the allosterically sensitive NHs in hemoglobin.[25] To cover most of the oxyHb HX curve, exchange-in times used in different experiments ranged from 1 min at pH 7.4, 0° to 2 days at 37°, pH 9. Exchange-out times were adjusted accordingly. Results show that about one-quarter of the peptide NHs in hemoglobin are sensitive to its allosteric form. These always exchange more rapidly in the oxy form. Thus selective labeling experiments start by exchange-in with

[25] E. L. Malin and S. W. Englander, *J. Biol. Chem.* **255**, 10695 (1980).

oxyHb, then switch to the slow-exchanging deoxyHb form. Rate ratios (oxy rate/deoxy rate) found for the different sensitive NHs range from about 15 to 10^4. The sensitive NHs detected in these experiments appeared to occur in kinetically distinct groupings. The NHs in each set exhibit roughly similar rates in the oxy form, and all move more or less in unison to a new rate, slower by some common factor, in the deoxy form. This behavior was interpreted in terms of sets of NHs with HX rate determined by cooperative local unfolding reactions.[6]

These general parameters are interesting, even provocative, but unfortunately incomplete. The experiments define the behavior of functionally sensitive NHs in any allosteric or otherwise modified protein form only at low resolution. The results are silent concerning the specific structural sites that carry the tritium label and the intimate details of allosteric mechanism that seem to be hidden in the HX patterns and rates.

Locating Allosterically Sensitive Segments

In principle, one might locate the tritium-labeled sites by proteolyzing the protein, then separating the fragments, and analyzing them for carried tritium. The exchangeable tritium need not be lost in this process because the HX half-time for even freely exposed peptide NHs can be longer than 1 hr at pH ~3 and 0°. Under these conditions, the approach can be made to succeed.

A fragment separation analysis was first reported by Rosa and Richards[26] and then by others[27–32] and is described in detail in Englander *et al.*[24,33] Results of a fragment separation analysis that locates the faster exchanging sites seen in Fig. 1B are shown in Fig. 2. These data place a set of four allosterically sensitive NHs on a COOH-terminal fragment of the hemoglobin β chain.

The experiments in Fig. 2 started with an oxyHb sample selectively labeled for 35 min. The longer exchange-in time is used in order to locate and study, in the same experiment, sites that are labeled in the shorter 1-min period and also a set of sensitive sites that exchange in oxyHb with a half-time of 15 min and are found on the α chain N terminus.[22,23,32]

[26] J. J. Rosa and F. M. Richards, *J. Mol. Biol.* **133**, 399 (1979).
[27] S. W. Englander, D. B. Calhoun, J. J. Englander, N. R. Kallenbach, R. K. H. Liem, E. L. Malin, and J. R. Rogero, *Biophys. J.* **32**, 577 (1980).
[28] A. M. Beasty and C. R. Matthews, *Biochemistry* **24**, 3547 (1985).
[29] D. S. Mallikarachichi, D. S. Burz, and N. M. Allewell, *Biochemistry* **28**, 5386 (1989).
[30] S. M. Kaminsky and R. M. Richards, *Protein Sci.* **1**, 10 (1992).
[31] J. J. Englander, J. R. Rogero, and S. W. Englander, *J. Mol. Biol.* **169**, 325 (1983).
[32] J. Ray and S. W. Englander, *Biochemistry* **25**, 3000 (1986).
[33] J. R. Rogero, J. J. Englander, and S. W. Englander, this series, Vol. 131L, p. 508.

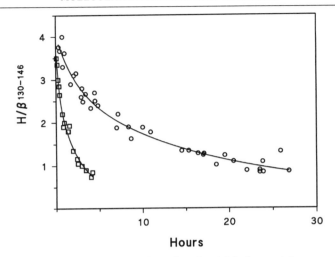

FIG. 2. An example of HX results from functional labeling and fragment separation experiments. OxyHb was labeled for 25 min, then exchanged out as deoxyHb. After the exchange-out times indicated, samples were subjected to the fragment separation analysis. The results (○) reveal four allosterically sensitive NHs on the fragment β130–146, exchanging in deoxyHb on an ~5-hr time scale. Subfragmentation experiments place these NHs between residues 140 and 146, at the very COOH terminus. Similar experiments with NES–Hb (□), in which two salt links at the β-chain COOH terminus have been cleaved, show that the allosterically sensitive NHs move in unison to a new rate faster by eightfold. We conclude that the exchange of this set of NHs is controlled by a cooperative transient unfolding reaction at the β-chain COOH terminus that cleaves the protecting H bonds. When salt links stabilizing this segment are prebroken in the NES–deoxyHb derivative, the equilibrium constant for unfolding is enhanced by eightfold, corresponding to 1.1 kcal/mol.

In these experiments, oxyHb is exposed to tritium labeling for 35 min, then the sample is deoxygenated, gel filtered to remove solvent tritium, and allowed to exchange out in the deoxyHb form. To locate the tritium still bound after various exchange-out times, a sample is plunged into slow HX conditions, namely pH ~2.7 at 0°, and briefly fragmented using an acid protease (pepsin). The proteolyzed sample is resolved by a fast HPLC separation and the eluant peaks are collected and analyzed for carried tritium. The tritium level in each fraction is determined by liquid scintillation counting. The concentration of each fragment collected is read directly from the area of the HPLC trace, using an absorbance coefficient previously determined for each peptide fragment on the basis of quantitative amino acid analysis.

Samples taken in time during exchange out of a selectively labeled sample and analyzed in this way provide a hydrogen exchange curve for the labeled, allosterically sensitive set of NHs in any well-resolved fragment. As in the low-resolution method described above, this can be done for Hb in either allosteric form or for mutationally or chemically modified hemoglobins. Figure 2 shows results for unmodified Hb A (upper curve) and for NES–hemoglobin (lower curve) in which the reactive sulfhydryl at Cys-93 in the β chain was reacted with N-ethylmaleimide so that two bonds at the β-chain COOH terminus are broken, namely a salt link that bridges the $\alpha_1\beta_2$ interface and an internal β-subunit H bond.[34]

As before, a background curve can be generated and used to remove possible contributions due to allosterically insensitive NHs, but the need for this is much reduced because only the NHs in the fragment of interest can now contribute.

The quality of the data obtained in fragment separation experiments depends on the yield of the fragments of interest and on the quality of the HPLC separations obtained. The yield of any fragment is always well below 100%. This is because acid proteases are notoriously unspecific. Thus the population of a given fragment in the hydrolysate will rise in time for a while and then decrease again as further breaks decompose it into a heterogeneous collection of smaller pieces. In the best cases we have obtained yields in the range of 15 to 30%. Some positions of a protein are more accessible to the fragment separation analysis than others. NHs near to a chain terminus tend to be found on terminal fragments that appear early in the cleavage process and in relatively high yield because only a single break is required to produce a terminal fragment. Sites that are more internal appear on fragments that require two breaks. These are produced with a kinetic lag and are split into a heterogeneous collection of fragments by further breaks. In the case of hemoglobin, it has turned out that important allosteric behavior happens to involve chain termini. The generality of this result remains to be seen.

Experience shows that it is important to keep tritium losses during the fragment separation analysis below about 50%, or else results can become seriously misleading. Thus each of the three steps in the procedure should be carefully optimized to reduce exchange losses. All operations are performed at 0° and at pH 2.5 to 3 to minimize the HX loss rate. The selectively labeled hemoglobin sample being processed can be quickly dropped to low pH by directly adding a predetermined volume of acid phosphate and/or by passing the sample through a short (3-cm) Sephadex column with adequate low-pH buffering (0.1 M phosphate). The proteolysis step initiated by added pepsin requires 5 min. Although pepsin and other acid

---

[34] M. F. Perutz, H. Muirhead, L. Mazzarella, R. A. Crowther, J. Greer, and J. V. Kilmarten, *Nature (London)* **222**, 1240 (1969).

proteases in general are relatively nonspecific, one takes advantage of a kinetic specificity—some bonds are broken much more rapidly than others—to reduce the diversity and increase the yield of fragments produced. The proteolyzed sample can then be injected into the HPLC column and resolved. Separation time and quality can be optimized by using a solvent gradient to elute the predetermined peptide of interest in perhaps 15 min. The presence of nonaqueous solvent (acetonitrile) during the HPLC separation helps to reduce HX rates.[24,31,33]

In our experience peptides can be separated and captured with a loss of only 30% of the carried tritium label. This can be corrected by determining a loss curve[22,24] and extrapolating it to zero time to obtain a loss correction factor. Figure 3 shows a loss curve obtained for the allosterically sensitive set of NHs at the Hb β-chain COOH terminus.

The effectiveness of the overall fragment separation technology can be tested by processing hemoglobin that has initially been fully tritium labeled by exchange-in at elevated pH and temperature, for example, at pH 9 and 40° for 2 days. The value of H/fragment recovered for each

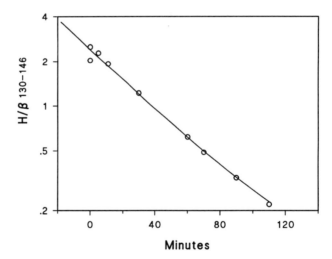

Minutes

FIG. 3. The loss curve for β130–146. Hemoglobin was selectively labeled as in Fig. 2. Samples were quenched at pH 2.7, 0° as before, held for increasing times to allow tritium loss to continue at its normal loss rate, then proteolyzed, subjected to the HPLC separation, and analyzed for remaining carried tritium. The curve shows the loss rate. Back extrapolation to the time at which the sample was taken from the exchanging native deoxyHb (shown as −20 min) indicates the loss correction factor to be used in the analysis of fragment separation data for this peptide. The loss factor is taken as the ratio of the tritium level in the native protein to the level obtained in the HPLC analysis with no excess waiting time (zero time). The loss factor indicated for this fragment is 1.5.

fragment should be close to the known number of amino acids in the fragment minus one. An example can be seen in Fig. 1 of Englander *et al.*[24]

Toward Higher Resolution

The fragment isolation method used as just described specifies the location of sets of allosterically sensitive NHs only to within the sequence represented by the peptide fragment on which it is found. Somewhat better resolution can be obtained by further subfragmentation. One may consider the use of different acid proteases, which might present a different but overlapping library of fragments. An alternative approach is to isolate a given fragment by HPLC as just described, then subfragment it with another protease, reisolate the subfragments, and determine the carried tritium on each. This has been done[22,32] to specify more closely the positions of two sets of allosterically sensitive NHs. A set of five NHs initially isolated on a high-yield α-chain N-terminal fragment, $\alpha1$–29, could be localized to $\alpha1$–12, when the pepsin fragment was subfragmented with another protease. The set of four NHs found on $\beta130$–146 (Fig. 2) was localized to $\beta140$–146. The method suffers, of course, from increased loss of label during the double-separation process and thus becomes only semiquantitative.

Local Unfolding Model

The local unfolding model for protein hydrogen exchange suggests that slow exchange generally reflects H bonding, that exchange requires transient H-bond breakage, and that H bonds in units of regular secondary structure tend to break in a cooperative way. According to the statistical mechanical theory of helix–coil transitions,[35,36] cooperative unfolding will appear as fraying behavior at the ends of helices and as a concerted unfolding of sets of internal NHs (referred to as *unfoldons*). Experience with the hydrogen exchange of peptides[35] and proteins[7,19] now indicates the essential correctness of this view.[7,19,37] When expressed quantitatively, the unfolding model indicates that the HX rate of cooperatively exchanging NH sets can be connected to an equilibrium constant for transient unfolding of a region of structure.[6,19] For example, if a concerted set of NHs in a turn or two of helical structure exchanges more slowly than the free

[35] B. H. Zimm and J. K. Bragg, *J. Chem. Phys.* **31**, 526 (1959).
[36] S. Lifson and A. Roig, *J. Chem. Phys.* **34**, 1963 (1961).
[37] C. A. Rohl, J. M. Scholtz, E. J. York, J. M. Stewart, and R. L. Baldwin, *Biochemistry* **31**, 1263 (1992).

peptide rate by 10^4-fold, one can conclude that the NHs are exposed to exchange by a transient structural opening that exists 10^{-4} of the time. That is, the equilibrium constant, K_{op}, for this local unfolding reaction is 10^{-4}. The free energy, $\Delta G°$, for the opening reaction can then be evaluated as $\Delta G° = -RT \ln K_{op} = -RT \ln (10^{-4}) = 5.6$ kcal at room temperature. It is promising that change in HX rate varies exponentially with change in structural stabilization energy. Thus measured changes in HX rate can be expected to sensitively mark segments that actively participate in the allosteric process.

The capability for localizing and measuring changes in structural energy is essential for understanding allosteric mechanism because, as suggested before, the mechanism is based fundamentally on ligand-induced changes in structural stabilization free energy. The local unfolding model indicates that if a segment of structure is destabilized in an allosteric event, for example by the ligand-induced breakage of a stabilizing salt link, then transient unfolding reactions that normally break that salt link in a hydrogen exchange event will be promoted by an amount determined by the loss in the salt link stabilization free energy. For example, the removal of two salt links at the hemoglobin β-chain COOH terminus in the NES modification studied in Fig. 2 causes a set of four adjacent NHs to exchange faster by eightfold, corresponding to a loss of 1.1 kcal of stabilization at this position.[7]

Thus, from HX results of this kind one can hope to read out, at a structurally resolved level, the effects in real free-energy terms of ligand binding, allosteric transition, mutations, defined chemical modifications, Bohr effect titrations, and the like. (The quantitative free-energy analysis for stability change assumes that the open state is effectively unchanged in the transition being studied.) With these approaches, the structurally resolved free-energy changes resulting from a number of allosterically effective modifications in hemoglobin have been measured and compared with the changes in cooperative free energy measured for the same samples by global methods, namely by oxygen-binding and subunit dissociation experiments.[7] Good agreement was found, indicating the essential correctness of these approaches. More discussion on these issues can be found in Englander *et al.*[7,19,22,38]

Some Problems

It is an oversimplification to suppose that the NHs in a cooperative set should all exchange at the same rate. At the level of HX chemistry,

[38] R. E. McKinnie, J. J. Englander, and S. W. Englander, *J. Chem. Phys.* **158**, 283 (1991).

one now knows that even fully exposed peptide NHs can exchange at somewhat different rates because different side chains impose somewhat different inductive and steric blocking effects on their nearest neighbor peptide NHs.[39,40] In addition, the amino acids that participate in a cooperative unfolding reaction may experience different degrees of exposure and therefore somewhat different HX rates in the open state (work in progress). These chemical and physical factors together operate to produce some dispersion of HX rates among the residues that share a common unfoldon.

Because of this rate dispersion, functional labeling protocols do not work as efficiently as they otherwise might. In attempting to label the NHs in a cooperative unit, one must lengthen the exchange-in time in order to pick up the stragglers or else accept their loss. A lengthened exchange-in time produces more background labeling of insensitive NHs. The natural dispersion in HX rate also works to reduce the difference, for a given unfoldon, between the two functional states being tested (e.g., oxy- and deoxyHb). It is the size of this difference that governs the ability to achieve selective labeling. On the positive side, small free-energy changes produce large HX rate changes, and results so far obtained show that, at least for some available well-characterized examples,[22,31,32] allosterically sensitive sets of NHs can be isolated and located, and their relative exchange rates can be measured with good accuracy.

Summation

Hydrogen exchange-labeling methods can in principle show which parts of hemoglobin are actively involved in the allosteric process and which are not. The approach derives from the local unfolding model. The model leads one to expect that local bonding changes of the kind thought to underly the allosteric transition[41-43] will produce sizeable HX rate changes, and that these changes will show up among concerted, structurally related sets of exchanging peptide NHs. The present state of a selective labeling methodology designed to locate and quantify such changes was outlined here. Results available indicate that only a fraction of the hemoglobin molecule is actively involved in allosteric function and point to some of the loci of activity. The methods described have made it possible to locate some of the important interactions and measure their

---

[39] R. S. Molday, S. W. Englander, and R. G. Kallen, *Biochemistry* **11**, 150 (1972).
[40] Y. Bai, J. S. Milne, L. Mayne, and S. W. Englander, *Proteins: Struct. Funct. Genet.* **17**, 75 (1993).
[41] M. F. Perutz, *Nature (London)* **228**, 726 (1970).
[42] M. F. Perutz, *Q. Rev. Biophys.* **22**, 130 (1989).
[43] J. Baldwin and C. Chothia, *J. Mol. Biol.* **129**, 17 (1979).

free-energy contribution to the overall allosteric transition. Results have demonstrated a quantitative relationship between structural free energy measured locally by these methods and globally by other established methods that involve the analysis of ligand-binding curves and the measurement of subunit dissociation equilibria.[7] These results appear to validate the overall hydrogen exchange approach and the specific methodology described here.

[4] High-Pressure Fluorescence Methods for Observing Subunit Dissociation in Hemoglobin

By SERGE PIN and CATHERINE A. ROYER

In this chapter, we present detailed experimental procedures used to investigate subunit interactions in human oxyhemoglobin by fluorescence and high-pressure techniques.[1] With a covalent extrinsic fluorescence probe having a long lifetime, the measurements of fluorescence emission depolarization due to the Brownian tumbling of the protein allow the direct observation of oligomer size variations on dissociation. Because the dissociation constant is in the micromolar range, the study of tetramer dissociation can be made by fluorescence measurements as a function of the protein concentration. The application of high pressure is used to dissociate the dimer. Despite the high affinity between the monomers (dissociation constant in the nanomolar range), the high sensitivity of fluorescence and the destabilizing effect of pressure on the interactions between subunits permit the observation of complete dimer dissociation and the determination of the dissociation constant.

The time scale and sensitivity of fluorescence spectroscopy make this a powerful technique for the study of equilibrium interactions between biological molecules. Quenching, energy transfer, solvent relaxation, Brownian tumbling, and segmental motions of biopolymers generally occur in the fluorescence time scale (10^{-12}–10^{-7} sec) and thus can be used to monitor complex formation between protein subunits, proteins and ligands, proteins and nucleic acids, and antibodies and antigens in solution. Depending on the probe employed, these measurements can be made at concentrations reaching the subnanomolar range, such that equilibrium binding titrations often can be performed.

[1] S. Pin, C. A. Royer, E. Gratton, B. Alpert, and G. Weber, *Biochemistry* **29**, 9194 (1990).

In spectroscopic studies, pressure has been used widely as a means of physical perturbation of biological molecules.[2-6] In particular, to monitor extremely high-affinity interactions under reasonable conditions of fluorophore concentration, fluorescence spectroscopic techniques may be coupled with high hydrostatic pressure. It has been demonstrated that complexes between subunits are destabilized by the application of pressure.[5,7] This destabilization is based on Le Chatelier's principle that the application of pressure to a system in equilibrium will shift the equilibrium to the side that occupies the least volume. Because of exposure of subunit interfaces to the solvent, the dissociated form of an oligomeric protein system occupies less volume than the associated oligomer. This reversible dissociation usually occurs in a pressure range below 3 kbar and is generally not accompanied by protein denaturation.[5,7]

In a general sense, the dependence of the fluorescence polarization on protein concentration constitutes a sensitive experimental confirmation for oligomerization equilibria. However, as we shall demonstrate, it is the basic underlying assumption of the homogeneity of the population that allows extraction of a dissociation constant from a dilution/polarization profile. In the case of hemoglobin, the particular relationship between quenching of the fluorophore by the heme and protein conformation results in a fluorescence depolarization signal that reports both on the state of oligomerization and on the intrinsic heterogeneity of the preparation. Heterogeneity in intrinsic fluorescence lifetimes of hemoglobin samples, which could be diminished by chromatographic procedures, has been reported.[8-10]

In oligomeric proteins, multiple equilibria are superimposed and energetically coupled.[11-15] Ligation effects are modulated through free-energy

[2] R. Jaenicke, *Annu. Rev. Biophys. Bioeng.* **10**, 1 (1981).

[3] E. Morild, *Adv. Protein Chem.* **34**, 93 (1981).

[4] K. A. H. Heremans, *Annu. Rev. Biophys. Bioeng.* **11**, 1 (1982).

[5] G. Weber and H. G. Drickamer, *Q. Rev. Biophys.* **16**, 89 (1983).

[6] P. T. T. Wong, *Annu. Rev. Biophys. Bioeng.* **13**, 1 (1984).

[7] G. Weber, *in* "High Pressure Chemistry and Biochemistry" (R. van Eldik and J. Jonas, eds.), p. 401. Reidel Publ., Dordrecht, The Netherlands, 1987.

[8] E. Bucci, H. Malak, C. Fronticelli, I. Gryczynski, and J. R. Lakowicz, *J. Biol. Chem.* **263**, 6972 (1988).

[9] E. Bucci, H. Malak, C. Fronticelli, I. Gryczynski, G. Laczko, and J. R. Lakowicz, *Biophys. Chem.* **32**, 187 (1988).

[10] A. G. Szabo, K. J. Willis, D. T. Krajcarski, and B. Alpert, *Chem. Phys. Lett.* **163**, 565 (1989).

[11] J. Wyman, *Adv. Protein Chem.* **4**, 407 (1948).

[12] J. Wyman, *Adv. Protein Chem.* **19**, 223 (1964).

[13] G. Weber, *Biochemistry* **11**, 864 (1972).

couplings of ligand-binding and subunit interactions. Changes in the ligand affinity are correlated to changes in the subunit affinity. Thus, the influence of allosteric effectors (protons[16] and inositol hexaphosphate or IHP[17]) on the oxyhemoglobin subunit interactions has been investigated. The experimental conditions selected for these fluorescence measurements are pH 7 and pH 9, and with IHP, pH 7.

Labeling Reactions

Adult human oxyhemoglobin can be purified from fresh blood by the method of Perutz,[18] and 2,3-diphosphoglycerate can be eliminated by the method of Jelkmann and Bauer.[19] The oxyhemoglobin sample, prepared following the standard procedures (Hb sample), is then either labeled with the fluorescent probe for fluorescent measurements or fractionated, using modified methods.[8-10] The Hb sample is applied on a fast protein liquid chromatography (FPLC) DE-52 Whatman (Clifton, NJ) anion-exchange column, and oxyhemoglobin fractions are eluted with a discontinuous gradient of ionic strength.[1] Figure 1 displays a typical elution profile. Four elution peaks are visible and the fourth contains two different components.

The number and the reproducibility of oxyhemoglobin components eluted by FPLC have excluded the possibility that they correspond uniquely to the A_0, A_1, and A_2 hemoglobins or to abnormal hemoglobins. Rather, the heterogeneity in the oxyhemoglobin elution profile appears to be subject to different conformational states of molecules. The major oxyhemoglobin component that elutes from this column, designated Hb_{III}, corresponds to the central 30% of the volume of the third peak in the chromatogram (hatch marks in Fig. 1). The Hb_{III} sample is concentrated under vacuum to carry out the labeling reaction with the same Hb and Hb_{III} concentrations.

The Hb and Hb_{III} samples are labeled with dansyl chloride (5-dimethylaminonaphthalene-1-sulfonyl chloride, DNS) from Molecular Probes, Inc. (Eugene, OR). The fluorescence lifetime of dansyl is sufficiently long for the observation of the global tumbling of proteins.[20] This is a covalent

[14] G. K. Ackers, M. L. Johnson, F. C. Mills, and S. H. C. Ip, *Biochem. Biophys. Res. Commun.* **69**, 135 (1976).
[15] C. A. Royer, J. M. Beechem, and W. R. Smith, *Anal. Biochem.* **191**, 287 (1990).
[16] C. Bohr, K. A. Hasselbalch, and A. Krogh, *Skand. Arch. Physiol.* **16**, 402 (1904).
[17] R. Benesch, R. E. Benesch, and C. I. Yu, *Proc. Natl. Acad. Sci. U.S.A.* **59**, 526 (1968).
[18] M. F. Perutz, *J. Cryst. Growth* **2**, 54 (1968).
[19] W. Jelkmann and C. Bauer, *Anal. Biochem.* **75**, 382 (1976).
[20] G. Weber, *Biochem. J.* **51**, 155 (1952).

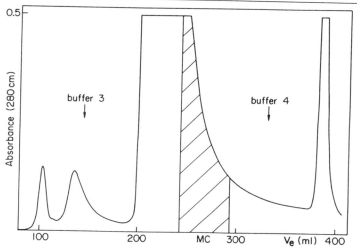

FIG. 1. Elution profile of human oxyhemoglobin on an FPLC DE-52 Whatman anion-exchange column. Buffers 3 and 4 correspond to 20 mM Tris-HCl at pH 8.4 with 60 and 200 mM sodium acetate, respectively. The fourth peak was found to contain two different oxyhemoglobin fractions. The hatched area of the third peak corresponds to the Hb_{III} sample. V_e, Elution volume. [Reprinted with permission from S. Pin *et al.*, *Biochemistry* **29**, 9194 (1990). Copyright 1990 American Chemical Society.]

probe that forms highly stable sulfonamide derivatives with amines.[21-23] The concentration of reactive sites is not only determined by the amino acid composition, but also by the degree of protonation of the residues. The modification of the ε-amino group of the lysine residues in proteins occurs readily at pH values greater than 8, at which the concentration of unprotonated amine becomes significant. The protonated aliphatic amine is devoid of reactivity and modification is slow below this pH.[24] The hemoglobin tetramer contains 44 lysine residues,[25,26] so the labeling reaction should be carried out at pH 8 to obtain a probe/protein ratio of less than one DNS/heme to minimize the effects of the chemical modification on subunit interactions. The oxyhemoglobin solutions for the labeling

[21] W. R. Gray, this series, Vol. 11, p. 139.
[22] N. Seiler, *Methods Biochem. Anal.* **18**, 259 (1970).
[23] B. A. Davis, *J. Chromatogr.* **151**, 252 (1978).
[24] R. P. Haugland, *in* "Excited States of Biopolymers" (R. F. Steiner, ed.), p. 29. Plenum, New York and London, 1983.
[25] G. Braunitzer, R. Gehring-Muller, N. Hilschmann, K. Hilse, G. Hobom, V. Rudloff, and B. Wittmann-Liebold, *Hoppe-Seyler's Z. Physiol. Chem.* **325**, 283 (1961).
[26] W. Konigsberg and R. J. Hill, *J. Biol. Chem.* **237**, 3157 (1962).

reaction are 0.1–0.2 mM Hb and Hb$_{III}$ tetramers in 240 mM sodium borate at pH 8.

Dansyl chloride is insoluble in aqueous solution; to add the probe to the oxyhemoglobin samples, it can be dissolved in N,N-dimethylformamide (DMF). This is a highly polar aprotic solvent, compatible with many biological molecules.[24] The dansyl chloride solution should be used immediately, because it begins to degrade after 1 or 2 min in DMF. The 20 mM dansyl chloride/DMF solution is diluted in the Hb and Hb$_{III}$ samples to a final concentration of 1 mM. The duration of the labeling reaction is 4 hr for the Hb sample (bulk preparation) and 2 hr for the Hb$_{III}$ sample (major component obtained from FPLC) to obtain similar labeling ratios.

Chromatography on a Pharmacia (Piscataway, NJ) Sephadex G-25 Superfine column is used to remove excess probe. The Tris-HCl chromatography buffer, pH 7, consequently stops the labeling reaction. The eluted Hb and Hb$_{III}$ samples are dialyzed three times against the appropriate buffers to be used for the fluorescence measurements: 100 mM Tris-HCl, pH 7, 100 mM Tris-HCl plus 1 mM IHP, pH 7, and 100 mM Tris-HCl, pH 9. Absorption and fluorescence emission measurements are recorded to confirm the absence of free dansyl chloride in the last dialysis solutions.

Control Parameters of Preparations

The dansyl/hemoglobin tetramer ratios are determined from the oxyhemoglobin absorption spectra before and after the labeling reaction. The wavelength of maximum absorption of dansyl chloride is 340 nm.[24] It may be assumed that, after labeling, the absorbance at 340 nm (A) is the sum of the oxyhemoglobin absorbance (A_h) and the dansyl absorbance (A_d)

$$A = A_h + A_d \tag{1}$$

The contribution of oxyhemoglobin absorbance A_h is deduced from the ratio (R), before labeling, of the absorbance at 576 nm ($A_{b,576}$) to the absorbance at 340 nm ($A_{b,340}$)

$$R = A_{b,576}/A_{b,340} \tag{2}$$

and from the absorbance at 576 nm of the oxyhemoglobin after labeling ($A_{a,576}$)

$$A_h = A_{a,576}/R \tag{3}$$

The dansyl absorbance at 340 nm of the labeled oxyhemoglobin is given by

$$A_d = A - (A_{a,576}/R) \tag{4}$$

TABLE I
RATIOS OF DANSYL MOLECULES PER HEMOGLOBIN TETRAMER[a]

Preparation	Dansyl/Hb ratio		
	Tris-HCl, pH 7	Tris-HCl + IHP, pH 7	Tris-HCl, pH 9
Hb	1.56 ± 0.43	1.98 ± 0.40	1.68 ± 0.47
Hb_{III}	2.00 ± 0.41	2.38 ± 0.49	2.76 ± 0.58

[a] The ratios were measured for 13 different preparations (7 Hb and 6 Hb_{III} preparations). (From Pin et al.[1])

A molar extinction coefficient of 4500 mol^{-1} liter cm^{-1} is used to obtain the dansyl concentration.[24] The absorption at 576 nm and the extinction coefficient of 15,150 mol^{-1} liter cm^{-1} are used to calculate the oxyhemoglobin concentration.[27]

The average dansyl/hemoglobin tetramer ratios of the Hb (bulk preparation) and Hb_{III} (major component obtained from FPLC) samples are given in Table I. The apparent labeling range is between one and three dansyl molecules per oxyhemoglobin tetramer. The difference in the ratios for each particular preparation may result from small changes in the extinction coefficient of the probe and/or the heme when the external conditions are changed. Moreover, the systematic order of calculated ratios (in the presence of IHP at pH 7 > pH 9 > pH 7 for the Hb preparations; pH 9 > in presence of IHP at pH 7 > pH 7 for the Hb_{III} preparations) serves as a control for the reproducibility of preparations.

The reproducibility of preparations before labeling can be determined by the value of ratio R [see Eq. (2)]. The average value is 0.561 for the Hb preparations and 0.554 for Hb_{III} preparations. For both, the standard deviation is extremely low, ± 006. After labeling, the ratio of the absorbance at 576 nm to the absorbance at 542 nm is calculated as a control for the formation of methemoglobin. The values, which are between 1.030 ± 0.009 for the Hb_{III} preparations in presence of IHP at pH 7 and 1.052 ± 0.004 for Hb preparations at pH 9, are indicative of the absence of methemoglobin. The value of the ratio of the two visible peaks for Hb_{III} preparations is lower than the value for Hb preparations. This difference could be indicative of subtle differences in the heme environment of the oxyhemoglobin molecules.

Fluorescence Experiments

Steady state fluorescence polarization measurements on dansyl-labeled oxyhemoglobin samples have been made with a Greg PC fluorimeter (L-

[27] R. Banerjee, Y. Alpert, F. Leterrier, and R. J. P. Williams, Biochemistry 8, 2862 (1969).

format) from ISS (Champaign, IL). The excitation and emission wavelengths are 340 and 500 nm, respectively. The average standard deviation of the polarization measurements is $\pm$ 0.005. Polarization values, p, are converted to emission anisotropy values, A, by

$$A = 2p/(3 - p) \tag{5}$$

The average molecular volume (V_m) is calculated from the Perrin equation,[28]

$$(A_0/A) - 1 = RT\langle\tau\rangle/\eta V_m \tag{6}$$

where A_0 is the limiting anisotropy of the fluorophore, $\langle\tau\rangle$ is the average lifetime, R is the gas constant, T is the temperature in degrees Kelvin, and η is the viscosity of the solution. The average molecular volumes for Hb and Hb_{III} samples are obtained using an A_0 value of 0.31[20] and 0.35,[1] respectively.

The dansyl emission lifetime measurements have been made with a multifrequency phase and modulation fluorimeter.[29] The 325-nm exciting light from a Liconix HeCd (helium–cadmium) continuous-wave laser is modulated with an acoustooptic modulator system.[30] The standard fluorophore, 1,4-bis(5-phenyl-2-oxazolyl)benzene, or POPOP, with a lifetime of 1.32 nsec in ethanol, is used to correct the instrumental phase delay and demodulation. For each sample, 10–12 frequencies are monitored between 2 and 100 MHz. The standard phase error is $\pm$ 0.2 degrees of phase and the error in the modulation ratio is $\pm$ 0.004. Multifrequency phase and modulation data are analyzed with the Globals Unlimited software (LFD, University of Illinois, Urbana, IL),[31] which allows the simultaneous analysis of multiple data sets.

The measurements of the frequency response of the dansyl–oxyhemoglobin samples are analyzed globally, assuming a ground state that is perturbed by dilution (or pressure). Dilution (or pressure) results in a change of the preexponential factor for an unvarying lifetime value. The $\langle\tau\rangle$ average lifetime values are obtained from the linear combination of the lifetime values from the global fits weighted for their fractional contribution to the lifetime.

[28] F. Perrin, *J. Phys. Radium* **1**, 390 (1926).
[29] E. Gratton and M. Limkerman, *Biophys. J.* **44**, 315 (1983).
[30] D. W. Piston, G. Marriott, T. Radivoyevich, R. M. Clegg, T. M. Jovin, and E. Gratton, *Rev. Sci. Instrum.* **60**, 2596 (1989).
[31] J. M. Beechem, E. Gratton, M. A. Ameloot, J. R. Knutson, and L. Brand, *in* "Fluorescence Spectroscopy" (J. R. Lakowicz, ed.), Vol. 2. Plenum, New York, 1991.

Fig. 2. Photograph of the pressure cell with four quartz windows and the internal quartz pressure cuvette.

High-Pressure Fluorescence Experiments

High-pressure fluorescence measurements are carried out using a high-pressure cell similar to that described by Paladini and Weber.[32] Figure 2 shows a high-pressure cell: the cell has one inlet of the solvent from the high-pressure generator and four quartz windows in a plane placed at 90° with respect to each other for absorption and fluorescence measurements. The quartz bottle-shaped cuvette (pictured beside the cell in Fig. 2) has the capacity for 1.2 ml of sample. A polyethylene flexible tube, sealed at one end, caps the cuvette and permits the pressure transmission inside the cuvette. In the cell, the cuvette rests in a holder. Thermal regulation of the cell can be provided by a thermostatic jacket. The solvent of compression is ethanol of the highest purity.

The high-pressure polarization measurements are corrected for the slight birefringence of cell windows induced by pressure. The correction factors are obtained with a modification of the original method[33] by measuring the polarization on a scattering glycogen solution as a function of pressure and wavelength. For the lifetime measurements, the reference lifetime used to correct for the instrument response is the zero value for

[32] A. A. Paladini and G. Weber, *Biochemistry* **20**, 2587 (1981).
[33] A. A. Paladini and G. Weber, *Rev. Sci. Instrum.* **53**, 419 (1981).

the scatter of the exciting light from the protein particles. The pressure is applied up to 2.4 kbar in 200-bar increments for the polarization measurements and up to 2 kbar in 500-bar increments for the lifetime measurements. The data are acquired after 5 min of equilibration at each pressure. From a comparison of the polarization values and absorption spectra before and after application of the pressure, we have found that pressure transitions are $\geq 95\%$ reversible.

The dissociation constant for each dimer–monomer equilibrium is determined from the changes in the calculated average molecular volumes under pressure. The degree of dimer dissociation (α) at each pressure $[\alpha_{(P)}]$ is obtained by

$$\alpha_{(P)} = [V_{mD} - V_{m(P)}]/(V_{mD} - V_{mM}) \tag{7}$$

where $V_{m(P)}$ is the average molecular volume at pressure P, V_{mD} is the average molecular volume of hemoglobin dimer taken from the dilution curve as the plateau value at low concentration, and V_{mM} is the average molecular volume of hemoglobin monomer taken as the plateau value at high pressure. For the simplest case of association of two monomers into a dimer, the dissociation constant (K_d) is related at each pressure $[K_{d(P)}]$ to the degree of dimer dissociation, and the total protein concentration is expressed in dimer (C_0) by

$$K_{d(P)} = 4C_0\alpha_{(P)}^2/(1 - \alpha_{(P)}) \tag{8}$$

Because the derivative of the free-energy change with respect to pressure corresponds to the volume change at constant temperature, it is possible to relate the change in dissociation constant from atmospheric pressure (K_{do}) to the experimental pressure P:

$$\ln K_{d(P)} - \ln K_{do} = P \Delta V/RT \tag{9}$$

Using Eqs. (8) and (9), the dimer dissociation constant and the negative volume change associated with dimer dissociation are obtained from the plot of $\ln[\alpha^2/(1 - \alpha)]$ versus pressure. The dissociation constant at atmospheric pressure is then calculated from the y intercept, and the volume change from the slope.

Interpretation of Results

The emission anisotropies of Hb and Hb$_{III}$ dansyl–oxyhemoglobin samples exhibit a protein concentration dependence. The anisotropy variation is a result of changes both in the rotational rate of the probe as well as in the rotation of the dansyl–oxyhemoglobin molecule (correlation time). The global analysis of measurements of frequency response of samples has yielded three lifetime values. The very low value (between 120 and 340 psec) of the short lifetime is indicative of efficient energy transfer of the dansyl emission to the heme. The intermediate and long lifetime values are between 2 and 4 nsec and between 12 and 15 nsec, respectively. The values of preexponential factors and fractional intensities obtained by the global fits point out a critical observation: between 30 and 80% of the total intensity originates from 1 to 10% of the molecular population. To illustrate, Fig. 3 shows the concentration dependence of fractional intensities and preexponential factors for the three lifetime components of the dansyl–Hb samples at pH 9: 64, 70, 73, and 78% of the total intensity came from 2.8, 3.7, 4.7, and 7.6% of the molecular population, respectively. The percentage of observed molecules would not be important if the oxyhemoglobin molecular populations were homogeneous. However, chromatographic techniques yielded fractions with different quenching and dissociation properties, demonstrating that the population was heterogeneous both in terms of fluorescence lifetime properties as well as in terms of subunit affinity. The fluorescence measurements were thus selective for particular subset(s) of this population. Such heterogeneity has already been observed by high-pressure fluorescence on the subunit affinity of the yeast glyceraldehyde dehydrogenase.[34]

The concentration dependence of calculated average molecular volumes (Fig. 4) was ascribed to tetramer–dimer dissociation. Compared to the values determined by gel filtration, kinetic techniques, osmotic pressure, or sedimentation studies,[35,56] the dissociation constant of Hb tetramers at pH 7 (estimated from the midpoint of the dilution curve) is one order of magnitude higher, whereas the constant for Hb_{III} dissociation at pH 7 is of the same order of magnitude. This demonstrated that the chromatographic separation, which yielded the Hb_{III} fraction, eliminated a subset of low-affinity tetramers. The observation of dimer dissociation by dilution necessitated the use of low concentrations, even for the fluorescence measurements.

High pressure is used to destabilize the dimer and to reach its complete dissociation. The anisotropy and average lifetime of dansyl–Hb and Hb_{III} have been measured under pressure. For all samples, the average lifetime increases linearly with pressure (1.3 nsec/2 kbar), probably due to a decrease in intersubunit energy transfer. Figures 5 and 6 show the concentration dependence under pressure of the average molecular volume of the Hb and Hb_{III} samples, respectively. At high concentrations (>1 μM in

[34] K. Ruan and G. Weber, *Biochemistry* **28**, 2144 (1989).
[35] E. Antonini and E. Chiancone, *Annu. Rev. Biophys. Bioeng.* **6**, 239 (1977).
[36] A. H. Chu and G. K. Ackers, *J. Biol. Chem.* **256**, 1199 (1981).

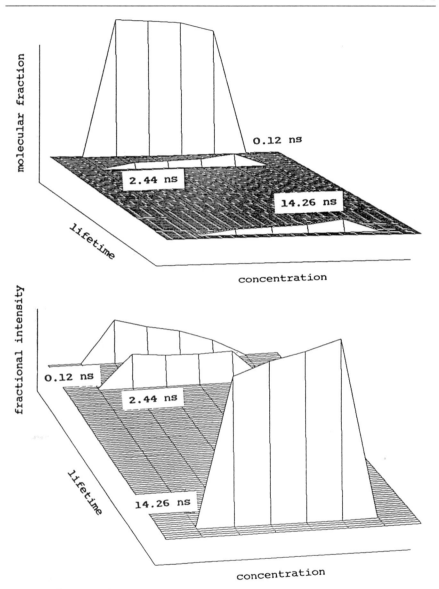

FIG. 3. Concentration dependence of molecular fraction and fractional intensities for the three lifetime components of dansyl-labeled oxyhemoglobin for the Hb samples at pH 9. The concentrations are 30, 13, 4, and 1 μM Hb tetramer (from left to right).

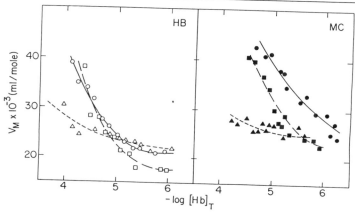

FIG. 4. Concentration dependence of average molecular volumes calculated from steady state fluorescence anisotropy and average dansyl–hemoglobin lifetime values, using Eq. (6). $[Hb]_T$ is the concentration (molar) of oxyhemoglobin tetramer. *Left*: Hb samples. *Right*: Hb_{III} samples. Solid lines (○, ●), pH 7; large dashed lines (□, ■), in the presence of IHP at pH 7; medium dashed lines (△, ▲), pH 9. [Reprinted with permission from S. Pin *et al.*, *Biochemistry* **29**, 9194 (1990). Copyright 1990 American Chemical Society.]

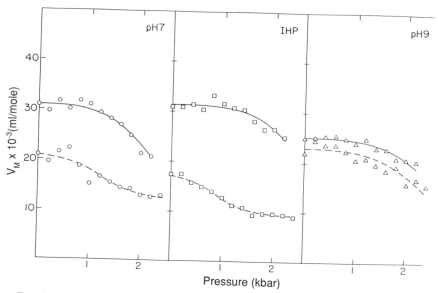

FIG. 5. Pressure dependence of the average molecular volumes for the Hb samples. Solid lines, 32 μM; dashed lines, 1 μM oxyhemoglobin tetramer. [Reprinted with permission from S. Pin *et al.*, *Biochemistry* **29**, 9194 (1990). Copyright 1990 American Chemical Society.]

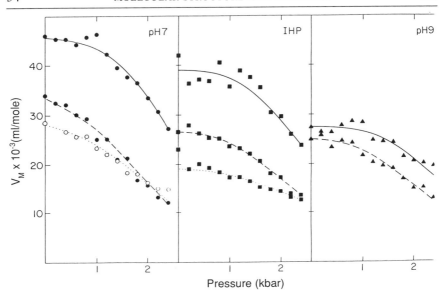

FIG. 6. Pressure dependence of the average molecular volumes for the Hb_{III} samples. Solid lines, 32 μM; dashed lines, 4 μM; dotted lines, 1 μM oxyhemoglobin tetramer. [Reprinted with permission from S. Pin *et al.*, *Biochemistry* **29**, 9194 (1990). Copyright 1990 American Chemical Society.]

tetramer), the calculated molecular volumes are indicative of the superposition of tetramer and dimer dissociations under pressure. At low concentration (1 μM) and high pressure, complete dimer dissociation is detected by the presence of a plateau for the Hb samples or by a convergence to the same average molecular volume value for the Hb_{III} samples. The dimer dissociation constants and the negative volume changes of Hb and Hb_{III} samples in the different experimental conditions can be extracted from these curves and are found to be in the nanomolar range.

High-pressure dissociation profiles yield dimer–monomer affinities (approximately nanomolar) and volume changes ($\approx$80–100 ml/mol) that are in the range of those observed for a number of proteins previously studied by this method.[32,37–41] A nanomolar value for the dimer dissociation constant is consistent with the 30 nM upper limit estimated from sedimentation

[37] G.-J. Xu and G. Weber, *Proc. Natl. Acad. Sci. U.S.A.* **79**, 5268 (1982).
[38] D. F. Senear and G. K. Ackers, *Biophys. J.* **53**, 104a (1988).
[39] J. L. Silva, E. W. Miles, and G. Weber, *Biochemistry* **25**, 5781 (1986).
[40] J. B. A. Ross, C. A. Royer, and P. H. Petra, *Biophys. J.* **55**, 517a (1989).
[41] K. Ruan and G. Weber, *Biochemistry* **27**, 3295 (1988).

experiments,[42] although higher than the subpicomolar value inferred from exchange kinetics.[43]

Weber and Drickamer[5,7] offered the following explanation for the volume decrease of the protein–water system on protein dissociation. The first assumption, previously given by Paladini and Weber,[32] is the existence of "free volumes" or "dead spaces" at the intersubunit surfaces in proteins because of uneven packing. The irregularity of protein surfaces suggests that surface texture may be a factor influencing the molecular interactions. The second assumption derives from the nature of interactions between subunits. The newly exposed charges (from electrostatic interactions) could undergo hydration (electrostriction) on dissociation, and interactions between nonpolar groups could be replaced by interactions of nonpolar groups with water molecules.[44] To sum up, the dissociated protein–water system presents a smaller volume because protein hydration should be more efficient. All of these effects, filling of void volumes, electrostriction of dissociated ion pairs, and exposure of hydrophobic interfacial residues, would lead to a destabilizing effect of pressure on subunit interactions. The relative contributions, however, of these three factors remain to be determined.

Hemoglobin presents an interesting application of fluorescence polarization to the study of oligomeric interactions. On the one hand, the quenching action of the heme results in the actual observation of only a small percentage of the molecules. On the other, this has allowed us to assess the underlying heterogeneity of the preparations. Further applications of the technique toward testing both affinities and heterogeneities of mutant hemoglobins or those of other than human origin could yield more information concerning the thermodynamic and structural basis for different functional properties.

[42] G. L. Kellet and H. K. Schachman, *J. Mol. Biol.* **59,** 387 (1971).
[43] N. T. Mrabet, J. R. Shaeffer, M. J. McDonald, and H. F. Bunn, *J. Biol. Chem.* **261,** 1111 (1986).
[44] L. Boje and A. Hvidt, *Biopolymers* **11,** 2357 (1971).

[5] Optical Measurements of Quaternary Structural Changes in Hemoglobin

By ANDREA BELLELLI and MAURIZIO BRUNORI

Introduction

The cooperative ligand binding by hemoglobin (Hb) demands that (at least) two functionally different conformational states be accessible to the protein; indeed, two quaternary structures of Hb have been discovered and solved by Perutz and co-workers,[1] using crystallographic methods. The classic allosteric model proposed by Monod et al.,[2] which postulates two quaternary conformational states (T and R) in equilibrium at every level of ligation, provides a general framework for quantitative analysis of the structure–function relationships in Hb. Ligand-linked quaternary structural changes have been documented either indirectly, by modeling of the equilibrium and kinetics of ligand binding and the ligand-linked tetramer–dimer dissociation, or directly, mainly by X-ray crystallography and spectroscopy.

The physical and chemical methods applied to the characterization of the two quaternary states of Hb are dealt with in other chapters (see [2], [8], and [11] of this volume). Here we briefly allude to some of these, as an introduction to the main thrust of this chapter, namely the use of static and dynamic optical spectroscopy in the characterization of the quaternary structural changes in Hb. We therefore limit our contribution to the optical changes of the heme and of the globin associated either with the assembly process involved in the formation of tetrameric Hb or with the ligand-linked quaternary change of the tetramer.

Crystallography

The most direct evidence for the ligand-dependent structural transition has been provided by Perutz[3] and Perutz and Fermi,[3,4] using X-ray crystallography. Because the tetramer is assembled from two symmetrical $\alpha_1\beta_1$ dimers (whose quaternary structure changes very little on ligand binding),

[1] M. F. Perutz, G. Fermi, B. Luisi, B. Shaanan, and R. C. Liddington, *Acc. Chem. Res.* **20,** 309 (1987).

[2] J. Monod, J. Wyman, and J. P. Changeux, *J. Mol. Biol.* **12,** 88 (1965).

[3] M. F. Perutz, *Nature (London)* **228,** 726 (1970).

[4] M. F. Perutz and G. Fermi, "Hemoglobin and Myoglobin. Atlas of Protein Sequence and Structure," Vol. 2. Oxford Univ. Press, Oxford, 1981.

the ligand-linked quaternary transition has been described as the relative rotation of the two symmetrical dimers around a pivot centered between the α chains, associated with a significant reorganization of the heme site and of the molecular contacts at the $\alpha_1\beta_2$ interface (the sliding interface[5]). The quaternary transition, as described by X-ray crystallography, is an "all-or-none" phenomenon, as the $\alpha_1\beta_2$ contacts seem to allow only the R- or T-like interfaces.[5] However, crystallized Hb Ypsilanti [β99 (G1) Asp $\rightarrow$ Tyr] has been shown to yield a different quaternary structure[6]; the impact of this new finding on our thinking about Hb (although still unassessed) may be considerable.

Reactivity

Ligand binding to Hb is known to have large effects on the reactivity of the protein matrix as well as the heme iron.[7] Among the wealth of available results are the following: (1) the $-SH$ reagent p-hydroxymercuribenzoate specifically reacts with the sulfhydryl group of β93 (F9) Cys; the reaction time course is strongly influenced by the quaternary state of Hb, the reactivity with deoxyHb being 50- to 80-fold slower than that of HbO_2 (or $HbCO^8$); (2) dissociation of tetramers into $\alpha_1\beta_1$ dimers is favored in liganded Hb; this has been correlated to the number of weak bonds that stabilize the two quaternary states. This property (known for a long time[7]) has been exploited by Ackers and co-workers[9] to test the applicability of the two-state allosteric model to adult human hemoglobin (Hb A). It should be recalled that the dissociation constant of at least one of the two allosteric states depends on the extent of ligation; to a first approximation, dissociation into dimers depends on the number of ligands bound in the T state but not in the R state.[10]

Spectroscopy

The changes in the porphyrin optical spectrum caused by ligand binding to the sixth coordination position usually overwhelm those associated with the quaternary transition(s); hence the definition of the T to R optical changes relies on careful experiments; some of these are detailed in the next section. Spectroscopic techniques that have been extensively em-

[5] J. Baldwin and C. Chothia, *J. Mol. Biol.* **129,** 175 (1979).
[6] F. R. Smith, E. E. Lattman, and C. W. Carter, *Proteins* **10,** 81 (1991).
[7] E. Antonini and M. Brunori, "Hemoglobin and Myoglobin in Their Reactions with Ligands." North-Holland Publ., Amsterdam, 1971.
[8] E. Antonini and M. Brunori, *J. Biol. Chem.* **244,** 3909 (1969).
[9] G. K. Ackers, M. L. Doyle, D. Myers, and M. A. Daugherty, *Science* **255,** 54 (1992).
[10] S. J. Edelstein and J. T. Edsall, *Proc. Natl. Acad. Sci. U.S.A.* **83,** 3796 (1986).

ployed to probe the quaternary conformational changes in hemoglobin are nuclear magnetic resonance (NMR) and resonance Raman.

Nuclear Magnetic Resonance. The ^{1}H NMR spectrum of hemoglobin has not been resolved completely; however, many resonances have been assigned to specific residues, and at least three of those are sensitive to the quaternary conformation, namely (1) a line at -14 ppm from the reference peak of 3-trimethylsilylpropane sulfonate, characteristic of T-state derivatives irrespective of the ligation state, and attributed to the exchangeable proton of α122 (H5) His,[11,12] and (2) the lines of the two hydrogen bonds α42 (C7) Tyr–β99 (G1) Asp (characteristic of the T state, with a resonance at -9.4 ppm from water) and α94 (G1) Asp–β102 (G4) Asn (typical of the R state, -5.8 ppm from water[13]; see also Inubushi *et al.*[14] The methodologies of high-resolution NMR as applied to Hb are presented elsewhere in this volume.[15]

Resonance Raman Spectroscopy. Resonance Raman spectroscopy offers some unequivocal probes of the conformation and coordination of Hb, such as the stretching frequency of the Fe–His(F8) bond[16,17] and the resonance frequency of the peaks at 1550 and 1615 cm^{-1} (attributed to α42 (C7) Tyr and β37 (C3) Trp at the $\alpha_1\beta_2$ interface[18,19]). It is of particular interest for dynamic studies that the resonance Raman spectrum can be followed after photodissociation of the ligand over a time range starting from picoseconds, both in the Soret[20,21] and ultraviolet (UV) regions.[18,19] This technique is dealt with in Friedman.[22]

In the following sections we focus on the results obtained by means of optical spectroscopy in the study of quaternary structural changes of Hb.

[11] D. J. Patel, L. Kampa, R. G. Shulman, T. Yamane, and M. Fujiwara, *Biochem. Biophys. Res. Commun.* **40,** 1224 (1970).

[12] R. G. Shulman, J. J. Hopfield, and S. Ogawa, *Q. Rev. Biophys.* **8,** 325 (1975).

[13] L. W. M. Fung and C. Ho, *Biochemistry* **14,** 2526 (1975).

[14] T. Inubushi, C. D'Ambrosio, M. Ikeda Saito, and T. Yonetani, *J. Am. Chem. Soc.* **108,** 3799 (1986).

[15] C. Ho and J. R. Perussi, this volume [8].

[16] K. Nagai, T. Kitagawa, and H. Morimoto, *J. Mol. Biol.* **136,** 271 (1980).

[17] M. R. Ondrias, D. L. Rousseau, T. Kitagawa, M. Ikeda Saito, T. Inubushi, and T. Yonetani, *J. Biol. Chem.* **257,** 8766 (1982).

[18] C. Su, Y. D. Park, G. Y. Liu, and T. G. Spiro, *J. Am. Chem. Soc.* **111,** 3457 (1989).

[19] S. Kaminaka, T. Ogura, and T. Kitagawa, *J. Am. Chem. Soc.* **112,** 23 (1990).

[20] J. M. Friedman, D. L. Rousseau, M. R. Ondrias, and R. A. Stepnoski, *Science* **218,** 1244 (1982).

[21] W. Findsen, J. M. Friedman, M. R. Ondrias, and S. R. Simon, *Science* **229,** 661 (1985).

[22] J. M. Friedman, this volume [11].

Static Optical Spectroscopy

Effects of Quaternary Structure on Spectrum of Heme

It has been clearly proved that the optical absorption of the heme is affected by the quaternary structure of the globin, but in fact spectral perturbations of the deoxyhemes similar to those illustrated below have been associated with tertiary structural changes as well. Nonetheless, optical spectroscopy represents a powerful tool to monitor directly the assembly of the tetramer and the ligand-linked quaternary conformational change in deoxygenated and liganded Hb.

Deoxygenated Hemoglobin. The first and fully documented quaternary-linked optical changes were obtained with deoxygenated Hb. The $T_0 - R_0$ (T_0, unliganded T state of Hb; R_0, unliganded R state of Hb) difference spectrum has been recorded after rapid removal of the bound ligand by photolysis, before the unliganded R_0 state formed immediately after photolysis decays to the predominant species, T_0. It should be remarked that when photolysis of HbCO is obtained with a photographic flash (with 2- to 200-μsec duration), the $R_0 \rightarrow T_0$ transition cannot be recorded because its half-life ($\sim$20 μsec) overlaps with the light pulse; nevertheless, the first evidence for a $T_0 - R_0$ spectroscopic difference was reported by Gibson[23] in his classic work on the quickly reacting form of Hb. We now know that this was due to the presence of a large fraction of R-state $\alpha\beta$ dimers in the dilute HbCO solution at alkaline pH.[24,25] The $T_0 - R_0$ difference spectrum for human Hb A obtained from laser photolysis experiments is described in Dynamics of Quaternary Change by Optical Spectroscopy, below.

The spectrum of unliganded isolated α and β chains corresponds to that of R_0 and differs from that of T_0 in Hb A[26,27]; thus a prototype difference spectrum (corresponding in shape and amplitude to $T_0 - R_0$) has been obtained by mixing deoxygenated α and β chains, and recorded either statically or kinetically (by stopped flow) over the range 400–650 nm.[27] Figure 1 reports this difference spectrum in the Soret region, where the quaternary-linked optical change is maximal (at pH 7.0, $\Delta\varepsilon = 17$ mM^{-1}

[23] Q. H. Gibson, *Biochem. J.* **71**, 293 (1959).

[24] Q. H. Gibson and E. Antonini, *J. Biol. Chem.* **242**, 4678 (1967).

[25] S. J. Edelstein, M. J. Rehmar, J. S. Olson, and Q. H. Gibson, *J. Biol. Chem.* **245**, 4372 (1970).

[26] E. Antonini, E. Bucci, C. Fronticelli, J. Wyman, and A. Rossi Fanelli, *J. Mol. Biol.* **12**, 375 (1965).

[27] M. Brunori, E. Antonini, J. Wyman, and S. R. Anderson, *J. Mol. Biol.* **34**, 357 (1968).

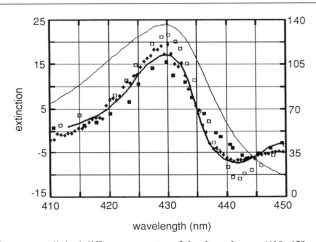

FIG. 1. Quaternary-linked difference spectra of the deoxyhemes (410–450 nm). *Continuous thin line:* The absolute absorption spectrum of deoxygenated human Hb A, with a maximum at 430 nm and an extinction of 133 mM^{-1} cm^{-1} (referred to the right ordinate). This spectrum was recorded in a Cary 14/Olis (Olis, Athens, GA) reconverted spectrophotometer, in a 1-cm cuvette, at [Hb] = 4 μM in 0.1 M Bis–Tris/HCl buffer, pH 7, containing 0.1 M NaCl, at 20°. *Continuous bold line:* The difference spectrum [deoxygenated Hb] − [deoxygenated, isolated α and β chains in equimolar amounts]. The isolated chains were prepared by the p-chloromercuribenzoate (PMB) procedure[28] and freed from PMB according to De Renzo *et al.*[29] A baseline (corresponding to the R_0 state) was recorded in a two-compartment cuvette (total path length, 1 cm), each compartment containing the same amount of each type of chain. The contents of the two compartments were mixed and reassociation of deoxygenated chains to deoxygenated tetramers was allowed to proceed for 5–10 min before recording a new spectrum, corresponding to deoxyHb T_0. Experimental conditions: 0.1 M phosphate buffer, pH 7, containing 1 mM Na$_2$S$_2$O$_4$, at 20°; protein concentration, 2.5 μM (heme, for each chain, before mixing); light path, 1 cm. An identical difference spectrum was also recorded by stopped flow. (Data from Brunori *et al.*[27]) *Filled diamonds:* The kinetic difference spectrum recorded with a photodiode array spectrophotometer (Tracor Northern TN6500, Middleton, WI) coupled to a Gibson-Durrum stopped-flow apparatus, during the course of the reaction leading from HbO$_2$ to Hb. As detailed in text, this difference spectrum corresponds to the optical density change associated with recombination of deoxygenated $\alpha\beta$ dimers (R_0) to yield deoxygenated tetramers (T_0). Experimental conditions: a 3 μM (heme) solution of HbO$_2$ in 0.1 M phosphate buffer, pH 7, was mixed with 50 mM Na$_2$S$_2$O$_4$ in the same buffer, and the time course was followed from 10 msec to 8 sec over a 150-nm wavelength range. Sixty spectra were deconvoluted, applying the singular value decomposition (SVD) algorithm,[30] and the first two columns of the V matrix were fitted to two exponentials, because the time course is biphasic.[31,32] Deoxygenation was complete 400 msec after mixing, but reassociation of deoxygenated $\alpha\beta$ dimers was minimal; thus the spectrum at 400 msec after mixing, reconstructed from the first 3 U columns and multiplied by −1 to obtain the difference T_0 − R_0, was chosen for presentation. The corresponding millimolar extinction was calculated from the fraction of $\alpha\beta$ dimers, taking a tetramer–dimer dissociation constant for HbO$_2$ of K_d = 2 μM heme.[7] *Squares:* The T_0 − R_0 difference spectrum obtained after laser photolysis of HbCO at two pH values, that is, 6.5 (closed

cm^{-1} at 430 nm, i.e., 12.8% of $\varepsilon = 133$ mM^{-1} cm^{-1}). This difference spectrum has a characteristic isosbestic point at 436 nm. The detailed experimental protocol to obtain this difference spectrum is given in Fig. 1.[28–33]

The deoxygenated $\alpha_1\beta_1$ dimers, like the isolated chains, display the chemical reactivity and spectroscopic properties of R-state Hb; thus the same spectroscopic transition has been recorded when dimers of deoxyHb associate into tetramers, or when tetrameric deoxyHb dissociates into dimers (e.g., on addition of an excess of haptoglobin[34]). In the former case the optical difference spectrum may be obtained as follows: a solution of HbO_2 at micromolar concentration (which contains a significant fraction of oxygenated dimers) is mixed with excess dithionite in a stopped-flow apparatus; the time course recorded at 430 nm is strongly biphasic[31,32] and can be described by the following scheme:

$$2[\alpha\beta(O_2)_2] \rightarrow 2[\alpha\beta] \rightarrow [\alpha\beta]_2$$

The first phase, corresponding to the time course of oxygen dissociation, has an overall rate constant of ~25/sec at pH 7 and is independent of Hb concentration; the second phase is dependent in rate and amplitude on the concentration of Hb and corresponds to the reassociation of the unliganded dimers.[32] The difference spectrum obtained from the latter process (shown in Fig. 1 with the details of the experimental protocol) agrees satisfactorily with that obtained by mixing the deoxygenated α and β chains.

Liganded Hemoglobin. Experimental conditions must be selected such that the quaternary state of the protein can be switched without a change

[28] E. Bucci and C. Fronticelli, *J. Biol. Chem.* **240**, PC551 (1965).

[29] E. C. De Renzo, C. Ioppolo, G. Amiconi, E. Antonini, and J. Wyman, *J. Biol. Chem.* **242**, 4850 (1967).

[30] E. R. Henry and J. Hofrichter, this series, Vol. 210, p. 129.

[31] E. Antonini, M. Brunori, and S. R. Anderson, *J. Biol. Chem.* **243**, 1816 (1968).

[32] G. L. Kellet and H. Gutfreund, *Nature (London)* **227**, 921 (1970).

[33] C. A. Sawicki and Q. H. Gibson, *J. Biol. Chem.* **251**, 1533 (1976).

[34] E. Chiancone, J. B. Wittenberg, B. A. Wittenberg, E. Antonini, and J. Wyman, *Biochim. Biophys. Acta* **117**, 379 (1966).

symbols) and 7.0 (open symbols). Experimental conditions: [Hb] = 52 μM (heme) in 0.05 M phosphate buffer, 20°, path length = 1 mm. Photolysis was promoted by the 1-μsec light pulse from a dye laser (Electrophotonics system 100, New Durham, NH; peak wavelength = 540 nm; total energy delivered, 0.15 J/pulse). The difference spectrum shown here was recorded 2 μsec after the laser pulse, and thus represents deoxyHb that has already partially undergone the tertiary conformational relaxation but not yet the quaternary one (see the text and Fig. 5 for comparison). (Data modified from Sawicki and Gibson.[33])

in the number of ligands bound. Although, of course, this is no problem with fully deoxygenated Hb (see above), it is more difficult for (partially or fully) liganded Hb. It has been achieved with the CO and the azide derivatives of fish hemoglobins, such as trout HbIV, or carp Hb displaying the Root effect,[35,36] with the NO derivative of ferrous Hb A and with the fluoride and water derivatives of metHb A, whose allosteric conformation is switched from R to T in the presence of inositol hexakisphosphate (IHP)[37-41] and with the CO derivative of Hb Kansas $\beta102$ (G4) Asn $\rightarrow$ Thr.[11,42]

Fish hemoglobins displaying the Root effect (i.e., a stabilization of the T state at low pH, even with fully liganded Hb[43,44]) lent themselves to the spectroscopic characterization of the T_4 state using primarily the CO derivative. The quaternary equilibrium was shifted by changing pH and/or organic phosphate concentration (e.g., IHP), and the spectral changes recorded statically. The $T_4 - R_4$ difference spectrum obtained by Giardina et al.[35] using trout HbIV is shown in Fig. 2 (and the pertinent experimental conditions are reported in the caption). The T state of the CO and O_2 derivatives also was studied in the case of Hb Kansas [$\beta102$ (G4) Asp $\rightarrow$ Thr], a naturally occurring mutant of human hemoglobin.[41]

In some of the liganded ferrous derivatives (notably with ferrous HbNO), addition of IHP at pH 6.5 leads to rupture of the axial bond between His(F8) and the heme iron, mainly in the α chains; thus the globin is stabilized in the T state, while the heme iron remains pentacoordi-

[35] B. Giardina, F. Ascoli, and M. Brunori, Nature (London) 256, 761 (1975).
[36] M. F. Perutz, J. K. M. Sanders, D. H. Chenery, R. W. Noble, R. R. Pennelly, L. W. M. Fung, C. Ho, I. Giannini, D. Porschke, and H. Winkler, Biochemistry 17, 3640 (1978).
[37] M. R. Adams and T. M. Schuster, Biochem. Biophys. Res. Commun. 58, 525 (1974).
[38] J. M. Salhany, S. Ogawa, and R. G. Shulman, Proc. Natl. Acad. Sci. U.S.A. 71, 3359 (1974).
[39] M. F. Perutz, A. R. Fersht, S. R. Simon, and G. C. K. Roberts, Biochemistry 13, 2174 (1974).
[40] M. F. Perutz, E. J. Heidner, J. E. Ladner, J. C. Beetlestone, C. Ho, and E. F. Slade, Biochemistry 13, 2187 (1974).
[41] M. F. Perutz, J. V. Kilmartin, K. Nagai, A. Szabo, and S. R. Simon, Biochemistry 15, 378 (1976).
[42] S. Ogawa, A. Mayer, and R. G. Shulman, Biochem. Biophys. Res. Commun. 49, 1485 (1972).
[43] R. W. Noble, L. J. Parkhurst, and Q. H. Gibson, J. Biol. Chem. 245, 6628 (1970).
[44] M. Brunori, Curr. Top. Cell. Regul. 9, 1 (1975).

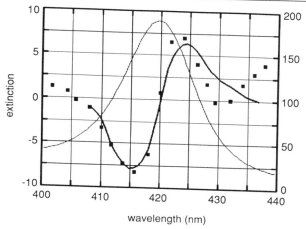

wavelength (nm)

FIG. 2. Quaternary-linked difference spectra for carbonmonoxyhemes. *Continuous thin line:* Spectrum of carbonmonoxyhemoglobin with a maximum at 419 nm and an extinction of 191 mM^{-1} cm^{-1} (referred to the right ordinate). Experimental conditions as in Fig. 1. *Continuous bold line:* Difference spectrum of trout HbIV-CO at pH 6.2 minus trout HbIV-CO at pH 8. Because of the Root effect, trout HbIV-CO switches from the R_4 to the T_4 state as the pH is lowered to 6.2, while remaining liganded due to the sufficiently high affinity of CO for both allosteric states.[35] Experimental conditions: [Hb] = 6 μM, in 0.1 M Bis–Tris buffer, 20°. *Squares:* Spectrum of the second out-of-phase component recorded by Martino and Ferrone,[45] using the intensity-modulated photoexcitation method; it is attributed to a quaternary-linked perturbation of the CO heme. Experimental conditions: [Hb] = 1 mM in 0.05 M Bis–Tris buffer, pH 7, containing 2 mM $Na_2S_2O_4$ and 1 atm CO; the 1 to 1.3% photolysis was promoted by the intensity modulated beam of a Coherent 599 (Coherent, Palo Alto, CA) jet stream dye laser, with emission wavelength of 573 nm. The original points were scaled using as a reference the larger difference spectrum of the first out-of-phase component, which corresponds to the $T_0 - R_0$ transition, whose extinction is available (see Fig. 1 and references therein).

nated.[41,46] This condition implies an abnormal T state, and the porphyrin is presumably domed in the opposite direction with respect to unliganded Hb (i.e., toward the distal site); the spectral changes of the porphyrin are large, as shown in Fig. 3.

Of interest also are some derivatives of metHb (such as F$^-$, H$_2$O, and cyanide) whose optical spectra were shown to be sensitive to the quaternary structure of the protein, as discussed by Perutz *et al.*[39,40]

Optical Intermediates along Binding Curve. The high cooperativity of human Hb implies that partially saturated intermediates are scarcely

[45] A. J. Martino and F. A. Ferrone, *Biophys. J.* **56**, 781 (1989).
[46] A. Szabo and M. F. Perutz, *Biochemistry* **15**, 4427 (1976).

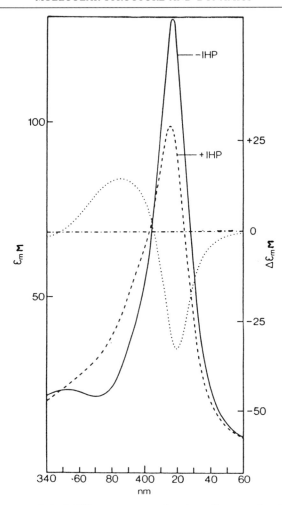

FIG. 3. Quaternary-linked difference spectra for nitrosylhemes, showing the effect of IHP on the optical spectrum of nitrosyl Hb A (HbNO), from 340 to 460 nm. (Reprinted with permission from Perutz *et al.*[41] Copyright 1976 American Chemical Society.) A solution of human HbO_2 at a concentration of 3 mM in heme, in 0.1 M NaCl, was deoxygenated by alternated equilibrations with vacuum and nitrogen, 1 atm, under shaking; a 0.05 M Bis–Tris buffer solution, pH 6.5, containing 0.1 M NaCl, was saturated with pure NO from a cylinder (after washing the gas with concentrated NaOH). The two solutions were mixed in a nitrogen-filled glove box to give a solution 30 μM in heme; 1.5 ml of this solution was filled into each of two stoppered cuvettes with path length 4 mm. Twenty microliters of 5 mM IHP neutralized at pH 6.5 was added to one cuvette and 20 μl of water to the other; 20°. —, absolute spectrum in the absence of IHP; – – –, the same in the presence of 65 μM IHP, both referred to the left ordinate; ···, difference spectrum referred to the right ordinate (extinction coefficients in mM^{-1} cm^{-1}).

populated at equilibrium;[47-50] therefore, identification of the optical changes associated with the quaternary transition along the titration curve proved to be difficult. Static titration of Hb A with CO in solution[51] showed no evidence of deviation from linearity, as may be expected if T state-liganded intermediates (T_1, T_2), characterized by distinct absorption properties, were not significantly populated.

In the case of oxygen, the lower affinity makes the analysis of the equilibrium curve more complex but more informative.[47,52] No clear-cut evidence for a significant contribution of quaternary spectral changes from the intermediates has been suggested by experiments carried out monitoring the oxygen saturation at single wavelength. However, a careful study carried out by Ownby and Gill,[53] recording the Soret spectrum at each saturation in the thin-layer dilution apparatus and analyzing the results by the singular value decomposition (SVD) algorithm, seems to suggest that more than a single spectroscopic transition must be considered. Whether this result points to spectroscopic and functional heterogeneity of the two chains or to the existence of spectroscopically distinct ligation intermediates (i.e., liganded T state or unliganded R state) is still unclear.

Perturbation of Aromatic Side Chains at $\alpha_1\beta_2$ Interface

The near-UV difference spectrum between the R and T state of Hb displays (on top of a broad band due to the porphyrin) a fine structure centered around 290 nm. This spectral feature, first observed in the late 1960s,[54,55] has been attributed to a perturbation of the optical spectrum of aromatic residues at the $\alpha_1\beta_2$ interface, namely α42 (C7) Tyr, β37 (C3) Trp, and β41 (C7) Phe.[56] A careful scrutiny reveals four sharp peaks at 279, 287, 294, and 302 nm in the difference spectrum of HbO_2 minus Hb and in that obtained by addition of one (or more) molar equivalent(s) of

[47] K. Imai, "Allosteric Effects in Hemoglobin." Cambridge Univ. Press, Cambridge, 1982.
[48] M. Perrella, L. Benazzi, L. Cremonesi, S. Vesely, G. Viggiano, and L. Rossi-Bernardi, *J. Biol. Chem.* **258,** 4511 (1983).
[49] S. J. Gill and J. Wyman, "Binding and Linkage." University Science Books, Mill Valley, CA, 1990.
[50] S. J. Gill, E. Di Cera, M. L. Doyle, G. A. Bishop, and C. H. Robert, *Biochemistry* **26,** 3995 (1987).
[51] S. R. Anderson and E. Antonini, *J. Biol. Chem.* **243,** 2918 (1968).
[52] G. K. Ackers and M. L. Johnson, *J. Mol. Biol.* **147,** 559 (1981).
[53] D. W. Ownby and S. J. Gill, *Biophys. Chem.* **37,** 395 (1990).
[54] Y. Enoki and I. Tyuma, *Jpn. J. Physiol.* **14,** 280 (1964).
[55] R. W. Briehl and J. F. Hobbs, *J. Biol. Chem.* **245,** 544 (1970).
[56] M. F. Perutz, J. E. Ladner, S. R. Simon, and C. Ho, *Biochemistry* **13,** 2163 (1974).

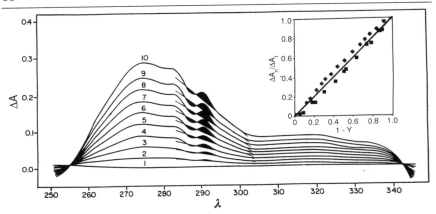

FIG. 4. Difference spectra recorded in UV region. The sample cell contains HbO_2; the reference cell, successively deoxygenated Hb, starting from HbO_2 (spectrum 1) and ending with Hb (spectrum 10). The broad difference spectrum is due to the contribution of the porphyrin and the sharp difference spectrum around 290 nm to perturbation of the optical absorption of β37 (C3) Trp. Straight lines extrapolated from a linear region between 295 and 300 nm were used as baselines for assessing the magnitude of the fine structure around 290 nm (filled) and were used for calculation of the oxygen saturation of hemoglobin. Experimental conditions: 0.05 M Bis–Tris buffer, pH 7.4, containing 1.7 mM IHP; [Hb] = 58 μM (heme). *Inset:* The relationship between the fraction of deoxyHb and the amplitude of the fine structure at 290 nm. Experimental conditions: 60 μM (heme) hemoglobin in 0.05 M Bis–Tris buffer, pH 7.4, either stripped (■) or in the presence of 1.7 mM IHP (∗). The line indicates the behavior expected for a linear correlation. (Modified from Imai and Yonetani.[58])

IHP to aquo- or fluorometHb (Hb^+ and Hb^+F^-), to nitrosyl ferrous Hb (HbNO), to deoxyHb Kempsey [β99 (G1) Asp $\rightarrow$ Asn], or to oxy- and carbonmonoxyHb Kansas β102 (G4) Asn $\rightarrow$ Thr].[39,41,56] These optical bands are paralleled by a number of negative CD peaks, among which the most prominent is the band at 287 nm.[57]

The tryptophan residue at position β37 (C3) participates to the $\alpha_1\beta_2$ interface, where the quaternary transition is associated with the largest reorganization of contacts,[5] and contributes two of the four sharp peaks, that is, those at 294 and 302 nm. Figure 4[58] reports the difference spectrum Hb minus HbO_2 in the region 250–350 nm; the experimental details are given in the caption. Assignment of the difference peaks at 294 and 302 nm to β37 (C3) Trp was first made on the basis of structural considerations; experiments carried out with the mutant Hb Hirose, β37 (C3) Trp $\rightarrow$ Ser,[59]

[57] S. R. Simon and C. R. Cantor, *Proc. Natl. Acad. Sci. U.S.A.* **63**, 205 (1969).

[58] K. Imai and T. Yonetani, *Biochem. Biophys. Res. Commun.* **50**, 1055 (1973).

[59] K. Yamaoka, *Blood* **38**, 730 (1971).

supported this assignment.[39,55] An additional optical probe at the $\alpha_1\beta_2$ interface has been documented by difference spectroscopy of a site-directed mutant of the α chains, α38 (C3) Thr $\rightarrow$ Trp[60]; with this mutant the amplitude of the sharp difference spectrum around 290 nm is approximately doubled, indicating that the two tryptophans at the $\alpha_1\beta_2$ interface are perturbed upon the quaternary transition.

The relevance of the UV difference spectrum becomes obvious when one considers that this is the only optical probe of the quaternary transition directly monitoring the globin. Indeed, it is not affected by the ligation state of the heme iron, because the same spectral perturbation has been detected with several fully saturated liganded Hb derivatives that were shifted from R_4 to T_4 by allosteric effectors (pH, IHP); among these we recall the carbon monoxide derivative of Hb Kansas[41] and the carbon monoxide derivative of trout HbIV.[35] Because the UV spectral perturbation is far from an isosbestic point, and because several spectroscopic contributions (from the heme and the globin) superimpose, the relationships of the amplitude of the difference band around 290 nm' with the fractional oxygen saturation and with the population of the R state (calculated according to the two-state model[58]) deviate (slightly but significantly) from linearity (Fig. 4).

Dynamics of Quaternary Change by Optical Spectroscopy

Analysis of the time course of absorbance changes following photolysis of the bound ligand by a very short laser pulse (i.e., subpicoseconds to nanoseconds in duration) has been informative, in agreement with the expectation that intermediates are largely populated in a kinetic experiment if the photolysis occurs randomly and the dissociation rate of the ligand in the dark is negligible compared to the rebinding rate (for references on the photochemical properties of hemoglobin and the technicalities of laser photolysis experiments, see Hofrichter *et al.*[61]

Sawicki and Gibson[33,62] employed a 1-μsec laser pulse of sufficient power to dissociate oxygen or carbon monoxide from Hb A. The long length of the pulse yielded ~100% photolysis, because the geminate component was pumped out by multiple photodissociations. These authors recorded the kinetic difference spectrum of equilibrium deoxyHb and the 2-μsec photoproduct (i.e., R_0 deoxyHb); on the basis of this spectrum

[60] B. Vallone, P. Vecchini, V. Cavalli, and M. Brunori, *FEBS Lett.* **324**, 117 (1993).
[61] J. Hofrichter, E. Henry, A. Ansari, C. M. Jones, R. M. Deutsch, and J. Sommer, this volume [18].
[62] C. A. Sawicki and Q. H. Gibson, *J. Biol. Chem.* **252**, 5783 (1977).

A

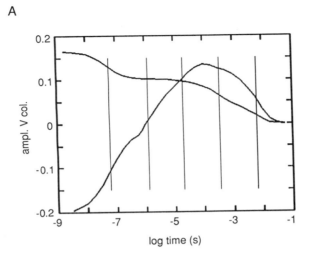

B

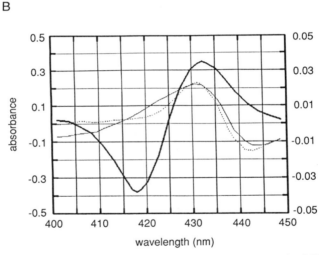

FIG. 5. Time course of spectral changes in Hb A after laser photolysis of CO. (A) Time evolution of the amplitude of the first two columns from the V matrix, obtained by applying the SVD algorithm to the rebinding of CO to Hb A after laser photolysis (Hofrichter *et al.*,[65] modified). The columns of the V matrix represent the time courses of the evolution of the spectroscopic components described by the corresponding columns of matrix U, and can be fitted using five independent exponential relaxations. These, denoted I through V, correspond (from left to right) to: (I) geminate ligand rebinding ($1/e$ time $= 57$ nsec); (II) tertiary rearrangement of the unliganded chains ($1/e$ time $= 1.3$ μsec); (III) quaternary transition

they selected appropriate wavelengths to follow the time course of either $R_0 \rightarrow T_0$ (425 nm) or $T_0 \rightarrow R_4$ (436 nm, isosbestic for $T_0 \rightleftharpoons R_0$; see Fig. 1). The $T_0 - R_0$ difference spectrum recorded by Sawicki and Gibson[33] is compared in Fig. 1 with that obtained from association of the isolated deoxygenated α and β chains or the deoxy $\alpha\beta$ dimers. The excellent agreement provides strong evidence for the general validity of the quaternary-linked spectral perturbation of the deoxyhemes. It is important to notice that the amplitude of the $T_0 - R_0$ difference spectrum is pH dependent, being smaller at pH 6.5 and larger at pH 7.0–9.2, for hitherto unknown reasons. Moreover, its amplitude was found to be much smaller for the $R_0 \rightarrow T_0$ transition in trout HbI, which was investigated extensively by Hofrichter et al.[63]

A different experimental setup and a more refined data analysis were applied by Eaton and co-workers to the study of transient spectroscopy of photodissociated Hb[64,65] (see also Hofrichter et al.[61]). These authors used two synchronized laser pulses, one to photolyze and the other to probe the absorbance of the sample; the spectra recorded at different times after photolysis over the wavelength range 400–460 nm were combined into a matrix and analyzed using the SVD algorithm.[30]

This analysis yielded a quantitative description of the observations in terms of three basic spectral components, as shown in Fig. 5; all the experimental details are given in the caption. The largest spectral component was assigned to ligand rebinding, whereas the second largest had the

[63] J. Hofrichter, E. R. Henry, A. Szabo, L. P. Murray, A. Ansari, C. M. Jones, M. Coletta, G. Falcioni, M. Brunori, and W. A. Eaton, *Biochemistry* **26**, 6583 (1991).

[64] J. Hofrichter, J. H. Sommer, E. R. Henry, and W. A. Eaton, *Proc. Natl. Acad. Sci. U.S.A.* **80**, 2235 (1983).

[65] J. Hofrichter, E. R. Henry, J. H. Sommer, R. Deutsch, M. Ikeda Saito, T. Yonetani, and W. A. Eaton, *Biochemistry* **24**, 2667 (1985).

(1/e time = 20 μsec); (IV) and (V) second-order ligand rebinding to the R and T states of Hb, respectively. The 1/e times characteristic of each relaxation are indicated by thin vertical lines. (B) Normalized difference spectra of carbonmonoxy- and deoxyhemes (Hofrichter et al.,[65] modified). Bold line and left ordinate: deoxyHb minus carbonmonoxyHb (corresponds to the static difference spectrum); thin line and right ordinate: difference spectrum associated with relaxation II of the deoxy hemes (tertiary conformational change); dotted line and right ordinate: difference spectrum associated with relaxation III in deoxyhemes (quaternary conformational change). In both cases the baseline is the deoxy species populated at the end of each relaxation. The two difference spectra are similar in shape, with isosbestic points at 438 and 437 nm. (A) and (B) Experimental conditions: [Hb] = 143 μM (heme); path length = 0.34 mm; pCO = 1 atm; 0.1 M phosphate buffer, pH 7, containing 10 mM $Na_2S_2O_4$. The instrument employed is described in Hofrichter et al.[61]

spectroscopic properties of the $T_0 - R_0$ difference spectrum (with an isosbestic point at 437 nm, to be compared with 436 nm found by Sawicki and Gibson[33] and Fig. 1). The rate constant of the $R_0 \rightarrow T_0$ quaternary transition is 50,000/sec at pH 7 in 0.1 M phosphate buffer, much faster than that of Sawicki and Gibson at pH 9.2, that is, 6400/sec [in agreement with the larger value of the allosteric constant L_0 (equilibrium constant of the reaction $R_0 \rightleftharpoons T_0$) at pH 7]. The experimental approach employed by Eaton and co-workers[64,65] has made it possible to assign with certainty the microsecond process to the $R_0 \rightarrow T_0$ quaternary transition on the basis of the following: (1) the dependence of the amplitude on the extent of photolysis, (2) the independence of the rate constant on CO and Hb concentration, and (3) the features of the difference spectrum. The time course of the spectroscopic transitions that follow the photodissociation of carbon monoxide is described by five exponential relaxations over the time scale of 10 nsec to 1 sec. These, on the basis of their spectroscopic features and the dependence on ligand concentration and on extent of photolysis, were attributed to: (I) the geminate rebinding of ligand molecules from inside the protein matrix, (II) the tertiary relaxation of unliganded subunits, (III) the quaternary transition, and (IV and V) the bimolecular rebinding of ligand from the solvent (see Fig. 5A).

It is interesting to remark that the intermediate associated with a relaxation of the tertiary structure of the globin ($t \approx 1\ \mu sec$) is characterized by a difference spectrum similar (if not identical) in shape to the $T_0 - R_0$ difference spectrum, with an isosbestic point at 438 nm (as shown in Fig. 5B).

A different experimental approach to characterize the rate and spectral properties of the quaternary transition in partially liganded Hb has been introduced by Ferrone and Hopfield[45,66] (see also Ferrone[67]). These authors made use of an intensity-modulated light source to photolyze HbCO (and HbO_2) partially in order to populate preferentially a triply liganded intermediate. The absorbance changes induced by the modulated photolytic beam were analyzed by SVD to yield (1) one "in-phase" spectral component, due to the variable levels of photolysis, corresponding to the HbCO − Hb (or HbO_2 − Hb) difference spectrum and (2) two "out-of-phase" spectral components, corresponding to processes that lag behind the modulation frequency. The fine tuning of the excitation beam frequency and the possibility of electronically compensating for the "in-phase" components make this a powerful technique. Analysis of the results starting from HbCO shows that the value of L_3 is 0.33 (pH 7, 0.1 M phosphate buffer,

[66] F. A. Ferrone and J. J. Hopfield, *Proc. Natl. Acad. Sci. U.S.A.* **73,** 4497 (1976).
[67] F. A. Ferrone, this volume [15].

19°), with $k_3(RT) = 1000$ sec^{-1} and $k_3(TR) = 3000$ sec^{-1}; interestingly, somewhat different values were obtained for O$_2$ ($L_3 = 1.5$, $k_3(RT) = 3000$ sec^{-1}, and $k_3(TR) = 2000$ sec^{-1}). Two observations are relevant to the present chapter: that the larger of the two "out-of-phase" spectral components corresponds to the $T_0 - R_0$ difference spectrum shown in Fig. 1 (i.e., a quaternary-linked perturbation of the deoxyhemes), and that the smaller "out-of-phase" spectral component, which has been attributed to a quaternary-linked spectral perturbation of the CO–hemes, corresponds to that obtained with trout HbIV-CO by Giardina et al.,[35] as compared in Fig. 2. It is rewarding that the two difference spectra display a remarkable quantitative agreement even though they have been obtained on two different hemoglobins (human Hb A and trout HbIV) by totally different approaches. Thus, it may be concluded that the difference spectrum shown in Fig. 2 truly represents the T_4-R_4 quaternary-linked spectral perturbation of the CO hemes, with an isobestic at 420 nm, a trough at 415 nm ($\Delta\varepsilon = -8$ mM^{-1} cm^{-1}) and a positive peak at 424.5 nm ($\Delta\varepsilon = +6.3$ mM^{-1} cm^{-1}).

Kinetics of the quaternary-linked structural change occurring at the $\alpha_1\beta_2$ interface were obtained by UV resonance Raman spectroscopy after photolysis of human HbCO[18,19]; the UV resonance Raman bands were assigned to $\alpha42$ (C7) Tyr and $\beta37$ (C3) Trp (see above). The probes of the interface structural transition yield, remarkably but consistently, an independent estimate of 10 to 20 μsec for the half-time of the $R_0 \rightarrow T_0$ quaternary transition. This may be taken as evidence that the dynamic pathway of the quaternary change in fully deoxygenated Hb is rate limited by a single barrier, which controls the interface structural transition and the heme perturbation.

Acknowledgments

The authors express their appreciation to the MURST of Italy for financial support.

[6] Allosteric Equilibrium Measurements with Hemoglobin Valency Hybrids

By MICHAEL C. MARDEN, JEAN KISTER, and CLAUDE POYART

The cooperative nature of hemoglobin (Hb) tetramers makes observation of the properties of the partially ligated species difficult. A cooperative system such as Hb (with maximum Hill coefficient near 3) favors oxygenation of species that already have ligands bound, relative to the deoxy species, thus inducing the formation of fully liganded forms. Instead of forming 37% doubly liganded Hb species at half-saturation for a random distribution of four identical sites, the amount of this species based on a cooperative two-state model[1] is typically less than 10%. A similar result occurs for a triply liganded hemoglobin species, which has a maximum relative population of less than 5%. Although different models predict specific properties of these intermediates, it is difficult to test the models for lack of precise knowledge of the properties and relative populations of the substates. It is also difficult to determine whether the allosteric mechanism is operational at each ligation level. To advance to the next level of resolution, it will be necessary to characterize the partially liganded species.

Symmetrical valency hybrids, in which one type of chain (α or β) is oxidized, are useful forms providing large populations of a specific intermediate. These symmetrical hybrids allow studies of the doubly liganded forms and have been reviewed previously.[2] Methods are also being developed to physically separate the partially liganded forms by low-temperature isoelectric focusing.[3] The mapping of the energy levels of the substates, from studies of the dimer–tetramer equilibria, also provides information about the partially liganded forms.[4]

Another method to enhance the population of certain substates is by perturbation of the system. Photodissociation of ligands (from four identical sites) would produce the binomial distribution of partially liganded species; however, the distribution for the ligand-rebinding signal requires a weighting for the number of deoxy sites (Fig. 1). Low levels of photolysis can be used to observe recombination to triply liganded tetramers, a form difficult to populate at equilibrium; for example, at 5% photodissociation of identical subunits, the relative populations are 0, 0.05, 1.4, 17.1, and 81.5% for the forms with zero, one, two, three, and four ligands, which provides ligand-rebinding signals (Fig. 1) of 0, 1, 13, 86, and 0%, respectively. Variation of the percentage photodissociation was used by Gibson to demonstrate the difference in the rebinding rate of CO to Hb in the deoxy (T) and triply liganded (R) conformations.[5] The combination of equilibrium and kinetic methods provides a more accurate description of this complex system.

In an analogous method, triply liganded tetramers can be studied by using samples that are nearly completely oxidized. Oxygen- or CO-binding

[1] J. Monod, J. Wyman, and J.-P. Changeux, *J. Mol. Biol.* **12**, 88 (1965).
[2] R. Cassoly, this series, Vol. 76, p. 106.
[3] M. Perrella and L. Rossi-Bernardi, this series, Vol. 76, p. 133.
[4] G. K. Ackers, M. L. Doyle, D. Myers, and M. A. Daugherty, *Science* **255**, 54 (1992).
[5] Q. Gibson, *J. Physiol.* (*London*) **134**, 123 (1956).

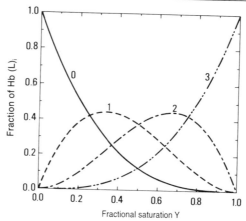

FIG. 1. Distribution of partially liganded substates contributing to the ligand recombination signal after photodissociation. The fractional saturation represents the fraction not photolyzed. Assuming identical subunits the partial photolysis produces the binomial distribution of partially liganded tetramers. These species must then be weighted by the number of deoxy subunits; for example, tetramers with four ligands will not show a rebinding signal and have a weight of zero. At low levels of photolysis (high saturation), the main contribution is from triply liganded tetramers (curve 3). The same distribution can be used for the partially oxidized samples; the fractional saturation then refers to the fraction oxidized. Assuming complete photolysis of the ferrous subunits, the plots show the relative contribution to the (ferrous) ligand-rebinding signal. Thus curve 3 refers to the fraction of the photodissociation signal due to triply oxidized tetramers.

experiments for a highly oxidized sample will essentially probe tetramers with a single ferrous subunit. One can then study how different ferric ligands will affect the overall allosteric equilibrium, using the well-known difference in oxygen affinity as a probe.

The scope of this chapter is limited to partially oxidized Hb samples. The goal is twofold: (1) There is a need to correct for the influence of a slightly oxidized sample. Although the oxidation levels can usually be maintained below 5%, this level is worse than it seems when dealing with a tetrameric system; in the worst case 20% of the tetramers would possess one ferric subunit (18.5% for a random distribution). Because the ferric subunits are liganded, they can potentially shift the allosteric equilibrium by the same order of magnitude (factor of 100) as a ferrous ligand. This problem may be aggravated in Hb mutants, natural or artificial, in which the rate of autoxidation is increased. Thus an understanding of how ferric subunits affect their neighbors within a Hb tetramer is important in order to compare the ligand-binding parameters of different samples. (2) The use of highly oxidized solutions provides another method to study the triply liganded form. With different ferric ligands and the use of external effectors, one can determine whether the two-state model is still operational at the triply liganded level. In addition, by comparing different ferric ligands with varying amounts of high- and low-spin ferric hemes, the coupling of the spin state and the allosteric equilibrium can be studied.

Preparation of Partially Oxidized Hemoglobin

Autoxidation is a natural process that, in the absence of the natural reducing enzymes, requires about 24 hr at 25° for 50% oxidation of oxyHb.[6] This rate is accelerated at higher temperatures, lower Hb concentrations, under conditions of partial oxygenation, and with effectors favoring the low-affinity conformation. For high levels of oxidation, which might require 48 hr, chloramphenicol (20 μg/ml) should be added to prevent bacterial growth.

An alternate method is by addition of oxidizing agents such as ferricyanide. In this case the sample must be stripped of any by-products of the reaction.[6] This method is faster than autoxidation and may be preferable for hemoglobins that are less stable. Adjusting the percentage metHb by reducing agents such as sodium dithionite may produce adverse side effects.[7]

The percentage oxidation may be calculated using the known absorption spectra at a given pH for the oxy, deoxy, and met forms.[8] A second measure of the percentage ferrous is the change in absorption on oxygen or CO binding. The most commonly used ferric ligands (and their observed affinities) are fluoride (10 mM), cyanide (2 μM), nitrite (1 mM), azide (10 μM), and imidazole (4 mM).[6] Without the addition of specific ferric ligands, there will be a water molecule or OH$^-$ bound depending on the pH of the solution.

Methemoglobin is slowly reduced by carbon monoxide[9]; approximately 1 hr is required for a change from 98 to 90% oxidized. This makes equilibrium studies with CO at a fixed percentage of metHb difficult. However, there may be certain advantages for rapid studies, such as flash photolysis: a completely oxidized sample can be prepared and then exposed to CO; the sample at different levels of oxidation may then be studied by making observations at various times during the reduction process. NO is a ligand

[6] E. Antonini and M. Brunori, "Hemoglobin and Myoglobin in Their Reactions with Ligands." North-Holland Publ., Amsterdam, 1971.

[7] J. S. Olson, this series, Vol. 76, p. 631.

[8] R. E. Benesch, R. Benesch, and S. Yung, *Anal. Biochem.* **65**, 245 (1973).

[9] D. Bickar, C. Bonaventura, and J. Bonaventura, *J. Biol. Chem.* **259**, 10777 (1984).

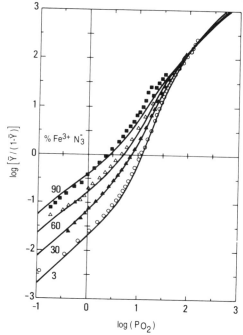

FIG. 2. Oxygen equilibrium curves (OECs) for the percentage azidometHb shown, 50 mM Bis–Tris buffer at pH 6.5, 0.1 M NaCl, 2 mM sodium azide, 25°. The solid lines are simulations using a modified two-state model in which ferrous ligands (c) and ferric-azide hemes (m) have a similar effect ($m/c = 1.5$) on the allosteric equilibrium ($L = 1.3 \times 10^6$, $c = 0.0066$, $K_R = 0.33$ mmHg); the fitting was constrained to a single set of values of the allosteric parameters for the family of curves at different oxidation levels.

of both ferrous and ferric Hb; the reduction of metHb by NO is faster than by CO.

Equilibrium Measurements

The methods for oxygen equilibrium measurements have been extensively reviewed.[10] Comments are limited here to the features relevant to the study of partially oxidized samples. Because the absorbance change for oxygen binding decreases with increasing oxidation level, it is necessary to increase the sample concentration to maintain a large signal.

Oxygen equilibrium curves (OECs) at various levels of oxidation, with azide bound to the ferric hemes, are shown in Fig. 2. The fractional

[10] K. Imai, this series, Vol. 76, p. 438.

saturation Y refers only to the ferrous hemes: for example, at 90% overall oxidation, $Y = 0.5$ represents 5% oxygen bound and 5% deoxy hemes (and 90% azide bound). The measurements were made with a Hemox analyzer (TCS Medical Products Co., Huntington Valley, PA)[11,12]; sample concentrations were 60 μM (total heme) for ferrous Hb samples, but progressively higher concentrations were used for the partially oxidized samples (300 μM at 90% oxidation) in order to maintain a sufficiently large oxygen-binding signal.

Equilibrium Simulations

The original two-state model[1] provides a compact form for the allosteric equilibrium: $T_i/R_i = Lc^i$, where L is the allosteric equilibrium coefficient for the deoxy form and i is the number of ferrous ligands bound. Each ferrous ligand shifts the equilibrium by a factor $c = K_R/K_T$, with K the ligand dissociation constant (mmHg). An analogous parameter m for the ferric (met) hemes can be used; for example, tetramers with two ferric hemes will have allosteric equilibria Lm^2, Lm^2c, and Lm^2c^2 for the forms with zero, one, and two ferrous ligands (oxygen or CO) bound.[12]

For identical subunits, the fraction of each partially oxidized species (zero to four ferric subunits) is given by the binomial distribution

$$F_j = 4!(1 - f)^{(4-j)} \times f^j/[j!(4 - j)!] \tag{1}$$

with f the fraction of hemes that are oxidized and j the number of ferric hemes per tetramer.

The fraction of ferrous ligand sites saturated can then be calculated for each partially met species, using the two-state formalism with $\alpha = [\text{ligand}]/K_R$:

$$Y_j = \frac{\alpha(1 + \alpha)^{(3-j)} + L(m)^j(\alpha c)(1 + \alpha c)^{(3-j)}}{(1 + \alpha)^{(4-j)} + L(m)^j(1 + \alpha c)^{(4-j)}} \tag{2}$$

The overall fraction saturation of ferrous sites is the sum (from $j = 0$ to 4) of the components, weighted by the fraction of each species and by the number of ferrous hemes per tetramer:

$$Y = \sum_{j=0}^{4} (4 - j)F_j Y_j \tag{3}$$

Simulations should be made with a single set of parameters for a series of curves at different oxidation levels. When oxygenation curves do not

[11] M. C. Marden, J. Kister, B. Bohn, and C. Poyart, *J. Mol. Biol.* **217**, 383 (1991).
[12] M. C. Marden, L. Kiger, J. Kister, B. Bohn, and C. Poyart, *Biophys. J.* **60**, 770 (1991).

show the full allosteric transition, such as at high oxidation levels where oxygen binds mainly to tetramers with a single ferrous subunit, it is not possible to separate the affinity and allosteric parameters. This compensation between the affinity and allosteric parameters also occurs for Hb in the presence of strong effectors.[13,14]

Simulations for the series of partially azidometHb data (solid lines in Fig. 2) yielded values of $m/c = 1.5$. With CN as ferric ligand, $m/c = 1$, meaning that each CN ion makes the same contribution as an oxygen molecule when calculating the allosteric equilibrium.[12] For example, tetramers with three CN ligands will show the same oxygen affinity (mainly R state) as for the fourth ligand in a completely ferrous tetramer. This leads to a left shift in the OEC relative to ferrous Hb, because with CN as ligand there is always a larger fraction R state for the partially oxidized tetramers.

Oxygen equilibrium curves for partially aquometHb samples are shown in Fig. 3. The left shift at low levels of oxygenation is smaller than for the case with azide or CN as ferric ligand, and the series of curves at different levels of oxidation intersect. Simulations required $m/c = 3$, indicating that triply aquometHb tetramers have an allosteric equilibrium shifted (by a factor of 27) toward the T state relative to triply oxygenated Hb; this leads to the cross-over of the OEC near 75% oxygenation due to the increased T-state contribution in the region of the upper asymptote.

For all metHb derivatives, there is a shift toward lower oxygen affinity on addition of effectors. The combination of inositol hexaphosphate (IHP) and L345, 2-[4(3,4,5-trichlorophenylureido)phenoxy]-2-methylpropionic acid, a more potent derivative of bezafibrate,[15] induces additive shifts in the oxygen affinity. For triply cyanometHb samples, the addition of both effectors shifts the oxygen affinity to nearly normal (ferrous) T-state values.[12]

A more general model has been described, which accounts for dimers and uses a value of c_j that depends on the number (j) of ferric subunits.[16] A factor of about 4.8 was reported for m/c_0 for partially aquometHb (their parameter d/c_0), again indicating a smaller shift toward the R state for the binding of a high-spin ferric ligand relative to the binding of an oxygen molecule. Note that the distribution of substates may also be affected by the dimer–tetramer equilibrium.[16] This distribution may change during the

[13] M. C. Marden, J. Kister, B. Bohn, and C. Poyart, *Biophys. J.* **57,** 397 (1990).

[14] C. Poyart, M. C. Marden, and J. Kister, this volume [24].

[15] I. Lalezari, P. Lalezari, C. Poyart, M. C. Marden, J. Kister, B. Bohn, G. Fermi, and M. F. Perutz, *Biochemistry* **29,** 1515 (1990).

[16] L. Cordone, A. Cupane, M. Leone, V. Militello, and E. Vitrano, *Biophys. Chem.* **37,** 171 (1990).

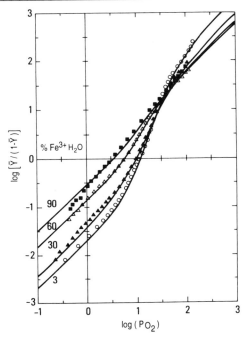

FIG. 3. Oxygen equilibrium curve for different percentages of aquometHb at pH 6.5, 25°. Each high-spin ferric ligand causes a weaker shift (factor of $1/m$) toward the R state, as compared to oxygen ($1/c$). Simulations for the series of OEC were made with a generalized two-state model, with $c = 0.0066$, $m/c = 3.0$, $L = 1.3 \times 10^6$, $K_R = 0.33$ mmHg.

oxygenation process, and the rates involved may become important; for example, several hours may be required to obtain the equilibrium distribution at low saturation levels.

Ligand Rebinding after Photodissociation

The method of flash photolysis is an excellent complement to equilibrium measurements. In addition to providing the ligand association rates, the distribution of partially liganded substates after photodissociation is different from that at equilibrium; thus there is an enhancement of the resolution for certain substates (Fig. 1).

The experimental system consists of two convergent optical beams. A pulsed light source is used to break the iron–ligand bond, and a second beam detects the subsequent absorption changes of the protein. This method and the instrumentation have been described previously.[17] Advances in digital oscilloscopes and analog/digital interface cards for microcomputers have facilitated data collection and storage.

Geminate Recombination

Just after photodissociation, the iron–ligand bond is broken, but the ligand is still close to the binding site. The ligand may rebind directly to the same iron atom (the geminate phase, which occurs typically on the picosecond–nanosecond time scale in aqueous solvents) or migrate to the surface of the protein and escape to the solvent. The relative rates for direct rebinding versus dissociation from the heme pocket will determine which fraction of the ligands rebinds via the geminate phase; the geminate fraction depends on the type of ligand, the protein and the solvent viscosity.[18,19] The ligands that escape into the solvent will rebind on a slower time scale (microsecond–millisecond); this bimolecular phase is characterized by a linear dependence on the solvent ligand concentration. Although we concentrate exclusively on the bimolecular phase in this discussion, it is important to take into account the amplitude of the geminate phase to obtain the correct initial distribution of substates for the bimolecular phase. Rather than measuring the geminate phase, it is more practical to compare the amplitude of the observed bimolecular signal with that for the static difference between the liganded and deoxy forms, also measured in the kinetic apparatus. This measurement should be made at a wavelength at which the ligand-binding signal dominates, such as 436 nm.

Photolysis Pulse Length

Generally, a shorter laser pulse permits a greater variety of experimental results. However, the presence of a large geminate recombination fraction may prevent observation of certain bimolecular phenomena. It is important in Hb studies to be able to vary the laser energy to dissociate from zero to four ligands. For short laser pulses, the maximum bimolecular fraction is independent of laser energy beyond one photon per heme. For example, if each heme is dissociated by a 1-psec pulse, a sample of HbCO at 25° will show about 60% geminate recombination on a nanosecond time scale. More photolysis energy will simply be absorbed by deoxyhemes. One way to increase the bimolecular amplitude is to flash a second time

[17] C. A. Sawicki and R. J. Morris, this series, Vol. 76, p. 667.

[18] R. H. Austin, K. W. Beeson, L. Eisenstein, H. Frauenfelder, and I. C. Gunsalus, *Biochemistry* **14**, 5355 (1975).

[19] D. Beece, L. Eisenstein, H. Frauenfelder, D. Good, M. C. Marden, L. Reinisch, A. H. Reynolds, L. B. Sorensen, and K. T. Yue, *Biochemistry* **19**, 5147 (1980).

just after the geminate phase (about 1 μsec); alternatively a pulse length of about 1 μsec would allow multiple ligand dissociation of hemes initially involved in the geminate phase. Thus while the 10-nsec yttrium–aluminium–garnet (YAG) lasers have advantages in stability and repetition rate, the older dye lasers[17] with pulse lengths about 600 nsec may be more useful for certain bimolecular studies.

Allosteric Reequilibration

A two-state model implies 10 tetrameric forms: R and T, each with 0 to 4 ligands bound. After partial photolysis of liganded (R-state) Hb tetramers, there will be a distribution of partially liganded forms. These forms will reequilibrate with their T-state counterparts, in parallel with ligand rebinding. The rapid (R-state) and slow (T-state) bimolecular recombination fraction after photodissociation provides information on the allosteric equilibrium. To observe this phenomenon, the ligand recombination should be slower than the R-to-T transition. Consider the complete photolysis of Hb (from R_4 to R_0): this species will relax to the more energetically favorable T_0, but the transition R_0 to T_0 is not instantaneous. Direct measurements[20] at isosbestic points for ligand dissociation indicate that the R-to-T transition requires approximately 100 μsec. If ligand recombination is rapid compared to this rate, one will simply observe recombination to the liganded (R-state) conformation. Lower ligand concentrations can be used to permit completion of the allosteric transition; however, a low ligand affinity (as for oxygen) might not permit a large range of useful concentrations.

Choice of Ligand

The rebinding rate of NO after photolysis is rapid; it is therefore useful for picosecond geminate studies,[21] but provides practically no bimolecular signal. Oxygen binding also shows a large geminate fraction (>90%). For both oxygen and NO, the difference of a factor of 100 in ligand affinity between the R and T states is due mainly to the difference in the ligand dissociation rate. The situation is reversed for CO, which shows a T-state association rate about 30 times slower than for the R state. CO is the best choice for bimolecular measurements, because (1) the slower rebinding rates permit completion of the allosteric transition, (2) the high affinity of Hb for CO allows a wide range of ligand concentrations (0.01 to 1 atm CO) to probe the competition between R-state rebinding and the allosteric

[20] C. A. Sawicki and Q. H. Gibson, *J. Biol. Chem.* **251**, 1533 (1976).
[21] J. L. Martin and M. H. Vos, this volume [19].

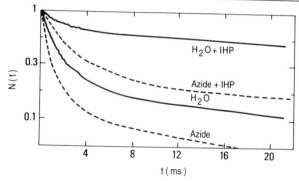

FIG. 4. Recombination kinetics of CO to 95% aquomet- or azidometHb, pH 6.5, 25°, 200 μM total heme, 2 mM azide, 0.1 atm CO (100 μM free CO). With (1 mM) or without the effector IHP, the high-spin ferric ligand (water) shows more slow phase, characteristic of rebinding to T-state Hb, than the low-spin ferric ligand (azide).

transition, (3) the large difference in the R and T recombination rates results in a better analysis of the two phases, and (4) the higher bimolecular quantum yield (0.4 for CO at 25°) permits a larger range of photolysis.

Kinetics for the recombination of CO to 95% metHb, (200 μM total heme) equilibrated with 0.1 atm CO, 50 mM phosphate buffer at pH 6.5, are shown in Fig. 4. The biphasic kinetics are characteristic of the Hb tetramer, with the slow phase corresponding to recombination to T-state Hb. For samples equilibrated with 0.1 atm CO, the R–T transition is rapid compared to ligand rebinding; the allosteric equilibrium is thus established after the photodissociation and the slow fraction is a measure of the amount of T state. Tetramers with the high-spin ligands fluoride or water bound to the ferric subunits showed the most T state behavior.

A series of measurements consists of recording the kinetics for the partially aquometHb sample, first without and then with effector (IHP); finally the alternate ferric ligand was added. Because all three measurements could be made within a few minutes, the oxidation level is little changed. In another series, the second ferric ligand can be added before the effector. Note that the ferric ligands are not photodissociated.

Inositol hexaphosphate induces a decrease in the recombination rates and an increase in the T-state behavior.[22] With or without IHP, the high-spin ligands (water and fluoride) show the most T-state behavior. As observed for the equilibrium studies, addition of both IHP and L345 produced the maximum amount of T-state behavior. With both effectors

[22] R. D. Gray and Q. H. Gibson, *J. Biol. Chem.* **246,** 7168 (1971).

present, the CO recombination kinetics of aquo- and fluorometHb samples showed over 95% slow phase; cyanometHb samples showed 55% slow phase.[12]

Kinetic Simulations

Unlike for the equilibrium simulations, there are no simple formulas for the kinetic simulations. The proper initial distribution of substates is the starting point; as mentioned above, this may involve two types of distributions: oxidation and photodissociation. The kinetic curves are then generated by numerical integration, taking into account all substates and their interconversion rates. For simulations of kinetic data, the dimer–tetramer reequilibration is slow compared to the ligand-rebinding kinetics, and the dimers can be treated as a static fraction of "R-like" Hb.

Distribution of States

Identical subunits were assumed for the calculation of the distribution of partially met species. It has been reported that within Hb tetramers the α chains oxidized more rapidly than the β chains[23]; however, studies have shown little difference in the oxidation rates.[15,24] One can test for a large deviation from the supposed random distribution, such as a mixture of completely ferrous or ferric tetramers. The fraction of slow ligand recombination of completely ferrous tetramers shows a dependence on the number of ligands photodissociated. Variation of the laser energy showed little change in the kinetics for 95% oxidized samples, as expected for tetramers with a single ferrous subunit.

Double Distribution

In addition to the distribution of partially met species, there may be problems in calculating the exact amount of partially liganded ferrous species after photolysis. For identical sites, the binomial distribution can be used. There is evidence that the α and β chains might have different quantum yields. The compound errors of both distributions involved in photolysis experiments of samples near 50% oxidized may lead to a hopeless analysis. Two experimental parameters are associated with this problem.

The first concerns the properties of the photolysis beam. For a typical sample concentration (100 μM in a 1-mm cuvette) with photodissociation in the visible absorption bands (532 nm for a frequency-doubled YAG

[23] A. Mansouri and K. H. Winterhalter, *Biochemistry* **12**, 4946 (1973).
[24] A. Tomoda, Y. Yoneyama, and A. Tsuji, *Biochem. J.* **195**, 485 (1981).

laser), the absorption is about 0.15, meaning that the front side (incoming laser pulse) of the sample receives 30% more energy than the far side. Ideally all molecules should receive the same intensity. Photolysis at a wavelength of lower sample absorption would provide a more uniform beam, but a higher laser energy would then be needed to provide the same photolysis levels. Flashing the sample from both sides also helps provide a more uniform intensity across the sample. In addition it is necessary to consider the homogeneity of the beam intensity (hot spots), polarization effects, and the centering of the detection beam within the photolyzed region (1-mm optical path length cuvettes are better that 1-cm cuvettes). Note that this is one example in which signal averaging with a large jitter in the photolysis energy may not yield a better data set.

The second experimental parameter is the detection light intensity, usually a continuous source for bimolecular studies. As with the principal photolysis pulse, the detection beam is capable of ligand dissociation. If this induced dissociation rate is high enough, it may depopulate the bound state and change the initial (preflash) distribution of substates. Such an effect can usually be observed by a loss of signal size at progressively higher detection light levels. However, some subtle effects may occur. Consider an HbCO sample with effectors bound that favor the T state, and a partial photolysis by the detection beam that induces a steady state of 90% ligand bound; one would expect a 10% loss of signal, but if there is a preflash shift toward the T state, the higher (bimolecular) quantum yield might more than offset this effect. A simple test is to lower the light level until there is no further change in the shape of the kinetic trace. Samples such as symmetrical valency hybrids, or the extreme conditions such as 5 or 95% metHb, provide conditions that are more easily interpretable.

It is also necessary to take into account the changing ligand concentration. Low CO levels are often used to permit completion of the R-to-T transition after the flash. High heme concentrations (100 μM) are used to provide a large signal and to avoid a large dimer contribution. Note that the amount of CO photodissociated may be equivalent to that in solution before the flash (about 100 μM for equilibration with 0.1 atm CO); thus for 100 μM (in heme) samples equilibrated with 1 and 0.1 atm CO, there may be only a relative change in the initial CO-rebinding rate of about a factor of 5.

Coupling of Spin and Allosteric Equilibria

The influence of the spin state of the heme–ligand complex on the allosteric equilibrium in ferric Hb was suggested by Perutz, based on IHP-induced changes in the absorption spectra[25] and oxidation potentials,[26] and supported by the fact that low-spin cyanometHb crystallizes in the liganded (R-state) conformation,[27] whereas high-spin fluorometHb with IHP shows a ferrous deoxy (T-state) structure.[28] Studies on completely ferric Hb do not permit a comparison with the usual oxygen-binding properties. By varying the ferric ligand with triply oxidized tetramers, one can determine whether the heme–heme interaction is still operational at the triply liganded level and study the influence of spin state on the allosteric equilibrium, using the oxygen- or CO-binding properties as a probe of the fourth (ferrous) chain.

Relative to the (ferrous) unliganded form, all ligands shift the allosteric equilibrium toward the R state. A value of $m/c = 3$, with aquometHb, for example, indicates that the shift toward the R state is three times less for water binding than for the binding of an oxygen molecule. Thus values of $m/c > 1$ indicate a shift toward the T state relative to the oxygenated form. A value of $m/c = 3.5$ was obtained for fluorometHb relative to cyanometHb ($m/c = 1$), which should represent the total difference between the high-spin and low-spin forms.

Note that the high-spin fraction and the T-state fraction are not necessarily the same, but rather there are four possible states in equilibrium: the R and T conformations, each of which may be high spin or low spin. The overall correlation shows that the spin state is an important parameter in determining the overall R–T equilibrium. The percentage T state for fully metHb ($j = 4$) can be calculated assuming that the fourth ferric ligand produces the same shift in the allosteric equilibrium as observed for the first three ($T_4/R_4 = Lm^4$). The extrapolated values are consistent with aquometHb (without IHP) being mainly R state (18% T state) and fluorometHb plus IHP as T state (72%), but show that the R-to-T transition ($\pm$IHP) is not complete.

With completely oxidized Hb there is no way to calibrate the IHP-induced change in the absorption spectra to obtain the initial and final T-state fraction. Because the crystal formation will apparently not allow for a mixture of the two allosteric conformations, the crystallographic studies give the misleading impression of an all-or-nothing transition induced by IHP. The extrapolation of the simulations for partially aquometHb solutions indicates that the transition is from 18 to 52% T state on addition of IHP. Using the spin data of Philo and Dreyer[29] and the results

[25] M. F. Perutz, A. R. Fersht, S. R. Simon, and C. K. Roberts, *Biochemistry* **13**, 2174 (1974).

[26] J. V. Kilmartin, *Biochem. J.* **133**, 725 (1973).

[27] J. F. Deatherage, R. S. Loe, C. M. Anderson, and K. Moffat, *J. Mol. Biol.* **104**, 687 (1976).

[28] G. Fermi and M. F. Perutz, *J. Mol. Biol.* **114**, 421 (1977).

[29] J. S. Philo and U. Dreyer, *Biochemistry* **24**, 2985 (1985).

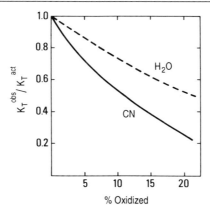

FIG. 5. Relative error for simulations of data for partially metHb samples when using a model (MWC) that does not account for the ferric ligands. Plotted is the ratio of the observed T-state affinity to the actual (0% metHb) value versus the percentage metHb. The low-spin ferric ligands (such as CN) show larger deviations than for the partially aquometHb samples.

for the partially oxidized Hb solutions,[11,12] the spin and allosteric equilibria do appear to be coupled.

For aquometHb, there is a sharp drop in the T-state fraction kinetics above pH 7.2 (data not shown); this results from a combination of changes: (1) the ferric ligand (OH) at alkaline pH is low spin; (2) the IHP affinity decreases at higher pH values; and (3) the allosteric equilibrium is also pH dependent, with less T state at high pH.

The use of multiple effectors can be used to confirm that the IHP-induced transition from R to T is not complete. For aquometHb, the calculation of 18 to 52% T state indicates that a much larger change of absorption for the R-to-T transition is potentially observable. The absorbance changes of metHb due to both IHP and bezafibrate are larger than those of IHP alone.[30] Similarly the changes in the absorption spectra induced by both IHP and L345 were nearly twice those for IHP alone[12]; the kinetic results also show a larger shift towards the T state when both effectors are present, again confirming that the R-to-T transition is not complete with IHP alone.

The fraction of Hb in the T state, as evidenced by the low oxygen affinity or slow CO bimolecular rebinding rate, depends on the nature of the ligands bound. Ferrous ligands such as oxygen or CO, and low-spin ferric ligands such as CN, show a similar contribution toward the allosteric

---

[30] R. W. Noble, A. DeYoung, S. Vitale, M. Cerdonio, and E. E. DiIorio, *Biochemistry* **28,** 5288 (1989).

equilibrium. The binding of high-spin ferric ligands also shifts the allosteric equilibrium toward the R state, but by a factor three to four times less than for the low-spin ligands; ferric ligands involving a mixture of high- and low-spin states (azide, nitrite, or imidazole) show intermediate values. The observed allosteric equilibrium depends on the ferric ligand and is generally correlated with the ferric spin state.

One can estimate the error involved in simulating OECs for samples with a known percentage metHb by using the simple two-state model, that is, by comparing the two-state parameters to those for the model that accounts for the contribution of the ferric subunits. Simulations were made by varying only the affinity parameters; the apparent error in affinity depends on the ferric ligand, as shown in Fig. 5. High-spin ligands such as water lead to a smaller error in the T-state affinity than for low-spin ligands such as CN, because the high-spin ligands cause a smaller shift from the T to the R state. This is also true for the R-state affinity, even though there is little change in the upper asymptote of the equilibrium curves for partially metHb with low-spin ligands; this is because after correction in the T-state affinity a subsequent correction is needed for the R-state affinity. Note that for data measured near pH 7, there will be a mixture of high-spin (water) and low-spin (OH) forms and the correction will be intermediate. Ideally a model accounting for the ferric ligands should be used before comparing specific parameters for different Hb samples when a significant percentage metHb is involved.

[7] Electron-Transfer Reactions of Hemoglobin with Small Molecules: A Potential Probe of Conformational Dynamics

By GEORGE MCLENDON and JEHUDAH FEITELSON

Introduction

Conformational fluctuations clearly play a central role in the chemistry of hemoglobin (Hb), in facilitating the necessary ligand binding and release. They also may play a critical role in the (less desirable) redox chemistry of hemoglobin, including redox-induced methemoglobinemia. In this latter context, we have been interested in studying how hemoglobin redox chemistry is modulated by conformational fluctuations.

All electron transfer reactions, whether involving simple organic molecules or more complex redox-active proteins, are sensitive to the distance that separates the electron donor and acceptor. In general, biological redox-active chomophores, like the heme group in hemoglobin, are somewhat buried within an insulating protein matrix, which, at least partially, prevents close approach of the reactants, thereby greatly diminishing their redox activity *vis-à-vis* analogous isolated (protein-free) chromophores. As a rule of thumb, each 2-Å separation between the oxidant and reductant (separated by the protein) results in a 10-fold drop in the rate constant for electron transfer.[1] Therefore, any structural fluctuations that allow closer approach to a buried redox chromophore can result in a significant increase in rate. If such increases are sufficiently large, the overall rate of reaction will be limited by the rate of these conformational fluctuations: the reaction becomes "conformationally gated."

As an approach to the study of such processes, we here present data on photoinduced electron transfer reactions of hemoglobin with a variety of small molecules. By using a filled shell, photoactive zinc-substituted heme, the electron transfer can be studied without complications arising from ligand substitution effects. As is appropriate for this series, we focus primarily on the methodology of such studies: how the measurements are made and interpreted.

Reactants

The preparation and characterization of zinc-substituted hemoglobins have been discussed in detail elsewhere.[3] Of present interest is the observation that Zn^{II} serves as a stereochemical equivalent to deoxy-Fe^{II}. Thus, tetrazinc-substituted hemoglobin adopts a limiting T-state quaternary structure. The bisligated hybrid hemoglobin $(\alpha FeCN)_2 \beta(Zn)_2$ has already switched to the R quaternary state. Thus, it is possible to investigate reactions of the same chromophore within two different quaternary states of Hb. A specific protocol for the solutions of interest is reproduced below.

$\alpha Zn\beta Fe^{3+}CNHb$ is prepared as described previously.[2] It is fully oxidized by reacting it with $K_3Fe(CN)_6$ and removing the surplus ferricyanide by chromatography on a Sephadex G-25 column. Zn_4Hb is prepared as previously described by Hoffman.[2] The proteins are stored in the form of frozen pellets in liquid nitrogen. For each experiment, a small amount of the hemoglobins is dissolved in pH 7.2, 0.02 M phosphate buffer to a concentration of about 8×10^{-5} M (heme). The $\alpha Zn\beta FeHb$ is cyanide ligated by adding an equimolar amount of aqueous KCN to the solution.

[1] G. McLendon, *Acc. Chem. Res.* **21**, 160 (1988).
[2] B. Hoffman and A. Ratner, *J. Am. Chem. Soc.* **109**, 6237 (1987).
[3] K. Simolo, G. Stucky, B. Chen, M. Bailey, C. Scholes, and G. McLendon, *J. Am. Chem. Soc.* **107**, 2865 (1985).

The ligand exchange is followed by the disappearance of the 408-nm shoulder in the Soret absorption band. Subsequently, the protein and the quencher solutions are deoxygenated by flushing with high-purity nitrogen in septum-stoppered Erlenmeyer flasks and transferred to a glove box.

For quenching with AQS (anthraquinone sulfonate) and MV (methyl viologen dichloride) a 1-ml portion of protein solutions plus 3 ml of deoxygenated buffer and the appropriate volume of quencher solution are anaerobically transferred to an optical cell and stoppered with subseal septa. The cells containing the solution are illuminated in a water-jacketed cell holder by a DCR2 neodymium:yttrium–aluminum–garnet (Nd:YAG) laser at 532 nm and the transient absorption is followed by an R-928 (Hamamatsu) photomultiplier. The signal is digitized by a Tektronix 7912 digitizer and transferred to an IBM PC for data analysis.

For quenching with oxygen the solution contained in a deoxygenation vessel with an optical cell as a side arm[4] is freed of oxygen by flushing with prepurified nitrogen. A measured amount of air is injected with a syringe into the apparatus and the solution is equilibrated with the atmosphere in the vessel for at least half an hour by stirring it at the temperature of the experiment. The oxygen concentration in solution is determined from the solubility of oxygen at the given temperature.

Oxygen-containing porphyrin solutions readily undergo irreversible photochemical changes producing products that absorb light in the absorption range of the triplet and thus interfere with the determination of the decay constants. Therefore, the quenching rate by oxygen is determined by illuminating the solution by a weak laser flash from a nitrogen laser-pumped dye laser (Molectron DL220) at 550 nm. Both the decay of the E-type delayed fluorescence from the zinc protoporphyrin in hemoglobin and its triplet absorption are determined as a function of temperature.

Photochemistry of Zinc Porphyrins and Zinc-Substituted Heme Proteins

The ground states of filled shell porphyrins, like zinc protoporphyrins, are hard to oxidize. However, on photoexcitation, the excitation energy promotes an electron to a high energy level, so that electron transfer becomes both thermodynamically and kinetically favorable. The energy available for redox photochemistry can thus be calculated by summing the excitation energy and the redox energy:

$$E^o_{(\text{excited state redox})} = E^{o-o}_{\text{excitation}} - E^o_{(\text{oxidation ground state})}$$

[4] N. Barboy and J. Feitelson, *Biochemistry* **26**, 3240 (1987).

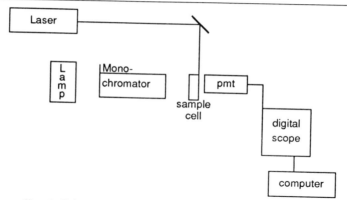

FIG. 1. Schematic diagram of the flash spectroscopy apparatus.

Both a singlet excited state ($^1E_{o-o}$ = 2.05 eV) and a triplet excited state ($^3E_{o-o}$ = 1.74 eV) are available for reaction. In practice, electron transfer reactivity proceeds almost exclusively through the triplet state because it has a much longer lifetime than the singlet, thus allowing sufficient time for a diffusional reaction to occur ($^1\tau \approx$ 10 nsec, $^3\tau \approx$ 10 msec).

When a redox-active center is brought close enough to react, the excited state becomes deactivated by electron transfer and the observed lifetime is thereby decreased:

$$1/\tau_{obs} = 1/^3\tau + 1/\tau_{electron\ transfer} \quad \text{or} \quad k_{obs} = {}^3k + k_{et}$$

where the observed electron transfer rate constant, k_{et}, may include those conformational fluctuations that facilitate electron transfer.

Measurement of Rates

Flash photolysis provides the most convenient method for measuring the photochemical properties of excited state hemes. In our experiments, reaction is initiated by a 10-nsec laser flash either from a frequency-doubled Nd:YAG laser (λ_{max} 532 nm) or by a dye laser, tuned to an absorption feature of the Zn^{II} heme. A basic block diagram of a "standard" instrument is provided in Fig. 1.

In principle, the rates of interest can be measured either by monitoring the disappearance of the reactant (triplet excited state, $\lambda_{max} \approx$ 450 nm) or by the appearance of products. In practice, it is often more precise to measure reactant loss, because the net yield of products can be low, with a correspondingly low signal. This low net yield is due primarily to charge recombination of the initial (high-yield) redox products as summarized in the following equations:

Excitation: $(ZnHb) \xrightarrow{h\nu} {}^{3}(ZnHb)*$

Reaction: ${}^{3}(ZnHb)* + Q \xrightarrow{k_{et}} {}^{3}(ZnHb)*/Q \rightarrow (ZnHb^{+}/Q^{-})$

Prompt recombination: $ZnHb^{+}/Q^{-} \xrightarrow{k_{b}} ZnHb/Q$

Slow recombination: $ZnHb^{+}/Q^{-} \xrightarrow{k_{escape}} ZnHb^{+} + Q^{-}$

$ZnHb^{+} + Q^{-} \xrightarrow{k_{diff}} ZnHb + Q$

The net observable yield of redox products therefore is given by $k_{esc}/(k_{esc} + k_{b})$. In general, $k_{b} > k_{esc}$, therefore the product yield, and associated signal, can be rather low. This is why the best signal-to-noise (S/N) ratio is obtained by monitoring the triplet state reactant (at 450 nm) (Fig. 2). It is important, however, to demonstrate that the observed quenching indeed occurs by electron transfer, with production of the expected redox products. For the reaction of ${}^{3}(ZnHb)*$ with methyl viologen studied here, the redox products have indeed been observed by Magner and McLendon.[5]

Rates and Activation Parameters

With this background, we can briefly discuss some relevant results. The rate of electron transfer quenching of photoexcited zinc protoporphyrin by methyl viologen, anthraquinone, or O_2, occurs at the diffusion controlled limit: $k_{obs} \approx 3 \times 10^9 \ M^{-1} \ sec^{-1}$. For zinc-substituted hemoglobins, this rate is reduced by over an order of magnitude, $k_{obs} \approx 0.8\text{–}2.0 \times 10^8 \ sec^{-1}$. Such a decrease is not surprising, given the fact that the reactive chromophore is insulated from reaction by the hemoglobin matrix.

There are two possible explanations for this rate decrease. The first is based on a static structural perspective. The reaction always occurs via direct collisions between the quencher and the small portion of the heme in hemoglobin that is exposed to solvent (Fig. 3). In this simplest case, the observed rate constant would simply be $k_{obs} = k_p(f)$, where k_p is the rate constant for the reaction with porphyrin and f is the fraction of heme "exposed" for direct reaction with the quencher. Such a "static" model would predict about the right magnitude for the observed decrease in the rate constant. However, it does not predict any change in activation

[5] E. Magner and G. McLendon, *Biochem. Biophys. Res. Commun.* **159,** 472 (1989).

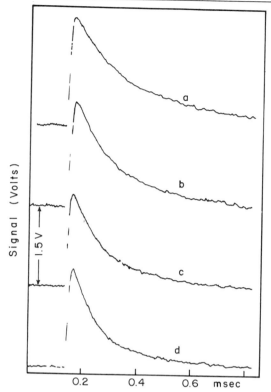

FIG. 2. Transient absorption signal for the first-order decay of the Zn_4Hb triplet in the presence of 385 μM AQS at (a) 7°, (b) 17.5°, (c) 25°, and (d) 36.8°. The ordinate shows the decrease in light transmittance (in volts) as measured by the digitizer.

parameters. Because f is only a statistical factor, it can affect only the prefactor and not the activation energy. Thus, for a static model, one predicts $\Delta G^{\ddagger}_{Hb} = \Delta G^{\ddagger}_{Porph}$. This is not observed: $\Delta G^{\ddagger}_{Hb} \approx 5$ kcal/M > $\Delta G^{\ddagger}_{Porph} \approx 3$ kcal/M. Furthermore, in a direct collision, the closest approach of the quencher will be modulated by the size and charge of that quencher: a small O_2 molecule can more readily penetrate the side chains that surround the heme than can a bulky charged molecule such as MV^{2+}. However, no significant rate differences are observed between a small uncharged quencher (O_2) and the larger cationic MV^{2+}, or anionic AQS^-.

These observations suggest that some structural fluctuation(s) (for which $\Delta G^{\ddagger} \approx 6$ kcal/M) must occur that make the heme more readily accessible to an external reactant. In this case, the rate could be controlled

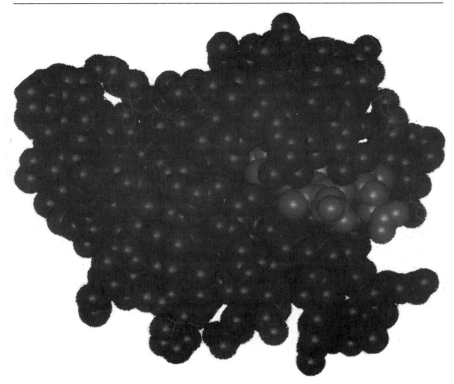

FIG. 3. Model of hemoglobin, showing exposed heme groups.

not by simple diffusion, nor even by the electron transfer step, per se, but by the rate(s) of fluctuation(s) that produce the active species. This more reactive conformational state would serve as a gateway to reaction, and the observed electron transfer reaction rate would be gated by the rate of conformational change.

Conformational Gating

A simple approach to treating gated reactions is summarized below. A gating mechanism[6-8] was previously suggested for ZnPPN-substituted myoglobin.[4] Adopting the theoretical approach of Northrup and McCammon, a process in a restricted medium, such as a protein, is described by the reaction coordinate for the electron transfer, and by an additional

[6] S. H. Northrup and J. A. McCammon, *J. Am. Chem. Soc.* **106,** 930 (1984).
[7] N. Agmon and R. Kosloff, *J. Phys. Chem.* **91,** 1988 (1987).
[8] D. Shoup and A. J. Szabo, *Biophys. J.* **40,** 33 (1982).

auxiliary coordinate that is related to changes in the protein conformation. The potential energy surface is depicted as a topographic map in which the x axis is the reaction coordinate and the y axis is the (conformational) auxiliary coordinate. If the direct passage from location A_1 to location B_1 in the protein is blocked by a higher potential barrier, ΔG ($A_1 \rightarrow B_1$), that is, a bottleneck in the passageway, there might exist a different protein conformation for which the free energy of activation for the passage from A to B is lower, ΔG ($A_2 \rightarrow B_2$). The subscripts 1 and 2 denote protein conformation such that ΔG ($A_2 \rightarrow B_2$) < ΔG ($A_1 \rightarrow B_1$). The theory describes two limiting cases. If the conformational fluctuations in the protein occur on a time scale that is short in comparison to the migration of the quencher, the protein conformation adapts to the movement of the latter and the activation energy of the process represents the lowered energy barrier ΔG ($A_2 \rightarrow B_2$). The picosecond fluctuations of the amino acid side chains described by molecular dynamics calculations would constitute such fluctuations. These small, local fluctuations cannot be expected to facilitate the passage of the small O_2 molecule and the much large AQS^- and MV^{2+} molecules to the same extent. Therefore, the energy barrier experienced by these different size molecules would not be the same in contrast to the observed similarities in rate constants and activation energies. In the other limiting case, the conformational changes are slow with respect to the time it takes the quencher to pass the constriction. In this case, the activation energy measures the energetics of the protein dynamics, that is, the opening of a gate. This is denoted on the above two-dimensional energy surface as a change along the auxiliary coordinate from point A_1 to point A_2. If we approximate the free energy of activation by the enthalpy of activation, then the similarity of the activation energies for all three quenchers indicates that the gating mechanism operates in hemoglobin and that the measured E_a values represent the energy needed to effect the above conformational changes. Once this conformation change, the opening of the gate, has taken place, small and large quencher molecules alike can pass the above constricted region with equal ease on their way from location A to location B. This suggested mechanism necessarily implies that a rather large gate, that is, a passage of >6-Å radius, is required to accommodate the large AQS^- or MV^{2+} quenchers. These data are particularly germane in the context of the interest in electron transfer in proteins. Conformational gating has been proposed to play an important role in protein redox reactions,[9] but few examples exist of such gating.[10] The present data provide a possible example of one type of conformational gating in electron transfer in which relatively large

[9] D. M. Scholler, M. R. Warig, and B. Hoffman, this series, Vol. 52C, p. 487.
[10] G. McLendon, K. Pardue, and P. Bak, *J. Am. Chem. Soc.* **109**, 7540 (1987).

structural rearrangements allow a close approach between the reactants. These data also point out a possible difficulty in other studies in interpreting the activation parameters observed for protein electron transfer. In the present case, the observed activation energy bears no relationship to the fundamental electron transfer step but rather to a preceding conformational rearrangement. Similar results can be predicted for other proteins for which gating is involved in the rate-determining step.

Acknowledgments

This work was supported by the National Institutes of Health: NIH GM 33881 (G.McL.).

Section II

Spectroscopy

[8] Proton Nuclear Magnetic Resonance Studies of Hemoglobin

By CHIEN HO and JANICE R. PERUSSI

Introduction

Proton nuclear magnetic resonance (^{1}H NMR) has proved to be a powerful technique for investigating the structure–function relationship in hemoglobin (Hb). ^{1}H NMR spectroscopy has played an important role in our understanding of the structural changes associated with the cooperative oxygenation of Hb and the structural features of partially ligated Hb species,[1] and of the molecular basis of the Bohr effect.[2-4] New methods for multinuclear, multidimensional NMR, combined with recombinant DNA techniques, which make possible the production of any desired mutant Hb and the incorporation of isotopic labels using cloned genes in suitable expression systems, promise even greater advances in our understanding of Hb.

Hemoglobin is the oxygen carrier of the blood. Human normal adult hemoglobin (Hb A) is a tetrameric protein consisting of two α chains of 141 amino acid residues each and two β chains of 146 amino acid residues each. Each chain contains a heme group, an iron complex of protoporphyrin IX. Under physiological conditions, the heme-iron atoms of Hb normally remain in the ferrous state. In the absence of oxygen, the four heme-iron atoms in Hb A are in the high-spin ferrous state (Fe^{II}) with four unpaired electrons each. In the presence of oxygen, each of the four heme-iron atoms in Hb A can combine with an O_2 molecule and is then converted to a low-spin, diamagnetic ferrous state. For details, see Ho[1] and the references therein.

Research efforts on Hb during the past 60 years have been concentrated in two main areas, namely the molecular basis for the cooperative oxygenation of Hb A and the molecular basis for the Bohr effect. The oxygen dissociation curve for Hb exhibits sigmoidal behavior, with an overall association constant expression giving a greater than first-power dependence on the concentration of O_2. Thus, the oxygenation of Hb is a cooperative process, such that when one O_2 is bound, succeeding O_2

[1] C. Ho, *Adv. Protein Chem.* **43**, 153 (1992).
[2] C. Ho and I. M. Russu, *Biochemistry* **26**, 6299 (1987).
[3] M. R. Busch and C. Ho, *Biophys. Chem.* **37**, 313 (1990).
[4] M. R. Busch, J. E. Mace, N. T. Ho, and C. Ho, *Biochemistry* **30**, 1865 (1991).

molecules are bound more readily. The oxygen affinity of Hb depends on pH.[5,6] At pH values between 6.1 and 9.0, the O_2 dissociation curve is shifted to higher oxygen pressure with decreasing pH, that is, Hb has a lower affinity for O_2 at lower pH values. This effect is known as the alkaline Bohr effect and is physiologically important in helping to unload O_2 from oxyhemoglobin (HbO_2) when muscle acidity indicates that more O_2 is needed for metabolic reactions. On the other hand, at pH values between 4.5 and 6.1, deoxyhemoglobin (deoxyHb) is a stronger acid than HbO_2.[5,6] This effect is known as the acid Bohr effect.

Hemoglobin is an allosteric protein, that is, its functional properties are regulated by a number of metabolites other than its physiological ligand, O_2. Hemoglobins of vertebrates are among the most extensively studied allosteric proteins, as their allosteric properties are physiologically and medically important in optimizing O_2 transport by erythrocytes. The large number of mutant forms available provides an array of structural alterations with which to correlate effects on function. The functional properties of the Hb molecule are the result of homotropic interactions between its oxygen-binding sites as well as of heterotropic interactions between individual amino acid residues and solvent components. Heterotropic or allosteric effectors include hydrogen ions, chlorides, carbon dioxide, 2,3-diphosphoglycerate (2,3-DPG), and inositol hexaphosphate (IHP). For details, see Ho[1] and the references cited therein.

In many aspects, one-dimensional proton NMR spectroscopy is an excellent technique for investigating the structure–function relationship in Hb. Because of the presence of the unpaired electrons in the high-spin ferrous atoms in deoxyHb and of the presence of the highly conjugated porphyrins in the Hb molecule, the proton chemical shifts of various Hb derivatives cover a wide range (Fig. 1). Resonances vary from about 20 ppm upfield from H_2O to about 90 ppm downfield from H_2O, depending on the spin state of the iron atoms and the nature of ligands attached to the heme groups. This unusually large spread of proton chemical shifts for deoxyHb A provides the selectivity and the resolution necessary to investigate specific regions of the Hb molecule.

^{1}H NMR

^{1}H NMR spectroscopy has played an important role in increasing our understanding of how the structure of the Hb molecule relates to the

[5] C. Bohr, K. Hasselbalch, and A. Krogh, *Skand. Arch. Physiol.* **16,** 402 (1904).
[6] B. German and J. Wyman, Jr., *J. Biol. Chem.* **117,** 533 (1937).

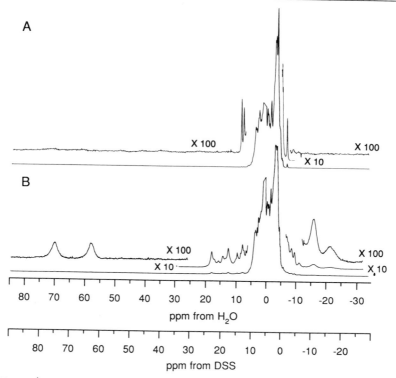

FIG. 1. ^{1}H NMR spectra (300 MHz) of 1.4 mM human normal adult hemoglobin in 0.1 M phosphate at pH 7.1 in H_2O at 29°: (A) oxyHb A; (B) deoxyHb A. ×10 and ×100, expansion of the spectrum to 10 or 100 times, respectively, the original spectrum. [Reprinted with permission from C. Ho, *Adv. Protein Chem.* **43**, 153 (1992).]

cooperative oxygenation of Hb and the Bohr effect. For a general description of NMR, refer to references 7–10.

The NMR parameters determine the appearance of the NMR spectra and also contain information on the molecules under investigation. A resonance in NMR spectroscopy is generally characterized by five parame-

[7] J. A. Pople, W. G. Schneider, and H. J. Bernstein, "High Resolution Nuclear Magnetic Resonance." McGraw-Hill, New York, 1959.

[8] A. Carrington and A. D. McLachlan, "Introduction to Magnetic Resonance." Harper & Row, New York, 1967.

[9] E. D. Becker, "High Resolution NMR: Theory and Chemical Applications," 2nd ed. Academic Press, New York, 1980.

[10] C. P. Slichter, "Principles of Magnetic Resonance," 3rd ed. Springer-Verlag, Berlin, 1990.

ters: the chemical shift (δ), the intensity or the area of the resonance (proportional to concentration), the multiplet structure [related to the spin–spin coupling constant (J)], and two relaxation times, that is, the spin–lattice (or longitudinal) relaxation time (T_1) and the spin–spin (or transverse) relaxation time (T_2). These parameters can be used to derive structural and dynamic information about the Hb molecule under different experimental conditions. Several articles give theoretical background needed for the application of ^{1}H NMR spectroscopy to the investigation of Hb, see references 1, 11–15.

Advances in multidimensional NMR spectroscopy offer unique opportunities for making ^{1}H resonance assignments and for determining the three-dimensional structures of proteins with molecular weights up to approximately 15,000.[16,17] For larger proteins such as Hb, the problems of overlapping cross-peaks and ambiguous connectivities in two-dimensional (2D) ^{1}H NMR spectra pose severe difficulties in making precise resonance identification. J-correlated spectroscopy (COSY) has also been used to identify the resonances of various amino acid residues in protein molecules with molecular weights less than 15,000.[16] For larger proteins, most of the proton linewidths become comparable to, or in some cases exceed, the ^{1}H–^{1}H coupling constants, so that spin–spin multiplet structure cannot be observed.[16] However, in large proteins at extreme sample conditions (high temperature, low pH) in D_2O, some linewidths are narrow enough for coupling to be observed. Two-dimensional nuclear Overhauser and exchange spectroscopy (NOESY) has been used to provide interproton distances in protein molecules with molecular weights less than 15,000. In larger molecules, the frequencies of the rotational motions tend to be too low to allow efficient coupling with the nuclear spin transitions. Hence, 2D ^{1}H NMR spectroscopy has so far had only limited applications in the case of Hb. Nevertheless, some 2D ^{1}H NMR experiments have been carried out and assignments made.[18–20] These techniques have limitations, but undoubtedly will become increasingly useful as new methods for multi-

[11] C. Ho, L. W.-M. Fung, K. J. Wiechelman, G. Pifat, and M. E. Johnson, *in* "Erythrocyte Structure and Function" (G. J. Brewer, ed.), p. 43. Alan R. Liss, New York, 1975.

[12] C. Ho, L. W.-M. Fung, and K. J. Wiechelman, this series, Vol. 54, p. 192.

[13] R. G. Shulman, J. J. Hopfield, and S. Ogawa, *Q. Rev. Biophys.* **8,** 325 (1975).

[14] C. Ho and I. M. Russu, this series, Vol. 76, p. 75.

[15] C. Ho and I. M. Russu, *in* "New Methodologies in Studies of Protein Configuration" (T. T. Wu, ed.), p. 1. Van Nostrand-Reinhold, New York, 1985.

[16] K. Wüthrich, "NMR of Proteins and Nucleic Acids." Wiley, New York, 1986.

[17] A. Bax, *Annu. Rev. Biochem.* **58,** 223 (1989).

[18] C. T. Craescu and J. Mispelter, *Eur. J. Biochem.* **176,** 171 (1988).

[19] C. T. Craescu and J. Mispelter, *Eur. J. Biochem.* **181,** 87 (1989).

[20] C. T. Craescu, J. Mispelter, and Y. Blouquit, *Biochemistry* **29,** 3953 (1990).

nuclear, multidimensional NMR spectroscopy of large molecules are developed.

Sample Preparation

Human Hb A is isolated from freshly drawn red blood cells lysed by the freeze-thaw method.[21] The samples can be prepared in either H_2O or D_2O media. Hemoglobin solutions ($\sim$10%) are exchanged with D_2O by pressure ultrafiltration or dialysis. The pH values reported in this article for Hb samples in D_2O are direct pH meter readings, because the deuterium effect on the glass electrode (i.e., pD = pH + 0.4)[22] is compensated by its effect on the pK value of the imidazole.

Deoxygenation of samples is accomplished by conversion of HbCO to HbO_2 in a rotatory evaporator at 4°, followed by deoxygenation under nitrogen.[23] The sample is transferred to NMR tubes previously flushed with N_2 under N_2 pressure, and the use of sodium dithionite is not necessary, even at high pH.[24]

Hemoglobin samples at partial O_2 saturations can be prepared by mixing approximate amounts of HbO_2 and deoxyHb solutions. The percentage of oxygenation of the sample can be measured directly in a specially constructed NMR sample tube by monitoring the optical densities at 540, 560, and 577 nm.[25] Another method for determining the percentage of oxygenation is to calculate it from the mixed volumes of oxy- and deoxyHb solutions, corrected for dissolved oxygen.[26] Significant amounts of metHb can be formed (especially from high-affinity mutant or modified Hbs) during the course of ^{1}H NMR measurements of HbO_2 or Hb samples at partial O_2 saturation. A metHb reductase system[27] can be added to the Hb samples to prevent or to greatly reduce the formation of metHb. This reductase system does not appear to affect either the oxygenation properties or the ^{1}H NMR spectra of Hb.[27,28]

NMR Techniques

Any modern Fourier transform NMR spectrometer manufactured in the 1980s by major instrument companies is capable of performing various

[21] M. K. Purcell, G. M. Still, T. Rodman, and H. P. Close, (Winston-Salem, N.C.) Clin. Chem. 7, 536 (1961).
[22] P. K. Glasoe and F. A. Long, J. Phys. Chem. 64, 188 (1960).
[23] T. R. Lindström and C. Ho, Proc. Natl. Acad. Sci. U.S.A. 69, 1707 (1972).
[24] J. R. Perussi, V. Simplaceanu, and C. Ho, unpublished results (1990).
[25] T.-H. Huang and A. G. Redfield, J. Biol. Chem. 251, 7114 (1976).
[26] G. Viggiano, N. T. Ho, and C. Ho, Biochemistry 18, 5238 (1979).
[27] A. Hayashi, T. Suzuki, and M. Shin, Biochim. Biophys. Acta 310, 309 (1973).
[28] K. J. Wiechelman, S. Charache, and C. Ho, Biochemistry 13, 4772 (1974).

types of ^{1}H NMR experiments needed for studies of hemoglobin. With a modern 7.0-T high-resolution NMR spectrometer operating at 300 MHz for ^{1}H, a satisfactory ^{1}H NMR spectrum (with a signal-to-noise ratio of ~20 or better) of 0.3–0.5 ml of Hb in millimolar concentration contained in a 5-mm sample tube, can be obtained in a few minutes.

^{1}H NMR studies of proteins are carried out either in deuterated or normal aqueous media. When proteins for NMR studies are dissolved in H_2O, the concentration of protons is about 110 M, whereas that of the protein is around 1 mM. Thus, there is a severe dynamic range problem. A traditional way to overcome the dynamic range problem is to dissolve proteins in D_2O rather than in H_2O. When the Hb samples are dissolved in D_2O media, standard pulse sequences available in a modern Fourier transform NMR spectrometer are capable of providing excellent ^{1}H NMR spectra. A pulse sequence in which one selects for observation the fastest relaxing resonances and suppresses the slow relaxing ones (i.e., the dia-magnetic proton resonances) is quite useful for observing the hyperfine-shifted (hfs) proton resonances of Hb.[29]

When proteins are in D_2O media, solvent-accessible exchangeable pro-tons are then replaced by deuterons. Hence, these protons that have been replaced by deuterons will not be observable in an ^{1}H NMR spectrum. If one would like to detect these exchangeable proton resonances in the presence of H_2O, solvent suppression techniques to reduce the dynamic range problem are needed. We have found that the jump-and-return pulse sequence[30] is quite useful for the spectral range from 8 ppm upfield from H_2O to 22 ppm downfield from H_2O. For the very low-field hfs exchange-able proton resonances (50 to 90 ppm downfield from H_2O), the $1\bar{2}1$ soft pulse sequence developed in our laboratory[31] works quite well. For details on various ^{1}H NMR techniques used to obtain Hb spectra, refer to our publications and those of other investigators mentioned in our publica-tions.

Is there an optimal magnetic field strength for ^{1}H NMR studies of Hb? For hfs proton resonances of deoxy- and metHb, the linewidths of these resonances increase with the square of the resonance frequency because of the paramagnetic contribution to the spin–spin relaxation, whereas the resolution between the resonances increases only linearly with fre-quency.[32] Thus, there is a magnetic field at which both optimal sensitivity and optimal resolution are obtained. For the hfs proton resonances of deoxyHb A, the optimal magnetic field appears to be around 7.0 T (or

[29] J. Hochmann and H. Kellerhals, *J. Magn. Reson.* **38,** 23 (1980).

[30] P. Plateau, C. Dumas, and M. Guéron, *J. Magn. Reson.* **54,** 46 (1983).

[31] C. Yao, V. Simplaceanu, A. K.-L. C. Lin, and C. Ho, *J. Magn. Reson.* **66,** 43 (1986).

[32] M. E. Johnson, L. W.-M. Fung, and C. Ho, *J. Am. Chem. Soc.* **99,** 1245 (1977).

300 MHz). Because of the chemical exchange contribution to the spin–spin relaxation, the linewidths of the C-2 proton resonances of histidyl residues also increase with the square of the resonance frequency. There is no improvement in resolution in going from 300 to 600 MHz, and thus the His–C-2 proton resonances of Hb can be readily obtained in a 300-MHz NMR spectrometer.[33] On the other hand, for the diamagnetic proton resonances (such as exchangeable and ring current-shifted proton resonances), it appears that a 500- or 600-MHz instrument gives better sensitivity and resolution than an instrument operating at 300 MHz.

A commonly used proton chemical shift standard in ^1H NMR studies of proteins in aqueous solution is the proton resonance of the methyl group of the sodium salt of 2,2-dimethyl-2-silapentane 5-sulfonate (DSS). This resonance is +4.73 ppm upfield from the proton resonance of water at 29°. Because all of our Hb samples contain H_2O (in the case of the samples in D_2O, there is residual HDO present), we have found that it is quite useful to use the proton resonance of H_2O in each sample as the internal reference. The ^1H chemical shift of H_2O varies with temperature. As long as we know its variation as a function of temperature, we can always refer the ^1H chemical shift of H_2O to that of DSS. In our ^1H NMR studies, we have used both H_2O and DSS as proton chemical shift references. At the recommendation of the International Union of Pure and Applied Chemistry,[33a] the chemical shift scale has been defined as positive in the region of the ^1H NMR spectrum at a lower field than the resonance of a standard and negative at higher fields. It should be noted that in some of our earlier publications, we used the negative sign to indicate a chemical shift that was downfield from that of the reference.

^1H NMR Spectrum of Hemoglobin

The ^1H NMR spectrum of Hb can be divided into the following spectral regions that have been used to monitor spectral (or structural) changes associated with the ligation of Hb A (refer to Fig. 1 and Ho[1]).

1. Resonances in the region +50 to +80 ppm downfield from H_2O arise from the $N_\delta H$-exchangeable protons of the proximal histidyl residues of the α and β chains of deoxyHb A. These resonances have been shifted more than 50 ppm downfield from their normal diamagnetic resonance regions due to the hyperfine interactions between the unpaired electrons of the high-spin ferrous iron atoms of the heme group and the $N_\delta H$ protons of the proximal histidyl residues of Hb A. They are markers for the

[33] M. Madrid, V. Simplaceanu, N. T. Ho, and C. Ho, *J. Magn. Reson.* **88,** 42 (1990).
[33a] IUPAC, No. 38, August 1974.

proximal histidyl residues of the α and β chains of deoxyHb A and can be used to monitor the binding of O_2 to the α- and β-hemes of Hb A.

2. Resonances in the region $+10$ to $+90$ ppm downfield from HDO arise from the protons of the heme groups and their nearby amino acid residues due to the hyperfine interactions between these protons and unpaired electrons of the high-spin ferric heme-iron (Fe^{III}) atoms of metHb A.

3. Resonances in the region $+5$ to $+25$ ppm downfield from HDO arise from the protons of the heme groups and their nearby amino acid residues due to hyperfine interactions of these protons and the unpaired electrons of the low-spin ferric heme-iron atoms of cyanometHb or azidometHb.

4. Resonances in the region $+6$ to $+22$ ppm downfield from H_2O of deoxyHb A are due to two sources. First, they arise from the protons on the heme groups and their nearby amino acid residues due to the hyperfine interactions between these protons and unpaired electrons of Fe^{II} in the heme-iron atoms. Second, they arise from the exchangeable protons due to intra- and intermolecular H bonds in Hb A. Exchangeable and hfs proton resonances are excellent tertiary and quaternary structural markers of deoxyHb.

5. Resonances in the region $+5$ to $+8.5$ ppm from H_2O arise from the exchangeable proton resonances of oxyHb A. They are excellent markers for the subunit interfaces and oxy quaternary structure.

6. Resonances in the region $+2$ to $+6$ ppm downfield from HDO arise from the protons of aromatic amino acid residues of oxy- and deoxyHb A. The region from $+2.8$ to $+3.8$ ppm contains the C-2 protons of the histidyl residues, whereas the region from $+1.8$ to $+2.8$ ppm contains resonances from the C-4 protons of histidines as well as aromatic resonances from tryptophan, tyrosine, and phenylalanine. The C-2 proton resonances have been used to investigate the molecular basis of the Bohr effect of Hb A.

7. Resonances in the region -1 to -5 ppm upfield from HDO arise from the protons of aliphatic amino acid residues of oxy- and deoxyHb A. This is a very crowded spectral region due to the large number of CH_2 and CH_3 groups in the amino acid residues.

8. Resonances in the region -5 to -8 ppm from HDO arise from the ring current-shifted protons due to those protons located above or below the aromatic amino acid residues and the porphyrins of oxyHb A.

9. Resonances in the region -6 to -20 ppm from HDO arise from ring current-shifted protons and the protons on the heme groups and their nearby amino acid residues due to the hyperfine interactions between these protons and the unpaired electrons of the high-spin ferrous heme-iron atoms of deoxyHb A.

Assignment of Proton Resonances of Hemoglobin

The first step in any NMR investigation of the structural and dynamic properties of a protein molecule is the assignment of resonances to specific amino acid residues. Table I summarizes the present state of the assignment of proton resonances of Hb A.[2,4,19,34–49] Here we describe briefly how some of these have been determined.

For a protein the size of Hb, one of the most widely used methods for assigning proton resonances to specific amino acid residues is to compare the ¹H NMR spectra of normal and appropriate mutant or chemically modified Hb molecules. This method is shown in Fig. 2, where the aromatic proton resonance region of the ¹H NMR spectrum of HbCO is compared to those of enzymatically prepared des(β146 His)HbCO and mutant Hbs Cowtown (β146 His → Leu) and York (β146 His → Pro). Figure 2 shows that a single resonance is consistently missing in the ¹H NMR spectra of these Hb variants in 0.2 M phosphate and 0.2 M chloride at pH 7.1. In fact, the assignment of this resonance has been highly controversial. Figure 3 gives some indication of the difficulties, in that resonances shift positions under different experimental conditions and in modified Hbs. The assignment of the C-2 of β146 His to peak Y has been confirmed by an incremental titration from conditions in which the assignment of this resonance is unambiguous (see Fig. 2) to other conditions of interest. This bridging of

[34] S. Takahashi, A. K.-L. C. Lin, and C. Ho, *Biochemistry* **19**, 5196 (1980).

[35] C. Ho, C.-H. J. Lam, S. Takahashi, and G. Viggiano, *in* "Hemoglobin and Oxygen Binding" (C. Ho, W. A. Eaton, J. P. Collman, Q. H. Gibson, J. S. Leigh, Jr., E. Margoliash, K. Moffat, and W. R. Scheidt, eds.), p. 141. Elsevier/North-Holland, New York, 1982.

[36] D. G. Davis, T. R. Lindström, N. H. Mock, J. J. Baldassare, S. Charache, R. T. Jones, and C. Ho, *J. Mol. Biol.* **60**, 101 (1971).

[37] T. R. Lindström, C. Ho, and A. V. Pisciotta, *Nature (London), New Biol.* **237**, 263 (1972).

[38] L. W.-M. Fung and C. Ho, *Biochemistry* **14**, 2526 (1975).

[39] I. M. Russu, N. T. Ho, and C. Ho, *Biochim. Biophys. Acta* **914**, 40 (1987).

[40] T. Asakura, K. Adachi, J. S. Wiley, L. W.-M. Fung, C. Ho, J. V. Kilmartin, and M. F. Perutz, *J. Mol. Biol.* **104**, 185 (1976).

[41] K. Ishimori, K. Imai, G. Miyazaki, T. Kitagawa, Y. Wada, H. Morimoto, and I. Morishima, *Biochemistry* **31**, 3256 (1992).

[42] G. Viggiano, K. J. Wiechelman, P. A. Chervenick, and C. Ho, *Biochemistry* **17**, 795 (1978).

[43] C. Dalvit and C. Ho, *Biochemistry* **24**, 3398 (1985).

[44] I. M. Russu, N. T. Ho, and C. Ho, *Biochemistry* **21**, 5031 (1982).

[45] I. M. Russu, N. T. Ho, and C. Ho, *Biochemistry* **19**, 1043 (1980).

[46] J. J. Kilmartin, J. J. Breen, G. C. K. Roberts, and C. Ho, *Proc. Natl. Acad. Sci. U.S.A.* **70**, 1246 (1973).

[47] L. W.-M. Fung, C. Ho, G. F. Roth, Jr., and R. L. Nagel, *J. Biol. Chem.* **250**, 4786 (1975).

[48] I. M. Russu, S.-S. Wu, N. T. Ho, G. W. Kellogg, and C. Ho, *Biochemistry* **28**, 5298 (1989).

[49] I. M. Russu, A. K.-L. Lin, S. Ferro-Dosch, and C. Ho, *Biochim. Biophys. Acta* **785**, 123 (1984).

TABLE I
ASSIGNMENTS OF PROTON RESONANCES OF HUMAN ADULT HEMOGLOBIN[a]

Resonance position (ppm from H_2O or HDO)	Hb derivative	Experimental conditions[b]	Assignment	Refs.
+71.0	DeoxyHb	0.1 M Bis–Tris in H_2O, pH 6.7, 27°	β92 (F8) His $N_\delta H$ proton, exchangeable hfs	34
+58.5	DeoxyHb	0.1 M Bis–Tris in H_2O, pH 6.7, 27°	α87 (F8) His $N_\delta H$ proton, exchangeable hfs	34
+17.5	DeoxyHb	0.1 M Bis–Tris, pH 6.7, 27°	β chain, hfs	34–37
+16.8	DeoxyHb	0.1 M Bis–Tris, pH 6.7, 27°	β chain, hfs	34, 35
+15.6	DeoxyHb	0.1 M Bis–Tris, pH 6.7, 27°	α chain, hfs	34, 35
+14.2	DeoxyHb	0.1 M Bis–Tris, pH 6.7, 27°	β chain, hfs	34, 35
+13.2	DeoxyHb	0.1 M Bis–Tris, pH 6.7, 27°	α chain, hfs	34, 35
+12.1	DeoxyHb	0.1 M Bis–Tris, pH 6.7, 27°	α chain, hfs	34–37
+9.5	DeoxyHb	0.1 M Bis–Tris pH 6.7, 27°	β chain, hfs	34
+9.4	DeoxyHb	0.1 M Bis–Tris in H_2O, pH 6.6, 27°	H bond between α42 Tyr and β99 Asp, $\alpha_1\beta_2$ interface	38, 39
+8.2	DeoxyHb	0.1 M phosphate in H_2O, pH 7.0, 29°	H bond between α126 Asp and β35 Tyr, $\alpha_1\beta_1$ interface	39, 40
+8.1	HbO_2 and HbCO	0.1 M phosphate in H_2O, pH 7.0, 29°	H bond between α126 Asp and β35 Tyr, $\alpha_1\beta_1$ interface	39
+7.8	DeoxyHb	0.1 M Bis–Tris, pH 6.7, 27°	α chain, hfs	34, 36, 37
+7.5	DeoxyHb	0.1 M phosphate in H_2O, pH 7.0, 29°	H bond between α103 His and β108 Asp, $\alpha_1\beta_1$ interface	39
+7.4	HbO_2 and HbCO	0.1 M phosphate, 29°	H bond between α103 His and β108 Asp, $\alpha_1\beta_1$ interface	39
+7.2	DeoxyHb	0.1 M Bis–Tris, pH 6.7, 27°	β chain, hfs	34
+6.4	DeoxyHb	0.1 M Bis–Tris in H_2O, pH 6.8, 27°	H bond between α94 Asp and β37 Trp, $\alpha_1\beta_2$ interface	38, 41[c]

TABLE I (*continued*)

Resonance position (ppm from H₂O or HDO)	Hb derivative	Experimental conditions[b]	Assignment	Refs.
+5.9	HbO₂	0.1 M Bis–Tris in H₂O, pH 6.6, 27°	H bond between β94 Asp and β102 Asn, $\alpha_1\beta_2$ interface	38
+5.5	HbCO	0.1 M Bis–Tris in H₂O, pH 6.6, 27°	H bond between α94 Asp and β102 Asn, $\alpha_1\beta_2$ interface	38
+5.69	HbCO	0.1 M phosphate, pH 5.6, 37°	Meso proton γ of the heme of α chain	19
+5.59	HbCO	0.1 M phosphate, pH 5.6, 37°	Meso proton γ of the heme of β chain	19
+5.42	HbCO	0.1 M phosphate, pH 5.6, 37°	Meso proton α of the heme of β chain	19
+5.09	HbCO	0.1 M phosphate, pH 5.6, 37°	Meso proton δ of the heme of α and β chains	19
+4.97	HbCO	0.1 M phosphate, pH 7.2, 29°	Meso proton δ of the hemes of α and β chains	43
+4.95	HbCO	0.1 M phosphate, pH 5.6, 37°	Meso proton α of the heme of α chain	19
+4.76	HbO₂	0.1 M phosphate, pH 7.2, 29°	Meso proton δ of hemes of α and β chains	43
+4.73	HbCO	0.1 M phosphate, pH 5.6, 37°	Meso proton β of the heme of β chain	19
+4.71	HbCO	0.1 M phosphate, pH 5.6, 37°	Meso proton β of the heme of α chain	19
+3.95	HbCO	0.1 M Bis–Tris, pH 6.3, 27°	β97 His, C-2 proton	2, 44
+3.80	DeoxyHb	0.1 M Bis–Tris pH 6.3, 27° or 0.1 M phosphate + 0.2 M NaCl, pH 6.2, 30°	β146 His, C-2 proton	45, 46
+3.68	HbCO	0.2 M phosphate + 0.2 M NaCl, pH 6.2, 27°	β2 His, C-2 proton	47
+3.60	DeoxyHb	0.1 M HEPES, pH 7.5, 29°	β146 His, C-2 proton	4
+3.47	HbCO	0.2 M phosphate + 0.1 M HEPES, pH 6.7, 29°	β146 His, C-2 proton	4
+3.40	HbCO	0.1 M Bis–Tris, pH 6.3, 29°	β2 His, C-2 proton	48

(*continued*)

TABLE I (*continued*)

Resonance position (ppm from H₂O or HDO)	Hb derivative	Experimental conditions[b]	Assignment	Refs.
+3.38	HbCO	0.1 *M* HEPES + 0.2 *M* phosphate, pH 6.7, 29°	β2 His, C-2 proton	48
+3.32	DeoxyHb	0.1 *M* Bis–Tris, pH 6.3, 27°	β2 His, C-2 proton	48
+3.31	HbCO	0.1 *M* Bis–Tris, pH 6.3, 29°	β146 His, C-2 proton	4
+3.27	HbCO	0.2 *M* phosphate + 0.2 *M* NaCl, pH 7.1, 29°	β146 His, C-2 proton	46
+3.20	HbCO	0.1 *M* Bis–Tris, pH 6.6, 27°	β2 His, C-2 proton	47
+3.23⎱ +3.17⎰	HbCO	0.1 *M* Bis–Tris, pH 6.3, 29°	β116 His or β117 His, C-2 proton	4, 49
+3.21	HbCO	0.2 *M* phosphate + 0.2 *M* NaCl, pH 7.1, 29°	β2 His, C-2 proton	48
+3.18⎱ +3.09⎰	DeoxyHb	0.1 *M* HEPES, pH 7.0, 29°	β116 His or β117 His, C-2 proton	4, 49
+3.13⎱ +3.02⎰	HbCO	0.1 *M* HEPES, pH 6.8, 29°	β116 His or β117 His, C-2 proton	4, 49
+3.11	HbCO	0.1 *M* HEPES, pH 6.8, 29°	β2 His, C-2 proton	48
+3.05	HbCO	0.1 *M* HEPES, pH 6.8, 29°	β146 His, C-2 proton	4
+2.93	DeoxyHb	0.1 *M* HEPES, pH 7.5, 29°	β2 His, C-2 proton	48
+2.93⎱ +2.90⎰	HbCO	0.2 *M* phosphate + 0.2 *M* NaCl, pH 7.1, 29°	β116 His or β117 His, C-2 proton	4, 49
−0.96	HbCO	0.1 *M* phosphate, pH 5.6, 37°	α62 (E11) Val, α-CH	19
−1.06	HbCO	0.1 *M* phosphate, pH 5.6, 37°	3-CH₃ of the heme of β chain; 8-CH₃ of the heme of α chain; β67 (E11) Val, α-CH	19
−1.09	HbCO	0.1 *M* phosphate, pH 5.6, 37°	8-CH₃ of the heme of β chain	19
−1.14	HbCO	0.1 *M* phosphate, pH 5.6, 37°	1-CH₃ of the heme of α chain	19
−1.21⎱ −1.56⎰	HbCO	0.1 *M* phosphate, pH 7.2, 29°	1- and 8-CH₃ of the hemes of α and β chains	43
−1.27⎱ −1.54⎰	HbO₂	0.1 *M* phosphate, pH 7.2, 29°	1- and 8-CH₃ of the hemes of α and β chains	43

TABLE I (*continued*)

Resonance position (ppm from H_2O or HDO)	Hb derivative	Experimental conditions[b]	Assignment	Refs.
−1.38	HbCO	0.1 M phosphate, pH 5.6, 37°	1-CH$_3$ of the heme of β chain	19
−2.00	HbCO	0.1 M phosphate, pH 5.6, 37°	5-CH$_3$ of the heme of α chain	19
−3.15	HbCO	0.1 M phosphate, pH 5.7, 37°	β67 (E11) Val, β-CH	19
−3.28	HbCO	0.1 M phosphate, pH 7.2, 29°	β67 (E11) Val, β-CH	43
−3.35	HbCO	0.1 M phosphate, pH 5.7, 37°	α62 (E11) Val, β-CH	19
−3.51	HbCO	0.1 M phosphate, pH 7.2, 29°	α62 (E11) Val, β-CH	43
−4.47	HbCO	0.1 M phosphate, pH 5.7, 37°	(E11) Val, γ-CH$_3$ of α and β chains	19
−4.54	HbCO	0.1 M phosphate, pH 7.2, 29°	β67 (E11) Val, γ$_1$-CH$_3$	43
−4.60	HbCO	0.1 M phosphate, pH 7.2, 29°	α62 (E11) Val, γ$_1$-CH$_3$	43
−4.80	HbO$_2$	0.1 M phosphate, pH 7.2, 29°	(E11) Val, γ$_1$-CH$_3$ of α and β chains	43
−6.32	HbCO	0.1 M phosphate, pH 5.6, 37°	α62 (E11) Val, γ-CH$_3$	19
−6.46	HbCO	0.1 M phosphate, pH 5.6, 37°	β67 (E11) Val, γ-CH$_3$	19
−6.51	HbCO	0.1 M phosphate, pH 7.2, 29°	(E11) Val, γ$_2$-CH$_3$ of α and β chains	43
−7.11	HbO$_2$	0.1 M phosphate, pH 7.2, 29°	(E11) Val, γ$_2$-CH$_3$ of α and β chains	43

[a] The proton chemical shift of H_2O is 4.73 ppm downfield from that of the methyl group of 2,2-dimethyl-2-silapentane 5-sulfonate (DSS) at 29°. (Modified and updated from Ho and Russu[14] and Ho.[1])

[b] The particular resonance can be observed in both H_2O and D_2O, unless specifically stated otherwise.

[c] Viggiano *et al.*[42] assigned the +6.4 resonance to the hydrogen bond between β98 Val and β145 Tyr. It has been pointed out by Ho and co-workers that this assignment is tentative and needs further verification. Ishimori *et al.*[41] reported their spectroscopic and biochemical studies on two recombinant hemoglobins Hb(β37 Trp → Phe) and Hb(β145 Tyr → Phe). They found that the exchangeable proton resonance at +6.4 ppm is missing in deoxyHb(β37 Trp → Phe), but is present in deoxyHb(β145 Tyr → Phe). On the basis of their ^{1}H NMR results on deoxyHb(β145 Tyr → Phe), they concluded that the +6.4 ppm resonance cannot originate from the hydrogen bond between β98 Val and β145 Tyr. Because this resonance is missing in deoxyHb(β37 Trp → Phe), they concluded that the +6.4 ppm resonance is due to the intersubunit hydrogen bond

(*continued*)

buffer conditions has confirmed the identification of peak Y in chloride-free 0.1 M N-2-hydroxyethylpiperazine-N'-2-ethanesulfonic acid (HEPES) and in 0.1 M Bis–Tris at pH 6.36, and made possible the determination of the contribution of β146 His to the Bohr effect under different solvent conditions[3,4] (see below).

The technique of comparing the spectra of normal and modified Hbs is weakened when a single mutation or chemical modification produces significant conformational alterations in other regions of the protein molecule. Such effects can produce changes in the ^{1}H NMR spectrum that make spectral assignments ambiguous. One approach to solving this problem is to use several mutant or chemically modified Hbs with a change in the same amino acid residue to clarify the assignment.[38,42,50] Another approach is to use independent techniques to confirm the resonance assignment made by the mutant Hbs. One-dimensional NOE and 2D NMR techniques can be combined with X-ray structural information to make and confirm resonance assignments of HbCO A.

The exchangeable proton resonances from $+5.0$ to $+10.0$ ppm downfield from the water proton resonance provide an example of assignments based on mutant Hbs and confirmed by NOE measurements correlated with X-ray crystal structures. These exchangeable proton resonances arise from amino acid residues that are located at the subunit interfaces of the Hb tetramer and are involved in the hydrogen-bonding interactions responsible for the quaternary structures of deoxyHb and ligated Hb. They are thus of special interest for understanding the quaternary structural transitions in Hb. Figure 4 shows the exchangeable proton resonances of Hb A in 0.1 M phosphate in H_2O at pH 7.0 and 29° in deoxy, oxy, and CO forms. The resonances at $+6.4$, $+7.5$, $+8.2$, and $+9.4$ ppm are absent in the spectra of isolated α and β chains, suggesting that they are specific

---

[50] T. R. Lindström, I. B. E. Norén, S. Charache, H. Lehman, and C. Ho, *Biochemistry* **11**, 1677 (1972).

---

between α94 Asp and β37 Trp. It should be mentioned that there are inconsistencies between their spectroscopic and biochemical results on these two recombinant Hbs. Ishimori *et al.*[41] need to ascertain the nature of the proton resonances over the spectral region from $+5$ to $+10$ ppm from H_2O, that is, which ones are due to the hfs proton resonances and which ones are due to the exchangeable proton resonances, or both. Without a knowledge of the nature of these resonances, they cannot make conclusions regarding the tertiary and quaternary structures of these two mutant Hbs from their ^{1}H NMR results and cannot make definitive resonance assignments. It should be noted that Fung and Ho[38] first suggested that the $+6.4$ ppm resonance could arise from the hydrogen bond between α94 Asp and β37 Trp. On the basis of available information, the origin of the $+6.4$ ppm resonance needs additional verification.

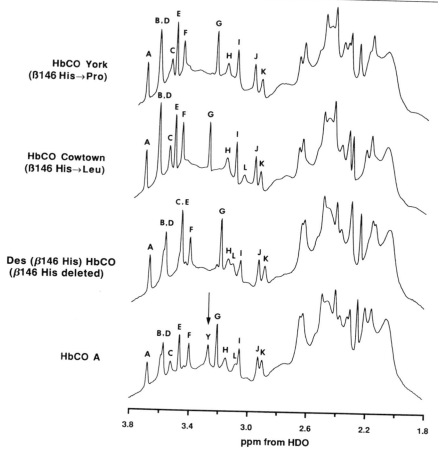

FIG. 2. ^{1}H NMR spectra (300 MHz) of 8–10% solutions of HbCO York ($\beta146$ His $\rightarrow$ Pro), HbCO Cowtown ($\beta146$ His $\rightarrow$ Leu), des($\beta146$ His)HbCO, and HbCO A in D$_2$O in 0.2 M phosphate plus 0.2 M NaCl at pH 7.1 and 29°. Resonances from C-2 protons of histidyl residues are seen from +2.8 to +3.8 ppm. [Reprinted with permission from M. R. Busch et al., Biochemistry **30**, 1865 (1991). Copyright 1991 American Chemical Society.]

markers for the quaternary structure of the Hb tetramer.[39] The +6.4 and +9.4 ppm resonances disappear on the binding of ligand to deoxyHb A,[38] whereas the resonances at +8.2 and +7.5 ppm are shifted by only about 0.1 ppm on ligand binding.[39] Using appropriate mutant Hbs, the resonances at +6.4, +8.2, and +9.4 ppm have been assigned by our laboratory to specific amino acid residues. The +9.4 ppm resonance has been assigned to the intersubunit H bond between $\alpha42$ (C7) Tyr and $\beta99$ (G1) Asp (an

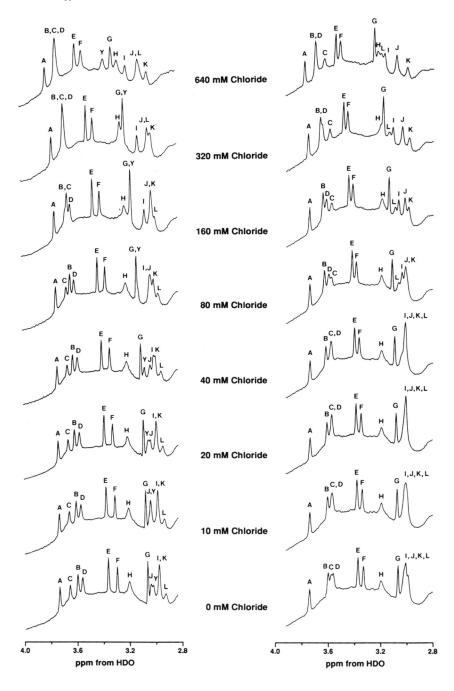

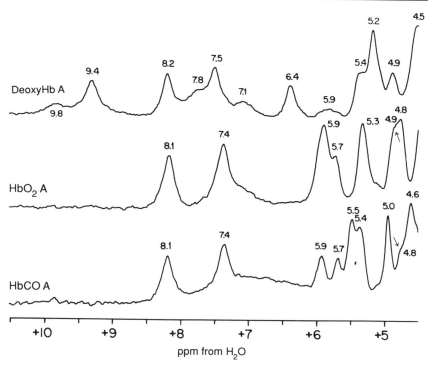

FIG. 4. ^{1}H NMR spectra (300 MHz) of deoxyHb A, HbO$_2$ A, and HbCO A in 0.1 M phosphate in H$_2$O at pH 7.0 and 29°: effects of ligation on the exchangeable proton resonances. Arrows indicate the positions of chemical shift. [Reprinted with permission from I. M. Russu *et al.*, *Biochim. Biophys. Acta* **914**, 40 (1987).]

important deoxy quaternary feature in the $\alpha_1\beta_2$ subunit interface), the +8.2 ppm resonance has been assigned to the intersubunit H bond between α126 (H9) Asp and β35 (C1) Tyr in the $\alpha_1\beta_1$ subunit interface, and, with less certainty, the +6.4 ppm resonance has been tentatively assigned to the H bond between α94 (G1) Asp and β37 (C3) Trp.[38,40–42] Also see footnote *c* in Table I.

By using the initial NOE build-up rates on the exchangeable proton resonances at +7.5, +8.2, and +9.4 ppm from H$_2$O, we have observed

---

FIG. 3. ^{1}H NMR spectra (300 MHz) of 8–10% solutions of (A) HbCO A and (B) des(β146 His)HbCO in D$_2$O in 0.1 M HEPES and 0–640 mM chloride at pH 6.95 and 29°. The histidyl C-2 resonances in HbCO are labeled A–L and Y. [Reprinted with permission from M. R. Busch *et al.*, *Biochemistry* **30**, 1865 (1991). Copyright 1991 American Chemical Society.]

specific NOEs for each of these exchangeable proton resonances in deoxyHb A (Fig. 5). To gain insight into the molecular origin of the observed NOEs, we have attempted to correlate the patterns of the NOEs to the predictions made on the basis of the X-ray crystal structures of deoxyHb A and HbCO A,[51-53] as well as on the basis of the assignments previously proposed by this laboratory for the +9.4 and +8.2 ppm resonances.[38,40] From these studies, we have confirmed the assignment of the exchangeable proton resonances at +9.4 and +8.2 ppm and have assigned the exchangeable proton resonance at +7.5 ppm to the intermolecular H bond between α103 His and β108 Asn in the $\alpha_1\beta_1$ subunit interface.[39]

Another example of assignments made by a combination of techniques is represented by the ring current-shifted resonances from -5 to -8 ppm upfield from HDO. The ring current fields of the porphyrins in Hb cause large shifts in the resonances of nearby protons. Changes in these shifts reflect structural changes in the heme pocket, thus giving information about changes in the tertiary structure of the active center of the Hb molecule. Much effort has, therefore, gone into the attempt to assign these ring current-shifted resonances in the region -5 to -8 ppm upfield from HDO (or -0.3 to -3.3 ppm upfield from DSS). The closest amino acid residues to the hemes are the methyl groups of E11 valine (distal valine) and the C-2 protons of E7 histidine (distal histidine). It has been suggested that the conformations of the heme pockets (as manifested by the ring current-shifted proton resonances) of the isolated α and β chains in the CO form are quite similar to those in intact HbCO A.[50] Thus, for both convenience and ease of spectral assignments, we have investigated the ring current-shifted proton resonances of isolated α and β chains and of Hb A in both CO and oxy forms.

Figure 6A, spectrum a, shows the 300-MHz ring current-shifted proton resonances of isolated α chains in the CO form. The resonance at -1.78 ppm from DSS (or -6.51 ppm from HDO) was previously assigned to the γ_2-CH$_3$ group of the α62 (E11) Val.[50] Figure 6A, spectrum b, shows the truncated-driven NOE difference spectrum of isolated α chains obtained on preirradiation of the resonance at -1.78 ppm from DSS with a radio-frequency (rf) pulse of 100-msec duration.[43] Only a few resonances appear in both aliphatic and aromatic resonance regions in the difference spectrum, indicating that the observed NOEs are extremely selective. The resonances that are present in Fig. 6A, spectrum b, originate from the

[51] G. Fermi, *J. Mol. Biol.* **97**, 237 (1975).

[52] J. M. Baldwin, *J. Mol. Biol.* **136**, 103 (1980).

[53] G. Fermi and M. F. Perutz, *in* "Hemoglobin and Myoglobin. Atlas of Molecular Structures in Biology" (D. C. Phillips and F. M. Richards, eds.), p. 3. Oxford Univ. Press (Clarendon), Oxford, 1981.

A. Control Spectrum

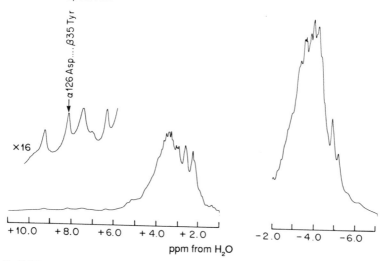

B. NOE on the Resonance at 8.2 ppm

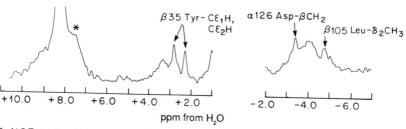

C. NOE on the Resonance at 7.5 ppm

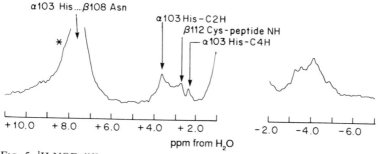

Fig. 5. ^{1}H NOE difference spectra (300 MHz) for the exchangeable proton resonances of deoxyHb A in 0.1 M phosphate in H_2O at pH 7.0 and 29°. The irradiation time at +8.2 or +7.5 ppm was 50 msec. Asterisks denote off-resonance spillage. [Reprinted with permission from I. M. Russu *et al.*, *Biochim. Biophys. Acta* **914**, 40 (1987).]

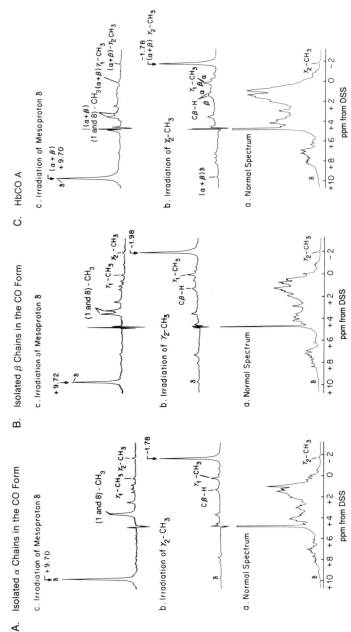

FIG. 6. ¹H NMR normal and NOE difference spectra (300 MHz) of isolated α and β chains and of HbCO A in the CO form in 0.1 *M* phosphate in D₂O at pH 7.3 and 29°: (A) isolated α chains; (B) isolated β chains; (C) HbCO A. The NOE difference spectra were obtained with a preirradiation pulse of 100 msec at the positions indicated by arrows. γ_1-CH₃, γ_2-CH₃, and Cβ-H are protons of E11 Val. [Reprinted with permission from C. Dalvit and C. Ho, *Biochemistry* **24**, 3398 (1985). Copyright 1985 American Chemical Society.]

protons located very close to the γ_2-CH$_3$ of α62 (E11) Val, and the resonance intensity reflects to a first approximation the proton–proton distance. The protons closest to the γ_2-CH$_3$ group of the E11 Val are the protons of the same amino acid residue, namely, γ_1-CH$_3$ and Cβ-H. From the relative intensities of these two resonances, we have assigned the resonances at $+0.13$ and $+1.21$ ppm from DSS (or -4.60 and -3.52 ppm from HDO), respectively, to the γ_1-CH$_3$ and Cβ-H of E11 Val. The assignment of the resonance at $+9.70$ ppm from DSS (or $+4.97$ ppm from HDO) to the mesoproton δ has been confirmed by the truncated-driven NOE difference spectrum of Fig. 6A, spectrum c.

Another method used to confirm our spectral assignment of E11 Val of isolated α chains of Hb A in the CO form is the COSY technique.[43] A typical COSY experiment can, in principle, provide a complete map of all ^{1}H–^{1}H J connectivities, thus avoiding the lack of selectivity of irradiation in crowded spectral regions. Figure 7 shows the COSY contour plot in the region between $+2$ and -2.2 ppm from DSS with the connectivities for E11 Val. These results are in agreement with the NOE experiments discussed above. Theoretical ring current calculations are also consistent with our assignment of the γ_1-CH$_3$ of E11 Val.[43] Thus, the present assignments for E11 Val of the α chain in the CO form may be considered definitive.

Figure 6B, spectrum a, shows the 300-MHz ^{1}H NMR spectrum of isolated β chains of Hb A in the CO form. The resonances in Fig. 6B, spectrum a, are not as sharp as those for isolated α chains shown in Fig. 6A, spectrum a. This is because isolated β chains exist in solution as tetramers whereas isolated α chains exist as either monomers or dimers.[54,55] The upfield resonance at -1.98 ppm from DSS (or -6.71 ppm from HDO) was assigned to the γ_2-CH$_3$ of β67 (E11) Val.[50] The NOE difference spectrum with preirradiation on this resonance for 100 msec is shown in Fig. 6B, spectrum b. The observed spectral features are similar to those shown in Fig. 6A, spectrum b. Following the same procedures and considerations used for the α chain, we have assigned the resonances at -0.10 and $+1.36$ ppm from DSS (or -4.83 and -3.37 ppm from HDO) to the γ_1-CH$_3$ and Cβ-^{1}H of β67 (E11) Val, respectively.

The present assignment of the resonance at -0.10 ppm from DSS to the γ_1-CH$_3$ of β67 E11 is different from that previously suggested,[50] -1.1 ppm upfield from DSS (or -5.83 ppm from HDO). We have observed only a small NOE on the resonance at -1.1 ppm on irradiating the resonance at -1.98 ppm, suggesting that this resonance comes from protons that are

[54] R. Benesch and R. E. Benesch, *Nature (London)* **202,** 773 (1964).
[55] R. Valdes, Jr. and G. K. Ackers, *J. Biol. Chem.* **252,** 74 (1977).

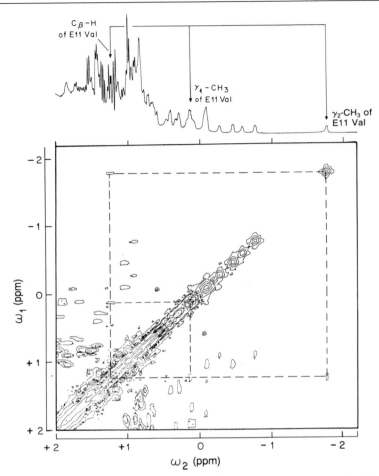

FIG. 7. ^{1}H NMR spectrum (300 MHz) (top) and ^{1}H COSY spectrum (bottom) of isolated α chains of Hb A in the CO form in 0.1 M phosphate in D_2O at pH 5.3 and 29°. The ^{1}H chemical shift scale is referred to DSS. [Reprinted with permission from C. Dalvit and C. Ho, *Biochemistry* **24**, 3398 (1985). Copyright 1985 American Chemical Society.]

relatively far from the γ_2-CH_3 of β67 (E11) Val. We believe that the present assignment is more reliable than the previous work, and it is also confirmed by other studies.[43]

Figure 6C, spectrum a, shows the 300-MHz ^{1}H NMR spectrum of HbCO A. The resonance at -1.78 ppm from DSS (or -6.51 ppm from HDO) corresponds to the γ_2-CH_3 groups of E11 Val from both the α and β chains.[50] Figure 6C, spectrum b, shows the truncated-driven NOE difference spectrum with preirradiation at -1.78 ppm upfield from DSS for 100 msec. Comparing the difference spectra of isolated α and β chains with that of HbCO A, we have assigned several resonances to specific groups in the α and β subunits. We have similarly carried out spectral assignments of the heme substituents and of the distal histidyl and valyl residues of the heme pockets of the isolated α and β chains and of Hb A, all in the oxy form, by using the same techniques that we have used in our spectral assignments for these three samples in the CO form (for details, see Table I and Dalvit and Ho[43]).

Cooperative Oxygenation of Hemoglobin

Hemoglobin has been a favorite protein molecule for designing and testing various models of the molecular basis of subunit interactions in multisubunit proteins. In the 1960s, two mechanisms were advanced to account for the sigmoidal nature of the oxygen-binding curve of Hb A, one proposed by Monod $et\ al.,$[56] based on a concerted mechanism, and the other proposed by Koshland $et\ al.,$[57] based on a sequential mechanism. The concerted mechanism postulates that there is an equilibrium between the low-affinity deoxy conformation (T form) and the high-affinity oxy conformation (R form) and that this transition is a concerted process, that is, an intermediate (or hybrid) conformation RT cannot exist. On the other hand, the sequential model does not assume the existence of an equilibrium between the T and R forms in the absence of ligand. The transition from the T form to the R form is induced by the binding of ligand, that is, a sequential process. Thus, the hybrid species RT plays a prominent role in the sequential model. Even though there are conceptual differences in these two mechanisms (in terms of the nature of conformational transitions induced by ligand binding), they both can account for the O_2-binding curve of Hb A quite satisfactorily. During the past three decades, extensive efforts by a large number of researchers have been devoted to attempting to understand the molecular basis for the oxygenation of Hb.

As a result of comparing the atomic models, based on X-ray crystallographic structural investigations, of human deoxyHb A and horse oxy-like metHb, Perutz proposed a stereochemical mechanism for the cooperative oxygenation of Hb.[58] In its original form, Perutz's model emphasizes the link between the cooperativity and the transition between two quaternary structures, that is, two different arrangements of the four subunits, the

[56] J. Monod, J. Wyman, and J.-P. Changeux, $J.\ Mol.\ Biol.$ **12,** 88 (1965).

[57] D. E. Koshland, Jr., G. Nemethy, and G. Filmer, $Biochemistry$ **5,** 365 (1966).

[58] M. F. Perutz, $Nature\ (London)$ **228,** 726 (1970).

deoxy quaternary structure, T, and the oxy quaternary structure, R. Movement of the heme-iron atoms and the sliding motion of the $\alpha_1\beta_2$ or $\alpha_2\beta_1$ subunit interface as well as the breaking of the intra- and intermolecular salt bridges and hydrogen bonds as a result of the ligation of the Hb molecule are among the central features of Perutz's stereochemical mechanism for the cooperative oxygenation of Hb. Perutz's mechanism allows tertiary structural changes, that is, changes in the conformation of the subunits, to take place each time a subunit is oxygenated, but a single concerted quaternary structural transition (i.e., T → R) is responsible for the cooperativity of the oxygenation process. Thus, the affinity for O_2 of an unligated subunit is not affected by the state of ligation of its neighbors within a given quaternary structure. The basic conceptual framework of Perutz's mechanism shares many features of a two-state allosteric model, such as the one proposed by Monod et al.[56]

Because of the highly cooperative nature of the oxygenation of Hb A, there are few partially oxygenated species [$Hb(O_2)_1$, $Hb(O_2)_2$, and $Hb(O_2)_3$] present during the oxygenation process, and thus the majority of the Hb species are either fully deoxy- or fully oxyHb. This fact, together with the lack of suitable techniques for investigating structural and functional properties of transient, partially ligated species, has led to the prevalence of two-state descriptions of the cooperative oxygenation process. Nevertheless, there are by now a number of reports of experimental results on Hb that suggest the existence of intermediate forms, and are thus not consistent with a two-state or two-structure type of model for hemoglobin. For details, see Ho.[1]

Selected Experimental Results

We shall give two examples of how ^1H NMR has contributed to our understanding of the cooperative oxygenation of Hb, one involving the hyperfine shifted resonances from 10 to 20 ppm from HDO, and the other making use of the exchangeable proton resonances from 5 to 10 ppm downfield from H_2O and the two exchangeable proton resonances at 58.5 and 71.0 ppm downfield from H_2O.

Hyperfine Shifted Resonances

^1H NMR and X-ray crystallography results have clearly shown that the environments of the heme pockets of the α and β chains of Hb as manifested by the conformations of proximal histidine (F8), distal histidine (E7), and distal valine (E11), as well as the conformation of several porphyrin protons and the electronic structure of the heme group in both deoxy and ligated states, are not equivalent.[34,43,59,60] It has not been clear what this difference in structure between the α- and β-heme pockets means in terms of differences in function. Because an understanding of the functional properties of the α and β chains of an intact Hb A molecule is important for our eventual understanding of the detailed mechanism for the cooperative oxygenation of Hb A, we have carried out extensive ^{1}H NMR investigations of ligand binding to the α and β chains of Hb A.

To monitor the structural changes associated with the cooperative oxygenation process for Hb A, we need a probe that allows a direct observation of α and β chains of an intact tetrameric Hb molecule at different stages of ligation. The ferrous hfs proton resonances of deoxyHb A in the spectral region from +10 to +20 ppm from HDO offer such a method. The spectra of the isolated deoxy-α and -β chains of Hb A do not show resonances at +18 or +12 ppm from H_2O, but on mixing of stoichiometric amounts of α and β chains, the ^{1}H NMR spectrum of the mixture is identical to that of Hb A (see Perutz et al., [61] Fig. 9). By comparing the hfs proton resonances of deoxyHb A and appropriate mutant Hbs, we have assigned the 12 ppm resonance to the α chain and the 18 ppm resonance to the β chain. When deoxyHb A combines with O_2 or CO, it becomes diamagnetic, causing these hfs proton resonances to disappear and making this spectral region flat (see Fig. 1). Thus, by following the intensity of the hfs resonances as a function of ligation, we can monitor the binding of a ligand to the α and β chains of Hb A.

To calibrate the intensity of each resonance over the spectral region from +10 to +20 ppm from HDO to the number of protons per heme for each resonance, we have used as a reference standard an NMR shift reagent Tris (6,6,7,7,8,8,8-heptafluoro-2,2-dimethyl-3,5-octanedionate) europium, complexed with the OH of tert-butanol, which gives a single proton resonance at about +27 ppm downfield from HDO.[35] There is no other observable proton signal from +10 to +30 ppm from HDO (see Fig. 8). By using this intensity standard, we have calculated the number of protons in each of the Hb resonances between +10 and +27 ppm from HDO. The total number of protons per heme over this spectral region is 22, with an accuracy of approximately ±10%.[35] A computer program based on the algorithm formulated by Marquardt[62] for the least-squares estimation of nonlinear parameters was used to simulate the ferrous hfs proton resonances of Hb A.[35] The parameters obtained from the least-

[59] B. Shanaan, J. Mol. Biol. 171, 31 (1983).

[60] Z. Derewenda, G. Dodson, P. Emsley, P. Harris, K. Nagai, M. Perutz, and J.-P. Renaud, J. Mol. Biol. 211, 515 (1990).

[61] M. F. Perutz, J. E. Ladner, S. R. Simon, and C. Ho, Biochemistry 13, 2163 (1974).

[62] D. W. Marquardt, J. Soc. Ind. Appl. Math. 11, 431 (1963).

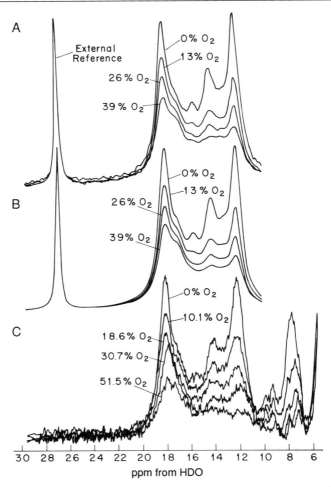

FIG. 8. ^{1}H NMR spectra (250 MHz) of Hb A in 0.1 M Bis–Tris plus 10 mM IHP in D_2O as a function of oxygenation at 27°: (A) experimental spectra of 10% Hb A at pH 6.6; (B) computer-simulated spectra; (C) experimental spectra of 11% Hb A at pH 6.42 (from Viggiano et al.[26]). (Reprinted with permission from C. Ho et al., in "Hemoglobin and Oxygen Binding," p. 141. Elsevier/North-Holland, New York, 1982.)

squares program were then used to generate a computer-simulated spectrum of deoxyHb A over the spectral region from +10 to +30 ppm from HDO. The difference spectrum obtained by subtracting the experimental spectrum from the simulated spectrum is only slightly above the noise level of the experimental spectrum, suggesting that the spectral parameters

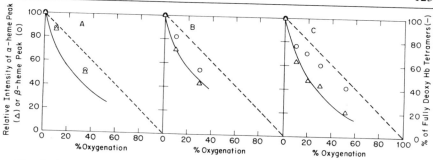

FIG. 9. A comparison of the relative intensities of the α-heme resonance at 12 ppm ($\triangle$) and the β-heme resonance at 18 ppm ($\bigcirc$) with the percent of fully deoxyHb tetramers (—) as a function of oxygenation: (A) in 0.1 M Bis–Tris; (B) in 0.1 M Bis–Tris plus 35 mM 2,3-DPG; (C) in 0.1 M Bis–Tris plus 10 mM IHP. The fraction of fully deoxyHb tetramers was calculated from the data of Tyuma et al.[63] [Reprinted with permission from G. Viggiano et al., Biochemistry 18, 5238 (1979). Copyright 1979 American Chemical Society.]

obtained by the fitting routine explain the experimentally observed spectrum reasonably well. The overlapping peaks in the experimental ^{1}H NMR spectra are resolved using the computer-generated spectra. Figure 8A and C gives a series of ^{1}H NMR spectra of Hb A in 0.1 M Bis–Tris plus 10 mM IHP as a function of O_2 saturations at pH ~6.6 and 27°. Figure 8B gives the corresponding least-squares simulated spectra. There is excellent agreement among several independent sets of experimental results on the variation of the ferrous hfs proton resonances as a function of oxygenation of Hb A.[35]

Figure 9 gives a comparison of the relative intensities of the α-heme resonance at +12 ppm and the β-heme resonance at +18 ppm with the percentage of fully deoxyHb tetramers as a function of oxygenation in the absence and presence of organic phosphates.[26] In the absence of organic phosphates, the areas of the 12 and 18 ppm resonance peaks decrease at the same rate. In the presence of organic phosphates, the area of the α peak at +12 ppm follows the behavior of the fully deoxyHb A populations, while the area of the +18 ppm β peak remains significantly larger than that of the +12 ppm α peak on oxygenation (Fig. 9B and C). This means that the presence of even one O_2 molecule in the Hb tetramer is enough to cause a decrease in the area of the +12 ppm α resonance, but not the +18 ppm β resonance.

The experimental results allow us not only to calculate various populations of the partially oxygenated species, namely Hb, Hb$(O_2)_1$, Hb$(O_2)_2$,

[63] I. Tyuma, K. Imai, and K. Shimizu, Biochemistry 12, 1491 (1973).

$Hb(O_2)_3$, and $Hb(O_2)_4$, but also the fractions of the α and β chains that are oxygenated in the Hb tetramer as a function of oxygenation. In the presence of IHP, it is clear that the α chains have a much higher O_2 affinity than the β chains. For example, the ratios of oxygenated α chains over β chains of Hb A are 31.5, 10.8, and 8.3 at 13, 26, and 39% of total oxygenation, respectively. As a first approximation, one can assume that at low O_2 saturations, both the singly and doubly oxygenated species all have O_2 molecules bound on the α chains. The behavior of the $+18$ ppm β-chain resonance, which is sensitive to both ligation and structural changes, shows that the number of modified β chains is less than the number of modified α chains.[26,35] The relative intensity of the $+12$ ppm α-chain resonance follows the population of the fully deoxyHb tetramers, suggesting that all of the oxygenated species either have O_2 bound to the α chains or have some structural modifications in the unligated α chains. Thus, our ^{1}H NMR results suggest that a strong cooperativity must exist in the oxygenation of the α chains and that the oxygenation of one α chain affects the ligand affinity of the other unligated α chain within a tetrameric Hb molecule.[26,35,64] These and similar experimental results have confirmed our earlier conclusions[23,26,64-66]: (1) there is no preference between the α and β chains of deoxyHb A for the binding of CO; (2) in the absence of organic phosphate, the α and β chains have similar affinities for O_2; and (3) in the presence of the allosteric effector 2,3-DPG, the α chains have a higher affinity for O_2 than do the β chains, and the difference is enhanced in the presence of IHP.

These results suggest that there are more than two affinity states in the Hb molecule on ligand binding. Thus, the structural changes at partial O_2 saturations in the presence of organic phosphates as seen by ^{1}H NMR spectroscopy cannot be concerted as required by two-state allosteric models, even if extended to take into account the difference in the affinity between the α and β chains as proposed by Ogata and McConnell.[67]

The structural information about partially ligated species derived from this approach is indirect. Because of rapid dimer exchange, intermediate species (such as singly, doubly, and triply ligated species) cannot be isolated. However, the exchange of two dissociative dimers of Hb can be prevented by introduction of the cross-linking reagent bis(3,5-dibromosalicyl)fumarate, which cross-links the Lys-82 residues of the two β subunits under oxygenated (or CO) conditions. In this way, we have prepared

[64] C. Ho and T. R. Lindström, *Adv. Exp. Med. Biol.* **28,** 65 (1972).

[65] M. E. Johnson and C. Ho, *Biochemistry* **13,** 3653 (1974).

[66] G. Viggiano and C. Ho, *Proc. Natl. Acad. Sci. U.S.A.* **76,** 3673 (1979).

[67] R. T. Ogata and H. M. McConnell, *Cold Spring Harbor Symp. Quant. Biol.* **36,** 325 (1971).

various asymmetrical cyanomet mixed valency hybrid Hbs. Because cyanomet heme serves as a model of a ligated heme, it is thus possible to prepare stable Hb molecules that can serve as models of singly and doubly ligated species. We have found that the intensity of the resonance at +9.2 ppm downfield from H_2O is greatly reduced in the spectrum of valency hybrid Hbs with one cyanomet chain compared with that in fully deoxyHb, suggesting that the $\alpha_1\beta_2$ subunit interface in these partially ligated hybrid Hbs is altered. Thus, these results also support the conclusion that there are at least three functionally important structures in going from the deoxy to the ligated state. For details, see Miura and Ho.[68]

Exchangeable Proton Resonances

An important feature in Perutz's stereochemical mechanism for the oxygenation of Hb is the critical roles played by the salt bridges.[58,69] The salt bridges of the α chain involve the α-carboxyl of Arg(141α_1) with both the α-amino group of Val(1α_2) and the ε-amino group of Lys(127α_1) and the guanidinium group of Arg(141α_2) with the carboxyl group of Asp(126α_1). The β-chain salt bridges involve the α-carboxyl of His(146β_2) with the ε-amino group of Lys(40α_1) and the imidazole group of His(146β_2) with the carboxyl of Asp(94β_2). To assess the roles of the salt bridges in the tertiary and quaternary structures of Hb, we have prepared the cross-linked asymmetrically modified Hbs [α(des-Arg)β]$_A$($\alpha\beta$)$_C$XL and [α(des-Arg-Tyr)β]$_A$($\alpha\beta$)$_C$XL, where the subscript A or C denotes that the $\alpha\beta$ dimer is from Hb A or Hb C, respectively, and XL indicates a cross-linked Hb.[70] We have used the same bifunctional reagent, bis(3,5-dibromo-salicyl)fumarate, for cross-linking as in our studies on cross-linked mixed valency hybrid Hbs,[68] in order to prevent exchange between dimers. ^1H NMR spectroscopy was used to investigate these modified Hbs and thus to assess the influence of the salt bridges located at the carboxy terminals of both the α and β chains on the tertiary and quaternary structures of Hb. For details, see Miura and Ho.[70]

^1H NMR spectra of Hb solutions in H_2O show several proton resonances between +5 and +10 ppm downfield from H_2O that arise from exchangeable protons and vanish when the samples are in D_2O. As shown in the section on assignment of resonances, the peaks at +9.4 and +6.4 ppm, like the +12.0 and +18.0 ppm hfs proton resonances discussed in the last section, disappear on ligation, and are thus specific probes for the deoxy structure. These exchangeable proton resonances originate from

[68] S. Miura and C. Ho, *Biochemistry* **21**, 6280 (1982).
[69] M. F. Perutz and L. F. Ten Eyck, *Cold Spring Harbor Symp. Quant. Biol.* **36**, 295 (1971).
[70] S. Miura and C. Ho, *Biochemistry* **23**, 2492 (1984).

hydrogen bonds that are thought to play a crucial role in the structure–function relationship in the Hb molecule. The exchangeable proton resonance at about +9.4 ppm downfield from H_2O is an indicator of the deoxy (T) quaternary structure, whereas the resonance at about +6.4 ppm from H_2O is a characteristic of the deoxy structure.

Two additional exchangeable proton resonances occur in the spectrum of deoxyHb A at about +58.5 and about +71.0 ppm. They originate from the exchangeable NH protons of the proximal histidyl residues of the α and β chains. The large chemical shifts result from the hyperfine interactions between the NH protons of the proximal histidyl residues and the unpaired electrons in the iron atoms of deoxyHb A. The resonance at +71 ppm has been assigned to the β chain and the resonance at +58.5 ppm to the α chain.[34,71,72]

It has been reported that the removal of the carboxy-terminal Arg(α141) can expose the carboxy group of Tyr(α140), which can then form a salt bridge with Val(α1).[69] The newly formed salt bridge between the carboxy group of tyrosine on the α_1 subunit and the amino-terminal amino group of valine on the α_2 subunit is sensitive to pH in the physiological range. Figure 10 gives the ^1H NMR spectra of deoxy[des-Arg(α141)]Hb A in 0.1 M Bis–Tris, 0.1 M Tris, and 0.2 M chloride. It may be seen that breaking of the newly formed salt bridge by increasing the pH can cause a downfield shift of the proximal histidyl $N_\delta H$ exchangeable proton resonance of the α subunit. It has been reported that at low pH, deoxy[des-Arg(α141)]Hb A is in a T-like (deoxy) quaternary structure, whereas at high pH it is converted into an R-like (oxy) quaternary structure.[73] The ^1H NMR spectra of deoxy[des-Arg(α141)]Hb A show pH-dependent changes in both the proximal histidyl $N_\delta H$ exchangeable proton resonances (the α-subunit resonance at +59 ppm is shifted downfield by about 12 ppm on raising the pH) and the H-bonded exchangeable proton resonances (the +9.3 ppm resonance is lost on raising the pH). Thus, these NMR results are in agreement with the finding of Kilmartin et al.[73]

The ^1H NMR spectra of deoxy[α(des-Arg)β]$_A$($\alpha\beta$)$_C$XL at pH 6.0 and 8.6 are also illustrated in Fig. 10. At low pH, the exchangeable proton resonances in the region +5 to +10 ppm from H_2O are essentially the same for deoxy[des-Arg]Hb A and deoxy[α(des-Arg)β]$_A$(aβ)$_C$XL. The resonances due to the α-subunit proximal histidyl $N_\delta H$ proton of deoxy[α(des-Arg)β]$_A$($\alpha\beta$)$_C$XL split into two peaks at 42°. The high-field peak shows

[71] G. N. La Mar, D. L. Budd, and H. Goff, Biochem. Biophys. Res. Commun. 77, 104 (1977).

[72] G. N. La Mar, K. Nagai, T. Jue, D. L. Budd, K. Gersonde, H. Sick, T. Kagimoto, A. Hayashi, and F. Taketa, Biochem. Biophys. Res. Commun. 96, 1172 (1980).

[73] J. V. Kilmartin, J. A. Hewitt, and J. F. Wootton, J. Mol. Biol. 93, 203 (1975).

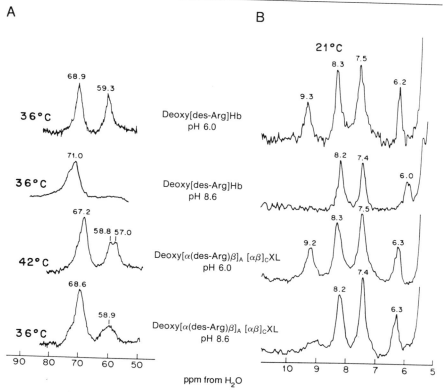

FIG. 10. ^{1}H NMR spectra of deoxy[des-Arg(α141)]Hb A and deoxy[α(des-Arg)β]$_A$ ($\alpha\beta$)$_C$XL in 0.1 M Bis–Tris, 0.1 M Tris, and 0.2 M chloride in 95% H$_2$O and 5% D$_2$O: (A) 300-MHz ^{1}H NMR spectra of hyperfine shifted exchangeable N$_\delta$H of proximal histidyl residues; (B) 600-MHz ^{1}H NMR spectra of exchangeable proton resonances. [Reprinted with permission from S. Miura and C. Ho, *Biochemistry* **23**, 2492 (1984). Copyright 1984 American Chemical Society.]

the same chemical shift as that of the α-subunit proximal histidyl N$_\delta$H exchangeable proton resonance of deoxyHb A (+56.8 ppm at 42°), presumably due to the normal α chain in the $\alpha\beta$ dimer from Hb C. The low-field peak shows the same chemical shift as that of deoxy[des-Arg(α141)]Hb A, and is presumably due to the modified α chain in the α(des-Arg) dimer from Hb A. The integrated intensity of the resonances due to the α-subunit proximal histidyl N$_\delta$H exchangeable protons is similar to that of the β-subunit proximal histidyl N$_\delta$H exchangeable proton signal around +70 ppm downfield from H$_2$O. Raising the pH to 8.6 causes about half the intensity of the α-subunit proximal histidyl N$_\delta$H proton resonance to

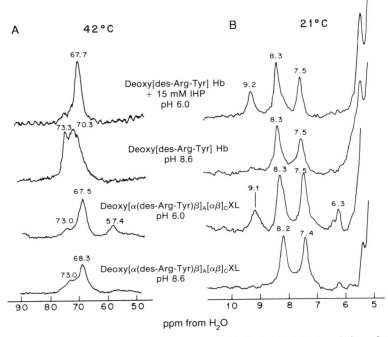

FIG. 11. ^{1}H NMR spectra of deoxy[des-Arg(α141)-Tyr(α140)]Hb A and deoxy[α(des-Arg-Tyr)β]$_A$($\alpha\beta$)$_C$XL in 0.1 M Bis–Tris, 0.1 M Tris, and 0.2 M chloride in 95% H_2O and 5% D_2O: (A) 300-MHz ^{1}H NMR spectra of hyperfine shifted exchangeable N$_\delta$H of proximal histidyl residues; (B) 600-MHz ^{1}H NMR spectra of exchangeable proton resonances. [Reprinted with permission from S. Miura and C. Ho, *Biochemistry* **23**, 2492 (1984). Copyright 1984 American Chemical Society.]

shift by about +10 ppm downfield, as does that of deoxy[des-Arg(α141)]Hb A on raising the pH. Thus, the tertiary structural change due to the destruction of the salt bridge is localized within the modified subunit, and the other intact α subunit in [α(des-Arg)β]$_A$($\alpha\beta$)$_C$XL remains unaffected. The $\alpha_1\beta_2$ (or $\alpha_2\beta_1$) subunit interface is altered on raising the pH as a result of the tertiary structural alteration of the α(des-Arg) subunit. The intensity of the exchangeable proton resonance at +9.2 ppm is reduced on increasing the pH, but the resonance at +6.3 ppm remains unchanged.

Deoxy[des-Arg(α141)-Tyr(α140)]Hb A is believed to exist in an R-like (oxy) structure in the absence of IHP and can be converted to a T-like structure in the presence of IHP.[73] The ^{1}H NMR spectra of deoxy[des-Arg(α141)-Tyr(α140)]Hb A in the regions from +5 to +10 ppm and from +50 to +80 ppm from H_2O with and without IHP are shown in Fig. 11.

In the absence of IHP, the exchangeable proton resonances in the spectral region from +5 to +10 ppm from H_2O show the spectral features of an R-like structure, that is, lacking the +9.3 and +6.3 ppm resonances. Corresponding to this, the hyperfine shifted proximal histidyl $N_\delta H$ exchangeable proton resonances from the α subunit are shifted downfield by more than 10 ppm from the position expected for deoxyHb A (Fig. 11A). In the presence of IHP, the deoxy quaternary structure marker resonance at +9.2 ppm reappears and the proximal histidyl $N_\delta H$ exchangeable proton signals merge into one resonance at +68 ppm. It should be noted that despite the appearance of the +9.2 ppm resonance on addition of IHP, the exchangeable proton resonance expected at +6.3 ppm from H_2O, also a deoxy structural marker, is absent.

As also shown in Fig. 11A, for deoxy[α(des-Arg-Tyr)β]$_A$($\alpha\beta$)$_C$XL at pH 6.0, about half the intensity of the hyperfine shifted proximal histidyl $N_\delta H$ exchangeable proton resonances has already shifted downfield by about 15 ppm, but the remaining resonance remains at a position close to that of the α-subunit proximal histidyl $N_\delta H$ proton resonance in intact Hb A. It is reasonable to assign the resonance that remains unchanged to the normal intact α subunit of ($\alpha\beta$)$_C$ in deoxy[α(des-Arg-Tyr)β]$_A$($\alpha\beta$)$_C$XL. In correspondence with the shifted $N_\delta H$ proton resonance of α(des-Arg-Tyr), both resonances at +9.2 and +6.3 ppm from H_2O have lost about half of their respective intensities, and they disappear when the pH is raised to 8.6. The increase in pH also causes the α-subunit histidyl $N_\delta H$ exchangeable proton resonance, normally occurring around +57 ppm from H_2O at low pH, to shift downfield by about 14 ppm and the broad resonance at +73 ppm from H_2O to increase in intensity (Fig. 11A).

The ^1H NMR results indicate that the effects on the hyperfine shifted proximal histidyl $N_\delta H$ exchangeable proton resonances at pH 6.0 of removing Arg(α141) or Arg(α141)Tyr(α140) from one of the two α subunits are limited to within the α subunit from which the carboxy-terminal amino acids are specifically removed. The two asymmetrically modified Hbs have the exchangeable proton resonance at +9.3 ppm from H_2O, which has been assigned to the H bond between α42 Tyr and β99 Asp located at the $\alpha_1\beta_2$ subunit interface. This suggests that these asymmetrically modified Hbs preserve the deoxy-like quaternary structure in the $\alpha_1\beta_2$ subunit interface as manifested by the presence of this intersubunit H bond. On the other hand, the ^1H NMR spectrum of deoxy[α(des-Arg-Tyr)β]$_A$($\alpha\beta$)$_C$XL at high pH shows no resonance at about +57 ppm from H_2O in the spectral region expected for the proximal histidyl $N_\delta H$ exchangeable proton resonance of the unmodified α subunit in the $\alpha\beta$ dimer from Hb C (Fig. 11A). Thus, these results clearly show that several features of the ^1H NMR spectra of these asymmetrically modified Hbs cannot be accounted for as a spectral sum of the intact deoxyHb C and chemically modified deoxyHb A. These results suggest that intermediate structures exist in which the tertiary and the quaternary structural transitions occur asymmetrically about the diad axis of the Hb molecule during the course of the successive removal of the salt bridges. These results show that within the tetrameric Hb molecule in the deoxy form, the conformation at the two intersubunit interfaces $(\alpha_1\beta_2)$ and $(\alpha_2\beta_1)$ is different in these modified Hbs. This implies that there is at least one additional structure other than the T and R structures that can exist in the Hb molecule in going from the deoxy to the oxy state.

Molecular Mechanism of the Bohr Effect

The Bohr effect is the change in the oxygen affinity of hemoglobin with pH.[5] It is due to the change in pK of ionizing groups in hemoglobin on oxygenation.[74,75] Above pH 6.0, HbO$_2$ is more negatively charged than deoxyHb because of an oxygen-linked ionizing group. The alkaline Bohr effect is the release of protons in the Hb molecule during the transition from the deoxy to the oxy state at a pH above 6.0.[5] Below pH 6.0 the reverse occurs. The O$_2$ affinity increases with decreasing pH and the Hb molecule absorbs H$^+$ ions on oxygenation (the acid Bohr effect). Besides the α-amino terminal, the imidazole groups from the histidyl residues, because of their pK values, which are in the physiologically relevant range, are the most likely amino acid residues of Hb for which changes in pK values on oxygenation could contribute significantly to the alkaline Bohr effect.

Identification of Bohr Groups

Several histidyl residues of Hb A have been proposed, on the basis of X-ray diffraction data and the results on mutant and chemically modified Hbs, to play an important role in the variation of the oxygen affinity with pH.[58,76] The involvement of these surface histidine residues in the Bohr effect originates from the changes in their electrostatic environments accompanying the change in the conformation of the Hb molecule on ligation. According to the crystallographic results,[58] β146 His is an important contributor to the Bohr effect. The imidazole of β146 His forms an intrachain salt bridge with the carboxyl group of

[74] J. Christiansen, C. C. Douglas, and J. S. Haldane, *J. Physiol. (London)* **48**, 244 (1914).

[75] L. J. Hendersen, *J. Biol. Chem.* **41**, 401 (1920).

[76] M. F. Perutz, J. V. Kilmartin, K. Nishikura, J. H. Fogg, P. J. G. Buttler, and H. S. Rollema, *J. Mol. Biol.* **138**, 649 (1980).

β94 Asp in deoxyHb crystals but not in oxy-like metHb crystals.[51,58] This local electrostatic interaction would greatly increase the pK values of the β146 His residue. In ligated Hb, β146 His is found to move ~10 Å away from the β94 Asp,[52,58,59] and thus the pK value of this residue should revert to a lower value.

To investigate the contribution of β146 His to the alkaline Bohr effect, functional studies have used molecules in which this histidyl residue is altered either by enzymatic or by chemical modification or replaced by mutation. In all the Hb molecules in which this histidyl residue is altered, the alkaline Bohr effect is reduced compared to that of human normal adult Hb.[76–78] The amount of this reduction, however, depends on the nature of the alteration or of the replacing amino acid residue as well as on the experimental conditions. Saroff[79] first pointed out that the identification of the Bohr groups with mutant or chemically modified Hb requires a complete analysis of the pH dependence of the number of H$^+$ ions released on oxygenation. This is because replacements or chemical modifications of single amino acids can induce long-range conformational effects in the Hb molecule and thus can affect the contributions of other groups to the Bohr effect.

^{1}H NMR spectroscopy is the only technique presently available that can monitor the conformation and environment of individual amino acid residues in the Hb molecule in solution.[2,14] As seen in Fig. 2, the C-2 proton resonances of the histidyl residues of Hb A are well separated from those of other protons. According to the convention in our laboratory, the histidyl C-2 resonances in HbCO A are given letters A–L and Y. The C-2 proton resonances of 11–13 histidyl residues per Hb dimer can be resolved and titrated individually by ^{1}H NMR in both deoxy and CO forms in the solution state and under a wide range of experimental conditions relevant to Hb function.[3,4,44,45,80] The pK values are determined by a nonlinear least-squares fit of the ^{1}H chemical shift, δ, as a function of [H$^+$] according to the following equation:

$$\delta = (\delta^+[H^+]^n + \delta^0 K^n)/([H^+]^n + K^n) \tag{1}$$

where δ^+ and δ^0 are the chemical shifts in the protonated and unprotonated forms of the histidyl residue, respectively, K is the H$^+$ dissociation equilibrium constant of the histidyl residue, and n is the titration coefficient for the ^{1}H NMR titration of the histidyl residue.

[77] J. V. Kilmartin, J. H. Fogg, and M. F. Perutz, *Biochemistry* **19**, 3189 (1980).

[78] S. Matsuakawa, Y. Itatani, K. Mawatari, Y. Shimokawa, and Y. Yoneyama, *J. Biol. Chem.* **259**, 11479 (1984).

[79] H. A. Saroff, *Physiol. Chem. Phys.* **4**, 23 (1972).

[80] I. M. Russu and C. Ho, *Biochemistry* **25**, 1706 (1986).

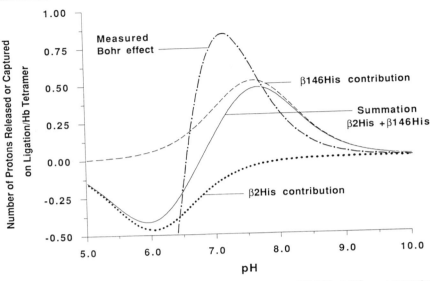

FIG. 12. Alkaline Bohr effect of Hb A in D_2O in 0.1 M HEPES at 29°, reconstructed from Eq. (2), and the contributions of $\beta146$ His and $\beta2$ His as calculated from the pK values determined by ^{1}H NMR spectroscopy. The summation of the $\beta146$ His and $\beta2$ His contributions is also shown. Curve symbols: measured Bohr effect (–·–), $\beta146$ His contribution (– – –), summation of $\beta146$ His and $\beta2$ His (—), and $\beta2$ His contribution (···). [Reprinted with permission from M. R. Busch *et al., Biochemistry* **30**, 1865 (1991). Copyright 1991 American Chemical Society.]

From the obtained pK values and the experimental values of P_{50} obtained for a wide range of pH, the H^+ ions released per Hb can be calculated using the following equation and under the assumption that the same 22–26 histidyl residues are observed in the deoxy and in the CO form:

$$\Delta H^+ = 4\left(\frac{K_1''}{[H^+] + K_1''} + \frac{K_2''}{[H^+] + K_2''} - \frac{K_1'}{[H^+] + K_1'} - \frac{K_2'}{[H^+] + K_2'}\right) \quad (2)$$

Where K_1 and K_2 are the H^+ dissociation equilibrium constants of two hypothetical oxygen-linked groups, and the single prime refers to deoxy Hb and the double prime to HbCO. The total number of H^+ calculated using this method is comparable to that observed experimentally.

The Bohr effect of Hb A in chloride-free 0.1 M HEPES at 29° as estimated by this method is shown in Fig. 12. The participation of $\beta2$ and $\beta146$ histidyl residues in the Bohr effect as a function of pH can be calculated using the pK values of $\beta2$ His of Hb A in 0.1 M HEPES buffer in HbCO A and deoxyHb A determined by Russu *et al.*[48] and the pK

values of $\beta146$ His of Hb A in the same conditions obtained by Busch *et al.*[4] As seen in the results for the $\beta2$ His, it is not correct to assume that all sites have pK changes that result in a positive contribution to the macroscopically observed effect. The microscopic behavior of a specific group will be defined by its site-specific chemical and electrostatic environments, and its resulting microscopic behavior may oppose the macroscopic behavior seen for the Hb molecule.

Effect of Anions

High-resolution ^{1}H NMR spectroscopy has also been used to investigate the molecular mechanism of the Bohr effect of Hb A in the presence of allosteric effectors such as chloride, inorganic phosphate, or 2,3-DPG. Various allosteric effectors can compete for the same binding sites on the Hb molecule, and a given solvent component can affect the affinity of an individual site for other allosteric effectors. The number of H$^+$ ions released on oxygenation of Hb is strongly dependent on the concentration of Cl$^-$ ions in solution. It has been shown that the release of about half of the Bohr protons is due to a difference in Cl$^-$ binding to the oxy and deoxy forms.[81] In this model, the Cl$^-$ ions bind to positively charged groups whose pK values are around neutral pH such as histidine residues and the NH$_2$ group of $\alpha1$ Val. The binding of Cl$^-$ ions to an individual site should clearly result in an uptake of H$^+$ ions, which, in turn, should depend on the affinity of that site for the anion. Thus, the release of Cl$^-$ ions on oxygenation of Hb should be accompanied by a release of H$^+$ ions and an enhancement in the Bohr effect as observed experimentally. The part of the Bohr effect that originates from the oxygen-linked interactions of the Hb molecule with solvent anions is generally known as the anion Bohr effect.

A recent ^{1}H NMR study in our laboratory indicates that $\beta2$ histidyl residues are strong binding sites for chloride and inorganic phosphate ions in Hb.[48] The affinity of $\beta2$ histidyl residues for these anions is larger in the deoxy than in the CO form. The interactions of Cl$^-$ and inorganic phosphate ions with the Hb molecule also result in lower pK values of several surface histidyl residues and/or changes in the shapes of the H$^+$-binding curves. These results suggest that long-range electrostatic interactions between individual ionizable sites in Hb could play an important role in the molecular mechanism of the anion Bohr effect. In another study using ^{1}H and ^{31}P NMR, individual H$^+$ NMR titration curves were obtained for 22–26 histidyl residues of Hb and for each phosphate group

[81] G. G. M. Van Beek, E. R. P. Zuiderweg, and S. H. de Bruin, *Eur. J. Biochem.* **99**, 379 (1979).

of 2,3-DPG with Hb in both deoxy and CO forms. The addition of an equimolar concentration of 2,3-DPG to Hb A solutions specifically affects several His–C-2 proton resonances in both the deoxy and CO forms. The largest effect is observed for the $\beta2$ histidyl residue that originates from an increase in the pK value of this histidyl residue in the presence of 2,3-DPG (Fig. 13). The results suggest that 2,3-DPG binds to deoxyHb at the central cavity between the two β chains and the binding involves the $\beta2$ histidyl residues. In addition, the results suggest that the binding site of 2,3-DPG to HbCO A involves at least some of the same amino acid residues in deoxyHb. Under the experimental conditions studied in this work, the $\beta2$ His makes a significant contribution to the alkaline Bohr effect (up to 0.5 H$^+$/Hb tetramer) because of the specific interactions with 2,3-DPG. These findings gave the first experimental demonstration that long-range electrostatic and/or conformational effects of the binding could play an important role in the allosteric effect of 2,3-DPG. Figure 13 shows the Bohr effect of $\beta2$ His and $\beta146$ His under different experimental conditions. It is clear that the molecular mechanism of the Bohr effect is dependent on the solvent composition.

Studies with Lysed Cells

To gain insights into the nature of the Bohr effect inside red blood cells (RBCs), we have compared the titration properties of the C-2 proton resonances of histidyl residues of purified Hb A, lysed RBCs and intact RBCs. The aromatic proton resonances of HbCO A, lysed RBCs, and RBCs in 0.1 M HEPES in the presence of 0.154 M NaCl in D$_2$O at 29° are shown in Fig. 14. It can be seen that the resonances from lysed cells are much better resolved than those of the whole cells. In the spectra from lysed cells, the resonances are narrower and it is much easier to follow them over the entire pH range. Because of the broad lines, it is difficult to resolve the individual resonances in RBCs, but under the conditions in which we can, the chemical shifts are essentially the same. The spectrum of lysed cells is essentially identical to that of Hb in solution.

The pH titration curves of the histidines in Hb solutions and lysed cells are superimposable for most of the resonances (several examples are shown in Fig. 15A), except for resonance L (Fig. 15B). Thus, most of the resonances have the same pK in purified Hb and lysed cells in the same conditions. These results suggest that the study of lysed cells can provide useful information about the Bohr effect and how the Hb molecule works inside the RBCs.

A

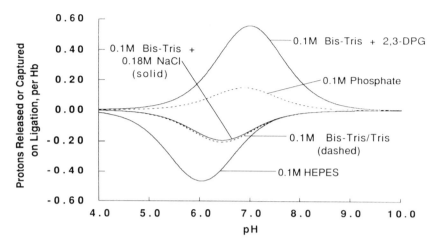

B

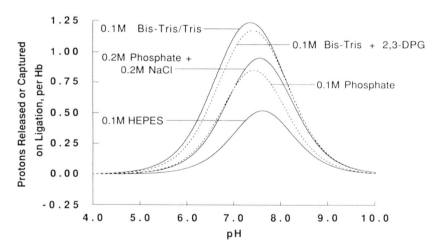

FIG. 13. Contributions of $\beta2$ His (A) and $\beta146$ His (B) residues to the Bohr effect of Hb A under different conditions. [Reprinted with permission from M. R. Busch and C. Ho, *Biophys. Chem.* **37**, 313 (1990).] For details, see Busch and Ho.[3]

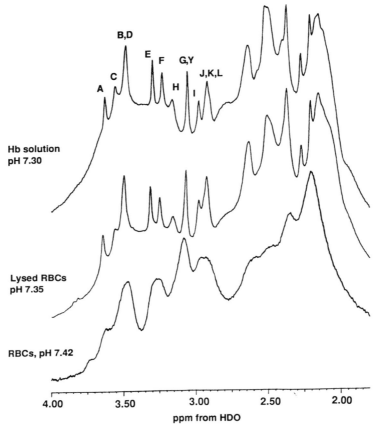

FIG. 14. ^{1}H NMR spectra (300 MHz) of the histidyl residues of HbCO A, pH 7.30; lysed RBCs, pH 7.35; and RBCs, pH 7.42 in 0.1 M HEPES plus 0.154 M NaCl in D_2O at 29°.[24]

Concluding Remarks

From our NMR studies of the molecular basis for the cooperative oxygenation of Hb, we have reached the following major conclusions: (1) The α and β chains of Hb A are nonequivalent, both structurally and functionally. The relative ligand affinities of the α and β chains of Hb A depend not only on the nature of heme ligands, but also on the nature of allosteric effectors. For example, the α and β chains have equal affinity for O_2 in the absence of organic phosphates. However, in the presence of 2,3-DPG or IHP, the α chains have a higher affinity for O_2 than the β

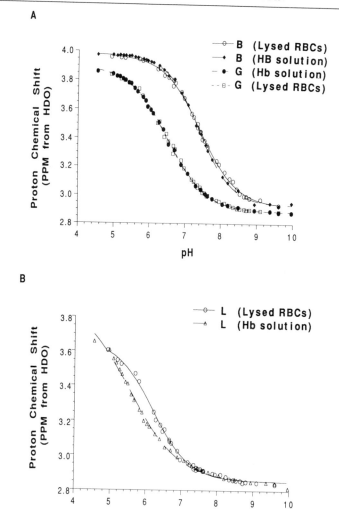

FIG. 15. ^{1}H NMR titration curves of the same resonances in HbCO A in solution and lysed RBCs in 0.1 M HEPES plus 0.154 M NaCl in D_2O at 29°.[24] Resonances B and G (A) and L (B) are the same ones as those shown in Fig. 14.

chains. (2) Structural changes that occur in Hb on oxygenation are not concerted. Ligation of one subunit can affect the structure of unligated subunits within a tetrameric Hb molecule, thus altering the ligand affinity of the unligated subunits. Some cooperativity must be present within the deoxy quaternary structural state during the oxygenation process. (3) Strong evidence exists that the structure of the singly ligated Hb species is likely to be different from that of the fully deoxy- or oxyHb (especially in the $\alpha_1\beta_2$ subunit interface). These results indicate that there are more than two quaternary structures during the transition from the deoxy to the oxy state and that two-structure allosteric models do not adequately describe the cooperative oxygenation of Hb A. These results are more consistent with a sequential (or induced-fit) description for the oxygenation of hemoglobin.

Our ^{1}H NMR studies of the alkaline Bohr effect clearly illustrate that the microscopic behavior of a specific group in Hb, which results from the site-specific chemical and electrostatic environment of the group, is determined both by the Hb structure and by the environment in which the Hb molecule resides. The variable role of a given residue under different solvent conditions has been demonstrated for $\beta2$ His and $\beta146$ His, and implies that the detailed molecular mechanisms of the Bohr effect are not unique, but are adaptive to solvent conditions.

The hemoglobin problem is certainly not yet solved and there are more exciting discoveries to be made by researchers in the years ahead. A complete description of ligand binding to Hb A must include detailed analyses of both intermediate structural states of Hb and their functional properties. A challenge to hemoglobin researchers is to elucidate the structures of singly, doubly, and triply oxygenated Hb species and to determine their respective functional properties. We can then compare these structures to those of the deoxy (or T) quaternary and oxy (or R) quaternary structures and correlate these structural properties with their energetic and functional properties. These partially ligated Hb species may exist only transiently, but they play an essential role in making possible the transition from the deoxy to the oxy state of Hb A. These new findings will lead us not only to an understanding of hemoglobin cooperativity, but also to new knowledge of the molecular basis for the regulation of metabolic reactions by allosteric enzymes and multimeric proteins. Thus, we will come an important step closer to our eventual understanding of biological information transfer and signaling phenomena at the atomic level. We believe that ^{1}H NMR spectroscopy in conjunction with appropriate engineered recombinant hemoglobins will provide new insights into or answers to the above-mentioned challenging problems in the field of hemoglobin.

Acknowledgments

We wish to extend our greatest appreciation to Dr. E. A. Pratt for indispensable help in getting this paper in order. Our hemoglobin research is supported by a research grant from the National Institutes of Health (HL-24525). J.R.P. is supported by Conselho Nacional de Desenvolvimento Científico e Tecnológico (CNPq-200771/90-6).

[9] Infrared Methods for Study of Hemoglobin Reactions and Structures

By Aichun Dong and Winslow S. Caughey

Introduction

Methods using infrared (IR) spectroscopy provide important, often unique, means for the study of the reactions and structures of hemoglobins (Hb). The infrared bands due to vibrations of infrared-active ligands (e.g., O_2, CO, NO, CN^-, and N_3^-) bound to heme iron not only permit direct measurements of ligand binding and insight into the nature of bonding between iron and ligand, but also reflect effects of differences in protein structure due to factors such as subunit type, mutation, interconversion among dynamic conformers, and perturbations of pH, temperature, and effector molecules.[1-4] The band due to the S–H vibration of the thiol group in a cysteine residue reflects directly the nature of interactions, such as hydrogen bonding, of the –SH group with the environment provided by protein and/or medium.[5,6] The amide I bands, due primarily to the carbonyl groups in the peptide bonds that constitute the linkages between the amino acid residues of the hemoglobin molecule, can be utilized for the qualitative and quantitative determination of α-helix, random, β-sheet, and turn sec-

---

[1] W. S. Caughey, R. A. Houtchens, A. Lanir, J. C. Maxwell, and S. Charache, in "Biochemical and Clinical Aspects of Hemoglobin Abnormalities" (W. S. Caughey, ed.), p. 29. Academic Press, New York, 1978.

[2] J. C. Maxwell and W. S. Caughey, this series, Vol. 54, p. 302.

[3] W. S. Caughey, in "Methods for Determining Metal Ion Environments in Proteins: Structure and Function of Metalloproteins" (D. W. Darnall and R. G. Wilkins, eds.), p. 95. Elsevier/North-Holland, New York, 1980.

[4] W. T. Potter, J. H. Hazzard, M. G. Choc, M. P. Tucker, and W. S. Caughey, Biochemistry 29, 6283 (1990).

[5] J. O. Alben, G. H. Bare, and P. A. Bromberg, Nature (London) 252, 736 (1974).

[6] J. O. Alben and G. H. Bare, J. Biol. Chem. 255, 3892 (1980).

ondary structures.[7–11] Infrared spectroscopy has also been used to detect and characterize sites occupied within hemoglobin by infrared active molecules of the anesthetic nitrous oxide.[12,13]

Infrared spectra of hemoglobin can be obtained under a wide variety of conditions. Intact tissue, whole blood,[14] packed cells,[15] or, with an infrared microscope, a single red cell,[16] as well as isolated protein may be studied.[2,3] Furthermore, the isolated protein may be in solution,[2,3] an amorphous solid, or crystals[17]; crystallization of adult human HbCO (HbCO A) causes a shift in CO infrared spectra that demonstrates an alteration in the ligand environment of β subunits.[17] Conditions of temperature, pH, and medium may be the same or widely different from physiological conditions. Water may be replaced with D_2O or organic cosolvents may be present. Some infrared bands are more readily measured in D_2O than H_2O. Both H_2O and D_2O absorb so strongly in certain regions of the spectrum as to limit severely the detection of solute bands in these regions (Fig. 1). Fortunately, the regions of relatively low absorbance ("windows") for H_2O and D_2O usually appear at sufficiently different wavenumbers to permit the use of one solvent where the other cannot be used effectively. Thus, the accurate measurement of an infrared band at a given wavelength depends on the absorption of the medium as well as on protein concentration, intrinsic band intensity, and the distance through the sample traversed by the infrared radiation. Improvements in infrared instrumentation and in data analysis have greatly enhanced the sensitivity of the infrared method applied to aqueous solutions. In this chapter experimental approaches and interpretations of data for ligand spectra, microspectroscopy, thiol spectra, and amide I spectra are considered.

Infrared Spectra of Exogenous Ligands Bound to Heme Iron

The most extensive use of infrared spectroscopy in hemoglobin studies has involved ligand infrared spectra.[2,3] Several exogenous sixth ligands

[7] T. Miyazawa and E. R. Blout, *J. Am. Chem. Soc.* **83**, 712 (1961).
[8] S. Krimm and J. Bandekar, *Adv. Protein Chem.* **38**, 181 (1986).
[9] H. Susi and D. M. Byler, this series, Vol. 130, p. 290.
[10] A. Dong, P. Huang, and W. S. Caughey, *Biochemistry* **29**, 3303 (1990).
[11] A. Dong, P. Huang, and W. S. Caughey, *Biochemistry* **31**, 182 (1992).
[12] J. C. Gorga, J. H. Hazzard, and W. S. Caughey, *Arch. Biochem. Biophys.* **240**, 734 (1985).
[13] J. H. Hazzard, J. C. Gorga, and W. S. Caughey, *Arch. Biochem. Biophys.* **240**, 747 (1985).
[14] J. C. Maxwell, C. H. Barlow, J. E. Spallholz, and W. S. Caughey, *Biochem. Biophys. Res. Commun.* **61**, 230 (1974).
[15] W. T. Potter, J. H. Hazzard, S. Kawanishi, and W. S. Caughey, *Biochem. Biophy. Res. Commun.* **116**, 719 (1983).
[16] A. Dong, R. G. Messerschmidt, J. A. Reffner, and W. S. Caughey, *Biochem. Biophys. Res. Commun.* **156**, 752 (1988).
[17] W. T. Potter, R. A. Houtchens, and W. S. Caughey, *J. Am. Chem. Soc.* **107**, 3350 (1985).

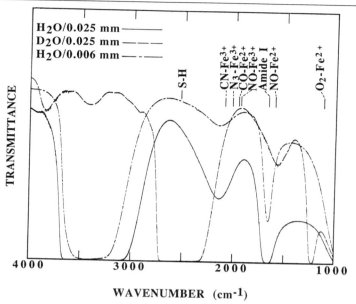

FIG. 1. Infrared spectra of H_2O vs air and D_2O vs air recorded in transmission mode. Approximate wavenumber locations of absorption bands due to some ligands bound to heme iron in hemoglobins and to cysteine –SH and amide I groups are indicated. The 0.025- and 0.006-mm spectra shown designate infrared cell path lengths.

exhibit sufficiently intense vibrational bands to be observed in infrared spectra. These ligands include O_2, CO, and NO bound to Fe^{2+} and CN^-, N_3^-, OCN^-, SCN^-, $SeCN^-$, and NO bound to Fe^{3+}. It is the vibration within the ligand itself that is observed. The iron–ligand bond vibrational bands appear at lower energies and have not been studied by infrared spectroscopy. Carbon monoxide is the first ligand to be observed and remains the ligand for which the most infrared data are available. Our discussion here is confined mainly to CO infrared spectra. More detailed discussions can be found elsewhere[2,3,18–20] for O_2, NO, N_3^-, and CN^- ligand infrared spectra.

Measurement of Ligand Infrared Spectra

As discussed above, the presence of water in biological materials limits measurement of solute infrared bands that occur in regions where water

[18] W. T. Potter, M. P. Tucker, R. A. Houtchens, and W. S. Caughey, *Biochemistry* **26**, 4699 (1987).
[19] S. Yoshikawa, D. H. O'Keeffe, and W. S. Caughey, *J. Biol. Chem.* **260**, 3518 (1985).
[20] J. C. Maxwell and W. S. Caughey, *Biochemistry* **15**, 388 (1976).

TABLE I
FREQUENCIES AND INTEGRATED INTENSITIES FOR MAJOR BANDS
OF LIGANDS BOUND TO HEME IRON IN ADULT
HUMAN HEMOGLOBIN[a]

Ligand	Frequency (cm^{-1})	Integrated intensity (B) $(M^{-1}\ cm^{-2})$	Refs.
CO	1951 (Fe^{2+})	3.4×10^4	4, 21
O_2	1107 (Fe^{2+})	0.4×10^4	18, 22
N_3^-	2025 (Fe^{3+})	2.0×10^4	23
CN^-	2122 (Fe^{3+})	0.5×10^4	19, 23
NO	1617 (Fe^{2+})		20, 24
	1925 (Fe^{3+})		24
$SeCN^-$	2075 (Fe^{3+})		23
SCN^-	2063 (Fe^{3+})		23
OCN^-	2167 (Fe^{3+})		23

[a] Data from Potter et al.[4,18]; Yoshikawa et al.[19]; Alben and
Caughey[21]; Maxwell and Caughey[20]; Dong et al.[24]; Barlow et
al.[22]; McCoy and Caughey.[23]

absorptions are very strong (Fig. 1). These regions include 3800–3000
cm^{-1} due to O–H stretching and 1700–1600 cm^{-1} due to H–O–H bending
vibrations.[2,3] To obtain an accurate spectral measurement, sufficient radiation must pass through the material under observation. Infrared measurements on solutes are made more readily in regions of low water absorption
(the windows) between 4000–3800, 2800–1750, and 1500–900 cm^{-1} than
in regions of strong absorption. Many ligand bands occur within the 2800
to 1750-cm^{-1} window (Fig. 1; Table I).[21–24] By using a Fourier transform
infrared spectrophotometer with high sensitivity, however, the limitation
due to water absorption in the region between 1700 and 1600 cm^{-1} has
been greatly reduced by shortening the distance between cell windows to
6–10 μm.[10,11,25] Furthermore, if the solute has infrared bands of interest
in the region not readily accessible in the presence of water, it is frequently
possible to observe these bands by the exchange of H_2O with D_2O, which
has windows at frequencies where H_2O does not (Fig. 1).

[21] J. O. Alben and W. S. Caughey, Biochemistry 7, 175 (1968).
[22] C. H. Barlow, J. C., Maxwell, W. J. Wallace, and W. S. Caughey, Biochem. Biophys.
Res. Commun. 55, 91 (1973).
[23] S. McCoy and W. S. Caughey, Biochemistry 9, 2387 (1970).
[24] A. Dong, P. Huang, V. Sampath, and W. S. Caughey, unpublished data.
[25] J. L. Koenig and D. L. Tabb, in "Analytical Application of FT-IR to Molecular and
Biological systems" (J. R. During, ed.), p. 241. Reidel Publ., Dordrecht, The Netherlands, 1980.

The selection of window material for the infrared cell is also an important consideration. The solubility in water, the refractive index, absorption characteristics, inertness, and brittleness are important factors in the selection, as has been discussed previously.[2] For most studies over the 4000 to 1000-cm^{-1} region, CaF_2 is the window material of choice. It has a low refractive index (1.40 at 2000 cm^{-1}) and is sufficiently transparent in both infrared and ultraviolet–visible spectral regions to permit the measurement of ultraviolet–visible spectra on the same sample in the same cell as is used for measurement of infrared spectra. However, CaF_2 windows are not suitable for use with solutions containing ammonia, in which it is soluble.

The minimum protein concentration for an accurate spectral measurement depends on the extinction coefficient ($\varepsilon = A/cl$) or integrated intensity [$B = (1/cl)\int A dv$] of the selected ligand. For example, the C–O stretch band of the CO–ligand in human HbCO A can be detected at a concentration as low as 10 μM in heme and can be measured very accurately at 200 μM with a 0.1-mm path length cell. Several ligands commonly used in infrared spectroscopic studies of hemoglobins and their typical integrated intensities (B) are listed in Table I.

Adult HbCO solutions are typically prepared for infrared measurement in our laboratory in a Beckman (Fullerton, CA) FH-01 cell with CaF_2 windows and a 0.1-mm path length. Infrared spectra are recorded at 20° with a Perkin-Elmer (Norwalk, CT) model 1800 Fourier transform infrared spectrophotometer equipped with an Hg/Cd/Te detector and interfaced with a Perkin-Elmer 7700 computer. For each spectrum a 1000-scan interferogram is collected in a single-beam mode with a 2-cm^{-1} resolution and a 1-cm^{-1} interval. A reference spectrum is recorded under identical scan conditions with only buffer in the cell. A single-step subtraction of reference spectrum from the observed protein spectrum is carried out using the water vapor band at 1918 cm^{-1} to evaluate the effectiveness of the subtraction. The spectrum between 2000 and 1900 cm^{-1} is selected and baseline corrected. Curve-fitting deconvolutions with Gaussian or Gaussian–Lorentzian mixed functions are carried out, if needed, by using a CURVEFIT function (Galactic Industries Corp., Salem, NH) on a 386-based personal computer.

Assignment of Ligand Bands

Typical infrared band assignments for many hemoglobin ligands are listed in Table II.[26,27] The expected frequency, or range of frequencies,

[26] M. Nagai, A. Dong, Y. Yoneyama, and W. S. Caughey, unpublished data (1992).
[27] W. S. Caughey, J. O. Alben, S. McCoy, S. Charache, P. Hathaway, and S. Boyer, *Biochemistry* **8**, 59 (1969).

TABLE II
PARAMETERS FOR MAJOR INFRARED BAND OF LIGANDS BOUND TO
HUMAN HEMOGLOBINS

Hemoglobin	Ligand	Frequency (cm^{-1})	Isotope shift (cm^{-1})	Refs.[a]
Human Hb A	$^{12}C^{16}O$	1951		4, 21
(Fe^{2+})	$^{13}C^{16}O$	1907	44	4, 21
	$^{12}C^{18}O$	1907	44	4, 21
	$^{16}O_2$	1155, 1106[b]		18
	$^{18}O_2$	1094, 1065[b]		18
	$^{14}N^{16}O$	1615		20
	$^{15}N^{16}O$	1587	28	20
(Fe^{3+})	$^{12}C^{14}N$	2122		19, 23
	$^{13}C^{14}N$	2077.5	44.4	19
	$^{12}C^{15}N$	2092	30	19
	N_3^-	2025		23
Hb Zurich	$^{12}C^{16}O$	α1950, β1958		4, 17
	$^{16}O_2$	1156, 1107[b]		18
	$^{18}O_2$	1094, 1066[b]		18
Hb Sydney	$^{12}C^{16}O$	α1951, β1956		1
Hb M Boston	$^{12}C^{16}O$	α1970, β1952		26
	$^{13}C^{16}O$	α1926, β1908	44, 44	26
	$^{12}C^{18}O$	α1926, β1908	44, 44	26
Hb M Saskatoon	$^{12}C^{16}O$	α1951, β1970		26, 27
	$^{13}C^{16}O$	α1907, β1926	44, 44	26
	$^{12}C^{18}O$	α1907, β1926	44, 44	26
Hb M Milwaukee	$^{12}C^{16}O$	α1951, β1947		26
	$^{13}C^{16}O$	α1907, β1903	44, 44	26
	$^{12}C^{18}O$	α1907, β1904	44, 43	26
Horse Hb	$^{12}C^{16}O$	1950		4
Cow Hb	$^{12}C^{16}O$	1950		4
White rabbit II Hb	$^{12}C^{16}O$	α1928, β1952		4
		α1951		4
Guinea pig Hb	$^{12}C^{16}O$	1949		4
Rat Hb	$^{12}C^{16}O$	1950		4
Mouse Hb	$^{12}C^{16}O$	1951		4

[a] Data from Caughey et al.[1,27]; Potter et al.[4,17,18]; Yoshikawa et al.[19]; Alben and Caughey[21]; Maxwell and Caughey[20]; Nagai et al.[26]; McCoy and Caughey.[23]

[b] Band frequencies were measured from difference spectra of $^{16}O_2$ minus $^{18}O_2$.

for a given ligand usually can be deduced from assignments that can be clearly made for the ligand in small coordination compounds of known structure. Nevertheless, isotopic substitution is an extremely useful method to identify unequivocally the band for a given ligand. For example, the assignment of the C–O stretch band in human HbCO A[21] was obtained by the isotopic shifts from 1951 cm^{-1} for $^{12}C^{16}O$ to 1907 cm^{-1} for $^{13}C^{16}O$ and also to 1907 cm^{-1} for $^{12}C^{18}O$. The magnitude of the isotopic shift in each case is 44 cm^{-1}, which is nearly the shift expected for a simple diatomic molecule as computed from the reduced mass relationship

$$\nu = (k/\mu)^{1/2}/2\pi$$

in which $\mu = m_1 m_2/(m_1 + m_2)$, the reduced mass, where m_1 and m_2 represent the masses of the two atoms in the diatomic molecule and k is the harmonic force constant.[2,21] Because k is expected to be independent of the masses of the vibrators for an ideal isolated diatomic oscillator, the stretch frequency expected on isotopic substitution can be computed. The use of isotopic shifts proved even more important for the identification of bands due to the O–O stretch in HbO$_2$ A.[18,28] Although interconversions among carbonyl, oxy, and deoxy species may be carried out easily, the comparison of infrared spectra among these species does not permit a unique identification of the O–O stretch band(s) because of interference of bands from porphyrin and/or protein in the same region (near 1100 cm^{-1}) that also shift with a change in ligand or oxidation state. By use of computer-generated difference spectra between oxy species that differ only in terms of oxygen isotope ($^{16}O_2$, $^{17}O_2$, or $^{18}O_2$) it was possible to identify unique bands due to bound O$_2$ (Fig. 2).[18,28] Positive identification of the N–O stretch band for heme-bound NO in HbNO A is made difficult by its location near 1617 cm^{-1}, in a region where the strong absorptions from both the H–O–H bending band of H$_2$O (1645 cm^{-1}) and the C=O stretching of protein amide groups (1700–1620 cm^{-1}) are found (Fig. 1). Water absorption near 1645 cm^{-1} can be avoided by using a D$_2$O-based medium,[20] but interference from the amide C=O stretch band can be eliminated only by subtraction of the spectrum of HbA liganded with a ligand other than NO from the spectrum of HbNO A. Figure 3 presents the superimposed infrared spectra of HbNO A and HbCO A from 2000 to 1200 cm^{-1}. The N–O stretch band in the infrared spectrum of HbNO A, is indicated by a circle. Two discrete N–O stretch bands near 1617 cm^{-1} (CI band) and 1632 cm^{-1} (CII band) are observed. Isotope substitution in NO is also helpful[20] and shows only the CI band is due to NO.[28a]

The measurable parameters (frequency, bandwidth, and intensity) of heme-bound CO are sensitive to changes in the ligand-binding environment that result from changes in protein structure due to dynamic conforma-

[28] W. S. Caughey, M. G. Choc, and P. Hathaway, in "Biochemical and Clinical Aspects of Oxygen" (W. S. Caughey, ed.), p. 1. Academic Press, New York, 1979.

[28a] V. Sampath, X.-J. Zhao, and W. S. Caughey, Biochem. Biophys. Res. Commun., in press.

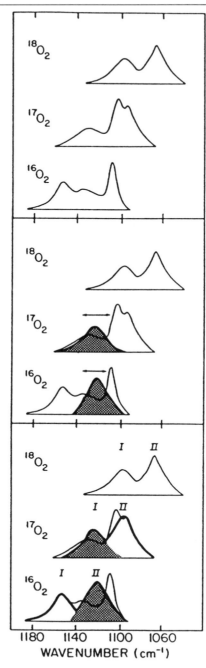

tions,[29] oxidation state,[3] amino acid sequence,[1,4,27,30,31] and external environment such as pH,[32,33] temperature,[4,29] and solvents.[4] Four C–O stretch bands (CI, CII, CIII, and CIV) have been observed in deconvoluted spectra of a variety of hemoglobins.[4,29,32,34] Each C–O stretch band reflects both the nature of the Fe–C bond and the interaction (steric and/or electrostatic) between CO and its protein environment within the heme pocket.[4] The relative intensities of these four C–O stretch bands vary with pH,[32] temperature,[4,29] and solvent composition.[4] The infrared evidence supports the presence of four rapidly interconverting structural conformers at the heme site.[4,33]

Mutation-Induced Environment Changes at Ligand-Binding Site

CO-infrared spectra have contributed significantly to studies of roles of E7 and E11 residues in hemoglobin structure and function. Mutations

[29] W. S. Caughey, H. Shimada, M. G. Choc, and M. P. Tucker, *Proc. Natl. Acad. Sci. U.S.A.* **78**, 2903 (1981).

[30] P. W. Tucker, S. E. V. Phillips, M. F. Perutz, R. A. Houtchens, and W. S. Caughey, *Proc. Natl. Acad. Sci. U.S.A.* **75**, 1076 (1978).

[31] D. Braunstein, A. Ansari, J. Berendzen, B. R. Cowen, K. D. Egeberg, H. Frauenfelder, M. K. Hong, P. Ormos, T. B. Sauke, R. Scholl, A. Schulte, S. G. Sligar, B. A. Springer, P. J. Steinbach, and R. D. Young, *Proc. Natl. Acad. Sci. U.S.A.* **85**, 8497 (1988).

[32] M. G. Choc and W. S. Caughey, *J. Biol. Chem.* **256**, 1831 (1981).

[33] H. Shimada and W. S. Caughey, *J. Biol. Chem.* **257**, 11893 (1982).

[34] W. E. Brown, J. W. Sutcliffe, and P. D. Pulsinelli, *Biochemistry* **22**, 2914 (1983).

FIG. 2. Schematic interpretation of dioxygen infrared spectra for Hbs liganded to $^{16}O_2$, $^{17}O_2$, and $^{18}O_2$. *Top*: Typical $^{16}O_2$, $^{17}O_2$, and $^{18}O_2$ stretch bands as deduced from oxygen isotope difference spectra. The spectrum for each isotope contains multiple bands and an overall shape that is distinctly different from the spectra for the other isotopes. *Middle*: A cross-hatched band corresponding to the postulated unperturbed O–O stretch band for $^{16}O_2$ and $^{17}O_2$ at ~1125 cm^{-1} has been superimposed on the observed spectra of the top panel. The horizontal double-headed arrows indicate splitting of the unperturbed band to a strong, lower frequency band and a weak, higher frequency band. *Bottom*: Postulated dioxygen stretch bands (I and II) that would be observed for each isotope if no perturbation from vibrational coupling occurred; these bands are represented by heavy lines for $^{17}O_2$ and $^{16}O_2$. For $^{18}O_2$ the observed spectrum is represented as equivalent to the unperturbed spectrum. For $^{17}O_2$, band I is split, whereas for $^{16}O_2$ band II is split. Such an analysis supports two major bands (I and II) for each isotope separated by about 30 cm^{-1}. The magnitude of the isotope shifts is comparable to that for the shifts observed experimentally for model systems. The band shapes do not conform precisely to shapes predicted theoretically for single vibrations, which suggests that additional minor bands are also present. [Reprinted with permission from W. T. Potter *et al.*, *Biochemistry* **26**, 4699 (1987). Copyright 1987 American Chemical Society.]

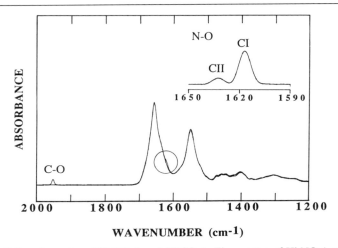

FIG. 3. Infrared spectra of HbNO A and HbCO A. The spectra of HbNO A and HbCO A are superimposed to show the C–O and N–O stretch vibrational bands. The circle encloses the area in which the two spectra differ due to bands for bound NO. The inset represents infrared bands obtained by subtracting the spectrum of HbCO A from the spectrum of HbNO A.

at E7 and E11, both natural[1,4,17,26,27] and recombinant site-directed mutations,[35,36] have been explored.

CO-infrared spectra first indicated that the coordination geometry of the Fe–ligand bond in hemoglobin differs markedly from those in free Fe^{II}–porphyrin model compounds.[1,2,37] X-Ray crystallography showed the Fe–C–O bonds in Fe^{II}(TPP)(N-MeIm)CO and Fe^{II}(TPP)(Py)CO were perpendicular to the porphyrin plane (TPP, tetraphenylporphin; N-MeIm, N-methylimidazole; Py, pyridine).[38,39] These complexes exhibit a single C–O stretch frequency at 1968 cm^{-1} for Fe^{II}(TPP)(N-MeIm)CO and 1975 cm^{-1} for Fe^{II}(TPP)(Py)CO,[40] whereas HbCO A, with steric constraints at the distal side, gives ν_{CO} at 1951 cm^{-1} for the major band.[4,21] The decrease in ν_{CO} of HbCO A compared to simple heme CO compounds was ascribed

[35] K. Nagai, B. Luisi, D. Shih, G. Miyazaki, K. Imai, C. Poyart, A. De Young, L. Kwiatkowsky, R. W. Nobel, S.-H. Lin, and N.-T. Yu, *Nature* (*London*) **329**, 858 (1987).

[36] S.-H. Lin, N.-T. Yu, J. Tame, D. Shih, J.-P. Renaud, J. Pagnier, and K. Nagai, *Biochemistry* **29**, 5562 (1990).

[37] N.-T. Yu and E. A. Kerr, *in* "Biological Applications of Raman Spectroscopy" (T. G. Spiro, ed.), Vol. 3, p. 39. Wiley, New York, 1988.

[38] S.-M. Peng and J. A. Ibers, *J. Am. Chem. Soc.* **98**, 8032 (1976).

[39] J. P. Collman, J. I. Brauman, and K. M. Doxsee, *Proc. Natl. Acad. Sci. U.S.A.* **76**, 6035 (1979).

[40] W. S. Caughey, *Ann N.Y. Acad. Sci.* **174**, 148 (1970).

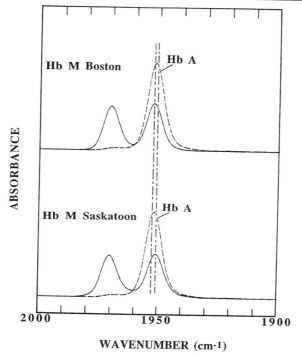

FIG. 4. The C–O stretch infrared spectra of Hb M Boston, Hb M Saskatoon, and Hb A carbonyls. The C–O stretch bands of abnormal subunits in both Hb M Boston (α58 His→Tyr) and Hb M Saskatoon (β63 His→Tyr) are shifted to 1970 cm^{-1}, while the C–O stretch bands of normal subunits remain unaffected. (From Nagai *et al.*[26])

to a bent and/or tilted mode of CO binding,[1] which was later proved correct by X-ray crystallographic studies. Replacement of the distal βE7 His of Hb A by Arg in Hb Zurich shifts the C–O stretch band to 1958.2 cm^{-1}, which is about 6 cm^{-1} blue shifted from that found in HbCO A (1951.9 cm^{-1} for the normal β subunit),[17,27] a shift consistent with an enlarged ligand-binding pocket as a result of the E7 His→Arg substitution. This explanation was confirmed by X-ray crystallography; the side chain of E7 Arg attaches itself to the propionate of the heme, opening the heme–ligand pocket.[30] Replacements of the distal E7 His by Tyr in both Hb M Boston (α58 His→Tyr) and Hb M Saskatoon (β63 His→Tyr) shift the C–O stretch bands of the abnormal subunits to 1970.2 cm^{-1}, while leaving the C–O stretch bands of normal subunits unaffected (Fig. 4).[26,27] These results suggest that ligand pockets also open as a result of E7 His→Tyr substitutions. A schematic presentation of effects of amino acid

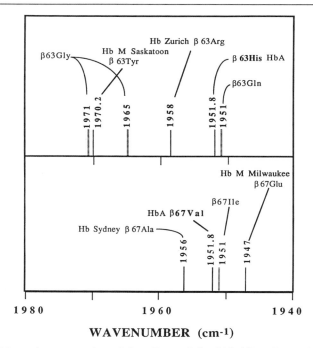

WAVENUMBER (cm⁻¹)

FIG. 5. Schematic presentation of the effects of distal histidine (E7) and valine (E11) substitutions on the frequency of C–O stretch bands of hemoglobins. Data obtained from Hb A and natural mutant Hbs[1,4,17,26,27] are presented as solid lines, and those from site-directed mutagenesis[35,36] are presented as broken lines.

replacements at E7 and E11 positions on the C–O stretch bands of hemoglobins is shown in Fig. 5. Results for mutations via site-directed mutagenesis were obtained from Raman spectroscopic studies.[35,36]

Asymmetric Ligand Binding at α and β Subunits of Hemoglobin

Interactions between the α and β subunits and their intrinsic differences in ligand affinity play extremely important roles in the physiological function of hemoglobin.[41,42] The ability of infrared spectroscopy to distinguish between ligands at α and β subunits provides a direct and clear means for determination of differences in subunit reactivities in the hemoglobin tetramer, even within intact erythrocytes.[15] By titrating HbO₂ with CO both in red cells and in solution, Potter et al.[15] used CO-IR spectroscopy to demonstrate that the replacement of O₂ ligands by CO in human packed

[41] G. Weber, Nature (London) 300, 603 (1982).
[42] F. Peller, Nature (London) 300, 661 (1982).

erythrocytes occurs preferentially at the β subunits (Fig. 6). The C–O stretch band maximum position shifts 0.6 cm^{-1} to lower frequency in the course of the CO titration. At very low CO levels, only the β subunit is liganded with CO. Infrared evidence of a much greater affinity for CO of the abnormal β subunits compared to the normal α subunits of Hb Zurich proved important clinically.[43]

pH-Induced Changes in Environment at Ligand-Binding Sites

The C–O stretch infrared spectra are highly sensitive to pH-induced changes at the ligand-binding sites, as shown in Fig. 7 over the pH range 3 to 12. CO-band parameters over this pH range obtained from deconvolution of the spectra, using a curve-fitting procedure, are presented in Table III. Only subtle spectral changes were observed from pH 6.0 to 10.5. The slight frequency shifts at absorption maxima indicate the immediate environment about the ligand in Hb A is relatively stable over this pH range. However, marked spectral changes, compared with pH 7.5, were observed under extreme pH conditions (pH 3.0, 4.5, and 12.0, especially at pH 3.0). The C–O stretch infrared spectral changes at pH extremes were characterized by an intensity increase at 1969 cm^{-1} (CIV band), accompanied by an intensity decrease at 1951 cm^{-1} (CIII band), whereas the bandwidth ($\Delta\nu_{1/2}$) of the CIV band increases dramatically from ~6 cm^{-1} at pH 7.5 to ~15 cm^{-1} at pH 12.0 and ~18 cm^{-1} at pH 3.0. These findings suggest that as the pH reaches the two extremes, the ligand pocket of hemoglobin gradually opens up, which permits the Fe–C–O bond to become perpendicular to the heme plane and causes the CO ligand to experience a much more mobile immediate environment.

Infrared Microspectroscopy of Single Erythrocytes

Use of a microscope with infrared optics coupled to a highly sensitive Fourier transform infrared spectrophotometer permits the measurement of an individual erythrocyte under physiologically relevant conditions.[16,44] The single-cell infrared (SC-IR) spectrum is obtained by digital subtraction of the spectrum of the medium adjacent to the target cell from the observed spectrum of a single erythrocyte plus surrounding medium. Even though a resolution as low as 8 cm^{-1} for an infrared spectrum of a single cell in H$_2$O media may be required in order to obtain a satisfactory signal-to-

[43] W. H. Zinkham, R. A. Houtchens, and W. S. Caughey, *Science* **209,** 406 (1980).
[44] R. G. Messerschmidt and J. A. Reffner, *Proc., Annu. Conf.—Microbeam Anal. Soc.* **23,** 215 (1988).

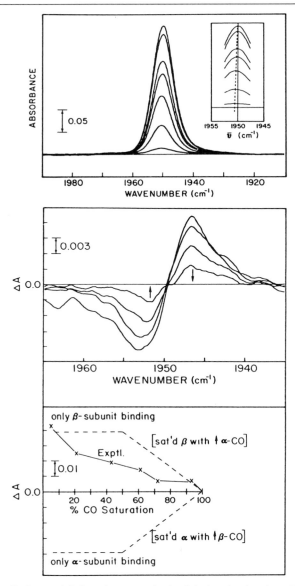

FIG. 6. The C–O stretch infrared spectra obtained during the titration of HbO$_2$ with CO in packed human erythrocytes at 4°. *Top*: Spectra were recorded at 5, 21, 43, 61, 72, 93, and 100% CO saturation. The remaining heme sites were liganded with O$_2$. The inset shows the position of the maximum absorbance on an expanded *x* axis. *Middle*: Difference spectra obtained by subtracting spectra at partial CO saturations from the spectrum of fully saturated HbCO A in packed erythrocytes. Difference spectra for 21, 43, 61, and 72% CO saturation

noise ratio, the microspectroscopic method provides a unique opportunity to acquire molecular information on the hemoglobins in an individual cell.[16]

Preparation of CO-Saturated Erythrocytes and Infrared Measurements

Fresh blood (1 ml) containing a small amount of anticoagulant [e.g., ethylenediaminetetraacetic acid (EDTA) or heparin] from a human donor is saturated with CO by passing CO gas over the sample surface for 30 min. Following the addition of 4 ml of Krebs–Ringer phosphate–dextrose buffer (119 mM NaCl, 4.7 mM KCl, 1.2 mM MgSO$_4$, 0.7 mM CaCl$_2$, 3.7 mM NaH$_2$PO$_4$, 11.2 mM NaHPO$_4$, and 5.5 mM dextrose), the mixture is centrifuged at 2000 g for 10 min. The supernatant is discarded and the packed cells suspended in buffer and collected by centrifugation an additional three times to remove plasma proteins from the cell suspension. The final cell suspension used for infrared studies contains packed cells and buffer in the volume ratio of 1 : 4, respectively.

The infrared spectrum is measured by placing the cell suspension on the stage of an infrared microscope IR-PLAN (Spectra-Tech, Stamford, CT) integrated with a Fourier transform (FT)-IR spectrophotometer [Perkin-Elmer (Norwalk, CT) model 1800 with an Hg/Cd/Te detector]. The microscope is equipped with a 15× magnification, NA 0.58 Reflachromat objective and a 10×, NA 0.71 Reflachromat condenser for visual and infrared use. The erythrocyte suspension is contained between two BaF$_2$ disks (1-mm thick, 13 mm in diameter). Two variable image-masking apertures are used to select and isolate the target cell and to reduce stray light and other spurious signals. The viewing area of about 10 × 10 μm, containing the single cell, is chosen. A similar area adjacent to the cell that contains only medium is chosen for recording the spectrum of the medium. The achievable resolution of SC-IR spectrum with a satisfactory signal-to-noise ratio is heavily affected by factors such as instrument alignment, air humidity, effectiveness of dry air purging, and sample preparation. The spectrum for the cell *per se* is obtained by an appropriate digital subtraction of the medium spectrum from the cell-plus-medium

are shown. *Bottom:* Plot of the total difference between the minimum and maximum absorbance of each difference spectrum vs the percentage CO saturation. Theoretical curves (– – –) show results expected if one subunit type (α or β) were to become completely saturated with CO before the other subunit begins to bind CO. The experimental curve (–×–) indicates the CO replacement of O$_2$ is preferentially at the β subunits, but some binding to the α subunit occurs prior to the complete replacement of O$_2$ at β subunits. [Reprinted with permission from W. T. Potter *et al., Biochem. Biophys. Res. Commun.* **116,** 719 (1983).]

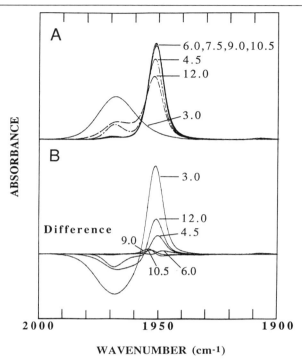

FIG. 7. The C–O stretch infrared spectra of HbCO A obtained under various pH conditions. (A) Primary spectra. (B) Difference spectra obtained by subtraction of the spectrum for pH 7.5 from the spectrum for the pH values indicated. Variations in pH for solutions of HbCO A were achieved by dialysis of a solution of HbCO A (8 mM in heme) against CO-saturated 50 mM sodium phosphate/50 mM sodium acetate buffers at pH values from 3.0 to 7.5 or against CO-saturated 50 mM sodium phosphate/50 mM boric acid buffers at pH 9.0 to 12.0 at 4° overnight. The final HbCO A concentrations were adjusted to 6 mM in heme. (From Dong et al.[24])

spectrum. The final difference spectrum is subjected to nine-point smoothing with a Savitsky–Golay[45] function.

Analysis of Single-Erythrocyte Spectra

Figure 8 presents an infrared spectrum of a single erythrocyte (saturated with CO gas) from a suspension of normal adult human erythrocytes that was obtained at 8-cm^{-1} resolution with an infrared microscope as described above. For comparison, the spectrum of purified HbCO A ob-

[45] A. Savitsky and J. E. Golay, Anal. Chem. **36**, 1628 (1964).

TABLE III

PARAMETERS OF DECONVOLUTED INFRARED C–O STRETCH BANDS FOR CARBOXYHEMOGLOBIN A[a]

	Band I		Band II		Band III		Band IV	
pH	ν_{CO} ($\Delta\nu_{1/2}$) (cm^{-1})	A%[b]	ν_{CO} ($\Delta\nu_{1/2}$) (cm^{-1})	A%	ν_{CO} ($\Delta\nu_{1/2}$) (cm^{-1})	A%	ν_{CO} ($\Delta\nu_{1/2}$) (cm^{-1})	A%
3.0	—		1945.5(9.5)	1.9	1953.5(9.6)	3.6	1967.6(18.1)	94.5
4.5	1634	c	1943.5(5.6)	1.1	1951.6(8.0)	85.1	1968.0(9.1)	13.8
6.0	1634	c	1945.2(7.6)	3.1	1951.3(7.5)	95.3	1969.5(6.6)	1.6
7.5	1634	c	1944.1(7.0)	3.0	1951.2(7.7)	95.9	1969.1(6.2)	1.1
9.0	1634	c	1944.7(8.4)	3.9	1951.1(7.6)	95.6	1969.3(6.0)	0.5
10.5	1634	c	1944.4(8.1)	3.0	1951.1(7.8)	96.1	1968.9(6.1)	0.9
12.0	1634	c	1942.2(5.3)	0.5	1951.7(8.7)	70.5	1966.3(15.1)	29.0

[a] At 20°.

[b] The individual band area is given by CURVEFIT deconvolution (see text).

[c] The band is too small to be measured accurately.

tained without a microscope at 2-cm^{-1} resolution is also presented in Fig. 8. Strong bands near 1656 and 1547 cm^{-1} are due to the amide I and amide II bands, respectively, and a small, but readily measured, band at 1951 cm^{-1} is the C–O stretch band of the heme-bound CO ligand. When ^{13}CO is used instead of ^{12}CO, the small band appears at 1907 cm^{-1} because of an isotopic shift (Fig. 9). An erythrocyte contains many components other than hemoglobin, for example, membrane lipids and proteins of the membrane skeleton as well as cytosolic proteins.[46] However, the hemoglobin concentration approaches 22 mM in the normal adult erythrocyte and represents about 97% of total protein and 34% of erythrocyte mass.[46] Thus the spectrum of an erythrocyte can be expected to resemble closely the spectrum of purified hemoglobin.

Figure 10a shows the spectrum obtained with an infrared microscope of a CO-saturated single human erythrocyte that contains Hb M Boston as well as Hb A. Also shown are the spectra of a suspension of erythrocytes containing Hb M Boston with and without reduction with sodium dithionite (Fig. 10B), as well as the spectra of purified, fully reduced Hb M Boston and Hb A carbonyls obtained without using an infrared microscope (Fig. 10C). The spectra of Hb M Boston-containing cells exhibit an unusual C–O stretch band at 1970 cm^{-1}, which can be assigned to the abnormal α subunits of Hb M Boston, and is clearly identified in the spectrum of a single cell. The band centered at 1951 cm^{-1} represents the superposition

[46] W. J. Williams, E. Beutler, A. Erslev, and M. A. Lichtman, "Hematology," 3rd ed., p. 281. McGraw-Hill, New York, 1983.

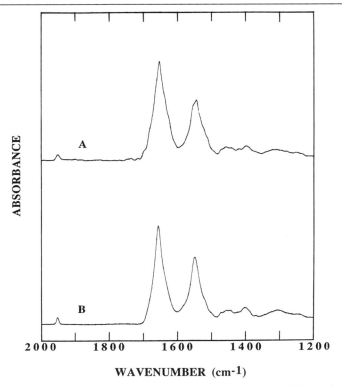

FIG. 8. Infrared spectra of a CO-saturated single erythrocyte and purified carbonyl Hb A. (A) Spectrum of a single adult human erythrocyte obtained with an infrared microscope at 8-cm^{-1} resolution. (B) Spectrum of a solution of purified human carbonyl Hb A (8 mM) obtained at 2-cm^{-1} resolution without the use of a microscope.

of bands for the CO bound to the α and β subunits of HbCO A and to the normal β subunits of Hb M Boston. By comparing the infrared spectra of single erythrocytes, the distribution of abnormal hemoglobin molecules in a cell population and the relative amounts of oxidized and reduced hemoglobin subunits under circulating conditions can be determined.[47]

Several technical difficulties are associated with single-cell infrared microscopic analysis. Particularly important is the interference from the strong absorptions between 2000 and 1300 cm^{-1} from both liquid water and water vapor. Especially bothersome is the spectral interference from water vapor, because it cannot be avoided by use of a D$_2$O medium. A plastic microscope cover combined with an extensive dry air purge may

[47] A. Dong, M. Nagai, Y. Yoneyama, and W. S. Caughey, unpublished data.

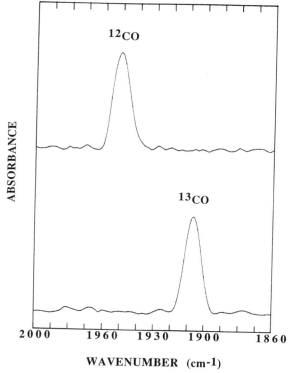

FIG. 9. Infrared spectra of ^{12}CO and ^{13}CO bound to heme iron within a single erythrocyte. *Top*: Spectrum with ^{12}CO exhibits a stretch band maximum at 1951 cm^{-1}. *Bottom*: Spectrum with ^{13}CO exhibits a stretch band maximum at 1907 cm^{-1}. Spectra were measured with 8-cm^{-1} resolution.

help reduce the amount of water vapor present in the radiation path. A proper dilution factor for the cell suspension is also important in order to isolate a single cell easily for infrared measurement, leaving enough area free of cells and immediately adjacent to the targeted cell to permit measurement of a background spectrum.

S–H Stretch Band of Cysteine Thiol Group

The infrared bands associated with the S–H stretch vibration of the cysteine thiol groups in hemoglobins provide a unique opportunity to probe changes in protein structure at these groups that may result from changes in amino acid sequence, oxidation state, ligand binding, pH,

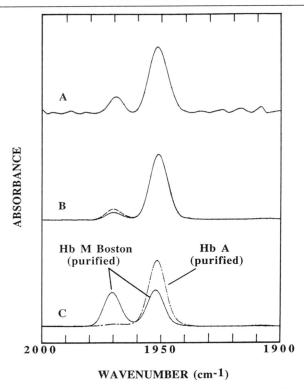

FIG. 10. Infrared C–O stretch spectra of a single erythrocyte and a cell suspension from a patient with Hb M Boston disease, as well as of purified Hb M Boston and Hb A carbonyls. (A) CO-IR spectrum of a single erythrocyte containing both Hb M Boston and Hb A saturated with CO from freshly drawn blood suspended in Krebs–Ringer phosphate–dextrose buffer. The spectrum was measured with an infrared microscope at 4-cm^{-1} resolution at room temperature. (B) Spectra of suspensions of CO-saturated erythrocytes that contain Hb M Boston as well as Hb A as freshly drawn (—) and after reduction by sodium dithionite (– – –). (C) Spectra of solutions of purified Hb M Boston [with 3 sodium dithionite (3 mg/ ml) present] and Hb A carbonyl. Both solutions contained about 4 mM heme and were in 5 mM sodium phosphate buffer, pH 7.0. (From Dong et al.[47])

temperature, and other factors. In 1974, Alben et al.[5] first identified the S–H stretch bands that arise from the sulfhydryl groups of cysteine residues in the infrared spectrum of human Hb A. Subsequent studies demonstrated the sensitivity of the S–H stretch frequency of α104 cysteine, which is located at the nonpolar $\alpha_1\beta_1$ interface, to changes in oxidation state and ligation,[6,48–50] as well as to the β109 Val→Met substitution in

48 G. H. Bare, J. O. Alben, and P. A. Bromberg, *Biochemistry* **14**, 1578 (1975).

Hb San Diego.[51] Refinements of this infrared method in our laboratory permit the extensions in the application and interpretation of the thiol infrared spectra discussed below.

Sample Preparation and Infrared Measurements

Because of the weakness of the S–H stretch bands, relatively high concentrations of protein must be used. Use of hemoglobin solutions at least 5 mM in heme in a 0.1-mm path length infrared cell is required to obtain a highly accurate measurement of the S–H bands. However, the extinction coefficients (ε) and integrated intensities (B) for the S–H stretch band of the cysteine residues in hemoglobins vary greatly. Band intensity depends heavily on the strength of hydrogen bonding, if any, between the SH group (as hydrogen donor) and a hydrogen acceptor, as well as on the polarity and other aspects of the immediate environment in which the SH group of the cysteine residue is located. Table IV lists some presently available extinction coefficients and integrated intensities for species of human Hb A.

The infrared spectra of hemoglobin solutions are measured and the protein-free buffer subtracted as described for ligand IR spectra above. The spectrum between the 2630- to 2500-cm^{-1} region is selected and baseline corrected. The wavenumbers of the S–H bands are fortunately within a water window (Fig. 1). The curve-fitting deconvolutions with Gaussian or Gaussian–Lorentzian mixed functions are carried out by use of a CURVEFIT function (Galactic Industries Corp., Salem, NH) on a 386-based personal computer.

S–H Stretch Band Assignments

The normal human adult hemoglobin A contains six cysteine residues, two at α104 (G11), two at β112 (G14), and two at β93 (F9) positions. On the basis of X-ray crystallographic studies, both α104 and β112 cysteines are located at the nonpolar $\alpha_1\beta_1$ interface,[5,48,52] whereas the β93 cysteine is located in the F–H pocket near the surface of the hemoglobin molecule.[52,53] The S–H stretch bands for the three types of cysteine are readily distinguished from each other in terms of frequency (ν_{SH}), half-bandwidth

[49] J. O. Alben, G. H. Bare, and P. P. Moh, in "Biochemical and Clinical Aspects of Hemoglobin Abnormalities" (W. S. Caughey, ed.), p. 607. Academic Press, New York, 1978.

[50] P. P. Moh, F. G. Fiamingo, and J. O. Alben, Biochemistry 26, 6243 (1989).

[51] S. El Antri, O. Sire, and B. Alpert, Chem. Phys. Lett. 161, 47 (1987).

[52] B. Shaanan, J. Mol. Biol. 171, 31 (1983).

[53] E. J. Heidner, R. C. Ladner, and M. F. Perutz, J. Mol. Biol. 104, 707 (1976).

TABLE IV
PARAMETERS FOR DECONVOLUTED S–H INFRARED STRETCH BANDS OF CYSTEINE RESIDUES IN HUMAN Hb A SPECIES[a]

S–H	Ligand	ν_{SH} (cm^{-1})	$\Delta\nu_{1/2}$ (cm^{-1})	ε (mM^{-1} cm^{-1})[b]	B (mM^{-1} cm^{-2})[c]	Ref.
α104	Aquomet	2553.7	12.6			24
	O_2	2553.1	12.2		2.05	24
	CO	2552.9	11.7		2.32	24
	CO	2552.6	13.5	0.17	2.43	48
	NO (Fe^{2+})	2552.5	11.8			24
β112	Aquomet	2566.3	11.3			24
	O_2	2566.1	12.0		0.90	24
	CO	2566.1	11.7		1.04	24
	CO	2566.3	12.5	0.055	0.80	48
	NO (Fe^{2+})	2566.4	11.8			24
β93	Aquomet	2590.1	14.7			24
	O_2	2588.8	12.4		0.18	24
	CO	2589.3	14.4		0.46	24
	CO	2591.5^d	19.5^d	0.046^d	0.88^d	50
	NO (Fe^{2+})	2580.7	15.6			24

[a] At neutral pH and 20°. Because of an extensive overlapping between S–H bands of α104 and β112, accurate data for the deoxy form cannot be obtained and, therefore, are not presented here. (Data from Bare et al.[48]; Dong et al.[24]; Moh et al.[50])

[b] $\varepsilon = A/cl$.

[c] $B = 1/(cl)\int A d\nu$.

[d] This value was calculated for HgCl$_2$-treated HbCO A sample.

($\Delta\nu_{1/2}$), and integrated intensity (B) in many, but not all, forms of Hb (Fig. 11). Comparison of the SH-IR spectra of Hb A, Hb F, and bovine hemoglobin carbonyls permit the unequivocal band assignment for each individual SH group; the β112 cysteine is absent in Hb F, and neither the β112 nor the α104 cysteine are found in bovine hemoglobin. The variations in the S–H stretch band parameters reflect the differences in local environment and hydrogen bonding experienced by the SH groups of the three types of cysteine. The frequency of the S–H stretch vibration is expected to increase with increasing S–H bond strength. Band intensity will increase as the strength of the dipole associated with the SH bond increases. For example, the stronger a hydrogen bond is, with the SH group serving as proton donor, the lower the frequency and greater the intensity expected. The bandwidth is related to the mobility of the environment about the SH group; the more mobile, that is, less rigid, the environment is, the greater the expected bandwidth is. Thus, the SH group of α104 cysteine, known to be located at the $\alpha_1\beta_1$ interface and strongly hydrogen bonded to the peptide carbonyl of α100 leucine,[5,48] on comparison with the locations of

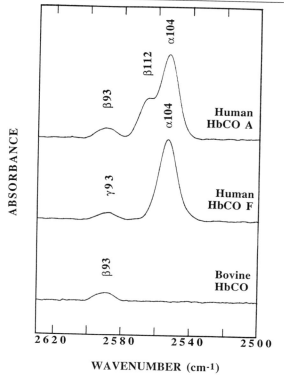

FIG. 11. The S–H stretch infrared spectra of the α104, β112, and β93 or γ93 cysteines of human HbCO A, human HbCO F, and bovine HbCO at pH 7.5, 20°. (From Dong et al.[24])

the β112 and β93 thiols, is reasonably expected to exhibit the lowest wavenumber and highest intensity. The α104 band is also expected to provide a particularly sensitive probe of the tertiary structure of the α-subunit G helix at the $\alpha_1\beta_1$ interface and the quaternary structure of the hemoglobin tetramer.

Ligand-Induced Changes in SH-IR Spectra

The S–H stretch vibrations of cysteine residues in Hb A are sensitive to the structural changes that result from ligand binding at heme iron.[6] The S–H stretch infrared spectra of deoxyHb A, metHb A, HbO$_2$ A, HbCO A, and HbNO A, are shown in Fig. 12. The S–H stretch bands of α104 cysteine of all liganded Hb A species are observed at lower frequencies than that for unliganded deoxy species (Table IV). Frequency and intensity differences between the S–H stretch bands of deoxyHb and

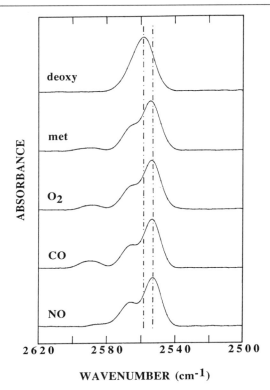

FIG. 12. The S–H stretch infrared spectra of deoxyHb A, aquomet Hb A HbO$_2$ A, HbCO A, and HbNO A at pH 7.5, 20°. The dashed lines indicate the positions of absorption maxima of deoxy and NO species (From Dong et al.[24])

HbCO A are shown in Fig. 13 (top) by superimposition of the spectra. However, the band parameters of the S–H stretch bands for α104 and β112 cysteines (and the β93 cysteine, if present) in deoxyHb cannot be established accurately by deconvolution analysis because of extensive overlapping between the S–H bands. Despite the location of the β93 cysteine near the surface of the hemoglobin molecule, the wide difference in frequency and intensity among liganded and unliganded species indicates the β93 S–H stretch is especially sensitive to protein conformational changes.

pH-Induced Changes in SH-IR Spectra

The S–H stretch vibrations of cysteine residues of Hb A are also sensitive to changes in pH.[54] We have found the changes in HbCO A

[54] S. El Antri, C. Zentz, and B. Alpert, Eur. J. Biochem. **179**, 165 (1989).

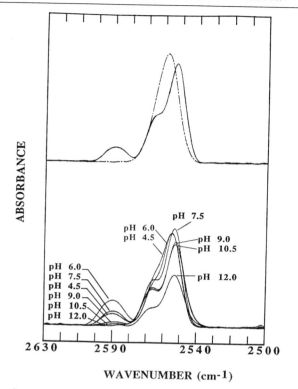

FIG. 13. The S–H stretch infrared spectra of deoxy and carbonyl species of Hb A at pH 7.5 and of HbCO A at pH values from 4.5 to 12.0 at 20° *Top*: Spectra of HbCO A (—) and unliganded (deoxy) Hb A (– · –) at pH 7.5. *Bottom*: Spectra of HbCO A under various pH conditions. Variations in pH were achieved by overnight dialysis at 4° against appropriate buffers: 50 m*M* sodium phosphate/50 m*M* sodium acetate for pH 4.5, 6.0, and 7.5; 50 m*M* sodium phosphate/50 m*M* boric acid for pH 9.0, 10.5, and 12.0. (From Dong et al.[24])

structure that accompany the changes in pH between 4.5 and 12.0 to be reflected in the S–H stretch infrared spectra as shown in Fig. 13. The integrated S–H stretch intensities *vs* pH and the S–H stretch frequencies vs pH are shown in Figs. 14 and 15, respectively. Because the hemoglobin samples used here were CO-liganded Fe^{2+} forms, any protein conformational changes that would result from changes in oxidation state or ligation can be ruled out. The changes in the S–H stretch vibration are therefore limited to pH-induced dynamic changes in protein structure.

The crystal structures of Hb A at neutral pH indicate that the $\alpha104$ cysteine SH groups are hydrogen bonded to the $\alpha100$ leucine carbonyl oxygen associated with a peptide bond.[5,48] This hydrogen bonding takes place in a nonpolar region of the $\alpha_1\beta_1$ interface. Therefore, under normal

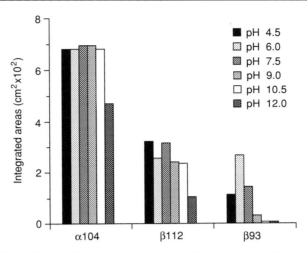

FIG. 14. The pH-induced changes in integrated S–H stretch band intensity of Hb A carbonyl for spectra of Fig. 13. The integrated intensities are calculated for hemoglobin at 6 mM in heme in a 0.1-mm path length cell. (From Dong *et al.*[24])

conditions, no direct interactions between this SH group and the external aqueous environment is likely. The frequency (ν_{SH}) values for α104 are the lowest among the three cysteines, as expected, because the crystal structures indicate the α104 SH is involved in the strongest hydrogen bonding. Also consistent with the α104 SH forming the strongest hydrogen bond is the fact that the α104 SH gives the most intense S–H stretch band. At pH 7.5 the half-bandwidth ($\Delta\nu_{1/2}$) values for α104 SH and β112 SH are comparable and smaller than is found for the β93 SH. The $\Delta\nu_{1/2}$ values at pH 7.5 indicate the mobility/rigidity of the immediate environments about the SH groups are comparable for the α104 and β112 groups, whereas the β93 SH environment is more mobile and less rigid.

The α104 SH band frequency is lowest at pH 9.0 and 10.5, a little higher at pH 7.5 and 12.0, and still higher at pH 4.5 and 6.0. Band intensities (B) are essentially the same from pH 4.5 to 10.5 but the intensity is about one-third at pH 12.0. The low intensity at pH 12.0 cannot be ascribed solely to weaker hydrogen bonding but is undoubtedly mainly due to partial deprotonation of the α104 SH at this high pH. In fact, one can infer from this data that the pK_a for the α104 SH is between 12.2 and 12.5. The $\Delta\nu_{1/2}$ values are lowest at pH 7.5 and increase slightly as pH increases or decreases from pH 7.5. A pH value of 12.0 produces the widest band and, therefore, the most mobile environment, consistent with partial denaturation (unfolding).

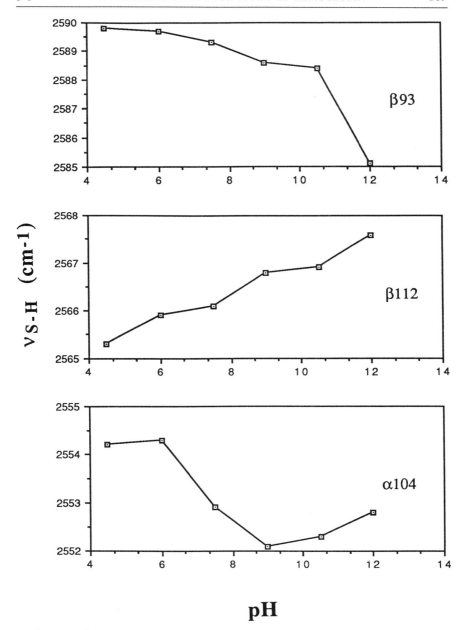

FIG. 15. The pH-induced changes in S–H stretch band frequency of Hb A carbonyl for spectra in Fig. 13. (The bands for β93 at pH 10.5 and 12.0 are possibly too weak to be significant.)

The SH group of β112 cysteine, which is also at a nonpolar site in the $\alpha_1\beta_1$ interface, appears to be involved in hydrogen bonding to the carbonyl oxygen of a peptide bond involving β108 asparagine or β109 valine.[5,48] The higher frequency and lower intensity of the S–H stretch band for β112 compared with the band for α104 SH is consistent with weaker hydrogen bonding at β112 than at α104. The frequency of the β112 SH band steadily increases as the pH increases from 4.5 to 12.0 (Fig. 15), a trend not seen with either α104 or β93. Although the origin of this pH effect is unclear, a simple explanation is that the SH group serves as a weaker hydrogen donor as the pH becomes greater. It is also of interest that the frequency difference between the β112 and α104 bands increases with pH: 11.1 cm^{-1} at pH 4.5, 13.2 cm^{-1} at pH 7.5, and 14.8 cm^{-1} at pH 12.0. The sharp drop in band intensity when the pH is increased from pH 10.5 to 12.0 is consistent with a pK_a for the β112 SH group a little less than 12, a lower value than is found for the α104 SH group. The lower band intensities at pH 4.5 to 10.5 compared with the α104 values are consistent with higher ν_{S-H} values. The basis for somewhat greater intensity at pH 4.5 and 7.5 compared with pH 6.0, 9.0, and 10.5 values is unclear. The bandwidths observed indicate the SH group environment is a little less mobile at β112 than at α104, but the mobility is affected very little by the changes in pH; there is no evidence of a significant increase in mobility at the pH extremes studied.

The effects on the SH stretch band for the β93 cysteine, caused by variations in pH, are quite different from the effects noted for the α104 and β112 bands. This is undoubtedly due to the location of the β93 SH group in the F–H pocket near the surface of the protein,[52,53] where it can be associated randomly with water molecules. In contrast, the other two SH groups are isolated in less polar regions at the $\alpha_1\beta_1$ interface. From pH 4.5 to 10.5 the band frequency decreases slightly, then drops markedly at pH 12.0. However, the bands detected at pH 10.5 and 12.0 are so weak that the significance of the reported band parameters is questionable; even the assignment of these weak bands to β93 SH may not be valid. The bandwidth goes through a maximum at pH 6.0; the larger width suggests the environment is more mobile than for the β112 or α104 SH groups. The band intensity also reaches a maximum at pH 6.0. The decrease in band intensity (B) from 2.59 at pH 6.0 to 1.37 at pH 7.5 to 0.29 at pH 9.0 is consistent with a pK_a for deprotonation of the SH group of about 7.5, which is clearly less than 8.0, the approximate pK_a of free cysteine. Although the sharp drop in band intensity from pH 6.0 to pH 4.5 without a significant change in band frequency might also be consistent with loss of a proton from the SH group, proton loss on acidification is not expected. A more likely explanation is a rearrangement of the aqueous environment with the decrease in pH from 6.0 to 4.5 that results in a decrease in band intensity without a change in frequency. The higher frequency at lower pH suggests only very weak, if any, donation of the thiol proton to an acceptor (e.g., the oxygen of water), so the decrease in pH is unlikely to have a substantial affect on hydrogen bonding. On the other hand the decrease in pH may reduce the strength of the dipole associated with the S–H bond by changing the polarity of the environment and, thereby, decrease the band intensity.

These studies support the view that SH-IR spectroscopy is a uniquely useful tool for probing the structures at the loci of cysteine SH groups in hemoglobins and other proteins. The infrared band observed represents the vibration of the S–H bond; if the bond is present, the proton must be bound to the sulfur atom. The method, therefore, also provides a unique means for determination of pK_a values of cysteine SH groups. In the case of HbCO A, the order of pK_a values for the cysteines is $\beta 93 \ll \beta 112 < \alpha 104$. The S–H stretch band parameters also reflect the interactions of the SH group with its immediate environment.

Determination of Secondary Structures by Amide I Infrared Spectra

Infrared spectroscopy, advanced by computerized Fourier transform instrumentation and spectral handling techniques, such as Fourier self-deconvolution (resolution enhancement) and second-derivative analysis, has become a widely used method for studying the secondary structures of polypeptides and proteins.[8–10,55,56] Nine characteristic vibrational bands or group frequencies that arise from the amide groups of protein have been identified.[8] Among them, the amide I band (1700–1620 cm^{-1}), which is due almost entirely to the C=O stretching vibration of the peptide linkages that constitute the backbone structure, is the most thoroughly studied. Another rather intense absorption, the amide II band (1600–1500 cm^{-1}), due to an out-of-phase combination of N–H in-plane bending and C–N stretching vibrations of peptide groups,[8] has been used very little for determination of protein structures. High sensitivity to small variations in molecular geometry and hydrogen bonding makes the amide I band uniquely useful for analyzing protein secondary structures and monitoring the conformational changes in proteins that involve secondary structure.[7–11,55,56] Each type of secondary structure gives rise, in principle, to a different C=O stretch frequency in the amide I region of the infrared spectrum. The basic theoretical calculations for these vibrations have

[55] W. K. Surewicz and H. H. Mantsch, *Biochim. Biophys. Acta* **952,** 115 (1988).
[56] D. M. Byler and H. Susi, *Biopolymers* **25,** 469 (1986).

TABLE V
DECONVOLUTED AMIDE I BAND FREQUENCIES AND ASSIGNMENTS TO SECONDARY
STRUCTURE FOR PROTEINS IN D_2O AND H_2O MEDIA

H_2O[a]		D_2O[b]	
Mean frequency (cm^{-1})	Assignment	Mean frequency (cm^{-1})	Assignment
1624 ± 1.0	β Strand	1624 ± 4.0	β Strand
1627 ± 2.0	β Strand		
1633 ± 2.0	β Strand	1631 ± 3.0	β Strand
1638 ± 2.0	β Strand	1637 ± 3.0	β Strand
1642 ± 1.0	β Strand	1641 ± 2.0	3_{10} Helix
1648 ± 2.0	Random	1645 ± 4.0	Random
1656 ± 2.0	α Helix	1653 ± 4.0	α Helix
1663 ± 3.0	3_{10} Helix, α_{II} helix, or type III turn	1663 ± 4.0	Turn
1667 ± 1.0	Turn	1671 ± 3.0	Turn
1675 ± 2.0	Turn	1675 ± 5.0	β Strand
1680 ± 2.0	Turn	1683 ± 2.0	Turn
1685 ± 2.0	Turn	1689 ± 2.0	Turn
1691 ± 2.0	β Strand	1694 ± 2.0	Turn
1696 ± 2.0	β Strand		

[a] Data from Dong et al.[10,58]
[b] Data from Susi and Byler[9]; Byler and Susi[56]; Prestrelski et al.[59]

been studied extensively by Miyazawa and Blout[7] and Krimm and Bande-kar.[8] The theoretical backgrounds of several infrared data-handling techniques, including Fourier self-deconvolution and second-derivative analysis, have been discussed in detail by Susi and Byler.[9,57] An improved method for carrying out second-derivative analysis has established the utility of the method for obtaining quantitative as well as qualitative determinations of α-helix, random and several turn and β-sheet structures.[10,11,58] Assignments of the amide I band components to protein secondary structure elements, such as α-helix, distorted α-helix, 3_{10}-helix, β-sheet, turn, and random structures, are available for proteins in both D_2O[9,56,59] and H_2O media[10,58] and are presented in Table V. Here, we focus our attention on the application of the second-derivative analysis method in monitoring conformational changes in hemoglobins.

[57] H. Susi and D. M. Byler, Biochem. Biophys. Res. Commun. 115, 391 (1983).
[58] A. Dong, P. Huang, B. Caughey, and W. S. Caughey, unpublished data.
[59] S. J. Prestrelski, D. M. Byler, and M. N. Liebman, Biochemistry 30, 133 (1991).

Measurement of Infrared Spectra

Subtraction of Infrared Absorption Due to Water. In the analysis of secondary structures of peptides and proteins in water, the region of the infrared spectrum that gives the greatest amount of information is the "secondary structural fingerprint" region from 1700 to 1600 cm^{-1}. Unfortunately, this immediately raises the problem of resolving the weak protein amide I bands from the much more intense absorption bands due to water in both liquid and gaseous states.[10,11] To overcome the "water problem," the majority of the infrared studies on protein secondary structures have been carried out in either solid state preparations[60] or deuterium oxide solutions.[9,59,61,62] The study of proteins in D_2O has some advantages over H_2O in that a larger path length (50-μm) infrared cell can be used, which can result in a higher signal-to-noise ratio, a factor of particular importance for proteins of low solubility. Nevertheless, H_2O-based media have the advantage of providing a more native environment. It is well known that the amide I infrared band frequencies are strongly affected by the hydrogen–deuterium exchanges in the peptide linkages.[9,55,56] The effects of these exchanges on protein structural properties are not fully understood, especially under incomplete hydrogen–deuterium exchange conditions. Furthermore, because the exchange of D for H can affect the strength and length of hydrogen bonds, it is possible that protein secondary structures may be altered by replacement of H_2O by D_2O.

The problem of water absorptions can be overcome by use of an infrared cell of sufficiently small path length (6–10 μm) to permit enough infrared radiation to pass through the material under observation.[10,25] By using the same cell to record both a blank water reference spectrum and the spectrum of the protein solution under identical scan conditions, the water contribution can be removed from the spectrum of the protein solution by digital subtraction of the water spectrum. Two criteria that are especially critical to the successful subtraction of absorption bands due to both liquid water and water vapor have been established in our previous studies.[10]

Criterion 1: The bands originating from water vapor must be subtracted accurately from the protein spectrum in the 1800 to 1500-cm^{-1} region regardless of the baseline.

[60] F. S. Parker, "Applications of Infrared, Raman and Resonance Raman Spectroscopy in Biochemistry." Plenum, New York, 1983.
[61] H. L. Casal, U. Korich, and H. H. Mantsch, *Biochim. Biophys. Acta* **957**, 11 (1988).
[62] P. W. Holloway and H. H. Mantsch, *Biochemistry* **27**, 7991 (1988).

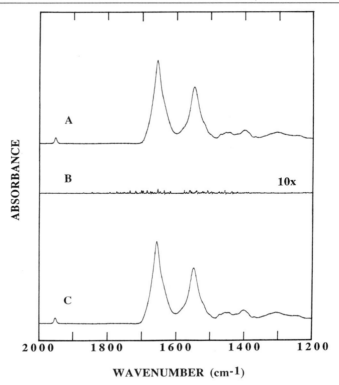

FIG. 16. Difference spectra of aqueous solutions of carbonyl Hb A from 2000 to 1200 cm^{-1} obtained during a double-subtraction analysis. (A) Difference spectrum of HbCO A obtained from the subtraction of spectrum of Ref-1 from the spectrum of the protein solution. (B) Difference spectrum of gaseous water obtained from the subtraction of the spectrum of Ref-2 from the spectrum of Ref-1. The intensity of this spectrum is expanded 10-fold. (C) Difference spectrum of HbCO A obtained by double subtraction, that is, spectrum A minus spectrum B (see text for explanation of Ref-1 and Ref-2 designations).

Criterion 2: A straight baseline must be obtained from 2000 to 1750 cm^{-1}.

The band intensities of liquid water and water vapor are superimposed but independent. The band intensity of liquid water is determined by the thickness of the cell (the path length), whereas the band intensities of water vapor are determined by the moisture in the atmosphere through which infrared radiation passes in the instrument. The latter continuously changes during the course of the experiment due to variations in the effectiveness in purging the instrument of water vapor. Satisfying both

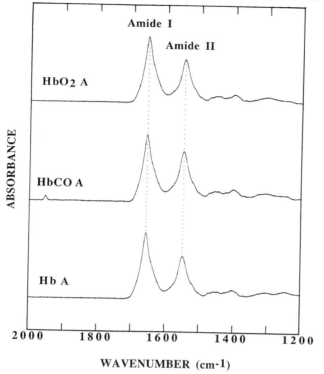

FIG. 17. Primary infrared spectra of oxyHb A, carbonylHb A, and deoxyHb A. The strong bands at 1656 and 1547 cm⁻¹ are the amide I and amide II bands, respectively. The weak band at 1951 cm⁻¹ in the carbonyl spectrum is the C–O stretch band of heme-bound CO.

criterion 1 and criterion 2 in one single step of water subtraction is usually not possible, whereas success is readily achieved by means of a double-subtraction procedure.[11] In the double-subtraction procedure, two reference spectra are first recorded from the same cell containing only buffer, thereby accumulating data twice under conditions identical to those used for the protein solution. These two reference spectra, Ref-1 and Ref-2, will be identical except for the time-dependent changes in water vapor intensity during continuous purging of the spectrophotometer with dry air. Therefore, a difference spectrum generated by subtraction of Ref-2 from Ref-1 represents the water vapor spectrum (Fig. 16B). Second, a spectrum of the protein solution is recorded using the same cell as used to obtain Ref-1 and Ref-2 after the cell has been vacuum dried. In all cases, caution must be taken to eliminate air bubbles in the infrared cells

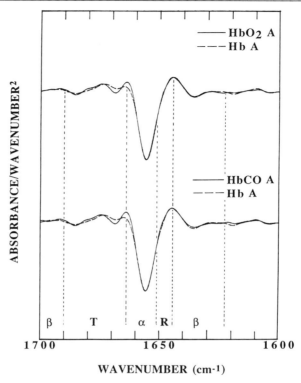

FIG. 18. Superimposed second-derivative amide I infrared spectra of solutions (6 mM in heme) of oxyHb A and deoxyHb A (top) and of carboxyHb A and deoxyHb A (bottom) in 50 mM phosphate buffer, pH 7.4, at 20°. α, α helix; β, β sheet; T, turn; and R, random structure.

completely, which can lead to a poor subtraction of the water vapor spectrum.

The spectrum of liquid water is subtracted from the observed spectrum of the protein solution to satisfy criterion 2 (a straight baseline must be obtained from 2000 to 1700 cm^{-1}); this is achieved by subtracting either Ref-1 or Ref-2 from the spectrum of the protein solution (Fig. 16A). After removal of the liquid water spectrum, the water vapor spectrum generated from the difference between Ref-1 and Ref-2 is subtracted, using the disappearance of the water vapor bands between 1850 and 1720 cm^{-1} (a window region) as an indicator of success in satisfying criterion 1 (the bands originating from water vapor must be subtracted accurately regardless of the baseline. (Fig. 16C). The final difference spectrum after double subtraction is smoothed with a nine-point Savitsky–Golay[45] function to

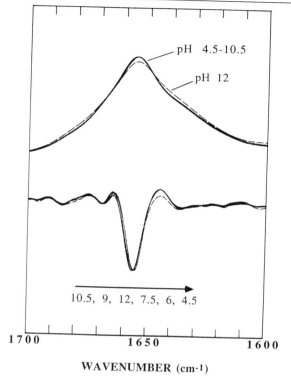

FIG. 19. Amide I infrared spectra of Hb A carbonyls from pH 4.5 to 12.0. *Top*: Primary spectra. *Bottom*: Second-derivative spectra. Spectra for pH 12.0 are shown as a dashed line (see text for further experimental details).

remove the possible white noise to give the primary spectrum. The second-derivative spectrum is obtained from the primary spectrum with Savit-sky–Golay derivative function software for a five-data point window.

Assignment of Amide I Bands to Secondary Structures. The frequency at maximum absorbance in the amide I band of the primary spectrum has been shown to indicate the predominant secondary structure in proteins.[10,25,56] Thus, the maximum absorbance of the amide I band of a protein with predominant α helix will occur near 1656 ± 2 cm^{-1}, and a protein with predominant β-sheet structure will be found between 1645 and 1632 cm^{-1}.[10] For human Hb A, which contains 75–85% α-helical structure on the basis of X-ray crystallographic analysis,[63,64] the amide I

[63] J. S. Richardson, *Adv. Protein Chem.* **34**, 167 (1981).
[64] M. Levitt and J. Greer, *J. Mol. Biol.* **114**, 181 (1977).

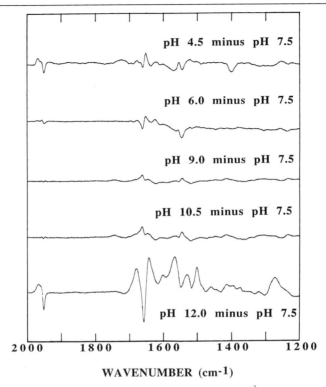

FIG. 20. pH-induced changes in primary infrared spectra of Hb À carbonyl. The difference spectra were obtained by subtraction of the spectrum for pH 7.5 from the spectrum for the pH indicated. Conditions are the same as for Fig. 19.

band maximum is found at 1656 cm^{-1} (Fig. 17). Much more detailed substructural information can be obtained by further deconvolution of the amide I band into its vibrational band components. Several spectral data-handling techniques, including curve-fitting, Fourier self-deconvolution, and second-derivative analysis, have been developed in the past decade to deconvolute the protein amide I band contour.[9,55] Of these techniques, second-derivative analysis provides the most detailed information on the different types of secondary structure present[10] and is especially useful for monitoring subtle changes in protein conformation.[11]

Ligand-induced changes in hemoglobin secondary structure are small but detected in second-derivative spectra (Fig. 18). The ligand-induced quaternary structure changes in hemoglobin are among the best known classic examples of allosteric protein conformation transitions and have been defined to atomic resolution by X-ray crystallographic analysis and thoroughly discussed.[65,66] Amide I infrared spectra reflect ligand-induced conformational changes at the secondary structure level; the changes in the corner regions (turns) between α helices and in the F–G corner regions of both α and β subunits[66] are the most readily observed. Figure 18 presents two pairs of superimposed second-derivative amide I infrared spectra: HbO_2 A/Hb A and HbCO A/Hb A. On going from deoxyHb (T state) to oxyHb (R state), the conformations of corner (turn) regions are altered due to relative movements of the α helices. In second-derivative amide I spectra, these secondary structure changes are represented mainly by small frequency and intensity changes of bands at 1667 cm^{-1}. These results demonstrate the high sensitivity of second-derivative analysis in monitoring the conformational changes of Hbs in H_2O solutions.

Both the primary and second-derivative amide I infrared spectra have proved useful for measurement of effects of changes in pH and temperature on hemoglobin conformations. Only the pH effects are discussed here. Figure 19 presents the superimposed primary and second-derivative spectra of HbCO A at pH 4.5 to 12.0. Compared with pH 7.5, only slight spectral changes, mainly a frequency shift near 1656 cm^{-1} (α helix), was observed for pH values from 4.5 to 10.5, suggesting that the secondary structures of HbCO A are remarkably stable over the pH range from 4.5 and 10.5. However, more dramatic changes were observed when the pH was raised from 10.5 to 12.0. Figure 20 presents a series of difference spectra from 2000 to 1200 cm^{-1} for HbCO A. Each difference spectrum is generated by subtraction of the primary spectrum obtained at a given pH from the primary spectrum obtained at pH 7.5. These difference spectra demonstrate the marked effect of increasing the pH from 10.5 to 12.0 in both amide I and II conformation-sensitive regions, with only small effects of pH change at the other pH values.

Acknowledgments

This work was supported in part by U.S. Public Health Services Grant HL-15980 and by a gift from Strohtech, Inc. We are grateful to Ping Huang for technical assistance in infrared spectra analysis and to Dr. Vijaya Sampath for supplying purified hemoglobins. We are also grateful to Professor Masako Nagai and Yoshimasa Yoneyama for supplying HbM samples. We thank the generosity of Spectra-Tech, Inc., for allowing us to use their IR-PLAN infrared microscope.

[65] M. F. Perutz, *Nature (London)* **228**, 726 (1970).
[66] J. Baldwin and C. Chothia, *J. Mol. Biol.* **129**, 175 (1979).

[10] Picosecond Infrared Spectroscopy of Hemoglobin and Myoglobin

By Robert H. Austin and Lewis J. Rothberg

General Philosophy behind Use of Transient Infrared Experiments to Study Protein Dynamics

Several central issues in understanding protein function are related to the dynamics of these remarkable macromolecules. It is apparent that the ability of a protein to assume a variety of configurations and to interconvert between these conformations is intimately related to the ability of a protein to perform biologically related functions. A central goal, therefore, in achieving an understanding of the mechanism of protein function is to identify the relevant configurations and the microscopic motions of importance for assuming these various geometries. A companion issue is to determine how energy, both thermal and chemical, flows in the macromolecule during these dynamics. In special cases, such as photosynthesis and vision, the coupling of optical energy into the protein and how it acts are also critical to the function.

The complexity of even the simplest of proteins makes this an extremely daunting task, which is exacerbated by a lack of detailed structural knowledge about most proteins at the atomic level and by a limited understanding of the effect of the surrounding environment (usually water).

It is our goal here to describe one particular experimental approach to obtaining substantive information regarding the protein dynamics relevant to biological function. We hope to provide a feeling for both the promise and limitations of transient infrared (IR) spectroscopy of proteins for advancing our microscopic understanding of mechanistic biochemistry. The vast majority of work has concentrated on heme proteins as model systems. The iron porphyrin group is a widespread component of proteins, even those performing a variety of functions such as ligand transport (myoglobin and hemoglobin), electron transfer (cytochrome oxidase), and enzymatic catalysis (cytochrome *c*). Although the heme group appears to be central to the function of all of these proteins, it is evident that the surrounding structure is essential in determining what these proteins actually do. This gross sensitivity of the function of heme proteins to their structure makes them excellent model systems to identify critical features that determine their behavior. Furthermore, the heme proteins are among the best characterized, having not only been sequenced but in most cases also crystallized. High-resolution neutron diffraction work[1] has led to crystallographic determination of myoglobin–CO (MbCO) at the atomic level along with determination of the water molecules of hydration.

The time scales for important motions in proteins span many orders of magnitude. They vary from normal vibrational motions (10^{-14} to 10^{-11} sec) of quasiequilibrium structures to large-scale structural rearrangements, which can occur over an enormous range of times from 10^{-12} sec (charge transfer in photosynthesis) to years (cataract formation in the lens of the eye). The intrinsic rapidity of these dynamics necessitates the use of optical probes (loosely defined here as 200–100,000 nm) with ultrafast measurement. In cases in which the methods are nondestructive or large amounts of sample are available, time-resolved optical studies can be powerful. When photolytic initiation is intrinsic to the protein function (e.g., vision), transient optical measurements are indispensable.

What, then, do we hope to gain from the picosecond IR avenue of experimentation? Ultimately, one would like a "movie" (or statistical set of movies) of the atomic motions of the protein to truly isolate the important structural features and to understand the molecular mechanism behind various functions. Naturally, what we are describing is essentially equivalent to molecular dynamics simulations of protein behavior. Because computer dynamics simulations appear to be directly relevant to the questions we posed above one might ask, why do dynamics experiments at all? Our answer is this: although we believe that molecular dynamics calculations are promising, there are both conceptual problems in modeling the forces in such a complex system and practical difficulties in carrying out such computations for time scales longer than 10^{-9} sec. The experimental work, then, can at present reach far beyond the time scales of molecular dynamics simulations and, of course, uses the "correct" potentials. Although someday the simulations will provide the detailed information experiments can provide, until then the theoretical calculations must rely on experiments to provide useful "boundary conditions" that dynamics simulations must at a minimum be able to reproduce.

For concreteness, several useful specifics obtainable by time-resolved infrared spectroscopy are as follows: (1) vibrational relaxation rates (no topological change in the protein), (2) orientational relaxation rates (minimal topological change), and, most importantly, (3) rates of changes in the global conformation (large topological changes).

This last point brings up a problem: it is not clear to us that picosecond IR spectroscopy of hemoglobin is going to be a useful technique with

[1] X. Cheng and B. P. Schoenborn, *J. Mol. Biol.* **220**, 381 (1991).

which to study conformational relaxation in hemoglobin. Nor in fact is it obvious that one should concentrate on hemoglobin. This chapter is restricted to hemoglobin and does not include myoglobin in the discussion. Myoglobin is a "simple" 18-kDa heme protein that binds a single oxygen. Hemoglobin is a more complex protein that binds 4 oxygens in a cooperative manner: the last oxygen binds with approximately 300 times the binding constant that the first oxygen has. Hemoglobin also shows allosterism in addition to cooperativity because certain effectors such as 2,3-diphosphoglycerate alter the binding of ligands. Thus, hemoglobin is particularly fascinating as a model system to discover the rules of nonlinear behavior in proteins. To our mind, the best reason for studying hemoglobin as opposed to myoglobin is to follow the conformational flow of the protein from the fully liganded state (the R state) to the fully unliganded state (the T state).

This is not to say that there are not many important problems that can be addressed in studying myoglobin; there are. In fact, myoglobin is the protein of choice if one wants to study the dynamics of a protein within the confines of a single state, the liganded state. Perhaps one can view those dynamics as the movement of a point in some complicated phase space landscape around a fixed point, the liganded state. Hemoglobin is the next step up to study conformational diffusion from one fixed point to another one, from the R to the T state.

However, few picosecond hemoglobin dynamics articles are concerned with the important question relevant to hemoglobin: how does the protein move from R to T? If one simply wants to study the initial picosecond dynamics of deligation, then by all means use myoglobin. If the conformational dynamics from one state to another state are of interest, then use hemoglobin, but be prepared to do dynamics over a much larger time scale then just picoseconds. The first picoseconds of the motion are not even the tip of the iceberg: the close-packed nature of the protein interior requires that there be concerted movements of the protein backbone before the ligand can get out.[2] These concerted motions might take many nanoseconds and are of great importance in understanding protein dynamics. Because our mission is to understand how proteins move, we also discuss nonpicosecond IR experiments and contrast them with the picosecond work in order to elucidate the real biological questions that ultimately are why we study biophysics.

Technical Problems in Transient Infrared Techniques

Energy states in the midinfrared portion of the electromagnetic spectrum (3–30 μm) are special and can be particularly valuable because

---

[2] D. A. Case and M. Karplus, *J. Mol. Biol.* **132**, 343 (1979).

absorption in this spectral region can be identified by specific molecular motions. Because there are a huge number of vibrational modes in even the smallest of proteins (e.g., for a protein of 1000 atoms there are over 3000 normal modes), this may seem at first to be futile. Nevertheless, it is often the case that there are a limited number of important functional groups with infrared absorptions that are spectrally isolated and can be probed or excited in a controlled fashion. The ability to monitor localized states in the protein not associated with the heme group promises to become increasingly useful when combined with the burgeoning power of biochemical techniques such as site-specific mutagenesis and atomically selected isotopic substitution. The opportunities are enormous. There is already a vast body of static difference infrared spectroscopy based on this approach, in which the difference spectra are interpreted in terms of structural differences associated with the altered group. Here, however, we restrict our consideration to transient dynamic studies.

There are many limitations associated with transient infrared studies. One of these is the poor compatibility of IR work with aqueous environments. Water absorbs strongly nearly everywhere in the midinfrared region. In principle, the broad IR absorption of water is not a problem because it is not spectrally structured (the absorption spectrum in the IR is broad and featureless), whereas the protein modes at issue are, by and large, sharp (linewidth δ, <20 cm^{-1}). In practice, however, the generally poor transmission of samples and small shifts of water spectra caused by transient heating can hamper quantitative experimentation and must be considered carefully, although in some cases these shifts can actually be put to good use.

Another problem is that in the IR, the absorption cross-sections and the corresponding extinction coefficients, ε, tend to be small (ε_{IR} for many transitions is on the order of 1 mM^{-1} cm^{-1} compared to optical extinction coefficients, ε_{VIS}, on the order of 10–100 mM^{-1} cm^{-1}). A fundamental reason for the low extinction coefficient lies in the Einstein coefficients: both the spontaneous emission terms and the absorption terms scale as the frequency ν^2, and hence the inherent transition probabilities scale as ν^2.[3] Because of the poor transmission due to water and small ε_{IR}, typically one must work with concentrated thin films of the protein to keep the water path length down and yet maintain a large amount of protein. Final protein amounts tend to be milligrams per square centimeter in the thin film, and the film is often no more than tens of microns thick.

Another difficulty with transient infrared absorption experiments is that infrared sources and detectors tend to be quite poor compared with

[3] J. T. Verdeyen, "Laser Electronics." Prentice-Hall, Englewood Cliffs, NJ, 1989.

optical ones: they often have poor quantum efficiency,[4] and they must compete with ambient thermal background radiation. As mentioned above, the absorption cross-sections are small and generate poor signal-to-noise ratios. In the case of IR sources, the low intensity of typical blackbody sources can be alleviated to some extent by using infrared lasers, but these are generally not spectrally tunable in a useful way. The poor quantum efficiency of IR detectors and the poor fluxes from continuous-wave IR sources mean that common techniques such as pulsed excitation followed by monitoring of the transmission with a continuous-wave (CW) beam typically cannot give adequate time resolution to explore the faster time scales cited above.

Thus, almost by necessity picosecond pump–probe methods have been the main form of subnanosecond time-resolved IR work. The complexity of generating picosecond IR pulses, given current technology, is a limitation on the utility of this approach.

The nature of vibrational spectroscopy is such that spectral resolution on the order of a few reciprocal centimeters is important, and yet the nature of picosecond pulses is to have intrinsically poor spectral resolution due to the uncertainty principle. Specifically, suppose we have a probe pulse with center frequency ν_0 and duration Δt. Fourier analysis tells us that the linewidth of the pulse (ignoring factors of π) is $\Delta\nu \approx 1/\Delta t$, and so the fractional linewidth of the pulse $\Delta\nu/\nu_0 \approx 1/\Delta t\nu_0$. It is more convenient in the IR to use units of inverse wavelength ($1/\lambda$, or cm^{-1}). Let the reciprocal centimeters of the center frequency be given as $\mathscr{E}_l$. The fractional linewidth is

$$\Delta\nu/\nu_0 = 1/\Delta t\mathscr{E}_l c \tag{1}$$

where c is the speed of light. If Δt is 10^{-11} sec (10 psec) and $\mathscr{E}_l = 2000$ cm^{-1} then the transform-limited linewidth is on the order of 3 cm^{-1}. If the pulse width is 1 psec then the linewidth is 30 cm^{-1}, uncomfortably large.

One approach by Hochstrasser and co-workers to surmount this problem was to use a narrow-band tunable infrared continuous-wave laser for the probe source and mix with an optical picosecond pulse.[5] A short-pulse visible laser was used both to photodeligate the protein and to gate the continuous infrared source, using a nonlinear mixing scheme that upconverts the IR beam to the visible, where it is detected by a conventional photoelectric effect detector. The highly nonlinear up-conversion process

[4] J. Jamieson et al., eds., "Infrared Physics and Engineering." McGraw-Hill, New York, 1963.
[5] J. N. Moore, P. A. Hansen, and R. M. Hochstrasser, Proc. Natl. Acad. Sci. U.S.A. 85, 5062 (1988).

in a crystal strongly peaks the optical output at the tuned center frequency ν_0 of the CW laser and rejects the transformed broadened tails of the pulse. We expect it to become a standard methodology for single-wavelength IR work on the picosecond time scale. Other possible approaches are to pass the transform-broadened pulse through a monochromator to regain frequency resolution. As long as the free induction decay time, T_2, of the sample is sufficiently long, high-resolution spectroscopy can be obtained. A review article by Stoutland et al.[6] summarizes these technical questions and has many valuable references.

As mentioned above, the transient IR absorption signals are small compared to transient optical signals, so one must optimize the signal-to-noise ratio as much as is possible. In IR absorbance experiments using picosecond lasers and time-delay techniques, shot noise is not usually a problem because each pulse contains many photons. For example, even a nanojoule pulse at 2000 cm^{-1} contains 3×10^{10} photons, so the variance expected per shot due to shot noise is only on the order of 10^{-5}, and even a 1% quantum efficiency detector would give a variance on the order of 10^{-4}. Because high-powered Nd:YAG (neodymium:yttrium–aluminum–garnet) systems are capable of running at high-power pulse repetition rates of 10^3 Hz, it is easy to imagine obtaining 10^{-6} relative variances in a few seconds; however, these are laser sources. The pulse energies even at the primary pump energy easily can have variances of 0.01, and these variances increase as a high power of pulse energy when nonlinear mixing techniques are used to create IR pulses.

As an example of this, consider what was observed when up-conversion techniques were explored (these observations have not been published). The main source was a Nd:YAG laser (pumped at 1 kHz) frequency doubled to 533 nm. The 533-nm radiation was used to pump a grating-tuned dye laser with primary wavelength of 600 nm. This 600-nm light was then mixed with the 532-nm radiation to produce tunable IR pulses with 70-psec duration. The probing IR was detected by up-conversion to the visible by sum frequency mixing with a 583-nm short pulse (3 psec) dye laser to the visible. There are many sensitive IR detectors around; the purpose of the up-conversion was to repeat what Moore et al.[5] did to obtain shorter time resolution.

The severe penalty that one pays for up-conversion is a big increase in pulse energy variance: It was found that the variance in the up-converted light is approximately 30%. Assuming 10^3-Hz pulse rates and random noise in the pulse energy, it takes on the order of 15 min per point to get the variance below 10^{-3}. This is not good, especially considering that

[6] P. O. Stoutland, R. B. Dyer, and W. H. Woodruff, *Science* **257**, 1913 (1992).

the sample remains under intense light illumination during this extended period, during which many photochemical process can degrade the sample.

The problem with the large variances in the IR probe pulse typically generated by nonlinear processes brings up the question of the advisability of using low- or high-pulse repetition rates. That is, suppose that one can arrange to have a fixed average power IR probe beam, I_{ir}. Should one opt for many small pulses, N_s, or a few large pulses, N_b, given that the energy/pulse, E_p, of the optical excitation energy roughly scales as $1/N$ for our fixed average power? The temptation is to use many small pulses, on the assumption that the variances can be averaged by $1/N^{1/2}$. However, the size of the signal created by the optical excitation usually scales as E_p, and hence the signal scales roughly as $1/N$, hence the signal-to-noise ratio will scale roughly as $1/N^{1/2}$, and it is therefore better to use a few, large pulses. Another important point in favor of using a few, large pulses is that often the biological molecule takes many milliseconds to recycle back to the initial state, and use of kilohertz repetition rates can lead to either pumping of the sample into out-of-equilibrium kinetic bottlenecks or even into photogenerated excited states. In the case of hemoglobin or myoglobin, the problem of pumping is particularly acute at low temperatures, where the bimolecular phase for ligand recombination can result in recombination times of many milliseconds to many hours. Use high repetition rates at your own risk!

Ligand Dissociation from Heme Proteins as Model Problem

The potential and limitations of transient infrared measurements in biophysics are well illustrated by reviewing the results of applying time-resolved infrared spectroscopy to a specific problem, the motion of small ligands (O_2, CO, and NO) in heme proteins. There are several underlying motivations for selecting this particular problem. The first is that it is not an artificial problem: ligand binding and unbinding are intrinsic to protein function for myoglobin, hemoglobin, cytochrome P-450, and cytochrome oxidase, for example. Moreover, ligand transport through the proteins can be modeled by molecular dynamics simulations, and for many known structures it is clear from those simulations that the protein can represent a very close-packed environment. Studies of ligand diffusion can tell us a great deal about the rate at which the protein structure relaxes. From a practical point of view, ligand unbinding can be accomplished efficiently by visible photolysis, thus the measurement can begin with a well-defined configuration, which facilitates synchronized time-resolved measurements. The result is the achievement of large enough concentrations of intermediate configurations to measure their properties and lifetimes. This experimental program is also exactly analogous to what is typically done in the theoretical simulations. A final reason (and most important) for choosing photodeligation of heme proteins as a model problem is that the infrared stretching absorptions of typical ligands (CO and NO) are spectrally well isolated and strongly absorbing, which makes these experiments relatively easy to perform and interpret. We note here that these ligand vibrational modes are not usually amenable to study using Raman spectroscopy, because of the lack of resonant Raman enhancement at wavelengths transparent to the protein. In general, however, Raman probing of protein dynamics is likely to have greater impact in many cases, because critical low-frequency motions are more accessible and interference with water absorption and scattering is much less of a problem than in infrared experiments.

Most of the IR work has been done on CO, and little has been done on NO. One important reason for this is that NO recombines much more quickly (on a subnanosecond time scale) than CO.[7] The process of internal recombination is called geminate recombination, and if the geminate phase is large, then there is no ligand left to probe the protein dynamics at longer times. Not all (in fact, very few) would agree, but in our opinion (R.H.A) the reason NO has such a large geminate phase is because it is a spin $\frac{1}{2}$ molecule and the normal selection rules of perturbation theory make the reaction rate very fast,[8-10] not because it is able to move physically in a different manner than CO. Oxygen also poses problems because the quantum yield for photolysis is considerably smaller than for CO, and the oxygen–iron bond is less stable chemically compared to CO–Fe. We also should note that the room-temperature fraction of geminate recombination is much larger in hemoglobin than in myoglobin, even if one restricts oneself to CO.[11] A careful comparison of the low-temperature activation energy distributions to the high-temperature kinetics in principle could be used to "predict" the size of the geminate phase if no other substates are activated above the glass transition.

Let us now explain how we can predict room-temperature kinetics from low-temperature kinetics. A seminal body of work in the area of heme

[7] J. W. Petrich, J. C. Lambry, K. Kuczera, M. Karplus, C. Poyart, and J. L. Martin, *Biochemistry* **30**, 3975 (1991).
[8] M. H. Redi, B. S. Gerstman, and J. J. Hopfield, *Biophys. J.* **35**, 471 (1981).
[9] B. Gerstman, R. H. Austin, and J. J. Hopfield, *Phys. Rev. Lett.* **47**, 1636 (1981).
[10] B. S. Gerstman and N. Sungar, *J. Chem. Phys.* **96**, 387 (1992).
[11] S. Pin, P. Valat, H. Tourbez, and B. Alpert, *Chem. Phys. Lett.* **128**, 79 (1986).

protein dynamics has been developed by Frauenfelder and co-workers.[12,13] These are principally temperature- and pressure-dependent optical studies of the iron porphyrin chromophore in myoglobin (Mb). Before discussing results of picosecond infrared measurements, we will review some of this work. This will help in both defining a framework in which to think about issues in protein dynamics and in motivating picosecond optical work.

The largest subset of studies can be traced back to Austin *et al.*,[12] who used an optical pulse to dissociate CO from the iron atom contained in the prosthetic heme group in Mb. The subsequent reformation of MbCO was monitored optically by recording the change of absorbance at 436 nm versus time after photolysis. The absorbance at this probe wavelength can be used to infer the (normalized) number of protein molecules, $N(t)$, that remain deligated as a function of time after dissociation. This work covers the temperature range from 40 to 320 K and some 10 decades in time (10^{-7} to 10^3 sec). All of this work has been done in glycerol–water mixtures with good optical quality at low temperature, but that also have glass transition temperatures at about 200 K.

If, as above, we refer to the carbon monoxide-bound form of Mb as a state, then a central concept arrived at in this work is recognition that there exists a large variety of conformational substates. Each conformational substate has a somewhat different barrier to rebinding, which explains the nonexponential rebinding kinetics observed at low temperatures, where thermal interconversion between these conformations is slow. That is, the total amount of unrebound hemes at time t after a photolyzing pulse can be written as

$$N(t) = N(0) \int_0^\infty g(k)\exp(-kt)\,dk \qquad (2)$$

where $N(0)$ is the number of photolyzed molecules at time $t = 0$, $g(k)$ is a distribution of rates k, which is determined by a distribution of conformational substates of the protein. The connection between the rate constant k and the temperature dependence of the rate is given by the Arrhenius relation, $k(H) = A \exp(H/k_B T)$, where H is the activation energy, k_B is Boltzmann's constant, and A is the preexponential term that contains the selection rule effects we discussed above. Thus, the rate distribution can be transformed to an activation energy distribution:

[12] R. H. Austin, K. W. Beeson, L. Eisenstein, H. Frauenfelder, and I. C. Gunsalus, *Biochemistry* **14**, 5355 (1975).
[13] G. U. Nienhaus, J. R. Mourant, and H. Frauenfelder, *Proc. Natl. Acad. Sci. U.S.A.* **89**, 2902 (1992).

$$N(t) = N(0) \int_0^\infty g(H) \exp[(-k(H)t] \, dH \qquad (3)$$

where $N(0)$ is the number of photolyzed molecules at $t = 0$ and $g(H)$ is the distribution of activation energies. Significantly, the distribution $g(H)$ is assumed to be both time (t) and temperature (T) independent below the glass transition temperature. Because these experiments were done at low temperatures, where presumably the ligand CO is photolyzed to a single nearby site, the distribution is presumed due to a frozen distribution of protein conformations. This distribution of conformations is to be distinguished from a distribution of binding sites within a single protein structure. A second crucial concept that came from this work was the idea of a glass transition temperature in the protein, T_g. Experimentally, this means that at temperatures above about 230 K, an increasing fraction of the recombination is nongeminate, that is, external ligand concentration dependent, indicating (1) that the protein "diffuses" among the various substates, which at low temperatures gave rise to the distribution $g(H)$ in the recombination kinetics, and (2) the CO can now escape from the immediate heme group and diffuse into the external solvent. The crossover in behavior is nearly coincident with the glycerol–water solvent glass transition, which may indicate that the protein flexibility is intimately coupled with that of its environment; technically the coupling of the two transitions is called a slaved-glass transition.

The final idea in Frauenfelder's work, perhaps not clearly expressed in the papers cited but nonetheless of primary biological importance, is that the conformational fluctuations at physiological conditions are scaled up from the dynamics that begin just after the glass transition, and that in order to understand these clearly biologically important motions, one must approach them from below to untangle the room-temperature dynamics. Doing measurements at room temperature alone because that is the temperature at which the molecules work misses the larger picture of protein dynamics and does not work anymore than restricting studies of semiconductors to room temperature because that is the temperature transistors run at. It is annoying to struggle with cryogenics, but in our opinion it must be done.

Probing these three concepts [(1) a complex distribution of conformational substates within a given protein state, (2) freezing of the dynamics at some critical temperature, and (3) dynamic exchange between these conformations at physiological temperatures] should lie at the heart of infrared experiments. Basically, we would like precise probes for both the presence of the conformation distribution within the protein and measurement of the rate of conformational interchange.

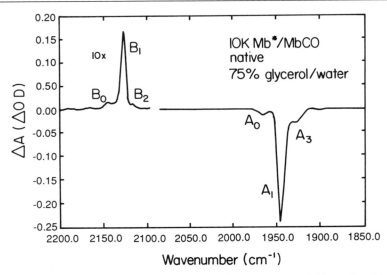

FIG. 1. A difference spectrum of photolyzed carbonmonoxymyoglobin at $T = 10$ K, in the region of the CO stretch, 1900 to 2200 cm^{-1}. Photolysis at this low temperature results in essentially an irreversible loss of bound CO, giving rise to the so-called A bands of iron-coordinated CO. The photolyzed CO, presumably now within the protein and near the iron, absorbs at an energy nearer the gas phase value of 2140 cm^{-1}. These bands are termed the B bands. The splitting of the bands gives rise to the subbands $A_{0,1,3}$ and $B_{0,1,2}$.

A second body of work by Frauenfelder and co-workers was aimed precisely at the nature of these metastable substates, in which the ligand remains in the protein but is not bound to the iron. They use Fourier-transform infrared spectroscopy (FT-IR) to look at CO in the protein matrix and work at cryogenic temperatures, low enough to keep quasipermanently the CO from recombining after photolysis. A typical result depicting the difference before and after photolysis is shown in Fig. 1.[14] The A bands represent the loss of CO bound to the iron and the B bands are the appearance of the CO elsewhere in the protein matrix. There are several features of these data worth noting. (1) Most importantly, the A bands and the B bands are, as stated, not single bands but instead are

[14] D. Braunstein, A. Ansari, J. Berendzen, B. B. Cowen, K. D. Egeberg, H. Frauenfelder, M. K. Hong, P. Ormos, T. B. Sauke, R. Scholl, A. Schulte, S. G. Sligar, B. A. Springer, P. J. Steinbach, and R. D. Young, *Proc. Natl. Acad. Sci. U.S.A.* **85,** 8497 (1988).

clearly split into a number of discrete broad lines.[15] The splitting of these bands apparently is due to a discrete population of protein conformations, and each band is broadened due to a distribution of substates, as we mentioned above. The population of a given line within the A band is a function of temperature and solvent pH, and line population can interconvert at high enough temperatures.[16] Time-resolved IR studies of the A bands offer a potentially important probe of protein dynamics. (2) In the metastable "B state," CO absorbs near its gas phase-stretching frequency. This indicates that the CO is weakly bound, although sufficiently constrained, to suppress all of its rotational side bands,[17] which are due to transitions between different rotational levels. (3) CO is not observable (or with great difficulty) when dissolved in a strongly hydrogen-bonding solvent such as water, presumably due to inhomogeneous broadening of the transition energy due to hydrogen bonding. Thus, at some point in time (or space) after photolysis, as the CO enters the solvent, the signal should disappear altogether even in the absence of recombination. This means that B-state dynamics can give some information concerning the true rate at which the CO leaves the molecule, a much more powerful probe than just monitoring rebinding kinetics.

Time-Resolved Ligand Infrared Studies in Myoglobin

Myoglobin stores and transports oxygen in muscle tissue. It is a globular protein consisting of 153 amino acids and the active iron porphyrin group. It has a molecular weight of around 18,000 and has been crystallized in both its liganded and unliganded forms. Because it is among the simplest and best characterized proteins, it serves as a prototype for a wide variety of experiments and theory. The binding and unbinding of ligands by Mb therefore have been studied extensively to achieve a molecular understanding of these relatively simple biochemical processes.

As we noted before, the visible absorption spectrum is characteristic of the iron porphyrin chromophore and is sensitive to ligation, forming the basis for the flash photolysis work and for a large number of nanosecond and picosecond studies, which are principally sensitive to reconfigu-

[15] M. K. Hong, D. Braunstein, B. R. Cowen, H. Frauenfelder, I. E. T. Iben, J. Mourant, P. Ormos, R. Scholl, A. Schulte, P. J. Steinbach, A. Xie, and R. D. Young, *Biophys. J.* **58,** 429 (1990).
[16] A. Ansari, J. Berendzen, D. Braunstein, B. Cowen, H. Frauenfelder, M. K. Hong, I. E. T. Iben, B. Johnson, P. Ormos, T. Sauke, R. Scholl, A. Schulte, P. J. Steinbach, J. Vittitow, and R. D. Young, *Biophys. Chem.* **26,** 337 (1987).
[17] G. Jiang, W. B. Person, and K. G. Brown, *J. Chem. Phys.* **62,** 1201 (1975).

ration of the heme group after photolysis and, along with time-resolved Raman experiments, indicate rapid out-of-plane motion of the iron on photodeligation. Here, however, we focus on the ligand behavior.

We know that the ligand leaves the iron rapidly. Petrich and Martin[18] have done photolysis experiments with subpicosecond resolution and observed rapid dynamic effects only on subpicosecond time scales. Although they report 20% bleaching recovery in MbCO in less than 3 psec, they ascribe the optical signal to an excited but associated species that recovers to the ground state in that time. They do not associate these dynamics with ligand recombination.

Because the CO ligand leaves the iron-bonding orbitals in under a picosecond, it is fair to ask, what is the subsequent path that the ligand takes as it moves away from the protein and out into the solvent? Only IR spectroscopy can answer that question.

The first picosecond time-resolved infrared experiments on MbCO were done in a set of seminal experiments by Hochstrasser and colleagues at the University of Pennsylvania. Moore et al.[5] made use of the pump-and-probe transient IR polarization to determine orientational information about the MbCO structure in solution at room temperature. The idea here is to use visible light, absorbed into the planar heme group, to photolyze the CO and monitor the polarization of the bleached A states. The orientation of the electronic absorption dipole for the dissociation pulse with respect to the protein frame is known.[19] Similarly, the absorption moment for the infrared probe at the CO stretching frequency is clearly along the bond direction. Hence, these authors were able to extract the angle of the CO with respect to the protein frame by looking at the size of bleaching under different excitation conditions. Let the ratio R of the absorption changes at 1944 cm^{-1} be defined as

$$R(t) = \frac{\Delta A_\perp}{\Delta A_\parallel} \tag{4}$$

where $\Delta A_\perp$ is the absorption change with IR polarization axis perpendicular to the visible pump polarization and $\Delta A_\parallel$ is the absorption change with IR polarization axis parallel to the visible pump polarization. Then it can be shown that[20]

$$R = \frac{4 - \sin^2\gamma}{2 + 2\sin^2\gamma} \tag{5}$$

[18] J. W. Petrich and J. L. Martin, *Chem. Phys.* **131**, 31 (1989).
[19] W. A. Eaton and J. Hofrichter, this series, Vol. 76, p. 175.
[20] P. A. Hansen, J. N. Moore, and R. M. Hochstrasser, *Chem. Phys.* **131**, 49 (1989).

where γ is the angle of the CO with respect to the normal to the porphyrin plane. It is important to bear in mind that under physiological conditions these measurements must be done with subnanosecond time resolution because rotation of the protein in solution erases orientational information on a 10-nsec time scale.

The results for CO angle γ in the iron-bound configuration (A state) turn out to be different for the different A substates in MbCO. For the dominant A_1 state (1943 cm^{-1}) the angle is 20°, whereas the A_3 state (1933 cm^{-1}) is inclined at 35° to the normal. The carboxyhemoglobin band analogous to A_1 at 1951 cm^{-1} is 18°, indicating a similar pocket structure. This work has shown that these binding angles clearly are influenced by the protein, as the CO behaves differently in the case of protoheme,[20] which can be considered to be the porphyrin chromophore without the associated protein.

Previous to these room-temperature experiments, low-temperature IR polarization experiments were done by Ormos et al.[21] on metastable photolyzed protein in a glycerol–water glass. Because the recombination was slow at the temperatures studied, static FT-IR techniques were used to monitor the states. The results of Moore et al.[5] turn out to be essentially the same as the low-temperature results, indicating that, at least at 100 psec at room temperature, the configuration of the CO in the A states is the same as the configuration in the low-temperature glasses.

Because the A band is converted to the B band as the CO enters the protein, the next step is to examine the polarization of the B state relative to the photolyzing flash polarization. This in principle can reveal how much the CO has rotated as it has moved in the protein. Such an experiment is clearly a picosecond experiment as one moves up in temperature, and it is extremely challenging because of the small oscillator strength of the B state.

Rothberg et al.[22] have done such a difficult experiment as a function of temperature at a fixed time delay of 100 psec. These experiments were carried out at pH 7.0 and, because of the extreme difficulty of obtaining adequate signal to noise, were not able to probe the question of the polarization of the different B-state lines. View these measurements as a determination of the average angle over the lines and as a pioneering first attempt.

Figure 2 shows B-state polarization data at two temperatures. The conclusion from these rather noisy data is that the CO molecule assumes

[21] P. Ormos, D. Braunstein, H. Frauenfelder, M. K. Hong, S.-L. Lin, T. Sauke, and R. D. Young, *Proc. Natl. Acad. Sci. U.S.A.* **85,** 8492 (1988).

[22] L. J. Rothberg, M. W. Roberson, and T. M. Jedju, *SPIE* **1599,** 309 (1991).

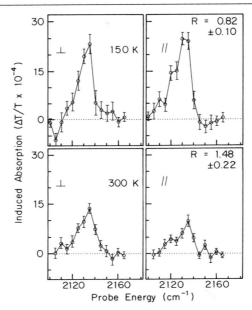

FIG. 2. Transient absorption spectra of photolyzed CO in myoglobin at 100-psec delay at $T = 150$ K and $T = 300$ K. The parameter R is defined in text. (From Rothberg et al.[22])

a different conformation in the protein as a function of temperature at 100 psec, even in the geminately combining state. Apparently the CO is able to shift its symmetry axis as it moves to different positions within the protein.

If we go back to the original model of Frauenfelder, then generally speaking, we can say that as we approach the glass transition, the ligand CO is able to access different conformations in the conformation distribution as the temperature increases, and the protein undergoes greater diffusion among the various conformations, be they the states that give rise to the different lines in the IR spectra or the substates within a given line.

There are several scenarios to which these altered physical orientations correspond. One possibility is that the protein reorganizes at higher temperatures, such that it is energetically favorable for CO to assume a different binding geometry in the metastable state, that is, the altered structures are locally restricted to the immediate iron-binding site. In fact, such a local distribution is certainly measured by Frauenfelder's group in their quasistatic A-state relaxation experiments carried out in the FT-IR apparatus.[21] There is no way to convince the reader at present that the polarization changes seen in the picosecond B state at higher temperatures are due to local changes in the environment, but let us speculate, guided by computer dynamics.

The second class of explanations is that the protein undergoes rapid conformational relaxation at higher temperatures, and the CO is able to reach within 100 psec an altogether different location in the protein, as opposed to the local rebinding pocket seen at low temperatures. This would indicate that the glass transition in a protein not only allows diffusion between different local states, but that the protein also may relax on a global scale and reach configurations that are not probed by the low-temperature distribution observed in the optical experiments. An example of this kind of global change in flexibility may be found in work on a totally unrelated protein, ribonuclease A.[23] It was found that the crystalline protein lost function below what could be interpreted as a glass transition around 220 K. Probably one of the most important problems facing us at present is in fact characterizing the dynamics of the entire protein structure above and below the glass transition over the time range from picoseconds to many seconds.

Rothberg et al.[22] have presented speculation about the dynamics, on the basis of picosecond B-state work and using molecular energetics calculations. These results seem to confirm that large-scale conformational changes occur near the glass transition.

It is instructive to review how these authors go about arguing for this hypothesis. First, they locate voids in the crystal structure of the protein, which are large enough to fit CO, after every atom is assigned to a hard sphere the size of its van der Waals radius. One finds three pockets fitting this description in the half-plane above the heme group. Not surprisingly, these correspond to pockets where xenon is found in X-ray diffraction studies of crystalline myoglobin equilibrated with overpressures of xenon.[24] If one assumes, on the basis of its near-gas phase stretch frequency, that the CO is bound only by van der Waals forces in these pockets, then it is expected that the CO axis will lie along the long axis off the pocket to minimize the repulsive forces. Given this, one can associate each pocket with an angle. A more sophisticated analysis, including the actual forces with molecular dynamics code Biograf and relaxation of the protein, essentially bears out these naive estimates. Two pockets were found, denoted M1 and M2. On the basis of the good correspondence of "theoretical" and experimental angles, it is tempting to identify pocket M1 with the CO in its low-temperature metastable state and pocket M2 with the CO in its higher-temperature metastable state. Rothberg et al.[22] go on to support this assignment with circumstantial evidence derived from the ligand behavior

[23] B. F. Rasmussen, A. M. Stock, D. Ringe, and G. A. Petsko, *Nature* (*London*) **357**, 423 (1992).

[24] J. Hermans and S. Shankar, *Isr. J. Chem.* **27**, 225 (1986).

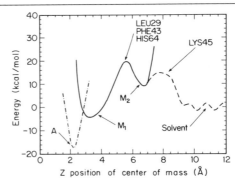

FIG. 3. Constrained minimization for CO ligand travel along the reaction coordinate (discussed in text). The dot-dashed line with a minimum at "A" represents the energy surface for bound CO. The solid line represents the energy surfaces for the photolyzed CO within the protein, and the dashed line represents the potential surface due to bound water molecules if Lys-45 moves to allow ligand escape. (From Rothberg *et al.*[22])

with temperature and site-specific replacement of several key amino acids. In particular, location M1 is directly above the heme pocket, and it makes sense that it would recombine from this location with high efficiency. Location M2, however, is in a pocket near the surface of the protein separated from the solvent by only one residue, Arg-45 in sperm whale Mb (or Lys-45 in horse Mb). Thus, at temperatures at which the CO reaches M2, it is reasonable to assume that it could escape the protein. The 60-nsec lifetime in M2 corresponds to an activation barrier of 5 to 7 kcal/mol, assuming Arrhenius prefactors of 10^{10} to 10^{12} sec^{-1}. This activation is plausible for breaking the salt bridge that supports the arginine "cap." Moreover, this residue is known from crystallography to exist in "open" and "closed" configurations, again suggesting this as a likely route of entry and escape for the ligand. Figure 3 shows a diagram of the potential surfaces for these sites.

The locations of these metastable pockets, therefore, argue for an unbinding trajectory at physiological temperatures that can be described as follows. The dissociated CO rapidly exits pocket M1 directly above the porphyrin in which it sticks at low temperature. Its excess energy allows it to push through residues Leu-29, Phe-43, and His-64, which separate it from the M2 pocket. It enters pocket M2 in less than 100 psec (probably less than 1 psec by analogy with hemoglobin) and remains trapped in this pocket for some time before existing to the solvent via the Arg-45 (or Lys-45) "door." The most sophisticated molecular dynamics calculations fail to predict this exit route for CO. Nevertheless, computations do suggest that the CO spends a great deal of time in the cavities identified by the xenon-saturated Mb experiments and that temperature is an important determinant in the diffusion of the CO in the protein. Advances in computational speed and methodology show great promise, but are beyond the scope of our discussion.

The above material is a kind of bottom-up approach to understanding what happens at the transition: that is, find all the parts and figure out in detail how they interconnect. It is also possible to imagine a top-down approach. That is, there may also exist general aspects of protein dynamics that involve collective motions as we discussed initially concerning the ideas of Frauenfelder. We can ask, what are the general modes that open up at the glass transition, what are the conformational relaxational rates of these modes at the glass transition, and how do the relaxation rates extrapolate to room temperature?

Unfortunately, the spectroscopic markers for collective modes in the protein are not known yet. We give here a brief example of how the CO IR absorption lines can begin to address this issue, but it is hardly satisfactory at present.

Suppose we observe the time dependence of the B-state absorbance spectrum. The time dependence of the B-state absorbance tells us how fast the protein relaxes and allows the CO to leave the protein, because we know that once the CO is in the solvent the B state disappears. If the loss of B-state absorbance was just due to diffusion of the ligand out of the protein then it would not be a picosecond process. To see this, imagine that the diffusion coefficient of CO in a protein at room temperature is 10^{-6} cm^2 s^{-1} and that the CO had to diffuse 10 Å to leave the protein. Then, using $\langle x^2 \rangle = 2Dt$, it would take at least 5 nsec for a ligand to diffuse out of the protein. This is not a picosecond experiment, but an important measure of protein dynamics.

To address this issue, Hong et al.[25] used a microsecond-duration flash lamp as an IR probe source and a 50-nsec response time IR detector to track the evolution of the B-state CO stretching mode absorption. In principle, such a low-cost experiment may not belong among a discussion of more costly picosecond experiments, but nanoseconds to microseconds may be in many cases relevant to biological action. It may be that picosecond spectroscopy is not the best way to answer some of the more basic questions.

The measurements spanned the temperature range between 160 and 230 K. Even at 230 K, far above where static FT-IR is viable, except in equilibrium situations in which a high CW flux can be used to pump

[25] M. K. Hong, E. Shyamsunder, R. H. Austin, B. S. Gerstman, and S. S. Chan, *Phys. Rev. Lett.* **66**, 2673 (1991).

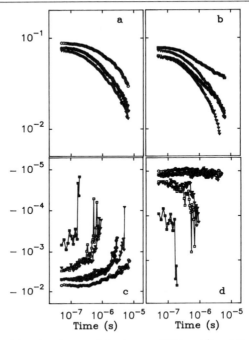

FIG. 4. The log of the absorbance of the A state (1950 cm^{-1}) and B state (2150 cm^{-1}) as a function of log time after photolysis. (a) A-state absorbance at 160 K (○), 190 K (●), and 200 K (▽). (b) A-state absorbance at 200 K (▽), 210 K (▼), 220 K (□), and 230 K (■). (c) The absorbance of the B state over the same temperature range as in (a) and (b), using the same symbols. (d) The ratio R (B-state population/A-state population) over the same temperature range, using the same symbols. (From Hong et al.[25])

into other long-lived states,[16] Hong et al.[25] were able to observe induced absorption in the 2130-cm^{-1} region as well as bleaching of the 1943-cm^{-1} iron-bound CO stretching mode. Figure 4 illustrates the dynamics of both the bleaching and induced absorption as a function of temperature. At 160 K the situation is relatively simple as these bands track each other. This reflects the fact that the disappearance of CO from the metastable B state results in the recovery of the heme-bound CO. At 220 K, the fall off of B diverges from the recovery of A, indicating loss of B is not accompanied by recombination. Hong et al.[25] point out, however, that there is still no concentration dependence to the recovery of A, which indicates that the recombination remains geminate. Thus, the CO leaving B is still associated with the same protein molecule from which it was photodissociated but resides in an environment where the CO stretching vibration has width substantially larger than is being monitored in the experiment (≥ 20 cm^{-1}). Note that Rothberg et al.[22] monitored the B state close to the binding pocket, whereas the experiments presently being discussed reach out to longer times and distances.

The loss of the B absorbance is speculated to be due to the trapping of CO in the hydration shell of the protein, where it experiences an inhomogeneous polar environment giving rise to a dramatic broadening of the stretching mode spectrum. Note that these results are in basic agreement with Rothberg et al.[22] in stating that the CO translates a considerable distance above the glass transition in nanosecond times, presumably due to the protein relaxing a great deal among different conformations not reached at low temperatures.

The 50-nsec time resolution of the data of Hong et al.[25] had the shortcoming that at temperatures greater than 240 K the disappearance of the B state was too rapid (<100 nsec) to observe in their apparatus. Actually, because Hong et al. were able to achieve signal-to-noise values on the order of 100 for the B state, the complete disappearance of the B-state signal at 240 K would seem to argue for a B-state lifetime of about 10–20 nsec at best at room temperature. Does this correspond to an expected protein dynamics relaxation time? No one has any direct measurement of this number, but Steinbach et al.[26] would argue that the relaxation seen here is due to protein relaxation and not ligand diffusion. Therefore further clarification is needed.

Hemoglobin

The experiments cited above concerned themselves with a relatively simple protein that lacks one fundamental aspect of biological action, namely the nonlinear response of the protein to multiple ligand binding. Structural rearrangement of proteins caused by ligand binding and subsequent alteration in biological activity is a key aspect of protein function. We have seen, however, from the transient IR work that the low-temperature distribution of substates probably does not represent the full spectrum of conformations that the protein can reach at room temperatures. We can now ask an even more difficult question. Suppose that the protein has a bimodal distribution of substates: how fast can the transition between a bimodal distribution of substates occur?

The protein hemoglobin is composed effectively of four Mb subunits and carries oxygen in the bloodstream. The thermodynamic cooperativity

[26] P. J. Steinbach, A. Ansari, J. Berendzen, D. Braunstein, K. Chu, B. R. Cowen, D. Ehrenstein, H. Frauenfelder, J. B. Johnson, D. C. Lamb, S. Luck, J. R. Mourant, G. U. Nienhaus, P. Omas, R. Philipp, A. Xie, and R. D. Young, Biochemistry 30, 3988 (1991).

between the subunits, the fact that the fully liganded R-state conformation lies lower in free energy than the fully liganded T state, is critical to hemoglobin function and adds additional richness to the protein dynamics. We do know that at the simplest level we can view the cooperativity of the protein as originating between switching of the Hb structure from a T state (for "tense") in the deoxy configuration to an R state (for "relaxed") in the four-ligand configuration. The binding constant of the protein for oxygen changes by a factor of 300 in these two states. The corresponding free energy change ΔG must then be:

$$\Delta G = kT \ln(300) = 0.14 \text{ eV} \qquad (6)$$

Studies on hemoglobin (Hb) have been the domain of many physical chemists and physicists. Historically, the major problem is to find out where the free energy is stored in the known R and T structures of hemoglobin.[27,28] A great deal of kinetics on the recombination of ligands with Hb has been done, presumably in the hope that somehow the dynamics of the recombination will lead us to insight into how structure dictates function. However, the recombination kinetics of Hb even at room temperature are very complicated.[29] Because of this additional complexity of Hb, most of the cryogenic recombination work has focused on myoglobin.

The most exciting possibilities for IR work in Hb would be (1) to map out the regions of the structure that are perturbed in the conformational relaxation and (2) to determine accurately the rate at which these conformational changes occur. This knowledge, coupled with advances in our understanding of the crucial role of the solvent (water) in the thermodynamics of the conformational equilibrium,[30] might allow us ultimately to say that the hemoglobin problem has been "solved." However, we are still far from that mark. Within the scope of this chapter we confine ourselves to picosecond direct IR work on hemoglobin. The reader will find that the work has not yet addressed some of the major issues.

The most complete work once again comes from Hochstrasser's group,[31] which used the infrared up-conversion gating technique previously described to study both the bleaching of the A bands and the absorbance of the B bands of HbCO with subpicosecond resolution. The solvent used was buffered D_2O, which has somewhat better transmission

[27] E. Antonini and M. Brunori, "Hemoglobin and Myoglobin in Their Reactions with Ligands." North-Holland Publ., New York, 1971.

[28] L. Stryer, "Biochemistry." Freeman, New York, 1988.

[29] L. P. Murray, J. Hofrichter, E. R. Henry, and W. A. Eaton, *Biophys. Chem.* **29**, 63 (1988).

[30] M. F. Colombo, D. C. Rau, and V. A. Parsegian, *Science* **256**, 655 (1992).

[31] P. A. Anfinrud, C. Han, and R. M. Hochstrasser, *Proc. Natl. Acad. Sci. U.S.A.* **86**, 8387 (1989).

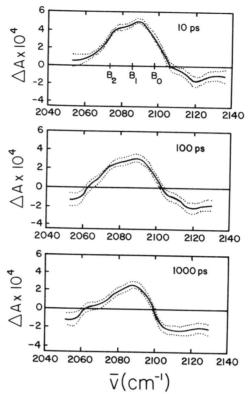

FIG. 5. Transient IR absorbance spectra in the free ^{13}CO region. The authors used ^{13}CO in these experiments. The dotted curves span 1 standard deviation on either side of the mean. The data were smoothed. The B_n are as described in text. (Taken from Anfinrud et al.[31])

than H_2O around 5 μm. Measurements of the absorption recovery of the 1951-cm^{-1} heme-bound CO band after photolysis yield a geminate recombination of around 3% of the ligands in the first nanosecond, consistent with transient electronic spectroscopy.[29] The infrared work would appear to be conclusive in this regard and is, in general, much more reliable than visible absorption spectroscopy simply because one knows precisely what molecular species and motion one is observing.

This group also observed transient absorption in the B state nearly free CO stretching region. The transient spectra thus obtained are smoothed with a 10-cm^{-1} filter and are presented in Fig. 5. There are several things to point out. First, the spectra are labeled $B_{0,1,2}$ with the corresponding absorptions from difference FT-IR measurements in low-temperature glycerol–water glasses for myoglobin. The correspondence between myoglobin and hemoglobin may, however, be fortuitous, and the high-temperature metastable location may be distinct from the low-temperature sites in Mb. Second, Anfinrud et al.[31] show that the loss of the A-band absorption appears faster than their instrumental resolution of 300 fsec. Calculations indicate that the ligand could travel ballistically (not diffusively) about 5 Å in this time, so the conclusion is that the ligand after photolysis resides no further than this distance from the iron. Third, one can see from Fig. 5 that there is a background rising on the tens of picoseconds time scale. This corresponds to shifts in the water spectrum from heating of the solvent around the protein, and the 10-psec response time of the water was interpreted as the time it takes "heat" to flow to the protein surface.

It is interesting to consider what happens to the absorbed energy in the protein once it has thermalized into the heme group. Two things will happen: acoustic transmission of energy and thermal diffusion due to thermodynamic heating of all the degrees of freedom. Imagine that the picosecond pulse results in a point source of heated material due to an absorbed photon. Because the absorption event is prompt, we can view the absorption event as giving rise first to the generation of an acoustic wave propagating out from the source. The attenuation coefficient $\alpha_{thermal}$ for an acoustic wave due to thermal diffusion is[32]

$$\alpha_{thermal} = \left(\frac{\gamma - 1}{\gamma}\right)\left(\frac{\omega^2 \kappa}{2\rho c^3 c_v}\right) \tag{7}$$

where γ is the ratio of specific heat at constant pressure to constant temperature, ω is the frequency of the wave, κ is the thermal conductivity, ρ is the density of the fluid, c_v is the specific heat per gram at constant volume, and c is the speed of sound in the medium. If the attenuation length is set to the diameter of the protein, and we assume that a protein macromolecule is approximately like water in its physical properties, we determine that the maximum frequency ω_{max} at which we can expect acoustic transmission of energy is roughly 2×10^{12} Hz. Thus, acoustic (underdamped) transmission occurs on the subpicosecond time scale. It is interesting to note that the acoustic wave will be phase coherent over the protein, and that such kinds of phase-coherent vibrational influences have been seen in reaction centers.[33]

[32] H. Anderson, ed., "Physics Vade Mecum." Am. Inst. Phys., New York, 1989.
[33] M. H. Vos, J.-C. Lambry, S. J. Robles, D. C. Youvan, J. Breton, and J. Martin, *Proc. Natl. Acad. Sci. U.S.A.* **88**, 8885 (1991).

Thermal diffusion is a slower, incoherent process, and again it is difficult to make anything more than a rough guess as to the relaxation rate. It is known from Raman scattering that the heme group cools in several picoseconds after transient heating by an absorbed photon,[34] giving rise to a transiently heated protein. This excess energy can then be transmitted via thermal diffusion to the surrounding water bath. If we assume that the protein can be modeled by a sphere of radius R, specific thermal conductivity κ, density ρ, and specific heat at constant volume c_v, then there is an analytical power-series solution to the cooling of the sphere.[35] The lowest order term with the largest amplitude has a time constant τ_1 given by

$$\tau_1 = R^2 \rho c_v / \kappa \pi^2 \tag{8}$$

Again, assuming that the protein has a radius of about 30 Å, density ρ, specific heat c_v, and specific thermal conductivity κ about the same as water (6×10^{-3} J/cm-K-sec), we determine that the characteristic cooling time is about 5 psec. The molecular dynamics simulations of Henry et al.[34] give nonexponential solutions to the cooling of the heme group with the characteristic time for the largest amplitude process on the order of about 4 psec and a long-time tail extending out to 20 psec or so, so our assumption appears fairly accurate.

At the same time as Anfinrud et al.[31] did their work, observation of a metastable B state in HbCO/D_2O was reported by Rothberg et al.[36] The results of that work are substantially the same as above, with one exception. These authors reported an additional transient absorption with 2-psec duration at around 2010 cm^{-1}, which they assigned to CO still bound to the heme but in an excited electronic state, HbCO*. This would appear to be at odds with most other work in the field; it is very unusual for an organometallic photodissociation reaction to have a barrier that would cause such a delay. It is possible that this band is due to a two-photon (infrared plus visible) absorption resonance and is not associated with an excited state. One must be careful in interpreting results at very short delays, where pump-and-probe pulses are temporally overlapped in the sample.

Neither of these experiments is able to address the much more important issue of the conformational relaxation of the quaternary structure of the Hb molecule from the T to the R state after photolysis. The structural

[34] E. R. Henry, W. A. Eaton, and R. M. Hochstrasser, *Proc. Natl. Acad. Sci. U.S.A.* **83**, 8982 (1986).

[35] V. S. Arpact, "Conduction Heat Transfer." Addison-Wesley, Reading, MA, 1966.

[36] L. Rothberg, T. Jedju, and R. H. Austin, *Biophys. J.* **57**, 369 (1990).

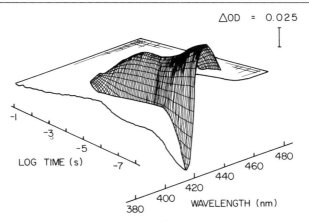

ΔOD = 0.025

LOG TIME (s)

WAVELENGTH (nm)

FIG. 6. Photoproduct minus HbCO difference spectra following 22% photolysis. Difference optical densities are shown at 1-nm intervals for spectra measured at 89 time delays between firing the photolysis and probe lasers. Spectra are shown after smoothing and removal of baseline offsets. (Reprinted with permission from Jones et al.[37] Copyright 1992 American Chemical Society.)

relaxations of the T to the R state are believed to occur in the 10- to 20-μsec time domain, whereas the IR pulse–probe techniques of both Hochstrasser and Rothberg are carried out in a 0- to 10-nsec regime. We are looking literally where the light is the brightest and not where the reaction is occurring.

For example, consider the optical work of Eaton and colleagues on the transient spectroscopy of hemoglobin after photolysis at room temperature.[37] They have utilized an electronically delayed pulse/probe technique, using broadband probe light to map the spectral changes that occur in the heme group as a function of time after photolysis from nanoseconds to milliseconds. Figure 6 is reproduced from one of their papers.[37] It is less important to consider the details here than it is to realize that there are significant changes occurring in the electronic spectrum of the heme group over many nanoseconds in time, unlike the "simple" molecule myoglobin. Presumably these electronic changes are due to changes in the heme environment, and hence the hemoglobin protein conformation. Transient IR spectroscopy could be used to map these changes.

Use of Infrared Spectroscopy in Protein Dynamics

The previous section leads naturally to the subject of this section: how can we exploit IR spectroscopy to study the dynamics of the protein? We

[37] C. M. Jones, A. Ansari, E. R. Henry, G. W. Christoph, J. Hofrichter, and W. A. Eaton, Biochemistry 31, 6692 (1992).

have seen from the above experiments that the CO experiments at the picosecond scale basically answer questions such as orientation of the CO and the immediate state of the CO after photolysis that do not really address broader issues of conformational relaxation of the protein. The nanosecond IR experiments of Hong et al.[25] tried to use the CO B-state absorbance as a probe of protein dynamics, but at the time of those experiments they were not able to extend the time domain to the nanosecond time scale necessary for room-temperature (active protein) work. In addition, the B-state absorbance is not well understood in the sense that we do not know how to calculate its width and oscillator strength as a function of position within myoglobin or hemoglobin, so that these experiments will probably be a good source for arguments among researchers, but little knowledge will be gained.

We wish now to discuss briefly some experiments that have tried to address questions of protein dynamics using IR techniques. There is a hint of this kind of work in the picosecond B-state dynamics presented by Anfinrud et al.[31] that we discussed earlier. They observed energy flow to the water molecules in about 20 psec, and we know that energy is stored in the protein. How could transient IR spectroscopy quantify these events?

Raman spectroscopy has traditionally offered a route to viewing protein dynamics while avoiding the water absorbance problem, but is hampered by a poor signal-to-noise ratio. Typically resonance Raman scattering has been used to enhance the signal, but the act of using resonance techniques limits the view to the absorbing group, typically the heme group.

Friedman and colleagues have been concerned for a number of years with the relationship between protein structure and reactivity, particularly in hemoglobin. In one paper utilizing resonance Raman techniques[38] they were able to show that there was a correlation between the iron–proximal histidine stretch frequency and the R–T state, indicating that there was significant strain in this bond that correlated with the conformation of the molecule. The idea that the energy difference between the R and the T states is stored both in a localized mode, such as the iron–proximal histidine bond, and delocalized over the protein structure has had a long history.

The initial idea of Perutz was of energy storage in salt bridges,[28] and Hopfield suggested delocalized strain[39] over the entire molecule. Then came the elaboration of Agmon and Hopfield to include structural flow of the strain in the protein,[40] and finally we have the synthesis of the views

---

[38] E. W. Findsen, J. M. Friedman, M. R. Ondrias, and S. R. Simon, *Science* **229**, 661 (1985).
[39] R. G. Shulman, J. J. Hopfield, and S. Ogawa, *Q. Rev. Biophys.* **8**, 325 (1975).
[40] N. Agmon and J. J. Hopfield, *J. Chem. Phys.* **79**, 2042 (1983).

by Sage and Champion involving both local strain and protein strain.[41] In all of these models, what is always missing is a direct experimental measurement of time-dependent displacement in the protein itself.

However, progress has been made in measurements of protein motion in hemoglobin between the R–T states, using ultraviolet (UV) Raman techniques.[42,43] Consider the results from the Spiro laboratory,[43] using transient UV Raman to observe the aromatic tryptophan and tyrosine groups in hemoglobin. These experiments indicate that there are both "prompt" changes in certain tryptophan residues remote from the heme at times less than 4 nsec after the photolysis event[43] and later evolution of the structure over the microsecond time scale. Miller and colleagues have utilized transient grating spectroscopy[44] to show that there are sub-nanosecond volume changes occurring in hemoglobin (and myoglobin) after photolysis.

Dryer et al.[45] at Los Alamos, New Mexico have begun a series of experiments using picosecond IR spectroscopy to probe the backbone dynamics of myoglobin after photolysis. We use the term *backbone* to refer to the α-carbon chain to which the amino acids are attached. Infrared studies of the backbone dynamics of proteins, such as the α helix, would be of great interest in probing fundamental modes of conformational relaxation. The amide I stretch band at 1650 cm^{-1} is a critical region for these kinds of experiments, because it participates in the hydrogen bonding that holds the α helix together. Raman work is of little use here because there is no resonance enhancement that could be used to pick out the amide transition.

A written description of the work of Dryer et al.[45] may be of some use, because we believe it strikes at the heart of the potential of picosecond IR spectroscopy. They have used mixing of the 1.06-μm light from an oscillator with the output from a dye laser pumped by doubled 0.532-μm light to generate IR pulses of a few picoseconds in duration at 1650 cm^{-1}. The 0.532-μm doubled light is used to photolyze the CO, and standard pump/probe delay lines with direct IR detectors are used to probe the absorbance change at the amide I versus time over time ranges of 0 to 100 psec. Surprisingly, there is a change in the amide I absorbance seen versus time, the absolute change in absorbance being approximately the same as is seen at the CO A-state bleach with photolysis, although the

[41] P. Li, J. T. Sage, and P. M. Champion, *J. Chem. Phys.* **97,** 3214 (1992).

[42] S. A. Asher and C. R. Johnson, *Rev. Sci. Instrum.* **54,** 1657 (1983).

[43] K. R. Rodgers, C. Su, S. Subramaniam, and T. G. Spiro, *J. Am. Chem. Soc.* **114,** 3697 (1992).

[44] L. Genberg, L. Richard, G. McLendon, and R. J. D. Miller, *Science* **251,** 1051 (1991).

[45] B. Dryer et al., unpublished observations (1993).

fractional change seen in the amide band is about 1/100 the A-state bleach. The time response of the amide I change at room temperature is somewhere in the ranges of tens of picoseconds. The point is that the technology is available to observe changes in protein structure, using IR spectroscopy.

Of course, the amide I stretch is a difficult area to study because the relatively infamous Davydov soliton was assumed to be connected to the amide I stretch.[46] The problem with the Davydov soliton, which was proposed to be the carrier of mechanical energy in a protein, is that there are 1000 theory papers for every experimental paper. Davydov solitons are inherently picosecond-lived phenomena and must be studied on the picosecond time scale, and probably in the IR domain.

The only picosecond IR experiment done to test for anomalous vibrational energy relaxation in hydrogen-bonded chains that we are aware of was done by Fann et al.,[47] using a free-electron laser. This was a primitive experiment, which simply looked at saturation recovery of an anomalous low-temperature side band that occurs in acetanilide (a molecule that resembles amino acids and forms a hydrogen-bonded molecular crystal). Although this experiment found that the saturation recovery time of the anomalous side band is short (approximately 20 psec) at low temperatures, it would be misleading to think that this experiment in any way is a definitive test of soliton models. For example, if the speed of the soliton is a standard value of 1000 m/sec (10 Å/psec) then in 20 psec energy could travel 200 Å, an interesting number biologically. Much more work needs to be done in this area if we are to take the theorists seriously.

Likewise, there is a great deal of theoretical speculation that low-frequency collective modes in proteins might play a role in the conformational dynamics of proteins.[48] The high-frequency IR modes discussed above are essentially localized over a few hydrogen bonds. One can also think of large-scale collective modes in which the energy is spread over many atoms, rather like Hopfield's model of how the R–T hemoglobin difference energy is stored over a sum of small displacements in the protein.[39] Because of the large effective mass of such a collective mode, its frequency can be low (tens to hundreds of wavenumbers)[49] in the far-infrared (FIR) part of the electromagnetic spectrum. The problem of a surfeit of theory simulations is not quite as severe in this field as for the

[46] P. L. Christiansen and A. C. Scott, eds., "Davydov's Soliton Revisited." Plenum, New York, 1990.

[47] W. Fann, L. Rothberg, M. Roberson, S. Benson, J. Madey, S. Etemad, and R. Austin, *Phys. Rev. Lett.* **64**, 607 (1990).

[48] B. Bialek and R. Goldstein, *in* "Protein Structure: Molecular and Electronic Reactivity" (R. Austin et al., eds.), p. 187, Springer Verlag, 1987.

[49] R. H. Austin, M. K. Hong, C. Moser, and J. Plombon, *Chem. Phys.* **158**, 473 (1991).

soliton theory, but it is still fair to say that there has been almost no picosecond relaxation work done in the far-infrared portion of the spectrum and only a few experimental simulations.[50] Some work has been done to characterize the FIR spectrum and some experiments have attempted to see if the excitation in the FIR could possibly change reaction rates,[49,51] but outside of these few primitive experiments the field is still open. Possibly, some of the answers to how proteins transport and store energy may lie in the detailed chemical pictures, but hopefully more general pictures will emerge.

Conclusions

Picosecond transient IR absorption work in proteins is still in its infancy. At present, because laser technology is costly and technically difficult, the biological insight gained from these experiments is small. Although a fair amount is known about myoglobin structural relaxation, the basic steps involved in hemoglobin structural relaxation are not well known. Time-resolved UV Raman work seems to have been the most productive, in spite of the promise of IR spectroscopy. However, we hope that this chapter might entice scientists interested in the conformational dynamics of protein function to work with laser technologists to address this issue in a meaningful manner.

Acknowledgments

We would like to thank Dr. Kim Vandegriff for inviting us to write this paper. R.H.A. would like to thank Dr. Bill Eaton for useful discussions and Professor Hans Frauenfelder for going over the manuscript exhaustively and pointing out many grammatical and conceptual errors. The literature is so vast that we have attempted to cite only representative works.

[50] A. Garcia, *Phys. Rev. Lett.* **68**, 2696 (1992).
[51] R. H. Austin, M. W. Roberson, and P. Mansky, *Phys. Rev. Lett.* **62**, 1912 (1989).

[11] Time-Resolved Resonance Raman Spectroscopy as Probe of Structure, Dynamics, and Reactivity in Hemoglobin

By JOEL M. FRIEDMAN

Introduction

This chapter focuses on the use of time-resolved resonance Raman scattering as a tool to address specific questions pertaining to how hemoglobin functions. The chapter is divided into two sections: the first deals with the technology connected with this form of spectroscopy and the second with how this technology is being applied to specific hemoglobin-related problems.

Experimental Procedures

Laser Technology

The current questions being addressed with regard to how hemoglobin functions require that phenomena occurring over a wide range of time scales be probed. This multiplicity of time scales requires the use of a variety of different laser systems, making it difficult for any one investigator to cover the full range of temporal phenomena in hemoglobin. In addition, for each temporal regime there is often a requirement that the excitation or probe wavelengths be variable or tunable. The tunability requirement typically originates from the wavelength dependence of the resonance Raman cross-section. The Raman cross-section for a given chromophore is often dramatically enhanced when the excitation wavelength is tuned to a strong absorption of the chromophore. In addition, particular Raman spectral peaks associated with a given chromophore may be enhanced selectively by tuning to a particular absorption region of that chromophore.

On the fastest time scales, one is limited by the uncertainty principle. Raman spectra generated using pulses shorter than 1 or 2 psec display a broadening in energy that is a direct result of the temporal bounds imposed by the excitation pulse. Because many of the Raman bands associated with hemoglobin and other biological materials are relatively broad to start with (10 to 15 cm^{-1} full width at half-height), it is possible to use pulses on the order of 0.8 to 1 psec without losing spectral detail pertaining to many key features. To generate spectra with pulses on this time scale requires a balance between pulse energy and average power. To obtain a reasonable level of signal to noise typically requires at least a few milliwatts of average power. To initiate a sufficient level of photolysis with each pulse typically requires at least a few microjoules of energy per pulse. For these short pulses, high-pulse energies mean a very high concentration of photons per short time interval. This high density of photons can lead to nonlinear optical phenomena and sample destruction. The combination of repetition rate with pulse energy is a means of generating the appropriate compromise between pulse energy and average power. Until recently the obvious choice for an appropriate laser system almost always started with a mode-locked continuous wave (cw) Nd : YAG (neodymium : yttrium–aluminum–garnet) laser. This laser puts out 100-psec pulses at megahertz repetition rates. This laser is then used to synchronously pump single or dual jet dye lasers resulting in an output of tunable nanojoule pulses in the 0.2- to 5-psec regime. To use these pulses as a means of initiating photolysis requires amplification of the pulse energy.

Amplification of the output of the synchronously pumped dye laser can be achieved using a variety of different laser sources to pump a series of dye amplifier cells. Low repetition rate–high pulse energy nanosecond Q-switched YAG lasers have been used to amplify ultrashort pulses up to the range of tens of millijoules. This and other nanosecond amplifier lasers have the drawback of producing amplified stimulated emission (ASE), which can contribute a broad intense background to the Raman spectrum that dramatically reduces the signal to noise. Use of subnanosecond amplification pulses reduces the ASE problem. Reasonable success has been achieved using low and high repetition rate YAG regenerative amplifier lasers. These systems can go as high as several kilohertz with energies on the order of tens to hundreds of microjoules for the amplified pulses.

A tunable solid state system based on titanium-doped sapphire is now available. Passively mode-locked Ti : sapphire lasers produce nanojoule pulses at ~100 MHz in the tens of femtoseconds to several picosecond regime. These pulses are tunable over a region that covers 700 to 1000 nm. Solid state-based amplification schemes are actively being pursued by several companies and groups. Frequency doubling, tripling, and quadrupling of these short pulses should provide a wide range of wavelengths that complement the green-to-red range that is most directly obtained using the YAG-based synchronously pumped dye laser systems.

The most straightforward, relatively short-pulse system appropriate for many hemoglobin studies is the high-power active–passive mode-locked Nd : YAG laser. This 10- to 30-Hz system can be used to generate tens of millijoules in the second harmonic at 532 nm. The second harmonic can in turn be used to generate up to a few millijoules of Raman-shifted blue light at 436 nm by being passed through a high-pressure hydrogen cell. The green light can be used to photodissociate a sample and the blue can be delayed variably as a resonance Raman probe pulse for the heme.

Pulse–probe studies requiring time resolution of nanoseconds or longer can be conducted using a variety of laser combinations involving either Q-switched Nd : YAG or excimer lasers to pump either dye lasers or Raman shift cells. This approach allows for the generation of tunable laser excitations from ~200 to 800 nm. Time-resolved resonance Raman studies in the ultraviolet (UV) require higher repetition rates and lower pulse energies than for the heme studies conducted in the visible. This requirement stems from the relative ease with which the aromatic amino acids undergo photoinduced processes that interfere with or modify the normal Raman spectra. The availability of high repetition rate excimer and YAG-based systems is proving to be a great boon for these studies because they allow for high average powers with low pulse energies. Intracavity doubling of either cw argon ion lasers or ring dye lasers lasing in the blue may eventually yield cw UV sources. Ultraviolet resonance Raman spectra (UV-RR) of tryptophan using ~1 mW of quasi-cw UV have been generated using the frequency-doubled output of a blue, synchronously pumped dye laser that was pumped with the tripled output of a mode-locked cw YAG laser (S. Courtney and J. Friedman, unpublished results, 1989).

Detectors

Multichannel detectors have replaced photomultiplier tubes as the detectors of choice for most time-resolved Raman studies. These detectors allow for monitoring an entire segment of the Raman spectrum at a single setting of a spectrograph. Intensified diode array detectors can be gated on the 5-nsec and slower time scale. The gating capability allows for discrimination between two temporally distinct signals, including Raman or luminescence generated from the excitation pulse and the probe pulse. It appears that charge-coupled devices (CCDs) are being used more and more as a replacement to diode array detectors. The CCD's are two dimensional in the sense that multiple full spectra can be obtained at once. Thus a reference as well as multiple samples or delays can be accumulated simultaneously. The diode array detectors are sensitive from ~200 to 900 nm, whereas the CCD, which can be UV enhanced to 200 nm, has coverage to 1000 nm but with much higher quantum efficiency than any diode array. There are now spectrographs available that allow for accurate imaging of the entrance slit into a CCD. This feature enhances the use of CCDs as a means of simultaneously generating Raman spectra from different, spatially distinct regions of a sample (e.g., along the length of a flowing capillary cell).

Structure, Function, and Dynamics

Introduction

A full understanding of cooperative ligand binding in hemoglobin (Hb) requires a detailed account both of how the quaternary structure of the protein influences functionally significant structural elements at the heme and concomitantly of how the local events associated with ligand binding initiate the global changes associated with the T–R quaternary switch. Resonance Raman (RR) spectroscopy has been used successfully to address several aspects of the above questions. Resonance Raman studies on equilibrium forms of deoxygenated and ligand-bound hemoglobins show which structural degrees of freedom associated with the heme are responsive to the changes in the tertiary and quaternary structure of the protein for a given state of ligation.[1–3] Such studies do not expose the protein-induced changes at the heme that occur on change in ligation state. This limitation follows because the electronic changes that occur on changing ligation state are a significantly greater perturbation on the structure of the heme than are the induced changes in the protein. To expose the influence of ligation-induced changes in the protein on the heme requires that the heme be in the same electronic state and the protein be the only variable. This requirement is analogous to looking for solvent effects on the structure of a solute molecule: one does not change the solute at the same time that one changes the solvent. A convenient and appropriate reference state for the solute, that is, the heme, is the high-spin five-coordinate ferrous heme. This species is of course the heme chromophore found in equilibrium deoxy forms of Hb. It can also be generated by rapid photodissociation of ligand-bound six-coordinate ferrous hemoglobins. If probed on a fast enough time scale, the photogenerated "deoxy" heme is still surrounded by the as yet unrelaxed protein solvent that prior to photodissociation had been the equilibrium "solvent" structure for the starting ligand-bound species. This condition follows because the elec-

[1] J. M. Friedman, M. R. Ondrias, and D. L. Rousseau, *Annu. Rev. Phys. Chem.* **33**, 471 (1982).

[2] T. Kitagawa, *in* "Biological Application of Raman Spectroscopy" (T. G. Spiro, ed.), Vol. 3, p. 97. Wiley, New York, 1988.

[3] R. Schweitzer-Stenner, *Q. Rev. Biophys.* **22**, 381 (1989).

tronic rearrangements associated with photodissociation are orders of magnitude faster than those associated with the ensuing structural relaxation of the globin. Thus an RR comparison of equilibrium and photogenerated transient forms of deoxy hemoglobins in different quaternary states provides a means of deconvoluting which local tertiary changes at the heme are ligand induced and which are purely from differences in the quaternary state, independent of ligation. Such studies have set the stage for time-resolved studies that provide the sequence of structural relaxation events for the different ligation and R–T sensitive degrees of freedom at the heme. In addition, advances in UV-RR spectroscopy allow for similar types of studies of the aromatic amino acids within the globin. The combination of the two types of Raman studies should help expose the communication pathways between those structural elements of the globin that dictate quaternary structure and those at the heme that control ligand reactivity. The following is a progress report describing what has been accomplished to date, and possible future directions with respect to the use of time-resolved RR spectroscopy in the quest for a molecular level understanding of cooperative ligand binding in hemoglobins.

This chapter covers the use of time-resolved resonance Raman spectroscopy as a probe of (1) the physical phenomena that occur on laser excitation of Hb; (2) the sequence of structural events that occur on photodissociation; (3) the relationship between structure, structural dynamics, and reactivity; and (4) geminate recombination.

Heme Relaxation and Heme Heating: First 30 psec

Ultrafast absorption studies show that on visible excitation of HbCO, there is a loss of the ligand-bound spectrum within 100 fsec and the appearance of a deoxy-like heme spectrum within 350 fsec of the photoexcitation.[4] The rapid recovery of a deoxy-like heme species indicates a comparable rate for the nonradiative electronic relaxation of the optically accessed excited π electronic configuration. Indications of such nonradiative rates for the excited states associated with the Q band and Soret band absorptions were inferred from both emission quantum yield measurements and RR excitation profile measurements.[5-7] Time-resolved Raman studies of this subpicosecond process are almost impossible because of the uncertainty broadening of the Raman bands generated with ultrashort

[4] J. W. Petrich and J. L. Martin, *Chem. Phys.* **131,** 31 (1989).
[5] K. T. Schomacker and P. M. Champion, *J. Chem. Phys.* **84,** 5314 (1986).
[6] J. M. Friedman, D. L. Rousseau, and F. Adar, *Proc. Natl. Acad. Sci. U.S.A.* **74,** 2607 (1977); J. M. Friedman and D. L. Rousseau, *Chem. Phys. Lett.* **55,** 488 (1978).
[7] F. M. Adar, M. Gouterman, and S. Aronowitz, *J. Phys. Chem.* **80,** 2184 (1976).

pulses. Impulsive Raman scattering may be a means of overcoming this inherent limitation of making an energy measurement over a subpicosecond interval. Impulsive Raman scattering[8] is a means of probing a vibration state by monitoring in real time the evolution of the vibrational wavepacket created by using an ultrashort excitation. It is in effect the Fourier transform of the cw Raman experiment.

Subsequent to the near-instantaneous photodissociative event, there are several processes besides the nonradiative electronic decay that can both influence spectral evolution and be of significance in modulating the early dynamics of either a spontaneously dissociated or photodissociated ligand. These processes include local heating due to the dissociative event, rapid local structural relaxation, geminate pair formation and decay, and equilibrium fluctuations of the structure. Each of these processes is now considered.

Local Heating. The absorption of a visible photon on electronic excitation deposits ~50 kcal/mol of energy within the heme. In the case of a ligand-bound heme, a fraction of this energy may be dissipated in the dissociation process; however, even in this case there is anticipated to be an excess of energy, which on some time scale can create a nonequilibrium population of heme vibrational levels. Local heating could have a transient effect on thermal fluctuations that as is discussed below, can play a role in modulating geminate recombination.

The results of a classic molecular dynamics simulation indicate that the absorption of visible light can raise the temperature of a deoxyheme by 500–700 K in a solvent-free deoxymyoglobin.[9] The calculations also suggest that ~50% of this excess vibrational energy should be dissipated within a few picoseconds and the remainder within ~50 psec. These predictions are amendable to testing, using the anti-Stokes Raman bands from the heme to probe the population of excited heme vibrational levels.

Two time-resolved RR experiments bear directly on the issue of heat dissipation on the 1- to 10-psec time scale. In a pulse–probe protocol using pulses on the order of 0.8 psec, Petrich *et al.* observed systematic shifts in the ν_4 Stokes Raman band of a partially photodissociated HbCO sample over the first 10 psec subsequent to excitation.[10] These changes were interpreted as reflecting the cooling of the heme over this 10-psec interval. Although this interpretation is both plausible and consistent with

[8] X. Y. Yan, E. B. J. Gamble, and K. A. Nelson, *J. Chem. Phys.* **83,** 5391 (1985).
[9] E. R. Henry, W. A. Eaton, and R. M. Hochstrasser, *Proc. Natl. Acad. Sci. U.S.A.* **83,** 8982 (1986).
[10] J. W. Petrich, J. L. Martin, D. Houde, C. Poyart, and A. Orszag, *Biochemistry* **26,** 7914 (1987); J. L. Martin, A. Migus, C. Poyart, Y. Lecarpentier, R. Astier, and A. Antonetti, *Proc. Natl. Acad. Sci. U.S.A.* **80,** 173 (1983).

the results of an anti-Stokes experiment (*vide infra*), there is an alternative interpretation for the observed spectral shifts that is discussed in the section on conformational disorder. In a pulse–probe anti-Stokes Raman experiment,[11] evidence for a picosecond decay of a non-Boltzmann population of heme vibrational levels was obtained using the third and second harmonics of an Nd : YAG regenerative amplifier. The signal to noise in that experiment was in part limited by the less than optimal resonance conditions associated with the 355-nm probe wavelength.

Evidence for cooling within 30 psec of excitation has been obtained with both one and two pulse protocols using 30-psec probe pulses at 435 nm.[12] In this series of experiments, the low-frequency anti-Stokes Raman spectrum within 30 psec of excitation shows no indication of a non-Boltzmann distribution of intensities. At higher fluence levels of blue light, the ν_4 band and a few other high-frequency Raman bands display enhanced intensities in the anti-Stokes region of the spectrum.[13] High intensities of green light (second harmonic from an active–passive mode-locked Nd : YAG laser) in combination with low-intensity blue probe pulses did not result in any detectable increase in the anti-Stokes spectrum. The observations that it is primarily the resonantly enhanced ν_4 mode that displays the anti-Stokes enhancement, that the low-frequency modes do not seem to be affected, that Soret and not visible excitations are effective in generating the enhanced anti-Stokes Raman bands, and that not all high-frequency modes show the effect all suggest that the effect is not simply laser-induced heating. These observations are not consistent with but do not rule out the possibility that the enhanced anti-Stokes spectrum is due to resonance Raman from some bottleneck excited electronic state whose population is built up via optical pumping.[14] A more consistent explanation for the selective enhancement of the anti-Stokes for ν_4 is that with blue excitation, the ν_4 mode is pumped via a resonantly enhanced stimulated Raman process. Alternatively or in addition, the ν_4 mode could have a substantially slower vibrational relaxation time than other modes.

Geminate Recombination

On dissociation of the iron–ligand bond, the ligand is initially localized near the heme. We term this complex the proximate geminate pair (PGP),

[11] R. Lingle, X. Xu, H. Zhu, S. Yu, and J. B. Hopkins, *J. Am. Chem. Soc.* **113**, 3992 (1991); *J. Phys. Chem.* **95**, 9230 (1991).

[12] R. G. Alden, M. Chavez, M. Ondrias, S. H. Courtney, and J. M. Friedman, *J. Am. Chem. Soc.* **112**, 3241 (1990).

[13] R. G. Alden, M. C. Schneebeck, M. R. Ondrias, S. H. Courtney, and J. M. Friedman, *J. Raman Spectrosc.* **23**, 569 (1992).

[14] P. Li, J. T. Sage, and P. M. Champion, *J. Am. Chem. Soc.* **113**, 3992 (1991).

also known as the contact pair. The PGP can decay either through the reformation of the bound species or through diffusion of the ligand to a more distant locus within the protein. The further separated heme–ligand complex, which still consists of a ligand within the protein, is termed the separated geminate pair (SGP). The SGP can decay either by reforming the PGP or by having the ligand diffuse out of the protein. Obviously there is a distribution of possible or accessible SGPs. Recombination originating from either the PGP or the SGPs is termed geminate recombination (GR).

Geminate recombination plays a pivotal role in understanding how structure controls ligand binding in hemoglobin. This significance stems both from the relative simplicity of the process and from the indications that protein control of ligand binding occurs primarily at the level of the bond-forming (*vide infra*). Geminate recombination can provide a relatively unobstructed view of the bond-forming process. Whereas absorption spectroscopy is typically the probe of choice for accurately monitoring the kinetics associated with GR, RR spectroscopy can provide the details regarding the structure and the structural dynamics that dictate the energetics of the rebinding process. Furthermore, by using the technique of kinetic hole burning, which is discussed in a later section, it is possible to relate spectral features (and as a consequence, the structural determinant of that spectral feature) to the potential energy barrier controlling the rebinding process.

Geminate recombination within Hb has been reported to occur on several time scales under ambient solution phase conditions. Three groups almost concurrently first reported nanosecond GR of CO in Hb: two using transient absorption spectroscopy[15,16] and the other[17] using a pulse–probe nanosecond time-resolved RR protocol. Subsequently, a subnanosecond geminate phase for O_2 but not for CO was observed to occur within a few hundred picoseconds of photodissociation.[18,19] And finally, what appears to be an ultrafast geminate phase for O_2 and NO but not for CO occurs within a few picoseconds of photodissociation.[4,20] The following important

[15] D. A. Duddell, R. J. Morris, and J. T. Richards, *J. Chem. Soc., Chem. Commun.*, p. 75 (1979).

[16] B. Alpert, S. E. Mohsni, L. Lindqvist, and F. Tfibel, *Chem. Phys. Lett.* **64,** 11 (1979).

[17] J. M. Friedman and K. B. Lyons, *Nature (London)* **284,** 570 (1980).

[18] D. A. Chernoff, R. M. Hochstrasser, and W. A. Steele, *Proc. Natl. Acad. Sci. U.S.A.* **77,** 5606 (1980); P. A. Cornelius, W. A. Steele, D. A. Chernoff, and R. M. Hochstrasser, *ibid.* **78,** 7526 (1981).

[19] J. M. Friedman, T. W. Scott, G. Fisanick, S. Simon, E. W. Findsen, M. R. Ondrias, and V. Macdonald, *Science* **229,** 187 (1985).

[20] K. A. Jongeward, D. Magde, D. J. Taube, J. C. Marsters, T. G. Traylor, and V. S. Sharma, *J. Am. Chem. Soc.* **110,** 380 (1988).

questions concerning these geminate phases are being addressed: (1) What is the origin of the three geminate phases? (2) What is the origin of the ligand specificity of these three geminate phases? (3) How does the protein structure influence these processes? (4) To what extent do equilibrium fluctuations and structural relaxation play a role in modulating GR?

Origin of Geminate Phases. There are several possible explanations for the different time scales for the GR observed in Hb at ambient temperatures. It was initially suggested by Friedman and co-workers[19] that the 100-psec and 100-nsec geminate phases correspond to what Frauenfelder and co-workers called, respectively, process I and the matrix process in cryogenic samples.[21] Process I refers to the GR at cryogenic temperatures that originates from the ligand located within the heme pocket; the matrix process, on the other hand, was thought to arise from rebinding from within the bulk protein. There has been experimental evidence[22] indicating that the matrix process is process I occurring subsequent to or concurrent with structural relaxation as originally claimed by Agmon and Hopfield.[23]

Extensive femtosecond and picosecond transient absorption studies have led to an elaboration of the above idea regarding the origin of the geminate phases. The basic idea is still that the fastest recombination arises from proximate geminate pairs in which the ligand is only slightly separated from the heme, and the slower phases arise from separated geminate pairs in which the ligand is at the distal boundaries of the heme pocket.[20] Crystallographic and mutagenic evidence supports the concept of ligands localizing at the perimeter of the distal heme pocket.[24] Thus, in this model and similar models,[25] the different geminate phases are the consequence of a time- and temperature-dependent distribution of ligand–heme spatial separations subsequent to photodissociation.

Martin and co-workers have interpreted the fast 2- to 5-psec recombination and the reduced quantum yield of photodissociation for the oxygen ligand as resulting from a branching ratio in the excited electronic state decay subsequent to photon absorption by the ligand-bound species.[4] One decay pathway leads to the metastable deoxy-like photoproduct, whereas the other leads to reformation of the initial ligand-bound species. This

[21] H. Frauenfelder, F. Parak, and R. D. Young, *Annu. Rev. Biophys. Biophys. Biochem.* **17,** 451 (1986).

[22] G. U. Neihaus, J. R. Mourant, and H. Frauenfelder, *Proc. Natl. Acad. Sci. U.S.A.* **89,** 2902 (1992).

[23] N. Agmon and J. J. Hopfield, *J. Chem. Phys.* **79,** 2042 (1983).

[24] K. A. Johnson, J. S. Olson, and G. N. Phillips, *J. Mol. Biol.* **207,** 459 (1989).

[25] L. P. Murray, J. Hofrichter, E. R. Henry, M. Ikeda-Saito, K. Kitagishi, T. Yonetani, and W. A. Eaton, *Proc. Natl. Acad. Sci. U.S.A.* **85,** 2151 (1988).

description implies that repetitive photoexcitations should pump the system toward complete photodissociation. Photodissociation experiments[26] at cryogenic temperatures using pulses of 30-psec duration indicate that even at liquid helium temperatures, where process I slows down dramatically, the yield of photodissociation cannot be pumped noticeably. Approximately 60% of an HbO_2 sample remains ligand bound under these conditions. This result indicates that a simple branching ratio cannot be operative in determining the initial (<10 psec) ratio of photoproduct to reformed ligand-bound species.

An additional model for the different geminate phases has been proposed,[27] on the basis of the static and dynamic properties of the conformational disorder associated with the protein structure. It is based on the idea of conformational substates developed by Frauenfelder and co-workers,[21] which is discussed in more detail below. The basic idea as it relates to the issue of the different geminate phases is that on photodissociation of an equilibrium population of ligand-bound Hb, a population of photoproducts is generated whose members are energetically similar but with slightly different conformations. Each of these slightly different conformations is defined as a conformational substate. Each conformational substate has a slightly different rate of recombination. Thus, on photodissociation there is created instantaneously a distribution of photoproduct conformational substates.

In the new model, the 2 to 5-psec GR originates from those conformational substates that have no activation barrier for recombination. Calculations by Agmon[23,28] predict the existence of these zero barrier conformational substates. The nanosecond geminate phase is directly linked to process I and is hence determined at least in part by a distribution or an average of activation barriers. The geminate phase occurring over tens to hundreds of picoseconds is attributed to depletion of the photoproduct population due to thermal fluctuations that convert those conformational substates having a low activation barrier to conformational substates having no barrier. This model does not preclude ligand diffusion as an added determinant of the GR kinetics that is of potentially significant status. In addition, there is the added possibility that fast tertiary relaxations occurring over the range of time scales being discussed is causing a time-dependent change in the distribution of activation barriers controlling GR.

[26] M. R. Chance, S. H. Courtney, M. D. Chavez, M. R. Ondrias, and J. M. Friedman, *Biochemistry* **29**, 5537 (1990).

[27] A. M. Ahmed, B. F. Campbell, D. Caruso, M. R. Chance, M. D. Chavez, S. H. Courtney, J. M. Friedman, I. E. T. Iben, M. R. Ondrias, and M. Yang, *Chem. Phys.* **158**, 329 (1991).

[28] N. Agmon, *Jerusalem Symp. Quantum Chem. Biochem.: Tunneling* **19**, 373 (1986).

At ambient temperature it has been shown that it is the escape of the ligand that dictates the actual kinetics of the 100-nsec phase.[29]

Ligand Specificity. There is a substantial ligand dependence for each of the geminate phases. For adult human Hb (Hb A) the extent of GR decrease in the order of NO, O_2m and CO. Whereas both NO[30] and O_2[19] exhibit substantial GR on the subnanosecond time scale, CO is observed to rebind geminately only (>90%) on the nanosecond and longer time scales, except at low pH (<3.2), where even in myoglobin (Mb) the CO recombines on a picosecond time scale.[31]

Protein and Structure Dependence. The yield of GR for a given ligand varies with both different proteins and structure (for a given protein).[19,32] Whereas the yield of GR for Hb A is ~50% in the case of CO, for Mb it is less than 10% under ambient conditions.[17,33] In the case of O_2 the yield for the 100-psec geminate phase decreases with conditions that destabilize the R quaternary state.[19] Similar behavior is observed for CO in the nanosecond phase.[25,34,35] Systematic variation in the geminate yield in Hb is also observed as a function of vertebrate species.[19,32]

Time-Resolved Resonance Raman Scattering as Probe of Geminate Recombination. The resonance Raman spectrum of the heme can be used to monitor the yield and progress of GR. The ν_4 band clearly distinguishes between ferrous forms of five-coordinate (~1353 cm^{-1}) and six-coordinate species (~1375 cm^{-1}) of the kind that typically participate in the most commonly studied cases of GR. The relative sharpness of these well-separated Raman bands greatly facilitates the determination of whether recombination or some other process is occurring, a problem that can pose problems for time-resolved absorption studies. One- and two-pulse time-resolved Raman studies have been used to address several issues pertaining to the extent and time course of GR in Hb.

Using the ratio of the ν_4 bands for five- and six-coordinate heme, it was shown[34,35] that the extent of photodissociation over the duration of a single 10-nsec blue pulse, that functions both as a photolysis pulse and a probe pulse, depends on the tertiary and quaternary structures of a given Hb. The more T-like the structure (as determined by the iron-

[29] W. D. Tian, J. T. Sage, V. Srajer, and P. M. Champion, *Phys. Rev. Lett.* **68,** 408 (1992); P. M. Champion, *J. Raman Spectrosc.* **23,** 557 (1992).

[30] P. A. Cornelius, R. M. Hochstrasser, and W. A. Steel, *J. Mol. Biol.* **163,** 119 (1983).

[31] I. E. T. Iben, B. R. Cowen, R. Sanches, and J. Friedman, *Biophys. J.* **59,** 908 (1991).

[32] J. M. Friedman, S. A. Simon, and T. W. Scott, *Copeia* **3,** 679 (1985).

[33] E. R. Henry, J. H. Sommer, J. Hofrichter, and W. A. Eaton, *J. Mol. Biol.* **166,** 443 (1983).

[34] J. M. Friedman, T. W. Scott, R. A. Stepnoski, M. Ikeda-Saito, and T. Yonetani, *J. Biol. Chem.* **258,** 10564 (1983).

[35] T. W. Scott, J. M. Friedman, and V. W. Macdonald, *J. Am. Chem. Soc.* **106,** 5677 (1985).

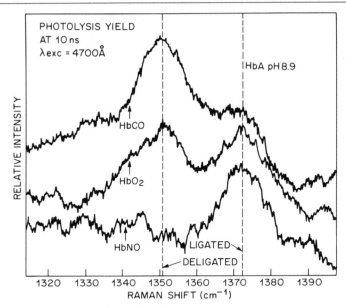

FIG. 1. The difference in the relative intensity of the ν_4 Raman band for five- and six-coordinate Hb A as a function of ligand under excitation conditions sufficient to photodissociate fully the HbCO sample within 10 nsec. All three samples are at same concentration of protein.

proximal histidine stretching frequency), the greater the ratio of five- to six-coordinate heme. The result was interpreted in terms of a convolution of both the ongoing photodissociation and concomitant GR occurring during the 10-nsec pulse. The major variable was assumed to be the yield of GR and not the intrinsic quantum yield for the actual subpicosecond photodissociative event. The ligand-specific differences in "effective" quantum yield at 10 nsec can be seen in Fig. 1. Figure 1 shows the ν_4 region of the spectrum for HbO_2, HbCO, and HbNO under conditions in which the excitation conditions are set to be just sufficient to photodissociate the HbCO sample with a single 10-nsec pulse. It can be seen that under this condition the oxy sample is partially photodissociated and the NO sample is still fully six coordinate. Similarity, ν_4 was used to compare the temperature dependence of the "photolysis yield" at 10 nsec for O_2 and CO derivatives of Hb A.[36] It was observed that as the temperature

[36] M. R. Ondrias, T. W. Scott, J. M. Friedman, and V. W. Macdonald, *Chem. Phys. Lett.* **112,** 351 (1984).

was lowered below 200 K, the yield of photoproduct for CO approached 100%, whereas for O_2 the ratio of photoproduct to ligand-bound heme leveled off at a point substantially below 100%. This result is consistent with subsequent picosecond studies that show that for O_2 the ratio of photoproduct to ligand-bound heme within a few picoseconds of photolysis is ~40% at temperatures below 70K.[26]

Geminate recombination is more directly studied in a pulse–probe protocol, in which an initial pulse photodissociates the sample, and a second delayed pulse monitors the relative population of five- and six-coordinate heme as a function of delay time between the pulses. The difficulty in using Raman for such studies is that the probe pulse needs to be tuned to the Soret band for maximum resonance enhancement and, at the same time, under these conditions of high optical density the pulse energy must be maintained at a level that minimizes rephotolysis of the sample. These conditions have been met in several studies. A pulse–probe protocol using nanosecond pulses was used first to demonstrate the existence of an ~50% and ~0% geminate yield in HbCO A and MbCO, respectively.[17] A similar methodology was used to follow the extent of GR in HbCO A as a function of solvent viscosity.[37] Similarly, pulses of 30-psec duration were used to follow the GR of CO in a cryogenic sample of MbCO at pH 3.[31] At pH 3 the geminate recombination rate increases dramatically.[38]

Similar one- and two-pulse protocols using other Raman bands also have been used to expose other aspects of GR. Using both ν_4 and the coordination-sensitive bands at 1638 and 1648 cm^{-1} as a function of pulse energy for a 10-nsec, 480-nm pulse, it was shown that for T-state HbNO A, the β subunits are more easily photodissociated than the α subunits.[34] The Raman spectrum can also be used to track the GR of subpopulations of CO by monitoring the iron–carbon stretch of the CO-bound heme. It has been shown using both Fourier transform-infrared spectroscopy (FT-IR)[39] and Raman[40] as well as X-ray crystallography[41] that there exist within a given heme protein different conformations for the CO–heme complex that have been termed A states. It has been shown that on photodissociation the different A states of MbCO at cryogenic tempera-

[37] E. W. Findsen, J. M. Friedman, and M. R. Ondrias, *Biochemistry* **27**, 8719 (1988).

[38] J. B. Miers, J. C. Postlewaite, B. R. Cowne, G. R. Roenig, I-Yin Sandy Lee, and D. D. Dlott, *J. Chem. Phys.* **94**, 1825 (1991).

[39] W. S. Caughey, H. Shimada, M. G. Choc, and M. P. Tucker, *Biochemistry* **29**, 4844 (1981).

[40] D. Morikis, P. M. Champion, B. A. Springer, and S. G. Sligar, *Biochemistry* **28**, 4791 (1989); J. Ramsden and T. Spiro, *ibid.*, p. 3125.

[41] J. Kuriyan, S. Wilz, M. Karplus, and G. A. Petsko, *J. Mol. Biol.* **192**, 133 (1986).

tures (≤ 180 K) do not interconvert and have different recombination rates.[42,43] The iron–carbon bond in HbCO A at pH 6 in solution at 300 K has been monitored with nanosecond resolution during the course of the 100-nsec geminate phase.[44] This study has the added complication, beyond the rephotolysis problem, of having a starting population of near zero for the Raman-active species, that is, the rebound CO–hemes. Nonetheless, within the limits of the signal to noise it appears that the initial population of rebound hemes is not the same as the equilibrium population; however, within a few hundred nanoseconds, the rebound population has regained the equilibrium distribution of A states. This result is consistent with a picture in which the A states rebind with different rates and have an interconversion time on the order of tens to hundreds of nanoseconds at ambient temperatures, as proposed from kinetic isolation experiments in MbCO.[29]

Correlation between Raman Spectrum and Geminate Yield. Vertebrate hemoglobins show a rough correlation between the frequency of the iron-proximal histidine Raman stretching frequency and the yield of GR.[19,45,46] In general those photoproducts having the lowest frequencies for this Raman band have the lowest yields for GR. For O_2 derivatives, this correlation is evident both for a large variety of hemoglobins derived from upper and lower vertebrates and for different tertiary and quaternary structures of a given Hb. It is observed that the yield of the 100-psec geminate phase diminishes as the Fe–His frequency of the photoproduct decreases. For Hb A, destabilizing the R-state photoproduct either by dropping the pH and adding inositol hexaphosphate (IHP) or by cross-linking the subunits via lysine $\alpha 99$ causes the Raman frequency to decrease and the yield of the subnanosecond recombination to decrease.[47]

In addition to correlating with the frequency of the Fe–His Raman band, the ligand–heme bond-forming tendency, reflected in the geminate yield, also appears to correlate with the intensity of this Raman band relative to the pure heme-associated bands. For a given frequency, the

[42] M. R. Chance, B. F. Campbell, R. Hoover, and J. M. Friedman, *J. Biol. Chem.* **262,** 6959 (1987).

[43] A. Ansari, J. Berendzen, D. Braunstein, B. R. Cown, H. Frauenfelder, M. K. Hong, I. E. T. Iben, J. B. Johnson, P. Ormos, T. B. Sauke, R. Scholl, A. Schulte, P. J. Steinbach, R. D. Vittitow, and R. D. Young, *Biophys. Chem.* **26,** 337 (1987).

[44] M. C. Schneebeck, L. E. Vigil, J. M. Friedman, M. D. Chavez, and M. R. Ondrias, *Biochemistry* **32,** 1318 (1993).

[45] J. M. Friedman, *Science* **228,** 1273 (1985).

[46] D. L. Rousseau and J. M. Friedman, *in* "Biological Applications of Raman Spectroscopy" (T. G. Spiro, ed.), Vol. 3, p. 133. Wiley, New York, 1988.

[47] R. W. Larsen, M. D. Chavez, M. R. Ondrias, S. H. Courtney, J. M. Friedman, M. J. Lin, and R. Hirsch, *J. Biol. Chem.* **265,** 4449 (1990).

lower the relative intensity of the Fe–His band, the greater are the bond-forming tendencies of the associated structure. This relationship is observed in two clear-cut cases. For MbCO at cryogenic temperatures below 100 K, extended illumination or optical pumping results in a slowing in rebinding kinetics that is attributed to an increase in the activation barrier for part of the distribution of conformational substates.[43] The primary effect of this optical pumping on the structure of the photoproduct population, as reflected in the blue wavelength-excited resonance Raman spectrum, is an increase in the relative intensity of the Fe–His band with continued illumination.[27,46] The other case in which the relative intensity of the Fe–His band appears to correlate with ligand reactivity is in a deep sea fish hemoglobin having both extremely low ligand affinities and marked subunit heterogeneity.[48] One of the subunits has an extraordinarily low O_2 on rate, allowing for the preparation of a half-saturated sample. Using a spinning cell, it was posible to demonstrate that the heme sites with the extraordinarily low affinity for O_2 have an Fe–His band with a frequency comparable to the β subunits in normal deoxyHb A but with the added feature of having an anomalously high relative intensity.

Raman, Geminate Recombination, Conformational Disorder, and Kinetic Hole Burning. The most direct spectroscopic means of relating structure to reactivity in a system such as hemoglobin is through a technique called kinetic hole burning (KHB). This approach is based on several fundamental concepts. The most essential of these concepts is the notion that proteins are conformationally disordered, that is, a given protein does not exist in a single equilibrium conformation; instead there is a distribution of slightly different conformations (conformational substates) having roughly the same average equilibrium structure.[21] Thus for a given equilibrium state of a protein, there is a distribution of conformational substates, each differing slightly in the details of the nuclear coordinates. Furthermore, each of these conformational substates has a slightly different functional property, such as rate of geminate rebinding. At ambient conditions thermal fluctuations cause the conformational substate to interconvert on an as yet undermined time scale that is much faster than the bimolecular combination of heme proteins with ligands originating in the solvent. At cryogenic temperatures below ~180 K, the distribution becomes frozen. Each molecule is locked into a specific conformational substate. In the cases of ligand-bound forms of Hb and Mb, each conformational substate has a well-defined rate constant for geminate recombination, which for the solution results in a distribution of geminate recombination times at a given temperature below 180 K. At the detailed molecular

[48] J. M. Friedman, B. F. Campbell, and R. W. Noble, *Biophys. Chem.* **37**, 43 (1990).

level of the conformational substates, the structure–function issues now focus conformational substates on which conformational substates-dependent parameter of structure correlates with the conformation substates-dependent variation in recombination rates. Thus both functionality and structure are distributed over the conformational substates. The research goal now becomes how to identify those distributed elements of structure that dictate the functional distribution. Kinetic hole burning provides a means of connecting these two distributed variables.

A given spectral band may have a spectral parameter that varies with each conformational substates. The spectral band may then reflect the conformational disorder associated with the protein solution. In the case in which each conformational substates has a slightly different absorption wavelength, an absorption band may be inhomogeneously broadened as a result of the distribution of conformational substates. If a spectral band associated with the photoproduct of ligand-bound Hb or Mb is inhomogeneously broadened by a functionally important structural variable, then as the ligand rebinds, the contribution to the spectral band originating from those conformational substates having the fastest recombination times will disappear first, causing an asymmetry or "hole" in the overall decreasing spectral band.

Kinetic hole burning has been observed in cryogenic samples of CO- and O_2-bound Hb and Mb.[49–53] It is observed that subsequent to photodissociation, as the ligand recombines, the red edge disappears first for the photoproduct charge transfer absorption band at ~760 nm (band lll). This result has been interpreted to mean that those conformational substates having the reddest band llls undergo the fastest geminate recombination. This relationship has been made quantitative by generating a direct linear relationship between the wavelength of band lll and the activation energies controlling the geminate process (process I) for the different conformational substates.[50,52,53] An important question is what structural parameter is responsible for the distribution of functionally linked absorption wavelengths. Two categories of time-resolved resonance Raman studies are being used to address this and related questions. The first involves correlat-

[49] B. F. Campbell, M. R. Chance, and J. M. Friedman, *Science* **238**, 373 (1987); D. A. Case and M. Karplus, *J. Mol. Biol.* **132**, 343 (1979).

[50] N. Agmon, *Biochemistry* **27**, 3507 (1988).

[51] M. D. Chavez, S. H. Courtney, M. R. Chance, D. Kuila, J. Nocek, B. M. Hoffman, J. M. Friedman, and M. R. Ondrias, *Biochemistry* **28**, 4844 (1990).

[52] V. Srajer and P. M. Champion, *Biochemistry* **30**, 7390 (1991).

[53] P. J. Steinbach, A. Ansari, J. Berendzen, D. Braunstein, K. Chu, B. R. Cowen, D. Ehrenstein, H. Frauenfelder, J. B. Johnson, D. C. Lamb, S. Luck, J. R. Mourant, G. U. Nienhaus, P. Ormos, R. Philipp, A. Xie, and R. D. Young, *Biochemistry* **30**, 3988 (1991).

ing the changes in the wavelength of band III with changes in the structurally more explicit Raman spectrum.[51] The second approach is to conduct KHB studies directly, using the Raman spectrum.[54] Both approaches are discussed below.

The wavelength of band III has been correlated to the displacement of the iron from the heme plane.[49–53] The bluest and reddest wavelengths correspond to the most out-of-plane and in-plane configurations, respectively. This relationship is derived in part from the iron–porphyrin charge transfer character of the electronic transition associated with this absorption band.[55] The key question is what distributed element of the protein structure gives rise to this distribution of displacements. It has been suggested that the distribution is primarily the result of the iron being forced out of the heme plane as a result of the low- to high-spin transition occurring on photodissociation.[52] It has been argued strongly both from theoretical and experimental grounds that the high-spin iron can be accommodated in the heme plane and consequently the displacement is a result of protein-induced factors.[56] As is discussed in the next section, it appears that the protein-modulated repulsive interaction between the heme and the proximal histidine plays a major role in the displacement process.

There have been several studies that relate the behavior of the Raman spectrum to that of band III. In particular both time-resolved comparisons at ambient conditions[57] and cryogenic comparisons[51] show that tertiary or quaternary structural changes that are accompanied by a change in the iron–His stretching frequency are also associated with a systematic change in the wavelength of band III. Structural changes that increase the iron–His frequency typically result in a red shift in the wavelength of band III. More recently it has been shown that on optically pumping the photoproduct of MbCO at temperatures below 40 K, the wavelength of band III shifts to the blue,[43,53] and the relative intensity of the iron–His band increases relative to the pure heme modes.[27,58] On the basis of these results, a model has been put forth relating structural parameters connected with both the intensity and the frequency of the iron–His band to both the wavelength of band III and the KHB properties of band III.[27,51] The protein-induced variations in the tilt of the proximal histidine (the imidazole ring), which appears to be intimately coupled to the quaternary structure, is associated

[54] J. M. Friedman and B. F. Campbell, *Spectroscopy* **1**, 34 (1986).

[55] W. A. Eaton and J. Hofrichter, this series, Vol. 76, p. 175.

[56] W. A. Goddard and B. D. Olafson, in "Biochemical and Clinical Aspects of Oxygen" (W. S. Caughey, ed.), p. 87. Academic Press, New York, 1979; B. D. Olafson and W. A. Goddard, *Proc. Natl. Acad. Sci. U.S.A.* **74**, 1315 (1977).

[57] M. Sassaroli and D. L. Rousseau, *Biochemistry* **26**, 3092 (1987).

[58] M. Sassoroli, S. Dasgupta, and D. L. Rousseau, *J. Biol. Chem.* **261**, 13704 (1986).

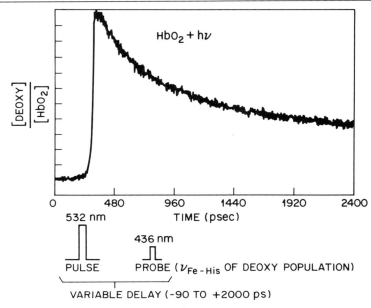

FIG. 2. Schematic for a picosecond Raman kinetic hole-burning experiment for HbO$_2$.

with structure-induced changes in the frequency of the iron–His stretching mode. The induced changes in the relative intensity of the iron–His Raman band are attributed to variations in the azimuthal angle of the imidazole ring with respect to the iron–pyrrole–nitrogen axis. Both of these protein-controlled structural parameters are important in determining the repulsive interaction that dictates the iron displacement. In this model, it is the repulsive interaction that determines both the iron displacement (and hence the wavelength of band III) and the bulk of the protein contribution to the activation energy controlling geminate rebinding at the level of the bond-forming step. Support for this model has come from Raman dispersion studies that separate different symmetry-breaking perturbations on the heme.[59]

Preliminary KHB studies using the iron–His Raman band are being pursued. The difficulties lie in maintaining a probe pulse that does not rephotolyze the sample, finding a fast enough geminate process at ambient temperatures, and dealing with slow sample recovery times at cryogenic temperatures. The 100-psec geminate recombination of O$_2$ in Hb A at room temperature was used in an attempt to detect KHB in the iron–His band[54] (Figs. 2–4). A green photolysis pulse–weak blue probe pulse proto-

[59] R. Schweitzer-Stenner and W. Dreybrodt, *J. Raman Spectrosc.* **23,** 529 (1992).

PULSE (532 nm) – PROBE (435 nm)
HbO$_2$

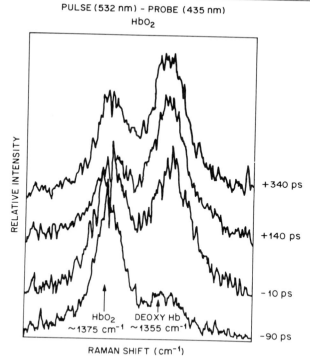

FIG. 3. Time evolution of the ν_4 Raman band prior to and subsequent to photodissociation of HbO$_2$ by a 30-ps laser pulse. The probe pulse (30 psec) is sufficiently weak so as not to photodissociate the sample as seen from the −90-psec spectrum.

col with 30 psec was used as shown in Fig. 2. An essential component of the experiment is to have a probe that does not rephotodissociate the sample. Figure 3 shows the evolution of ν_4 from an unphotodissociated sample at $t = -90$ psec through 340 psec after photolysis. The evolution of the iron–His band during the course of geminate recombination is shown in Fig. 4. Within the limits of only fair signal to noise there were no indications of frequency changes as ~70% of the photodissociated population recombined within 1 nsec of the initial excitation. In this early study, no attempt was made to monitor the relative intensity of this Raman band relative to the other bands. Results on cryogenic samples of Mb suggest that the fastest recombining conformational state are those with an iron–His Raman band having both the highest frequencies and lowest relative intensities.[27] Although it was not analyzed in terms of KHB, the results of the study by Martin and co-workers,[10] involving picosecond and

PULSE (520 nm) – PROBE (435 nm)
HbO$_2$

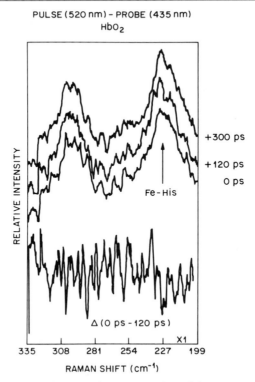

FIG. 4. Time evolution of the low-frequency region of the resonance Raman spectrum of HbO$_2$ under the same pulse–probe conditions as in Fig. 3.

subpicosecond time-resolved Raman of the ν_4 band, may reflect KHB. In this study, the ν_4 band of a partially photodissociated R-state HbCO sample showed a T-state transient value for the frequency of ν_4 before assuming, at several picoseconds subsequent to photodissociation, the more R-like values previously reported for the photoproduct at 30 psec[60] and 10 nsec.[61] A possible interpretation of this result is that under these partially photo-dissociating conditions it is the more T-like members of the distribution of conformational substates that are photodissociated. The subsequent evolution into the expected photoproduct population would then be a reflection of the equilibration time for the creation of the full distribution of conformational substates from the partial distribution of "T"-like con-

[60] E. W. Findsen, J. M. Friedman, M. R. Ondrias, and S. R. Simon, *Science* **229**, 661 (1985).
[61] K. B. Lyons and J. M. Friedman, *in* "Hemoglobin and Oxygen Binding" (C. Ho, ed.), p. 333. Elsevier, New York, 1982.

formational substates. Thus, this experiment may provide a direct mesurement of the time it takes a given conformational substate to sample the full distribution of accessible conformational substates.

Time-Resolved Resonance Raman, Tertiary Structure Changes, and Control of Geminate Recombination. To understand the structural basis for observed GR it is important to determine not only what are the structural parameters that contribute to the activation energy barrier controlling GR but also whether these structural elements are relaxing subsequent to photodissociation on a time scale that can influence the geminate process. Because the structural parameters associated with the iron–His band are implicated in the control of the activation energy barrier for GR, it is reasonable to pursue the question as to whether this Raman band changes over the time course of the different geminate processes. The results to date indicate that well within 30 psec of photodissociation, the frequency of this Raman band in Hb photoproducts has achieved a quasi steady state value that does not begin to relax further until tens of nanoseconds later.[60-62] Nothing is known about whether the relative intensity is changing during the first 10 nsec subsequent to photodissociation. Within the context of the above-discussed model relating this Raman band to structure, these results indicate that the proximal histidine assumes and maintains a relatively fixed tertiary and quaternary structure-sensitive tilt with respect to the heme from within a few picoseconds of dissociation, to tens of nanoseconds. This result also indicates that substantial displacement of the iron out of the heme plane occurs within at most a few picoseconds of photodissociation. Transient Raman studies focusing on Raman bands sensitive to the heme core diameter suggest that there may be some additional adjustment of the iron coordinate occurring on a time scale of hundreds of picoseconds in the case of Mb[63] and tens or hundreds of nanoseconds for Hb.[64] An intriguing possibility, which can be tested directly by following the time dependence of the relative intensity of the iron–His Raman band, is whether this additional slower displacement of the iron is coupled or in fact driven by a change in the azimuthal angle of the proximal histidine that is independent of the tertiary structural elements that dictate the quasistable tilt (iron–His frequency) angle. The picosecond Raman results have a direct and profound implication for the understanding of how the tertiary and quaternary structures control ligand dissociation. Continuous wave resonance Raman comparisons[46,65] of R-

[62] T. W. Scott and J. M. Friedman, *J. Am. Chem. Soc.* **106,** 5677 (1984).
[63] S. Dasgupta and T. G. Spiro, *Biochemistry* **26,** 5689 (1987).
[64] S. Dasgupta and T. G. Spiro, *Biochemistry* **25,** 5941 (1986).
[65] D. L. Rousseau and M. R. Ondrias, *Annu. Rev. Bioeng.* **12,** 357 (1983).

and T-state forms of ligand-bound hemoglobins reveal little or no structural differences at the iron and its immediate environment, raising the question of how the quaternary structure influences the bond-breaking and -reforming steps connected with dissociation.[1] The picosecond Raman results[60] indicate that R- and T-related differences become manifest at the iron within picoseconds or faster on iron–ligand bond dissociation. These results and conclusions[46] have led to a molecular level model[27,50] of how the quaternary structure controls local events at the ligand-binding site. The key elements of this model are as follows:

1. For a given tertiary/quaternary structure, both the relative stability of the ligand-bound state and the relative magnitude of the activation energy barrier for bond formation are directly related to the amount of work necessary to move the iron from its ligand-free position to the more in-plane position associated with the bound species.

2. Protein structure-specific variations in the amount of work required to move the iron to an in-plane conformation arise largely from protein-controlled variations in the heme–proximal histidine interaction.

3. The repulsive interaction between the heme and the imidazole ring of the proximal histidine contributes significantly to the force constant associated with the movement of the iron.

4. This repulsive interaction is modulated through tertiary and quaternary structure-induced variations in the orientation of the heme with respect to the proximal histidine.

5. The local conformation of the histidine with respect to the heme (e.g., tilt) is further modulated by the ligation-specific displacement of the iron.

6. When the iron is out of the heme plane, the repulsive interaction is minimized, and the tilt of the histidine with respect to the heme is set by the determinants of the quaternary structure (e.g., the $\alpha_1\beta_2$ interface), largely via the F helix.

7. When the iron is in the heme plane, the repulsive interactions increase to a degree such that this local interaction dictates the local geometry of the heme–histidine unit, thereby overriding the R- and T-specific influences.

8: The springlike coupling between the iron coordinate and the determinants of quaternary structure via the F helix and the proximal histidine results in quaternary influence materializing at the iron as soon as the iron is displaced from the heme plane.

9. This springlike coupling[37,46,52] ensures that as soon as the iron–ligand bond is either broken or stretched (due to a fluctuation moving the iron out of the heme plane), there is virtually instantaneous quaternary control

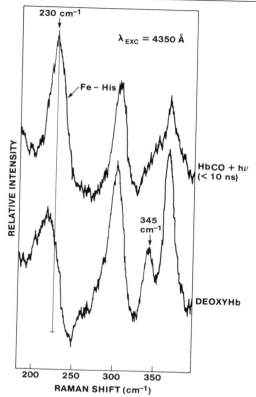

FIG. 5. A comparison of the low-frequency region of the resonance Raman spectrum of deoxyHbA and the photoproduct of HbCO A with a 10-nsec laser pulse.

of either the ultrafast geminate recombination subsequent to dissociation or the return to an in-plane six-coordinate conformation subsequent to a thermal fluctuation involving iron displacement.

10. The frequency and relative intensity of the iron–histidine stretching mode Raman band for a given photoproduct is a reflection of the effective globin-induced repulsive interaction between the heme and the histidine and, as a consequence, reflects the "ease" with which the iron can be restored to the in-plane configuration for that particular tertiary/ quaternary structure (ligand binding can induce structural changes that can result in a new set of "spring" parameters for the final ligand-bound structure, which would be reflected in the Raman spectrum of the photoproduct of the new ligand-bound species).

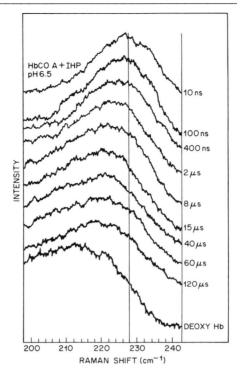

FIG. 6. The time evolution of the iron–proximal histidine stretching mode in photodissociated HbCO A. The behavior of the spectrum at longer times is influenced by ligand recombination from the solution.

Nanoseconds and Beyond: Time-Resolved Resonance Raman Studies of Tertiary Relaxations

Time-resolved RR has been useful in revealing the sequence of structural events that accompany the transition from a photodissociated R- or T-state ligand-bound species of Hb all the way to the equilibrium deoxy form. Until recently the focus has been exclusively on the heme and its environs. More recently progress has been made in using time-resolved UV resonance Raman to probe other regions of the globin, including the $\alpha_1\beta_2$ interface.[66,67]

The observed sequence of events associated with the heme in going

[66] K. R. Rodgers, C. Su, S. Subamanian, and T. G. Spiro, *J. Am. Chem. Soc.* **114**, 3697 (1992).
[67] T. Kitagawa, *Prog. Biophys. Mol. Biol.* **58**, 1 (1992).

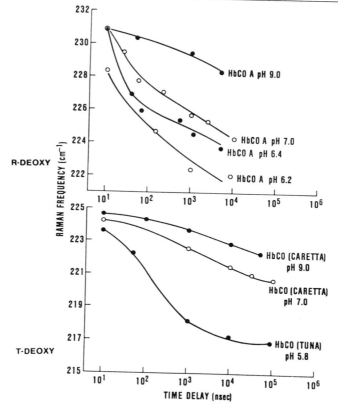

FIG. 7. The relaxation of the frequency of the iron–proximal histidine stretching mode as a function of solution conditions.

from the R-state photoproduct to the equilibrium deoxy T state of Hb A, as revealed in the Raman spectrum, can be summarized as follows.

1. The initial photoproduct structure at 10 nsec subsequent to photodissociation reflects the influence on the heme of the tertiary structure of the ligand-bound R-state protein (see Fig. 5).

2. Starting at a few tens of nanoseconds and ending a few microseconds subsequent to photodissociation, the Raman spectrum of the heme changes to a spectrum characteristic of a deoxy R-state Hb (see Fig. 6). The relaxation associated with these changes represents the loss by the protein of those tertiary signatures associated with ligand binding within the R structure. The spectrum at 5 to 10 μsec after photodissociation for

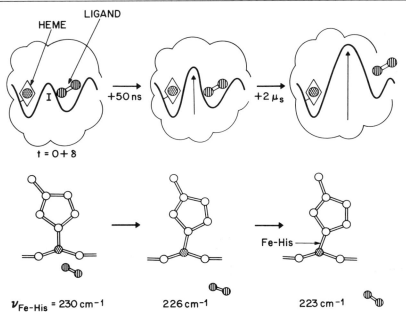

FIG. 8. A representation of how tertiary relaxation can influence the geminate rebinding process.

HbCO A at pH 7 is virtually identical to that of R-state deoxyHb Kempsey.[62]

3. Between 10 and 30 μsec subsequent to photodissociation for HbCO A at pH 7, the iron–His band assumes a frequency and line shape characteristic of the equilibrium deoxy T-state form of Hb A.[62]

4. The tertiary relaxations occurring on the nanosecond to microsecond time scale are dependent on solvent conditions. The rate of relaxation increases with those solution conditions that favor the T state,[62] as shown in Fig. 7.

5. In contrast to the springlike picosecond changes that occur on photodissociation, the nanosecond and slower tertiary relaxations are slowed on increasing the viscosity of the solvent.[37]

6. Because the tertiary relaxation involving the functionally important iron–His conformation partly overlaps the nanosecond geminate phase, it is possible that this solution-dependent relaxation represents a dynamic modulation of the geminate process within the R state.[62] As the structure relaxes, the barrier controlling GR increases, giving rise to a self-trapping phenomenon with respect to the ligand, as illustrated in Fig. 8. Such

behavior was predicted by Agmon and Hopfield[23] for Mb at cryogenic temperatures.

The UV time-resolved RR studies have focused on the tryptophans and tyrosines within Hb. Two studies[66,67] reveal that the R-to-T transition as reflected in the UV-RR occurs at 20 μsec subsequent to photodissociation. One of these studies[66] also reveals indications of a tertiary transition occurring on the time scale of a few microseconds. Comparisons between Hb A and Hb Rothschild (β37 Trp $\rightarrow$ Arg)[57] have allowed for the assignment of a particular shoulder of a tryptophan peak as arising from β37. This important assignment provides a direct window to the events occurring at the $\alpha_1\beta_2$ interface. As UV laser technology improves, it can be anticipated that both steady state and time-resolved UV-RR will provide views of the structure and structural dynamics associated with ligand binding in hemoglobins and related heme proteins.

[12] Front-Face Fluorescence Spectroscopy of Hemoglobins

By RHODA ELISON HIRSCH

When a molecule absorbs light, different forms of energy conversion may take place. An electron excited from the singlet ground state to the first or second singlet excited state may fall back to the ground state, emitting heat or emitting light known as *fluorescence*. Substances displaying significant fluorescence generally possess delocalized electrons such as those found in conjugated double bonds. Not all molecules that absorb light fluoresce. Heme (protoporphyrin IX with iron conjugated at its center) does not fluoresce. Other porphyrins, for example, protoporphyrin IX and zinc protoporphyrin, exhibit significant fluorescence. Excitation by visible light at about 420 nm results in characteristic emission spectra for these porphyrins, with emission maxima, respectively, at 632 and 588 nm in 80 : 20 (v/v) acetone–water.[1] The emission shift to longer wavelengths relative to the excitation (absorption) is known as the Stokes shift. This loss of energy between the excitation and emission is universal for fluorescing molecules in solution and dependent on the electronic characteristics of the molecule. The reader is referred to other works for an in-depth explanation of principles of fluorescence.[2–4]

[1] D. Hart and S. Piomelli, *Clin. Chem. (Winston-Salem, N.C.)* **27**, 220 (1981).
[2] J. R. Lakowicz, "Principles of Fluorescence Spectroscopy." Plenum, New York, 1983.

The observation that the aromatic amino acids fluoresce broadened the horizons for high-resolution and detailed solution-active conformational studies of proteins.[5,6] Tryptophan (Trp), tyrosine (Tyr), and phenylalanine (Phe) exhibit ultraviolet fluorescence. Each exhibits characteristic excitation and emission maxima. The quantum yield for Phe is low, and in proteins containing Tyr and/or Trp, Phe fluorescence is generally not detectable. Class A proteins contain only Tyr and, when excited by light at 275 nm, have been reported to yield a fluorescence emission with a maximum at 303–313 nm. Class B proteins contain both Trp and Tyr. It is generally assumed that excitation with light at 280 nm results in fluorescence emanating primarily from Trp because of resonance energy transfer from Tyr to Trp. The emission maxima of these proteins falls between 325 and 360 nm, depending on the microenvironment of the Trp. Burstein et al.[7] have shown that in proteins, Trp in a hydrophilic or exposed environment exhibits a maximum at 350–353 nm, whereas the maximum for a hydrophobic or buried Trp falls at 330–332 nm; Trp in limited contact with water exhibits a maximum at 340–342 nm. (For uncorrected spectra, the exact position of the maxima could vary by about 5 nm, depending on the spectrofluorometer employed.) In proteins containing both Trp and Tyr, Eisinger[8] has shown that the Trp fluorescence can be isolated from Tyr fluorescence by using an excitation wavelength of 296 nm. This has proved most useful, especially in the fluorescence of hemoglobins as described below.

In brief, the intrinsic fluorescence of proteins provides a powerful spectroscopic tool to learn about dynamic protein structure. The correlation of site-specific conformational changes, as detected by spectroscopy, with the known crystal structure provides a more complete understanding of protein structural behavior. Because crystal intermolecular contacts may place a constraint on attaining the dynamic conformational states in solution, high-resolution, site-specific spectroscopy provides critical insight into protein conformation in its reactive milieu.[9]

---

[3] R. F. Steiner and I. Weinryb, eds., "Excited States of Proteins and Nucleic Acids." Plenum, New York, 1971.

[4] R. F. Steiner, ed., "Excited States of Biopolymers." Plenum, New York, 1983.

[5] F. W. J. Teale and G. Weber, Biochem. J. **65,** 476 (1957).

[6] V. G. Shore and A. B. Pardee, Arch. Biochem. Biophys. **60,** 100 (1955).

[7] E. A. Burstein, N. S. Vedenkina, and M. N. Ivkova, Photochem. Photobiol. **18,** 263 (1973).

[8] J. Eisinger, Biochemistry **8,** 3902 (1969).

[9] D. G. Nettesheim, R. P. Edalji, K. M. Mollison, J. Greer, and E. R. P. Zuiderweg, Proc. Natl. Acad. Sci. U.S.A. **85,** 5036 (1988).

Specifically, intrinsic fluorescence spectroscopy of proteins is used to study conformational changes at or near tryptophans and tyrosines, whereas the binding of extrinsic probes allows for the study of conformational changes at other specific sites.[10] Extrinsic fluorescence (covalent binding of fluorescent probes to specific amino acid residues) probes the environment of nonaromatic amino acid residues. Extrinsic probes also serve in resonance energy transfer experiments as a "spectroscopic ruler" to measure internal and external distance, and may also be used to identify the magnitude of conformational change on protein–ligand binding and/or protein–protein interaction.[11]

Before 1980, characterization of intact heme proteins by fluorescence techniques had not been attempted, because it was generally assumed that tryptophan and/or tyrosine residues were effectively quenched by the nearby heme groups.[12-14] In 1980, the intrinsic fluorescence of hemoglobins was independently demonstrated by two laboratories. Hirsch et al.,[15] using front-face fluorometry, and Alpert et al.,[16] using photon-counting spectroscopy (and later with the use of synchrotron radiation[17]), first demonstrated significant intrinsic fluorescence from intact human hemoglobins. The term *intact hemoglobin* is used to emphasize that all four chains of the hemoglobin tetramer ($\alpha 2\beta 2$) contain the heme moiety, that the hemoglobin dimer does not predominate, and that there is no detectable apohemoglobin present. (This is discussed below in "Precautions to be Exercised".) The intrinsic fluorescence of hemoglobin and the lack of total quenching by the hemes may be explained in part by relative orientations of motions of groups involved in energy transfer mechanisms, which may dramatically reduce the transfer efficiency to result in the observed unquenched emission.[17]

There are multiple advantages of front-face fluorometry in measuring hemoglobin fluorescence or, for that matter, the fluorescence of any protein solution with a high extinction coefficient of absorption. One advantage is that it utilizes a simple and inexpensive modification of a standard spectrofluorometer. Fluorescence measurements of protein

[10] R. P. Haugland, in "Excited States of Biopolymers" (R. F. Steiner, ed.), p. 29. Plenum, New York, 1983.

[11] L. Stryer, Annu. Rev. Biochem. 47, 819 (1978).

[12] G. Weber and F. W. J. Teale, Discuss. Faraday Soc. 28, 134 (1959).

[13] F. W. J. Teale and G. Weber, Biochem. J. 72, 156 (1959).

[14] F. W. J. Teale, Biochem. J. 76, 381 (1960).

[15] R. E. Hirsch, R. S. Zukin, and R. L. Nagel, Biochem. Biophys. Res. Commun. 93, 432 (1980).

[16] B. Alpert, D. M. Jameson, and G. Weber, Photochem. Photobiol. 31, 1 (1980).

[17] M. P. Fontaine, D. M. Jameson, and B. Alpert, FEBS Lett. 116, 310 (1980).

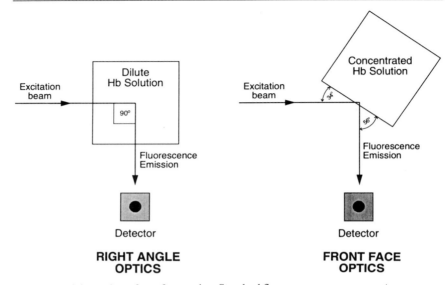

FIG. 1. Right-angle vs front-face optics. Standard fluorescence measurements are generally made using right-angle optics, by which the exciting light beam enters the solution through a relatively long light path, and fluorescence is detected at right angles from the exciting beam. Because of inner-filter effects, right angle-detected fluorescence emission from hemoglobin will be detected only at dilute concentrations, at which the equilibrium is shifted from the native tetramer to the dimer state. In front-face optics, inner-filter effects are overcome by the front face of the cell inclined at an angle other than 45° to the excitation beam.[18] For optimal measurements, the angle of incidence of the exciting light may be either 34° (as shown here) or 56°, depending on the orientation of the front-face cell adapter. This allows for the detection of fluorescence emission from concentrated solutions of hemoglobin (or any other molecule with a high extinction coefficient of absorption).

solutions are generally made using right-angle optics, with the detector positioned at 90° to the exciting light beam. The exciting light beam enters the solution through a relatively long light path. Measurements of fluorescence are then made by light emitted at right angles from the exciting beam. If a solution is strongly absorbing, such as hemoglobin, all excitation will take place near the front surface, and the detector will receive little or no emitted light. This is an "inner-filter" effect (i.e., when not all emitted light reaches the detector) and not true quenching. Inner-filter effects may be overcome with the use of front-face optics, from which the sample fluorescence emission is detected at an angle other than 45° to the incident beam (Fig. 1). It has been found that for optimal measurements, the incident light should make an angle of 34° with the normal to the cell face, or 56°, depending on the orientation of the front-face cell adapter.[18] The use of front-face optics results in the excitation of a relatively small volume of sample, with light penetrating to a depth of less than 100 μm. This in effect eliminates inner-filter effects caused by a molecule with a high molecular extinction coefficient of absorption. (For a detailed description of front-face fluorometry, see Eisinger and Flores.[18])

A further advantage of front-face fluorometry is that it allows one to work with concentrated solutions of hemoglobin, in which the subunit equilibrium is shifted to the tetrameric native form of the molecule. Right-angle optical fluorescence measurements are performed typically on protein solutions in the micromolar range. For a protein such as oxyhemoglobin, the dimer–tetramer equilibrium is shifted significantly to the dimer form at low concentrations.[19] (See Detection of Hemoglobin Aggregation States by Front-Face Fluorometry, and Precautions to Be Exercised, below).

An additional advantage of front-face optics is that, unlike right-angle optics, there is a certain concentration of protein at which the fluorescence intensity reaches a plateau and is not sensitive to small concentration increases (Fig. 2). This eliminates any artifactual intensity changes that occur as a result of small errors of pipetting.

Front-face measurements can be made by altering the sample compartment to front-face geometry. (The term *front surface* is sometimes used.) This is simple and can be done by obtaining a specially designed cuvette or purchasing a front-face adapter from the manufacturer of the standard fluorometer being used. Most companies offer adapters with a temperature-control jacket. In our initial studies, we were able to remove the standard cuvette holder in the sample compartment and design a holder that positioned the cuvette for optimal front-face measurements. Triangular cuvettes also may be used in a right-angle holder.[15,20,21] The disadvantage of triangular cuvettes is that the walls are weaker than typical 1 × 1 cm, four-walled cuvettes and may not hold up after repeated conversions to anaerobic (deoxy) conditions.

Some investigators have designed special cuvettes for front-face measurements. Recall that quartz cuvettes are required for ultraviolet excitation and emission detection. Eisinger and Flores[18] designed a front-face cell that can be inserted into a standard cuvette holder for 1 × 1 cm cells and results in an excitation at 34° to the normal to the cell face. Horiuchi

[18] J. Eisinger and J. Flores, *Anal. Biochem.* **94,** 15 (1979).
[19] G. Guidotti, *J. Biol. Chem.* **242,** 3685 (1967).
[20] E. Bucci, H. Malak, C. Fronticelli, I. Gryczynski, G. Laczko, and J. R. Lakowicz, *Biophys. Chem.* **32,** 187 (1988).
[21] S. Pin, C. A. Royer, E. Gratton, B. Alpert, and G. Weber, *Biochemistry* **29,** 9194 (1990).

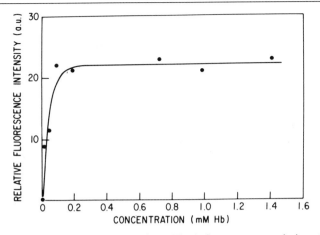

FIG. 2. Concentration dependence of oxyHb A fluorescence emission. Steady state fluorescence measurements were made on the Perkin-Elmer (Norwalk, CT) MPF-3 fluorescence spectrophotometer at 25°. Excitation wavelength was 280 nm. The hemoglobin solutions were buffered in 0.05 M potassium phosphate, pH 7.35. [Reprinted with permission from R. E. Hirsch *et al.*, *Biochem. Biophys. Res. Commun.* **93**, 432 (1980).]

and Asai[22] designed a front-face cell that they termed a "rhombiform optical cell" to measure simultaneously fluorescence and optical spectra of hemoglobins. This cell was designed to fit into a standard right-angle holder without any modification of a standard fluorometer. Bucci and colleagues[23] have designed a front-face cell with a free liquid surface so that the protein fluorescence is emitted from molecules not in contact with the quartz or glass face. This avoids any possible interference from surface effects (e.g., reflections, stray emissions, or denaturation).

Clinical Applications of Front-Face Fluorometry

It may be noted here that front-face fluorometry also has considerable applicability to the study of other macromolecules and cell organelles that have intense absorption as well as light scatter. In addition, specific clinical applications of front-face fluorometry are well proven as highly sensitive

[22] K. Horiuchi and H. Asai, *Biochem. Biophys. Res. Commun.* **97**, 811 (1980).
[23] E. Bucci, Z. Gryczynski, C. Fronticelli, G. Laczko, H. Malak, and J. R. Lakowicz, *Soc. Photo-Opt. Instrum. Eng.* (*SPIE*) **1024**, 813 (1990).

assays.[24-27] In particular, front-face fluorometry is the principal design of the hematofluorometer, an instrument used to determine bilirubin levels in whole blood[24] and also zinc protoporphyrin or protoporphyrin IX in whole blood, high levels of which are indicative of lead poisoning or iron-deficiency anemia.[25] Front-face fluorometry also may be used to determine the levels of fluorescent drugs circulating in whole blood, for example, tetracycline[26] and aminosalicylic acid derivatives.[27]

Applications of Front-Face Fluorometry in Studies of Hemoglobins

Source of Steady-State Fluorescence Emission

Adult human hemoglobin (Hb A) contains three Trp residues in each $\alpha\beta$ dimer, for a total of six in the tetramer: two $\alpha14$ Trp, two $\beta15$ Trp, and two $\beta37$ Trp. Hirsch et al.,[15] with the study of Trp$^+$ and Trp$^-$ hemoglobin mutants, suggested that the primary origin of the fluorescence signal is $\beta37$ Trp. Using steady state front-face fluorometry, Hirsch et al.[15] studied purified normal and variant hemoglobins such as Hb A ($\alpha_2\beta_2$), Hb F ($\alpha_2\gamma_2$), Hb H (β_4), and Hb Rothschild (RC) ($\beta37$ Trp $\rightarrow$ Arg) ($\alpha_2\beta_2$). Using an excitation of 296 nm to excite only Trp,[8,13,15] the hemoglobins were observed to fluoresce in increasing magnitude correlated to their tryptophan content: Hb H $\gg$ Hb F $>$ Hb A $>$ Hb RC (Fig. 3). Because the emission from Hb RC ($\beta37$ Trp $\rightarrow$ Arg) was similar to that of the buffer baseline with 296-nm excitation, $\beta37$ Trp was assigned as the possible origin of the emitted fluorescence in Hb A.[15] This later was confirmed by Itoh et al.,[28] using different hemoglobin mutants. The 325-nm fluorescence emission of oxyhemoglobins and the inability to quench this fluorescence with 1 M KI can be interpreted to mean that the fluorescent tryptophan is located in an internal, hydrophobic environment.[7,15] This study demonstrates that the fluorescence quenching of the aromatic amino acids by the hemes is incomplete, because the intrinsic fluorescence of hemoglobin remains detectable.

As discussed above, it has been accepted that a protein containing both Trp and Tyr, when excited by light at 280 nm, primarily emits a

[24] W. E. Blumberg, J. Eisinger, A. A. Lamola, and D. M. Zuckerman, Clin. Chem. (Winston-Salem, N.C.) 23, 270 (1977).

[25] A. A. Lamola, J. Eisinger, W. E. Blumberg, S. C. Patel, and J. Flores, Anal. Biochem. 100, 25 (1979).

[26] C. M. Park, U. R. Pulakhandam, and R. E. Hirsch, Clin. Res. 34, 466A (1986).

[27] R. E. Hirsch, M. J. Lin, and K. M. Das, J. Lab. Clin. Med. 116, 45 (1990).

[28] M. Itoh, H. Mizukoshi, K. Fuke, S. Matsukawa, K. Mawatori, Y. Yoneyana, M. Sumitani, and Y. Keitaro, Biochem. Biophys. Res. Commun. 100, 1259 (1981).

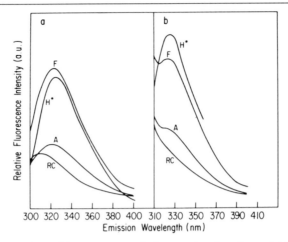

FIG. 3. A comparison of fluorescence emission spectra of oxyhemoglobin variants at (a) 280-nm excitation and (b) 296-nm excitation. The hemoglobins are designated as follows: Hb F (F), Hb H (H), Hb A (A), and Hb Rothschild (RC). H* denotes that the sensitivity of the recorder was one-third less than that recorded for the other hemoglobins. See Hirsch *et al.*[15] for an in-depth discussion of these observed differences. [Reprinted with permission from R. E. Hirsch *et al.*, *Biochem. Biophys. Res. Commun.* **93**, 432 (1980).]

signal dominated by Trp. As shown in Fig. 3, 280-nm excitation of these variant hemoglobins results in an emission maximum characteristic of Trp, with the exception of Hb RC. With 280-nm excitation, Hb RC exhibits an emission maximum at 310 nm, which is characteristic of Tyr in proteins. This is explained by the substitution of β37 Trp with Arg, which results in the loss of close-proximity resonance energy transfer of β35 Tyr to β37 Trp, thereby releasing the quenching constraint on β35 Tyr {Hb A [Tyr(β35)-Pro(β36)-Trp(β37)] vs Hb RC [Tyr(β35)-Pro(β36)-Arg(β37)]}. In this case β35 Tyr could serve as an intrinsic reporter group for Hb RC. The contribution of Tyr to the emission signal was demonstrated for myoglobin variants with different numbers of Tyr residues but with the Trp content and Trp residue position remaining invariant.[29] These studies indicate that tyrosine and tryptophan fluorophores can be isolated spectrally and utilized as site-specific probes in heme proteins.

Prospects for increasing the structural and dynamic informational content of the tryptophan fluorescence of hemoglobin are provided by Friedman and co-workers.[30] They have shown that time-resolved (<20 nsec) fluorescence of cryogenic samples (2–180 K) yields structured emission

[29] R. E. Hirsch and J. Peisach, *Biochim. Biophys. Acta* **872**, 147 (1986).
[30] T. W. Scott, B. F. Campbell, R. L. Cone, and J. M. Friedman, *Chem. Phys.* **131**, 63 (1989).

for several different proteins, including hemoglobin. Both line narrowing and site selectivity are observed. It also appears that the temperature and time-dependent ratio of structured to broad fluorescence can be used to evaluate the polarity and "flexibility" of the chromophore microenvironment.

Fluorescence Sensitivity to Ligand Binding and Quaternary Structure Changes in Hemoglobin

It is well established that the fluorescence emission is ligand dependent.[17,20,23,28,31–34] An 18–25% increase in the fluorescence emission intensity, as a result of the oxy(R)-to-deoxy (T) transition, has been reported for tetrameric Hb A.[28,31] This ligand dependency reflects the quaternary R-to-T conformational transition, as demonstrated by the use of hemoglobin variants, some of which are known to be independent of quaternary structural change on deoxygenation.[28,31] Some mutants exhibit different intensity changes, which suggest differences in R- or T-like structures.[28,31] The R-to-T transitional change is known to involve changes in the environment of the aromatic residues at the $\alpha_1\beta_2$ contact, where nearly all the movement between subunits takes place. The residues include β37 Trp (C3), α42 Tyr (C7), and β41 Phe (C7).[35,36] β37 Trp and α94 Asp are capable of H-bond formation in deoxy but not in liganded hemoglobin, and α140 Tyr (located in the microenvironment of β37 Trp), α42 Tyr, and β145 Tyr undergo conformational changes during the R-to-T transition.[35,36] Because β37 Trp is located at the $\alpha_1\beta_2$ interface and is assigned as the primary origin of the fluorescence emission, fluorescence spectroscopy is an important tool that can be used to understand conformational changes and perturbation at this important functional site.

Effect of pH and Organic Phosphates on Intrinsic Fluorescence of Hemoglobins

Carp hemoglobin is an interesting model to study quaternary structure changes, because at low pH and in the presence of inositol hexaphosphate (IHP) it remains in the T quaternary structure even when saturated with

[31] R. E. Hirsch and R. L. Nagel, *J. Biol. Chem.* **256,** 1080 (1981).

[32] R. E. Hirsch and R. W. Noble, *Biochim. Biophys. Acta* **914,** 213 (1987).

[33] A. G. Szabo, D. Krajcarski, M. Zuker, and B. Alpert, *Chem. Phys. Lett.* **108,** 145 (1984).

[34] E. Bucci, H. Malak, C. Fronticelli, I. Gryczynski, and J. R. Lakowicz, *J. Biol. Chem.* **263,** 6972 (1988).

[35] H. F. Bunn and B. G. Forget, "Hemoglobin: Molecular, Genetic and Clinical Aspects." Saunders, Philadelphia, 1985.

[36] J. Baldwin and C. Chothia, *J. Mol. Biol.* **129,** 175 (1979).

ligand.[37] A study of pH-dependent changes in the intrinsic fluorescence of carp hemoglobin in the presence and absence of IHP revealed that the exact nature of the R-to-T transition difference is dependent on the oxidation state and the specific ligand bound to carp Hb (i.e., CO, met, metazide), with α42 Trp contributing to the fluorescence emission.[32]

The fluorescence intensity behavior of normal, abnormal, and/or modified hemoglobins in the presence and absence of IHP suggests that IHP induces an altered T conformation (i.e., further modifies the T structure of the IHP-free species).[38] Studies of the fluorescence of HbA as a function of pH and in the presence of IHP demonstrate long-range communication of conformational change in the hemoglobin molecule from the diphosphoglycerate (DPG)-binding cavity to the microenvironment of the $\alpha_1\beta_2$ interface.[39–42] Perturbation by amidation of β43 Glu by glycine ethyl ester results in an alteration in the oxy-to-deoxy fluorescence intensity change in the presence of IHP, with a maximal response at pH 6.6. These studies suggest that the ionization status of β43 Glu plays a role in the intrinsic fluorescence of β37 Trp. In addition, titration studies of alkaline pH have been helpful in explaining the high pK for Tyr-146 in myoglobin and differences in the unfolding of hemoglobin and myoglobin.[29] It is important to note that all of these studies were conducted in a pH range that does not affect the fluorescence of Trp or Tyr.[43]

Binding of Extrinsic Fluorescent Probes to Hemoglobin

Fluorescent probes have been useful for the study of hemoglobin structure. Alfimova and Likhtenshtein[44] used 4(5)-(N-maleinisoimido)rhodamine B to bind to β93 Cys for the determination of intramolecular distances by resonance energy transfer measurements. Using right-angle optics, they were required to work at low hemoglobin concentrations in order to detect the emission from the bound fluorophore. With the use of front-face fluorometry, significant visible fluorescence emission emanating from

[37] A. L. Tan, R. W. Noble, and Q. H. Gibson, *J. Biol. Chem.* **248**, 2880 (1973).
[38] H. Mizukoshi, M. Itoh, M. Shigeru, K. Mawatari, and Y. Yoneyana, *Biochim. Biophys. Acta* **700**, 143 (1982).
[39] M. J. Lin, M. J. Rao, J. M. Friedman, A. S. Acharya, and R. E. Hirsch, *Biophys. J.* **59**, 290a (1991).
[40] R. E. Hirsch, M. J. Lin, M. J. Rao, and A. S. Acharya, *Blood* **78**(10), Suppl. 1, 86a (1991).
[41] R. E. Hirsch, M. J. Lin, M. J. Rao, and A. S. Acharya, *Biophys. J.* **61**(2), A57 (1992).
[42] R. E. Hirsch, M. J. Lin, M. J. Rao, R. L. Nagel, and A. S. Acharya, *Clin. Res.* **40**(2), 282A (1992).
[43] A. White, *Biochem. J.* **71**, 217 (1959).
[44] Y. Y. Alfimova and G. I. Likhtenshtein, *Biofizika* **17**, 49 (1972).

extrinsically bound probes is detected at high hemoglobin concentra-tions.[45] This again serves to demonstrate that the hemes do not totally quench fluorescence and that front-face fluorometry provides direct access to the signal.

The sensitivity of extrinsic probes to quaternary structure also is dem-onstrated: for example, the emission of the fluorescent probe, 5-iodoacetamidofluorescein (5-IAF), covalently bound to hemoglobin at β93, serves as a reporter group for the R-to-T transition.[45,46] These findings, along with the sensitivity of the intrinsic fluorescence to the R-to-T transi-tion, made it apparent that a stopped-flow front-face fluorometer could probe R-to-T transitions at specific sites, such as the aromatic amino acids and sites selectively binding extrinsic fluorophores. A prototype instrument using the core of a Gibson–Durrum stopped-flow apparatus on line with digital data acquisition was developed; the kinetic data were fit with the use of a modified Marquardt algorithm.[46] For the detection of fluorescein bound to Hb A, excitation and emission narrow-bandpass filters were used (470 and 520 nm, respectively). To study the R-to-T transition, a solution (1.0 g%) of purified oxyHb A covalently bound to the fluorescent probe 5-iodoacetamidofluorescein was mixed rapidly with deoxygenated buffer (pH 7.35, 0.05 M potassium phosphate) containing sodium dithionite (2 mg/ml). The hemoglobin, at a final concentration of 0.5g%, is primarily tetrameric. A pseudo-first-order reaction was observed with a rate constant near 8 sec^{-1}, which is similar to that reported for unmodified oxyHb A.[47] This is consistent with the findings of Taylor et al.,[48] showing that the oxygen dissociation rate constant did not differ for unmodified and iodoacetamide-modified hemoglobin. Hence, extrinsic fluorophores covalently bound to β93 of hemoglobin may be used as a reporter group of R-to-T transition kinetics. This instrument allows the investigation of R-to-T transition kinetics at sites previously inaccessible to direct study.

In summary, the binding of extrinsic fluorescent probes to intact hemo-globins serves (1) to probe site-specific residues/regions of hemoglobin that do not contain an intrinsic fluorophore for the study of structure–func-tion interrelationships and (2) as a spectroscopic tool to measure intramo-lecular and intermolecular distances.

[45] R. E. Hirsch, R. S. Zukin, and R. L. Nagel, Biochem. Biophys. Res. Commun. **138,** 489 (1986).
[46] R. E. Hirsch and R. L. Nagel, Anal. Biochem. **176,** 19 (1989).
[47] E. Antonini and M. Brunori, "Hemoglobin and Myoglobin in Their Reaction with Li-gands." North-Holland Publ., Amsterdam, 1971.
[48] J. F. Taylor, E. Antonini, M. Brunori, and J. Wyman, J. Biol. Chem. **241,** 241 (1966).

Detection of Hemoglobin Aggregation State by Front-Face Fluorometry

Front-face fluorometry of hemoglobins also proves useful in the study of hemoglobin dissociation. As stated above, it is established that the tetramer–dimer equilibrium of liganded hemoglobin is shifted to the dimer state by dilution and/or the addition of high salt concentrations.[19,47] In contrast, deoxyhemoglobin does not dissociate as readily.

Percent dimerization (α) may be calculated by the equation of Herskovits et al.[49]:

$$\alpha = \{K_D M_4 [1 + (16c/K_D M_4)]^{1/2} - K_D M_4\}/8c$$

where K_D is the dissociation constant in molarity, M_4 is the molecular weight of the tetramer, and c is the concentration in grams per liter.

The assignment of β37 Trp as the primary fluorescence-emitting residue of oxyHb A is based partially on the assumption that the mutation in Hb Rothschild (β37 Trp $\rightarrow$ Arg) does not involve conformational changes in α14 Trp and β15 Trp. Craik et al.[50] and Sharma et al.[51] independently concluded that Hb Rothschild is almost completely dissociated into dimers in its liganded state, whereas the unliganded state is essentially tetrameric. Investigations of the effect of dimerization on the fluorescence properties of hemoglobin show that unlike Hb Rothschild, hemoglobin dimers exhibit shifts to longer wavelengths compared with that of the tetramer.[52] Emission maximum shifts begin to be detected at about 0.3 mM heme for oxyHb A.[52] This shift is probably due to the increase in surface exposure of β37 Trp that occurs during dimerization.[53] It must be emphasized that dimeric hemoglobin predominantly forms when liganded hemoglobin is diluted to the micromolar range.[19,47,49] Hence, fluorescence characteristics of liganded hemoglobin at micromolar concentrations are not relevant for the native tetramer. This fact must be considered when interpreting studies done at low concentrations by other investigators.[16,17,33,34,54]

The emission maximum shift on dissociation is useful in that it may detect different aggregation states of hemoglobins for which the crystal structures are not known. For example, the major and minor hemoglobin

[49] T. T. Herskovits, S. M. Cavanagh, and R. C. San George, *Biochemistry* **16,** 5795 (1977).
[50] C. S. Craik, I. Vallette, S. Beychok, and M. Waks, *J. Biol. Chem.* **255,** 6219 (1980).
[51] V. S. Sharma, G. L. Newton, H. M. Ranney, F. Ahmed, J. W. Harris, and E. Danish, *J. Mol. Biol.* **144,** 267 (1980).
[52] R. E. Hirsch, N. A. Squires, C. Discepola, and R. L. Nagel, *Biochem. Biophys. Res. Commun.* **116,** 712 (1983).
[53] C. Chothia, S. Wodak, and J. Janin, *Proc. Natl. Acad. Sci. U.S.A.* **73,** 3793 (1976).
[54] A. G. Szabo, K. J. Willis, D. T. Krajcarski, and B. Alpert, *Chem. Phys. Lett.* **163,** 565 (1989).

components of the clam *Noetia ponderosa* and the minor hemoglobin component of *Anadara ovalis* exhibited shifts in the emission maxima relative to Hb A in a manner characteristic of dimers.[55] Light scattering and crystallography confirmed the dimer structure of these hemoglobins.[56,57]

Front-face fluorometry has been useful in studying acid vs alkaline dissociation in the giant hemoglobin of the earthworm, *Lumbricus terrestris*.[58] Studies at atmospheric pressure suggest that different subunit species result from acid and alkaline dissociation.[58] High-pressure studies of the fluorescence and light-scattering characteristics of *Lumbricus* hemoglobin indicate a ligand (CO vs oxy) dependency on subunit stabilization.[59] Comparative fluorescence and light-scattering studies of *Lumbricus* hemoglobin[59] and that of the giant earthworm hemoglobin *Glossoscolex paulistis*[60] indicate species structural differences. (For high-pressure studies of hemoglobins, see [4] in this volume.[61])

Time-Resolved Fluorescence Measurements

The above discussions concern the applications and insights that can be gained from the use of steady state front-face fluorometry of hemoglobins. Fluorescence lifetime measurements are of interest, because they provide detailed information about processes that occur on the picosecond to nanosecond time scale.[62] This will enrich our understanding of hemoglobin structural dynamics and its relationship to function. The feasibility of doing time-resolved fluorescence measurements of intact hemoglobin has been shown by several investigators,[17,20,28,33,34,54] but at present the interpretation of these measurements is complex. Multiexponential fluorescence decays for hemoglobins are reported. Hochstrasser and Negus[63] studied fluorescence lifetimes in the heme protein myoglobin and argue that the longer (nanosecond) component that they observe arises from a source other than the two tryptophans in myoglobin. However, Szabo and colleagues[33] have addressed this question directly and have shown that a single tryptophan residue may give rise to two and three fluorescence components, including a 2-nsec component, which suggests that hemoglo-

[55] R. E. Hirsch, R. C. San George, and R. L. Nagel, *Anal. Biochem.* **149,** 415 (1985).
[56] R. C. San George and R. L. Nagel, *J. Biol. Chem.* **260,** 4331 (1985).
[57] W. E. Royer, Jr., W. A. Henderson, and E. Chiancone, *J. Biol. Chem.* **264,** 21052 (1989).
[58] J. P. Harrington and R. E. Hirsch, *Biochim. Biophys. Acta* **1076,** 351 (1991).
[59] R. E. Hirsch, J. P. Harrington, and S. F. Scarlata, *Biochim. Biophys. Acta* **1161,** 285 (1993).
[60] J. Silva, M. Villa-Boas, C. F. Bonafe, and N. C. Meirelles, *J. Biol. Chem.* **264,** 15863 (1989).
[61] S. Pin and C. A. Royer, this volume [4].
[62] J. M. Beecham and L. Brand, *Annu. Rev. Biochem.* **54,** 43 (1985).
[63] R. M. Hochstrasser and D. K. Negus, *Proc. Natl. Acad. Sci. U.S.A.* **81,** 4399 (1984).

bins exist in solution as a mixture of at least three average conformations, having different tryptophan–heme orientations. They go on to explain that differing reports by different laboratories may result from differences in instrument resolution. Szabo et al.[33] obtained a higher degree of precision and statistical accuracy using time-correlated single-photon counting with high temporal resolution. Bucci and colleagues[23] use higher resolution techniques with a 10-GHz laser for time-resolved fluorescence measurements of hemoglobin.

Szabo and colleagues,[64] from fluorescence lifetime studies, suggested the possibility that $\alpha 14$ and $\beta 15$ Trp also may contribute to the signal. Further studies are necessary to determine the extent of the lifetime contribution of each of these residues.

Using dilute solutions of hemoglobin, Bucci et al.[34] concluded that the picosecond component is sensitive to the ligation state of hemoglobin, whereas the nanosecond components are probably due either to impurities or to hemoglobin molecules in conformations that do not permit energy transfer. One must keep in mind that dilute solutions of hemoglobin are primarily dimers as opposed to the functional tetramer. Such dilute solutions are prone to oxidation (metHb : Fe^{3+}) and denaturation. These problems may be avoided by the use of front-face fluorometry; with this technique hemoglobin may be studied at high concentrations, at which it remains a tetramer and is not prone to denaturation. Bucci and colleagues,[20,23] in their most recent investigations of hemoglobin, now employ front-face optics. At this point, the current understanding is that a picosecond lifetime component appears to be sensitive to the ligand state.[20,23] The significance and source of the nanosecond component(s) in hemoglobin remain controversial. The complexity of the issue is tied further to the observation that tryptophan itself in aqueous solution exhibits (depending on conditions) double and triple exponential fluorescence lifetime decays, including a nanosecond component.[65,66] Improvements in instrumentation resolution, agreement of decay computational analysis, criteria of hemoglobin purity (i.e., removal of all minor hemoglobin components such as Hb F, Hb A_2, Hb A_{1a1}, Hb A_{1a2}, Hb A_{1b}, and Hb A_{1c}[67,68]) (see Precautions to be Exercised, below), and a complete understanding of the fluorescence of tryptophan, tryptophans in peptide, and tryptophans in proteins are necessary to gain a clearer and more accurate understanding

[64] J. Albani, B. Alpert, D. T. Krajcarski, and A. G. Szabo, FEBS Lett. **182**, 302 (1985).

[65] A. G. Szabo and D. M. Rayner, J. Am. Chem. Soc. **102**, 554 (1980).

[66] E. Gudgin, R. Lopez-Delgado, and W. R. Ware, Can. J. Chem. **59**, 1037 (1981).

[67] M. J. McDonald, M. Bleichman, H. F. Bunn, and R. W. Noble, J. Biol. Chem. **254**, 702 (1979).

[68] L. M. Garrick, M. J. McDonald, R. Shapiro, M. Bleichman, M. McManus, and H. F. Bunn, Eur. J. Biochem. **106**, 353 (1980).

of the fluorescence lifetime components of heme proteins. However, given the same instrument and conditions, relative lifetime studies of different hemoglobins and perturbations of hemoglobin are informative and interpretable.

Precautions to Be Exercised

In general, because fluorescence analysis is a highly sensitive technique, highly purified material is required. Hemoglobin purity is critical for these studies.[20,21,34,54] Because an erythrocyte hemolysate primarily consists of several hemoglobins with a low percentage of erythrocyte enzymes,[35] it is necessary to isolate the hemoglobin to be studied. This usually involves cation- or anion-exchange procedures. Dialysis and gel filtration are also required, especially because the interaction of organic phosphates affects hemoglobin fluorescence.[38–42] Finally, the homogenity of the hemoglobin to be studied should be verified, possibly by fast protein or high-performance liquid chromatography (FPLC or HPLC). Different hemoglobins require rigorous, specific methods of isolation, which is discussed in several references.[69–71]

Many hemoglobin preparations call for the use of Tris buffer. It is important to eliminate the Tris by dialysis or by column exchange, because Tris reagents contain material with a fluorescent background that interferes with hemoglobin emission. However, some investigators[21,54] report the use of an "ultrapure" Tris product that may not exhibit interfering fluorescence.

Reducing agents such as sodium dithionite are often used to convert oxyhemoglobin to the deoxy form.[72] Sodium dithionite cannot be used for intrinsic fluorescence studies of hemoglobin because of its significant absorption in the ultraviolet (UV). However, in the visible range there is little interference, which allows its use in studies of extrinsic probes (which emit in the visible range) bound to hemoglobin.[46] The method of choice for deoxygenation of hemoglobins used in fluorescence studies is to blow nitrogen or helium gently over slowly rotating solutions or by alternating vacuum–nitrogen exposure cycles or helium replacement methods.

Denatured forms of hemoglobin exhibit emission spectra distinctly different from that of intact hemoglobin. For example, metHb A exhibits a greater fluorescence intensity than the oxy or deoxy forms.[31] Therefore, care must be taken to avoid the formation of significant amounts of methe-

[69] W. A. Schroeder and T. H. J. Huisman, "The Chromatography of Hemoglobin." Dekker, New York, 1980.
[70] This series, Vol. 76.
[71] This volume.
[72] R. L. Nagel and H. Chang, this series, Vol. 76, p. 760.

moglobin. The methemoglobin content may be determined from absorption spectroscopy by comparing the ratios at specific peaks.[73] Apohemoglobin (i.e., globin without the hemes) is also a product of denatured hemoglobin and may also affect the observed fluorescence emission. The characteristic fluorescence of apoglobin has an emission maximum that is shifted to longer wavelengths, and the emission intensity is greater compared to that of Hb A.[29] The fluorescence characteristics of apohemoglobin are useful in ascertaining and confirming its structural difference from Hb A: it is a dimer[47] and, in addition, appears to be more unfolded as indicated by its tryptophan emission maximum, which is shifted to longer wavelengths.[29] The presence of apohemoglobin can also be easily detected by 1-anilinonaphthalene-8-sulfonic acid titration as described by Alpert et al.[16]

Binding of extrinsic probes to hemoglobin must be done with great care. Because the binding of extrinsic fluorophores to hemoglobin usually involves chemical modification, the following must be ascertained: elimination of unbound fluorophore, determination of site specificity of binding, and alterations in hemoglobin function. (For details of protein chemical modification and its requirements, see Lundblad and Noyes.[74])

Heating, caused by intense lasers, can interfere with results and interpretation.[75] Intense lasers used in time-resolved studies may cause problems related to protein denaturation. This can be avoided by several techniques, such as replacing the sample after every second or third scan,[76] recording spectra in a closed flowing cell,[77] and cuvette rotation.[78]

In conclusion, the intrinsic fluorescence of hemoglobin is useful as a probe to assay dynamic, site-specific conformational changes in normal and mutant hemoglobins.

Acknowledgments

The author gratefully acknowledges the interest and collaboration of Dr. Ronald L. Nagel. I am also thankful to Dr. J. B. Alexander Ross for his critical review of the manuscript. The author wishes to acknowledge support by research grants from the American Heart Association, New York City Affiliate, and The National Institutes of Health (1RO1 DK-41253-01A1, HL-38655, and HL-21016).

[73] R. E. Benesch, R. Benesch, and S. Yung, *Anal. Biochem.* **55,** 245 (1973).
[74] R. L. Lundblad and C. M. Noyes, "Chemical Reagents for Protein Modification," Vols. I and II. CRC Press, Boca Raton, FL, 1984.
[75] E. R. Henry, W. A. Eaton, and R. M. Hochstrasser, *Proc. Natl. Acad. Sci. U.S.A.* **83,** 8982 (1986).
[76] R. A. Copeland and T. G. Spiro, *Biochemistry* **26,** 2134 (1987).
[77] C. Su, Y. D. Park, G.-Y. Liu, and T. G. Spiro, *J. Am. Chem. Soc.* **111,** 3457 (1989).
[78] J. M. Friedman, personal communication, 1993.

[13] Stationary and Time-Resolved Circular Dichroism of Hemoglobins

By CHRISTIAN ZENTZ, SERGE PIN, and BERNARD ALPERT

In this chapter we review and update the applications of circular dichroism (CD) spectroscopy to hemoglobin. Essentials of CD theory were presented by Geraci and Parkhurst in Volume 76 of this series.[1] Because many electronic states are involved in proteins, it is possible to investigate the molecule in different wavelength regions. Each region contains some information concerning a part of the three-dimensional (3D) organization of the macromolecule. By choosing a set of different wavelengths, different chromophores can be observed.

1. In the spectral region below 250 nm, the optical activity is dominated by the peptide backbone organization. Protein structural analysis is interpreted by the existence of a correlation between CD spectra and the organization of the secondary structure: α helix, β sheet, and random coil. The CD spectrum of each secondary structure is assumed to be independent of others, and the resulting CD spectrum is a simple summation. Two kinds of reference were used to analyze protein structure by CD. The first approach[2,3] utilized the experimental CD data from synthetic polypeptides of known structure; the second method[4-7] extracted the α-helix content from CD spectra of proteins in solution whose secondary structures were determined from X-ray diffraction of their crystals. Provencher and Glockner[8] compared the CD spectra of 16 proteins whose secondary structures are known from X-ray crystallography. The authors found a good correlation with the X-ray data for the α helix and β sheet. Hennesey and Johnson[9] refined the method by a multicomponent matrix analysis: 5 basic CD spectra allowed reconstruction of the 16 original CD spectra.

[1] G. Geraci and L. J. Parkhurst, this series, Vol. 76, p. 262.
[2] N. Greenfield, B. Davidson, and G. D. Fasman, *Biochemistry* **6**, 1630 (1967).
[3] N. Greenfield and G. D. Fasman, *Biochemistry* **8**, 4108 (1969).
[4] V. P. Saxena and D. B. Wetlaufer, *Proc. Natl. Acad. Sci. U.S.A.* **68**, 969 (1971).
[5] Y. H. Chen, J. T. Yang, and H. M. Martinez, *Biochemistry* **11**, 4120 (1972).
[6] Y. H. Chen, J. T. Yang, and K. H. Chau, *Biochemistry* **13**, 3350 (1974).
[7] C. T. Chang, C. S. Wu, and J. T. Yang, *Anal. Biochem.* **91**, 13 (1978).
[8] S. W. Provencher and J. Glockner, *Biochemistry* **20**, 33 (1981).
[9] J. P. Hennessey, Jr. and W. C. Johnson, Jr., *Biochemistry* **20**, 1085 (1981).

The influence of the length of the α-helix segment was empirically corrected.[6] The possible interactions among the different structural elements and the contributions from nonpeptide chromophores were neglected.[10,11] Most analyses have been restricted to a relatively narrow spectral range (185–250 nm), in which there are few electronic transitions of the polypeptide backbone. When the CD spectrum was extended to 165 nm, substantial improvements occurred in the precision of the amounts of different secondary structures.[12] Helical content of mammalian hemoglobin was investigated extensively by far-ultraviolet (UV) CD, and the analysis of the optical activity gave a good correlation with X-ray data.

2. In the near-ultraviolet (250- to 300-nm region), the CD bands arise from tryptophanyl, tyrosyl, and phenylalanyl residues, the S–S bridge, and heme groups.[13] The disulfide chromophore may be inherently optically active.[14] For aromatic rings having a plane of symmetry, or a center of inversion, optical activity occurs only if this chromophore is in interaction with nearby groups or with the amino acid moiety. The aromatic CD bands reflect essentially, up to 10–15 Å, some neighboring interactions.[15] Vibrational fine structure of the aromatic residues influences the CD spectrum.[15–18] The heme group lacks fine structure in the near-UV.[19] A tilt of the aromatic residue or an alteration in its asymmetric environment can produce a change of the sign and the magnitude of the optical activity.[13,20] The CD of aromatic residues also can be modified by the nonnegligible optical absorption of the heme in this UV region.[19]

3. Above 300 nm, the optical activity of the heme group (including the iron ligand) results from short- and long-distance interactions of the heme with the protein matrix. Theoretical calculations established that a coupled oscillator interaction between the transitions of the heme and those of the surrounding aromatic side chains could account for the observed CD bands in myoglobin and hemoglobin.[21] Hsu and Woody have carried out calculations of the heme rotational strength in heme proteins,[21]

[10] I. Tinoco, Jr. and A. W. Williams, Jr., *Annu. Rev. Phys. Chem.* **35,** 329 (1984).
[11] R. W. Woody, *Biopolymers* **17,** 1451 (1978).
[12] S. Brahms and J. Brahms, *J. Mol. Biol.* **138,** 149 (1980).
[13] S. N. Timasheff, *in* "The Enzymes" (P. D. Boyer, ed.), 3rd ed., Vol. 2, p. 371. Academic Press, New York, 1970.
[14] J. Lindenberg and J. Michl, *J. Am. Chem. Soc.* **92,** 2619 (1970).
[15] E. H. Strickland, *CRC Crit. Rev. Biochem.* **2,** 113 (1974).
[16] J. Horwitz, E. H. Strickland, and C. Billups, *J. Am. Chem. Soc.* **91,** 184 (1969).
[17] E. H. Strickland, J. Horwitz, and C. Billups, *Biochemistry* **8,** 3205 (1969).
[18] J. Horwitz, E. H. Strickland, and C. Billups, *J. Am. Chem. Soc.* **92,** 2119 (1970).
[19] D. W. Urry, *J. Biol. Chem.* **242,** 4441 (1967).
[20] S. Beychok, *Science* **154,** 1288 (1966).
[21] M. C. Hsu and R. W. Woody, *J. Am. Chem. Soc.* **93,** 3515 (1971).

based on the theory of Tinoco.[22] The sign and magnitude of the Cotton effect are determined by the distribution of this rotational strength among two nearby degenerated components, depending on the orientation of the transition moments in the heme plane. Therefore a simple rotation of the heme in the porphyrin plane results in a change of the sign of the Soret CD band. Various other possible mechanisms contribute to the visible optical activity as a mixing of $\pi \rightarrow \pi^*$ heme transition with $d \rightarrow d$ iron transition. A binding of an optically inactive ligand on the heme iron, in the fifth and/or sixth coordinated position, can perturb CD spectra in the visible domain (Q bands), reflecting symmetric properties of the heme iron-bound material.[23]

In summary, the focus of these approaches is their contribution to the knowledge of protein organization. Through interactions at short and long distances, CD helps in the conformational analysis of proteins, but the information is not always clearly visible without other, complementary investigations.

Stationary Circular Dichroism

Far-Ultraviolet Region

Human Hemoglobin. Circular dichroism measurements in the far-UV region (Fig. 1) have been carried out on freeze-dried human hemoglobin, in the absence and presence of protective compounds against denaturation (glucose, *meso*-inositol). The structural changes that unprotected hemoglobin undergoes during freeze-drying give a CD signal about 6% higher than that of the protected hemoglobin. Because the Soret CD does not change, the authors conclude that the absence of a protector increases only the helicity, from 75 to 81%, of the secondary structure.[24] Human apohemoglobin (apoHb) can be labeled at the β1 Val residue with pyridoxal 5'-phosphate (PLP), and it is known that the heme pocket binds 8-anilino-1-naphthalene sulfonate (ANS) with high affinity. The far-UV CD spectrum of apoHb–PLP, both in the absence and presence of ANS, is indistinguishable from that of normal untreated apoHb. Therefore labeling with PLP and addition of ANS do not modify the structure of apoHb.[25] However, the incorporation of Mn-phthalocyanine into apoHb

---

[22] I. Tinoco, Jr., *Adv. Chem. Phys.* **4**, 113 (1962).

[23] J. Bollard and A. Garnier, *Biochim. Biophys. Acta* **263**, 535 (1972).

[24] C. Thirion, D. Larcher, B. Chaillot, P. Labrude, and C. Vigneron, *Biopolymers* **22**, 2367 (1983).

[25] J. Kowalczyck and E. Bucci, *Biochemistry* **22**, 4805 (1983).

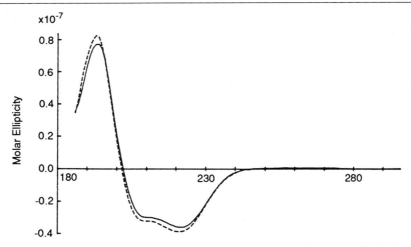

FIG. 1. Far-UV CD spectra of human hemoglobin, pH 7, before (—) and after (– – –) freeze-drying. [From C. Thirion *et al.*, *Biopolymers* **22**, 2367 (1983). Reprinted by permission of John Wiley & Sons, Inc. Copyright © 1983.]

increases the helical content of the protein, whereas that of Zn-phthalocyanine increases its unfolding.[26] Circular dichroism of human apohemoglobin α subunit (apoα) has been investigated previously.[27] Apoα labeled with N-iodoacetylaminoethyl-5-naphthylamine 1-sulfonate (AEDANS) shows an ellipticity at 222 nm about 20% lower than that of apoα.[28]

Limited proteolysis of human hemoglobin α chain with *Staphylococcus aureus* V8 protease (SAP) produces the complementary fragments 1–30 and 31–141. The far-UV CD spectra of the digest reflect the native-like structure of the noncovalent fragment system. The slightly lower helical content is presumably a result of local perturbation of the B helix due to the discontinuity introduced by the generation of the two complementing fragments.[29] The fragment α1–23 is obtained by BrCN cleavage in formic acid followed by fragmentation with SAP. Peptide(1–7) does not exhibit a helical structure, whereas peptide(1–23) forms a measurable population of α-helix structure that spans the same residues (namely residues 6 to 17) that are in the helical structure of the native protein.[30]

Native hemoglobin has been partially digested to find out whether

[26] H. Przywarska-Boniecka and H. Swirska, *J. Inorg. Biochem.* **13**, 283 (1980).

[27] M. Waks, Y. K. Yip, and S. Beychok, *J. Biol. Chem.* **248**, 6462 (1973).

[28] J. Oton, D. Franchi, R. F. Steiner, C. Fronticelli, A. Martinez, and E. Bucci, *Arch. Biochem. Biophys.* **228**, 519 (1984).

[29] R. Seetharam, A. Dean, K. S. Iyer, and A. S. Acharya, *Biochemistry* **25**, 5949 (1986).

[30] M. C. Pena, M. Rico, M. A. Jimenez, J. Herrans, J. Santoro, and J. L. Nieto, *Biochim. Biophys. Acta* **957**, 380 (1989).

and how thiol groups (glutathione or cysteine) participate in hemoglobin digestion by proteases. Far-UV CD spectra of hemoglobin before and after treatment with thiol do not exhibit great changes: α-helical structure is not unfolded by these agents. The authors consider that, because the absorbance in the Soret region decreases after treatment with thiol, the heme groups are dissociated from the globin moiety. Thus, the enhanced susceptibility of hemoglobin to proteases should be due to globin dissociation from its heme.[31]

Crustacean and Mollusk Hemoglobins. Adult hemoglobin of the crustacean *Artemia* is purified and cleaved into functional units by limited subtilisin digestion. The far-UV CD of fraction E is then compared to that of total purified hemoglobin. The α-helical content of the intact molecule and the separated fragments is rather low (25–35%) compared to that of most other invertebrate hemoglobins (35–75%).[32]

Dimeric hemoglobin Hb I (γ_2) and tetrameric hemoglobin Hb II $(\alpha_2\beta_2)$ of the mollusk *Anadara broughtonii* have no Bohr effect but exhibit cooperative ligand binding. Circular dichroism spectra in the far-UV region change significantly on ligand binding.[33] Circular dichroism of dimeric and tetrameric hemoglobins of the mollusk *Scapharca inaequivalvis* has been studied. The heme removal brings a larger conformational change in the tetrameric than in the dimeric protein.[34]

Annelid Hemoglobin. Far-UV CD of *Glycera dibranchiata* hemoglobin (both monomer and polymer) has been analyzed. The α-helix content for the single chain is in the expected range ($\approx$70%) as predicted by X-ray structure. The lower estimate of the α-helix content for the polymer ($\approx$50%) might reflect the differences in the primary structure and the polypeptide chain folding. The removal of the heme moiety from the monomer did result in a major decrease in its helix content, similar to the loss of heme from myoglobin.[35]

The ellipticity of *Nephtys incisa* hemoglobin has been found to be appreciably more negative than that of other extracellular annelid hemoglobins. The α-helical content of *N. incisa* oxyhemoglobin is estimated to be 50–60%.[36] *Tubifex tubifex* oxyhemoglobin has been dissociated into

[31] H. Kimura, H. Murata, and H. Uematsu, *J. Biochem.* (*Tokyo*) **88**, 395 (1980).
[32] L. Moens, D. Geelen, M. L. Van Hauwaert, G. Wolf, R. Blust, R. Witters, and R. Lontie, *Biochem. J.* **223**, 861 (1984).
[33] H. Furuta, M. Ohe, and A. Kajita, *Biochim. Biophys. Acta* **625**, 318 (1980).
[34] D. Verzili, N. Rosato, F. Ascoli, and E. Chiancone, *Biochim. Biophys. Acta* **954**, 108 (1988).
[35] E. Pandolfelli O'Connor, J. P. Harrington, and T. T. Herskovits, *Biochim. Biophys. Acta* **624**, 346 (1980).
[36] U. Messerschmidt, P. Wilhelm, I. Pilz, O. H. Kapp, and S. N. Vinogradov, *Biochim. Biophys. Acta* **742**, 366 (1983).

three types of subunits (13, 23, and 25 kDa) and reassociated. The α-helical content, estimated from CD at 222 nm, of the reassociated molecule is identical to that of the native hemoglobin.[37]

Near-Ultraviolet Region

Artificial Hemoglobin. Previous CD studies on human hemoglobin suggest that the region from 270 to 290 nm provides a sensitive probe of hemoglobin quaternary structure at the $\alpha_1\beta_2$ interface.[38,39] The 280- to 290-nm CD band should reflect the local environment of α42 Trp and β37 Trp,[40,41] which are located at the $\alpha_1\beta_2$ interface. In T-state hemoglobin, this band exhibits a negative ellipticity. In contrast, for R-state hemoglobin, a small positive ellipticity occurs in this region. When a 5-thio-2-nitrobenzoate group is anchored at the β93 Cys, the R–T transition can also be observed by a new 312-nm CD band.[42] Iron of hemoglobin A has been substituted by nickel ($Hb^{Fe \rightarrow Ni}$). The nickel is present in a mixture of different coordination states, because the Soret absorption exhibits two peaks (398 and 420 nm).[43] Near-UV CD spectra and X-ray crystallographic properties suggest that $Hb^{Fe \rightarrow Ni}$ exists in a structure similar to the deoxygenated T form of normal hemoglobin A.

Mixed metal hybrids of hemoglobin (one subunit contains zinc while the other subunit contains iron) have been investigated as models for partially ligated hemoglobins. Zinc(II) provides a particularly valuable probe because five-coordinate Zn^{II} porphyrins are virtually isostructural with five-coordinate, high-spin Fe^{II} porphyrins. Furthermore, Zn^{II} does not react with exogenous ligands. The iron subunit may then be liganded, or deliganded, without affecting the state of the Zn^{II} porphyrin-containing subunit. Ultraviolet CD spectra of $\alpha^{Fe \rightarrow Zn}\beta CO$, $\alpha CO\beta^{Fe \rightarrow Zn}$, and $\alpha^{Fe \rightarrow Zn}$ $\beta^{Fe \rightarrow Zn}$ show a significant negative ellipticity band centered at 280 nm, analogous to deoxyhemoglobin (and other T-state hemoglobins).[44,45] These

[37] G. Polidori, M. Mainwaring, T. Kosinski, C. Schwarz, R. Fingal, and S. N. Vinogradov, *Arch. Biochem. Biophys.* **233**, 800 (1984).

[38] M. F. Perutz, A. R. Fersht, S. R. Simon, and G. C. K. Roberts, *Biochemistry* **13**, 2174 (1974).

[39] M. F. Perutz, *Br. Med. Bull.* **32**, 195 (1976).

[40] C. F. Plese and E. L. Amma, *Biochem. Biophys. Res. Commun.* **76**, 691 (1977).

[41] C. F. Plese, E. L. Amma, and P. E. Rodesiler, *Biochem. Biophys. Res. Commun.* **77**, 837 (1977).

[42] I. Tabushi, T. Sasaki, and K. Yamamura, *Bioorg. Chem.* **12**, 242 (1984).

[43] K. Alston, A. N. Schechter, J. P. Arcoles, J. Greer, G. R. Parr, and F. K. Friedman, *Hemoglobin* **8**, 47 (1984).

[44] M. D. Fiechtner, G. McLendon, and M. W. Bailey, *Biochem. Biophys. Res. Commun.* **96**, 618 (1980).

[45] K. Simolo, G. Stucky, S. Chen, M. Bailey, C. Scholes, and G. McLendon, *J. Am. Chem. Soc.* **107**, 2865 (1985).

TABLE I

APPARENT QUATERNARY STRUCTURE OF ZINC-SUBSTITUTED HEMOGLOBINS
OBTAINED USING DIFFERENT TECHNIQUES[a]

Technique	$\alpha^{Zn}\beta CO$	$\alpha CO\beta^{Zn}$
CO off rate	T	T
CD_{280}	T	T
Val(E11) NMR resonance	R	R
Asp(β99)-Tyr(α42) exchangeable proton	R	R
Val(β98)-Tyr(β145) exchangeable proton	T	R
Asp(α94)-Asp(β102) exchangeable proton	T	R
His(β146) and other His resonances	T	T
ENDOR[b] of Fe^{III} C^--substituted species	R	R

[a] Reprinted with permission from K. Simolo et al., J. Am. Chem. Soc. **107**, 2865 (1985). Copyright 1985 American Chemical Society.
[b] Electron nuclear double resonance.

data suggest that the Zn/FeCO hybrid hemoglobins exist in the T state. However, it is possible that only the structural region probed by CD resembles the deoxygenated (T) structure, whereas other regions may resemble the oxygenated (R) structure. To obtain a more complete picture of the structure of $\alpha^{Fe \to Zn}$ and $\beta^{Fe \to Zn}$ hybrids, different spectroscopic techniques have been used. Table I lists all results that give different and apparent quaternary structure.[45]

Leghemoglobin resembles myoglobin in ligand-binding properties and optical absorption spectra. Artificial leghemoglobins have been reconstituted from apoleghemoglobin (prepared from soybean root nodules) with unnatural hemes: mesoheme, deuteroheme, and diacetyldeuteroheme.[46] The L band (in the vicinity of 260 nm[47]) is much smaller for leghemoglobin than for myoglobin. This L band must reflect the interactions between the heme and the surrounding apoprotein because many of the aromatic residues around the heme in myoglobin are substituted by nonaromatic residues in leghemoglobin. The ellipticity associated with the 260-nm band is strongly influenced by the attached ligand and, thereby, the spin state of the iron atom. The distinctly lower ellipticity of the tryptophanyl band at 289 nm of the aquoferric derivative of meso- and deuteroleghemoglobin, as compared to that of the native protein, probably indicates a different orientation of the heme (at least of its 2,4 side chains) with respect to Trp-128 producing that band. This study is in accordance with earlier

[46] U. Pertilla and G. Sievers, Biochim. Biophys. Acta **624**, 316 (1980).
[47] W. A. Eaton and J. Hofrichter, this series, Vol. 76, p. 175.

observations on artificial myoglobins[48] indicating no major changes in the protein structure on substitution of meso- or deuteroheme for protoheme.

Dromedary Hemoglobin. Near-UV CD spectra of dromedary and human globins have been compared. The L_b bands at 283 and 290 nm (typical of tryptophan residues[15]) are less intense in dromedary globin.[49] This can be ascribed to the substitution of $\alpha 14$ Trp→Phe. Therefore the simple Trp-14 residue present in the α chains strongly contributes to the dichroism of human apohemoglobin. Dichroic bands of deoxygenated and ligated forms of dromedary and human hemoglobins are due to the contribution of the aromatic side chains and to the induced optical activity of the heme group as well. In this region the dichroic bands are in large part affected by the nature of the heme ligand and the spin state of the iron atom.[50] Using the previous findings on quaternary structural changes,[51] the CD modifications suggest that, in contrast with human hemoglobin, the conformational changes induced by inositol hexaphosphate (IHP) are attributable mainly to a local change of the tertiary structure.[49,52–54]

Crocodilian Hemoglobin. Oxygen affinity of *Caiman crocodilus* hemoglobin is greatly lowered by CO_2. Therefore the effect of CO_2 on the near-UV CD spectra has been examined. A small fall in the spectrum is observed when CO_2 is added to deoxyhemoglobin solution. An addition of CO_2 in aquomethemoglobin solution produces a large fall around 287 nm, similar to the changes induced by IHP in human hemoglobin. The authors interpret these data in terms of R–T transition[55] (Fig. 2).

Fish Hemoglobin. The major hemoglobin component of the rainbow trout *Salmo irideus* in the carbon monoxide form has been studied under different experimental conditions. A decrease of the negative ellipticity at 287 nm occurs with decreasing pH; this change is enhanced by addition of IHP at low pH. From this result and previous work,[56] the authors assume that the CD optical changes are associated with the R–T transition.[57]

[48] M. Tamura, G. V. Woodrow, and T. Yonetani, *Biochim. Biophys. Acta* **317**, 34 (1973).

[49] R. Santucci, F. Ascoli, G. Amiconi, A. Bertollini, and M. Brunori, *Biochem. J.* **231**, 793 (1985).

[50] T. K. Li and B. P. Johnson, *Biochemistry* **8**, 3638 (1969).

[51] M. F. Perutz, J. E. Ladner, S. R. Simon, and C. Ho, *Biochemistry* **13**, 2163 (1974).

[52] R. Santucci, G. Amiconi, F. Ascoli, and M. Brunori, *Biochem. J.* **240**, 613 (1986).

[53] P. Ascenzi, R. Santucci, A. Desideri, and G. Amiconi, *J. Inorg. Biochem.* **32**, 225 (1988).

[54] G. Amiconi, R. Santucci, M. Coletta, A. Congiu Castellano, A. Giovannelli, M. Dell'Ariccia, S. Della Longa, M. Barteri, E. Burattini, and A. Bianconi, *Biochemistry* **28**, 8547 (1989).

[55] C. Bauer, M. Forster, G. Gros, A. Mosca, M. Perrella, H. S. Rollema, and D. Vogel, *J. Biol. Chem.* **256**, 8429 (1981).

[56] B. Giardina, F. Ascoli, and M. Brunori, *Nature (London)* **256**, 761 (1975).

[57] F. Ascoli, R. Santucci, G. Falcioni, and M. Brunori, *Biochim. Biophys. Acta* **742**, 565 (1983).

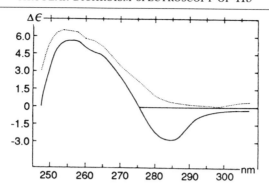

FIG. 2. Near-UV CD spectra of *Caiman* hemoglobin at pH 7.0 and 22–23°. Aquomethemoglobin (1.6–2.0 × 10^{-4} M heme) in the absence (···) and presence (—) of CO_2 (pCO_2 70 torr). [Reproduced with permission from C. Bauer *et al.*, *J. Biol. Chem.* **256**, 8429 (1981).]

All derivatives of *Cyprinus carpio* methemoglobin are reported to be switched from the R to the T quaternary structure when IHP is bound.[58] The intensity difference between 283 and 287 nm exhibited by human hemoglobin appears between 287 and 293 nm in carp hemoglobin. The similar size of the UV difference spectra of all carp derivatives suggests that this dip is a signature of a quaternary structure transition.[59] In contrast, the changes of the L band (around 260 nm) ellipticity of carp hemoglobin with ligation reflect tertiary structural alterations and bear no relationship to quaternary transition.[60]

The distributions between the T and R allosteric states in the bluefin tuna *Tunnus thynnus* and menhaden *Brevoortia tyrannus* hemoglobins have been studied as a function of pH. The CD spectra between 270 and 300 nm have been examined. As the pH increases from 5.5 to 7.5, the major negative peak at 285 nm largely disappears, and a triplet structure centered around 295 nm becomes prominent.[61]

Mollusk Hemoglobin. Dimeric (Hb I) and tetrameric (Hb II) hemoglobins from the mollusk *S. inaequivalvis* have been analyzed. In the deoxygenated derivative, both Hb I and Hb II exhibit broad negative bands at 285 to 295 nm. But in the case of Hb II, the negative band is split into

[58] M. F. Perutz, J. K. M. Sanders, D. H. Chenery, R. W. Noble, R. R. Pennelly, L. W.-M. Fung, C. Ho, I. Giannini, D. Porschke, and H. Winkler, *Biochemistry* **17**, 3640 (1978).

[59] E. R. Henry, D. L. Rousseau, J. J. Hopfield, R. W. Noble, and S. R. Simon, *Biochemistry* **24**, 5907 (1985).

[60] J. C. W. Chien, L. C. Dickinson, F. W. Snyder, Jr., and K. H. Mayo, *J. Mol. Biol.* **142**, 75 (1980).

[61] C. Greenwood and Q. H. Gibson, *J. Biol. Chem.* **258**, 4171 (1983).

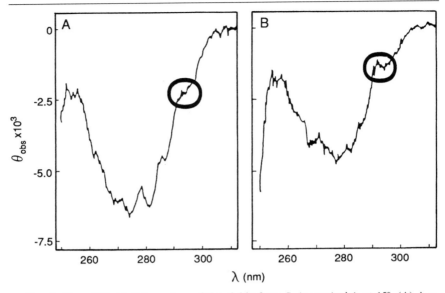

FIG. 3. Near-UV ellipticity values of the globin from *S. inaequivalvis* at 15°. (A) Apo-Hb I at 0.5 mg/ml in water; (B) Apo-Hb II at 1.4 mg/ml in water plus 1.5 mM dithiothreitol. The shoulder at 295 nm is encircled. Optical path, 1 cm; sensitivity, 0.5×10^{-3} deg cm^{-1}. [Reprinted with permission from D. Verzili *et al.*, *Biochim. Biophys. Acta* **954**, 108 (1988).]

two peaks. On ligation, the negative bands are conserved. The authors conclude that the tryptophan residues appear to be in a relatively rigid environment and that, for Hb II, this environment is insensitive to the quaternary structural change.[62] The heme removal brings about a larger conformational change in Hb II than in Hb I, and the shoulders at 295 nm seem to indicate that the tryptophan residues acquire a more rigid conformation in the tetramer[34] (Fig. 3).

Annelid Hemoglobin. The monomer and polymer components of *G. dibranchiata* hemoglobin have been investigated for various low-spin (O$_2$, CO, N$_3^-$, CN$^-$) and high-spin (deoxy, F$^-$, H$_2$O) complexes. For all liganded derivatives of monomer, the spectra exhibit only negative bands in the 250- to 300-nm domain. Although the polymer has negative bands in this domain, the low-spin derivatives of the polymer possess a positive transition, at 260 nm, attributed to a heme transition usually evidenced by vertebrate hemoglobins.[63] Interestingly, the monomer lacks the 260-

[62] E. Chiancone, P. Vecchini, D. Verzili, F. Ascoli, and E. Antonini, *J. Mol. Biol.* **152**, 577 (1981).
[63] S. Beychok, T. Tyuma, R. E. Benesch, and R. Benesch, *J. Biol. Chem.* **242**, 2460 (1967).

nm peak in all its liganded derivatives. These data indicate that the heme environment of the polymer more closely resembles human hemoglobin and myoglobin.[35]

Soret Region

Human Hemoglobin. Normal and modified human hemoglobin α and β subunits in which C-terminal residues have been removed enzymatically are prepared in the carbonmonoxy form. CO is removed in a rotary evaporator under strong illumination. The Soret CD spectra of the deoxygenated form of each chain and the reconstituted modified hemoglobins are measured.[64] The spectrum of reconstituted normal hemoglobin is different from the arithmetic mean of respective spectra of its constituent chains.[65] Thus, the Soret CD change reflects the subunit interaction in the $\alpha_1\beta_1$ dimer.[66,67] It is generally accepted that the interactions between α_1 and β_1 subunits (the $\alpha_1\beta_1$ contact) does not change during the quaternary T–R transition. Chemically modified hemoglobins, in which C-terminal residues are removed by carboxypeptidases A and B, are capable of undergoing the T–R structural interconversion within the deoxygenated state.[51,68,69] The CD spectra of the modified hemoglobins are also markedly different from the arithmetic means of respective spectra of their constituent chains. The authors show that the maxima of the difference spectra at about 437 ± 1 nm and 433 ± 1 nm are indicators of R and T structures, respectively (Table II). These results show that the $\alpha_1\beta_1$ contacts are different in R and T structures.[64]

Soret CD spectral changes on subunit association in valency hybrid hemoglobins have been examined.[67] Changes in the spectra on subunit association to form valency hybrid hemoglobins are also considered to result from the formation of the $\alpha_1\beta_1$ subunit contact. It should be accompanied by some changes in the spatial orientation of aromatic amino acids relative to the heme and changes in interaction between the heme iron and the proximal histidine. The structural properties in the heme vicinity (including the Fe–His bond of ferric as well as ferrous chains) in the deoxygenated $\alpha_2^+\beta_2$ are considerably different from those of $\alpha_2\beta_2^+$.

Theoretical studies on Soret CD spectra[21] show that contributions of intersubunit interactions between unlike chains such as $\alpha_1\beta_1$, $\alpha_2\beta_1$, and

[64] Y. Kawamura-Konishi and H. Suzuki, *Biochem. Biophys. Res. Commun.* **156**, 348 (1988).
[65] Y. Sugita, M. Nagai, and Y. Yoneyama, *J. Biol. Chem.* **246**, 383 (1971).
[66] Y. Kawamura, H. Hasumi, and S. Nakamura, *J. Biochem. (Tokyo)* **92**, 1227 (1982).
[67] K. Mawatari, S. Matsukawa, and Y. Yoneyama, *Biochim. Biophys. Acta* **748**, 381 (1983).
[68] J. Baldwin and C. Chothia, *J. Mol. Biol.* **129**, 175 (1979).
[69] K. Nagai, G. N. La Mar, T. Jue, and H. F. Bunn, *Biochemistry* **21**, 842 (1982).

TABLE II
WAVELENGTH OF MAXIMA OF DIFFERENCE CIRCULAR DICHROISM SPECTRA BETWEEN
CIRCULAR DICHROISM SPECTRA OF RECONSTITUTED MODIFIED HEMOGLOBINS AND
ARITHMETIC MEANS OF SPECTRA OF CONSTITUENT CHAINS[a]

α chain	β chain	pH 6		pH 9
		+ IHP	− IHP	
Normal	Normal	433 T	433 T	433 T
	Des-His	433 T	433 T	433 T
	Des-His-Tyr	433 T	433 R	433 (T)
Des-Arg	Normal	433 T	433 T	437 R
	Des-His	433 T	437 R	437 (R)
	Des-His-Tyr	433 (T)	437 (R)	437 (R)
Des-Arg-Tyr	Normal	433 T	437 (R)	437 R
	Des-His	433 (T)	437 R	437 (R)
	Des-His-Tyr	437 (R)	437 (R)	437 (R)

[a] Reprinted with permission from Y. Kawamura-Konishi and H. Suzuki, *Biochem. Biophys. Res. Commun.* **156**, 348 (1988). Each wavelength includes an error of ±1 nm. (R) and (T) are the quaternary structures predicted from Kawamura-Konishi and Suzuki.[64] R and T indicate the quaternary structures of the modified hemoglobin characterized previously.

$\alpha_1\beta_2$ to the Soret rotational strength are not negligible. Abnormal methemoglobin Hirose, β37 Trp→Ser, exists as an $\alpha\beta$ dimer. The loss of the heme (α_1) interactions with the β_2 Trp-37 is sufficient to account for the change of the CD spectrum of the hemoglobin Hirose compared with the normal hemoglobin tetramer.[70] Different dimer $\alpha_1\beta_1$ organization between normal and mutated San Diego (β109 Val→Met) hemoglobins in their liganded form has also been investigated. In each $\alpha\beta$ dimer of normal hemoglobin, the β109 Val residue interacts directly with the α103 His and α122 His residues.[71] Thus, the CD change in the Soret region between the two tetrameric proteins certainly arises from a perturbation in the coupling between the heme (β_2) and the histidine groups ($\alpha_1$103 and $\alpha_1$122) induced by a slight movement of the $\alpha_1\beta_1$ contact in hemoglobin San Diego.[72]

Interpretation of the CD, from gels and suspensions, is problematic because linear dichroism and birefringence complicate the real CD signals. Thus, observed spectra of suspensions (large aggregates or gels) are some-

[70] K. Mawatari, S. Matsukawa, and Y. Yoneyama, *Biochim. Biophys. Acta* **745**, 219 (1983).
[71] J. S. Sack, L. C. Andrews, K. A. Magnus, J. C. Hanson, J. Rubin, and W. E. Love, *Hemoglobin* **2**, 153 (1978).
[72] S. El Antri, C. Zentz, and B. Alpert, *Eur. J. Biochem.* **179**, 165 (1989).

times different from the spectrum of the same molecules in solution. For example, apparent CD spectra in the Soret region of sickle-cell deoxyhemoglobin (deoxyHb S) fiber gels are radically different from the CD of deoxyHb S in solution.[73] The observed spectra of Hb S reflect the superposition of the linear birefringence (associated with oriented domains of the gels), and the linear and the circular dichroisms from the heme. It has not been possible to determine the relative importance of each contribution. Thus, the apparent CD is not entirely associated with the molecular CD of the chromophore.

Mollusk Hemoglobin. The tetrameric (Hb II) and dimeric (Hb I) hemoglobins from the mollusk *S. inaequivalvis* show Soret CD spectra similar to those of vertebrate hemoglobins: a large positive band is present in all derivatives. However, the deoxygenated derivatives of both Hb I and Hb II exhibit an additional negative band at about 425 nm, which is absent in human hemoglobin. This fact can be interpreted by changes in the direction of the pair of transition moments in the heme plane.[62]

Annelid Hemoglobin. Circular dichroism spectra of several liganded derivatives of the monomer and polymer hemoglobins of *G. dibranchiata* have been investigated in the Soret region.[35,74–76] Interestingly, in contrast to vertebrate hemoglobins, all of the monomer-liganded forms of *G. dibranchiata* display predominantly negative transitions in the Soret region. Only two low-spin ligands, carbonmonoxy and azidomet, show an additional positive transition within this region. Nuclear magnetic resonance (NMR) studies on several heme proteins demonstrate that these molecules exist in two forms differing by 180° in the orientation of the heme group about the α–γ axis.[77,78] The A form (found in the sperm whale myoglobin) has the Phe residue CD1 located near the δ mesoproton and the pyrrole ring.[79] The reversed B form is found on various *G. dibranchiata* hemoglobin derivatives.[80] On the basis of these independent NMR studies, a correlation appears evident between heme orientation in hemoglobin and negative ellipticity in the Soret CD spectrum (Fig. 4). In contrast, the giant

[73] R. P. Hjelm, Jr., P. Thiyagarajan, and M. E. Johnson, *Biopolymers* **25**, 1359 (1986).

[74] A. W. Addison and J. J. Stephanos, *Biochemistry* **25**, 4104 (1986).

[75] R. Santucci, J. Mintorovitch, I. Constantinidis, J. D. Satterlee, and F. Ascoli, *Biochim. Biophys. Acta* **953**, 201 (1988).

[76] T. J. Difeo and A. W. Addison, *Biochem. J.* **260**, 863 (1989).

[77] G. N. La Mar, D. L. Budd, D. B. Viscio, K. M. Smith, and K. C. Lungry, *Proc. Natl. Acad. Sci. U.S.A.* **75**, 5755 (1978).

[78] G. N. La Mar, K. M. Smith, K. Gersonde, H. Sick, and M. Overkamp, *J. Biol. Chem.* **255**, 66 (1980).

[79] T. Takano, *J. Mol. Biol.* **110**, 537 (1977).

[80] R. L. Kandler, I. Constantinidis, and J. D. Satterlee, *Biochem. J.* **226**, 131 (1985).

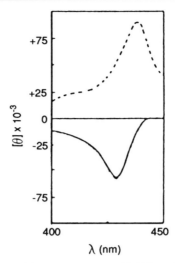

FIG. 4. Soret CD spectra of deoxy components II, III, and IV (—) of *G. dibranchiata* hemoglobin. Comparison with deoxygenated sperm whale myoglobin (---). Phosphate buffer, (0.1 *M*), pH 7, 25°. The molar ellipticity [θ] (deg · cm^2 · dmol^{-1}) is expressed on molar hemebasis. [Reprinted with permission from R. Santucci *et al., Biochim. Biophys. Acta* **953,** 201 (1988).]

hemoglobin of the polychaete annelid *Cirriformia tentaculata* presents major heme orientation identical with that of human hemoglobin.[81]

Visible Region

Human Hemoglobin. Amides and alcohols are known to increase and to decrease hemoglobin affinity for its ligands. X-ray absorption near-edge structure (XANES) and Soret CD measurements show that no core deformation or heme reorientation occurs with the affinity changes.[82] The cosolvents produce small shifts in the visible domain of the CD spectra. The absence of any cosolvent effect in the CD spectrum of the ligand-free hemoglobin shows that the iron–ligand bond angle is the sole parameter that depends on the external solvent. Thus, each cosolvent induces constraints that favor and stabilize a particular iron–ligand configuration.[82]

The effect of the replacement of the histidine, either at position E7 (distal histidine) or at the F8 position (proximal histidine), on the heme

[81] T. J. Difeo, A. W. Addison, and T. F. Kumosinski, *Comp. Biochem. Physiol. B* **97B,** 391 (1990).

[82] C. Zentz, S. El Antri, S. Pin, R. Cortes, A. Massat, M. Simon, and B. Alpert, *Biochemistry* **30,** 2804 (1991).

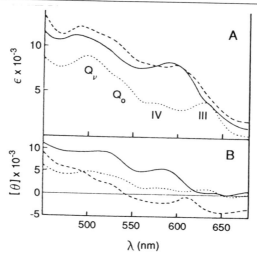

FIG. 5. Visible absorption (A) and CD (B) spectra of abnormal human hemoglobins M. MetHb M (Iwate (—) and metHb M Boston (– – –) are compared with normal metHb (···). Hemoglobins (100 μM in heme) are in 0.05 M Bis–Tris buffer, containing 0.1 M NaCl, pH 7.0. The letters Q_v and Q_0 are used for the $\pi \rightarrow \pi^*$ transition absorption bands and Roman numerals III and IV signify the charge transfer bands[47] of human metHb. [Reproduced with permission from M. Nagai et al., Biochem. Biophys. Res. Commun. 128, 689 (1985).]

structure of hemoglobin has been measured through visible CD. The hemoglobins examined are Hb M Boston [α (E7) His→Tyr], Hb M Iwate [α (F8) His→Tyr], Hb M Saskatoon [β (E7) His→Tyr], and Hb M Hyde Park [β (F8) His→Tyr].[83–85] Although these hemoglobins M have similar characteristic absorption spectra, their CD spectra exhibit great differences at long wavelength. The proximal histidine-replaced hemoglobins M show only positive CD in the visible region, whereas the distal histidine-replaced hemoglobins M display a negative CD band at 650 nm (Fig. 5). The changes probably originate from the charge transfer between the iron and the porphyrin or between the iron and the axial ligand.[47] The spectrum of another hemoglobin M, hemoglobin Milwaukee [β (E11) Val→Glu], has also been examined. The visible CD spectrum of methemoglobin Milwaukee is not very different from that of normal methemoglobin. The relatively distant amino acid substitution from iron could explain this observation by the absence of energy transfer.

[83] M. Nagai, S. Takama, and Y. Yoneyama, Biochem. Biophys. Res. Commun. 128, 689 (1985).
[84] M. Nagai, S. Takama, and Y. Yoneyama, Stud. Biophys. 116, 135 (1986).
[85] M. Nagai, S. Takama, and Y. Yoneyama, Acta Haematol. 78, 95 (1987).

Artificial Hemoglobins. Sheep apohemoglobin complexes with magnesium protoporphyrin ($Hb^{Fe \rightarrow Mg}$) and magnesium mesoporphyrin (MP-$Hb^{Fe \rightarrow Mg}$) have been studied by visible CD.[86] The Q_0 transition for $Hb^{Fe \rightarrow Mg}$ has a single band (589 nm), whereas the Q_1 transition is split into two bands (548 and 558 nm). The shapes of the bands (Q_0 band at 576 nm and Q_1 at 542 and 549 nm) of MP-$Hb^{Fe \rightarrow Mg}$ are essentially the same as those of $Hb^{Fe \rightarrow Mg}$. The authors compare these CD data to those obtained with analog myoglobin. They conclude that magnesium is five coordinated in hemoglobin, whereas magnesium is six coordinated in myoglobin (the sixth ligand being a water molecule). Visible CD spectra of artificial leghemoglobin are compared to natural protoleghemoglobin.[46] A striking split in the Q bands indicates that the geometry of the iron–ligand bond would be correlated with the electron affinity of the side chains, vinyl and acetyl groups at the 2,4-position of heme.

Fish Hemoglobin. Visible CD spectra have been made of the major hemoglobin component (Hb IV) of the rainbow trout *Salmo irideus.*[57] Significant differences are observed on the carbonmonoxy derivative between pH 6.2 and 7.8. The visible CD spectrum is also affected at pH 6.2 by the presence of inositol hexaphosphate. The authors correlate these changes with the physiological properties of Hb IV. Indeed, the liganded protein stabilization is effector dependent (protons and IHP). These optical CD changes, from 500 to 600 nm, imply a release of the heme asymmetry[23,87] due to the nonperpendicular iron–ligand bond above the *xy* plane of the heme group. The chromatographically slow component of swordfish HbCO also displays a marked reduction of the 570-nm band as the pH is lowered or IHP added. This component appears to have properties like that of trout Hb IV.[88]

Mollusk Hemoglobin. The heme environments of dimeric and tetrameric hemoglobins of the mollusc *S. inaequivalvis* have been studied by visible CD.[62] The spectra of both hemoglobins are almost identical and resemble that of human hemoglobin. The deoxygenated, oxygenated, and carbonmonoxy derivatives show two dichroic bands due to the Q transitions. In the liganded species the bands are centered at approximately the same wavelengths of the absorption maxima as in human hemoglobin. In the deoxygenated derivatives, at variance with human hemoglobin, there are two Q bands of similar amplitude at 565 and 586 nm. The unusually high ellipticity of the Q_0 band suggests a more constrained structure at the

[86] C. C. Ong and G. A. Rodley, *J. Inorg. Biochem.* **19,** 189 (1983).
[87] W. A. Eaton, L. K. Hanson, P. J. Stephens, J. C. Sutherland, and J. B. R. Bunn, *J. Am. Chem. Soc.* **100,** 4991 (1978).
[88] J. M. Friedman, S. R. Simon, and T. W. Scott, *Copeia* **3,** 679 (1985).

heme site in the deoxygenated conformation of the molluskan hemoglobins than in that of human hemoglobin. Such spectral changes should be due to the lengthening of the iron–N_ε bond[38] and/or the orientation and tilting of the proximal imidazole relative to the heme.[89]

Time-Resolved Circular Dichroism

Apparatus for Millisecond Time Range

Stationary CD measurements are performed through a piezooptical modulator, which gives the elements of circular polarization at the light. This type of modulation offers the advantage of a wide acceptance angle and an operating frequency enabled for fast CD measurements with time resolution on the order of milliseconds. A summary of the different instruments using the piezooptical modulation methods is given by Bayley.[90]

Luchins and Beychok[91] have constructed a stopped-flow CD instrument with a wavelength-accessible range from 200 to 750 nm. The instrument uses a stabilized xenon light source, a 50-kHz piezooptical birefringence modulator, and a phase-sensitive heterodyning lock-in amplification technique. The dead time of the apparatus is 12 msec. Kawamura *et al.*[66] have made CD stopped-flow measurements with a Union-Giken rapid-scan reaction analyzer (RA-1300) and a Union-Giken dichrograph III-j. Fronticelli and Bucci[92] have followed the time variations of the optical activity, using a modified Jasco-20 spectropolarimeter equipped with a Jasco stopped-flow apparatus. The Pockels cell is replaced by a modulated quarter-wave plate (50.3 kHz). A reference signal from the piezoelastic modulator is sent to a phase detector. Although the instrument has a 10-msec dead time, the analysis is started 100–150 msec after solution mixing.

Apparatus for Nanosecond and Microsecond Time Scale

The time resolution of CD measurements has been extended to the nanosecond scale by a new technique.[93] A linear polarized light beam is sent through a strained quartz plate. Depending on the orientation of the plate, right or left elliptical polarized light is produced. The light ellipticity

[89] C. M. Wang and W. S. Brinigar, *Biochemistry* **18,** 4960 (1979).
[90] P. M. Bayley, *Prog. Biophys. Mol. Biol.* **37,** 149 (1981).
[91] J. Luchins and S. Beychok, *Science* **199,** 425 (1978).
[92] C. Fronticelli and E. Bucci, *Biophys. Chem.* **23,** 125 (1985).
[93] J. W. Lewis, R. F. Tilton, C. M. Einterz, S. J. Milder, I. D. Kuntz, and D. S. Kliger, *J. Phys. Chem.* **89,** 289 (1985).

is modified by passing through the sample. The ellipticity change is monitored with a second polarizer perpendicularly oriented to the first one. This determines the magnitude of the minor axis from the polarization ellipse of the probe beam, which, in turn, is related to the CD of the sample. In this study a carbonmonoxyhemoglobin sample is photolyzed by a 7-nsec pulse (532 nm) from a neodymium : yttrium–aluminum–garnet (Nd : YAG) laser. The hemoglobin photolysis studies raise interesting new possibilities, but improvement in the apparatus is required.

Experimental Kinetic Data

Leutzinger and Beychok[94] have studied the kinetics of heme binding to the apohemoglobin-α chain, and the protein reorganization after this binding. By using three stopped-flow techniques they observed (1) the far-UV circular dichroism to gauge the recovery of secondary structure (Fig. 6), (2) the Soret absorption, whose position and intensity give information on the heme environment and the fraction of the heme protein associated, and (3) the tryptophan fluorescence quenching, to provide the distance between α14 Trp and the heme.

The kinetics of the tryptophan fluorescence quenching is biphasic: an initial second-order decay representing 80–85% of the total amplitude marks the binding of the heme into the heme pocket. Ultraviolet circular dichroism and Soret absorption kinetics are both multiphasic (Fig. 6). The first-order processes of the three kinetics studied by the different techniques have the same halftime of 25–40 sec. During this time the secondary and tertiary structures of the heme pocket are established. A slower CD time evolution, with a halftime of 116 sec, marks the full acquisition of the native subunit conformation. In summary, it seems that the heme pocket achieves its final three-dimensional structure well before the entire chain is completely refolded.[94]

Denaturation kinetics of hemoglobin with pH have been performed at 10° using 0.07-mg/ml protein solutions in 0.1 M NaCl. In the stopped-flow apparatus the protein solution is mixed with an equal volume of acid solution to obtain the desired pH. Total transient times range from 2 to 20 sec. The observed changes in CD at 225 nm can be simulated by two exponentials for all pH values. The large amount of secondary structure remaining 5 min after mixing suggests the presence of other kinetic components too slow to be detected on the time scale used. Thus, the acid denaturation is a complex event involving various structural domains, each characterized by their own rate.[92]

[94] Y. Leutzinger and S. Beychok, *Proc. Natl. Acad. Sci. U.S.A.* **78,** 780 (1981).

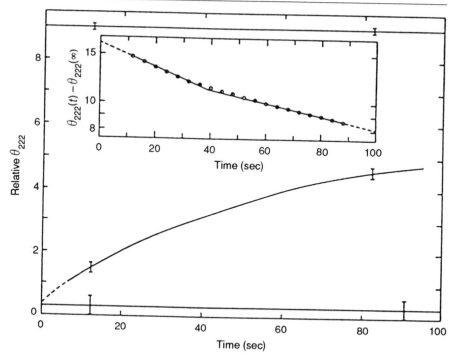

FIG. 6. Far-UV CD recovery for the reaction at 4° of an α subunit of apoglobin (in 20 mM potassium phosphate buffer, pH 5.7) with equimolecular hemin dicyanide (in 20 mM potassium phosphate buffer, pH 6.7). Final concentration of associated heme protein, 2.4 μM. Inset: First-order plot of the data. [Reproduced with permission from Y. Leutzinger and S. Beychok, Proc. Natl. Acad. Sci. U.S.A. **78**, 780 (1981).]

Kinetics of α- and β-chain association have been carried out at pH ≈ 7.5 by mixing equimolecular amounts of the α and β chains under three buffer conditions: 0.1 M Tris-HCl and 0.1 and 0.01 M phosphate buffers. The protein concentration range is 2×10^{-6} to 10^{-5} M. Total transient times for hemoglobin reconstitution range from milliseconds to seconds. Circular dichroism changes are measured at 285 nm and in the Soret region. From their experimental kinetic curves, the authors consider that the buffer conditions would affect the β tetramer dissociation rather than the β monomer combination with the α chain.[66,95,96] Dissociation of oxygenated $\beta_4 \rightarrow \beta$ would have a rate constant on the order of 2.8×10^{-3} sec^{-1}.

[95] Y. Kawamura and S. Nakamura, J. Biochem. (Tokyo) **93**, 1159 (1983).
[96] Y. Kawamura and S. Nakamura, J. Biochem. (Tokyo) **94**, 1851 (1983).

The $\alpha\beta$ association would have a rate constant of $7.5 \times 10^5 \, M^{-1} \sec^{-1}$ in the oxygenated form, and $6.4 \times 10^5 \, M^{-1} \sec^{-1}$ in the deoxy form. The dimer→tetramer formation in the deoxy form would have a rate constant of $1 \times 10^6 \, M^{-1} \sec^{-1}$.

[14] X-Ray Absorption Spectroscopy of Hemoglobin

By Serge Pin, Bernard Alpert, Agostina Congiu-Castellano, Stefano Della Longa, and Antonio Bianconi

Introduction

The local structure of the active site of hemoglobin and its variation on ligand binding is a key system for understanding the general aspects of the relationship between protein structure and function. X-Ray diffraction provides the standard structural probe to measure the atomic positions in a protein. The diffraction methods have several limitations that are of relevance for a biological system. The sample should be a protein crystal, and structural information on the Fe active site in the diffraction patterns is averaged over a large number of protein molecules within the diffraction correlation length. Therefore the diffraction methods fail to investigate some important problems such as (1) the interaction of the protein with the solvent and allosteric effectors or (2) the landscape of the protein conformations that modulate the dynamics of the protein. Several spectroscopic methods have been used to study proteins in solution and protein dynamics; however, generally the spectral features are only indirectly related to the atomic distribution in the space.

X-Ray absorption spectroscopy (XAS) of hemoglobin has provided an experimental method for probing the Fe site structural configurations in solution. The spectral features in the XAS spectra are determined by the scattering of the photoelectrons, emitted by the absorbing Fe ion of hemoglobin and returned to the Fe site by the neighboring atoms within a sphere of about 5 Å. Therefore the XAS method is a direct probe of atomic distribution via electron diffraction. It probes the spatial distribution of the atoms of the heme, the proximal histidine, and the ligand.

In the X-ray energy range between 7090 and 8000 eV the Fe ion in hemoglobin determines the absorption of X-ray photons. Photons of this energy can excite an electron from the iron $1s$ level to the unoccupied states, while the rest of the protein is transparent to X rays. The absorption spectrum exhibits an absorption edge at the energy required to excite the core electron to the lowest unoccupied molecular orbital (LUMO). The expected transitions to the unoccupied molecular orbitals of the Fe–heme complex from the core level are known as Kossel structure.[1] However, the X-ray absorption spectrum extends about 800 eV beyond the absorption edge, while the molecular orbitals are expected to extend about 5 eV above the LUMO. The absorption spectrum beyond 5 eV from the edge is due to transitions to high-energy states in the continuum, which can be described as due to the scattering of the high-energy photoelectrons excited in the continuum from the neighboring atoms.

The spectral features due to final states in the continuum over a low energy range, of about 60 eV, constitute the X-ray absorption near-edge structure (XANES)[2-4]; the spectral features in the higher energy region[5-7] from 60 to 800 eV, due to single scattering, constitute the extended X-ray absorption fine structure (EXAFS).[8,9] In the XANES region the high probability for scattering of low-energy (10–60 eV) photoelectrons by neighboring atoms requires that all the scattering pathways starting and ending at the Fe site contribute to the XANES pattern. Therefore in the XANES region the electrons are in the full multiple scattering regime. In the high-energy EXAFS region the low probability for scattering of photoelectrons with high kinetic energy, 60–800 eV, allows one to neglect the multiple scattering processes, and only single or triple scattering processes determine the measured spectral features. In Fig. 1, the multiple scattering (MS) pathways of the outgoing electron emitted at the Fe site, involving scattering by the ligand or by the heme nitrogens and ending at the same Fe site, are shown. The pathways can be classified according to the number (n) of neighboring atoms involved in the scattering.

Therefore the XANES probes the higher order correlations of the atomic distribution of the atoms within a sphere of 5 Å, with the Fe atom

[1] W. Kossel, Z. Phys. **1**, 119 (1920).

[2] A. Bianconi, Appl. Surf. Sci. **6**, 392 (1980).

[3] A. Bianconi, in "X-Ray Absorption: Principles, Applications, Techniques of EXAFS, SEXAFS and XANES" (D. C. Koningsberger and R. Prins, eds.), p. 573. Wiley, New York, 1988.

[4] P. J. Durham, in "X-Ray Absorption: Principles, Applications, Techniques of EXAFS, SEXAFS and XANES" (D. C. Koningsberger and R. Prins, eds.), p. 53. Wiley, New York, 1988.

[5] R. de L. Kronig, Z. Phys. **70**, 317 (1931).

[6] R. de L. Kronig, Z. Phys. **75**, 191 (1932).

[7] R. de L. Kronig, Z. Phys. **75**, 468 (1932).

[8] D. E. Sayers, F. W. Lytle, and E. A. Stern, Adv. X-Ray Anal. **13**, 248 (1970).

[9] E. A. Stern, in "X-Ray Absorption: Principles, Applications, Techniques of EXAFS, SEXAFS and XANES" (D. C. Konigsberger and R. Prins, eds.), p. 3. Wiley, New York, 1988.

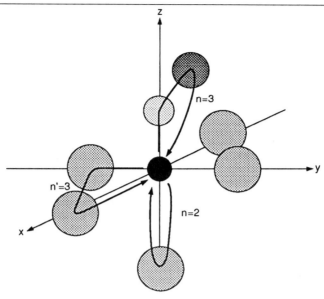

FIG. 1. Pictorial view of the single and multiple scattering processes of the excited photoelectron giving the XAS signals. The central sphere represents the absorbing iron atom and the other spheres indicate the neighboring atoms. The arrows represent the pathways of the excited photoelectron outgoing from the central atom and ending at the same iron site. The single scattering pathway, indicated by $n = 2$, contributes to the EXAFS modulation term $\chi_2(E)$, the two pathways $n = 3$ and $n' = 3$ contribute to the double scattering term $\chi_3(E)$. (From Bianconi[63] and reproduced with the permission of Springer-Verlag.)

at its center. It probes not only interatomic distances but also bond angles and the overall geometry of the Fe–heme complex. The EXAFS spectrum probes mainly the pair-correlation function, and therefore it is a direct probe of the distances of neighboring atoms from the Fe ion.

X-Ray absorption spectroscopy of metalloproteins was developed only after 1975, thanks to the use of synchrotron radiation, because of the lack of intense continuum X-ray sources. An intense X-ray beam is necessary because only a single iron atom per protein subunit contributes to the XAS spectrum. Hemoglobin samples with about 5 mM heme concentration have been studied by using an X-ray synchrotron radiation beam line on a bending magnet at a synchrotron radiation facility.[10–17] To study more

[10] J. Jaklevic, J. A. Kirby, M. P. Klein, A. S. Robertson, G. S. Brown, and P. Eisenberger, *Solid State Commun.* **23**, 679 (1977).

[11] J. E. Penner-Hahn and K. O. Hodgson, *in* "Iron Porphyrins" (A. B. P. Lever and H. B. Gray, eds.), part III, p. 235. VCH, Verlagslesellschaft, Weinheim, Germany.

dilute samples the X-ray intensity of the source has been increased and fluorescence detectors have been developed. The use of wiggler insertion devices in the storage rings provides higher intensity sources, and the X-ray fluorescence detectors[10] are used to study iron heme proteins.[11] In the near future the use of the new, dedicated high-brilliance synchrotron radiation X-ray sources, under construction at Grenoble and Argonne, and sensitive, fast, large-area X-ray detectors should allow the study of more diluted systems in the range 10–100 μM.

The X-ray absorption spectrum of hemoglobin can be divided into three regions. In the first region of about 4 eV, or edge region, only very weak absorption peaks (due the transitions to the molecular orbitals formed by mixing the Fe $3d$ with ligand orbitals) appear in the spectrum. The $1s \rightarrow 3d$ transition is dipole forbidden, following the dipole selection rule $\Delta l = \pm 1$; therefore only transitions from the Fe $1s$ level to the p-like components of the molecular orbitals increasing with the deviation from the square plane or octahedral Fe coordination contribute to the absorption. Therefore this part of the spectrum is sensitive to the chemical bonding. The increase of the Fe metal effective charge induces an increase in the Fe $1s$ level binding energy and therefore a shift of the absorption threshold E_0 and of the full absorption spectrum is expected.

The second part of the spectrum, the XANES region, beyond ~5 eV, exhibits the absorption rising edge that corresponds to the threshold of dipole-allowed transitions to continuum states with p symmetry, and strong peaks up to about 60 eV. The third part at higher energies, the EXAFS region, exhibits weak modulations of the atomic absorption. The measured absorption coefficient, $\mu(E)$, a function of the incident X-ray energy E, can be described both in the XANES and EXAFS part as the product of $\mu_0(E)[1 + \chi(E)]$, where $\mu_0(E)$ is the structureless atomic absorption to the states in the continuum, and $\chi(E)$ is the structural term due to the scattering of the photoelectron from neighboring atoms. The structural information in the XANES and EXAFS spectra is given by the

[12] R. G. Shulman, Y. Yafet, P. Eisenberger, and W. E. Blumberg, *Proc. Natl. Acad. Sci. U.S.A.* **73**, 1384 (1976).

[13] P. Eisenberger and B. M. Kincaid, *Science* **200**, 1441 (1978).

[14] R. G. Shulman, P. Eisenberger, and B. M. Kincaid, *Annu. Rev. Biophys. Bioeng.* **7**, 559 (1978).

[15] B. M. Kincaid, P. Eisenberger, K. O. Hodgson, and S. Doniach, *Proc. Natl. Acad. Sci. U.S.A.* **72**, 2340 (1975).

[16] P. Eisenberger, R. G. Shulman, G. S. Brown, and S. Ogawa, *Proc. Natl. Acad. Sci. U.S.A.* **73**, 491 (1976).

[17] P. Eisenberger, R. G. Shulman, B. M. Kincaid, G. S. Brown, and S. Ogawa, *Nature (London)* **274**, 30 (1978).

modulation term $\chi(E) = [\mu(E) - \mu_0(E)]/\mu_0(E)$. The Broglie wavelength, λ, of the propagated electron wave is defined by $\lambda = h/[2m_e(E - V_0)]^{1/2}$, where h is the Planck constant, m_e is the electron mass, and V_0 is the average interstitial potential, also called muffin-tin zero, of the potential seen by the photoelectron. Therefore the energy scale of the absorption spectra can be converted into a wavevector scale $\hbar\kappa = 1/\lambda$; the EXAFS region spans the wavevector range of 3–15 Å^{-1}, while the XANES region spans the range 1–3 Å^{-1}.

The first EXAFS data analysis of hemoglobin was carried out by using the plane wave approximation for the photoelectron of wavelength λ and considering only scattering pathways where the outgoing photoelectron is reflected by the neighboring atoms toward the absorbing center (single scattering regime). By using these two approximations the structural term $\chi(E)$ is given by a simple linear combination of $\sin(2kR + \phi)$ terms, where R values are the absorber–scatterer distances and ϕ is the backscattering phase function. Therefore the distances R of neighboring atoms from the iron atom can be obtained.

In this work we first discuss the results obtained by the XAS experimental method on the Fe site structure of hemoglobin and then we report the basic principles of the X-ray absorption spectroscopic method.

Structure Determination by EXAFS

The first EXAFS spectrum of hemoglobin, using very intense X rays from a high-energy electron storage ring, was measured in 1975 at Stanford.[15] The iron-to-porphyrin nitrogen (Fe–N$_p$) bond distances in the deoxygenated (deoxyHb), carbonmonoxy (HbCO), and oxygenated (HbO$_2$) forms of hemoglobin A were determined by fluorescence X-ray absorption instead of by traditional transmission studies. The average Fe–N$_p$ measured distances were 1.98 Å for both HbCO and HbO$_2$ and 2.055 Å for deoxyHb[16,17] (Fig. 2). The experimental accuracy was ±0.01 Å.

A comparison was made between the EXAFS spectra of the low oxygen affinity form of deoxyHb A and the high-affinity form of deoxyHb Kempsey [Asp(β99)Asn]. The Fe–N$_p$ distances led to the conclusion that the spacings are identical within $\Delta x = \pm 0.02$ Å. The absence of bond distance changes between the low- and high-affinity forms of deoxyhemoglobin has energetic consequences for the cooperative oxygenation. Indeed, the difference in oxygen-binding energies to the high- and low-affinity forms (without organic phosphates) is 3.4 kcal or 0.15 eV.[18] The strain energy at the iron, being proportional to Δx^2, reveals that only 2.7%

[18] I. Tyuma, K. Imai, and K. Shimizu, *Biochemistry* **12**, 1491 (1973).

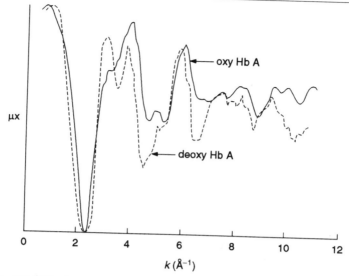

FIG. 2. EXAFS of adult human oxyhemoglobin (oxyHb A) and deoxyhemoglobin (deoxyHb A). The modulation of the X-ray absorption (μx) is plotted as a function of the wavevector scale (k) of the photoelectron. (From Eisenberger *et al.*[16] and reproduced with the permission of the authors.)

of the binding energy is stored in the Fe–N_p bonds, which is negligible. These data showed that the energies responsible for the increase of oxygen affinity are not localized at the heme iron.[16]

There is a small change in the Fe–N_p distance on deoxygenation (from 1.98 to 2.055 Å). The Perutz model estimated a distance of 2.18 Å in deoxyhemoglobin. This would place the iron atom at 0.83 Å out of the porphyrin plane.[19] This magnitude was sufficient to explain the movement of the F8 proximal histidine toward the heme plane during the transition from deoxy- to oxyhemoglobin.[19,20] Crystallographic studies of human deoxyhemoglobin at 2.5 Å allowed a measurement of the iron atom distance from the porphyrin plane. The distance, assuming a planar porphyrin, was 0.60 Å for the α-heme and 0.63 Å for the β-heme.[21] The Fe–C_t distance from the iron atom to the center of the porphyrin ring was also obtained by EXAFS studies. The Fe–C_t distance was calculated by triangulation, using a C_t–N_p distance of 2.045 Å measured with the 2-methylimidazole iron tetraphenylporphyrin model compound [Fe(TPP)(2-MeIm)].

[19] M. F. Perutz, *Nature (London)* **228**, 726 (1970).
[20] J.L. Hoard, *Science* **174**, 1295 (1971).
[21] G. Fermi, *J. Mol. Biol.* **97**, 237 (1975).

TABLE I

DISTANCES AROUND IRON ATOM IN DEOXYHEMOGLOBIN[a]

Method	Distance (Å)			Ref.
	$Fe-N_p$	C_t-N_p	$Fe-C_t$	
EXAFS	2.06 ± 0.01	2.04 ± 0.01	0.20 (+0.10, −0.20)	17
High-resolution X-ray diffraction	2.06 ± 0.02	2.02	0.38 ± 0.04	25

[a] From Fermi et al.[27] and reproduced with the permission of the authors.

The $Fe-C_t$ distance was 0.2 (+0.10, −0.20) Å.[17] This result was interpreted as showing that the iron atom is not forced out of the porphyrin plane during the deoxygenation.[22]

Because the heme distances obtained from EXAFS studies were shorter than those from crystallography, new X-ray absorption and diffraction experiments and analyses were undertaken on hemoglobin and model compounds. All studies[23–26] found exactly the same $Fe-N_p$ distance of 2.06 Å in deoxyhemoglobin as originally determined by Eisenberger et al.[17] The measurement of the $Fe-C_t$ distance was finally resolved:[27] the distance was shortened at first to 0.56 Å by a novel difference refinement procedure on X-ray data[23] and then to 0.38 Å ± 0.04 Å from a high 1.74-Å resolution crystallographic study.[25] This value is not far from the upper limit of the distance of 0.2 (+0.10, −0.20) Å calculated by triangulation from the first EXAFS experiments[17] (Table I).

The considerable discussion concerning the movement of the iron atom during the protein conformational changes on ligation was extended to a hemoglobin that exhibits a conformational change with pH. At pH 9, carp hemoglobin is stabilized in the high-affinity state, and at pH 6 it is locked into the low-affinity state.[28] When carp HbCO was switched from high to low affinity, it showed no differences in iron–ligand distances: 2.015 Å for the $Fe-N_p$ distance, 2.14 Å for Fe–N (of histidine), and 1.89 Å for Fe–C (of CO). This study suggested that the quaternary changes can

[22] R. G. Shulman, in "Transport by Proteins" (G. Blauer and H. Sund, eds.). p. 95. de Gruyter, Berlin, 1978.

[23] G. Fermi, M. F. Perutz, L. C. Dickinson, and J. C. W. Chien, J. Mol. Biol. 55, 495 (1982).

[24] M. F. Perutz, S. S. Hasnain, P. J. Duke, J. L. Sessler, and J. E. Hahn, Nature (London) 295, 535 (1982).

[25] G. Fermi, M. F. Perutz, B. Shaanan, and R. Fourme, J. Mol. Biol. 175, 159 (1984).

[26] R. G. Shulman, Proc. Natl. Acad. Sci. U.S.A. 84, 973 (1987).

[27] G. Fermi, M. F. Perutz, and R. G. Shulman, Proc. Natl. Acad. Sci. U.S.A. 84, 6167 (1987).

[28] R. Noble, L. Parkhurst, and Q. Gibson, J. Biol. Chem. 245, 6628 (1970).

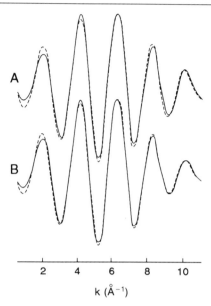

FIG. 3. Comparison of zinc EXAFS contributions. The solid curve is the spectrum of $(\alpha^{Fe\to Zn})_2(\beta^{Fe\to Zn})_2$ compared with $(\alpha CO)_2(\beta^{Fe\to Zn})_2$ (A) and with $(\alpha^{Fe\to Zn})_2(\beta CO)_2$ (B). (Reprinted with permission from Simolo *et al.*[31] Copyright [1986] American Chemical Society.)

induce a 0.1-Å motion of the iron atom out of the heme plane but not a large-scale movement of 0.4 Å, as on ligation.[29]

Metal distances were measured in hybrid hemoglobins in which Zn^{II} replaced Fe^{II}. These molecules mimic the intermediate ligation states of the hemoglobin molecule.[30] Because zinc and iron atoms have different K-edge energies, the structures of the ligated iron and unligated zinc sites were analyzed independently. The ligand-free $(\alpha^{Fe\to Zn})_2(\beta^{Fe\to Zn})_2$, the half-ligated $(\alpha CO)_2(\beta^{Fe\to Zn})_2$ and $(\alpha^{Fe\to Zn})_2(\beta CO)_2$, and the fully ligated $(\alpha CO)_2(\beta CO)_2$ molecules presented no appreciable strain in the metal–N_p distances[31] (Fig. 3). If the iron–imidazole bond is responsible for reducing the iron affinity,[32] then this "proximal strain" is not manifested in an Fe–N_p length change. An angular strain and/or a length variation of

[29] M. R. Chance, L. J. Parkhurst, L. S. Powers, and B. Chance, *J. Biol. Chem.* **261**, 5689 (1986).

[30] W. R. Scheidt, *Acc. Chem. Res.* **10**, 339 (1977).

[31] K. Simolo, Z. R. Korszun, G. Stuky, K. Moffat, G. McLendon, and G. Bunker, *Biochemistry* **25**, 3773 (1986).

[32] M. F. Perutz, *Annu. Rev. Biochem.* **48**, 327 (1979).

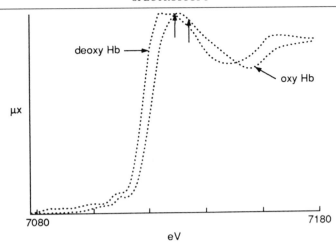

μx

7080   7180

eV

FIG. 4. X-Ray absorption spectra of oxyhemoglobin (oxyHb) and deoxyhemoglobin (deoxyHb). The arrows indicate the iron K-edge energy, which was 7130 eV for deoxyhemoglobin and 7135 eV for oxyhemoglobin. (From Eisenberger et al.[16] and reproduced with the permission of the authors.)

metal–sixth ligand bond could also explain the affinity changes.[33] The formation of long and short Fe–O (of O_2) bonded structures, associated with Fe–N (of histidine) distance variations, was observed in EXAFS studies of model compounds. These two structures offer a new, plausible interpretation of the cooperativity in oxygen binding.[34]

XANES Investigations

XANES Determination of Iron Electronic Arrangement

Eisenberger et al. observed in 1976 that the iron K-edge energy of oxyhemoglobin was 5 eV higher than that of deoxyhemoglobin[16] (Fig. 4). This result was confirmed by several independent XANES works[35–39] in

[33] K. Simolo, G. Stuky, S. Chen, M. Bailey, C. Scholes, and G. McLendon, *J. Am. Chem. Soc.* **107**, 2865 (1985).

[34] G. L. Woolery, M. A. Walters, K. S. Suslick, L. S. Powers, and T. G. Spiro, *J. Am. Chem. Soc.* **107**, 2370 (1985).

[35] S. Pin, B. Alpert, and A. Michalowicz, *FEBS Lett.* **147**, 106 (1982).

[36] A. Bianconi, A. Congiu-Castellano, M. Dell'Ariccia, A. Giovannelli, E. Burattini, M. Castagnola, and P. J. Durham, *Biochim. Biophys. Acta* **831**, 120 (1985).

[37] A. Bianconi, A. Congiu-Castellano, M. Dell'Ariccia, A. Giovannelli, E. Burattini, and P. J. Durham, *Biochem. Biophys. Res. Commun.* **131**, 98 (1985).

[38] S. Pin, P. Valat, R. Cortes, A. Michalowicz, and B. Alpert, *Biophys. J.* **48**, 997 (1985).

[39] A. Congiu-Castellano, *in* "Biophysics and Synchrotron Radiation" (A. Bianconi and A. Congiu-Castellano, eds.), p. 89. Springer-Verlag, Berlin, 1987.

which the energy shift was found to be from 3.5 to 6 eV. In fact, the differences between different reports depend on the definition of the energy at the iron K-edge. Several features appear on the rising absorption edge and at the main peak of spectra of deoxy- and oxyhemoglobin. These peaks are not identical because the XANES spectrum is modified by different $Fe-N_p$ distances (as discussed in the preceding section) and by the additional presence of the ligated oxygen molecule. In deoxyhemoglobin, the ferrous state of the iron determines the paramagnetic character of the molecule. In oxyhemoglobin, one oxygen molecule is associated with the iron atom and should give a ferric ion. However, the diamagnetic character of the oxyhemoglobin molecule[40] was interpreted more easily from a ferrous iron[41] than from an iron ionic complex, including electron redistribution.[42] This latter vision is, however, supported by infrared and Mössbauer spectroscopy studies that demonstrate a partial negative charge developed on the oxygen ligand. This is due to the formation of delocalized molecular orbitals with larger electron density on the distal oxygen.[43,44] Moreover, a hydrogen bond O_2-E7 histidine, proposed by crystallographic data[45-47] on the hemoglobin α subunit as on the myoglobin, confirmed the slight charge on the oxygen molecule, probably induced by the electronic redistribution inside the porphyrin–iron–oxygen complex. In the β subunit, if the hydrogen bond exists, it is a weak one.[45,46] The large energy shift of the deoxy to the oxyhemoglobin iron K-edge (including the effect of contraction of the $Fe-N_p$ distances and the effect of the presence of the oxygen molecule) cannot be explained without an increase of the iron effective charge as large as that found in chemical compounds going from Fe^{2+} to Fe^{3+}.[12] This is also confirmed by a shift of the oxyhemoglobin similar to that of aquo- and azidomethemoglobins[35,38,48] (Fig. 5).

The ionic state of the iron (Fe^{2+} or Fe^{3+}) can be separated from the effect due to the protein conformation. For each protein, the iron electronic states depend on the protein environment. For example, the differences in the local structure of the heme iron in the deoxygenated isolated α, β, and γ chains of adult and fetal hemoglobins induce small changes

[40] L. Pauling and C. D. Coryell, *Proc. Natl. Acad. Sci. U.S.A.* **22**, 210 (1936).

[41] L. Pauling, *Nature (London)* **203**, 182 (1964).

[42] J. J. Weiss, *Nature (London)* **202**, 83 (1964).

[43] C. H. Barlow, J. C. Maxwell, W. J. Wallace, and W. S. Caughey, *Biochem. Biophys. Res. Commun.* **55**, 91 (1973).

[44] G. R. Hoy, D. C. Cook, R. L. Berger, and F. K. Friedman, *Biophys. J.* **49**, 1009 (1986).

[45] B. Shaanan, *Nature (London)* **296**, 683 (1982).

[46] B. Shaanan, *J. Biol. Chem.* **171**, 31 (1983).

[47] S. E. V. Phillips and B. P. Schoenborn, *Nature (London)* **292**, 81 (1981).

[48] S. Morante, M. Cerdonio, S. Vitale, A. Congiu-Castellano, A. Vaciago, G. M. Giocometti, and L. Incoccia, *Springer Ser. Chem. Phys.* **27**, 352 (1983).

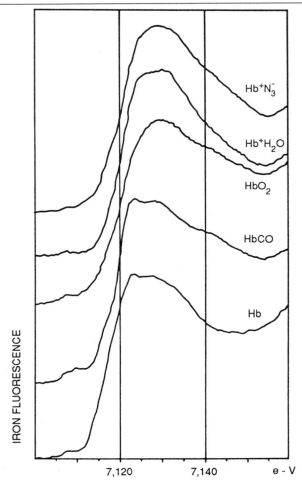

FIG. 5. Influence of the iron ligand nature on the position and shape of the hemoglobin iron XANES spectrum. From bottom to top, deoxy- (Hb), carbonmonoxy- (HbCO), oxy- (HbO$_2$), aquomet- (Hb$^+$H$_2$O), and azidomet- (Hb$^+$N$_3^-$) hemoglobin. (Reproduced from Pin et al.,[38] 1985, by copyright permission of the Biophysical Society.)

in the XANES spectra[35,49] (Fig. 6). By contrast, Fig. 5 shows large differences in the pattern of the XANES spectrum of hemoglobin with different ligands.[38,48]

In the same ligated ferric hemoglobin showing iron spin state equilib-

[49] A. Congiu-Castellano, M. Castagnola, E. Burattini, M. Dell'Ariccia, S. Della Longa, A. Giovannelli, P. J. Durham, and A. Bianconi, *Biochim. Biophys. Acta* **996**, 240 (1989).

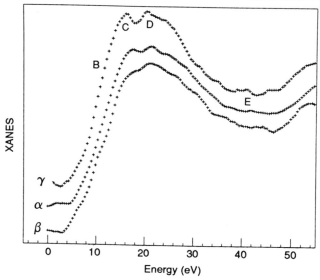

FIG. 6. Influence of the protein matrix nature on the iron XANES spectrum. From bottom to top, the isolated β, α, and γ chains in the deoxygenated form. The main spectral changes are indicated by B, C, D, and E. (From Congiu-Castellano et al.[49] and reproduced with the permission of Elsevier Science Publishers.)

rium, the small iron electronic reorganizations, caused by the spin state changes, were detected by XANES spectroscopy.[50,51] The iron atom is in equilibrium between two close electronic configurations ($^6A_1$ and $^2T_2$) characteristic of five and one unpaired electrons, respectively ($S = 5/2$ and $S = 1/2$). Thermal evolution of the spin equilibrium was studied in myoglobin. At 80 K, the iron of the hydroxymetmyoglobin was in a purely low-spin state and expected to be in the heme plane. The XANES spectrum exhibited a shoulder structure at 7122 eV. This shoulder gradually disappeared with the increase of high spin content, and at 300 K no shoulder structure was observed.[52] Carp azidomethemoglobin exhibits different

[50] S. Morante, A. Congiu-Castellano, M. Dell'Ariccia, P. J. Durham, A. Giovannelli, E. Burattini, and A. Bianconi, in "Biophysics and Synchrotron Radiation" (A. Bianconi and A. Congiu-Castellano, eds.), p. 107. Springer-Verlag, Berlin, 1987.

[51] S. Pin, V. Le Tilly, B. Alpert, and R. Cortes, FEBS Lett. 242, 401 (1989).

[52] H. Oyanagi, T. Iizuka, T. Matsushita, S. Saigo, R. Makino, and Y. Ishimura, in "Biophysics and Synchrotron Radiation" (A. Bianconi and A. Congiu-Castellano, eds.), p. 99. Springer-Verlag, Berlin, 1987.

iron spin state equilibria depending on external conditions.[53] In this compound, the iron spin change, at room temperature, is not induced by a significant out-of-plane iron motion but by a reorientation of the proximal histidine.[54,55] Thus, the iron d electron redistribution occurs without heme–iron geometric distortion. The pattern of the first derivative of XANES spectra revealed the fractions of high and low spin states between 7110 and 7130 eV[51] (Fig. 7). Thus, the iron K-edge is suitable to detect electronic differences as subtle as spin redistribution.

Therefore, the iron electronic reorganization, subsequent to a change in the affinity of hemoglobin for its ligands, should be detected in the K-edge position, or in the shapes of the XANES spectrum. Although XANES spectroscopy is sensitive, the Bohr effect was not found to perturb the electronic distribution of the hemoglobin iron.[38] This situation is analogous to that of the deoxygenated and ligated derivatives of human and carp hemoglobins in the absence and in the presence of the allosteric inositol hexaphosphate effector.[38,55–57] High signal-to-noise XANES spectra of carp hemoglobin showed that the R–T quaternary transition induced an Fe–N_p distance variation less than 0.01 Å.[55,56] All these results agree with the EXAFS data for the ligated form[29] and suggest that the hemoglobin oxygenation properties involve greater complexities than a simple iron–heme distance change. It appears that a reactivity change of the iron is improbable and that the high or low hemoglobin affinity can be associated with a change of the heme–ligand structure and/or a protein conformation modification.

Determination of Heme–Ligand Geometry

Although EXAFS spectroscopy gives accurate information regarding the distance of ligands surrounding the metal site, this technique cannot provide geometric information on the metal–ligand complex. Indeed, the EXAFS spectrum is due to a single scattering process with a spherical average.[58] Information on the coordination geometry is contained in the

[53] R. W. Noble, A. De Young, S. Vitale, S. Morante, and M. Cerdonio, *Eur. J. Biochem.* **168**, 563 (1987).

[54] E. R. Henry, D. L. Rousseau, J. J. Hopfield, R. W. Noble, and S. R. Simon, *Biochemistry* **24**, 5907 (1985).

[55] A. Bianconi, A. Congiu-Castellano, M. Dell'Ariccia, A. Giovannelli, S. Morante, *FEBS Lett.* **191**, 241 (1985).

[56] A. Bianconi, A. Congiu-Castellano, M. Dell'Ariccia, A. Giovannelli, S. Morante, E. Burattini, and P. J. Durham, *Proc. Natl. Acad. Sci. U.S.A.* **83**, 7736 (1986).

[57] S. Pin, R. Cortes, and B. Alpert, *FEBS Lett.* **208**, 325 (1986).

[58] P. A. Lee and J. B. Pendry, *Phys. Rev. B: Solid State* [3] **11**, 2795 (1975).

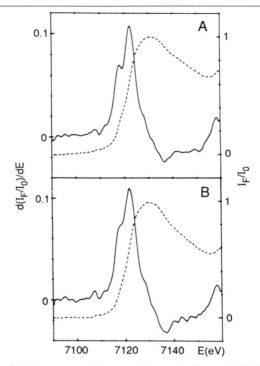

FIG. 7. Effect of the iron magnetic properties on the hemoglobin XANES spectrum. XANES spectra (– – –) and spectrum derivatives (—) of 9 mM carp azidomethemoglobin: (A) 16% high spin (200 mM Bis–Tris, pH 6.10); (B) 25% high spin (same buffer with 2.5 IHP/heme, pH 6.25). (From Pin et al.[51] and reproduced with the permission of Elsevier Science Publishers.)

XANES spectrum, where multiple scattering of the excited electron confers sensitivity to the angular distribution of the neighboring atoms.[59] To determine the ligand angle, XANES was first measured on oriented single crystals as a function of photon polarization.[60,61] The large dichroism of the X-ray absorption of the carbonmonoxymyoglobin crystal was fully interpreted by the multiple scattering theory. This allowed the determination of the variation of ligand bond angles of proteins in solution.

[59] P. Durham, A. Bianconi, A. Congiu-Castellano, A. Giovannelli, S. S. Hasnain, L. Incoccia, S. Morante, and J. B. Pendry, EMBO J. 2, 1441 (1983).
[60] R. S. Mulliken, J. Phys. Chem. 56, 801 (1952).
[61] A. Bianconi, A. Congiu-Castellano, P. J. Durham, S. S. Hasnain, and S. Phillips, Nature (London) 318, 685 (1985).

One can discriminate between electronic and steric contributions in the relative intensity of the iron XANES spectrum.[62] Without steric effect at the binding site, a bend of the iron–ligand bond can be induced by a small distortion of the heme plane. To identify the iron–ligand or porphyrin macrocycle deformations, the XANES theory used multiple scattering on a cluster of the first 30 atoms around the absorbing iron site.[63] The multiple scattering resonances of the photoelectron in the direction parallel to the normal (z) of the porphyrin plane reflect the configuration of the iron–ligand angle. It is revealed by the intensity value at the C_1 energy position of the XANES spectrum. The intensity value at the D energy position is mainly due to the multiple scattering in the heme plane (x, y) and reflects the porphyrin macrocycle geometry.[63,64]

XANES spectroscopy on hemoglobin in solution yields the iron–ligand angle averaged on the α and β subunits. Using XANES spectroscopy, the average Fe–C–O angle in the human carbonmonoxyhemoglobin in solution was found to be 165° ± 10°.[61] This result is in agreement with X-ray crystallography that reported, versus the heme normal, a tilted Fe–C–O configuration with an angle of 14° ± 7° in α chain and 13° ± 7° in β chain.[65] In oxyhemoglobin, it is interesting to note that the bonding angle value obtained by XANES in solution was different from that obtained by X-ray diffraction in crystal. The value of the Fe–O–O angle is 115° ± 10° in solution[66] and 156° in crystal.[45,46] This fact indicates that the lattice can produce some constraints on the iron–ligand geometry. In the isolated α, β, and γ chains of adult and fetal hemoglobins, the Fe–C–O angle decreases going from α to β and γ subunits[49] (Fig. 8). These angle modifications could be correlated with the affinity of these proteins. The dromedary hemoglobin is particularly suited to study this point and the interrelationships between heterotropic and homotropic sites.[67] It was shown under different conditions of allosteric effectors that the Fe–C–O bonding angle varied (between 153 and 159°) with a concomitant deformation of the porphyrin plane, but the porphyrin doming, the Fe–C–O angle change, and the iron movement relative to the heme plane seem to be

[62] A. Bianconi, A. Congiu-Castellano, M. Dell'Ariccia, A. Giovannelli, E. Burattini, P. J. Durham, G. M. Giacometti, and S. Morante, *Biochim. Biophys. Acta* **831**, 114 (1985).

[63] A. Bianconi, in "Biophysics and Synchrotron Radiation" (A. Bianconi and A. Congiu-Castellano, eds.), p. 81. Springer-Verlag, Berlin. 1987.

[64] C. Cartier, Ph.D. Thesis, Université Paris XI (1988).

[65] J. M. Baldwin, *J. Mol. Biol.* **136**, 103 (1980).

[66] A. Congiu-Castellano, A. Bianconi, M. Dell'Ariccia, S. Della Longa, A. Giovannelli, E. Burattini, and M. Castagnola, *Biochem. Biophys. Res. Commun.* **147**, 31 (1987).

[67] G. Amiconi, A. Bortollini, A. Bellelli, M. Coletta, S. G. Condo, and M. Brunori, *Eur. J. Biochem.* **150**, 387 (1985).

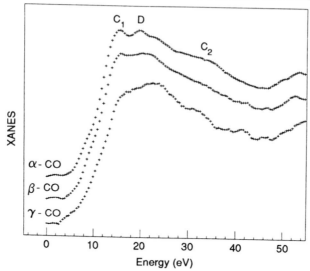

FIG. 8. Iron XANES spectra of the isolated γ, β, and α chains (from bottom to top) in the carbonmonoxy form. The variation of the Fe–C–O angle induces a change of the spectral features C_1 and C_2. Feature D is determined by multiple scattering in the heme plane. (From Congiu-Castellano et al.[49] and reproduced with the permission of Elsevier Science Publishers.)

uncoupled.[68,69] However, a correlation was found between the nonlinear Fe–C–O configuration and a longer Fe–N (of proximal histidine) distance (Fig. 9).[69a] This observation is in agreement with the conclusions from EXAFS investigations of oxygen bonding in picket fence iron porphyrins.[34] These results seem to indicate that the local iron coordination can be related to the ligand affinity and the tertiary structure of the subunit; it could be dissociated from the tetramer conformation.

Dynamic properties of the protein should be involved in the affinity changes. CO ligand binds the iron with different orientations.[70,71] The

[68] G. Amiconi, R. Santucci, M. Coletta, A. Congiu-Castellano, A. Giovannelli, M. Dell'Aricia, S. Della Longa, M. Barteri, E. Burattini, and A. Bianconi, *Biochemistry* **28**, 8547 (1989).

[69] A. Congiu-Castellano, S. Della Longa, A. Bianconi, M. Barteri, E. Burattini, P. Ascenzi, M. Coletta, R. Santucci, and G. Amiconi, *Biochim. Biophys. Acta* **1080**, 119 (1991).

[69a] A. Bianconi, A. Congiu Castellano, and S. Della Longa, in "Biologically Inspired Physics" (L. Peliti, ed.), p. 55. Plenum, New York, 1991.

[70] W. S. Caughey, H. Shimada, M. G. Choc, and M. P. Tucker, *Proc. Natl. Acad. Sci. U.S.A.* **78**, 2903 (1981).

[71] P. Ormos, D. Braunstein, H. Frauenfelder, M. K. Hong, S. L. Lin, T. B. Sauke, and R. D. Young, *Proc. Natl. Acad. Sci. U.S.A.* **85**, 8492 (1988).

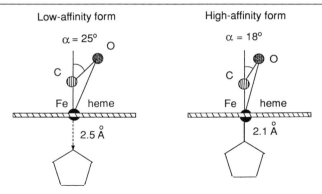

FIG. 9. Relationship between the Fe–C–O bonding angle and the Fe–N_ε distance. Data extracted from the dromedary carbonmonoxyhemoglobin XANES spectra. (From Bianconi et al.[69a] and reproduced with the permission of Plenum Press.)

multiple iron–ligand configurations cannot be calculated, and the iron–ligand bond angle extracted from the XANES data appears to result from an average of different orientations. By the modulation of the rapidly interconverting conformers of the heme pocket, particular organic solvents (alcohols or amides) induce a different distribution of the human hemoglobin conformers.[72] Moreover, the hemoglobin affinity for oxygen is markedly influenced by the presence of these cosolvents.[73] Using XANES spectroscopy, the effects of two amides and two alcohols on the possible alteration of the heme (core deformation) as well as the iron–ligand bonding geometry were investigated.[72] The D peak intensity did not change with external condition variations. Thus, there is a great stability of the heme doming with respect to different solution conditions. The intensity changes of the C_1 peak showed that the effects of amides and alcohols on the ligand bonding angle are opposite. The Fe–C–O and the Fe–O–O angle variations were estimated as between $-4°$ under the amide effect and $+3°$ under the alcohol effect. It appears that the ligand bonding angle decreases in the order alcohols > water > amides, and thus the angle is higher when the affinity is lower. The small rearrangement of the averaged iron–ligand configuration, subsequent to the solvent perturbation, demonstrates that a hierarchy of motions exists[74] and that the protein affinity can

[72] C. Zentz, S. El Antri, S. Pin, R. Cortes, A. Massat, M. Simon, and B. Alpert, *Biochemistry* **30**, 2804 (1991).

[73] L. Cordone, A. Cupane, P. L. San Biagio, and E. Vitrano, *Biopolymers* **18**, 1975 (1979).

[74] A. Ansari, J. Berendzen, D. Braunstein, B. R. Cowen, H. Frauenfelder, M. K. Hong, I. E. T. Iben, J. B. Johnson, P. Ormos, T. B. Sauke, R. Scholl, A. Schulte, P. J. Steinbach, J. Vittitow, and R. D. Young, *Biophys. Chem.* **26**, 337 (1987).

be associated to a hierarchy of subtle conformational and dynamics states.

Fast Measurements of XANES Spectra

An X-ray absorption spectrum is generally obtained with a monochromator scanning, step by step, the X-photon energy. Dispersive X-ray absorption spectroscopy[75] collects, at the same time, all the X photons, whatever their wavelength. The energy-position correlation on the 1024 sensing elements of the photodiode differentiates the X photons over 500 eV. Twelve milliseconds is needed to collect a full X-ray absorption spectrum. This can be repeated at 550 or 220 Hz according to the analog digital converter used. The measurement time for a XANES spectrum is determined by the overall number of X photons necessary to build one spectrum with a reasonable signal-to-noise ratio. This is determined by the product of the integration time for each frame by the number of summed frames. The spectra of HbCO can be obtained by summing up to 32 frames, each being collected in 2.2 sec; therefore the measurements take about 70 seconds for each HbCO spectrum. The progressive temporal change of the iron XANES spectrum during HbCO pH denaturation was measured by this method.[76] This technique permitted the observation of three transitions in the denaturation kinetics. The CO detachment at very low pH (2.6) is preceded by CO bending (pH 5.6) and by protein unfolding (pH 3.3).

Multiple Scattering Approach to X-Ray Absorption Spectrum Data Analysis

Classic Theory of X-Ray Absorption

In the early EXAFS experiments the EXAFS data analysis was performed using several approximations, such as the plane wave approximation and the single scattering approximation, and using nonaccurate phase shift, but more sophisticated data analysis is now available. In the early XANES experiments, the application of XANES for local geometry determination was limited to a fingerprint approach using model compounds, due to the lack of reliable theoretical analysis.[77] Important advances to-

[75] E. Dartyge, C. Depautex, J. M. Dubuisson, A. Fontaine, A. Jucha, P. Leboucher, and G. Tourillon, *Nucl. Instrum. Methods Phys. Res., Sect. A* **246,** 452 (1986).

[76] I. Ascone, A. Fontaine, A. Bianconi, A. Congiu-Castellano, A. Giovannelli, S. Della Longa, and M. Momenteau, *in* "Biophysics and Synchrotron Radiation" (A. Bianconi and A. Congiu-Castellano, eds.), p. 122. Springer-Verlag, Berlin, 1987.

[77] A. Bianconi, S. Alema, L. Castellani, P. Fasella, A. Giovannelli, S. Mobilio, and B. Oesh, *J. Mol. Biol.* **165,** 125 (1983).

ward quantitative analysis of the X-ray absorption spectral data have been made in describing the XANES and EXAFS in the same conceptual framework.

The Fe K-edge X-ray absorption of hemoglobin concerns the electronic transitions from the atomic Fe $1s$ inner shell to unoccupied states extending from the valence states to the high-energy states in the continuum. In classic quantum theory the absorption cross-section is given by many-body excitations of the N-electron system. Following the interaction with the photon beam of energy ω, the system is excited from the initial state i at energy E_i to a final state at energy E_f. In the dipole approximation the total absorption cross-section is given by

$$\sigma(\omega) \approx \omega \sum_f |M_{if}|^2 \delta(E_i - E_f + \omega) \tag{1}$$

where the sum is extended over all the possible final states f, and M_{if} is the matrix element involving the initial Ψ_i and final Ψ_f^* many-body radial wavefunctions

$$M_{if} = \int \psi_f^*(r_1, r_2, \ldots, r_n) \sum_n (\mathbf{r}_n \cdot \mathbf{e}) \Psi_i(r_1, r_2, \ldots, r_n) \, dr \tag{2}$$

where $\mathbf{e}$ is the unitary polarization vector of the electric field, and $\mathbf{r}_n$ is the vector describing the position of the nth electron.

In the one-electron approximation X-ray absorption is described by single-particle processes. The N-electron system is separated into two parts: the single electron in the $1s$ core level, which is excited into an unoccupied orbital, and the $N - 1$ passive electrons that do not participate directly in the electronic transition. Following the $1s$ core hole creation, the $N - 1$ passive electrons relax from the ground state to new excited states to screen the positive charge of the core hole. Many-body effects can arise because of the presence of different many-body final state configurations of the $N - 1$ passive electrons. These many-body effects can arise because of the partial relaxation of the $N - 1$ passive electrons in the time scale of the core excitation, or because of an excitation of a second core level or a valence electron. All these processes give different absorption peaks at different energies for the excitation of the primary photoelectron to the same final state, and the description of these lines requires the calculation of configuration interaction of many-body final states. Therefore the many-body satellites in the X-ray absorption spectra appear at higher energy than the one-electron transition and are often called "shake-up" satellites.

By definition the one-electron approximation considers only a single static configuration of the $N - 1$ passive electrons, the "fully relaxed configuration," defined as the final state many-body configuration with the lowest energy. In this configuration the core hole in the static final state potential is fully screened by the valence electrons. The wavefunction of the excited photoelectron is determined by the final state potential with the core hole and the fully relaxed $N - 1$ electrons, which is known as the von Barth and Grossmann final state rule.[78]

In the one-electron approximation the absorption coefficient $\sigma_c(\omega)$ is determined by the transitions from the core level c characterized by the quantum numbers n, l, and J with energy E_c and wavefunction ψ_c to the jth unoccupied state at energy E_j with wavefunction Ψ_j. Because the dipole transitions dominate the process of photoabsorption, an electron from a core level having angular momentum l is excited into the $l \pm 1$ final states. In the Fe K-edge spectrum only the $l = 1$ final states are allowed by the $\Delta l = +1$ dipole selection rule, therefore only the p-like final states can be reached in the core transition from the $l = 0$ ($1s$) core level.

The absorption coefficient is proportional to the partial oscillator strength $f_{c,l}(E)$ for transitions to final state energy E excited by the photon energy $\omega = E - E_c$, using atomic units where $h = 1$,[79]

$$f_{c,l}(E) = \rho_{c,l} N_l(E) \qquad \text{for} \qquad E > E_F \tag{3}$$

where E_F is the energy of the lowest unoccupied level, $N_l(E)$ is the angular projected density of states defined as

$$N_l(E) = 2 \sum_{\mathbf{k},j} \sum_m |\langle \mathbf{Y}_{lm} | \Psi_{kj} \rangle|^2 \delta(E - E_{kj}) \tag{4}$$

where the unoccupied states Ψ_{kj} are labeled by the reduced wavevector $\mathbf{k}$ of the excited electron and the unoccupied band index j. The effective matric element $\rho_{c,l}(E)$ is given by

$$\rho_{c,l}(E) = \langle \psi_c | \mathbf{r} | \phi_l(E) \rangle^2 / \langle \phi_l^2(E) \rangle \tag{5}$$

where the wavefunction $\phi_l(E, r)$ is a solution of the radial Schrödinger equation inside the muffin-tin (MT) sphere of radius $S (r < S)$ and outside the MT sphere $(r \geq S)$, the solution is given by

$$\phi_l(E, r) = [\cos \delta_l(E)] J_l[(E - V_0)^{1/2} r] - [\sin \delta_l(E)] n_l[(E - V_0)^{1/2} r] \tag{6}$$

where $\delta_l(E)$ is the lth phase shift of the muffin-tin potential, V_0, is the muffin-tin zero of the potential, and J_l and n_l are spherical Bessel functions. The X-ray absorption spectra are therefore interpreted in terms of the

[78] U. von Barth and G. Grossmann, Phys. Rev. B: Condens. Matter [3] 25, 5150 (1982).
[79] U. Fano and J. W. Cooper, Rev. Mod. Phys. 40, 441 (1968).

product of the partial (of selected $l = 1$ orbital momentum) density of states and of the matrix element.

Multiple Scattering Approach in Real Space

A long-standing discussion on the interpretation of the X-ray absorption spectra from the threshold to high energy was solved in the early 1980s. The X-ray absorption features in the first 5–10 eV were called a Kossel structure and assigned to atomic-like transitions or to transitions to unoccupied molecular levels of the molecule formed by the absorbing atom and its coordination shell. Following this old interpretation scheme, the X-ray absorption spectra of hemoglobin were first interpreted in terms of atomic transitions[12]. On the other hand, the Fourier transform of the EXAFS spectrum of hemoglobin was clearly probing the atomic distribution over a cluster of 5-Å radius, including all porphyrin atoms as indicated by the fourier transform.

In the early years of research on synchrotron radiation spectroscopy of biomolecules it was shown that the experimental features over a photon energy range of about 50 eV depend mainly on the geometric structure of a large cluster of atoms around the absorbing atom, and these features were called XANES.[78,80] An interpretation of the X-ray absorption near-edge structure was proposed in terms of multiple scattering resonances of the photoelectron in a large cluster.[80] This interpretation was based on an extention to condensed systems of the shape resonances, which are localized states in the continuum observed in the spectrum of molecules such as N_2,[81] and interpreted with the multiple scattering theory. The size of the cluster of atoms around the photoabsorbing atom, relevant for XANES, was found to depend on the electron mean free path for inelastic scattering and on the core hole lifetime. Only where the atomic distribution of neighboring atoms is highly disordered, such as for metal ions in solution, does the structural disorder reduce the number of neighbor shells to one or two.[3]

Several theoretical approaches have been developed to solve the absorption cross-section for core transitions in real space in the frame of multiple scattering theory solving the absorption coefficient both for the EXAFS and XANES energy region.[4,82] The extension of the bound state molecular scattering method of Johnson and co-workers[83] to determine the one-electron wavefunction for continuum states was formulated first

[80] A. Bianconi, S. Doniach, and D. Lublin, *Chem. Phys. Lett.* **59**, 121 (1978).

[81] A. Bianconi, H. Petersen, F. C. Brown, and R. Z. Bachrach, *Phys. Rev. A* **17**, 1907 (1978).

[82] C. R. Natoli, D. K. Misemer, S. Doniach, and F. W. Kutzler, *Phys. Rev. A* **22**, 1104 (1980).

[83] K. H. Johnson, *Adv. Quantum Chem.* **7**, 143 (1973).

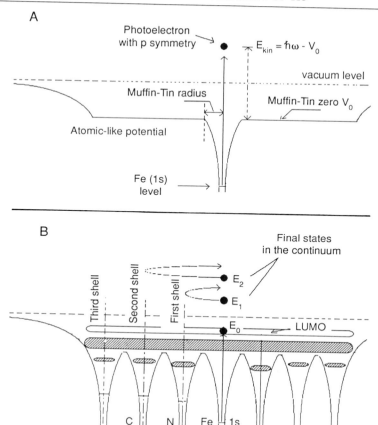

FIG. 10. (A) Potential used to calculate the atomic absorption of the central iron atom. The atomic-like potential is cut at the muffin tin zero below the vacuum level. (B) Muffin-tin potential for a cluster of three shells of atoms surrounding the absorbing iron atom. E_0 is the final state in the lowest unoccupied molecular orbital (LUMO). E_1 and E_2 are final states in the continuum above the vacuum level. In this region, the absorption coefficient is modulated by the scattering of the excited photoelectron on atoms of the neighbor shells.

by Dill and Dehmer.[84] The continuum wavefunction is matched to the proper asymptotic solution of the Coulomb scattering states, and in this way the multiple scattering problem is changed from a homogeneous eigenvalue problem (bound states) to an inhomogeneous one in which the continuum wavefunction is determined by an asymptotic T-matrix normalization condition. In this scheme the total potential is represented

[84] D. Dill and J. L. Dehmer, *J. Chem. Phys.* **61,** 692 (1974).

by a cluster of nonoverlapping spherical potentials centered on the atomic sites, and the molecule as a whole is enveloped by an "outer sphere." Three regions can be identified in this partitioning: atomic regions (spheres centered on nuclei, normally called region I), an extramolecular region (the space beyond the outer sphere radius, region III), and an interstitial region of complicated geometry in which the molecular potential is approximated by a constant "muffin-tin" potential (Fig. 10A and B). The Coulomb and exchange part of the input potential are calculated on the basis of a total charge density obtained by superimposing the atomic charge densities, calculated from Clementi and Roetti,[85] of the individual atoms constituting the cluster. For the exchange potential it is possible to use both the usual energy-independent Slater approximation[86] and the energy-dependent Hedin–Lundqvist[87] potential in order to incorporate the dynamic effect. Following this theory, the expression for the absorption coefficient for a cluster of atoms, for polarized light in the vector **e** direction, is given in the dipole approximation by

$$\mu_c(E) = N_c\sigma(E, \mathbf{e}) = N_c\sigma(k, \mathbf{e}) \tag{7}$$

$$\sigma(k, \mathbf{e}) = 4\pi\omega\alpha k\, \mathbf{Im}\left[\sum_{L,L'} \langle\psi_c|\mathbf{r}\cdot\mathbf{e}|\phi_l\mathbf{Y}_L\rangle\tau_{L,L'}\langle\phi_l\mathbf{Y}_L|\mathbf{r}\cdot\mathbf{e}|\psi_c\rangle] \tag{8}$$

where $\sigma(E, \mathbf{e})$ is the photoabsorbing cross-section, N_c is the density of atoms, $L = (l, m)$ is the angular momentum, α is the fine structure constant, k is the photoelectron wavenumber, and the spin dependence has been neglected. ϕ_l values are regular solutions of the radial Schrödinger equations in the photoabsorber muffin-tin sphere.[88] The structural information is contained in the quantity

$$\tau_{L,L'} = [\sin\delta_l(E) \sin\delta_{l'}(E)]^{-1}[(\mathcal{T}_a^{-1} - \mathcal{G})^{-1}]_{L,L'} \tag{9}$$

where δ_l is the lth phase shift of the absorbing atom, assumed to be located at site 0, $\mathcal{G} = G_{L,L'}$ is the matrix describing the free spherical wave propagation of the photoelectron from site i and angular momentum $L = (l, m)$ to site j and angular momentum $L' = (l', m')$ in the angular momentum representation, and $\mathcal{T}_a = \delta_{i,j}\delta_{L,L'}[\exp(i\delta_l^i) \sin\delta_l^i]$ is the diagonal matrix of atomic t-matrix elements describing the scattering process of the L spherical wave photoelectron by the atom located at site i with phase shift δ_l^i.

[85] E. Clementi and C. Roetti, *At. Data Nucl. Data Tables* **14**, 177 (1974).
[86] K. Schwartz, *Phys. Rev. B: Solid State* [3] **5**, 2466 (1972).
[87] L. Hedin and B. I. Lundqvist, *J. Phys. C* **4**, 2064 (1971).
[88] C. R. Natoli, M. Benfatto, and S. Doniach, *Phys. Rev. A* **34**, 4682 (1986).

Under certain conditions[89] it is possible to express the structural factor as an absolute convergent series and to expand the photoabsorption cross-section in partial contributions as follows:

$$\sigma(E, \mathbf{e}) = \sum_n \sigma_n(E, \mathbf{e}) \tag{10}$$

The $n = 0$ term represents the smoothly varying "atomic" cross-section, the $n = 1$ term is always zero, and the generic term n is the contribution to the photoabsorption cross-section from processes in which the photoelectron has been scattered $n - 1$ times by the surrounding atoms before returning to the photoabsorbing site. In particular, $\sigma_2(E, \mathbf{e})$, the EXAFS term in the spherical wave representation, can be written both for the K-edge and L_1-edge as an approximation to the general polarized EXAFS formula[90-92]

$$\sigma_2(E, \mathbf{e}) =$$
$$|M_{01}(E)|^2(-1) \sum_j \cos^2\theta_j \, \mathbf{Im}\{\exp[2i(\rho_j + \delta_l^0)/\rho_j^2] \sum_l (-1)_l(2l + 1) t_l^j f_l^2(\rho_j)\} \tag{11}$$

where $\rho_j = kR_j$, θ_j is the angle between the polarization vector $\mathbf{e}$ and the direction R_j joining the jth atom with the absorbing one, $|M_{01}(E)|^2$ is the radial matrix element between the initial $l = 0$ state and the final dipole-allowed R_1 radial wavefunction, and f_l takes into account the spherical correction to the free propagators. The total absorption coefficient can be written as

$$\mu(E, \mathbf{e}) = \mu_0[1 + \sum_{n\geq 2} \chi_n(E, \mathbf{e})] \tag{12}$$

where $\mu_0(E)$ is the structureless absorption coefficient of a central photoabsorbing atom, and $\chi_n(E)$ represents the contribution arising from all multiple scattering (MS) pathways beginning and ending at the central atom and involving $n - 1$ neighboring atoms. A pictorial view of the multiple scattering pathways contributing to the XANES is shown in Fig. 1. The muffin tin potential used to calculate μ_0 is shown in Fig. 10A.

The terms χ_n are usually plotted versus the wavevector k of the photoelectron. The EXAFS $\chi_2(k)$ term is the dominant term above wavevector values $k > \sim 3-4$ Å^{-1}. The $\chi_3(k)$ term is the double scattering signal arising

[89] M. Benfatto, C. R. Natoli, A. Bianconi, J. Garcia, A. Marcelli, M. Fanfoni, and I. Davoli, *Phys. Rev. B: Condens. Matter* [3] **34**, 5774 (1986).
[90] J. J. Boland, S. E. Crane, and J. D. Baldeschwieler, *J. Chem. Phys.* **77**, 142 (1982).
[91] M. Benfatto, C. R. Natoli, C. Brouder, R. F. Pettifer, and M. F. Ruiz Lopez, *Phys. Rev. B: Condens. Matter* [3] **39**, 1936 (1989).
[92] J. J. Rehr, R. C. Albers, C. R. Natoli, and E. A. Stern, *Phys. Rev. B: Condens. Matter* [3] **34**, 4350 (1986).

from all the triangular paths with one vertex on the central atom. The multiple scattering signal is defined as

$$\chi_{MS}(k) = \sum_{n \geq 3} \chi_n(k) = \{[\mu(k) - \mu_0(k)]/\mu_0(k)\} - \chi_2(k) \tag{13}$$

The general mathematical expression for $\chi_n(k)$, without taking into account the inelastic interactions of the photoelectron, that is, its mean free path and the structural Debye–Waller factor, can be written as

$$\chi_n^l(k) = \sum_{p_n} A_n^l(k, p_n) \sin[kRp_n + \phi_n^l(k, p_n) + 2\delta_l^0] \tag{14}$$

where δ_l^0 is the central atom lth phase shift, and the dependence of the amplitude and phase function A_n^l and ϕ_n^l on the particular path has been indicated symbolically by p_n. General expressions for calculating the quantities A_n^l and ϕ_n^l in terms of the atomic phase shifts and the geometry of the path p_n are provided by the MS theory. A substantial simplification of these expressions is achieved by means of a relatively simple approximation of the propagator of the photoelectron in the final state, the spherical wave approximation (SWA),[88] which preserves the spherical wave character of the propagation and is rather accurate even at very low wave vector values ($k \approx 1–2$ Å^{-1}).

Because of the general structure of the quantities $\chi_n^l(k)$, the amplitudes (A_n^l) decrease with increasing order n, so that usually $\chi_2^l(k)$ is the dominant term in the whole energy range where the series converges. Hence an analysis of the MS contribution beyond the first term is possible if the $\chi_2^l(k)$ contribution is subtracted from the experimental signal: $\chi_{MS}(k) = \chi(k) - \chi_2^l(k)$, provided a good estimate for $\mu_0(k)$ and $\chi_2^l(k)$ is used.

For K-edge, neglecting the angular dependence of the Hankel function in the free propagator, the usual EXAFS signal, which times the atomic part can be obtained for $n = 2$,

$$\mu_2 = \mu_0 \chi_2 = \mu_0 \sum_j \mathbf{Im}[f_j(k, \pi) \exp(2i\delta_1^0 + kr_j)/kr_j^2] \tag{15}$$

where r_j is the distance between the central atom and the neighboring atom j, and $f_j(k, \pi)$ is the usual backscattering amplitude. The first multiple scattering contribution is the μ_3 term, which can be written as

$$\mu_3 = \mu_0 \sum_{i \neq j} \mathbf{Im}\{P_1(\cos \phi) f_i(\omega) f_j(\theta) \exp[2i(\delta_1^0 + kR_{tot})]/kr_i r_{ij} r_j\} \tag{16}$$

where r_{ij} is the distance between atoms i and j, $f_i(\omega)$ and $f_j(\theta)$ are the scattering amplitudes that now depend on the angles [by Legendre polynomials $P_1(x)$] in the triangle that joins the absorbing atom to the neighboring atoms located at sites $\mathbf{r}_i$ and $\mathbf{r}_j$ and $R_{tot} = r_i + r_{ij} + r_j$. In this expression

$\cos \phi = -\mathbf{r}_i \cdot \mathbf{r}_j$, $\cos \omega = -\mathbf{r} \cdot \mathbf{r}_{ij}$, and $\cos \theta = \mathbf{r}_j \cdot \mathbf{r}_{ij}$. So the terms higher than two clearly contain information about the nth order correlation function. To conclude, it is possible to observe that because of $P_1(\cos \phi)$ $= \cos \phi$, there is a selection rule in the pathways that contribute to the μ_3 term. In fact, all the cases in which $\mathbf{r}_i$ is perpendicular to $\mathbf{r}_j$ do not contribute to this term because $\cos \rho = 0$.

EXAFS and XANES Regions

The generally strong scattering power of the low Z atoms such as N, C, and O atoms of condensed matter for low kinetic energy photoelectrons favors multiple scattering (MS) processes. At higher energies, such that the atomic scattering power becomes small, a single scattering (SS) regime is found. In an SS regime the modulation in the absorption coefficient is substantially due to interference of the outgoing photoelectron wave from the absorbing atom with the backscattered wave from each surrounding atom, yielding extended X-ray absorption fine structure (EXAFS).

EXAFS therefore provides information about the pair correlation function. By decreasing the photoelectron kinetic energy, a gradual transformation occurs from the EXAFS regime to the XANES regime.[93,94] Therefore EXAFS probes the first-order or pair-correlation function of the atomic distribution near the absorbing atom, whereas XANES probes the triplet and higher orders of the atomic distribution function. Interest in determination of higher order correlation functions of local atomic distributions in complex systems has stimulated the growth of XANES.

The absorbing Fe atom in hemoglobin has several neighboring atoms in colinear configurations. The high probability for forward scattering in this configuration enhances MS contributions in the EXAFS, therefore the $n = 3$ multiple scattering terms are also important in the EXAFS. For noncolinear configurations the probability of large angle scattering is low at high kinetic energies, and the single scattering approximation in the EXAFS region has been found to work.

Broadening Term

It is important to include in the XANES data analysis the lifetime of the photoelectron in order to obtain good agreement between calculated spectra and experiments. The lifetime of the photoelectron takes into account the inelastic scattering of the photoelectron by valence electrons,

[93] A. Bianconi, J. Garcia, A. Marcelli, M. Benfatto, C. R. Natoli, and I. Davoli, *J. Phys.* **46** (C9), 101 (1985).

[94] J. Garcia, M. Benfatto, C. R. Natoli, A. Bianconi, I. Davoli, and A. Marcelli, *Solid State Commun.* **58,** 595 (1986).

which is an essential physical aspect of the states at high energy in the conduction band, because the photoelectrons with energy $\varepsilon = E - E_F$ have enough energy to excite all valence electrons with binding energy smaller than ε.

The comparison between the calculated absorption coefficient and the experiments indicates that the theoretical spectra must be convoluted with an intrinsic Lorentzian broadening function $\Gamma(E)$ and with the instrumental bandwidth, in order to obtain good agreement. The intrinsic broadening of the excited states $\Gamma(E)$ is the convolution of two terms: the first is the core hole width Γ_h, and the second term is the energy bandwidth $\Gamma_e(E)$ of the excited electron of energy E. These are related to τ_h, the core hole lifetime, and to $\tau_e(E)$, the lifetime of the electron, which is a function of the mean free path of the excited electrons $\lambda(E)$. $\lambda(E)$ is determined by the inelastic scattering of the photoelectron with the passive electrons; it is energy dependent and varies for each material:

$$\Gamma_e(E) = (h/2\pi)/\tau_e(E) = 2(2E/m)^{1/2}/\lambda(E) \tag{17}$$

At low kinetic energies of the electron $\Gamma_h > \Gamma_e$, and the Γ_h term dominates, whereas at high energies $\Gamma_h < \Gamma_e$, therefore the $\Gamma_e(E)$ term dominates. It is possible to define an effective mean free path given by

$$\lambda_{eff}(E) = (h/2\pi)2(2E/m)^{1/2}/\Gamma_{tot}(E) \tag{18}$$

λ_{eff} is an estimation of the radius of the cluster probed by XANES, and Γ_{tot} is the energy broadening term due to the convolution of instrumental and lifetime effects. In the Fe K-edge XANES of Hb, the resulting radius is about 5 Å, therefore this spectroscopy probes the heme structure.

[15] Modulated Excitation Spectroscopy in Hemoglobin

By FRANK A. FERRONE

Modulated excitation is a kinetic perturbation method that uses frequency-domain techniques to observe small relaxations that may be partially masked by larger amplitude processes of lesser interest. Because it is basically a repetitive technique, it has found use thus far in reversible photolysis reactions, and there it has been used to study conformational change in hemoglobin (Hb).

Although the methodology does not require mathematics more difficult than complex algebra and simple differential equations, the lack of familiarity with this method and its principles requires an introduction of some length. In the next section we look carefully at pulse methods to set the scope of the kinetic problems that are addressed by modulated excitation. We then describe the theory necessary for understanding modulated excitation experiments, both in intuitive terms for the reader who wishes to obtain a general picture of the technique and in the detail needed for research work with this method. At that point we turn to the experimental aspects.

Central Kinetic Problem

The cooperative interaction between hemes that makes hemoglobin so interesting to study also serves to make such studies more difficult by keeping the equilibrium populations of intermediate species very low. This has added to the natural interest in the kinetic behavior of the molecule. In 1959, Q. Gibson used flash photolysis to demonstrate a transient species Gibson labeled as a quickly reacting form of hemoglobin.[1] This was observed under conditions in which a powerful flash of light removed all the ligands from each Hb tetramer in a time shorter than the observing technology resolved. It would be ideal if one could precisely titrate the different ligand levels by such methods, that is, if one could then flash three, two, and one ligand off. However, the intermediates are formed as members of a binomial distribution and cannot be seen independently, with one exception. A weak flash of light can serve to remove one ligand from a few molecules, and the likelihood that any tetramers are photolyzed at two or more heme sites can be made as small as desired. This type of excitation has been called "tickle" flash and is the basis of modulated excitation. This method uses weak, oscillatory light to excite the molecule. The oscillatory nature of the excitation allows certain strategies to be used that are impractical with pulse-and-probe methods, although all the information gathered by this method is in principle available in pulse excitation methods.

Hemoglobin is known to possess at least two alternative quaternary structures, labeled R and T for the structure found in the fully oxygenated and fully deoxygenated crystal structures, respectively. Spectroscopic signals have been correlated with these structures.[2] The simplest thermodynamic connection of structure to function is the MWC (Monod–Wyman–Changeux) model, which is shown schematically in Fig. 1 as a manifold of equally spaced free-energy levels in either structure.[3] More elaborate schemes are possible: for comparison, a proposal

[1] Q. H. Gibson, *Biochem. J.* **71**, 293 (1959).
[2] M. F. Perutz, J. E. Ladner, S. R. Simon, and C. Ho, *Biochemistry* **13**, 2163 (1974).
[3] J. C. Monod, J. Wyman, and J. P. Changeux, *J. Mol. Biol.* **12**, 88 (1965).

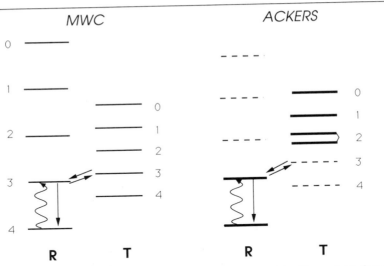

Fig. 1. Connection between structure and ligation for two models of hemoglobin function. Vertical separation represents differences in free energy; R and T represent quaternary structure. Number of ligands is indicated. In the MWC model, there is no cooperativity within a given structure (R or T). This is no longer true in the Ackers model for oxygen binding (in pH 7.4, 0.1 M Cl⁻ buffer).[4] Moreover, the doubly liganded state is split into one with lower energy when both ligands are on the same side of the $\alpha_1\beta_2$ interface and one with higher energy when one ligand is on either side of the interface. In the Ackers model, the states for which energy has been observed have been drawn in bold, while hypothetical states are indicated by dashed lines. In either model, a tickle excitation will probe equilibration processes between triply ligated states.

by Ackers and co-workers[4] is also shown in Fig. 1. The models differ in the equality of spacing within a given structure and in the way in which a given number of ligands specifies an affinity. In a tickle flash experiment, a single ligand is removed instantaneously within the R state. Subsequently there will be some change in equilibrium between R and T populations. Notice that the difference in details does not change the appearance of the basic kinetic scheme. Such an experiment is model independent.

Even if all rebinding is through the R state, the kinetics of ligand recombination will include $R \rightleftharpoons T$ rates. This idea is key to the modulated excitation method. First, however, we want to examine what happens if this system is probed by a pulse excitation method to see why the modulation method has certain experimental advantages for this specific problem.

The differential equations that govern CO recombination after a tickle flash to this two-state system are quite simple. If the three-liganded T-state

[4] G. K. Ackers, M. L. Doyle, D. Myers, and M. A. Daugherty, Science **235**, 54 (1992).

population is designated $T(t)$ and the three-ligated R-state population $R(t)$, then

$$dR(t)/dt = -k_R[CO]R(t) - k_{RT}R(t) + k_{TR}T(t) \tag{1}$$

and

$$dT(t)/dt = -k_T[CO]T(t) - k_{TR}T(t) + k_{RT}R(t) \tag{2}$$

where k_R and k_T are the bimolecular rate constants for binding a ligand to the R or T state, respectively, and k_{RT} and k_{TR} are the rate constants for conversion from R to T and vice versa. The above rate constants all refer to triply ligated species, and may differ for different ligand numbers. Such a system of equations will have solutions that are the sum of exponential decays; both R and T will have the same exponentials. Thus

$$R(t) = A \exp(-k_a t) + B \exp(-k_b t) \tag{3}$$

and

$$T(t) = C \exp(-k_a t) + D \exp(-k_b t) \tag{4}$$

The relaxation rates k_a and k_b are set by the rates in the problem. If we adopt the notation that

$$\langle k \rangle = \frac{1}{2}(k_R[CO] + k_T[CO] + k_{RT} + k_{TR}) \tag{5}$$

then we can write

$$k_{a,b} = \langle k \rangle \{1 \pm [1 - (k_R[CO]k_T[CO] + k_R[CO]k_{TR} + k_T[CO]k_{RT})/\langle k \rangle^2]^{1/2}\} \tag{6}$$

We shall associate k_a with the positive sign, so that $k_a > k_b$. Note that $\langle k \rangle = (k_a + k_b)/2$. The rate expression simplifies significantly when k_{RT} and k_{TR} are much greater than the ligand-binding rates. Then it is approximately true that

$$k_a = 2\langle k \rangle \tag{7}$$

and

$$k_b = k_R[CO]\frac{1}{L_3 + 1} + k_T[CO]\frac{L_3}{L_3 + 1} \tag{8}$$

where $L_3 = k_{RT}/k_{TR}$. However, even in rapid exchange two exponentials will appear in such a relaxation system.

The amplitudes will be set by the initial conditions. For example, if the initial state is all R_4, then at $t = 0$ there must be no population in

$T(t = 0)$, and hence $C = -D$. A and B are given in terms of the initial R population excited, which will be denoted $R(0)$. They are

$$A = \frac{k_R[CO] + k_{RT} - k_b}{k_a - k_b} R(0) \tag{9}$$

and

$$B = \frac{k_R[CO] + k_{RT} - k_a}{k_b - k_a} R(0) \tag{10}$$

Similarly, it is straightforward to show that

$$C = \frac{k_{RT}}{k_b - k_a} R(0) \tag{11}$$

and $D = -C$. $T(t)$ has a particularly simple expression, namely,

$$T(t) = \frac{k_{RT}}{k_b - k_a} R(0)(e^{-k_a t} - e^{-k_b t}) \tag{12}$$

The expression for the R + T populations is

$$R(t) + T(t) = \frac{R(0)}{k_a - k_b}[(k_R[CO] - k_b)e^{-k_a t} + (k_a - k_R[CO])e^{-k_b t}] \tag{13}$$

By following $T(t)$, say, by observation at an isosbestic point for ligated and unligated R states, one could determine the relaxation rates k_a and k_b. If $R(0)$ can be determined (say at an R–T isosbestic), and an appropriate extinction coefficient is known, then k_{RT} is directly found from $T(t)$. Likewise, with k_a and k_b known, the amplitude information from $R(t) + T(t)$ will directly allow $k_R[CO]$ to be found. This leaves two unknowns: k_{TR} and k_T. These can be solved directly from linear equations using the effective rates k_a and k_b. Thus, in principle, all the rates of the system are determined by observation of the decay of the system.

However, from rates generated by this technique[5,6] (as described below), a clean separation is difficult. For example, suppose $k_R = 6$ msec^{-1} mM^{-1}, $k_T = 0.1$ msec^{-1} mM^{-1}, $k_{RT} = 1$ msec^{-1}, and $k_{TR} = 3$ msec^{-1}. The two exponential rate constants and the ratio of the amplitudes of the two

[5] F. A. Ferrone, A. J. Martino, and S. Basak, *Biophys. J.* **48**, 269 (1985).
[6] N. Zhang, F. A. Ferrone, and A. J. Martino, *Biophys. J.* **58**, 333 (1990).

TABLE I

APPARENT RATE CONSTANTS AND AMPLITUDE RATIOS

[CO] (mM)	k_a (msec^{-1})	k_b (msec^{-1})	$\dfrac{A + C}{B + D}$
1	7.7	2.4	2.1
0.33	4.8	1.3	0.25
0.1	4.1	0.44	0.046

exponentials $(A + C)/(B + D)$ viewed at an R–T isosbestic are shown in Table I for three different CO concentrations.

The analysis that looks straightforward in theory is challenged by several experimental issues. The first of these is the noise level in the measurement. A successful resolution of two similar exponentials requires that noise be appropriately small. Figure 2 shows a double exponential as expected in the above example with [CO] = 1 mM, and the dashed line shows a fit by eye of a single exponential with an amplitude of 0.91 and an effective rate of 4.3 msec^{-1}. (This does not include a baseline offset, which would only improve the fit.) Although this is not representative of any of the true rates in the decay, it is actually rather close to the slow rate expected in rapid exchange [i.e., Eq. (8)]. In Fig. 2a, the average deviation between the single and double exponential is about 0.05% of the maximum excursion. If the maximum excursion is 10% of the direct current (DC) intensity (because it is a tickle experiment), then the two exponentials cannot begin to be resolved until the overall noise level is below 10^{-3} of DC. Observing the T-state signal (Fig. 2b) will allow easier resolution of the two exponentials. However, the maximum amplitude of that signal is only about 8% of the R + T signal because of the competing exponentials and small prefactor; moreover, the change in extinction coefficient due to R − T change is about an order of magnitude less than that seen for the ligand release (R + T) signal. Again, this produces a threshold of about 10^{-3} of the DC intensity before signal can be resolved. A noise level $<10^{-4}$ is thus sought for good resolution. This must be accomplished at reasonable bandwidth; for example, 10-μsec resolution is required. This necessitates about 10^9 photons/10 μsec or about 44 μW. A 150-W xenon arc and a fast (f 3.5) monochromator will give about 320 μW in a 1-nm bandpass. But when losses due to absorption by the sample and the optical train are taken into account, the flux at the detector is not far from the minimum 44 μW. That is, a 10% photolysis experiment has just enough signal to noise to resolve the above exponentials in single-shot fashion. This can be improved by signal averaging, of course.

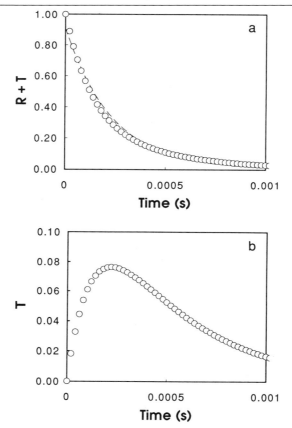

FIG. 2(a). Double-exponential decay based on 1 mM free CO and rate constants as follows[5]: $k_R = 6$ msec^{-1}mM^{-1}, $k_T = 0.1$ msec^{-1}mM^{-1}, $k_{RT} = 1$ msec^{-1}, and $k_{TR} = 3$ msec^{-1}. The amplitude is normalized to unity. Such an exponential would be expected if the system is observed at an R–T isosbestic, so that states R + T are observed. The dashed line shows a fit by eye of a single exponential with an amplitude of 0.91 and an effective rate of 4.3 msec^{-1}. The average deviation is 0.05% of maximum. If the raw data had such a noise level, the double exponential would not be resolved, and instead a single exponential could have been used to characterize the data. (b) Amplitude of the T-state signal based on the rates in (a) and Eq. (12). The same normalization applies here, so that the amplitude is shown relative to the excitation in (a).

A second difficulty is that, with a 10% excitation pulse, 1% of the sample will be doubly excited. This is 10% of the observed signal. If relaxation from a doubly ligated state is notably different from the relaxation with three ligands bound, it could obscure the second relaxation.

Thus a smaller pulse amplitude would be useful, although it would make signal to noise worse.

A third difficulty concerns interpretation of the results. In the case illustrated, the rate differences are in the range of those observed for $\alpha-\beta$ differences.[7-9] Although the amplitudes are quite different, the observed amplitudes depend on extinction coefficients. Hence it would be possible to rationalize away the amplitude difference by postulating equal amplitudes and different extinction. Of course, it is possible to distinguish chain differences from RT relaxations by varying the CO concentration. Another would be to collect decays at different wavelengths, because the spectral difference expected for the R–T change is known.

It is evident that some type of averaging system is highly desirable and that complete spectra are important. The averaging procedure naturally requires that the apparatus retain a high degree of repeatability and pulse-to-pulse stability.

Modulated excitation uses a beam that is repeatedly turned on and off, with observation of the accompanying optical changes. In this sense it is a means of performing the necessary averaging. Stability is assured by modulating a constant light output, rather than requiring pulse-to-pulse repetition. However, an unexpected bonus appears in the ability to adjust the phase of the detection system. We describe this in the next section.

Conceptual Framework for Method

Figure 3 shows the results of turning on and off a weak source of illumination at various rates. Notice that the approach to equilibrium is governed by the same kinetics in both on and off cases. When the periodicity is longer than the relaxation times, the response is almost a square wave. The faster the on-and-off rates are relative to the relaxation rate, the more triangular the wave looks. The peak-to-peak amplitude also falls as the modulation frequency becomes less than the relaxation rates.

To fit a sine wave to the response, the wave must be shifted further and further to the right, in addition to decreasing in amplitude as shown. If we fit a function $P(\omega) \sin(\omega t + \phi)$, the shift can be described as a shift in phase ϕ in the term $\sin(\omega t + \phi)$. ϕ will begin at 0 and move to 90° (or $\pi/2$ radians).

[7] C. A. Sawicki and Q. H. Gibson, *J. Biol. Chem.* **252,** 7538 (1977).
[8] J. S. Philo and J. W. Lary, *J. Biol. Chem.* **265,** 139 (1990).
[9] K. D. Vandegriff, Y. C. LeTellier, R. M. Winslow, R. J. Rohlfs, and J. S. Olson, *J. Biol. Chem.* **266,** 17049 (1991).

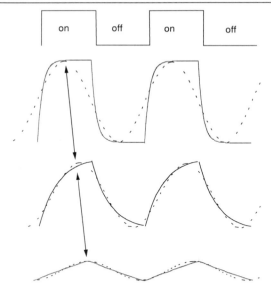

FIG. 3. The response of a system to a periodic driving force is shown. A single decay time governs the response of the system. As the period of the oscillation becomes shorter than the decay time, the shape becomes more and more triangular, and the peak-to-peak amplitude decreases. To fit the response with a sine wave requires that the sine wave be phase shifted. The phase shift depends on the ratio of the oscillation frequency to the relaxation rate. Thus the phase of the response can be used to recover the relaxation rate of the system.

If we are interested in the sine wave response, we can excite with a sine or cosine wave without loss of generality. And, as is discussed later, this has experimental advantages.

If we were to find the best phase angle empirically, we might move the trial sine wave across the signal, increasing the phase angle, until the overlap is maximal. It is interesting to note that there is not only a minimal overlap, but that the minimum is zero. This means that there is a phase angle at which a given signal will not be detected, even though it may be present. This ability to phase-tune away a given signal is key to the method.

To see how the frequency-domain picture is analyzed mathematically, consider photolysis and rebinding of a simple, single-binding site protein, such as the isolated chains of Hb, or myoglobin at room temperature. A pulse experiment would photolyze the ensemble of molecules at time $t = 0$, and the population P would then relax with rate k. In general k will depend on the free ligand concentration. In modulated excitation, the pulse is replaced by a continuous excitation that varies with frequency

ω. This excitation can be written as $A[\cos(\omega t) + 1]$. A is the rate of excitation, which includes the intensity of light and probability of its absorption. For samples optically thin in the spectral region of the exciting beam, $A = 2.3\, q\varepsilon z c_0 I$, where q is the quantum efficiency, ε is the extinction coefficient for the excitation light, z is the path length, c_0 is the concentration of ground state absorbers, and I is the average light beam intensity. Thus we can write

$$dP/dt = -kP + A[\cos(\omega t) + 1] \tag{14}$$

The addition of 1 to the cosine term keeps the light intensity (and ligand release rate) from becoming negative.

If we express the solution as $P(\omega)[\cos(\omega t) + \phi]$, then the phase angle ϕ is given by

$$\tan \phi = -\omega/k \tag{15}$$

The solution can be thought of as delayed by the phase angle. The amplitude of the oscillatory population is

$$P(\omega) = A/(k^2 + \omega^2)^{1/2} \tag{16}$$

From Eq. (16) it is clear that using too large an excitation frequency will cause a significant loss in signal, just as Fig. 3 showed.

The use of complex numbers simplifies the manipulations required, and offers insight into the modulation experiment. The only additional consideration is that the real part of the solution represents what is actually observed. If the excitation term in Eq. (14) is written as $A(e^{i\omega t} + 1)$ the solution is then succinctly given as

$$P = \frac{A}{k + i\omega}e^{i\omega t} + \frac{A}{k} \tag{17}$$

or

$$P = P(\omega)e^{i\omega t} + P(0) \tag{18}$$

where $i^2 = -1$. Taking $\mathrm{Re}(P)$ yields the solution of Eq. (16). The phase angle is determined from

$$\tan \phi = \mathrm{Im}[P(\omega)]/\mathrm{Re}[P(\omega)] \tag{19}$$

The amplitude of such a periodic signal, $P(\omega)$, is normally measured by a lock-in amplifier. Such an instrument detects the signal by multiplying the incoming signal by a reference signal oscillating at the same frequency. The reference signal can be represented as $\exp[-i(\omega_0 t + \phi_0)]$. The fre-

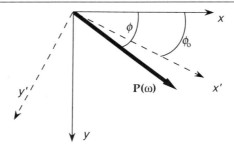

FIG. 4. Vector representation of the tuning procedure for a single relaxation process. Tuning the detector involves variation of the angle ϕ_0 until the population vector $P(\omega)$ lies along the x' axis. This is operationally achieved by minimizing the y' component of $P(\omega)$.

quency ω_0 of the detector is set equal to ω, the exciting frequency. Thus the lock-in output is the real part of $P(\omega)e^{-i\phi_0}$.

The phase of the amplifier ϕ_0 can be controlled by the user. Customarily, one would adjust the phase to obtain maximum signal. When $\phi_0 = -\phi$, the system is "in tune," and the real part is maximized. It is helpful to visualize this geometrically in the complex plane, as shown in Fig. 4. The x axis represents the real part of the signal, the y axis, the imaginary part $P(\omega)$ now looks like a vector (or phasor in electrical engineering terms) at an angle ϕ. It is in the fourth quadrant because the signal $P(\omega)$ lags the excitation, defined as purely real. Tuning the phase of the detector is represented by rotating the coordinate system (P is fixed) until the real part is maximal or, equivalently, until the imaginary part goes to zero. Rotated coordinates are designated x' and y'. Notice that at 90° to the signal there will be no component.

Theory of Modulated Excitation

Basic Theory

The value of modulated excitation comes from its ability to separate multiple relaxations. Consider excitations from a ground state G to state R, which can then relax to T, or rebind. In Fig. 1, the ground state is R_4 and R and T would be the triply liganded states. The equations that describe this process are

$$dR/dt = -k_R[X]R + A(1 + e^{i\omega t}) - k_{RT}R + k_{TR}T \tag{20}$$

and

$$dT/dt = -k_T[X]T - k_{TR}T + k_{RT}R \tag{21}$$

where we have denoted the free ligand concentration as [X]. So long as we take the viewpoint that k_R and k_T are rate constants for binding the last ligand, the analysis is independent of the exact model employed for the thermodynamic behavior of hemoglobin. This is simply because the models relate events at different ligation states; for example, k_T is assumed the same at the first and last step. Because these parameters can all be determined in the experiment, this analysis is truly model independent, and assumes only that there is an alternative structure with a known spectrum to which the molecule can switch with three ligands bound.

The solution to Eqs. (20) and (21) is a harmonic series, that is,

$$T = \sum_n T_n e^{in\omega t} \tag{22}$$

and likewise for R. Note that T_n (and R_n) will be functions of ω. Both T and T_n are also complex numbers. The first harmonic ($n = 1$) term relates R_1 and T_1 by simple substitution into Eq. (22). This creates a system of linear equations with complex coefficients that can be solved in their entirety for R and T.[5]

The signal V that is observed can be expressed as the product of a spectral weight, denoted $s(\lambda)$, which will depend on wavelength, times the population $R_1(\omega)$ or $T_1(\omega)$, which are wavelength independent but depend on excitation frequency. The spectral weight is the difference spectrum arising from removal of a ligand in the R or the T structure. Thus, dropping the subscript on R and T (because we will now exclusively consider the first harmonic terms),

$$V(\omega, \lambda) = s_R(\lambda)R(\omega) + s_T(\lambda)T(\omega)$$
$$= s_R(\lambda)[R(\omega) + T(\omega)] + [s_T(\lambda) - s_R(\lambda)]T(\omega) \tag{23}$$

The second equality emphasizes that the difference in spectra is more likely to be identifiable than either alone. The phase angle ϕ_0 is chosen to be that of $(R + T)$. Because we can write $R + T = |R + T| e^{i\phi_0}$ we can express the phase factor as $e^{i\phi_0} = (R + T)/|R + T|$. Then dividing by this factor gives

$$Ve^{-i\phi_0} = |R + T| \left[s_R + (s_T - s_R)\frac{T}{R + T} \right] \tag{24}$$

Multiplying V by the phase factor corresponds to the process of tuning the detector to the phase of the $R + T$ signal. This tuning process is equivalent to a rotation in vector notation, as shown in Fig. 5. The coefficient of s_R is clearly a real number in this rotated coordinate system. The imaginary part of the whole expression arises from the coefficient of $s_R - s_T$.

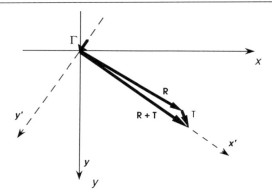

FIG. 5. Vector representation of R and T relaxations. Both R and T will vary in magnitude and phase with the excitation frequency; both populations are shown in equal weights. The observed signal weights these populations by their relative spectral contributions, so that R and T are illustrated at an isosbestic. T is drawn larger than normal for visibility. Note that in the rotated axes $(x'y')$ there is no contribution of $R + T$ projected onto the y' axis.

Whereas the coefficients do not depend on the observing wavelength, the spectral weights s_R and s_T do. In fact, if the wavelength is varied, a complete spectrum of s_R and $s_T - s_R$ can be obtained. We have

$$\text{In phase} = \text{Re}(Ve^{-i\phi_0}) = us_R + v(s_T - s_R) \qquad (25)$$
$$\text{Out of phase} = \text{Im}(Ve^{-i\omega_0}) = \qquad\quad w(s_T - s_R)$$

The in-phase channel contains a spectrum made up from s_R and $(s_R - s_T)$, and the out-of-phase channel contains a spectrum that is just $(s_R - s_T)$. In Eq. (25) u, v, and w are constants as λ is varied, but they will have different values at different excitation frequencies. The kinetic constants are contained in u, v, and w. We fit the in-phase channel spectrum to obtain the coefficient u of the s_R spectrum. Then the out-of-phase (rotated imaginary) spectrum can be fit to $s_R - s_T$ to give w. The fit coefficient w divided by the in-phase coefficient u gives Γ, where

$$\Gamma = \frac{w}{u} = -\text{Im}\left(\frac{T}{R + T}\right) \qquad (26)$$

We have not made use of v, the in-phase fit coefficient. This is not generally used because the average spectrum and the average population are both so much greater than the differences that v will be less well determined.

In terms of the rates involved, using Eqs. (20–22) it is easy to show that

$$\Gamma = \frac{k_{RT}\omega}{(k_{RT} + k_{TR} + k_T[X])^2 + \omega^2} \qquad (27)$$

This function peaks at frequency $\omega_{max} = k_{RT} + k_{TR} + k_T[X]$. Note that the product $2\omega_{max}\Gamma(\omega_{max}) = k_{RT}$. The maximum value for Γ [i.e., $\Gamma(\omega_{max})$] is simply related to the equilibrium constant between the R and T states if $k_T[X]$ is much smaller than the allosteric rates. If the equilibrium constant between R_3 and T_3 is denoted $L_3 \equiv k_{RT}/k_{TR}$, it is easy to see that $\Gamma(\omega_{max}) = L_3/2(L_3 + 1)$. As L_3 becomes large, $\Gamma(\omega_{max})$ approaches 0.5.

Note that k_R has not appeared in Γ, having been entirely removed by the tuning protocol. The out-of-phase signal is dominated by the results of the allosteric change. Consequently, as the monitoring wavelength is swept, the out-of-phase channel produces a spectrum of the allosteric change in real time. The value to this is that it allows the experimenter the ability to modify the experiment in progress. For example, if the out-of-phase spectrum has become too small to resolve at a given frequency, there is no reason to go to higher frequencies. Likewise, experimental problems (e.g., oxygen contamination) are viewed immediately. This is decidedly different from the way an experiment would be done if extensive transient averaging is required before the allosteric effects become visible.

To determine k_R, the tangent of absolute phase is measured at the R–T isosbestic. This is called absolute phase because it refers to the lag of the signal with respect to the driving term (photolysis beam), rather than a phase between different relaxation processes.

$$\tan \phi = \frac{\text{Im}(R + T)}{\text{Re}(R + T)} \tag{28}$$

The general expression is quite complicated because ligand rebinding is folded into the effective allosteric conversion rates. The general expression is easily derived and is given by

$$\tan \phi = -\frac{\omega[\omega^2 + (k_T + k_{TR} + k_{RT})^2 + k_{RT}(k_R - k_T)]}{k_R\omega^2 + (k_T + k_{TR} + k_{RT})[k_T(k_R + k_{RT}) + k_R k_{TR}]} \tag{29}$$

where we have suppressed the concentration of ligand for compactness. (Thus k_R in fact means $k_R[X]$, and similarly for k_T.) When ω becomes large compared to the various rates, this expression approaches a much simpler expression, namely,

$$\tan \phi \rightarrow -\omega/k_R(X) \tag{30}$$

Thus $\tan \phi$ approaches a straight line that passes through the origin and whose slope is the inverse of the ligand-rebinding rate. Although this limit is approached, it is difficult to collect much data at high frequencies

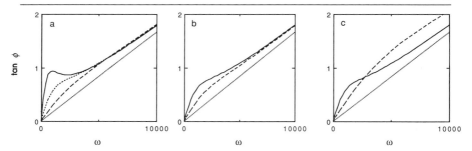

FIG. 6. Tan ϕ vs ω (in sec^{-1}) for various parameters showing deviation from linearity, plotted from Eq. (29). Usual experimental conditions will extend to frequencies at least twice as large as shown here; the restricted abscissa is to better illustrate the effects. $k_R[X] = 6000$ sec^{-1} for all the cases shown. Note that the straight lines in (a–c) show the asymptote, given by Eq. (30). (a) Variation of k_{TR}. $k_T[X] = 100$ sec^{-1}, $k_{RT} = 1000$ sec^{-1} for all curves. $k_{TR} = 3000$ sec^{-1} for the dashed curve, 1000 for the dotted curve, and 300 for the solid curve. (b) Variation of $k_T[X]$. $k_{RT} = k_{TR} = 1000$ sec^{-1} for both curves. $k_T[X] = 100$ sec^{-1} for the solid curve, 1000 sec^{-1} for the dashed curve. (c) Variation of k_{RT} and k_{TR} with fixed L_3. $k_T[X] = 1000$ sec^{-1} for both curves, $k_{RT} = k_{TR} = 1000$ sec^{-1} for the solid curve, and 3000 sec^{-1} for the dashed curve. (Note that for convenience the negative sign of the tangent is suppressed in customary usage.)

because the signal size decreases as ω^{-1} for high frequencies. Thus, in practice, this limit serves as a good starting estimate for an exact fit.

Tan ϕ must go through the origin, and it thus possesses a low-frequency linear asymptote. This is less easy to interpret, however. It is worth noting that large deviations from low-frequency linearity require substantial conversion to the T state and slow processes (either T-state ligand binding and allosteric conversion.) Figure 6 illustrates variation of the various parameters. Deviations from the asymptotic straight line can be viewed as the consequence of residency in the T state. Fast exchange, fast return, or fast rebinding minimize residence time and keep the tan ϕ plot closer to linear.

Subunit Inequivalences

It is interesting to ask what happens to the above analyses when there is inequivalence between the α and β subunits. It is easy to generalize the definitions of Γ and tan ϕ to include inequivalence. In general,

$$\Gamma = -\mathrm{Im}\left(\frac{T_\alpha + T_\beta}{R_\alpha + T_\alpha + R_\beta + T_\beta}\right) \qquad (31)$$

and

$$\tan \phi = \frac{\text{Im}(R_\alpha + R_\beta + T_\alpha + T_\beta)}{\text{Re}(R_\alpha + R_\beta + T_\alpha + T_\beta)} \qquad (32)$$

At the high-frequency limit, $\tan \phi$ has simple behavior:

$$\tan \phi \rightarrow -\frac{2\omega}{(k_{R\alpha} + k_{R\beta})[\text{X}]} \qquad (33)$$

That is, the inverse slope is just the average binding rate. If the quantum efficiency of the chains is different, this becomes

$$\tan \phi \rightarrow -\frac{2(q_\alpha + q_\beta)\omega}{(q_\alpha k_{R\alpha} + q_\beta k_{R\beta})[\text{X}]} \qquad (34)$$

If only the R-state binding rates differ, but not the T-state rates, then the equation for Γ assumes exactly the same form as Eq. (27) for simple rates. If the T-state rates differ, the equation is much more complicated, unless the T-state rates are small enough to be neglected. If the latter is true, however, Γ is again given by Eq. (27) despite binding rates that differ arbitrarily for α and β.

If the allosteric rates differ, Γ has a simple form at large excitation frequencies. We find that (assuming equal quantum efficiencies)

$$\Gamma \rightarrow \frac{k_{RT\alpha} + k_{RT\beta}}{2\omega} \qquad (35)$$

That is, the asymptotic value has the same functional dependence (ω^{-1}), and the average rate takes the place of the individual rates.

Inequivalence can arise in many different ways. Although the form of the resulting equations is not simple, it is possible to construct analytical expressions or to evaluate the various terms numerically to test a particular model. The difficulty in such cases is that of assigning α and β rates.

Experimental Realization

Figure 7 shows a diagram of the optical layout for modulated excitation with fluorescence detection.[10] This is one of the most elaborate possible setups, and has been shown to illustrate a system from which many simpler systems may be readily constructed. There are at least two beam lines in any experiment: the modulation beam, which provides the excitation, and the monitoring beam. We examine these in turn.

[10] A. J. Martino and F. A. Ferrone, *Biophys. J.* **56**, 781 (1989).

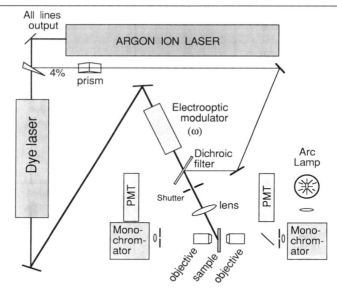

FIG. 7. An apparatus is shown for modulated excitation with optional fluorescence detection. The modulated excitation beam originates in the argon-pumped dye laser, which is modulated by an electrooptic modulator, and focused on the sample. The analysis beam for absorption originates in the arc lamp and passes through monochromator, mask, and ×15 objective before arriving at the sample. It then proceeds through a second ×15 objective, to a second matched mask and second monochromator, and into a photomultiplier tube (PMT). For absorbance, both monochromators are set to the same wavelength. For fluorescence, the analysis beam originates in the argon laser. A fraction of the beam is reflected, the lines are dispersed by a prism, and all but one are blocked. A dichroic filter combines the argon line, which is not modulated, with the dye laser line, which is modulated. Both beams are focused on the sample. The monochromators are set to different wavelengths; the final one is set to the fluorescence wavelength. For absorbance measurements, the laser beams are blocked by the shutter, and both monochromators scan in tandem. Then the photomultiplier nearest the arc lamp acts as a reference PMT.

Modulation Beam

Source. Our present arrangement (Fig. 7) utilizes laser excitation. A dye laser (Coherent CR 599) is pumped by an argon ion laser (Spectra Physics 2016-5). The dye laser offers a choice of wavelengths that are in the yellow–red region (or infrared). This can be important if a dye is used, because the dye will emit to the red of its excitation. If the modulation beam excites the dye, its fluorescence will have an unwanted AC component. When a dye is not required, an attenuated single line from the argon laser (514 or 488 nm) can be used by rearrangement of the mirrors. Laser sources are readily modulated, and this makes them desirable. On the other hand, the high intensity of the laser is emphatically not required because the excitation must be restricted to less than about 1%. In practice, we usually position the lens that focuses the laser beam so as to illuminate an area much larger than examined, so that the region of the sample that is probed is excited uniformly.

The first modulated excitation measurements were made with a 150-W tungsten lamp as the photolysis source, with suitable high-pass filters to separate the excitation light from the monitoring beam.[11] The beam was modulated by a mechanical chopper. It was subsequently discovered that the excitation was not maintained at the requisite low limits, despite the use of the incoherent source.[5]

Modulator. The simplest modulator is undoubtedly a mechanical chopper, of which a variety are marketed presently. The setup of a mechanical chopper and lamp must be done with care so that no phase shift arises from the motion of the blade image across the sample. If a mechanical chopper is used, the final lens should not form an image of the chopping wheel at the sample. If the sample is at a focus, this is ideal. In addition, mechanical choppers will typically produce a square wave pattern, and particular attention must be paid to suppression of the higher harmonics. Broadband lock-in amplifiers that demodulate by square wave mixing will allow harmonics to be detected. Either a front-end filter or harmonic-suppressing mixer is required.

Laser modulators do not have this spatial problem. Moreover, it is possible to drive the modulator with a sine wave so that harmonics are not generated that must be subsequently suppressed. Acoustooptic modulators offer broad bandwidth and less cost than electrooptic modulators.

The minimum amplitude of the modulated light should be as close to zero intensity as possible (making the average intensity about one-half the peak-to-peak value). At such low intensities, these modulators are typically nonlinear, and the nonlinearity implies harmonic generation. To circumvent such a problem, feedback schemes can be employed in which the modulator is driven so that its output is a pure sine wave. A commercial device using a stabilized detector for feedback and an electrooptic modulator is manufactured by Cambridge Research and Instrumentation (CRI, model LS 100, Cambridge, MA) and is the modulator in use in our appara-

[11] F. A. Ferrone and J. J. Hopfield, *Proc. Natl. Acad. Sci. U.S.A.* **73**, 4497 (1976).

TABLE II
CHARACTERISTICS OF VARIOUS MODULATORS

Type[a]	Bandwidth	Representative manufacturer and model number	Rough cost (1992)
Mechanical chopper	15–300 Hz 150–3000 Hz	EG&G PARC model 197[b]	$1500
Acoustooptic	DC-15 MHz	Isomet model 1205C-1/232A-1[c]	$1000
Electrooptic	DC-2 MHz	CRI model LS-PRO[d]	$8000

[a] Note that the electrooptic modulator involves a feedback system, and the acoustooptic modulator does not.
[b] Princeton, NJ.
[c] Springfield, VA.
[d] Cambridge, MA.

tus at present. Table II lists the performance of a sampling of various commercial modulators.

As mentioned above, it is desirable to maintain the light level sufficiently low that less than 1% of the hemes are excited. It is often convenient to use a neutral density filter after the modulator, rather than reduce the signal at the modulator itself, because the internal feedback system is somewhat more effective at higher light levels.

Signal Generator and Frequency Counter. A means is required for setting and measuring the modulation frequency. Depending on the instrumentation employed, a separate signal generator may or may not be required. Mechanical choppers, for example, have frequencies set by their own electronics. Furthermore, many lock-in amplifiers have excellent internal frequency generators, which when used eliminate the need for an external generator. Likewise, if the lock-in amplifier has a digital frequency readout, a separate frequency counter may be unnecessary.

Analysis Beam

To monitor the effect of modulation, a second optical train is required. The simplest system involves absorption, but fluorescence detection is also possible. In either case, the final detector is the same. The two methods differ in their source beams. For absorption, we employ a monochromator and 150-W xenon arc lamp. For fluorescence we have thus far used a piece of the argon beam. A thin wedge is inserted in the beam; each surface reflects about 4% of the light impinging on it. Because the wedge surfaces are not parallel, two weak divergent beams are generated and only one is used. (If a plate is used, both beams remain parallel and can create difficulties at the sample.) Because we run the laser multiline in this configuration, it is also necessary to select the wavelength required for excitation. This is done by a simple prism. Because the argon laser is polarized perpendicular to the optical table, it is important that the prism deflect the beam in the vertical direction. If the prism is used horizontally, a substantial fraction of the light will be lost due to reflection at the prism surface. Correct placement, however, can approximate Brewster's angle and minimize prism losses. The divergent lines can then be stopped except for that required for fluorescent excitation .The selected beam then is recombined after the modulator with the excitation beam by a dichroic filter selected to reflect the fluorescent exciting line and transmit the modulated beam.

Monochromator. Two identical monochromators to which we have added stepping motors are employed. For absorption measurements, they are run in tandem. However, they can be positioned independently, as is required for fluorescence detection. The independent positioning also allows some scattered photolysis light to be detected to set the zero of phase in absolute measurements.

The final monochromator has the important function of rejecting the photolysis light; the initial monochromator could be replaced by a wideband filter, but the current arrangement also reduces the net optical flux that continuously illuminates the sample. The two monochromators are more efficient than might at first appear, because the greatest losses occur due to polarization effects of the blazed grating.[12] Thus, a monochromator with 75% throughput most likely has almost 100% in one polarization, and 50% in the other. A second monochromator will then pass 100% of the good polarization, and a net 25% in the weak direction. The total transmission is therefore 63% rather than the 56% that might have been expected.

We have typically used different slits on the two monochromators. The intensity profile from a slit is triangular and, if equal slits are employed, slight differences in step size could lead to intensity losses. Instead, by maintaining one set wider, we limit the error that might result from such step mismatch.

[12] G. W. Stroke, *Phys. Lett.* **5,** 45 (1963).

For fluorescence measurements, we keep the final monochromator slits as the wider pair; for absorbance measurements without fluorescence we keep the final slits as the narrower pair, because it helps to reject the excitation beam.

Optics. Spatial resolution is also used to reject scattered excitation light. The first monochromator, lens, diaphragm, and condensing objective are arranged in Kohler illumination. The diaphragm is imaged at the sample and is positioned in the center of the excitation beam. The final objective, diaphragm, and lens are a mirror image of the first group, so that only light from the first mask goes through the second mask. This blocks a substantial amount of the excitation light that is collected by the final objective and might otherwise enter the monochromator.

The absorbance beam is constructed as a microspectrophotometer. This is useful in cases in which a high degree of photolysis is desired on the same sample. For example, photolysis of HbO_2 requires a large flux; the small size area monitored by the microscopic system allows the heat generated to dissipate with little rise in temperature. To maintain ultraviolet (UV) capacity, the objectives are Ealing $\times 15$ reflecting objectives with about 24-mm working distance (part No. 25-0506). The long working distance allows easy acess of the photolysis beam to the sample.

For simple (nonmodulated) absorbance measurements, a beam splitter diverts a fraction of the output of the initial monochromator into a reference phototube.

Detectors. Both detectors are S-20 photomultipliers (R453; Hamamatsu, Bridgewater, NJ). An important consideration is the maximum cathode current, because the detectors are typically exposed to substantial flux for a long time. We have found difficulties with bialkali-type tubes for this reason. Opaque rather than semitransparent photocathodes are another acceptable choice.

Although Figure 7 illustrates a tube adjacent to the monochromator, in practice we allow the beam exiting the monochromator to spread to a size that approaches that of the end window of the phototube. This decreases the effects of fatigue and saturation on the tube.

Electronics

Lock-In Amplifier. Although the original modulated excitation experiments used a single-phase lock-in with successive wavelength scans at orthogonal phases, it is fastest to collect in-phase and out-of-phase data simultaneously with a two-phase lock-in amplifier. The lock-in we presently use is a Stanford Research Systems model 530 two-phase lock-in (Sunnyvale, CA). This instrument allows computer control of all its main functions, most importantly time constant, phase, and gain. Critical to our selection was the good harmonic rejection, achieved in part by use of sine wave mixing in this device. Roughly comparable instruments are produced by EG&G PARC (model 5210) (Princeton, NJ) and Ithaco (model 3961B) (Ithaca, NY). Ithaco has produced an even lower cost alternative, which is constructed on a card that must be inserted into a PC-AT or clone personal computer and that performs amplification, two-phase demodulation, and filtering up to a maximum frequency of 10 kHz. Although this is 10 times less than the stand-alone units, we have not used frequencies higher than this at present.

In choice of an instrument, sensitivity is seldom an issue, because the typical signal levels are between 500 μV and 1 mV and only occasionally fall to the 200- to 500-μV range. Out-of-phase signals are 10 times smaller but are amplified at the output.

Converting Intensity to ΔA. We can write the intensity transmitted through the sample as I, given by

$$I = I_0 10^{-[A + \Delta A \cos(\omega t + \phi)]} \tag{36}$$

where A is the nonoscillatory absorbance, and ΔA is the magnitude of the oscillating absorbance.

$$\Delta A = z \sum_m \Delta \varepsilon_m \Delta c_m \tag{37}$$

where z is the path length, $\Delta \varepsilon_m$ is the change in extinction coefficient, and Δc_m is the change in concentration of each of the species that are varying with the modulation (e.g., R and T). In terms of our previous notation, $s_R = z \Delta \varepsilon_R$, and so on. Converting to natural logarithms and expanding the ΔA term,

$$I = I_0 e^{-2.3A}[1 - \Delta A \cos(\omega t + \phi)] \tag{38}$$

where terms of ΔA^2 or smaller have been discarded. There is a DC term, $I_0 e^{-2.3A}$, and a term with amplitude $I_0 e^{-2.3A} \Delta A$ oscillating at frequency ω. Thus ΔA is the ratio $AC(\omega)/DC$.

Photomultiplier Supplies. When a modulated excitation spectrum is scanned through a region of strong absorption, the light level drops considerably. Because the signal required is a ratio of AC to DC terms, the ratio may be large, although individual terms are small. In the lock-in amplifier, the smallness of the signal means that output noise can become significant.

What we have done to avoid this is to employ a programmable power supply on the final phototube. The programming is via the control computer, which adjusts the output to keep the DC voltage approximately constant. This is also valuable because the gain afforded by a photomultiplier has little noise associated with it. It is important, however, to assure that the maximum tube voltage not be exceeded in the attempt to achieve constancy of the DC output. Both AC and DC values are recorded, although for most experiments the DC remains constant.

Running Experiment

There are two types of measurements in the basic experiment. They differ most notably in how the lock-in is tuned. Measurement of Γ entails tuning to the ligand removal and rebinding signal and collecting complete spectra. Measurement of tan ϕ entails tuning to the laser excitation signal and measurement at a single wavelength. These are discussed in greater detail in the next section.

Γ *Determines Allosteric Rates.* In this measurement, phase tuning is performed at an R–T isosbestic, such as 436.5 nm. Both monochromators are set to this wavelength. Phase is adjusted until one channel is zero. It may be necessary to amplify the output of this channel (most lock-ins have a control to do so by a factor of 10). The channel that has been set to zero is the out-of-phase channel, whereas the channel that now has maximum signal at this wavelength is denoted the in-phase channel. Monochromators are then reset to one end of the range and scanned through the spectral range of interest. Once the spectra are recorded, a new excitation frequency is chosen, and tuning and scanning are repeated.

Dynamic Averaging. The principal noise source in these experiments is not electronic noise or laser noise but shot noise, that is, noise due to the fluctuations of the photoelectrons ejected by the light beam. This is approximately given by the square root of the number of photons in the beam. Hence in regions that absorb strongly, the signal-to-noise ratio drops. Amplification due to the gain in the photomultiplier tube (PMT) amplifies noise and signal together. Because of this effect, we employ a dynamic averaging procedure, in which we allow the control computer to perform all measurements to approximately equal levels of uncertainty. Thus a greater number of measurements is collected in the absorbing regions. A series of points is measured, and their average and standard deviation are computed. If the standard deviation is greater than a preset percentage of full scale, an additional set of points is taken, both are averaged together, and the deviation again compared. This proceeds until the deviations are sufficiently small. Note that it is important to take measurements beyond the time constant of the lock-in amplifier, otherwise the measurements are correlated intrinsically. For example, with the lock-in time constant set at 1 sec the computer should not record measurements taken less than 2 to 3 sec apart.

Determining Tan φ and Ligand-Binding Rates. In this measurement the lock-in is tuned to be in phase with the excitation. This is accomplished by setting only the detection monochromator to coincide with the excitation source wavelength. It will be necessary to reduce the high voltage somewhat for this part. Phase is again adjusted to make one channel zero, which is denoted the out-of-phase channel. Next, both monochromators are set to the R–T isosbestic (e.g., 436.5 nm). The ratio of the out-of-phase to in-phase signal at this wavelength and excitation frequency is called tan φ and is recorded and averaged over several measurements. The excitation frequency is then changed, and the process repeated. Tangent values are usually limited by noise to lie between 0.1 and 10.

Typical Data. Figure 8 shows typical spectra obtained for photolysis of the CO derivative. The filled circles show the most significant components in phase and out of phase, whereas the open circles show the next most significant components in the spectra. As can be seen, the data are well represented by the standards used in the fits. Figure 8c shows the three components used in fitting the most significant component of the out-of-phase spectrum. The solid curve represents the T − R difference spectrum taken from Perutz *et al.*[2] The dashed curve represents a spectrum that has been ascribed to a T − R transition of the CO-liganded hemes.[5] The dotted curve, just visible, is the deoxy-CO difference spectrum. The experiment whose spectra are described in Fig. 8 has been analyzed according to the procedures described above, as shown in Fig. 9. The low-frequency values of Γ are higher than the fit, due to multiple excitations.

Singular Value Decomposition. The previous method assumes that all the spectra that appear in the kinetic processes are known. A convenient way to summarize the accuracy of this assumption is provided by singular value decomposition (SVD) methods. Although these have been the subject of extensive exposition elsewhere,[13] our use is rather modest. We employ SVD to determine the number of independent spectra in the in-

---

[13] E. H. Henry and J. Hofrichter, this series, Vol. 210, p. 129.

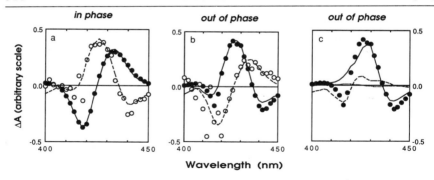

FIG. 8. Typical spectra obtained for photolysis of HbCO in pH 7.3, 0.15 M phosphate buffer. (a) In-phase spectrum. The filled circles show the first singular value component, which best represents the full data set of 13 spectra, whereas the open circles show the next most significant component in the set of spectra and has a singular value 1.7% of that of the first component. The third in-phase component, not shown, has a singular value of 1.0% of the first component. (b) Out-of-phase spectrum, which is 43 times smaller than the in-phase spectrum. The second singular value component, shown as open circles, is 30% as large as the first out-of-phase component; the third component, not shown, is 14% as large as the first out-of-phase component. As can be seen, the data are well represented by the standards used to generate the solid lines in (a) and (b). (c) Three components used in fitting the most significant component of the out-of-phase spectrum. The solid curve represents the T − R difference spectrum taken from Perutz et al.[2] The dashed curve represents a spectrum that has been ascribed to T − R transition of the CO-liganded hemes.[5] The dotted curve, just visible, is the deoxy-CO difference spectrum. (From M. Zhao and F. A. Ferrone, unpublished data, 1993.)

phase and out-of-phase spectra (also called the effective rank of the data matrix). It is necessary for the rank of significant spectra to be less than the number of standards employed and that the standards be capable of providing a satisfactory fit to all the significant spectra.

Other Applications

Probes Other Than Soret Absorbance. The procedure described has set the detector in tune with the excitation results. This provides a frame of reference from which other measurements also may be made. One such is fluorescence monitoring of the quaternary change.[10] Another application has been to observe the Bohr effect by dye absorbance. Tuning was performed in the Soret, while the dye absorbance was monitored in the visible.[14]

[14] M. Zhao, J. Jiang, and F. A. Ferrone, *Biophys. J.* **61,** A56 (1992).

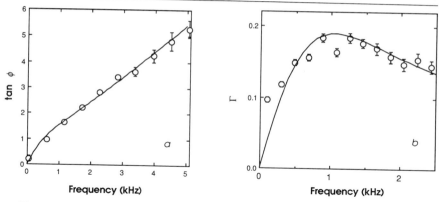

FIG. 9. The experiment whose spectra are described in Fig. 8 has been analyzed according to the procedures described in the text. (a) Tan ϕ is shown as a function of excitation frequency f. (Note that $\omega = 2\pi f$.) (b) Γ is shown as a function of excitation frequency f. Not as many high-frequency points are taken because the signal size relative to noise level is too small. Taking $k_T[CO] = 100$ sec^{-1}, (b) is analyzed to give $k_{RT} = 2500$ sec^{-1} and $k_{TR} = 4000$ sec^{-1}. This is shown as the solid line. Deviations at low frequencies are most likely due to multiple excitations. With these values and $k_R[CO] = 6000$ sec^{-1}, the tan ϕ curve in (a) is generated.

Polarization-Modulated Measurements. Modulated excitation also can be combined with a polarization-modulated analysis beam instead of a simple absorbance beam. We have done experiments using circular dichroism (unpublished data, 1976) and linear dichroism.[15] In these experiments, the averaging properties of the system are useful, because dichroism measurements are intrinsically noisier than absorbance measurements. Moreover, the combination of two modulation methods (polarization and excitation) is much more felicitous than the hybrid of pulse and modulation required in typical attempts to measure such transients.[16]

In such an experiment, the output of the polarization detecting lock-in must be the input to the modulation detection lock-in. Hence, the time constant must be set to its minimum to allow the maximum bandwidth for the second lock-in detector. Even so, there will be some phase shifts introduced into the second lock-in because of the final filter of the first lock-in. This makes absolute phase information more difficult to obtain.

[15] F. A. Ferrone, J. Hofrichter, and W. A. Eaton, *in* "Hemoglobin S Polymerization in the Photostationary State" (P. L. Dutton, J. Leigh, and A. Scarpa, eds.), p. 1085. Academic Press, New York, 1978.

[16] F. A. Ferrone, J. J. Hopfield, and S. E. Schnatterly, *Rev. Sci. Instrum.* **45,** 1392 (1974).

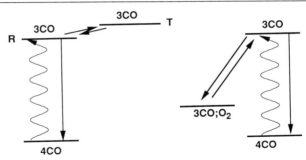

FIG. 10. Photolysis of CO followed by oxygen binding and release. The process of binding an O_2 molecule and spontaneously releasing it, as shown on the right, is formally equivalent to the R–T switch, shown on the left, following photolysis. Thus, with appropriate substitution, the allosteric analysis can be used directly for the replacement reaction.

Diagnosis and Repair of Experimental Problems

Oxygen Contamination in CO Samples

Modulated excitation is sensitive to small amounts of oxygen contamination in CO photolysis experiments because the O_2 will bind rapidly and subsequently be released. As can be seen in Fig. 10, the process of binding an O_2 molecule and spontaneously releasing it is formally equivalent to the R–T switch. Hence we can make the substitution of oxy–deoxy spectrum for the R–T spectrum and measure those coefficients. We can then construct a quantity analogous to Γ, which we shall denote Ω, given by

$$\Omega = \frac{k_{R, ox}[O_2]\omega}{(k_{R, ox}[O_2] + k_{off})^2 + \omega^2} \tag{39}$$

We have written the R-state oxygen binding rate as $k_{R, ox}$ and the off rate from the R state as k_{off}. Clearly the size of Ω will be governed by the concentration of O_2. With a mere 1 μM O_2, Ω reaches a maximum of 0.35. If the amount of oxygen is known, this provides a convenient means for determining the last ligand affinity. The equations unfortunately become somewhat more involved when there is T-state conversion as well. The presence of oxygen contamination is seen by the appearance of an oxygen related peak in the out-of-phase spectrum and the failure of the out-of-phase spectrum to zero at 400 nm.

Inaccurate Tuning

Although tuning is simple in principle, noise can make it difficult in practice to decide that a channel has been accurately nulled. This is especially true at higher frequencies. Imperfect tuning is not a fatal error and can be easily diagnosed and corrected. Proper tuning should not produce a ligand-loss spectrum in the out-of-phase channel. Hence the signature of mistuning is the presence of such a ligand-loss component when the out-of-phase spectrum is fit. Because the out-of-phase spectrum is normally enhanced 10-fold, the magnitude of the error is quite small and can be corrected by iteration.

In Eq. (24) above, mistuning corresponds to rotation of V by $\exp[-i(\phi + \delta\phi)]$. The real coefficient of s_R [denoted u in Eq. (25)] will be $\cos \delta\phi \, |R + T|$ instead of $|R + T|$, and the imaginary part containind the coefficient of $(s_T - s_R)$, called w above, will now be given by

$$w = \cos \delta\phi \, \mathrm{Im} \left(\frac{T}{R + T} \right) + \sin \delta\phi \, \mathrm{Re} \left(\frac{T}{R + T} \right) \qquad (40)$$

In addition, there is a new term, absent before, consisting of a coefficient of s_R in the out-of-phase channel, which we shall call u', and which is given by $\sin \delta\phi \, |R + T|$. Clearly $u/u' = \tan \delta\phi$, from which the mistuning angle $\delta\phi$ can be found. Then the following procedure is employed. The original data are used as a measure of $\Gamma(\omega)$, which generates the allosteric parameters. These parameters then generate both real and imaginary parts of $T/(R + T)$. Now with the correction and real and imaginary parts, a new $\Gamma(\omega)$ can be constructed.

Although at first the method of modulated excitation may seem to be critically dependent on the ability to precisely tune the ligand-binding spectrum away, the above discussion makes it clear that the tuning proto- col is more properly viewed as a tool for signal enhancement.

Overexcitation

To remove only one ligand, it is important that the fraction of the sample excited remains small. This can be surprisingly restrictive because of the problems of phase. The phase shift of T relative to R arises because of the delay in position of the population peaks between R and T. At low frequencies, there is not much shift between R and T [recall Eq. (27), which goes to zero as ω goes to zero]. However, multiple excitation to R_2, rapid conversion to T_2, and rather slow conversion to T_3 via T-state binding kinetics can produce more T_3 out of phase than does the direct cross-over conversion. To check the level of excitation requires a measure- ment of $\Delta A/A$ at low frequency. This is because, as ΔA depends on frequency, it will naturally fall at higher frequencies. Its amplitude is approximately given by Eq. (16), so that it is desirable to have $\omega \ll k_R[X]$. Alternatively, the fall in signal can be corrected from Eq. (16).

The multiple excitation error is most important at low frequencies, and a signature of multiple excitation is large out-of-phase signal at low frequencies and the failure of the $\Gamma(\omega)$ curve to have the expected shape. An approximate expression for the multiple excitation error δ is given by[5]

$$\frac{\delta}{\Gamma} = \frac{A}{k_{RT}} \frac{\omega^2 + k_{TR}^2 - k_{RT}^2}{\omega^2 + 4k_T^2[X]^2} \tag{41}$$

The excitation coefficient A is usually <10 sec^{-1}, so the leading factor is normally quite small. However, the T-state rebinding kinetics also may be small, in which case δ can be a sizable fraction of Γ.

Because multiple excitations can easily contaminate a small signal, excitation beam uniformity is important. For example, a tightly focused Gaussian laser beam could produce multiple excitations at its center, while the average monitored area retained the 1% or so criterion.

Stray Modulation Light

Because of the high sensitivity of this method, stray modulated light, being at the excitation frequency, can be seen at low levels. A good test for this is to block the light from the arc lamp, and set the high voltage to the highest value and the sensitivity of the lock-in to the lowest value used in the experiment. The amount of stray light can thus be observed and corrective measures taken.

A symptom of stray modulation light is the appearance of peaks in the out-of-phase spectrum that resemble the absorbance peaks. This is because the modulation spectrum is obtained by dividing AC by DC intensity. If the AC is augmented, we have $(I_0 e^{-2.33A} \Delta A + I_{leak})/(I_0 e^{-2.3A})$, which no longer produces the expected cancellation of the $I_0 e^{-2.3A}$ term. This produces almost a mirror image of the absorbance A under certain conditions.

A similar problem that occasionally arises is that of AC pickup via ground loops. With the detection PMT blocked, there should be no AC signal; anything seen is electronic, not optical. Usually this is too small a signal to create problems, but it can be important if small optical signals are sought. Because ground-loop problems depend critically on the electronic layout, there is no simple remedy, and if this problem appears, a good introductory electronics text should be consulted. (Note that optical tables can provide an unwanted electrical connection, and it may occasionally be necessary to insulate components electrically from the table top. A thin piece of plastic, found in commercial page protectors, works well.)

Limitations of Technique

Frequencies

Although in principle any reversible photodriven process can be examined by this means, the decrease in amplitude with increasing frequency [recall Fig. 3 and Eq. (16)] creates signal-to-noise problems. Not only does the driving (R-state) signal fall off as ω^{-1}, but $\Gamma(\omega)$ also falls as ω^{-1}, that is, the raw out-of-phase signal is falling as ω^{-2}. This limits the usable frequency range more than the technological aspects of modulators and lock-in detectors.

Spectral Accuracy

A second limit is posed by the fact that amplitude information is required to determine both k_{RT} and k_{TR}, in that this requires accurate spectra of the allosteric difference. There is some degree of ambiguity regarding the size of these spectra.[17] We have performed a study in which fluorescence and absorption probes were compared.[10] Error in the amplitude of the allosteric spectra would have shown up as differences between kinetics measured by the absorbance and fluorescence probes. The good agreement suggested that the difference spectrum derived from the model compound studied, namely, NES-des-Arg Hb $\pm$ inositol hexaphosphate (IHP),[2] provides an acceptable standard for these studies. On the other hand, this limit must be borne in mind when unusual conditions or hemoglobins are considered. This is especially true because, despite years of service to hemoglobin researchers, the exact nature of the allosteric difference spectrum remains unknown.[18]

[17] J. S. Olson, *Proc. Natl. Acad. Sci. U.S.A.* **73**, 1140 (1976).
[18] L. P. Murray, J. Hofrichter, E. R. Henry, and W. A. Eaton, *Biophys. Chem.* **29**, 63 (1988).

[16] Picosecond Phase Grating Spectroscopy: Applications to Bioenergetics and Protein Dynamics

By JOHN DEAK, LYNN RICHARD, MARCO PEREIRA, HUI-LING CHUI, and R. J. DWAYNE MILLER

Introduction

Picosecond phase grating spectroscopy (PGS) is a form of transient grating spectroscopy.[1-2] This form of spectroscopy is related to holography, in which the constructed image is a diffraction grating. In general, a diffraction grating is capable of light diffraction through spatially periodic modulations in optical absorption, index of refraction, or both. The different contributions to the light diffraction process can be manipulated through changes in optical wavelengths, choice of materials, or probing conditions. These manipulations change which mechanism dominates the diffraction process and has led to a number of different grating spectroscopies ranging from degenerate four-wave mixing (phase conjugation) to thermal grating descriptions of the grating formation. Phase grating spectroscopy can be distinguished from these other grating spectroscopies in that the probe wavelength is chosen to monitor periodic modulations in the index of refraction. For liquid-phase studies, this modulation is determined almost exclusively by the density dependence of the index of refraction. The material density is a rather simplistic probe of molecular processes. Yet there is a wealth of information pertaining to protein dynamical processes that cannot be attained readily by any other spectroscopy, as is highlighted below. In particular, this method is extremely sensitive to energy relaxation processes and photoinduced changes in molecular volume. In this regard, phase grating spectroscopy has been developed into a new probe of both protein motion and bioenergetics.

In terms of studying energetics, energy relaxation processes lead to thermal expansion, which modulates the density and correspondingly the index of refraction. Because of the interferometric nature of this technique, it is extremely sensitive, capable of detecting changes in energetics that raise the lattice temperature by less than $10^{-4}°$. The high sensitivity is essential as biologically significant amounts of energy correspond to 1–3

---

[1] R. J. D. Miller, *in* "Time Resolved Spectroscopy" (R. J. H. Clark and R. E. Hester, eds.), Vol. 18, p. 1. Wiley, Chichester, England, 1989; M. D. Fayer, *Annu. Rev. Phys. Chem.* **33**, 63 (1982).

[2] H. J. Eichler, P. Gunter, and D. W. Pohl, "Laser-Induced Dynamic Gratings." Springer-Verlag, Berlin, 1986.

kcal/mol—the amount of energy required to break a hydrogen bond. This level of energy detection is well within the limits of this spectroscopy. In this application, thermal phase grating spectroscopy is comparable to time-resolved photoacoustic spectroscopy,[3] except the time resolution is approximately three orders of magnitude better. The high time resolution of this technique, with respect to energetics, is an important feature of this spectroscopy. An operational time resolution of 10–20 psec has been attained. This time resolution is sufficient to follow even the primary events of vibrational energy relaxation. In the study of bioenergetics, the energy must show up in the aqueous phase to be properly credited as a relaxation process. Ten picoseconds is approximately the time frame needed for proteins to exchange energy with the water layer (see "Vibrational Energy Relaxation Studies") such that this method has achieved the fundamental upper limit with respect to time resolution in following energetics in biological systems. Questions concerning how energy exchanges within a protein and the time scale by which a protein accesses the stored energy within its structure can be addressed with this new methodology.

The high sensitivity of this spectroscopy is also ultimately responsible for its application in following protein motion. Even minute changes in protein structure that require net motion (material displacement) can be holographically recorded as a density grating to provide a real-time view of global protein motion. The two different contributions to the signal, energetics, and protein-driven density changes can be readily separated by the different temporal responses of the grating formation and by studies near the zero thermal expansion point of water. The unique sensitivity of phase grating spectroscopy to global structural relaxation is important as this feature of the spectroscopy can be used to determine the length scale over which the driving forces for the motion are distributed. By correlating the dynamics of the global relaxation to structural changes local to the focal point of the forces (the spatial location of the stimulus) the degree of coupling to collective modes of the protein can be determined. This correlation in turn provides information on the spatial distribution of the forces, as is discussed in the "Structural Relaxation Dynamics" section.

The global motion of the protein that leads to density changes must involve changes in the protein volume or strain ($\Delta V/V$). The concept of proteins undergoing strain changes in response to repulsive steric interactions is central to most discussions of biomechanics. In the study of functionally relevant protein motion, phase grating spectroscopy provides a direct probe of both protein strain and the released strain energy: the

[3] J. A. Westrick, J. L. Goodman, and K. S. Peters, *Biochemistry* **26**, 4497 (1987); J. A. Westrick, K. S. Peters, J. D. Ropp, and S. G. Sligar, *Biochemistry* **29**, 6741 (1990).

two fundamental parameters needed to understand the mechanics. These are important observables for problems ranging from allosteric regulation to protein folding.

Theory and Observables of Phase Grating Spectroscopy

The utility of phase grating spectroscopy has been demonstrated in many previous studies.[1-2] More than one photophysical process can contribute to the grating formation, making the signal complex in nature. The experimental conditions, however, can be manipulated to select only those processes of interest in a grating experiment. In a transient grating experiment, two time-coincident excitation pulses of the same wavelength are brought into the sample and crossed at an angle, θ_{ex}, to form an optical interference pattern (see Fig. 1). This pattern is holographically encoded in the sample through photoprocesses induced by the excitation pulses that change the complex index of refraction ($\tilde{n} = n + ik$) of the material. The spatial pattern, or fringe spacing, of the diffraction grating is dependent on the excitation beam geometry:[2,4]

$$\Lambda = \frac{\lambda_{ex}}{2 \sin(\theta_{ex}/2)} \tag{1}$$

where Λ is the grating fringe spacing, and λ_{ex} is the excitation wavelength. A variably time-delayed probe pulse is brought into the sample at the correct angle for Bragg diffraction from the grating image.

The diffracted probe intensity as a function of the time delay between the pump-and-probe pulses constitutes the signal. The diffraction efficiency of the probe has contributions from periodic modulations in the imaginary (amplitude) and real (phase) parts of the complex index of refraction induced in the material by the excitation pulses, that is,

$$\eta = \eta_a + \eta_p \tag{2}$$

where η is the diffraction efficiency, and the subscripts a and p refer to the diffraction from the amplitude and phase grating components, respectively. More explicitly, the amplitude and phase grating components are described in terms of the complex index of refraction by the relations

$$\eta_a = \exp - \left[\frac{\alpha(\omega)d}{\cos \theta}\right]\left(\frac{\pi d}{\lambda \cos \theta}\right)^2 \left[\Delta k(\omega)\right]^2 \tag{3}$$

$$\eta_p = \exp - \left[\frac{\alpha(\omega)d}{\cos \theta}\right]\left(\frac{\pi d}{\lambda \cos \theta}\right)^2 \left[\Delta n(\omega)\right]^2 \tag{4}$$

[4] A. E. Siegman, *J. Opt. Soc. Am.* **67**, 545 (1977).

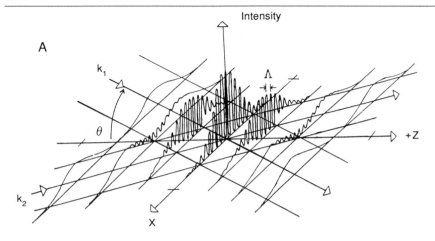

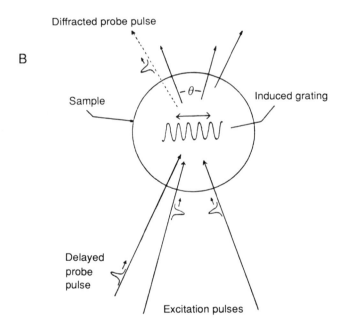

Fig. 1. (A) The effects of spatially Gaussian laser beams. The optical interference, with finite spatial width Gaussian beams, forms a Gaussian-shaped interference pattern of fringe spacing Λ, where k_1 and k_2 are the wavevectors of the excitation beams. The grating thickness is defined by the beam intersection region (reproduced from Miller,[1] Copyright 1989, by permission of John Wiley & Sons, Ltd.). (B) Transient phase grating pulse sequence. Two excitation pulses are used to write the grating, and a nonresonant probe is brought in at its respective Bragg angle for diffraction off of the induced grating.

for $\eta \ll 1$.[5,6] In this expression, α is the material absorptivity, d is the grating thickness, and ω, λ, and θ refer to the probe radial frequency, wavelength, and Bragg angle for diffraction, respectively. The term Δk is related to the induced peak–null variations in the material absorptivity or optical density ($k = 2.303\alpha\lambda/4\pi d$). The term Δn defines the induced peak–null changes in the real part of the complex index of refraction. The above expression is for plane waves. This expression needs to be modified by a geometric correction factor (G) for Gaussian excitation and probe beams,[4] that is,

$$\eta(\omega) = [\eta_a(\omega) + \eta_p(\omega)]G(\Omega_{ex}, \Omega_p) \tag{5}$$

where the correction factor is a multiplicative constant that depends on the excitation (Ω_{ex}) and probe (Ω_p) spatial beam waists.

An optically induced change in the material density is the observable of interest. Density-dependent changes in the absorption (i.e., spectral shifts) are negligible compared to the corresponding change in the index of refraction (n).[1,5] The optically triggered changes, due to either protein structural changes or energy deposition, contribute to the signal predominantly through the phase grating component. For probe wavelengths on-resonance to either ground or excited state transitions, there will be contributions to the diffracted signal from a population grating, with both amplitude and phase grating components, and a density-modulated phase grating. In this case, the diffraction efficiency is from a mixed grating such that

$$\eta(\omega, t) = \exp -\left[\frac{\alpha(\omega)d}{\cos\theta}\right]\left(\frac{\pi d}{\lambda\cos\theta}\right)^2\{[\Delta k_{ex}(\omega, t)]^2 + [\Delta n_{ex}(\omega, t) + \Delta n_s(t)]^2\}G(\Omega_{ex}, \omega_p) \tag{6}$$

The density-induced changes in the index of refraction, Δn_s, appear in superposition to the population grating components Δk_{ex} and Δn_{ex}. The density-modulated index, Δn_s, forms a cross-term with the population phase grating component, Δn_{ex}.

For resonant probe conditions the grating can be used to follow population dynamics through the time evolution of the population phase and amplitude grating terms [Δn_{ex} and Δk_{ex} in Eq. (6)]. However, this information can be obtained through more simple pump–probe experiments. With respect to heme proteins, the novel information contained in the grating experiment is the density modulation of the phase grating component (Δn_s)

[5] K. A. Nelson, R. Casalegno, R. J. D. Miller, and M. D. Fayer, *J. Chem. Phys.* **77**, 1144 (1982).
[6] H. Kogelnik, *Bell Syst. Tech. J.* **48**, 2909 (1969).

that arises from either energy or structural relaxation processes. To make
the phase grating as sensitive as possible to the ensuing density modulations and to avoid cross-term complications, it is desirable to choose probe
wavelengths in which the population phase grating contribution is at a
minimum. This condition normally can be accomplished by using a probe
wavelength well red of the lowest energy electronic transition. However,
the probe must also be well off resonance of any excited state or transient
absorption features. Excited electronic states normally absorb at wavelengths extending well out into the infrared, which may make this condition
difficult to achieve. The relative magnitudes of the population phase and
amplitude grating components at different wavelengths can be determined
by a Kramers–Kronig analysis of the transient absorption difference spectrum. The difference absorption spectrum $\Delta\alpha(\omega)$ between the final and
initial states of the excited molecules can be transformed into the corresponding population modulated index, Δn_{ex}, according to

$$\Delta n_{\text{ex}}(\omega) = \frac{c}{2\pi} P \int_{-\infty}^{\infty} \frac{\Delta\alpha(\omega')d\omega'}{\omega'(\omega' - \omega)} \tag{7}$$

where P indicates the Cauchy principal value, c is the speed of light, and
ω is the radial frequency of the radiation. Because the optical absorption
properties of heme proteins are well characterized, this analysis provides
an accurate determination of the relative contribution of population phase
and amplitude processes to the signal.

Even with the above caveat, finding a suitable probe at which the
density modulations dominate the grating diffraction is not usually a problem for most systems. In this case, an essentially pure phase grating
develops in which

$$\eta(t) = \left(\frac{\pi d}{\lambda \cos \theta}\right)^2 \Delta n_s^2(t) \tag{8}$$

where

$$\Delta n_s(t) = \left(\frac{\partial n}{\partial \rho}\right)\delta\rho(t); \qquad \delta\rho(t) = \rho\Delta S(t) \tag{9}$$

In this expression, the term ΔS refers to peak–null modulation in the
material strain, which is the parameter typically used to describe acoustic
fields and density changes. The main point is that for probes sufficiently
off resonance, the induced density changes dominate the grating signal.

The optically initiated density changes are holographically imaged by
the excitation pulse interference pattern into a standing acoustic wave in
which the acoustic wavelength is defined by the optical fringe spacing

(Λ). As discussed, the ensuing density changes can arise through either energy relaxation and thermal expansion or changes in the molecular volume of the protein with optically induced structural changes. If the density changes are rapid relative to the acoustic period of the grating image, the density changes coherently excite a standing acoustic wave, that is, two counterpropagating waves are launched that beat in and out of phase with the acoustic period, τ_{ac} ($\tau_{ac} = \Lambda/v_s$; v_s is the material speed of sound). For thermally induced density changes, this limit of the coupling to the acoustics can be solved analytically. In this case, the diffracted signal is given by[7,8]

$$\eta(t) = \left(\frac{\pi d}{\lambda \cos \theta}\right)^2 [A(1 - \cos \omega_{ac} t)]^2 \tag{10}$$

where ω_{ac} is the radial acoustic frequency ($2\pi/\tau_{ac}$), and A is the amplitude of the acoustic wave as defined by the relation

$$A = \alpha_{th} \Delta T/T = \alpha_{th} q/\rho V C_V \tag{11}$$

In this expression, α_{th} is the thermal expansion coefficient, ΔT is the temperature change induced in the material, which can be calculated from the amount of energy (q) deposited as heat into the irradiated volume (V) from the heat capacity (C_V). The dependence of the signal amplitude on the deposited heat forms the basis for this spectroscopy in the determination of reaction energetics.

The impulsive limit to the signal response defined above is depicted in Fig. 2. This signal response can be decomposed into an acoustic and a nonpropagating thermal grating component. The static thermal density modulation results from the static temperature differences induced in the material. It is the acoustics beating against the static thermal grating offset that leads to the $(1 - \cos \omega_{ac} t)^2$ dependence in the signal. The effect of finite molecular relaxation times on the thermal acoustic grating formation has been treated analytically.[8] These results are also reproduced in Fig. 2 to clarify this discussion. Dynamic processes that do not deposit energy impulsively into the lattice, relative to the acoustic period, augment the thermal grating component while diminishing the acoustic amplitude through interference effects. For typical signal-to-noise ratios, molecular relaxation processes on the order of one-tenth of an acoustic period can be resolved as a phase shift in the acoustics and a decrease in the depth of modulation of the grating signal. With appropriate excitation beam geometries, the acoustic period can be made as short as a few tens of

[7] K. A. Nelson and M. D. Fayer, *J. Chem. Phys.* **72**, 5202 (1980).
[8] L. Genberg, Q. Bao, S. Gracewski, and R. J. D. Miller, *Chem. Phys.* **131**, 81 (1989).

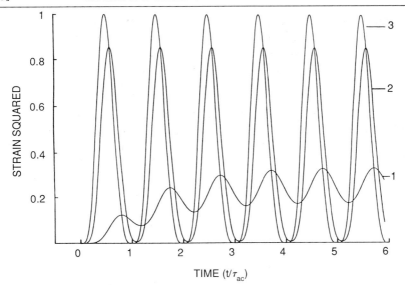

FIG. 2. Theoretical calculations of the effects of finite energy relaxation times $(1/k)$ on the grating formation dynamics. The normalized difference in the strain squared is plotted as a function of nondimensional time, t/τ_{ac}. Impulsive driven acoustic waves are shown in curve 3, for which $1/k \ll \tau_{ac}$. Curve 2 shows the case for which $1/k = 0.1\tau_{ac}$, and $1/k = \tau_{ac}$ is shown in curve 1. Within typical signal-to-noise ratios, it is possible to resolve energy relaxation processes that occur with time constants of one-tenth the acoustic period as phase shifts in the acoustics from the impulse response.[8]

picoseconds. Thus, this method is capable of picosecond time resolution in the study of the energetics of the protein response. The fast time resolution of phase grating spectroscopy is important in following the energetics of the initial tertiary structure changes. This high time resolution and sensitivity has been exploited to study vibrational energy relaxation,[8,9] the energetics of photoreactions,[10] and weak optical transitions.[1,1a]

In general, it is difficult to excite a system exactly at the electronic origin such that there will always be a rapid thermalization component that arises from vibrational relaxation of the excess energy. This general phenomenon of a rapid energy relaxation phase will impulsively drive acoustics of the form shown in Fig. 2. In certain cases it is desirable to eliminate the acoustic oscillations, as the ringing tends to obscure the underlying, slower energy relaxation dynamics that may be of interest. The acoustics can be eliminated completely by the proper choice of excita-

[9] L. Genberg, F. Heisel, G. McLendon, and R. J. D. Miller, *J. Phys. Chem.* **91**, 5521 (1987).
[10] M. B. Zimmt, *Chem. Phys. Lett.* **160**, 564 (1989).

tion pulse width. The dependence of the acoustics on pulse width and shape has been treated in detail theoretically and demonstrated experimentally.[8] By either manipulating the grating fringe spacing or the excitation pulse width to make it the same duration as one-half an acoustic period, it is possible to cancel the acoustic effect. In this case, the leading edge of the pulse generates an acoustic response that is 180° out of phase with respect to the trailing edge of the excitation pulse. The acoustics cancel, leaving the underlying nonpropagating thermal phase grating component. The slower relaxation components then add quadratically in signal amplitude to the initial rapid thermalization process. In this manner, a 10% slow thermalization component results in a signal amplitude 100 times larger than the signal would be in the absence of the fast thermalization component.

The most important aspect of this spectroscopy to the study of protein dynamics is the density change induced by the optically triggered changes in molecular volume or structure.[11] This component to the signal can be separated from thermal contributions by studying the protein response at the zero thermal expansion point of water. In addition, the well-defined thermal phase grating formation dynamics shown in Fig. 2 are important in distinguishing protein-driven acoustics from a thermal response. In the study of heme proteins, the protein structural changes make a significant contribution to the signal and have a different temporal dependence from thermal effects such that the separation of thermal contributions from protein structural changes is relatively straightforward. In the case of carboxymyoglobin (MbCO), this protein volume change dominates the density grating signal.[11] The coupling of the protein motion to the acoustics is found to approximate a $t = 0$ sinusoidal strain boundary condition rather than a stress boundary condition as in a pure thermal grating case (see "Structural Relaxation Dynamics"). The density changes arise from the material displacement that occurs during the course of the optically triggered heme protein structural transition. In this case, the dynamics and the amplitude of the grating image replicate the protein motion and yield a direct measure of this process.

Experimental Methods

Phase grating spectroscopy can provide information on the global motion of proteins and the corresponding energetics ranging in time from subpicoseconds to seconds. The different dynamic ranges under study require different experimental setups and varying degrees of complexity

[11] L. Genberg, L. Richard, G. McLendon, and R. J. D. Miller, *Science* **251**, 1051 (1991).

to generate the requisite pulse widths and wavelengths for the excitation pulse and probe sequence. For example, our previous work used a picosecond system designed with the capability to probe resonant and nonresonant processes by employing the second harmonic and the fundamental of an Nd:YAG (neodymium:yttrium–aluminum–garnet) laser, respectively. Exchanging a 100-mW continuous wave (cw) laser for the probe allowed longer time domains to be explored as well. For subpicosecond time resolution it is slightly more difficult to generate the infrared to near-infrared pulses required to study selectively the phase grating component. To achieve this end, the diffracted grating signal is upconverted with an amplified pulse from a synchronously pumped dye laser, which also serves to generate the grating pattern. These different experimental setups are shown in Figs. 3 and 4.

The experimental setup of the dual-probe arrangement is shown in Fig. 3. A single pulse selected from the output of a Q-switched and mode-locked Nd:YAG laser is frequency doubled and then spectrally separated with a dispersing prism. The 532-nm probe pulse is split from the excitation

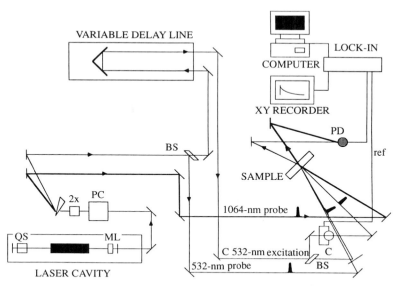

FIG. 3. Experimental setup of the dual-probe arrangement. A Q-switched (QS) and mode-locked (ML) Nd³⁺:YAG laser is used as the source for the excitation and probe pulses. The excitation pulses are variably delayed in time with respect to the two probe pulses so that only one delay line is needed. The $t = 0$ positions for the two probe wavelengths were determined independently, using cross-correlations with the excitation pulses. BS, Beam splitter; C, chopper; PC, pockel cell pulse selector; PD, photodiode; 2×, doubler crystal.

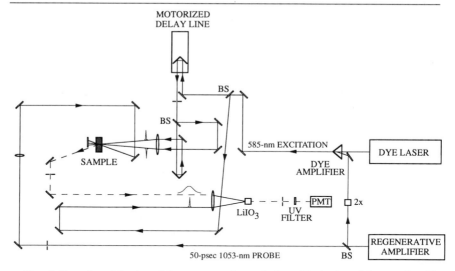

MOTORIZED
DELAY LINE

BS

BS

585-nm EXCITATION

DYE LASER

SAMPLE

DYE
AMPLIFIER

LiIO₃   UV
FILTER   PMT   2x

REGENERATIVE
AMPLIFIER

50-psec 1053-nm PROBE   BS

FIG. 4. Experimental setup of the upconversion technique. Excitation of the sample with 585-nm, picosecond pulses causes the dissociation of the Fe–CO bond. A 50-psec probe pulse at 1.053 μm, left over from the Nd^{3+}:YLF regenerative amplifier, is used to probe the grating. The diffracted infrared light is summed with a split-off of the dye laser to conserve picosecond time resolution.

pulses, using a 30% beam splitter to provide maximum excitation energies of 17 μJ. The excitation pulses are sent to a motorized optical delay line to provide the time resolution between the excitation and the two probe pulses. The excitation pulses are split with a 50% beam splitter and focused into the sample to give spot sizes of 200 μm. The excitation angle is varied from 2 to 11°. Both a 532-nm and 1.064-μm probe are focused and sent into the sample at the proper angle for Bragg diffraction. This setup is designed so that it is possible to measure simultaneously the resonant probe (532 nm) and an off-resonant probe (1.064 μm) diffraction efficiency under identical excitation conditions. The diffracted signal is collected and processed by a lock-in amplifier (EG & G 5209, Princeton, NJ) and then sent simultaneously to an X–Y recorder (Bausch and Lomb 2000, Rochester, NY) and a computer with a Data Translation 2801-A A/D board (Marlboro, MA) for data acquisition. The time base is provided by the voltage across a 10-turn potentiometer connected to the motorized delay. The excitation pulse energy is typically less than 500 nJ for the lowest excitation studies to ensure that only one photon process contributes to the signal. This excitation level corresponds to less than 10% bleach of the ground state population for the focusing conditions used. Absolute diffraction efficiencies are determined by measuring the probe intensity before the sample and the diffracted signal with the same photodiode. To correct for differences in optimized Bragg angle and beam parameters for the two wavelengths studied, the absolute diffraction efficiencies are referenced to the signal from a pure acoustic grating provided by the deoxy form of the protein under study.

The experimental setup of the upconversion technique is shown in Fig. 4. Subpicosecond pulses (750 fsec) are generated by a synchronously pumped dye laser that is pumped by the frequency-doubled output of an Nd:YLF (Neodymium:yttrium-lithium-fluoride) mode locked laser. Two dye amplifier stages, pumped by the second harmonic of an Nd:YLF regenerative amplifier, increases the dye laser output to the microjoule level at a repetition rate of 500 Hz. A 50% beam splitter creates two equally intense excitation beams from the amplified dye laser. These beams are crossed at an angle of 10° to write the grating pattern in a flowing solution of protein. A 50-psec, 1.053-μm pulse from the regenerative amplifier is then overlapped at the appropriate angle for Bragg diffraction. The time delay between the excitation and probe pulses is controlled using a stepper motor-driven delay stage (Daedal, Harrison City, PA) with an optical encoder. The phase grating signal is then spatially isolated and upconverted in a $LiIO_3$ crystal with a fraction of the amplified dye laser pulse, obtained before the delay, to produce <1 psec, 376-nm light. This ultraviolet (UV) signal is then detected on a photomultiplier tube (Hamamatsu R928, Bridgewater, NJ) and the data recorded on an IBM 386 computer. In the upconversion technique, the time resolution is defined by the sampling window created in the sum-frequency mixing process. For the experimental setup described above, a window of <1 psec of the diffracted infrared probe is sampled, which makes subpicosecond resolution possible.

The grating alignment is more demanding than a conventional pump–probe experiment in that it requires the spatial and temporal alignment of three beams rather than two. The spatial overlap of the beams on the sample can be ensured by crossing the pump and probe beams through a pinhole whose diameter is slightly smaller than the beam waist. The timing between the excitation beams can then be maximized by adjusting the relative beam lengths while monitoring the second harmonic produced by the overlap of the excitation pulses on a thin crystal. The conditions in which the pump-and-probe beams are all time coincident can then be determined in an analogous manner. The most difficult task in performing a grating experiment, however, is isolating the signal from background scatter. This is best accomplished by allowing the signal to travel a long distance through a series of irises before detection to spatially filter the signal from the scattered light. A reference sample that provides a strong grating signal is convenient for initial beam alignment. Typically, high concentrations of dyes, such as malachite green in ethanol, are used to produce intense thermal gratings so that the diffracted signal can be seen by eye. All the grating parameters can then be optimized by maximizing the reference signal at a fixed time delay.

The hemoglobin samples, used in the studies discussed below, are prepared from fresh whole blood, using standard techniques.[12] This method prepares the oxy derivative. If the CO form is desired, CO is passed over the solution until the UV–visible spectra indicate complete conversion. The preparation of deoxyHb involves extensive flushing of metHb with nitrogen and subsequent reduction with a molar excess of sodium dithionite. The myoglobin (horse heart, type III; Sigma, St. Louis, MO) is dissolved in 100 mM Tris buffer, pH 7, and is reduced anaerobically using a slight molar excess of sodium dithionite. If the CO form is desired, CO is passed over the sample, which readily converts. For the deoxy samples, special care is taken to ensure that oxygen is removed from the sample cells. The procedure includes the use of sealed flasks and a nitrogen-purged sample cell fitted with septums that enables sample transfer under nitrogen.

For the optical studies using high-repetition laser sources, the protein must be circulated rapidly enough to ensure that each laser shot encounters a fresh sample of the protein. The rapid circulation must be accomplished without denaturing the protein and without the formation of excessive bubbles in the protein solution, which strongly scatter light. Two different methods are used. For room-temperature studies, the protein samples are housed in an airtight rotating cell. Short path lengths are used, on the order of 250 μm, to avoid excess scatter from bubbles that otherwise are spun into the beam-sampling region with longer path lengths. The small path lengths necessitate the use of concentrations on the order of 1 to 3 mM in order to ensure adequate signal-to-noise ratios. For the temperature-dependence studies, a flowing cell design with a path length of 1 mm is used, which is cooled with a 50% (v/v) mixture of ethylene glycol and water from a temperature-controlled bath (RTE-100; Neslab Endocal, Newington, NH). The flow is obtained by using a Rainin Rabbit peristaltic pump with a flow rate of approximately 10 ml/min. The width of the cell is reduced to 1 mm so that this flow is sufficient to ensure an adequate sample turnover between laser shots without degrading the sample. The temperature is measured with a thermocouple readout combination (DP81-T; Omega, Stamford, CT) that is fitted with a 20-gauge needle. This needle is fitted through a septum port adapter in the cell so that the

[12] E. Antonini and M. Brunori, "Hemoglobin and Myoglobin in Their Reactions with Ligands." North-Holland Publ., Amsterdam, 1971.

temperature can be measured directly. Because of the longer path lengths of this cell, sample concentrations are on the order of 1 mM.

The energetics are crucial to a detailed understanding of the biomechanics behind the protein motion. As is detailed in "Structural Relaxation Dynamics" the protein-driven density changes have been found to dominate the signal. It has been found, however, that the energetics can be determined by going to a 75%/25% (v/v) glycerol–buffer solvent in which the thermal grating dominates. The protein samples are prepared as previously described, with the exception that the solutions are made up from stock solvent, and the conversion to the carboxy form of the protein takes several hours of stirring under CO to ensure that the protein has completely converted. The samples and stock solvent are always kept in sealed vessels, and transfers are made via syringe to avoid the picking up of water from the atmosphere. The deoxy form of the protein is used as the reference for 100% deposition of the absorbed photon energy. The optical density of the deoxy sample is adjusted to be as close as possible to the carboxyheme. Calibration curves are made as a function of optical density to correct for any small differences in the optical density between the sample and reference. The optical densities are measured directly in the cell used to make the measurements and the signal scaled according to the calibration curve. For most of the measurements, the optical densities are essentially identical (less than 5% difference), so that this procedure becomes unnecessary. The excitation conditions are monitored and kept to within less than 5% difference in the excitation pulse energies used to determine the signal level from the deoxy reference and the carboxyheme. A minimum of five independent measurements are made in the evaluation of the energetics, and a minimum of three measurements are made for all other reported measurements.

Applications

One of the essential issues in biophysics is the underlying mechanism of protein motion and structural transitions that are integral to the functionality of biological molecules. These motions often involve large amplitude-correlated motions that require the concerted action of a large number of atomic degrees of freedom. In this regard, heme proteins are one of the most important protein systems exhibiting large amplitude-correlated structural changes.

For hemoglobin and myoglobin, the changes in conformation are triggered by the binding and dissociation of small ligands, such as O_2, NO, and CO.[12] The CO-ligated form is particularly valuable in that it exhibits essentially unit quantum efficiency for photodissociation,[13] with recombination occurring much slower than the ensuing ligated (oxy) to deoxy tertiary structure changes.[14-16] The ability to trigger carbon monoxide dissociation optically and to monitor the subsequent structural changes toward the deoxy state has provided a wealth of information on protein dynamics as related to structure and function. A number of spectroscopies ranging from transient absorption,[17-19] time-resolved infrared,[15,20] and Raman[21-24] have been used to follow the optically triggered structure changes. From these studies and X-ray crystal structure determinations, it is currently believed that the displacement of the iron from the heme plane and the doming of the porphyrin ring that follow ligand dissociation are the primary events driving the system to the deoxy tertiary structure.[25,26] The changes in the tertiary structure of the monomer heme proteins, in turn, alter the forces at the $\alpha\beta$ interface of the hemoglobin tetramer. The cumulative action of the tertiary structure changes ultimately lead to the R $\rightarrow$ T quaternary structure change that controls the O_2-binding efficiency of hemoglobin. The state of ligand occupancy is communicated to the adjacent subunits through structure changes. This communication link is the heart of the mechanism of allosteric control of protein function. Thus, understanding the correlated structure changes that occur in the deoxy structure transition of heme proteins is important in understanding both protein dynamical responses and molecular cooperativity.

[13] Q. H. Gibson and S. Ainsworth, *Nature (London)* **180,** 1416 (1957).

[14] L. P. Murray, J. Hofrichter, E. R. Henry, and W. A. Eaton, *Biophys. Chem.* **29,** 62 (1988); L. P. Murray, J. Hofrichter, E. R. Henry, M. Ikeda-Saito, K. Kitagashi, T. Yonetani, and W. A. Eaton, *Proc. Natl. Acad. Sci. U.S.A.* **85,** 2151 (1988).

[15] P. A. Anfinrud and R. M. Hochstrasser, *Proc. Natl. Acad. Sci. U.S.A.* **86,** 8387 (1989).

[16] J. W. Petrich, J. C. Lambry, K. Kuczera, M. Karplus, C. Poyart, and J. L. Martin, *Biochemistry* **30,** 3975 (1991).

[17] R. H. Austin, K. W. Beeson, L. Eisenstein, H. Frauenfelder, and I. C. Gunsalas, *Biochemistry* **14,** 5355 (1975).

[18] T. E. Carver, R. J. Rohlfs, J. S. Olson, Q. H. Gibson, R. S. Blackmore, B. A. Springer, and S. G. Sligar, *J. Biol. Chem.* **265,** 20007 (1990).

[19] W. D. Tian, J. T. Sage, V. Srajer, and P. M. Champion, *Phys. Rev. Lett.* **68,** 408 (1992).

[20] L. Rothberg, T. M. Jedju, and R. H. Austin, *Biophys. J.* **57,** 369 (1990).

[21] T. G. Spiro, G. Smulevich, and S. Ogawa, *Q. Rev. Biophys.* **8,** 325 (1975).

[22] S. Kaminaka, T. Ogura, and T. Kitagawa, *J. Am. Chem. Soc.* **112,** 23 (1990).

[23] E. W. Findsen, J. M. Friedman, M. R. Ondrias, and S. R. Simon, *Science* **229,** 661 (1985).

[24] J. W. Petrich, J. L. Martin, D. Houde, C. Poyart, and A. Orszag, *Biochemistry* **26,** 7914 (1987).

[25] M. F. Perutz, *Nature (London)* **228,** 726 (1970).

[26] M. F. Perutz, G. Fermi, B. Luisi, B. Shaanan, and R. C. Liddington, *Acc. Chem. Res.* **20,** 309 (1987).

There have been numerous models proposed for the mechanics of the heme protein structure changes. In principle, the changes in structure could occur through correlated vibrational modes. However, the number of vibrational modes that can be coupled in a structural transition is limited by the spatial dispersion of vibrational energy. Vibrational energy relaxation and dispersion are intimately tied to the discussion of large amplitude-correlated motions and have been addressed elsewhere.[8,27] This work, which is summarized below, has shown that vibrational energy relaxation and dispersion among the protein modes is extremely fast, occurring on picosecond time scales. The rapid randomization of vibrational energy limits the coupling of vibrational modes to correlated motions. These findings raise questions concerning how the dynamics of conformational changes are triggered and directed to specific parts of a protein. The mechanism of energy transduction from a stimulus, such as the dissociation of a ligand, to the specific protein motions associated with the function of the molecule and the dynamics of the protein motion remains an enigma.

Most models for the structure changes rely on energy transduction processes involving potential energy and do not consider vibrational energy relaxation and dispersion processes. Fundamentally, these models rely on the forces that develop from changes in the potential energy surface of the protein through the interaction with a stimulus to displace the atoms to the new structure. In the case of heme proteins, the mechanisms all invoke the release of stored potential energy with the heme ligation site as the focal point. The major distinguishing features of these various models are the length scale over which the forces act and the multiplicity of pathways along the structure coordinate.

One of the most detailed models for the structure changes was developed by Perutz and is referred to, herein, as the localized strain model.[25,26] On the basis of static differences in structure between oxy- and deoxyhemoglobin, a specific sequence of events was envisaged. The breaking of the $Fe-O_2$ bond (or $Fe-CO$ bond) leads to the doming of the heme. The puckering of the porphyrin ring in this process reduces the repulsive potential at the proximal histidine site. The proximal histidine then tilts, relieving the strain, and creates a new strain line along the FG helix that ultimately is pushed into place by this force. Although not specifically stated in this model, this sequence of events views the potential energy gradient, or acting force, as localized over length scales comparable to one amino acid residue (~5 Å). First, one portion of the protein responds, creating a potential energy gradient in a new location, and then the next

[27] R. J. D. Miller, *Annu. Rev. Phys. Chem.* **42**, 581 (1991).

segment of motion occurs, and so on. This spatial propagation of the potential energy gradient, along with the subsequent response of the protein in that region, is analogous to the sequential falling of dominos. An important feature of this model is that the whole process is orchestrated by the specific structure of the protein. When a stimulus acts on the system, the energy released during the course of the protein response to the stimulus can cause motion in only one direction, as the other motions are sterically hindered.

A paradigm that incorporates the localized strain model but includes a dynamic sampling of the complex potential energy surface of the protein has been proposed by Frauenfelder and co-workers.[17,28,29] Within the framework of this model, a protein in a particular state can assume a large number of conformational substates, which have the same overall structure and function but differ in the details of the structure and rates of those functions. A protein reaction is paralleled to an earthquake in which a stress is relieved at the focus. The released strain energy is dissipated in the various modes of the protein and through the propagation of a deformation. On photodissociation of CO from myoglobin, the stress is relieved, and the protein evolves toward the deoxy structure through a series of "protein quakes." The protein quakes progress through various conformational substates until equilibrium is attained. This relaxation is sequential and nonexponential in time. The main distinction of this conformational substate model from the localized strain model is that it includes multiple pathways for the propagation of the structure changes.

In the connection to structural relaxation, the existence of conformational substates in heme proteins has been demonstrated mainly by temperature dependencies of the nonexponential recombination kinetics of geminate CO recombination. These studies provide details of the barrier distribution for different conformational states of the protein. For a system with as many degrees of freedom as heme proteins, there will undoubtedly be nearly degenerate conformational substates separated by small barriers. However, there is no direct connection between the barrier distribution that affects CO recombination and the potential energy surface that directs the structural relaxation of the entire protein. The use of CO recombination as a probe is potentially too restricted in the motions it probes to yield sufficient information on the initial moments of tertiary structure relaxation. Thus, even though conformational intermediates have been identified, it is unclear whether conformational substates represent the dominant

[28] A. Ansari, J. Berendzen, S. F. Bowne, H. Frauenfelder, I. E. T. Iben, T. B. Sauke, E. Shyamsunder, and R. D. Young, *Proc. Natl. Acad. Sci. U.S.A.* **82**, 5000 (1985).
[29] H. Frauenfelder, S. G. Sligar, and P. G. Wolynes, *Science* **254**, 1598 (1991).

path for the propagation of the atomic displacements as the protein evolves to its new structure.

The above models assume that the potential energy gradients that drive the system response are localized over some length scale. This assumption requires that the protein be flexible over the same length scale. At the opposite extreme, the protein may respond as a rigid body (on the relevant time scale of the structure transition). In this case, the forces are redistributed over the entire protein structure. In this event, the doming of the heme following ligand dissociation would lead to the collective displacement of a large number of atoms. The propagation of the structure changes would then best be described by a superposition of low-frequency collective modes of the protein. There is experimental evidence that low-frequency collective modes exist in proteins from Raman studies[30] and from neutron-scattering studies.[31] These types of motions also have been found in molecular dynamics studies and normal-mode analyses of proteins.[32,33] The frequency range available for a structural rearrangement, propagated by collective modes in heme proteins, is 1–50 cm^{-1}. These frequency components correspond to picosecond time scales for the atomic displacements associated with these modes. For a net change in structure to occur through collective modes, the segments of the protein at the focus of the forces responsible for the ensuing motion must be more rigid than the adjacent areas. In the case of heme proteins, the α-helical sections forming the heme pocket are more rigid than the corners.[33] If these sections are sufficiently more rigid than the corners, then the doming of the heme accompanied by ligand dissociation could lead to a collective displacement of the atoms onto the global energy minimum corresponding to the deoxy structure. Such a collective displacement of atoms would represent the most efficient mechanism possible for the functionally important motions of the protein.

The tertiary and quaternary structure changes of heme proteins, activated by changes in the state of ligation at the heme, offer an ideal response function for understanding deterministic protein motion. The above models for the heme protein response function vary in the length scale over which the forces act and the multiplicity of the functionally relevant pathways. To understand the dynamic response of the protein, the length scale over which the induced potential energy gradient is delocalized needs to be determined. This can be accomplished by studying the protein dynamics

[30] P. C. Painter, L. E. Mosher, and C. Rhoads, *Biopolymers* **21**, 1469 (1982).
[31] S. Cusack and W. Doster, *Biophys. J.* **58**, 243 (1990).
[32] B. Brooks and M. Karplus, *Proc. Natl. Acad. Sci. U.S.A.* **82**, 4995 (1985).
[33] Y. Seno and N. Go, *J. Mol. Biol.* **216**, 111 (1990).

with probes that are sensitive to different length scales of the protein motion. In addition, the number of pathways and functionally distinct conformations along the structural transition coordinate can be determined from a study of the energetics. As the protein transcends into its lower energy configuration, the time evolution of the energy release replicates the barrier distribution among the different structural intermediates along the reaction coordinate. The study of the protein motion with probes sensitive to different length scales, along with the energetics for the motion, should enable a distinction between the conformational substate (with various degrees of localized strain) and collective displacement models for the heme protein structure transition. Both descriptions of the atomic displacements operate to some degree. The issue is which mechanism is dominant. This issue can be resolved by determining which mechanism leads to the greatest change in the energetics of the system, that is, the greatest downhill energy excursion along the deoxy structure coordinate.

Overall, a detailed understanding of protein response functions requires information about (1) how the energy exchanges within the protein, (2) the spatial distribution of the driving forces, and (3) the energetics or amplitudes of the driving forces for the motion. Each of these aspects of the problem is discussed below.

Vibrational Energy Relaxation Studies

The initial PGS studies focused on determining how vibrational energy spatially disperses among the enormous number of mechanical degrees of freedom in a biological molecule. This information provides information on how energy exchanges within proteins. The problem of vibrational energy redistribution is particularly complex in biological systems, as there are mechanical degrees of freedom that are similar to solid state phonons (i.e., spatially extended low-frequency modes of proteins) and liquid-state collisional exchange processes (energy exchange through modulation of van der Waals contacts in the protein). The treatment and understanding of vibrational energy flow in biological systems requires a synthesis of liquid state and solid state concepts. This is an important problem as the exchange of vibrational energy among the system modes is related to fluctuation and dissipation processes that govern the motion of the protein.

In this regard, heme proteins represent ideal systems. The excited state lifetimes of deoxy ferrous heme proteins and ferric forms are extremely short, on the order of a couple of picoseconds to subpicoseconds. By optically exciting the heme chromophore, it is possible to selectively deposit large amounts of vibrational energy at the heme site (10,000–20,000 cm^{-1} in the ground state following internal conversion) and then, by using different probes, monitor the spatial propagation of the vibrational energy to the aqueous interface.[27] The vibrational energy relaxation pathway is shown schematically in Fig. 5. The trickling of vibrational energy from the initially prepared Franck–Condon modes into adjacent modes depends on the anharmonicities of the interatomic potentials and the density of states that conserve energy and momentum (i.e., the number of spatially adjacent modes that can accept energy). This can be expressed in terms of the mode density for the system as a function of energy, as shown in Fig. 5A. A simple hydrodynamic model for the energy flow, as an analogy, is shown in Fig. 5B. There is both an energy transfer and transport component to the spatial dispersion or spreading of this energy. In the large-molecule limit for intramolecular vibrational energy redistribution (IVR), $V_{vib}^2 \rho_{vib} \gg 1$, the redistribution of energy is fast, occurring on the picosecond to the subpicosecond time scale for hydrocarbons. The large-molecule limit would be expected to apply to biological molecules. However, one needs to consider the density of states specific to a particular site, that is, the local density of states.[27] As a good example, the heme chromophore is primarily held in place, within the protein globular structure, through van der Waals contacts. Thus, one needs to consider the modes local to the heme that act as conduits for the first transfer step from the heme to the covalently bonded network of the surrounding protein. Prior to the phase grating studies, this step alone was expected to occur on a 20- to 40-psec time scale.[34,35]

The PGS method selectively follows energy relaxation processes through relaxation of the initially excited vibrational modes into translational degrees of freedom of the bath—the final sink for the excess energy. It is through the increase in the translational energy of the molecules that the lattice thermally expands and changes the index of refraction. The index of refraction is determined largely by the water density, as these are by far the largest number of oscillators that interact with the probe and determine the optical dispersion. Thus, this method gives a fairly selective means of monitoring the energy transfer from the protein to the aqueous bath.

In the determination of energy relaxation processes, the time resolution depends on the grating fringe spacing [Eq. (1)]. The thermal expansion due to increases in translation energy content must create a density modulation of the correct spatial period to meet the Bragg condition for light diffraction. As discussed above, this limit is defined by the time it takes for material expansion to occur over one-half the grating fringe spacing

[34] D. D. Dlott, *J. Opt. Soc. Am. B* **7**, 1638 (1990).
[35] E. R. Henry, W. A. Eaton, and R. M. Hochstrasser, *Proc. Natl. Acad. Sci. U.S.A.* **83**, 8982 (1986).

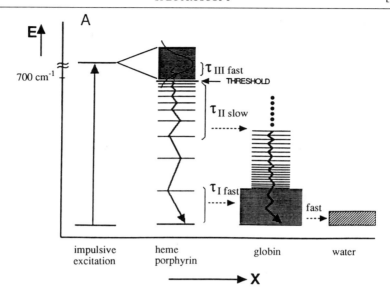

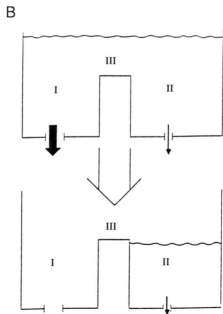

FIG. 5. (A) Energy-level diagram schematically representing the local density of state dependence for vibrational energy relaxation as a function of spatial position from the heme center. Impulsive excitation creates a nonstationary state centered on the heme porphyrin.

for full grating formation. The highest time resolution is achieved by making the fringe spacing as narrow as possible, which is accomplished by using counterpropagating beam geometries [$\theta = 90°$ in Eq. (1)]. Representative results are shown in Fig. 6 for counterpropagating 366-nm excitation and probe pulses. The time resolution for this grating geometry is limited to 20 psec by these speed of sound considerations.

The grating formation dynamics for the excitation of all heme proteins studied to date show a speed of sound-limited rise time of approximately 20 psec, as shown in Fig. 6. This finding demonstrates that most of the energy is dissipated from the protein into the aqueous bath in less than 20 psec. This observation also means that the excess vibrational energy leaves the initial heme site to the surrounding protein in less than 20 psec and becomes highly dispersed. The conclusion regarding the highly spatially dispersed nature of the energy within the protein matrix is based on studies of collisional exchange processes in liquids. This process takes significantly longer than 20 psec to relax if vibrational energies on the order of 10,000 cm^{-1} are localized on a specific site.[36] The fast energy dissipation processes have been confirmed by time-resolved infrared studies that show a heating up of the water following the optical excitation of deoxy heme proteins with a time constant of 16–19 psec.[15,37]

From the initial grating studies and related works,[24,38–40] we now have

[36] P. O. Scherer, A. Seilmeier, and W. Kaiser, *J. Chem. Phys.* **83**, 3948 (1985); A. Seilmeier and W. Kaiser, *Top. Appl. Phys. Ultrashort Laser Pulse Appl.* **60**, 279 (1988).

[37] R. M. Hochstrasser, R. Diller, S. Maiti, T. Lian, B. Locke, C. Moser, P. L. Dutton, B. R. Cowen, and G. C. Walker, *in* "Ultrafast Phenomena VIII" (J. L. Martin, A. Migus, G. A. Mourou, and A. H. Zewail, eds.), p. 517. Springer-Verlag, Berlin, Heidelberg, 1993.

[38] R. Lingle, X. B. Xu, H. P. Zhu, S. C. Yu, J. B. Hopkins, and K. D. Straub, *J. Am. Chem. Soc.* **113**, 3992 (1991).

[39] R. G. Alden, M. D. Chavez, M. R. Ondrias, S. H. Courtney, and J. M. Friedman, *J. Am. Chem. Soc.* **112**, 3241 (1990).

[40] H. Hayashi, T. L. Brack, T. Noguchi, M. Tasumi, and G. H. Atkinson, *J. Phys. Chem.* **95**, 6797 (1991).

There is a maximum in the acceptor mode density for transfer from the heme to the surrounding globin at 100 cm^{-1}, which extends past 400 cm^{-1}.[33] The most efficient energy transfer and transport channel is to these extended modes of the protein. The fastest relaxation rates are for energy regimes above the intramolecular vibrational energy redistribution (IVR) threshold (III) and the low-frequency regime (I) in which these short-lived modes are populated and act as doorway states to the globin and surrounding water.[34] (Reproduced, with permission, from the *Annual Review of Physical Chemistry*, Volume 42, © 1991, by Annual Reviews Inc.). (B) Hydrodynamic model of vibrational energy transfer. Transfer in region I is mediated by coupling to a large number of acceptor modes (large hole), whereas region II is coupled to fewer states (smaller hole). The barrier represents the two-phonon barrier in the protein energy reservoir between the low-frequency modes (I) and intermediate-frequency modes (II).

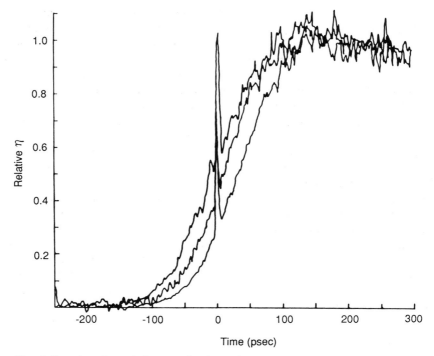

Time (psec)

FIG. 6. Transient thermal phase grating data using counterpropagating beam geometries at 366-nm excitation (Λ = 1300 Å). *Upper curve:* The instrument response. *Middle curve:* Cyanomethemoglobin in an aqueous buffer. *Lower curve:* Malachite green in ethanol. The rise time of the cyanohemoglobin thermal phase grating is less than 22 psec. The slower rise time ($1/e$ point) of the thermal phase grating in ethanol reflects the slower speed of sound of ethanol relative to water. Reprinted with permission from Ref. 9, Copyright 1987 American Chemical Society.

a fairly comprehensive pathway mapped out for the vibrational energy flow. Optical excitation of the heme places it in the ground state manifold on picosecond to subpicosecond time scales. This highly excited vibrational energy state is well above the threshold for rapid IVR, and the energy redistributes to produce essentially a Boltzmann distribution corresponding to an internal temperature of 700 K at the heme site in less than 2–5 psec. These latter dynamics have been determined from time-resolved anti-Stokes Raman studies of heme proteins and related systems.[38,39] The Raman studies have by and large focused on vibrational modes that lie above the IVR threshold of the heme and are expected to be short lived and serve as source terms for regime I and II modes (see Fig. 5). From thermal phase grating studies, even the regime II modes are short lived.

There appears to be no "two-phonon" bottleneck for heme proteins in contrast to observations of this effect in the vibrational relaxation of chromophores embedded in molecular crystals. If one takes into account that the IVR threshold for rapid IVR for the heme porphyrin moiety should be ~700 cm^{-1} and the fact that the acceptor mode distribution in the surrounding heme protein extends out to 400 cm^{-1}, there should not be a two-phonon bottleneck.[27] In contrast, for molecular crystals, there is normally a maximum phonon frequency of 200 cm^{-1}. These modes are unable to act as accepting modes and conserve energy for the energy stored in modes between 400 and 700 cm^{-1} except through higher-order interactions. This normally leads to a dramatic retardation of the rate of vibrational cooling.[34] However, the discrete size of the protein relative to typical phonon wavelengths in the solid state and the covalently bonded network throughout the protein structure quantizes the analogous collective modes of the protein and pushes them to higher frequencies so that no bottleneck occurs in the mode distribution.

The bottom line in all these studies is that the vibrational energy exchange and redistribution and the exchange with the aqueous interface is extremely efficient. This statement holds true even for isolated regions where the local density of states is relatively small compared to the rest of the protein, and the energy exchange is expected to go primarily through van der Waals contacts. These studies give direct experimental insight into how energy *sloshes* about in biological systems. In addition, the observed dynamics are in fairly good agreement with molecular dynamics simulations of this process,[35] although not perfect. It is expected that continued studies along these lines will provide rigorous bench marks for improving molecular dynamics simulations.

Structural Relaxation Dynamics

The next application of phase grating spectroscopy was directed to determining the energetics for the stored energy in the protein structure. This energy directs the specific structure changes unique to the oxy to deoxy tertiary changes of heme proteins. These studies were to be accomplished by following the formation dynamics of a thermal phase grating following the photodissociation and ensuing structural relaxation of MbCO and HbCO. However, during the course of these studies it was discovered that the protein's own optically triggered motion toward the deoxy structure produced a large change in the material density that dominated the signal.[11] These results are shown in Fig. 7 for MbCO, in which a direct comparison is made to deoxyMb, which undergoes no significant structural changes and serves as the control for a pure thermal phase grating re-

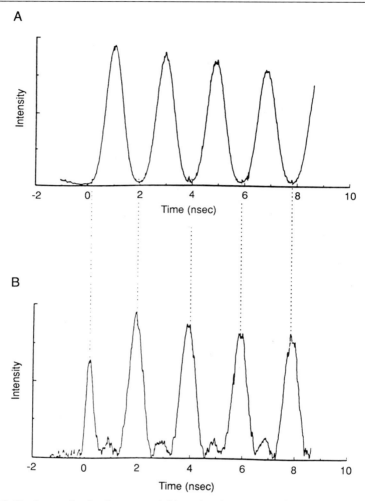

FIG. 7. Grating studies for deoxymyoglobin and carboxymyoglobin using 532-nm excitation and a 1.064-μm probe. (A) Deoxymyoglobin results with excitation pulse conditions of 0.8 mJ/cm^2. (B) Carboxymyoglobin results with excitation pulse conditions of 0.8 mJ/cm^2. The dotted lines are to illustrate that the acoustics driven by the protein tertiary structural changes following CO dissociation are 180° out of phase with the pure thermal acoustic response exhibited by the deoxymyoglobin.

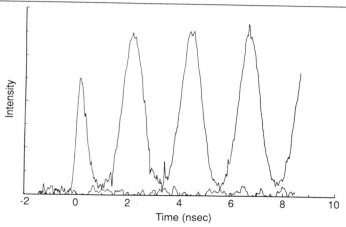

FIG. 8. Grating studies near the zero thermal expansion point of water. The data for both MbCO and deoxyMb were collected at $-0.3°$, using 532-nm excitation conditions of 15 mJ/cm² and a 1.064-μm probe. The top data set is for MbCO and the lower, noisy data is for deoxyMb under identical conditions. The deoxyMb results show that the thermally driven acoustics vanish at this temperature.

sponse. The most important observation is that the density grating for MbCO develops much quicker than the thermal phase grating. This is a result of the impulsive material displacement, driven by the motion of the protein, coherently exciting a standing acoustic wave that is 180° out of phase with the pure thermal acoustic response. The signal correlated to the carboxy-to-deoxy tertiary structural changes is much larger than the thermal acoustic response as well. If the density changes arise from optically triggered changes in the protein structure, the generated acoustics should be nonthermal in nature. This point was demonstrated by studying the phase grating formation dynamics at the zero thermal expansion point of water. These results are shown in Fig. 8. The thermal phase grating is reduced by more than two orders of magnitude, whereas the protein-driven density changes remain unchanged in amplitude. This study unambiguously demonstrated that the acoustics are generated by the nonthermal tertiary structural changes.

The form of the protein-driven acoustics could also be explained by an acoustic grating beating against an excited state phase grating component [see Eq. (6)]. This possibility needs to be considered as no matter how far off resonance the probe is, there will always be a small population grating contribution to the signal. The important issue is the relative magnitude of the population phase grating component to the induced density changes at the given probe wavelength. To explain the observed signal, the acoustic grating would have to result in an index of refraction change that coincidentally has the same amplitude as the excited state phase grating contribution. This is unlikely. However, to check this possibility a wavelength dependence was conducted in which the absolute diffraction efficiencies were measured for both resonant and nonresonant probe conditions with respect to the heme transitions, along with a complete Kramers–Kronig (K–K) analysis for the expected amplitude and phase grating components at the different probe wavelengths. By measuring the diffraction efficiency for resonant conditions, in which the population grating contribution should dominate the phase grating signal, the relative contribution of the population grating for off-resonant probe conditions can be determined.

For MbCO the absorption difference spectrum is essentially the same as the MbCO/deoxyMb equilibrium difference spectrum 30 psec after photoexcitation.[41] The absorption spectra for MbCO and deoxyMb and the corresponding population phase grating amplitudes are shown in Fig. 9. This analysis shows that the population phase grating component should lead to more than a two orders of magnitude decrease in diffraction efficiency in going from resonant probe conditions at 532 nm to nonresonant probe conditions at 1.064 μm [Eqs. (4) and (6)]. In contrast, absolute diffraction efficiency studies found that the signal from the protein-driven acoustics was only nine times less at 1.064 μm relative to 532 nm. These results are shown in Fig. 10. This ratio includes the amplitude grating contribution, which is present at 532 nm but absent at 1.064 μm. Taking the amplitude grating component into account at 532 nm, the change in diffraction efficiency at the two different probe wavelengths is only 2 : 1 [$\eta(532) : \eta(1.064)$]. This result is essentially in quantitative agreement with that expected for a pure acoustic phase grating. In fact, the acoustics even dominate the phase grating component for resonant probe conditions at 532 nm. There is less than a 5% population phase grating component at 1.064 μm. This finding illustrates that the phase grating dynamics observed for MbCO and HbCO at 1.064 μm are essentially from a pure acoustic response.

The most important observation from these data is the rapid rise time in the protein-driven density changes relative to the thermal grating response. The mechanism of the coupling of the protein motion to the density changes needs to be considered. Previously, it has been proposed that nonthermal density changes in optically excited carboxyheme proteins arise from electrostriction effects.[3] For example, the breaking of a salt bridge in the protein structure changes the electric fields in the material.

[41] J. W. Petrich, C. Poyart, and J. L. Martin, *Biochemistry* **27,** 4049 (1988).

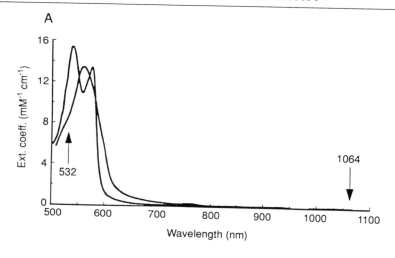

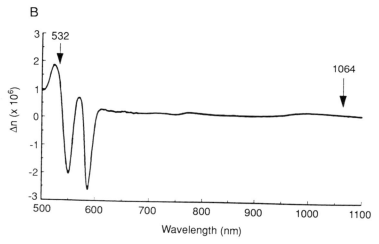

FIG. 9. (A) The absorption spectra of both deoxyMb and MbCO. The arrows indicate the positions of the two probes used in these studies. (B) The Kramers–Kronig calculation of the population phase grating component (Δn_{ex}) for MbCO as a function of wavelength, based on the absorption difference spectrum of MbCO and deoxyMb. The calculation uses changes in the fractions of MbCO and deoxyMb, determined by experimental conditions of Fig. 3. The population phase grating component is at least seven times larger at 532 nm relative to 1.064 μm. Reprinted with permission from *Biochemistry* **31,** 10703 (1992), Copyright 1992 American Chemical Society.

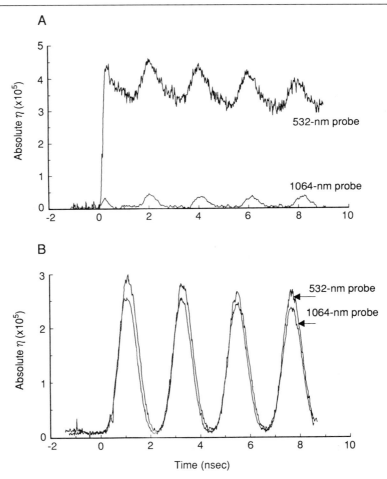

FIG. 10. Probe wavelength dependence of the absolute diffraction efficiencies. (A) Data give the results for MbCO. The upper curve is for the 532-nm probe, and the lower curve is for the 1.064-μm probe. The ratio for the absolute diffraction efficiencies at the maxima is 9 : 1 for 532 nm relative to 1.064 μm under identical excitation conditions. The ratio of the acoustically modulated signal is 2 : 4 : 1. (B) Studies of deoxyMb at the same excitation conditions and optical density as in (A). The ratio of the diffracted signal at 532 nm relative to 1.064 μm is 1.2. These data give the scaling with wavelength for a pure acoustic response with this grating configuration and optical density at 532 nm. The acoustic modulated signal for MbCO is virtually the same as the pure thermal acoustic response for deoxyMb, which illustrates that the protein-driven signal modulation arises from density changes. Reprinted with permission from *Biochemistry* **31**, 10703 (1992), Copyright 1992 American Chemical Society.

The polar water molecules reorientate in the new fields and cause a net material contraction. Electrostrictively driven sound fields, however, exhibit approximately the same grating formation dynamics as thermally excited acoustics.[5,42] The rise time for both thermal and electrostriction mechanisms of sound generation is determined by the speed of sound for material displacement occurring over the dimensions of the grating fringe spacing. The thermal grating rise times are on the order of one-half an acoustic cycle, which for the grating geometries used in the studies shown in Figs. 7 and 8 are ~500 psec and not less than 30 psec.

Both thermal and electrostriction mechanisms involve material displacement in response to $t = 0$ sinusoidal stress boundary conditions that match the optical interference pattern. Instead, the material displacement driven by the protein structure changes is responding to forces local to the heme. If the protein structure changes involve a change in volume of the protein, the material displacement will mimic the dynamics of this motion. The motion of the protein will displace both the surrounding water and sections of the protein matrix. The material displacement is arrested once the protein has undergone its full range of motion. This change in material density would be analogous to a time-dependent expansion coefficient in the thermal grating case in which there is initially a large value that rapidly goes to zero. If the material displacement driven by the initial protein motion goes to completion much faster than the grating acoustic period, then the acoustics will approximate a $t = 0$ sinusoidal strain boundary condition. In this case, the nonpropagating static component to the density grating would develop first, and the acoustics would respond to the strain-induced stress in the material. This condition would lead to a 180° phase shift in the acoustics relative to the thermal case, as observed. This model for the coupling of the protein motion to the hydrodynamics explains the general form of the observed acoustics. However, a detailed analysis of the mechanical coupling that includes the aqueous interface and the effect of the motion on the index of refraction needs to be developed to obtain quantifiable parameters.

The most important point of the above discussion is that the coupling of the motion of the protein to the acoustic modes of the water can occur only through global changes in the protein structure, that is, there must be a net displacement of the exterior amino acid residues relative to the interior to give a net volume change. Essentially, the material displaced by the global motion of the protein is imaged into a density hologram that can be probed to give a real-time view of the protein motion. The rise time of the density change (the amount of material displaced by the protein

[42] R. J. D. Miller, M. Pierre, T. Rose, and M. D. Fayer, *J. Phys. Chem.* **88,** 3021 (1984).

motion) is determined by the protein dynamics and not by the speed of sound over the grating fringe spacing. Thus, we now have pulse width-limited resolution to the global motion of the protein. The global relaxation was found to occur with a time constant of less than 30 psec.[11] In addition, we found that the amplitude of the material displacement was larger for MbCO than HbCO. This latter result is consistent with the long-held concept that the quaternary structure acts as a mechanical constraint to the tertiary structure changes.[23] This observation also demonstrates that the grating signal depends on the surrounding globin.

We have since reexamined the global relaxation dynamics using a kilohertz (kHz) regeneratively amplified subpicosecond dye laser system. As discussed above, the experiment consists of four laser pulses: two grating writing pulses, a long infrared pulse to probe the nonresonant phase grating amplitude, and an amplified dye pulse to upconvert the diffracted probe signal. The upconversion detection is needed to conserve the picosecond time resolution with infrared pulses. Representative results are shown in Fig. 11. There is an optical Kerr effect from the water that contributes to the signal for the first 2–5 psec. Continued study has revealed a protein excited state grating contribution to the signal for the first few picoseconds as well. These two contributions account for the pulse width-limited rise in the signal and the fast initial decay of approximately one-half the signal amplitude. This rapid rise and decay of the grating signal obscures the signal associated with the global structure change. Regardless of this complication, it is evident from Fig. 11 that the global relaxation is occurring within 5 psec. Within the observation window of Fig. 11, this signal component appears as the long-lived component that develops within the first few picoseconds. The observed dynamics for the global protein relaxation are similar to the best experimental estimates of the local motion in the vicinity of the heme as probed by the change in orientation of the proximal histidine. This correlation over two length scales provides evidence for the involvement of collective modes in propagating the initial tertiary structure changes. In addition, the observed dynamics are in agreement with Seno and Go.[33] From a normal-mode analysis, they observed that the globin deoxy tertiary structure could be reconstructed from the oxy structure primarily through the displacement of five collective modes ranging in frequency from 1 to 15 cm^{-1}. A highly damped half-period oscillation of these modes would be in the time range observed.

The above results represent an important observation, as the collective displacement of atoms is the most efficient mechanism possible for propagating protein structural changes. This mechanism is distinct from either the localized strain model or the conformational substate model in that

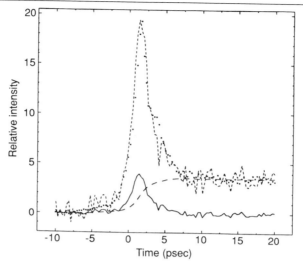

Fig. 11. (■) MbCO transient grating signal, using the experimental setup shown in Fig. 4. *Solid line:* Instrument function derived from a Kerr grating component measured using water alone. *Long-dash line:* Convolution of the instrument function with an infinitely fast system response (step function), scaled to match the long-time signal. *Short-dash line:* Conveniently scaled sum of the last two plots, showing that the rise time of the protein-driven density changes occur within 5 psec. Reprinted with permission from R. J. D. Miller *et al., in* "Ultrafast Phenomena VIII" (J. L. Martin *et al.,* eds.), p. 525, Springer-Verlag, New York, 1993.

the forces are extensively distributed, and the motion occurs in phase without segmental sampling of intermediates. In our work, we find no evidence as yet for slower relaxation components that would support structurally important substates. However, these substates may not involve as large a volume change as the initial collective phase of the global relaxation, such that they go undetected. Whether conformational substates are significant with respect to the structural relaxation coordinate needs to be decided through a direct determination of the energetics for each phase of the relaxation.

Energetics of Protein Motion

A direct measurement of the energetics relevant to the protein relaxation is needed to understand the effective driving forces for the protein motion. The energetics of the system response to ligand dissociation also provide an obvious criterion for determining the dominant mechanism for the structural relaxation. The mechanism that propagates the system furthest downhill in energy toward the global energy minimum corresponding to the final equilibrium structure should be classified as the dominant

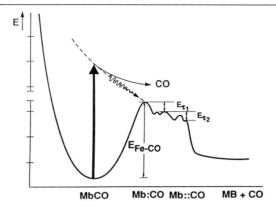

FIG. 12. The energetics for optically triggered protein motion. This diagram shows schematically a one-dimensional slice through a complex multidimensional energy surface of the protein. Optical excitation prepares the system on a dissociative surface in which the Fe–CO bond is broken. The initial optically prepared state is a highly excited deoxy state and a CO physically trapped near the heme. The excited deoxy state rapidly dissipates the excess energy above that needed to break the Fe–CO bond (E_{Fe-CO}), and the protein undergoes relaxation toward its deoxy tertiary structure, which also dissipates energy (E_{τ_1}). The next phase of the protein relaxation involves the diffusion of the CO out of the protein. Energy is released as the interactions between residues disrupted by the CO are reformed (E_{τ_2}), and the CO becomes solvated in the aqueous phase (E_s) to give the fully equilibrated deoxy heme protein. A determination of the energetics and dynamics gives the driving force for the protein relaxation at each phase as well as information on the barrier distribution. Reprinted with permission from *Biochemistry* **31**, 10703 (1992), Copyright 1992 American Chemical Society.

mechanism. The simultaneous measurement of the energetics and dynamics is unique in this regard as it should enable an unambiguous characterization of the relative importance of the various phases of the protein relaxation in the attainment of its final structure. If conformational substates play an important role in the overall evolution of the system, the energetics should show a distribution of relaxation rates. In contrast, if the structure changes are propagated largely through the impulsive excitation of collective modes, then the energetics should show the majority of the energy dissipated during the rapid collective phase of the motion.

The energetics of the optically induced structure change are shown schematically in Fig. 12. The energetics include the Fe–CO bond enthalpy (E_{Fe-CO}), the amount of absorbed photon energy above the bond energy ($E_{h\nu} - E_{Fe-CO}$), and the energetics of the triggered protein relaxation (E_{τ_1}). Optical excitation of 532 nm places the heme on a dissociative electronic surface that breaks the Fe–CO bond in less than 300 fsec.[15,41] A certain fraction of the absorbed photon energy goes toward breaking the Fe–CO bond, and the excess energy above this amount will be dissipated rapidly into the bath. The rapid dissipation of this excess energy is equivalent to the nonradiative relaxation of the short-lived deoxyheme protein excited states and contributes to the rapid rise time in the thermal grating. The subsequent relaxation of the protein structure toward the deoxy configuration will also dissipate energy into the bath. The amount of energy dissipated and the dynamics for this energy relaxation process will depend on the energy difference between the ligated and deoxy tertiary structure and the barrier distribution between conformational substates that may act as real intermediates in the structural relaxation. To use the system energetics as a criterion for protein relaxation, the temporal resolution of the probe determining the energetics must be sufficient to distinguish the various relaxation phases of the motion of the protein. As previously demonstrated, thermal phase grating spectroscopy is capable of attaining approximately 10-psec resolution to the energetics with appropriate grating beam geometries. This resolution is sufficient to resolve virtually all phases of the protein motion and to determine which phase of the motion has the largest driving force.

As can be appreciated from Figs. 7 and 8, the problem in attaining information on the heme protein energetics in water is that the grating is dominated by the protein volume changes and not by thermal expansion processes, especially for MbCO. This problem was overcome by going to a 75%/25% (v/v) glycerol–water solvent system. The advantage of the glycerol–water is that the thermal expansion coefficient is an order of magnitude larger than water. With this solvent system, the thermal phase grating components dominate the signal, allowing a fairly straightforward determination of the energetics. In addition, this is the same solvent system used by Frauenfelder and co-workers to establish the existence of conformation substates in heme proteins, and a great deal of effort has been made in making a correlation between the protein dynamics in this medium and physiological aqueous conditions.[28]

Representative results from grating studies of protein energetics are shown in Fig. 13 for MbCO and HbCO, using 532-nm excitation and a 1.064-μm probe. The results shown in Fig. 13 make a direct comparison of the thermal grating dynamics for MbCO and HbCO with their deoxy analogs under the same conditions. The deoxy heme proteins serve as a reference for the case in which all the absorbed photon energy is dissipated into the aqueous bath with a time constant less than 22 psec. As can be seen from Fig. 13B, the protein wave is still present for MbCO, but the thermal grating response now dominates the signal. In the case of MbCO, the amplitude of the thermal phase grating is significantly less than the deoxymyoglobin because some of the energy must go into breaking the Fe–CO bond. The difference in energy can be determined from the differ-

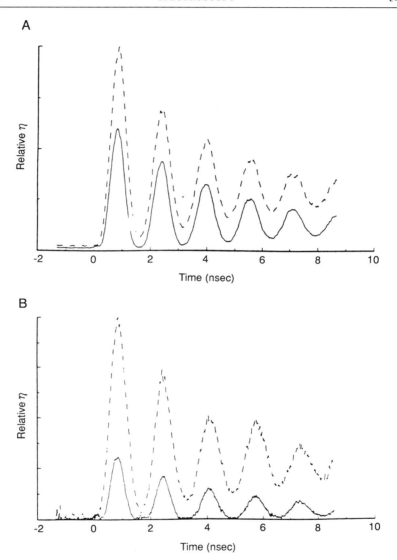

FIG. 13. Phase grating studies of the protein relaxation energetics. (A) HbCO (—) in 75%/25% (v/v) glycerol–water, using 532-nm excitation at 2 mJ/cm^2 and a 1.064-μm probe. Comparison is made to deoxyHb (– – –) under identical conditions, which provides a reference for complete dissipation of the photon energy in less than 20 psec. Carboxyhemoglobin is found to be endothermic by 11 kcal/mol relative to the deoxyHb reference. (B) Carboxymyoglobin under the same experimental conditions. Carboxymyoglobin (—) is found to be endothermic relative to deoxyMb (– – –) by 21 ± 2 kcal/mol. The decay of the acoustic modulation is due to acoustic attenuation in this viscous medium.

ence in signal amplitudes of the Mb reference and the MbCO results, that is,

$$\Delta E = [1 - (\eta_{MbCO}/\eta_{Mb})^{1/2}]E_{h\nu} \tag{12}$$

A detailed comparison of several samples in which the experimental conditions were kept constant and using calibration curves to correct for optical density differences between the deoxymyoglobin reference and MbCO found that the difference in energy is 21 ± 2 kcal/mol. This result can be inferred from Fig. 13B in which the thermal grating amplitude of MbCO is slightly more than one-quarter the Mb reference. This means that just more than one-half the 532-nm photon energy (53 kcal/mol) is dissipated.

This amount of energy corresponds closely to the expected value for the Fe–CO bond enthalpy. Previous studies have estimated the bond energy to be 25 kcal/mol.[12,43] The initial protein relaxation is found to be endothermic by 21 kcal/mol, which would mean at most a few kilocalories is stored in the protein structure of myoglobin and released on relaxation to the deoxy structure. The maximum barrier to CO recombination imposed by the protein relaxation would be approximately 3 kcal/mol, which is in good agreement with the temperature-dependent studies of the kinetics for geminate CO recombination.[28,29] These results have determined that the energetics of the initial protein relaxation for MbCO are virtually identical to the Fe–CO bond enthalpy. This result means that there is little energy stored in the protein structure. The surrounding globin acts like an elastic basket for the prosthetic heme group.

The dynamics for the energy relaxation are equally important. Within the signal to noise, the protein completes its energetic relaxation within 200 psec. The energetics are essentially identical to the calorimetric determinations of the enthalpy for the formation of MbCO, which has a reported value of 21.4 ± 0.3 kcal/mol.[44] This thermodynamic value includes the Fe–CO bond enthalpy and the amount of energy needed to change the protein structure to the fully relaxed deoxy structure. The close agreement between the measured energetics and the thermodynamic value suggests that the driving force for the deoxy tertiary structural changes is essentially exhausted in less than 200 psec. However, there are variations in the reported thermodynamic enthalpy measurements between 17 kcal/mol[45] and 21.4 kcal/mol.[44] The dynamic range of the grating experiment needs to be extended to follow the CO diffusion out of the protein to correspond completely to the thermodynamic measurements.

[43] S. A. Asher and J. Murtaugh, *J. Am. Chem. Soc.* **105**, 7244 (1983).
[44] M. H. Keyes, M. Falley, and R. J. Lumry, *J. Am. Chem. Soc.* **93**, 2035 (1971).
[45] S. A. Rudolf, S. O. Boyle, D. F. Dresden, and S. J. Gill, *Biochemistry* **11**, 1098 (1972).

The grating results should be compared to time-resolved photoacoustic studies of the protein energetics for myoglobin. The grating determinations appear to give endothermic values for the energetics that are approximately 8 kcal/mol higher than the corresponding photoacoustic results. It must be borne in mind that the photoacoustic studies are sensitive to different time scales than in the grating studies. The piezoelectric transducers (acoustic detectors) used in these studies have typically a 1-MHz fundamental resonant frequency with a 10% bandwidth. This frequency response corresponds to a dynamic window that primarily covers relaxation processes from 100 nsec to 10 μsec. The fact that the grating and the photoacoustic measurements do not completely agree most likely reflects that the two different approaches are measuring different phases of the protein relaxation. The measured energetics from the grating studies are in agreement with previous estimates of the Fe–CO bond enthalpy, whereas the photoacoustic measurements seem to give anomalous values for the protein relaxation in this regard. On the basis of the better time resolution of the grating relative to the photoacoustic studies, the thermal phase grating approach is expected to give a more accurate determination of the energetics of the initial relaxation processes.

In contrast to the myoglobin studies, HbCO was found to be endothermic for CO photodissociation by 11 ± 3 kcal/mol. This result is significant in that the structural relaxation of the heme protein tertiary structure involves the dissipation of approximately 10 kcal/mol extra energy relative to myoglobin. From the magnitude of the protein-driven material displacement for hemoglobin relative to myoglobin, it appears that the mechanical motion of hemoglobin is constrained by the quaternary structure relative to that of myoglobin. This observation is consistent with current models for the allosteric regulation of hemoglobin. The energy of cooperativity is stored as potential energy through interactions between the subunits at the $\alpha\beta$ interface between the subunits. These interactions become modified by the changes in the tertiary structure of the composite subunits which lead to changes in the number and location of hydrogen bonds and electrostatic interactions in the interface region. As a subunit undergoes its tertiary structure change to the deoxy form, the reduction in the repulsive forces between the proximal histidine and the heme, with the doming of the heme, has been postulated to release the strain in the structure.[25,26,46,47] Using 22 kcal/mol as the amount of energy required to break the Fe–CO bond, it appears that the strain energy released as a hemoglobin subunit

[46] J. M. Friedman, T. W. Scott, G. J. Fisanick, S. R. Simon, E. W. Findsen, M. R. Ondrias, and V. W. MacDonald, *Science* **229**, 187 (1985).
[47] B. R. Gelin and M. Karplus, *Proc. Natl. Acad. Sci. U.S.A.* **74**, 801 (1977).

undergoes its tertiary structure changes is on the order of 10 kcal/mol. This energy is approximately the amount of energy required to hold a heme protein subunit rigidly as it undergoes the tertiary structure changes with changes in ligation. Using rigid boundary conditions for an energy-minimized structure of MbCO, it was calculated that it costs 7–8 kcal/mol to constrain the heme protein from undergoing its full range of motion from the ligated to deoxy tertiary structure.[33] If the quaternary structure acts to constrain the subunit tertiary structure changes, then the amount of stored strain energy should be on this order.

As in the case of myoglobin, it is interesting that the amount of energy involved in the initial tertiary structural relaxation of hemoglobin is close to the thermodynamic values. Previous determinations of the enthalpy change from the fully relaxed deoxyHb to fully ligated Hb found enthalpy changes of approximately 13 kcal/mol.[44,48] This value is close to the measured value of 11 ± 3 kcal/mol. It appears that most of the driving force for the tertiary structural changes is relieved in less than 200 psec. This is an intriguing finding as it indicates that the protein accesses the energetics or driving force for its structural changes on an extremely fast time scale. Further, this rapid relaxation in the system energetics is consistent with a collective mode response as discussed above for MbCO. It is the tertiary structure changes that affect the quaternary structure force balance at the $\alpha\beta$ subunit interface. If the collective mode response is found to be the dominant mechanism for the propagation of tertiary structural changes, then the difference in energetics associated with the quaternary structure involves a relaxation in structure redistributed over many atomic degrees of freedom. In this event, the energy of cooperativity involved in the allosteric control of hemoglobin is best described within the distributed energy model of Hopfield rather than stored in specific regions in the protein structure.[49] However, before this evidence for quaternary structure effects can be taken as conclusive, the overall energetics need to be studies further, using isolated subunits rather than myoglobin as a reference. Such studies would eliminate potential complications from tertiary structure differences in fully assessing the effect of the quaternary structure on the energetics.

Concluding Remarks

The picosecond phase grating technique has some unique features. As demonstrated, it is capable of following global protein relaxation processes

[48] F. W. Roughton, *Biochem. J.* **29**, 2604 (1934).
[49] J. J. Hopfield, *J. Mol. Biol.* **77**, 207 (1973).

as well as determining the energetics for the different phases of the protein relaxation. With respect to the energetics, the time resolution is sufficient to study the dynamics of the vibrational energy relaxation and exchange with the water interface. Thus, the time resolution with respect to following protein energetics is at the fundamental limit. The dynamic range can be extended from picoseconds to milliseconds with changes in beam geometries. It will be interesting to extend the above studies of heme proteins to the microsecond range to follow the R-to-T quaternary structure changes. Numerous questions can be answered regarding the cumulative effect of ligand loss on the energetics and the driving forces for the R-to-T switch for O_2 affinity. In this regard, picosecond phase grating spectroscopy provides a direct determination of the protein strain and released energy that is at the heart of most models for allosteric regulation and molecular cooperativity. We are now at a point where we can address which mechanism dominates and which motions direct the structural relaxation important to protein function.

The combined ability to measure the dynamics of the global motion and energetics makes this technique an important new probe of protein dynamics. The only caveat is that the relaxation process must be optically initiated. Despite this limitation, there are numerous examples of protein systems exhibiting photoactivity that are amenable to study as model systems. These include photosynthetic reaction centers, rhodopsins, DNA photolyase, as well as heme proteins. There are also protein-unfolding processes that can be optically stimulated. It is hoped that the above applications and experimental details will lead to further applications of this general methodology to problems of biological interest.

Acknowledgments

The above research was supported by the NIH (1 R01 GM41909-01A1) and an NSF Presidential Young Investigator Award (R.J.D.M.). R.J.D.M. is the recipient of a John Simon Guggenheim Fellowship which contributed to this work. This research was the result of a number of fruitful collaborations over the years. The authors would like to further acknowledge the numerous contributions of Dr. Laura Genberg, and Professor George McLendon that led to the research discussed above.

Section III

Ligand Binding

[17] Assignment of Rate Constants for O_2 and CO Binding to α and β Subunits within R- and T-State Human Hemoglobin

By ANTONY J. MATHEWS and JOHN S. OLSON

In 1959, Gibson[1] and Gibson and Roughton[2] reviewed in detail how rapid mixing and flash-photolysis techniques can be used to assign rate constants to the first and last steps in ligand binding to mammalian hemoglobins. Over the next 20 years several new methods were applied to this problem and included the isolation of functionally active α and β chains, the recognition of intrinsic differences between the subunits, the preparation of valency and metal hybrid hemoglobins, the application of single and double mixing rapid-freezing electron paramagnetic resonance (EPR) methods to resolve these differences, and the use of laser flash-photolysis instruments to examine the rates of NO, CO, and O_2 rebinding to partially saturated intermediates. These techniques were reviewed in 1981.[3,4]

Several new approaches have since been developed. Perrella *et al.* have adapted cryoscopic isoelectric focusing techniques to the analysis of both kinetic and equilibrium intermediates during CO binding to human hemoglobin.[5,6] Sharma *et al.* have extended double mixing and high-performance liquid chromatography (HPLC) techniques to measure rate constants for CO and methyl isocyanide binding to specific mono-, di-, and triliganded intermediates.[7-9] Martino and Ferrone[10] and Zhang *et al.*[11] constructed a modulated excitation instrument that is designed to resolve ligand-binding spectral changes from those involved in quaternary conformational changes during partial photolysis of fully and partially saturated intermediates. Nagai and co-workers constructed a bacterial system for the expression of individual α- and β-globins and then used site-directed

[1] Q. H. Gibson, *Biochem. J.* **71**, 293 (1959).
[2] Q. H. Gibson and F. J. W. Roughton, *J. Physiol.* (*London*) **145**, 32P (1959).
[3] J. S. Olson, this series, vol. 76, p. 631.
[4] C. A. Sawicki and R. J. Morris, this series, vol. 76, p. 667.
[5] M. Samaja, E. Rovida, M. Niggeler, M. Perrella, and L. Rossi-Bernardi, *J. Biol. Chem.* **262**, 4528 (1987).
[6] M. Perrella, N. Davids, and L. Rossi-Bernardi, *J. Biol. Chem.* **267**, 8744 (1992).
[7] M. Berjis, D. Bandyopadhyay, and V. S. Sharma, *Biochemistry* **29**, 10106 (1990).
[8] V. S. Sharma, *J. Biol. Chem.* **264**, 10582 (1989).
[9] V. S. Sharma, D. Bandyopadhyay, M. Berjis, J. Rifkind, and G. R. Boss, *J. Biol. Chem.* **266**, 24492 (1991).
[10] A. J. Martino and F. A. Ferrone, *Biophys. J.* **56**, 781 (1989).
[11] N. Q. Zhang, F. A. Ferrone, and A. J. Martino, *Biophys. J.* **58**, 333 (1990).

mutagenesis to study the influence of distal amino acid residues on the rate and equilibrium constants for ligand binding to hybrid hemoglobins containing one native and one mutated subunit.[12-15] Hoffman et al.[16] and Wagenbach et al.[17] have described bacterial and yeast systems, respectively, that express both subunits simultaneously, producing soluble and fully functional hemoglobins. This chapter concentrates on the use of mutant hemoglobins to assign specific rate and equilibrium constants to the α and β subunits within R- and T-state hemoglobin in rapid mixing and partial photolysis kinetic experiments.

Experimental Definitions of T- and R-State Rate Constants

The complete, unambiguous assignment of rate constants to the expanded Adair scheme has proved to be a difficult task, requiring the adoption of specific models with limited numbers of conformational states. Consequently, it has been useful to make the following empirical definitions in order to collate and compare various experimental results. The functional properties of the low-affinity or T-state quaternary conformation are defined empirically by the rate constants for the first step in ligand binding:

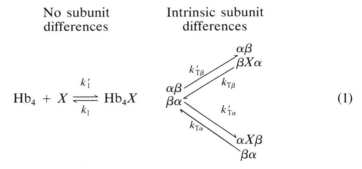

[12] K. Nagai, B. Luisi, D. T.-B. Shih, G. Miyazaki, K. Imai, C. Poyart, A. De Young, L. Kwiatowsky, R. W. Noble, S.-H. Lin, and N.-T. Yu, *Nature (London)* **329**, 858 (1987).

[13] J. S. Olson, A. J. Mathews, R. J. Rohlfs, B. A. Springer, K. D. Egeberg, S. G. Sligar, J. Tame, J.-P. Renaud, and K. Nagai, *Nature (London)* **336**, 265 (1988).

[14] A. J. Mathews, R. J. Rohlfs, J. S. Olson, J. Tame, J.-P. Renaud, and K. Nagai, *J. Biol. Chem.* **264**, 16573 (1989).

[15] A. J. Mathews, R. J. Rohlfs, J. S. Olson, J.-P. Renaud, J. Tame, and K. Nagai, *J. Biol. Chem.* **266**, 21631 (1991).

[16] S. J. Hoffman, D. L. Looker, J. M. Roehrich, P. E. Cozart, S. L. Durfee, J. L. Tedesco, and G. L. Stetler, *Proc. Natl. Acad. Sci. U.S.A.* **87**, 8521 (1990).

[17] M. Wagenbach, K. O'Rourke, L. Vitez, A. Wieczorek, S. Hoffman, S. Durfee, J. Tedesco, and G. Stetler, *Bio/Technology* **9**, 57 (1991).

In terms of simple Adair parameters, the association and dissociation rate constants are designated k_1' and k_1, respectively. When discussing subunit differences using the right-hand scheme in Eq. (1), it is more convenient to designate the α and β rate parameters as $k_{T\alpha}'$, $k_{T\beta}'$, $k_{T\alpha}$, and $k_{T\beta}$ than to use numbers for the individual reactions as was done originally by Olson[13] and Olson and Gibson.[18] Whether the Hb_4 and Hb_4X species are in the T quaternary conformation must be ascertained by independent experiments and depends on experimental conditions. These intermediates are in the T state for human hemoglobin at pH 7.0 and 20° in the presence or absence of organic phosphates.[3,4,18]

The functional properties of the high affinity or R-state quaternary conformation are defined empirically by the rate parameters for the last or fourth step in ligand binding:

No subunit   Intrinsic subunit
differences   differences

$$Hb_4X_3 + X \underset{k_4}{\overset{k_4'}{\rightleftharpoons}} Hb_4X_4 \qquad\qquad
\begin{array}{c}
\alpha X\beta \\
\beta X\alpha X
\end{array}
\begin{array}{c}
k_{R\beta} \\
\nearrow \;\; k_{R\beta}' \\
\searrow \;\; \beta X\alpha X \\
k_{R\alpha}' \;\; \nearrow \;\; \alpha X\beta X \\
\alpha\beta X \;\; \nwarrow \\
\beta X\alpha X \qquad k_{R\alpha}
\end{array} \qquad (2)$$

In terms of simple Adair parameters, the association and dissociation rate constants are designated k_4' and k_4, respectively, and again it is convenient to designate the α and β rate parameters as $k_{R\alpha}'$, $k_{R\beta}'$, $k_{R\alpha}$, and $k_{R\beta}$. These designations are more equivocal than those for the first step in ligand binding because the Hb_4X_4 species often has a certain amount of T-state character, particularly in the presence of organic phosphates or at low pH. However, the assumption of R-state character for the last step in ligand binding does appear to be valid for human hemoglobin at pH values ≥ 7.0 and in the absence of organic phosphates.[3,4,18]

Resolution of α- and β-Subunit Reactivity Differences in R-State Hemoglobin

Previous Experimental Work

Kinetic differences between R-state α and β subunits within tetramers were first shown for the reaction of aquomethemoglobin with azide and

[18] J. S. Olson and Q. H. Gibson, *J. Biol. Chem.* **247**, 3662 (1972).

nitrite.[19] The overall time courses for these reactions are biphasic, and the fast- and slow-reacting components can be resolved near the static isosbestic points in the Soret and visible wavelength region. The assignment of the fast- and slow-reacting components to β and α subunits, respectively, was based on correlations with the relative rates and spectral changes associated with isolated chains. A short time later subunit differences were established for reactions of alkyl isocyanides with the ferrous form of native human hemoglobin.[20,21] The β subunits have 2- to 10-fold higher association and dissociation rate constants for the last step in ligand binding compared to those for α subunits, but the affinities of the two subunits, $K_{R\alpha}$ and $K_{R\beta}$, for isocyanides are roughly equivalent regardless of size.[22–24] Again, the assignment of faster reacting components to β subunits was based on analogy with the relative rates and spectral properties of the isolated chains. These assignments were verified by rapid-freezing EPR methods, analyses of the ligand-binding properties of mixed heme and valency hybrids, and nuclear magnetic resonance (NMR) techniques.[3]

Gibson, Olson, Ho, Sugita, and others have extended these types of analyses to the reactions with O_2, CO, and NO.[3,25] However, in contrast to the isocyanide results, the differences between the α- and β-subunit rate constants are much smaller for the diatomic gases and often do not correlate in relative order with those observed for the isolated chains. As a result, some of these assignments were equivocal, and new experimental approaches were needed. The bacterial expression system developed by Nagai and co-workers has provided the methodology for producing hybrid hemoglobin tetramers in which an amino acid substitution specifically alters ligand binding to the mutated subunits without changing the functional properties of the native subunit partners.[12] In collaboration with Nagai's group, we measured the effects of E7 and E11 substitutions on the kinetic properties of hybrid mutant hemoglobins.[14,15] These studies had two objectives. The first was to determine the roles of the individual distal pocket residues in regulating ligand binding by measuring the effects of mutagenesis, and the second was to establish the rates of ligand binding to the native subunits. The effects of distal pocket mutations have been

[19] Q. H. Gibson, L. J. Parkhurst, and G. Geraci, *J. Biol. Chem.* **244**, 4668 (1969).

[20] J. S. Olson and Q. H. Gibson, *J. Biol. Chem.* **246**, 5241 (1971).

[21] J. S. Olson and Q. H. Gibson, *J. Biol. Chem.* **247**, 1713 (1972).

[22] P. I. Reisberg and J. S. Olson, *J. Biol. Chem.* **255**, 4144 (1980).

[23] P. I. Reisberg and J. S. Olson, *J. Biol. Chem.* **255**, 4151 (1980).

[24] P. I. Reisberg and J. S. Olson, *J. Biol. Chem.* **255**, 4159 (1980).

[25] C. Ho, *Adv. Protein Chem.* **43**, 153 (1992).

discussed extensively,[12-15,26] and this chapter deals primarily with the problem of assigning rates to the native subunits.

O_2 Displacement by CO

For a simple monomeric heme protein, this reaction can be written as

$$HbO_2 \underset{k'_{O_2}}{\overset{k_{O_2}}{\rightleftharpoons}} O_2 + Hb + CO \underset{k_{CO}}{\overset{k'_{CO}}{\rightleftharpoons}} HbCO \tag{3}$$

where k'_{O_2} and k'_{CO} are the association rate constants for O_2 and CO binding, respectively, and k_{O_2} and k_{CO} are the corresponding dissociation rate constants. If $[O_2]$ and $[CO]$ are high enough to ensure complete saturation and much greater than [heme], a steady state assumption can be invoked for [Hb], and the resultant rate equation predicts a time course that can be described by a single exponential expression. The observed, first-order replacement rate constant, r_{obs}, is given by the following general expression:

$$r_{obs} = \frac{k_{O_2}k'_{CO}[CO] + k_{CO}k'_{O_2}[O_2] + k_{CO}k_{O_2}}{k'_{O_2}[O_2] + k'_{CO}[CO] + k_{CO}} \tag{4}$$

In the case of oxygen displacement by carbon monoxide, k_{CO} (~ 0.01 sec^{-1}) $\ll k_{O_2}$ (10–25 sec^{-1}), so that at high [CO] all the terms containing k_{CO} can be neglected, and Eq. (4) reduces to

$$r_{obs} = \frac{k_{O_2}k'_{CO}[CO]}{k'_{O_2}[O_2] + k'_{CO}[CO]} \quad \text{or} \quad \frac{k_{O_2}}{1 + k'_{O_2}[O_2]/k'_{CO}[CO]} \tag{5}$$

In the case of tetrameric hemoglobin, all of the species $Hb_4(O_2)_{4-i}(CO)_i$, $i = 0$ to 4, are assumed to have equivalent R-state character because the hemes appear to react independently, and the replacement reaction does not exhibit cooperativity. The rate parameters in Eq. (5) apply to the last step in ligand binding because all the intermediates generated during the replacement reaction have either three or four bound ligands. Values of r_{obs} are measured as a function of [CO]/[O$_2$], and the results are fitted to the hyperbolic expressions in Eq. (5) to obtain values of k_{O_2} and the ratio of k'_{O_2}/k'_{CO}. These values are then assigned to k_4 for oxygen dissociation and the ratio of the last association rate constants, k'_4, for O_2 and CO binding. In terms of the definition in Eq. (2), these are considered to

[26] J. Tame, D. T.-B. Shih, J. Pagnier, G. Fermi, and K. Nagai, *J. Mol. Biol.* **218**, 761 (1991).

be R-state parameters that describe ligand binding to the high-affinity quaternary conformation.

In 1971, Olson et al.[27] examined the reaction of human oxyhemoglobin with CO at pH 7.0 and observed two kinetic components, a fast species with $k_{O_2} \approx 20$ sec^{-1} and a slow species with $k_{O_2} \approx 10$ sec^{-1}. Although O_2 − CO kinetic difference spectra suggested that the fast component represented oxygen dissociation from β subunits within the tetramer, comparisons with dissociation rate constants for the isolated chains indicated an opposite assignment, because $k_{O_2} = 28$ sec^{-1} for isolated α chains and 16 sec^{-1} for isolated β chains. This dilemma was resolved by specific chemical modification of β93 Cys with p-hydroxymercuribenzoate (pMB), which increased ~fivefold the faster rate of oxygen dissociation observed for intact hemoglobin and had little effect on the slower rate.[27-29] This result, together with the spectral correlation with isolated β chains, suggested assignment of the faster phase to oxygen dissociation from β subunits in native tetrameric hemoglobin.

This approach of selective modification has been extended and generalized by genetic engineering techniques. Sample time courses for the reactions of two oxyhemoglobins containing mutated α E7 residues and native β subunits are shown in Fig. 1. The α E7 His $\rightarrow$ Gly mutation produces a markedly biphasic time course, which is fitted readily to a 2-exponential expression with rate constants that differ by a factor of 30. Two phases are also apparent in the trace for the α E7 His $\rightarrow$ Gln mutant tetramer, whereas the time course for native hemoglobin looks monophasic without quantitative analysis. A complete set of results for mutant α subunits is presented in Table I. Absorbance changes were collected at 423 nm, at which the spectral changes for isolated α and β chains are equal, and the resultant time courses were fitted to a two-exponential expression with equal amplitudes.[27] The dependence of the fast and slow replacement rates on [CO]/[O_2] was measured by reacting buffer equlibrated with 1 atm of CO (1×10^{-3} M) with hemoglobin solutions equilibrated with $N_2 : O_2$ mixtures varying from 10 to 100% oxygen. Two types of analyses can be carried out. The observed fast and slow replacement rates can be fitted to Eq. (5), allowing both k_{O_2} and k'_{O_2}/k'_{CO} to vary; or the ratio of the association rate constants can be taken from independent partial photolysis experiments and k_{O_2} calculated for each set of ligand conditions. In our work with E7 and E11 mutants both types of analyses were used, and

[27] J. S. Olson, M. E. Anderson, and Q. H. Gibson, *J. Biol. Chem.* **246**, 5919 (1971).

[28] J. S. Olson, Ph.D. Dissertation, Cornell University, Ithaca, NY (1972).

[29] K. D. Vandegriff, Y. C. Le Tellier, R. M. Winslow, R. J. Rohlfs, and J. S. Olson, *J. Biol. Chem.* **266**, 17049 (1991).

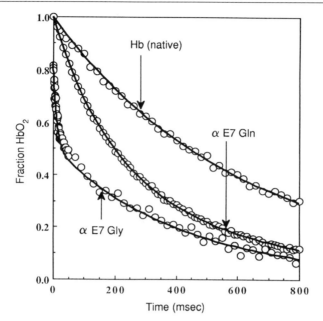

FIG. 1. Normalized time courses for the replacement of oxygen by carbon monoxide following rapid mixing of native and mutant hybrid oxyhemoglobins with carbon monoxide in a stopped-flow apparatus. Concentrations after mixing were 2–5 μM heme, 0.63 mM oxygen, and 0.5 mM carbon monoxide. The reactions were carried out at 20° and pH 7.0 in 0.1 M Bis–Tris, 0.1 M KCl buffer. The solid lines represent the fits to the sum of two exponentials with unequal amplitudes (α E7 Gly) and equal amplitudes (α E7 Gln and native). For α E7 Gly, a portion of the α-hemes had autoxidized. (Data are from Mathews *et al.*[14])

then the results for k_{O_2} were averaged for each independent experiment.[14] A detailed description of these analyses for native human hemoglobin has been presented by Vandegriff *et al.*[29]

Computed fast and slow k_{O_2} values for native, α wild-type, and five α-subunit mutants are presented in the second and third columns in Table I. These rate constants were assigned to α and β subunits by assuming that (1) the rate for the native subunits in the mutant protein should correspond to one of the rates observed for native hemoglobin; (2) if both rates are altered, the biggest change should be assigned to the mutated subunit; and (3) there should be some correspondence between the effects observed in hemoglobin with those observed for the same E7 and E11

TABLE I

ASSIGNMENT OF NATIVE β-SUBUNIT O_2 DISSOCIATION RATE CONSTANTS FROM CO
DISPLACEMENT REACTIONS USING DISTAL POCKET MUTATIONS IN α SUBUNITS[a]

| | Empirical fit | | Assignment | |
| | Fast k_{O_2} (sec^{-1}) | Slow k_{O_2} (sec^{-1}) | $k_{R\beta O_2}$ (sec^{-1}) | $k_{R\alpha O_2}$ (sec^{-1}) |
Protein				
α(native)$_2\beta$(native)$_2$	21 ± 4	11 ± 3	21	11
α(*Escherichia coli*)$_2\beta$(native)$_2$	24 ± 8	12 ± 5	24	12
α(*E7 Gly*)$_2\beta$(native)$_2$	620 ± 50	22 ± 16	22	620
α(*E7 Gln*)$_2\beta$(native)$_2$	53 ± 8	12 ± 2	12	53
α(*E11 Ala*)$_2\beta$(native)$_2$	67 ± 20	24 ± 20	24	67
α(*E11 Leu*)$_2\beta$(native)$_2$	15 ± 9	4.0 ± 2.3	15	4.0
α(*E11 Ile*)$_2\beta$(native)$_2$	31 ± 24	6.8 ± 3.2	31	6.8
Average for β(native):			22 ± 8	

[a] Conditions: 0.1 M Bis–Tris (2-[bis(2-hydroxyethyl)amino]-2-(hydroxymethyl)propane-1,3-diol), 0.1 mM KCl, pH 7.0, 20°. (Data from Mathews *et al.*[14])

mutations in myoglobin, in which no assignment problem occurs (Table II). On the basis of the initial work of Olson *et al.*,[27] $k_{R\beta O_2}$ and $k_{R\alpha O_2}$ were tentatively assigned values of ~20 and ~10 sec^{-1}, respectively. The result for the α E7 His → Gly mutant is the most definitive. The fast phase clearly represents rapid oxygen dissociation from the mutated α subunit

TABLE II

EFFECTS OF E7 AND E11 SUBSTITUTIONS ON OXYGEN
DISSOCIATION RATE CONSTANTS OF HUMAN HEMOGLOBIN AND
SPERM WHALE MYOGLOBIN[a]

Mutation	$k_{R\alpha O_2}$ (sec^{-1})	k_{MbO_2} (sec^{-1})	$k_{R\beta O_2}$ (sec^{-1})
Native (E7 His, E11 Val)	12 ± 3	14 ± 4	22 ± 8
E7 Gly	620 ± 50	1600	37 ± 19
E7 Gln	53 ± 14	130	31 ± 14
E11 Ala	67 ± 20	18	27 ± 12
E11 Leu	4.0 ± 2.3	6.0	20 ± 6
E11 Ile	6.8 ± 3.2	14	28 ± 16

[a] Conditions: R-state hemoglobin, 0.1 M Bis–Tris, 0.1 mM KCl, pH 7.0, 20° (data from Mathews *et al.*[14]); sperm whale myoglobin, 0.1 M phosphate, pH 7.0 (data from Rohlfs *et al.*,[30] Egeberg *et al.*,[31] and unpublished data for E11 Leu sperm whale myoglobin by T. Li and J. S. Olson, 1993).

due to loss of hydrogen bonding between the bound O_2 and the distal histidine. A similar, large increase in k_{O_2} was observed when E7 His was replaced with glycine in sperm whale myoglobin (Table II). The slow rate observed for the α E7 Gly mutant corresponds to the faster rate observed for native hemoglobin and confirms its assignment to native β subunits.

The results in Tables I and II also point out three major problems with this analysis. First, in the case of the α E7 His $\rightarrow$ Gln mutant, the assignment of the native subunit rate is equivocal. For this protein, the fast and slow rate constants were 53 and 12 sec^{-1}, respectively. By analogy with the myoglobin results, the rate of oxygen dissociation from the mutated α subunit is expected to increase significantly, and $k_{R\alpha}$ is probably 53 sec^{-1}. This would give the native β subunits a value of 12 sec^{-1} for $k_{R\beta}$, which is half the previously assigned value. However, if the assignments for the mutant protein are reversed, the native β subunit would have a value of k_{O_2} 2.5-fold greater than expected and there would be no change in the α subunit rate even though E7 His was replaced with glutamine. Similar ambiguities occur for α E11 Val $\rightarrow$ Ile and Leu mutants. Second, none of the β-subunit mutations produces significant changes in the observed time courses or fitted rate constants for oxygen dissociation (Table II). Thus, with the exception of pMB modification of β93 Cys, it has not been possible to increase or decrease selectively the rate of oxygen dissociation from β subunits and then assign unequivocally the native α-subunit rate constant. Third, the errors in the analyses are large if they are determined as the standard deviation from the mean of the average of at least three completely independent sets of experiments. In our analyses, $k_{R\beta O_2}$ for the native subunit was computed to be 22 $\pm$ 8 from the average of the values assigned to the native β subunits in the seven different hemoglobin tetramers listed in Table I. Thus, the twofold difference between the native α and β k_{RO_2} values is barely outside the error limits of ±30–40%.

R-State CO Dissociation Rate Constants

The rate of CO dissociation from fully liganded hemoglobin can be measured by mixing $Hb_4(CO)_4$ with a high concentration of NO. In this case, $k'_{NO} \gg k'_{CO}$ and, if $[CO] \leq [NO]$, Eq. (5) reduces to $r_{obs} = k_{CO}$.[3] When analyses similar to those described in Fig. 1 and Table I were carried out, the resultant values of $k_{R\beta CO}$ and $k_{R\alpha CO}$ were 0.0072 $\pm$ 0.0028 and 0.0046 $\pm$ 0.0015 sec^{-1}, respectively.[14] Thus the differences between the subunits are not statistically significant. The effects of mutagenesis on the R-state rate of CO dissociation were also much smaller than those observed for O_2 dissociation. The largest effect was observed for the β E7

His → Gly mutation, which increases $k_{R\beta CO}$ only 80%. A similar trend was observed for myoglobin mutants. E7 and E11 substitutions produce at most two- to threefold changes in the rate constant for CO dissociation, whereas the rate constant for O_2 dissociation varies from 2 to 15,000 sec^{-1} for the same set of mutants in myoglobin.[30,31]

Partial Photolysis of Hb$_4$(CO)$_4$ and Hb$_4$(O$_2$)$_4$

The distribution of partially liganded intermediates at different fractions of photolysis cannot be calculated readily because the quantum yields for the R and T states appear to be significantly different.[32] Fortunately, Brunori, Gibson, and co-workers have shown that the differences between quantum yields for the isolated α and β chains are less than 20–30%, and thus it is reasonable to assume equal quantum yields for the subunits within tetrameric R-state hemoglobin.[33] In practice, partial photolysis time courses are collected at decreasing levels of breakdown by putting neutral density filters in the excitation beam until no further change in rate or character of the time course is observed (usually at ≤10% breakdown). The resultant changes are then assigned to the rebinding of ligands to R-state or Hb$_4X_3$ hemoglobin.

Higher levels of photolysis that produce Hb$_4$, Hb$_4X_1$, and Hb$_4X_2$ also introduce complications due to unliganded heme absorbance changes associated with the R-to-T quaternary conformational transition. At pH 7 and in the absence of organic phosphates or cross-linking, these conformational changes occur at rates of $\sim 10^4$ sec^{-1}. When bimolecular ligand recombination is slow, as with CO complexes ($k'_{RCO}[CO] \leq 10^4$ sec^{-1}), these effects can be avoided only by low levels of photolysis. Monitoring ligand binding at 436 nm, the isosbestic wavelength for the R-to-T unliganded heme absorbance change, does not alleviate this problem because the observed time courses still reflect rates of binding to non-R-state intermediates. When bimolecular rebinding is fast, as with oxyhemoglobin at 1 atm O_2 ($k'_{RO_2}[O_2] \approx 10^5$ sec^{-1}), the rebinding rate is roughly 10-fold greater than that for the conformational transitions. Under these conditions, R-state rebinding parameters are observed at higher photolysis levels with a subsequent improvement in the signal-to-noise level.[14,29,34]

[30] R. J. Rohlfs, A. J. Mathews, T. E. Carver, J. S. Olson, B. A. Springer, K. D. Egeberg, and S. G. Sligar, *J. Biol. Chem.* **265**, 3168 (1990).
[31] K. D. Egeberg, B. A. Springer, S. G. Sligar, T. E. Carver, R. J. Rohlfs, and J. S. Olson, *J. Biol. Chem.* **265**, 11788 (1990).
[32] R. J. Morris and Q. H. Gibson, *J. Biol. Chem.* **259**, 365 (1984).
[33] J. S. Olson, R. J. Rohlfs, and Q. H. Gibson, *J. Biol. Chem.* **262**, 12930 (1987).
[34] C. A. Sawicki and Q. H. Gibson, *J. Biol. Chem.* **252**, 7538 (1977).

Experiments with CO complexes normally are carried out with conventional photographic strobes equipped with thyristor quenching devices that cause the excitation pulse to decay in ≤ 0.3 msec. Carbonmonoxyhemoglobin has a quantum yield of about 0.5 and is readily photolyzed. The observed pseudo-first-order rates can be manipulated by varying the ligand concentration, because the P_{50} for CO binding is $\leq 0.01 \mu M$. Consequently, partial photolysis time courses for CO complexes can be measured on millisecond time scales. The oxygen-rebinding reactions require much shorter and more intense excitation pulses due to a 10-fold lower apparent quantum yield and a 10-fold greater association rate constant. In addition, if the concentration of O_2 is lowered to less than about $200 \mu M$, hemoglobin becomes partially saturated at equilibrium. Thus, short ($< 1 \mu$sec) laser excitation pulses are required for studies with oxyhemoglobin complexes.

Until 1989, most analyses of partial photolysis experiments assumed that the α and β subunits within tetrameric hemoglobin have equal bimolecular rate constants for the binding of the fourth O_2 or CO molecule. This was due principally to the relatively poor signal-to-noise ratios of partial photolysis data, which causes difficulty in resolving multiple kinetic phases. An example of this analysis problem is shown in Fig. 2 for O_2 rebinding at $\sim 10\%$ breakdown. The poor signal-to-noise ratio causes difficulties in distinguishing whether the data are best fitted to a single exponential ($\chi^2 = 4.0 \times 10^{-6}$, Fig. 2A) or to a two-exponential expression with equal amplitudes ($\chi^2 = 3.6 \times 10^{-6}$, Fig. 2B). Attempts to fit these same data to two exponentials with variable amplitudes did not converge well because of noise in the trace. The difference between χ^2 values for the one- and two-exponential fits can be increased for O_2-binding data by increasing the number of traces averaged or by allowing more photolysis to increase the absolute value of the absorbance change. In the latter case, the protein remains in the R state even when $Hb_2(O_2)_2$ species are generated because the rate of R-to-T conformational transitions are roughly 10-fold slower than rebinding. In the case of CO binding, the ligand concentration is kept small so that rebinding occurs on millisecond time scales, and the noise can be decreased by electronic filtering (Fig. 3A). Even with these improvements, the observed time courses still fit fairly well to single-exponential expressions.

Three independent observations suggest that the postulated differences between the k_R' values for native α and β subunits are real. First, fits of partial photolysis time courses to a single-exponential function do show small but systematic deviations from the observed data (Fig. 2A). Second, the fast and slow components can be resolved near static isosbestic points. An example of this is shown in Fig. 3B for CO rebinding, where the fast component predominates on the short-wavelength side of the isosbestic

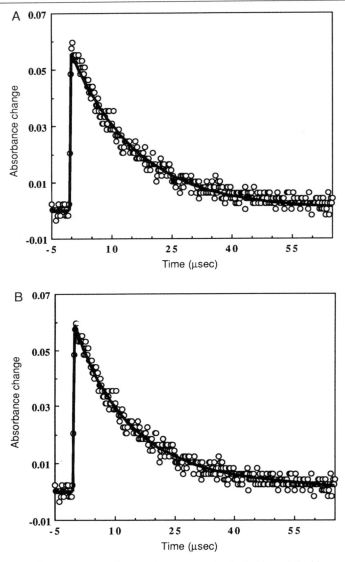

Fig. 2. Fits of oxygen recombination time courses to single (A) and double exponentials (B). The data show the recombination of 1.25 mM oxygen with R-state hemoglobin following 10% photolysis by a 300-nsec laser flash. The solid lines show the fit of the data (the average of eight traces) to one exponential (A), with $k = 60{,}000$ sec^{-1}, $\chi^2 = 4.0 \times 10^{-6}$, and to two exponentials with equal amplitudes (B), with $k_{fast} = 105{,}000$ sec^{-1}, $k_{slow} = 42{,}000$ sec^{-1}, $\chi^2 = 3.6 \times 10^{-6}$. The heme concentration was ~50 μM, and the reaction was monitored at 436 nm using a 1-mm path length cell. Conditions were 20° and pH 7.0 in 0.1 M Bis–Tris, 0.1 M KCl buffer.

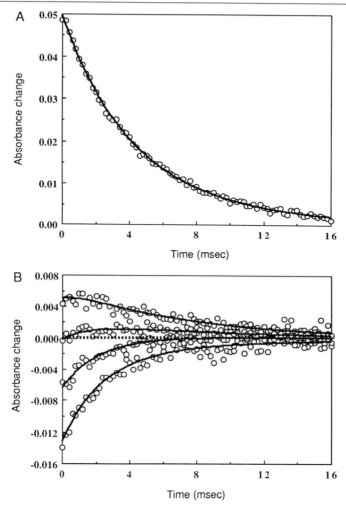

FIG. 3. Resolution of R-state α- and β-subunit reactivity toward carbon monoxide. (A) CO recombination time course monitored at 430 nm for the reaction of 50 μM CO with R-state hemoglobin following $\leq$10% photolysis by a 0.5-msec excitation pulse from a conventional flash-photolysis apparatus. The solid line shows the best fit of the data (symbols) to the sum of two exponentials with equal spectral amplitudes. The computed rates were 300 and 160 sec^{-1}. (B) Resolution of these components by observing recombination at wavelengths close to the isosbestic point for CO binding. From upper to lower trace these wavelengths are 426.0, 425.5, 425.0, and 424.5 nm. Conditions were 20° and pH 7.0 in 0.1 M Bis–Tris, 0.1 M KCl buffer. (Data are from Mathews et al.[14])

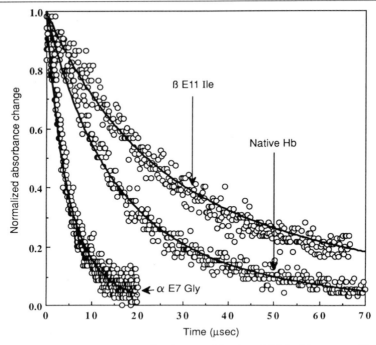

FIG. 4. Normalized time courses for the recombination of 1.25 mM oxygen with R-state hemoglobin after a 300-nsec excitation flash. The solid lines represent fits to the sum of two exponentials with equal spectral amplitudes. Conditions were 20° and pH 7.0 in 0.1 M Bis–Tris, 0.1 M KCl buffer. (Data are from Mathews *et al.*[14])

point at ~425 nm and the slow component predominates on the long-wavelength side. On the basis of previous spectral and kinetic studies with isolated chains, cyanomet hybrids, and hemoglobin Bethesda, Mathews *et al.*[14] were able to assign the fast and slow components for CO binding to R-state β and α subunits, respectively. Third, analysis of partial photolysis data for E7 and E11 mutant hybrid hemoglobins also indicated that O_2 and CO react more rapidly with native β subunits than with α subunits.

Qualitative proof that the native R-state subunits react with oxygen at different rates is shown in Fig. 4. Fits to the native hemoglobin time course indicate fast and slow association rate constants equal to ~100 × 10^6 and ~30 × 10^6 M^{-1} sec^{-1}, respectively. When α E7 His is replaced with glycine, O_2 binding occurs much more rapidly (lower trace, Fig. 4). There is no evidence of the slow component ($k'_{RO_2} \approx 30 \times 10^6 M^{-1}$ sec^{-1}) that is observed with native hemoglobin, and for this mutant the slow phase has a fitted rate constant that corresponds to that for the fast component in

TABLE III

ASSIGNMENT OF R-STATE, NATIVE β-SUBUNIT O_2 ASSOCIATION RATE CONSTANT FROM PARTIAL PHOTOLYSIS EXPERIMENTS USING DISTAL POCKET MUTATIONS IN α SUBUNITS[a]

| | Empirical fit | | Assignment | |
| | Fast k'_{O_2} (μM^{-1} sec^{-1}) | Slow k'_{O_2} (μM^{-1} sec^{-1}) | $k'_{\beta O_2}$ (μM^{-1} sec^{-1}) | $k'_{\alpha O_2}$ (μM^{-1} sec^{-1}) |
Protein				
α(native)$_2\beta$(native)$_2$	100 ± 24	29 ± 5	100	29
α(E7 Gly)$_2\beta$(native)$_2$	220	89	89	220
α(E7 Gln)$_2\beta$(native)$_2$	110	41	110	41
α(E11 Ala)$_2\beta$(native)$_2$	210	82	82	210
α(E11 Leu)$_2\beta$(native)$_2$	98	24	98	24
α(E11 Ile)$_2\beta$(native)$_2$	120	16	120	16
Average for β(native):			100 ± 13	

[a] Conditions: 0.1 M Bis–Tris, 0.1 M KCl, pH 7.0, 20°. (Data from Mathews et al.[14])

native hemoglobin ($k'_{RO_2} \approx 100 \times 10^6 M^{-1}$ sec^{-1}). On the basis of the X-ray structure of the mutant and previous results for the same mutation in myoglobin,[30] the E7 His → Gly substitution is expected to increase markedly the rate of oxygen binding to α subunits. Thus, the fast and slow components for α E7 Gly hemoglobin can be assigned to mutant α and native β subunits, respectively. When β E11 Val is replaced with isoleucine, the opposite result is observed. Rebinding is now slower (upper trace, Fig. 4), and there is no evidence for the fast component observed with native hemoglobin. For this mutant, the fast and slow phases have rate constants equal to ~30 × 10^6 and ~10 × 10^6 M^{-1} sec^{-1}, respectively. Again, on the basis of both X-ray structures and studies of the same mutation in myoglobin,[31] the large side chain of E11 Ile is expected to block access to the β-heme iron atom and decrease the rate of ligand binding. Thus, in the case of β E11 Ile hemoglobin, the fast and slow components can be assigned to native α and mutant β subunits, respectively.

Fitted results for O_2 binding to the complete series of α E7 and E11 mutant oxyhemoglobins are shown in Table III, along with the assignments to the individual subunits. The logic was the same as that used to assign dissociation rate constants in Table I. k'_{RO_2} for the native subunit in the mutant tetramer should be close to one of the rate constants observed for native hemoglobin; the largest change with respect to the native rate constants should be assigned to the mutant subunit; and there should be some correspondence between the mutational effect observed for hemoglobin with that observed for myoglobin, particularly in the case of altered

TABLE IV

OXYGEN ASSOCIATION AND DISSOCIATION RATE CONSTANTS FOR α AND β SUBUNITS WITHIN
NATIVE R-STATE HUMAN HEMOGLOBIN[a]

| Protein | k'_{RO_2} (μM^{-1} sec^{-1}) | | k_{RO_2} (sec^{-1}) | | K_{RO_2} or K_4 (μM^{-1}) | Ref. |
	α subunits	β subunits	α subunits	β subunits	Hb$_4$	
Hb A$_0$	—	—	13 ± 0.9	22 ± 2	—	27
Hb A$_0$	28 ± 9	100 ± 13	12 ± 3	22 ± 8	3.1	14
Hb A$_0$	—	—	6.3 ± 1.2	19 ± 1	—	35
Hb A$_0$	42 ± 8	150 ± 30	12 ± 2	40 ± 8	3.6	29
Hb A$_0$	28 ± 1	64 ± 1	9.0 ± 0.2	23 ± 0.7	2.9	36
Hb A$_0$	34	87	13	34	2.6	37
Isolated chains	50	60	28	16	2.4	33

[a] Conditions for each entry (from top to bottom): pH 7, 100 mM KCl, 20° (Olson et al.[27]); pH 7.0, 100 mM KCl, 20° (Mathews et al.[14]); pH 7.0, 20° (Bonaventura et al.[35]); pH 7.4, 100 mM NaCl, 25° (Vandegriff et al.[29]); pH 7.4, 100 mM NaCl, 21.5° (Philo and Lary[36]); pH 7.0, 100 mM KCl, 20° (Fronticelli et al.[37]); pH 7.0, 0.1 M potassium phosphate, 20° (Olson et al.[33]).

α subunits. As before, native rate constants were computed from the average of the values determined for the native and the appropriate set of mutant hybrid hemoglobins (i.e., Table III, column 4).

The same problems associated with the assignment of dissociation rate constants also apply to the analysis of partial photolysis experiments. (1) In some cases the fitted fast and slow rate constants do not correspond well with either native value; (2) with the exception of the E11 Ile substitution, the β E7 and E11 mutations have little effect on the observed time courses for O_2 and CO rebinding; and (3) the errors in the analyses are large, in this case due to low signal-to-noise ratios. These problems are compounded in the case of partial photolysis experiments by the possible occurrence of spectral changes associated with quaternary conformational changes on time scales similar to those for ligand rebinding.

R-State Parameters, Dimerization, and Quaternary Enhancement

A summary of published rate constants for the last step in oxygen binding to native human hemoglobin is given in Table IV.[35-37] Despite the

[35] C. Bonaventura, R. Cashon, J. Bonaventura, M. F. Perutz, G. Fermi, and B. T.-B. Shih, J. Biol. Chem. **266,** 23033 (1991).

[36] J. S. Philo and J. W. Lary, J. Biol. Chem. **265,** 139 (1990).

[37] C. Fronticelli, W. S. Briniger, J. S. Olson, Z. Gryczynski, J. K. O'Donnell, J. Kowalczyk, and E. Bucci, Biochemistry **32,** 1235 (1993).

large estimated errors, there is good agreement among all six independent studies. Using the definitions in Eq. (2), the intrinsic Adair equilibrium constant for the last step in ligand binding is given by

$$K_4 = \frac{2K_{R\alpha}K_{R\beta}}{K_{R\alpha} + K_{R\beta}} \tag{6}$$

Again, there is good agreement among the published results. K_4 or K_{RO_2} is equal to 2.6–3.6 μM^{-1} at pH 7.0–7.4, room temperature, and in the absence of phosphates.

To obtain better fits to equilibrium curves collected at differing heme concentrations, Chu et al.[38] and Di Cera et al.[39] postulated that Hb_4X_3 tetramers show a significantly higher oxygen affinity than dimers and called this effect "quaternary enhancement." Their reported values of K_4 were ~fivefold higher than those computed from partial photolysis and ligand replacement reactions. In most of the kinetic experiments, low heme concentrations are used, and an appreciable fraction of dimers is present, which could bias kinetically derived K_4 values. Gibson and Edelstein[40] and Philo and Lary[36] have questioned the existence of quaternary enhancement. The latter workers demonstrated unequivocally that the k'_{RO_2} and K_{RO_2} values given in Table IV are independent of heme concentration and that the kinetically derived value of K_R applies equally well to monoliganded dimers and triliganded tetramers at pH 7.0 in the absence of phosphates. Even if the parameters for isolated chains are used, the resultant K_R value (2.4 μM^{-1}) is comparable to that obtained at high heme concentrations for the $Hb_4(O_2)_3 + O_2 \rightarrow Hb_4(O_2)_4$ reaction. Thus, the assumption that tetrameric R-state parameters can be estimated, within a factor of two, from the properties of dimers and isolated chains appears to be valid. However, Ackers and Johnson[41] have carefully reanalyzed their equilibrium data and claim that quaternary enhancement cannot be dismissed without further experimentation.

Care must be taken when applying the kinetically derived value of K_R to the analysis of equilibrium binding curves. If the kinetically derived equilibrium constant was measured under conditions in which extensive dimerization occurred, it is readily defined as K_R in the two-state model of Monod, Wyman, and Changeux.[42] However, K_R may be greater than the value of K_4 because the extent of T-state character present in the

[38] A. H. Chu, B. W. Turner, and G. K. Ackers, *Biochemistry* **23**, 604 (1984).
[39] E. Di Cera, C. H. Robert, and S. J. Gill, *Biochemistry* **26**, 4003 (1987).
[40] Q. H. Gibson and S. J. Edelstein, *J. Biol. Chem.* **262**, 516 (1987).
[41] G. K. Ackers and M. L. Johnson, *Biophys. Chem.* **37**, 265 (1990).
[42] J. Monod, J. Wyman, and J.-P. Changeux, *J. Mol. Biol.* **12**, 22 (1965).

Hb_4X_3 and Hb_4X_4 intermediates depends on the corresponding allosteric isomerization constants (i.e., Lc^3 and Lc^4, respectively, where $L = [T_0]/[R_0]$, $c = K_T/K_R$, $Lc^3 = [T_3]/[R_3]$, and $Lc^4 = [T_4]/[R_4]$). Similarly, if the partial photolysis and ligand replacement reactions were carried out at high heme concentration in the absence of dimers, the observed equilibrium constant will equal K_4 but may be less than the true value of K_R if L and c are large.

The distinctions between K_R and K_4 are minimal for human hemoglobin at pH values ≥ 7.0, 20°, and in the absence of phosphates. Vandegriff et al.[29] have shown that the kinetically determined K_4 values for human hemoglobin and α G6 Lys–α G6 Lys cross-linked hemoglobin at pH 7.4 and 25° are comparable to those determined from fitting O_2 equilibrium curves to an unconstrained Adair equation. They then employed the kinetically derived parameter to reduce considerably the standard error of the third Adair coefficient, a_3. Large uncertainties in the assignment of a_3 have plagued equilibrium analyses and led to the publishing of negative values for this parameter even though a_3 represents the product $K_1K_2K_3$ and physically cannot be either 0 or negative. This tendency of a_3 to become small results in abnormally high values of a_4 in order for the theoretical curve to maintain complete binding and a high level of positive cooperativity. Considering the success of the analyses of Vandegriff et al.,[29] the results of Philo and Lary,[36] and the correspondence between the various rate constants listed in Table IV, it seems reasonable to estimate K_4 kinetically and then to use the resultant value to constrain the analysis of equilibrium binding data.

Assignment of Association Rate Constants for CO Binding to T-State Deoxyhemoglobin

Most published ligand-binding data with native, valence, and mixed metal hybrid hemoglobins suggest that the subunits within deoxyhemoglobin have similar affinities for O_2 or CO in the absence of organic phosphates.[3,25] Nagai et al.[12] and Tame et al.[26] have reported O_2 equilibrium binding curves for mutant hemoglobins that also suggest that the native α and β subunits have similar T-state affinities. However, in all of these experiments interpretations are greatly complicated by cooperativity, which precludes the build-up of partially liganded intermediates, and by the lack of simple spectral signals that discriminate between binding to the α- and β-heme groups.

The kinetics for O_2 binding to T-state deoxyhemoglobin are even more difficult to measure and analyze. The rate constants for oxygen dissociation from monoliganded species are 500–2000 sec^{-1}.[34] Consequently, the halftimes for oxygen binding to T-state hemoglobin are ≤ 0.001 sec, which precludes measurement by stopped-flow rapid mixing techniques. Using laser photolysis techniques, Sawicki and Gibson[34] examined oxygen rebinding to 10% saturated solutions of human hemoglobin. The resultant time courses were resolved into two kinetic components. The fast component exhibited association and dissociation rate constants equal to $11.8 \times 10^6 \, M^{-1} \, sec^{-1}$ and 2500 sec^{-1}, respectively, and the slow-component values were equal to $2.9 \times 10^6 \, M^{-1} \, sec^{-1}$ and 180 sec^{-1}, respectively. These rate constants were originally assigned to T-state β and α subunits. However, this assignment has been questioned because laser photolysis experiments with partially saturated, equilibrium mixtures of hemoglobin and ligands are complicated by the slow appearance of rapidly reacting species and by the rates of the R-to-T quaternary transition.[34,43–45]

The first step in binding to deoxyhemoglobin can be examined in mixing experiments when NO and CO are used as ligands. Under ordinary conditions, the NO and CO reactions are effectively irreversible because of their low dissociation rate constants (≤ 0.001 to 0.2 sec^{-1}).[9,46] Using rapid-freezing EPR techniques, Hille et al.[47,48] have shown unequivocally that the α and β subunits within Hb_4 react at equal rates with NO, confirming the earlier absorbance measurements by Cassoly and Gibson.[49] However, these same workers have shown that slow conformational changes also occur on second to minute time scales following the formation of Hb_4NO species.[48]

In general, carbon monoxide is the preferred ligand for studies with deoxyhemoglobin because this ligand is chemically stable, its reactions are effectively irreversible even at micromolar concentrations of protein and ligand, and its association rate constants are 2- to 100-fold smaller than those for NO binding. Thus, our first attempts to use site-directed mutants to assign T-state rate constants involved measurements of CO binding by rapid mixing techniques.

Measurement of First Step in CO Binding to Deoxyhemoglobin

To examine the binding of only one ligand molecule, deoxyhemoglobin must be present in great excess. As in the case of the partial photolysis

[43] C. A. Sawicki and Q. H. Gibson, J. Biol. Chem. 251, 1533 (1976).
[44] M. A. Khaleque and C. A. Sawicki, Photochem. Photobiophys. 13, 155 (1986).
[45] I. A. Zahroon and C. A. Sawicki, Biophys. J. 56, 947 (1989).
[46] E. G. Moore and Q. H. Gibson, J. Biol. Chem. 251, 2788 (1976).
[47] R. Hille, G. Palmer, and J. S. Olson, J. Biol. Chem. 252, 403 (1977).
[48] R. Hille, J. S. Olson, and G. Palmer, J. Biol. Chem. 254, 12110 (1979).
[49] R. Cassoly and Q. H. Gibson, J. Mol. Biol. 91, 301 (1975).

experiments, the exact ratio of heme to CO is determined empirically by mixing hemoglobin with decreasing amounts of ligand until the half-time and shape of the observed time course no longer change. In our experiments at pH 7.0 in the absence of phosphates, this occurred when native or wild-type human deoxyhemoglobin was mixed with only enough CO to achieve 5% saturation (usually 0.5 μM CO plus 10 μM heme). If the heme concentration is lowered much below 10 μM, the observed rates will be $\leq 1-2$ sec^{-1} and begin to contain significant contributions from the first dissociation rate constant, k_1 in Eq. (1), which is ~ 0.1 sec^{-1}. At higher heme concentrations the reaction can be considered irreversible, and the time course is described by

$$\frac{d[CO]}{dt} = -k'_{T(obs)}[Hb]_0[CO] = -(k'_{T\alpha}[\alpha]_0 + k'_{T\beta}[\beta]_0)[CO]$$

$$= -0.5(k'_{T\alpha} + k'_{T\beta})[Hb]_0[CO] \quad (7)$$

where $[Hb]_0$, $[\alpha]_0$, and $[\beta]_0$ are the initial ($t = 0$) concentrations of total heme, α subunits, and β subunits and do not change with time. As shown, the observed pseudo-first-order rate constant is determined by the concentration of free heme groups, which is effectively the total concentration of deoxyhemoglobin present, $[Hb]_0$. The reaction should always be described by a simple exponential decay with an apparent bimolecular rate constant, $k'_T = k_{obs}/[Hb]_0$, given by the average of the α and β T-state rate constants.

In this case, determination of the native subunit rate constant requires the construction of recombinant hybrid hemoglobin in which ligand binding to the mutated subunit is effectively blocked. For such a protein, the limited amount of CO present will bind exclusively to the native subunit, and the observed bimolecular rate constant will be equal to $0.5k'_{Tnative}$, because only half the hemes are available for reaction with CO [i.e., Eq. (7) with $k'_{T\alpha}$ or $k'_{T\beta} = 0$]. If the mutated subunit reacts much more rapidly than the native partner, the observed rate will equal 0.5 times the rate constant for the genetically engineered subunit. Thus, the strategy is to block binding to one of the subunits by site-directed mutagenesis and then determine the association rate constant of the native subunit partner by reacting the hybrid tetramer with a very small (5%) amount of CO.

Inhibition of Binding to β Subunits by E11 Val $\rightarrow$ Ile Mutations

Three previous experimental observations suggested strongly that β E11 Ile markedly inhibits ligand binding to β subunits. Nagai *et al.*[12] showed that the δ methyl group of the β E11 Ile side chain is directly above the β-heme iron atom in the crystal structure of [α(native)β(E11

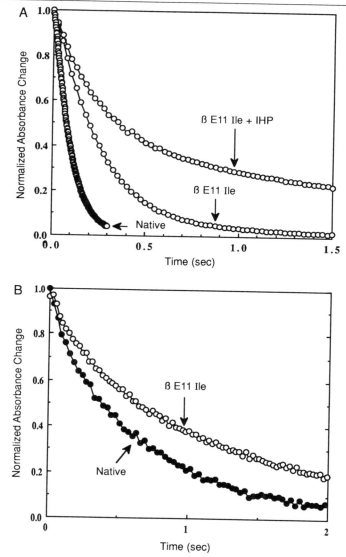

FIG. 5. Inhibition of CO binding to T-state deoxyhemoglobin by the β E11 Ile mutation. (A) Time courses for the reaction of excess CO with native deoxyhemoglobin and with deoxyHb (β E11 Ile) in the presence and absence of inositol hexaphosphate (IHP). In the presence of IHP the full effect of the mutation is observed with ~50% of the reaction occurring in a slow phase with a rate constant of 0.0082 μM^{-1} sec^{-1}. (B) Time courses for the reaction of excess deoxyhemoglobin with limiting (5%) CO. Conditions were 20° and pH 7.0 in 0.1 M Bis–Tris, 0.1 M KCl buffer. (Data are from Mathews *et al.*[15])

TABLE V
ASSIGNMENT OF ASSOCIATION RATE CONSTANTS FOR CO BINDING TO T-STATE
DEOXYHEMOGLOBIN[a]

	Reaction with 500% CO			Reaction with 5% CO
Protein	k'_{fast} ($\mu M^{-1} sec^{-1}$)	Percent fast	k'_{slow} ($\mu M^{-1} sec^{-1}$)	[k'_T($\mu M^{-1} sec^{-1}$)]
[α(native)β(native)]$_2$	0.19 ± 0.02	100		0.15 ± 0.02
+ IHP	0.10 ± 0.02	100		0.12 ± 0.02
[α(native)β(E11 Ile)]$_2$	0.089	100		0.066 ± 0.01
+ IHP	0.068 ± 0.020	63 ± 10	0.008 ± 0.003	0.055

[a] Data from Mathews et al.[15]

Ile)]$_2$ deoxyhemoglobin; these same workers reported that the P_{50} for oxygen binding to this mutant is increased roughly 2-fold, and finally, Mathews et al.[14] measured a 23-fold lower association rate constant for CO binding to R-state β E11 Ile subunits. Thus, we examined CO binding to deoxy [α(native)β(E11 Ile)]$_2$ hybrid tetramers in order to assign a value to $k'_{T\alpha}$ for native T-state α subunits.

Sample time courses for the reaction of the β E11 Ile mutant hemoglobin with excess and limiting amounts of CO are shown in Fig. 5, and fitted kinetic parameters are given in Table V. These results confirm that substitution of isoleucine for E11 Val greatly inhibits CO binding to T-state β subunits. This mutation causes a twofold reduction in the overall rate constant for the binding of excess CO and a loss of acceleration in the time course (Fig. 5A). There is a similar ~twofold reduction in the apparent bimolecular rate constant for the binding of only 5% CO (Fig. 5B). The latter twofold change is the maximum possible decrease if the native subunits initially exhibited equal rates [i.e., if $k'_{T\alpha(native)} = k'_{T\beta(native)}$ and $k'_{T\beta(E11\ Ile)} = 0$ in Eq. (7)].

When inositol hexaphosphate is added to [α(native)β(E11 Ile)]$_2$, the time course for CO binding in the presence of excess ligand becomes markedly biphasic (Fig. 5A). Roughly half the absorbance change occurs at the "normal" rate of ~1 × 10^5 M^{-1} sec^{-1}, whereas the remainder occurs at a much slower rate of ~1 × 10^4 M^{-1} sec^{-1} (Table V). The simplest interpretation of this biphasic time course is that the initial fast phase represents the binding of CO to native T-state α subunits, whereas the slow phase represents CO binding to T-state β E11 Ile subunits. The $\alpha_2(CO)_2\beta_2$ intermediate is maintained in the low-affinity quaternary conformation by the presence of bound inositol hexaphosphate. In the absence of the effector molecule, the $\alpha_2(CO)_2\beta_2$ and $\alpha_2(CO)_2\beta_2(CO)$ species

TABLE VI
EFFECTS OF E7 AND E11 MUTATIONS ON ASSOCIATION RATE CONSTANTS FOR CO BINDING TO R- AND T-STATE HUMAN HEMOGLOBIN AT pH 7.0, 20°[a]

Mutated residue	α subunit			β subunit		
	k'_{RCO} ($\mu M^{-1}\ sec^{-1}$)	k'_{TCO} ($\mu M^{-1}\ sec^{-1}$)	Ratio	k'_{RCO} ($\mu M^{-1}\ sec^{-1}$)	k'_{TCO} ($\mu M^{-1}\ sec^{-1}$)	Ratio
Native	2.9 ± 0.5	0.12 ± 0.03	24 ± 7	7.1 ± 2.4	0.18 ± 0.05	39 ± 17
E7 Gly	19 ± 3	9.2 ± 0.8	2.1 ± 0.4	5.0 ± 1.7	6.5 ± 0.8	0.8 ± 0.3
E11 Ala	32 ± 6	0.32 ± 0.09	100 ± 34	7.0 ± 2.4	1.4 ± 0.1	5.0 ± 1.8
E11 Ile	0.9 ± 0.2	0.04 ± 0.06	23 ± 35	0.3 ± 0.1	0.008 ± 0.003	36 ± 19

[a] Data from Mathews et al.[14,15]

switch over to the R quaternary conformation. As a result the β subunits in these phosphate-free intermediates react at higher rates [i.e., $k'_{R\beta(E11\ Ile)} = 3 \times 10^5\ M^{-1}\ sec^{-1}$; Table VI],[14] and the overall time course with excess CO appears monophasic and exhibits only a twofold reduction in the overall rate constant [Fig. 5A (middle curve) and Table V (row 3)]. This latter phenomenon is a good experimental example of how cooperativity can obscure large differences between the intrinsic reactivities of the α and β subunits.

One objection to using site-directed mutants concerns whether the hybrid tetramers exhibit native T-state conformations. Nagai and co-workers have shown that almost all of the distal pocket mutants listed in Tables I–VI can be crystallized in the deoxy T-state quaternary conformation.[12,26,50] We have confirmed the ability of these mutant tetramers to form the low-affinity T state in solution by measuring the reaction of p-hydroxymercuribenzoate with β93 Cys in the presence and absence of saturating amounts of ligand. In all cases, the deoxygenated tetramers showed ~100-fold lower rate constants for mercurial binding than the fully liganded forms, and the absolute rates were, within a factor of two, equal to those observed for native human deoxyhemoglobin.[15]

Computation of Native and Mutant Subunit k'_{TCO} Values

The starting point for the assignment of $k'_{T\alpha}$ and $k'_{T\beta}$ values to native hemoglobin are the results for $[\alpha(native)\beta(E11\ Ile)]_2$ shown in Table V. $k'_{T\beta(E11\ Ile)}$ was assigned a value of $8.2 \times 10^3\ M^{-1}\ sec^{-1}$ on the basis of the rate of the slow phase observed when this mutant protein was reacted with excess CO in the presence of inositol hexaphosphate (Fig. 5A).

[50] B. F. Luisi and K. Nagai, Nature (London) 320, 555 (1986).

$k'_{T\alpha(\text{native})}$ was computed as $[2k'_{T(\text{obs})} - k'_{T\beta(\text{E11 Ile})}]$, where $k'_{T(\text{obs})}$ is the apparent T-state association rate constant for the binding of 5% CO to the $[\alpha(\text{native})\beta(\text{E11 Ile})]_2$ mutant (Table V). $k'_{T\beta(\text{native})}$ was then calculated using this value of $k'_{T\alpha(\text{native})}$ and $k'_{T(\text{obs})}$ for completely native hemoglobin. The resultant values of $k'_{T\alpha(\text{native})}$ and $k'_{T\beta(\text{native})}$ were then used to assign T-state, CO association rate constants for several other mutant subunits, using Eq. (7), and the results are summarized in Table VI. It is important to note that the absolute value of $k'_{T\beta(\text{E11 Ile})}$ is not crucial for accurate assignments of the other T-state association rate constants. As long as $k'_{T\alpha(\text{native})} \geq 10 \times k'_{T\beta(\text{E11 Ile})}$, the value of $k'_{T\alpha(\text{native})}$ must be ~$2k'_{T(\text{obs})}$ for the reaction of $[\alpha(\text{native})\beta(\text{E11 Ile})]_2$ with limiting (5%) CO.

The results in Table VI provide three important conclusions. First, the α and β subunits of native T-state deoxyhemoglobin react at roughly equal rates with CO, in agreement with previous experimental results. Second, the absolute values, $k'_{T\alpha} = 0.12 \pm 0.03 \ \mu M^{-1} \ \text{sec}^{-1}$ and $k'_{T\beta} = 0.18 \pm 0.05 \ \mu M^{-1} \ \text{sec}^{-1}$, are remarkably similar to the corresponding rates reported by Perrella et al.[6] These workers used cryoscopic isoelectric focusing to separate the various intermediates formed when substoichiometric amounts of CO were mixed with high concentrations of deoxyhemoglobin and reported values of $k'_{T\alpha} = 0.10 \pm 0.04 \ \mu M^{-1} \ \text{sec}^{-1}$ and $k'_{T\beta} = 0.15 \pm 0.06 \ \mu M^{-1} \ \text{sec}^{-1}$. Third, the E7 His $\rightarrow$ Gly and E11 Val $\rightarrow$ Ala mutations cause 10- to 30-fold increases in the rates of CO binding to T-state β subunits, whereas the same substitutions are without effect on R-state β subunits (Table VI). This demonstrates rather unambiguously that the distal portion of the β-heme pocket is much more tightly packed in the T quaternary conformation than in the R conformation, as was observed by Perutz in 1970.[51] In contrast, the distal portion of the α-heme pocket appears to be sterically restricted in both quaternary conformations. A more detailed discussion of the mutagenesis effects in presented by Mathews et al.[15] The remaining task is to apply these mutagenesis techniques to the case of O_2 binding to T-state deoxyhemoglobin, using flow-flash methodology.

[51] M. F. Perutz, Nature (London) 228, 726 (1970).

[18] Ligand Binding and Conformational Changes Measured by Time-Resolved Absorption Spectroscopy

By JAMES HOFRICHTER, ANJUM ANSARI, COLLEEN M. JONES, ROBERT M. DEUTSCH, JOSEPH H. SOMMER, and ERIC R. HENRY

Introduction

In the original application of flash photolysis to kinetic measurements, Porter and co-workers used a short pulse of white light from a photographic flashlamp to measure the spectrum of a sample in which a reaction had previously been initiated by a pulse of exciting light. The probe light was passed through a spectrograph and the spectrum recorded on film.[1,2] We refer to this general experimental approach as time-resolved absorption spectroscopy (TRAS). Although TRAS continued to be applied to both flash photolysis and pulsed radiolysis studies, its use was, for many years, complicated by the lack of an efficient and quantitative method for recording and retrieving the data. The more commonly used technique for monitoring chemical kinetics was to measure transmitted intensity at a single wavelength using a fast detector such as a phototube or photodiode.[3-6] For kinetic experiments in which the signals could be time resolved directly, this method was simpler to execute, more quantitative, and less expensive to set up.

With the development of sensitive and quantitative detectors to replace the photographic emulsion in the classic photolysis experiment, TRAS has become a much more convenient and powerful technique. Electronic controllers that could quantitate the charge stored on a vidicon were introduced in the late 1970s; these have been supplanted in the last decade by conceptually similar detectors, which utilize linear arrays of photodiodes or two-dimensional arrays of charge transfer devices. The temporal dynamic range of the technique has also been vastly expanded by the availability of extremely short light pulses from Q-switched and mode-locked lasers. An explosion in the application of TRAS to kinetic measure-

[1] G. Porter and M. Topp, *Proc. R. Soc. London, Ser. A* **315**, 163 (1970).

[2] M. R. Topp, *Appl. Spectrosc. Rev.* **14**, 1 (1978).

[3] Q. H. Gibson, *Prog. Biophys. Chem.* **9**, (1959).

[4] G. Porter and M. A. West, *in* "Investigation of Rates and Mechanisms of Reactions" (G. Hammes, ed.), 3rd ed., Part II, p. 367. Wiley (Interscience), New York, 1974.

[5] R. H. Austin, K. W. Beeson, L. Eisenstein, H. Frauenfelder, and I. C. Gunsalus, *Biochemistry* **14**, 5355 (1975).

[6] C. A. Sawicki and R. J. Morris, this series, Vol. 76, p. 667 (1981).

ments was, in fact, triggered by the need to carry out measurements on the picosecond and subpicosecond time scales. Except by employing streak cameras, a costly alternative, conventional kinetic methods cannot be used to time resolve such experiments. As a result, TRAS, coupled with the use of picosecond continua[7] as probe sources, became a standard approach to such measurements.[8] These technologies led to the development of a number of spectrometers designed to measure complete time-resolved spectra over a temporal dynamic range extending from a few femtoseconds to a few nanoseconds.[9-15] There have been fewer applications of this technique to measurements at longer times, for which the alternative approach of directly time-resolving kinetic signals at a fixed wavelength was both more easily implemented and less expensive. In the past decade, however, a number of such instruments for measuring time-resolved absorption spectra have been developed.[16-24]

[7] The continuum, first described by Alfano and Shapiro in 1970, is generated by focusing a short, intense pulse of monochromatic laser light into a medium having a large dielectric constant, such as H_2O or D_2O and mixtures of these solvents with alcohols [cf R. R. Alfano and S. L. Shapiro, *Phys. Rev. Lett.* **24**, 584 (1970); A. G. Doukas, J. Buchert, and R. R. Alfano, *in* "Biological Events Probed by Ultrafast Laser Spectroscopy" (A. A. Alfano, ed.), p. 387. Academic Press, New York, 1982].

[8] G. R. Fleming, "Chemical Applications of Ultrafast Spectroscopy." Oxford Univ. Press, New York, 1986.

[9] C. V. Shank and B. I. Greene, *in* "Biological Events Probed by Ultrafast Laser Spectroscopy" (A. A. Alfano, ed.), p. 387. Academic Press, New York, 1982.

[10] B. I. Greene, R. M. Hochstrasser, R. B. Weisman, and W. A. Eaton, *Proc. Natl. Acad. Sci. U.S.A.* **75**, 5255 (1978).

[11] B. J. Greene, R. M. Hochstrasser, and R. B. Weisman, *J. Chem. Phys.* **70**, 1247 (1979).

[12] J. L. Martin, A. Migus, C. Poyart, Y. Lecarpentier, R. Astier, and A. Antonetti, *Proc. Natl. Acad. Sci. U.S.A.* **80**, 173 (1983).

[13] J. A. Hutchinson and L. J. Noe, *IEEE J. Quantum Electron.* **QE-20**, 1353 (1984).

[14] K. A. Jongeward, D. Magde, D. J. Taube, J. C. Marsters, T. G. Traylor, and V. S. Sharma, *J. Am. Chem. Soc.* **110**, 380 (1988).

[15] S. M. Janes, G. A. Dalickas, W. A. Eaton, and R. M. Hochstrasser, *Biophys. J.* **54**, 545 (1988).

[16] R. B. Weisman, J. I. Selco, P. L. Holt, and P. A. Cahill, *Rev. Sci. Instrum.* **54**, 284 (1983).

[17] J. Hofrichter, J. H. Sommer, E. R. Henry, and W. A. Eaton, *Proc. Natl. Acad. Sci. U.S.A.* **80**, 2235 (1983).

[18] J. Hofrichter, E. R. Henry, J. H. Sommer, R. Deutsch, M. Ikeda-Saito, T. Yonetani, and W. A. Eaton, *Biochemistry* **24**, 2667 (1985).

[19] E. P. L. Hunter, M. G. Simic, and B. D. Michael, *Rev. Sci. Instrum.* **56**, 2199 (1985).

[20] J. Sedlmair, S. G. Ballard, and D. C. Mauzerall, *Rev. Sci. Instrum.* **57**, 2995 (1986).

[21] L. P. Murray, J. Hofrichter, E. R. Henry, M. Ikeda-Saito, K. Kitagishi, T. Yonetani, and W. A. Eaton, *Proc. Natl. Acad. Sci. U.S.A.* **85**, 2151 (1988).

[22] S. J. Milder, T. E. Thorgeirsson, L. J. W. Miercke, R. M. Stroud, and D. S. Kliger, *Biochemistry* **30**, 1751 (1991).

For reactions in which the spectra of the kinetic intermediates are known in advance, the most compact and efficient means of measuring the time dependence of the composition of the system is by monitoring the kinetics at a set of wavelengths preselected to optimize the spectral differences among the species (and hence the signal to noise).[25] A problem routinely encountered when the lifetimes of the intermediate species are short, however, is that their spectra must be determined from the kinetic data because these intermediates do not exist as stable species. When the number of components or the spectra of the intermediates are not known in advance, characterization of the spectra by conventional multiple-wavelength techniques requires that the set of wavelengths be expanded significantly. When extended to its logical limit, this approach provides a set of data that is similar to that obtained by TRAS, characterizing the changes in absorbance as a function of both (log) time and wavelength. When TRAS is used as a routine probe, data at all wavelengths are collected as a matter of course, so complete spectra of kinetic intermediates are derived from analysis of the data, permitting a more straightforward determination of the spectra of these species. Another advantage of TRAS is that the spectral data are usually of much higher quality than those that can be obtained by sequential kinetic measurements of the time course of the reaction at a variety of wavelengths.[26] The improved quality of the spectra permits small, time-dependent changes in the spectra to be discerned even in the presence of simultaneous large changes in the sample composition, thus providing a much more sensitive probe of the kinetics of complex systems.[17,18,21,23]

In this chapter we describe and characterize the performance of a time-resolved absorption spectrometer with a time resolution of 10 nsec. Our spectrometer is capable of collecting spectral data over selected wavelength intervals throughout the visible and near-ultraviolet spectral region and provides performance and stability comparable to conventional double-beam recording spectrophotometers. Our objective in the development

---

[23] T. E. Thorgeirsson, S. J. Milder, L. J. W. Miercke, M. C. Betlach, R. F. Shand, R. M. Stroud, and D. S. Kliger, *Biochemistry* **30,** 9133 (1991).

[24] J. Hofrichter, E. R. Henry, A. Szabo, L. P. Murray, A. Ansari, C. M. Jones, M. Coletta, G. Falcioni, M. Brunori, and W. A. Eaton, *Biochemistry* **30,** 6583 (1991).

[25] J. F. Nagel, *Biophys. J.* **59,** 476 (1991).

[26] Careful automation of the more conventional technique has also proved to produce data of extremely high accuracy and precision. The reader is referred to D. G. Lambright, S. Balasubramanian, and S. G. Boxer, *Chem. Phys.* **158,** 249 (1991) for an elegant demonstration of this point.

of this instrument has been to time resolve changes in the optical absorption spectra of biological macromolecules. In this chapter we focus on hemoglobin (Hb), a system in which the subtle structural changes that take place during the relaxation of the deoxy photoproduct produce only small changes in their optical absorption spectra.[27] In designing an instrument that is able to resolve these spectral changes, optimization of stability and sensitivity have been primary considerations. A more comprehensive review of the technology and alternatives for such measurements is presented in a review by Chen and Chance.[28]

We also describe some techniques that we have developed to determine accurately the fraction of hemes photolyzed and to utilize this information in kinetic analyses. These techniques are based on characterizing the distribution of orientations of the photoproduct molecules generated by partial photolysis of the sample with a linearly polarized excitation pulse. Systematic variation of the photolysis level is of particular use in investigating hemoglobin kinetics, because different degrees of photolysis produce different distributions of tetramer ligation states. Analysis of the dependence of the kinetics on the extent of photolysis permits tertiary and quaternary conformational changes to be distinguished experimentally.[17,18,24,29–31]

At the wavelengths used in photolysis experiments, hemes behave like circular absorbers of linearly polarized light[32]; the hemes absorb light polarized parallel to any direction in the porphyrin plane equally and do not absorb light polarized perpendicular to the plane. The probability of photodissociation is, therefore, largest when the porphyrin planes of the heme–ligand complexes are parallel to the electric vector of a linearly polarized excitation pulse and smallest when they are perpendicular. Photoselection has two major effects. The first is that it produces a photoproduct sample that is optically anisotropic; the sample exhibits linear dichroism because the distribution of the photolyzed and unphotolyzed species is no longer random. Reorientation of the photoselected sample can interfere with the accurate determination of the kinetics of ligand rebinding. The second is to produce a nonrandom distribution of ligands in the photoproduct. Although the effects of photoselection have been considered in the

[27] M. F. Perutz, J. E. Ladner, S. R. Simon, and C. Ho, *Biochemistry* **13,** 2163 (1974).

[28] E. Chen and M. R. Chance, this series (submitted for publication).

[29] C. A. Sawicki and Q. H. Gibson, *J. Biol. Chem.* **251,** 1533 (1976).

[30] M. C. Marden, E. S. Hazard, and Q. H. Gibson, *Biochemistry* **25,** 7591 (1986).

[31] C. M. Jones, A. Ansari, E. R. Henry, G. W. Christoph, J. Hofrichter, and W. A. Eaton, *Biochemistry* **31,** 6692 (1992).

[32] W. A. Eaton and J. Hofrichter, this series, Vol. 76, p. 175.

design of experiments on retinal proteins for some time,[33,34] the substantial effect on the apparent geminate ligand rebinding to hemoglobin has been largely ignored until recently.[31]

The contribution of the decay of the sample dichroism to the apparent ligand-binding curves must be eliminated if the geminate rebinding of oxygen and carbon monoxide on the picosecond to microsecond time scale is to be characterized accurately. One way in which this can be accomplished is by measuring optical densities in both polarizations and calculating the isotropically averaged value at each wavelength from the conservation relation

$$\Delta OD_{iso} = \tfrac{1}{3}(\Delta OD_\| + 2\Delta OD_\perp) \tag{1}$$

We have used the notation ΔOD in Eq. (1) because, in most experiments, it is the difference in optical density between the photodissociated and unphotodissociated sample that is measured. In addition to producing accurate values for ΔOD_{iso}, this approach provides important dynamic information regarding the reorientational dynamics of the heme and the protein from the decay of the anisotropy, $r(\lambda, t, x)$, defined at each degree of photolysis (x), wavelength (λ), and time delay (t) as[31,35]

$$r(x, \lambda, t) = \frac{\Delta OD_\| - \Delta OD_\perp}{\Delta OD_\| + 2\Delta OD_\perp} \tag{2}$$

An alternative approach, useful at low degrees of photolysis where the linear dichroism is small, is to orient the polarization of the probe light at the so-called "magic angle" (54.7°) relative to the excitation polarization. Under these conditions, measurements at the magic angle are equivalent to measuring the isotropically averaged optical density. When the linear dichroism of the sample becomes large, however, the polarization direction is rotated by the anisotropic absorption of the sample toward the more weakly absorbing direction, and the isotropically averaged optical density is no longer measured.[36] It is important to note that removal of the contribution of the linear dichroism to the measurement in no way alters the effect of photoselection on the distribution of ligands in the photoproduct.

[33] J. F. Nagel, S. M. Bhattacharjee, L. A. Parodi, and R. H. Lozier, *Photochem. Photobiol.* **38**, 331 (1983).
[34] J. F. Nagle, L. A. Parodi, and R. H. Lozier, *Biophys. J.* **38**, 161 (1982).
[35] A. Ansari, C. M. Jones, E. R. Henry, J. Hofrichter, and W. A. Eaton, *Biophys. J.* **64**, 852 (1993).
[36] J. W. Lewis and D. S. Kliger, *Photochem. Photobiol.* **54**, 963 (1991).

The analysis of large sets of transient absorption spectra, such as those produced when sets of time-resolved spectra are measured in two polarizations for a number of different photolysis levels, presents a number of technical problems. The first is the requirement for a robust technique by which to extract and condense the spectral information in the data set. We have utilized the technique of singular value decomposition (SVD) for this purpose. This rank-reduction procedure permits us to represent the measured set of spectra in terms of a minimal set of basis spectra together with the associated time-dependent amplitudes of each of these spectra. A detailed discussion of SVD has been presented,[37] and we will touch only briefly on this aspect of the problem in this chapter. We use sets of data in which the degree of photolysis is varied systematically to obtain ligand-rebinding curves and conformational dynamics for each subspecies in the photoproduct population. This approach provides a model-independent method for unpacking the kinetic information to produce kinetic curves that characterize the individual species in the photoproduct population.

To illustrate these techniques, we present data on the recombination of carbon monoxide (CO) with the $\alpha_2(Co)\beta_2(FeCO)$ hybrid of human hemoglobin subsequent to laser photolysis. These experiments illustrate the sensitivity, stability, and dynamic range of the instrument and provide insight into the pathway by which ligands dissociate from heme proteins. We are able to characterize the deoxyHb photoproduct at times as short as 10 nsec after photolysis. The sensitivity of the spectrometer, together with SVD analysis, permit us to detect three distinct deoxy spectral species that evolve subsequent to the photolysis of HbCO and the hybrid.[18]

Transient Spectrometer

The transient spectrometer, shown in Fig. 1, is constructed using two Nd : YAG (neodymium : yttrium–aluminum–garnet) lasers. The first laser, referred to as the photolysis laser, produces an excitation pulse at 532 nm that is used to photolyze the sample and thereby initiate the experiment. The second laser, referred to as the probe laser, is delayed electronically before it produces a pulse at 355 nm that is used to excite a fluorescent laser dye in a flow cell. The spontaneous emission of the dye is used as a short-lived flashlamp to measure the absorption spectrum of the sample. The fluorescent emission is focused onto the photolyzed area of the sample, which in turn is imaged onto the slit of a spectrograph. The spectrograph disperses the light and the spectrum is measured by a vidicon

[37] E. R. Henry and J. Hofrichter, this series, Vol. 210, p. 129.

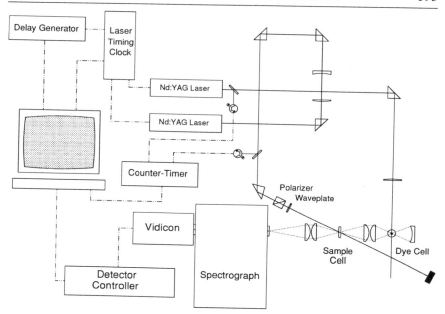

FIG. 1. A spectrometer for measuring time-resolved spectra with a time resolution of 10 nsec. The spectrometer is described in detail in text.

detector. The reference beam of the spectrometer is generated by focusing a second portion of the image of the fluorescent dye just below the photolyzed portion of the sample cell. This image is then focused on the slit by the same optics that focus the sample beam. The reference beam is dispersed by the spectrograph and measured as an adjacent track of intensity by the vidicon tube. In a more modern implementation, the vidicon would be replaced by a dual photodiode array or by an array of charge transfer detectors, such as a charge-coupled device (CCD) array. These newer systems can be read accurately by a single scan and, because they are cooled, accumulate a much smaller number of dark counts. These improvements remove many of the problems encountered in reading the vidicon, which are discussed below.

The photolysis beam is generated by a DCR-1A Nd : YAG laser, consisting of an oscillator stage and an amplifier stage (Spectra Physics, Inc., Mountain View, CA). The 1064-nm fundamental is doubled by a type II KDP crystal and separated from the fundamental using a PHS-1 dispersing prism harmonic separator (Spectra Physics). The 532-nm second harmonic is focused to a slit by a cylindrical telescope consisting of a 250-mm focal length cylindrical lens and a -100-mm focal length cylindrical lens. The telescope produces a diffraction-limited image on the sample cell with a total width at the $1/e$ points of 0.5 mm. The beam profile is roughly Gaussian, suggesting that the incident intensity is constant to within 10% over a total width of 0.1 mm. The height of the photolysis beam is masked to about 4 mm by a horizontal slit placed approximately 40 cm in front of the sample cell. The photolysis beam is folded and directed by three prisms placed between the pair of lenses and the slit. The beam can be translated across the face of the cell by adjustment of the final prism. For polarization experiments, a Glan-Laser polarizing prism (Melles Griot, Irvine, CA) with an extinction ratio of $<5 \times 10^{-5}$ is inserted into the beam to ensure the purity of the linear polarization of the photolysis beam. The polarization direction of the 532-nm beam is controlled by rotation of a $\lambda/2$ waveplate. Typical photolysis beam power measured at the cell under conditions of maximum photolysis is about 10 mJ/pulse, of which only about 1 mJ is absorbed by the sample. The absorbed energy is equivalent to approximately 20 photons/heme and is calculated to produce a temperature rise of less than 0.2° at sample concentrations of about 0.1 mM (heme). The intensity of the photolysis beam is modulated by adjusting the lamp energy of the amplifier stage of the laser and/or by insertion of neutral density filters (Rolyn Optics, Covina, CA) into the beam path.

The probe laser optical train is simpler than that used in the photolysis beam. A slit-shaped image of the probe laser beam is focused on the dye cell by a 300-mm cylindrical lens. For the experiments reported here, stilbene 420 (Exciton, Inc., Dayton, OH), a laser dye that emits over the range of 400–470 nm in methanol, is used as a source because of its favorable overlap with the Soret bands of HbCO and deoxyHb (Fig. 2).

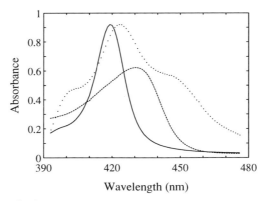

FIG. 2. The overlap between the emission spectrum of stilbene 420 in methanol and the Soret spectra of HbCO and the 10-nsec Hb photoproduct. (—) Spectrum of HbCO; (– – –) spectrum of the 10-nsec Hb photoproduct; (···) emission spectrum of stilbene 420 in methanol.

The dye is excited by the third harmonic of a Nd : YAG laser (DCR-1; Spectra Physics), using only 10 mJ/pulse. The dye cell is a piece of precision-bore quartz capillary tubing (Wilmad, Inc., Buena, NJ), through which the dye solution is circulated by a peristaltic pump (Pharmacia Biosystems, Inc., Piscataway, NJ). The high instantaneous energy at the dye cell produces stimulated emission in the dye, which is spatially focused along the vertical axis of the cylindrical tube and is the major source of energy loss in the dye cell. The ratio of spontaneous to stimulated emission is maximized by using vertical polarization for the 355-nm probe laser beam used to excite the dye. The fluorescence lifetime of this dye in ethanol is reported to be 0.8 nsec.[38] The short lifetime, coupled with the fact that the excited state population of the dye is depleted rapidly by stimulated emission directed along the slit image, causes the temporal profile of the spontaneous dye emission to follow that of the 355-nm excitation pulse closely. The unattenuated dye spectrum produces a peak intensity of approximately 5000 counts/channel in a single shot, equivalent to roughly 1.5×10^7 photons in a bandwidth of 0.2 nm when focused on a slit 0.1 mm in width. A dichroic sheet polarizer positioned after the dye cell is used to polarize the probe source. The extinction ratio of the probe light, measured by placing a polarization analyzer at the sample, is 0.005.

Laser timing is controlled using a programmable digital delay. Timing pulses for firing each laser are produced by a specially designed digital clock circuit that produces a set of three pulses for each laser. The first of these pulses activates a relay that applies high voltage to the laser Q-switch, the second fires the flashlamp, and the third fires the laser Q-switch. The pulses are generated by a set of decade counters, which count 10^6 cycles from a 10-MHz clock, and thereby produce a repetition rate for firing the lasers of 10 Hz. By logically AND-ing the appropriate outputs from this counter, and using these signals to set and reset flip-flops, pulses of any desired duration can be produced at any point in the 0.1-sec duration of the counter cycle. For each of the lasers, the required output pulses are (1) a pulse of 5-msec duration that closes the relay that applies high voltage to the Q-switch and opens an electromechanical shutter that blocks any intensity generated by the non-Q-switched output of the lasers, (2) a trigger pulse of 100-μsec duration, begining 3 msec after closing the relay, which fires the flashlamp, and (3) a trigger pulse of 1-μsec duration, delayed 250 μsec from the flashlamp pulse, which fires the Q-switch. This clock circuit is designed to work together with the delays obtained from a Hewlett-Packard (Santa Clara, CA) 5359A time synthesizer to control the laser timing in two modes. The synthesizer

[38] M. Rinke and H. Gusten, *Ber. Bunsenges. Phys. Chem.* **90**, 439 (1986).

produces two pulses of 20-nsec duration at an arbitrary interval after an arbitrary delay. Both the pulse interval and the delay can be varied over a range from 0.1 nsec to 160 msec. For long delays, all pulses are generated by the counters, and the phasing of the counters is controlled by the synthesizer, which is triggered on completion of the count cycle of the photolysis laser. An immediate output pulse from the synthesizer retriggers the counter for the photolysis laser, while the delayed pulse retriggers the counter for the probe laser. This mode permits delays to be set at precise 100-nsec intervals from about 100 nsec to 100 msec, and is used for all delays greater than 5 μsec. For short delays, the two clocks are triggered synchronously, but are used only to control the relays and flashlamps. The flashlamp trigger pulse is used to trigger the synthesizer, and the conditioned output pulses from the synthesizer are used to trigger both Q-switches. The root-mean-squared (RMS) variation in the time interval between the output pulses arising from all sources in either delay mode is less than 2 nsec.

Also incorporated in the timing control box are two counters. One determines the number of laser pulses fired in response to each trigger pulse received by the timing box, and the second determines the number of flashlamp pulses between each Q-switch firing. These two counters have been incorporated to permit multiple exposures to be collected on the vidicon prior to reading the signal from the target and to permit delays longer than the 100-msec delay between flashlamp firings. When data are collected using multiple exposures, the first counter determines the number of exposures and the second determines the interval between pulses to a resolution of 0.1 sec.

The actual timing of the laser output pulses is monitored by a pair of fast photodiodes (Cat. No. 5082-4220; Hewlett-Packard) that detect the output pulses from each laser. The diode outputs are adjusted by attenuating the input light intensity until the peak output is 20–30 mV into an impedance of 50 Ω. The two signals are then used to trigger an interval timer (Hewlett-Packard model 5335A) for which the trigger levels are adjusted to half the peak intensity. The second harmonic of the photolysis pulse has a duration at half-height of 10 nsec when averaged over a large number of shots (320). The third harmonic from the probe laser, which is used to pump the dye, produces a slightly shorter pulse (8 nsec). The duration of the dye "flash" is increased slightly relative to the pump source, producing an output pulse having a full width at half-maximum (fwhm) of 10 nsec, essentially identical to that of the photolysis pulse. Individual pulses have pulse shapes that exhibit two to four temporal lobes spaced by approximately 3 nsec and hence are distinctly different from the averaged pulse profile. The complete averaging observed in 320 shots suggests that these lobes occur with random phasing relative to the 50% point on the rising edge of the pulse on which both the timer and the oscilloscope are triggered. Shorter averages performed over 10 and 30 shots showed that RMS variations in the averaged photolysis pulse profile are less than 3% when 10 shots are averaged and 1% when 30 shots are averaged.

The spectrometer portion of the instrument is quite simple. The dye cell, together with its reflected image formed by a spherical mirror, acts as the light source. The source is imaged on the sample cell by a symmetric quartz condenser with high collection efficiency, and an identical condenser lens is used to image the sample cell on the spectrograph slit. The condenser lenses are 1.5 in. in diameter with a focal length of 35 mm (Melles Griot, Inc.). One portion of the beam passes through the photolyzed region of the cell, and a second portion passes through an unphotolyzed region, which is separated from the photolyzed region by a vertical distance of about 1 mm. The light that passes through the slit is dispersed by a 0.25-m spectrograph (M-20; J-Y Optical Systems, Metuchen, NJ) and imaged on the face of a silicon vidicon tube (1256E; EG&G Princeton Applied Research, Inc., Princeton, NJ). The spectrograph is most often used with a 600-groove/mm holographic grating and 100-μm slits. In this configuration the image of the slit covers 4 resolution elements of the vidicon, permitting measurement of 125 individually resolved intensities across a total bandwidth of 100 nm. Because the resolution of the instrument in this configuration is less by a factor of four than the inherent resolution of the vidicon, smoothing over four resolution elements is performed before spectra are processed or displayed. The information that results from a single experiment, then, is two dispersed tracks of intensities stored on the silicon faceplate of the vidicon tube. The incident intensities are read from the target and digitized by a detector controller (model 1216; EG&G Princeton Applied Research).

The spectrometer is controlled by a Hewlett-Packard model 9826 desktop computer (later renamed Series 300 model 226). The software for instrument control and data communication was written in-house using the Multi-FORTH system (Creative Solutions, Rockville, MD). The use of FORTH permitted the modular development of the various control functions, while allowing the use of assembler coding for time-critical routines. The computer is connected to the model 1216 detector controller by a 16-bit parallel interface, to the timing control box described above by a second 16-bit parallel interface, and to the model 5359A time synthesizer and the model 5335A countertimer by an IEEE-488 interface. Prior to each experiment, the time delay is set by programming the mode of operation of the control box, together with the delay and interpulse intervals of the time synthesizer. The computer then issues a set of commands to the detector controller to configure the reading sequence. These instructions specify the scan pattern, the number of scans used to prepare the target, the exposure interval, as well as the details of the reading process.

The reading sequence is then initiated. First, the target is prepared. Next, scanning of the electron beam is halted and the detector controller triggers the timing box to generate the set of laser pulses to expose the vidicon; finally, the target is scanned to remove the stored charge. During the reading scans, the channel-by-channel counts digitized by the model 1216 detector controller during each pass of the electron beam over the target are sent in real time to the computer, where they are accumulated to produce complete profiles of incident intensity vs channel number. During the exposure of the vidicon, the time intervals between the excitation and probe pulses are measured to a precision of about ± 1 nsec by the model 5335A counter and transmitted to the computer in real time. For applications in which actual pulse energies are required (e.g., partial photolysis studies), the output of a J3-05DW pyroelectric joulemeter optical detector (Molectron, Inc., Sunnyvale, CA) in the path of the transmitted excitation light is read following each incident pulse by a 12-bit analog-to-digital (A/D) converter (ADC 98640A; Hewlett-Packard) installed in the backplane of the computer.

The details of the reading sequence are critical to accurate spectrophotometry. To obtain dark count levels that are stable and independent of the time between scans, the signal tracks must be prepared by performing at least 10 scans. Scanning must then be halted while the charge produced by exposure to the measuring beam is stored on the silicon faceplate. Exposure of the vidicon face to a pulsed source while it is being scanned introduces output spikes at the channel being read and, because the firing of the lasers is not synchronized with the scan, these spikes occur at arbitrary points in the output spectrum. Finally, to ensure that the digitized output is linearly proportional to the intensity stored on the faceplate, the exposed target must be completely cleaned of charge during the reading process. Because the electron beam reads the charge deposited on the vidicon faceplate more efficiently when the impressed charge is high than when it is low,[39] incomplete scanning of the signal from the vidicon target produces nonlinearities that increase with the optical density of the sample.[40] Although an alternative reading procedure has been developed to deal with this problem,[39] we have found that cleaning the vidicon by

---

[39] G. W. Liesegang and P. D. Smith, *Appl. Opt.* **20**, 2604 (1981).
[40] Because high optical densities produce low light levels that are read less efficiently, the absorption maxima are artifactually high when scanning is incomplete.

multiple sequential scans provides satisfactory linearity for accurate spectroscopy. For a channel height of 100 pixels, 15 scans at a scan speed of 120 μsec/channel are required to remove 98% of the charge generated by the exposure of the target.

The charge accumulated on the vidicon target during the exposure and read phases of the measurement cycle results from both exposure of the vidicon target to the probe light pulses ("light counts") and thermally generated electrons ("dark counts"). The latter must be subtracted from the total accumulated charge to obtain an accurate measure of the incident light intensity. The dark counts are measured by blanking the Q-switch trigger outputs to both the excitation and probe lasers with the laser timing clock and performing the identical reading sequence. A reading sequence performed with only the excitation laser blanked produces a baseline spectrum containing intensities measured when neither the sample nor the reference track is photolyzed. Both dark counts and baselines are measured periodically throughout the course of an experiment, usually once for every eight spectra. In addition, absolute spectra of the two regions of the sample cell through which the sample and reference beams pass can be measured before and after a set of experiments by replacing the sample cuvette with a blank cuvette filled with solvent and calculating the logarithm of the intensity transmitted by the solvent to that transmitted with the sample in place. A criterion of spectrometer alignment is that these two spectra be identical to within 2% at all wavelengths.

With the vidicon at room temperature, approximately 8000 dark counts/channel are accumulated in the course of a typical reading cycle. Saturation effects limit the total counts to below 20,000. If the light counts are to be determined with an accuracy comparable to the reading noise (± 1 count), it is critical that the dark count levels be stable over the course of data acquisition. Because the rate of accumulation of dark counts is highly temperature dependent, we found it necessary to control the temperature of the vidicon by using a circulating water jacket and a foam blanket. Cooling the vidicon to approximately 5° reduces the dark counts to about 5000, making it possible to detect approximately 13,000–14,000 light counts/exposure. Although dark counts can be reduced further by lowering the temperature of the vidicon faceplate to below 0°, this both increases the contrast between maximally and minimally exposed regions of the vidicon and decreases the efficiency with which the target is cleaned by a single scan of the electron beam. Because both of these effects increase the nonlinearity of the reading process, we have not explored this approach.

A measured spectrum, in its raw form, consists of intensity-vs-channel profiles for the sample and reference tracks of the vidicon. A typical experiment consists of approximately 100 of these spectra measured at time intervals ranging from about 20 nsec prior to the photolysis pulse to about 80 msec after the pulse, together with time delays, periodic dark count and baseline intensity measurements, and a record of the intensities of the photolysis pulses. For times longer than 20 nsec, the delays are chosen on a logarithmically spaced grid. At the end of each experiment, a record of these measurements is transmitted by the computer to a local-area network of Unix workstations via a 19.2-kbps (kilobits per second) RS-232 connection to a networked communication server. A background process, running on a workstation on the network, monitors this port for incoming data and stores the incoming data set in compact form as a single Unix file. For subsequent analysis, the contents of this file, which serves as the basic archive of an experiment, are converted into a form that can be read directly by data analysis programs [e.g., MATLAB (Mathworks, Inc., Natick, MA)].

The sample cell, the cell holder, and sample thermostat are shown in Fig. 3. The fused silica cell, an EPR flat cell (WG-814; Wilmad, Inc.) is supported by an aluminum yoke that supports each of the tubular arms of the cell in a V-shaped groove, where each is clamped by a retaining fixture. The temperature of the sample is controlled by a recirculating bath (model RTE-110; Neslab, Inc., Nashua, NH) and a specially constructed copper sandwich that encases the sample cell. The copper plates are connected to a brass heat transfer block/thermocouple assembly, which is supported by a pair of rods connected to the yoke that holds the cell. Nitrogen, equilibrated at the temperature of the block, is passed over the exposed surface on both sides of the sample cuvette to minimize temperature gradients.

To prepare samples for kinetic experiments, CO-saturated samples of hemoglobin (HbCO) or of the hybrid $\alpha_2(Co)\beta_2(FeCO)$, stored as beads in liquid nitrogen, are thawed and diluted into 0.1 M potassium phosphate buffer (pH 7.0). The diluted solution is equilibrated exhaustively with water-saturated carbon monoxide (CO) in a vial covered with a perforatable septum (Pierce, Rockford, IL). A concentrated solution of sodium dithionite is then added anaerobically to a final concentration of 10 mM. The sample is then anaerobically transferred into an EPR flat cell, which has previously been flushed with CO. The cell is then sealed under nitrogen, using dental wax and glyptal varnish, providing a robust and anaerobically sealed container in which samples are stable for periods of up to 1 year. The data reported here were obtained from a sample having a heme concentration of 100 μM and a path length of 340 μm.

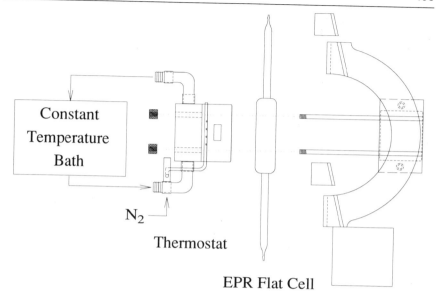

Thermostat

EPR Flat Cell

Cell Holder

FIG. 3. Schematic of sample cell, cell holder, and temperature-controlled cell shroud. The cell is clamped in the cell holder by two friction clamps that compress the upper and lower cylindrical arms of the cell into a V-shaped groove. The shroud, which consists of two copper plates mounted to a brass block, slides over the front and back surfaces of the cell and is supported by two threaded rods. Both photolysis and probe beams pass through the rectangular opening in the center of the shroud. Nitrogen, forced through narrow lengths of copper tubing affixed to the copper plates of the thermostat, minimizes temperature gradients on the exposed faces of the cell and prevents condensation at low temperatures.

Data Analysis

The data that we discuss here consist of sets of time-resolved spectra collected as described above under different experimental conditions (polarization and photolysis level). Typically, 10–20 photolysis levels, ranging from about 10% to over 90%, are measured with the polarization of the photolysis beam both parallel and perpendicular to that of the probe beam. The first steps in the analysis are to correct the measured spectra and baseline intensities for dark counts and to calculate difference spectra between the photolyzed and unphotolyzed tracks by computing, channel by channel, the log ratio of these pairs of intensities. The difference spectra are then corrected by subtracting the appropriate baseline in order to remove from the measured spectra any differences between the intensities from the two tracks that are not produced by the excitation pulse. An example of the output obtained from these procedures for an experiment carried out at about 30% photolysis using parallel polarization for the excitation and probe lasers is shown in Fig. 4a. The size of the plotted matrix of data has been reduced by carrying out a Gaussian average over a bandwidth of 4 pixels and interpolating the resulting data onto a grid of wavelengths separated by 1 nm. The plotted data, in particular the spectra measured at long times, show the characteristic baseline offsets that constitute a major source of noise in our experiments.

The next step in the analysis is to process all of the spectra obtained from each experiment by calculating its singular value decomposition.[37,41] To carry out the SVD, the data are assembled as a matrix A, which is arranged so that each column contains the spectrum measured at a given time delay; each column of A then corresponds to a specific time delay, and each row corresponds to a single wavelength. The SVD of this matrix rewrites A as the product of three matrices U, S, and V, having well-defined mathematical properties

$$A = USV^T \tag{3}$$

The $m \times n$ elements of A are rewritten in terms of $m \times n$ (elements of U) $+ n \times n$ (elements of V) $+ n$ (diagonal elements of S) $= (m + n + 1)n$ numbers. The effective use of SVD as a data reduction tool requires some method for selecting subsets of the columns of U and V and corresponding singular values that provide an essentially complete representation of the data set. This selection then specifies an "effective rank" of the matrix A. In practice, a reasonable selection procedure produces an effective rank that is much less than the actual number of columns of A, effecting a drastic reduction in the number of parameters required to describe the original data set. When the singular values are ordered by size, the useful information from each experiment is contained in the first three to six columns of the U and V matrices. The SVD of each set of data is therefore truncated to include the first 12 columns of U and V, and this compacted representation of the data is used as the starting point for further analysis. The truncated SVD output provides a compact and efficient format in which to store and examine the data.[37]

In the next step of the analysis, all of the experiments that comprise the data set are analyzed simultaneously using a more extensive singular value decomposition procedure.[24,31] Because as mentioned above, the

[41] G. Golub and C. VanLoan, "Matrix Computations." 2nd Ed. Johns Hopkins Univ. Press, Baltimore, MD, 1989.

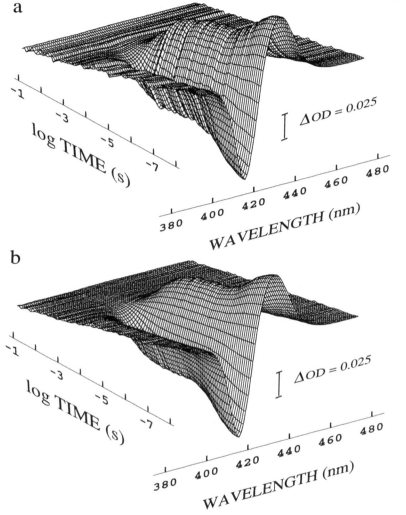

Fig. 4. A polarized absorption experiment. The set of 100 measured spectra has been smoothed by calculating the average value at each point as a weighted average using Gaussian weights having a $1/e$ width of 4 pixels. After smoothing, the data were interpolated onto a 1-nm wavelength grid for display. (a) Measured data prior to removal of baseline offsets. (b) Data after baseline offsets have been removed as described in text.

spectral resolution is determined by the 100-μm slit width of the monochromator, and the width of a single pixel is 25 μm, the data are smoothed over 4 pixels using a Gaussian filter with a full width at $1/e$ of 4 pixels. After smoothing, the polarized difference spectra at all times and levels of photolysis are linearly interpolated onto a common wavelength grid spaced at 1-nm intervals. This is identical to the representation of the experimental data used in Fig. 4a. The data also are mapped onto a standard time delay grid by linear interpolation after determining the initial times of the kinetic traces for each experiment; because the grid is chosen to coincide with the longer measured time delays, this interpolation affects only data within the first few hundred nanoseconds. To align the $t = 0$ points for each measured data set, the ligand-rebinding curves from about 20 nsec prior to the photolysis pulse to about 80 nsec after the photolysis pulse are averaged, and the deviations of the individual curves from the average kinetic trace are minimized by a linear correction of the times describing each experiment. This operation is carried out efficiently by first calculating the SVD of the data for the first 100 nsec of all the experiments, after interpolation onto a grid of times that extends from the shortest measured time to 80 nsec in 2-nsec intervals. The first column of the resulting V matrix is then rearranged so that the data from each experiment form a column of the reconstructed matrix. To ensure that the second component of the resulting matrix will reflect a temporal offset of the data, and to calibrate the temporal offset associated with the amplitude of this component, a copy of the same values, offset by three rows (i.e., a temporal offset of 6 nsec) is appended to this matrix. The first column of the U matrix generated by the SVD of this matrix is the averaged time dependence of the amplitudes of the first basis spectrum obtained from the original SVD. By design, the second component is a derivative of this curve and its amplitudes can be used to calculate the time by which the amplitudes from each experiment in the data set are offset from the average. The times corresponding to each experiment are then corrected for these offsets. In applying this procedure, the calculation is repeated until the starting times change less than 0.5 nsec, which generally requires only two or three iterations. Simulations in which the photolysis and probe pulses were modeled as Gaussians (fwhm = 10 nsec) have shown that the time at which the absorption signal is largest does not correspond to the time at which there is maximum temporal overlap between the photolysis and probe pulses.[35] The point at which the absorption signal is largest depends on both the short-time kinetics of the system and the incident laser intensity. For the conditions of our experiments, the simulations showed that the largest absorption signal occurs when the probe pulse is delayed by approximately 10 nsec from maximum overlap. On this basis we assign our initial point to 10 nsec. After alignment, the data prior to 10 nsec are discarded.

The data are then assembled into a global data matrix, $\mathbf{D}$, which consists of difference optical densities measured as a function of four variables: the wavelength of the probe beam, the time delay between the photolysis and probe beams, the degree of photolysis, and the relative polarization directions for the electric field vectors of the photolysis and probe beams. The singular value decomposition of $\mathbf{D}$ can be written as

$$\mathbf{D} = \mathbf{U}_D \mathbf{S}_D \mathbf{V}_D^T \tag{4}$$

where the columns of $\mathbf{U}_D$ are a set of linearly independent, orthonormal basis spectra, the columns of $\mathbf{V}_D$ describe the time-, polarization-, and photolysis-dependent amplitudes of these basis spectra, and the matrix $\mathbf{S}_D$ is a diagonal matrix of nonnegative singular values that describe the magnitudes of the contributions of each of the outer products of the ith column vectors, $U_{Di}V_{Di}^T$, to the data matrix, $\mathbf{D}$. The portion of the SVD of one such data set is shown in Fig. 5. Although the basis spectra, $U_{Di}S_{Di}$, are obtained from the entire data set, the time-dependent amplitudes are those measured at only 4 of the 17 different levels of photolysis, with parallel polarizations of the photolysis and probe pulses. Again, a truncated representation of the data, consisting of only the 12 components that have the largest singular values, is retained for further analysis; most of the higher-order components of the SVD contain no spectral information and their time-dependent amplitudes have random time dependence (i.e., they are only noise).

A major advantage of this approach is that all of the data are described by a relatively small set of basis spectra (the columns of $\mathbf{U}_D\mathbf{S}_D$). This representation permits both facile inspection of the data and significant advantages in its manipulation. One example is the removal of baseline offsets, which constitute the largest source of noise in the data collected from our instrument.[42] These offsets, which appear both in the difference spectra and in the baselines, originate from instabilities in the spatial distribution of the intensity produced by the dye cell used as a light source, which change the relative intensities of the source for the reference and sample tracks of the spectrometer. These instabilities arise from shot-to-shot variations in the intensity distribution of the output of the probe laser. There are, at present, two different procedures by which the contri-

[42] Another advantage of such a representation is that systematic changes in spectra collected over a range of experimental conditions, such as temperature, can sometimes be removed from a set of kinetic data. One such example is presented by A. Ansari, C. M. Jones, E. R. Henry, J. Hofrichter, and W. A. Eaton, *Biochemistry* (submitted for publication).

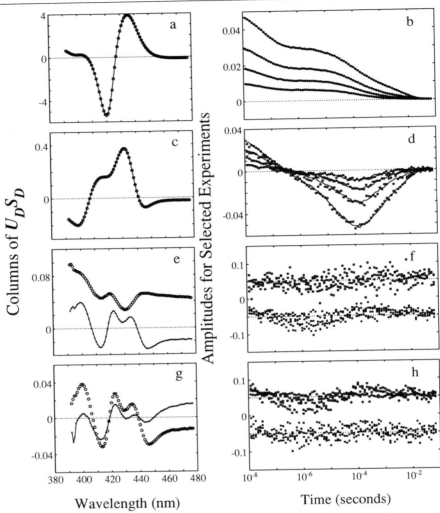

FIG. 5. Basis spectra obtained from analysis of a data set, using the procedure for removal of offsets described in text[31] and rotation of components 3–12 of the resulting SVD[37] and their amplitudes for four experiments. (○) SVD components prior to removal of baseline offsets; (□) SVD components obtained after removal of baseline offsets and rotation as described in text. (a) The first column of $U_D S_D$; (b) a portion of the first column of V_D associated with the parallel polarization at four different levels of photolysis. (c and d) The second columns of $U_D S_D$ and V_D; (e and f) the third columns of $U_D S_D$ and V_D; (g and h) the fourth columns of $U_D S_D$ and V_D. In (f) and (h) the values prior to offset removal (○) have been offset by 0.05 and those after offset removal and rotation (□) have been offset by −0.05 for clarity.

bution of these baseline offsets and other wavelength-correlated noise to the retained data may be reduced. The first is the rotation procedure used by Hofrichter and co-workers.[24] In this procedure, linear combinations of the columns of $\mathbf{V}_D$ that optimize the autocorrelations are calculated by diagonalizing the cross-correlation matrix of a selected group of components as described by Henry and Hofrichter.[37] An alternative is to discriminate against such random sources of experimental noise by postulating the form of the wavelength dependence of the noise component and explicitly removing contributions having this shape from a selected subset of the basis spectra.[31] We illustrate the latter procedure with the data presented here.

Note that the offsets appear mainly in the third basis spectrum (U_{D3}) produced by the SVD (Fig. 5), which appears to be composed of an offset spectrum (i.e., a constant optical density difference at every wavelength) mixed with contributions from real spectral changes. To remove the offset associated with this component, we assume that this basis spectrum can be described as a linear combination of an offset spectrum and the remaining 11 basis spectra. The values of the coefficients, c_j, can then be determined by a least-squares fit, that is,

$$s_{D3} U_{D3} \cong c_0 U_{D0} + \sum_{i \neq 3} c_i s_{Di} U_{Di} \tag{5}$$

where s_{Di} is the singular value of the ith component, U_{D0} is the normalized offset spectrum, and the sum runs over the remainder of the first 12 SVD components. After discarding the pure offset term, $c_0 U_{D0} V_{D3}^T$, the data matrix can be described by

$$\mathbf{D} \cong \sum_{i \neq 3} s_{Di} U_{Di} (V_{Di}^T + c_i V_{D3}^T) \tag{6}$$

In this representation of the data matrix the largest random fluctuations originally associated with the offset have been removed. Figure 4b illustrates the effectiveness of this procedure. Note, in particular, the dramatic decrease in the offsets in the data at long times. After applying this procedure and recalculating the SVD of the offset-corrected data set presented here, the third and fourth components of the SVD were similar in magnitude. To optimize the signal-to-noise ratio of the third component and to minimize the number of components retained for final analysis, the rotation procedure described by Henry and Hofrichter was applied to components 3–12 of this representation to obtain the offset-corrected results shown in Fig. 5.[37]

The next step in the analysis is to calculate the absorption anisotropy. To carry out this calculation, the data measured in the two polarizations must be corrected for small differences in the degree of photolysis. These differences arise both as the result of differential reflection losses at the sample cuvette surface (which is oriented at an angle relative to the propagation direction) and from slow changes in laser intensity arising from drifts in the beam position. The correction procedure is based on the experimental observation that, at time delays between 200 nsec and 1 μsec, the spectra measured in both polarizations are proportional to within a scale factor. At times less than 200 nsec the spectra measured in different polarizations differ because of photoselection effects, while at times longer than 1 μsec the measured spectra differ because the kinetics depend on the degree of photolysis. Accordingly, the optical densities measured for delays that fall between 200 nsec and 1 μsec are averaged for each experiment to produce an amplitude that is proportional to the extent of photolysis. The data measured in each polarization are then linearly interpolated to a photolysis level intermediate between each pair of experiments. The interpolated amplitudes of the components of V_{D1} for both polarizations at the four photolysis levels shown in Fig. 5 are plotted in Fig. 6a.

These polarized absorption data are then used to calculate the optical anisotropy, $f(x, \lambda, t)$, defined by Eq. (2). Although both geminate rebinding of CO and randomization of the molecular orientations by rotational diffusion decrease the linear dichroism, $\Delta OD_\parallel - \Delta OD_\perp$, the effects of the former process are removed by calculating the anisotropy. The average anisotropy, $\bar{r}(x, t)$, can then be calculated as the weighted average of the anisotropy at each wavelength,

$$\bar{r}(x, t) = \frac{\sum_\lambda r(x, \lambda, t) \, \Delta OD_{iso}(x, \lambda, t)^2}{\sum_\lambda \Delta OD_{iso}(x, \lambda, t)^2} \tag{7}$$

where $\Delta OD_{iso}(x, \lambda, t)$ is given by Eq. (1). Weighting by the square of the isotropic optical density difference is required because the noise in the anisotropy is large at wavelengths where the measured optical density in the isotropic spectrum is small. The anisotropy decay curves calculated from the polarized absorption data in Fig. 6a are shown in Fig. 6b. The curves at all levels of photolysis were fitted with a single exponential, yielding a rotational correlation time, τ_r of 35 ± 3 nsec. Results obtained from the anisotropy decay of hemoglobin and myoglobin are discussed in more detail by Hofrichter et al.[43]

In addition to the kinetics of its decay, the absorption anisotropy, $F(x, t)$, when examined as a function of the degree of photolysis (x), provides an independent and sensitive measure of the degree of photolysis, which

[43] J. Hofrichter, C. M. Jones, A. Ansari, E. R. Henry, O. Schaad, and W. A. Eaton, *Biophys. J.* (in preparation).

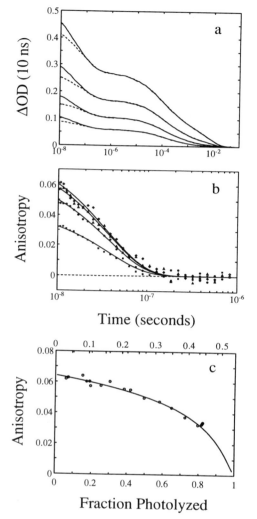

FIG. 6. Polarized absorption and anisotropy data obtained from partial photolysis data. (a) Amplitudes of $\mathbf{U_D S_D}$ for parallel (—) and perpendicular (---) polarizations of the photolysis and probe pulses measured at four levels of photolysis. The plotted vectors were obtained by interpolating the measured data onto a photolysis level intermediate between those obtained in the experiments carried out in the parallel and perpendicular polarizations. (b) Average anisotropy obtained from Eqs. (2) and (7). The points are the anisotropies calculated from the polarized absorption data, which produced the amplitudes plotted in (a). (—) Fits to an exponential decay, using the correlation time obtained from a simultaneous fit to all of the anisotropy decay data, $\tau_r = 35 \pm 3$ nsec. (c) Dependence of the anisotropy at 10 nsec on the fraction photolyzed. The anisotropies ($\bigcirc$), obtained from the exponential fits shown in (b), are plotted against the amplitude of the difference spectra at 10 nsec, obtained directly from the data. The fraction photolyzed, which is proportional to $\Delta\Delta OD$, is obtained by the fitted curve (—). The fitting procedure is described in text.

does not require any advance knowledge of the spectrum of the photoproduct. Theoretical expressions for the optical density changes measured with polarized light and for the absorption anisotropy as a function of time and laser intensity have been derived by Ansari and Szabo.[44] The theory shows that changes in the observed optical densities that result from rotational diffusion of the photoselected population can produce large deviations from the true ligand-binding curves. These effects have been analyzed quantitatively by Ansari *et al.* from data obtained on myoglobin in glycerol–water solvents.[35]

Figure 6c shows the average anisotropy at 10 nsec plotted as a function of the isotropic spectral amplitude of the photoproduct. These values can be analyzed using the theoretical results for a planar absorber by calculating the anisotropy at the end of a square pulse of width, t_p, as a function of the excitation intensity using the theory of Ansari and Szabo.[44] The anisotropy after the pulse is given by

$$r(t > t_p) = \frac{S}{10}\left[\frac{a_2(t_p)}{1 - a_0(t_p)}\right]\exp[-6D_r(t - t_p)] \tag{8}$$

where $a_0(t_p)$ and $a_2(t_p)$ depend on the excitation intensity and the rotational diffusion coefficient of the hemoglobin molecule, D_r.[45] The parameter $a_0(t_p)$ represents the fraction of hemes that have a ligand bound at the end of the pulse, and $a_2(t_p)$ is proportional to the linear dichroism in the sample. S is the generalized order parameter of the heme as defined by Lipari and Szabo[46] and is a measure of the amplitude of the internal motions of the heme that are much faster than rotational diffusion.

The application of Eq. (8) to experimental data has been described by Jones *et al.*[31] Because it is difficult to measure accurately the excitation intensity in each experiment, we take the following approach to fitting

---

[44] A. Ansari and A. Szabo, *Biophys. J.* **64**, 838 (1993).

[45] The parameters a_0 and a_2 are related by a system of coupled differential equations,

$$\frac{da_l(t)}{dt} = -l(l + 1)D_r a_l(t) - \frac{2l + 1}{2}\varepsilon I \int_0^\pi [1 - SP_2(\cos\theta)]\sum_{l'} a_{l'}P_l(\cos\theta)P_{l'}(\cos\theta)\sin\theta\, d\theta$$

where P_l are the Legendre polynomials, ε is the extinction coefficient of the liganded heme at the excitation wavelength, I is the excitation intensity, θ is the angle between the transition moment direction of the heme and the polarization direction of the photolysis laser, and a_l are the coefficients that describe the anisotropic distribution of the liganded molecules as a sum of Legendre polynomials.[44] Prior to excitation, the distribution is assumed to be isotropic with all hemes having a ligand bound; the initial values of the coefficients are therefore $a_0 = 1$, $a_l = 0$ for $l > 0$. The coefficients a_0 and a_2 are evaluated as a function of the excitation intensity using Eq. (1) after truncating the Legendre polynomial series to $l = 8$.

[46] G. Lipari and A. Szabo, *Biophys. J.* **30**, 489 (1980).

the data. First, $a_0(t_p)$ and $a_2(t_p)$ are obtained by solving equations given in footnote 44 for various values of the excitation intensity, using the value of D_r obtained from the measured rotational correlation time, $D_r = 1/6\tau_r$. The fraction of hemes photolyzed in each experiment, which is equal to $1 - a_0$, can be defined as the ratio of the amplitudes of the isotropic spectra, $\Delta\Delta OD_{iso}$, to the amplitude at infinite laser power with all the hemes photolyzed, $\Delta\Delta OD_{max}$:

$$1 - a_0 = \frac{\Delta\Delta OD_{iso}}{\Delta\Delta OD_{max}} \qquad (9)$$

The notation $\Delta\Delta OD$ denotes the wavelength-independent value of the amplitude of the difference spectrum, calculated as $\Delta OD_{436} - \Delta OD_{420}$. To fit the measured anisotropies, the parameters $\Delta\Delta OD_{max}$ and S are then varied simultaneously in order to minimize the residuals between the theoretically calculated values of the anisotropy at 10 nsec and the experimental values. In each iteration, the values of $a_0(t_p)$ are obtained from Eq. (9), and the corresponding values of $a_2(t_p)$ are obtained by interpolating on the tabulated values. The best fit to the data in Fig. 6c gives an order parameter S of 0.93 ± 0.03 and a spectral amplitude $\Delta\Delta OD_{max}$ of 0.54. Using these values, the degree of photolysis for each experiment (i.e., the fraction of hemes photolyzed in each of the calculated isotropic data sets) can be determined to a precision of better than $\pm 2\%$.

In the final step of the analysis, the degrees of photolysis obtained from the fit are used to calculate the populations of the photoproduct species that are present in the sample at the end of the photolysis pulse. These populations are then used to extract the amplitude vectors that describe the kinetics of ligand rebinding and conformational relaxation for these species. In the case of the $\alpha_2(Co)\beta_2(FeCO)$ hybrid, the example used in this chapter, there are only two photoproduct species: unliganded $\alpha_2(Co)\beta_2(Fe)$ and singly liganded $\alpha_2(Co)\beta(FeCO)\beta(Fe)$. To begin this analysis, we generate tables of the photoproduct populations that would be present in two limiting cases. In the first, the probability of photolysis of each heme is uncorrelated; this would be the case if rotational diffusion were rapid compared to the temporal duration of the excitation pulse and all hemes were photodissociated with the same (average) probability. In this case the distribution of photoproduct species is binomial and the fraction of dimers with i ligands bound, f_i, is given by the coefficient of λ^i in the expression

$$\sum_{i=0}^{k} \lambda^i f_i = (1 - y_p + \lambda y_p)^k, \qquad k = 2, 4 \qquad (10)$$

where y_p is the fractional saturation produced by the photolysis pulse, k denotes the number of sites that can be dissociated, and the subscript i denotes the number of ligands bound to the tetramer. The second limiting case is obtained when the temporal duration of the photolysis pulse is short compared to the rotational correlation time. In this case, the distribution of photoproduct species is nonrandom because absorption of the first photon generates a photoselected distribution. The probability that a second photon is absorbed by the remaining hemes is therefore either greater than or less than that for the absorption of the first photon, depending on the relative orientation of the heme planes. The distribution can be obtained from expressions having the same form as Eq. (10),

$$\sum_{i=0}^{n} \lambda_i f_i = \frac{1}{4\pi} \int_0^{2\pi} d\phi \int_0^{\pi} \sin \theta \, d\theta \prod_{n=1}^{k} (1 - P_n + \lambda P_n) \qquad (11)$$

where P_n, the probability that the nth heme–CO complex is not photodissociated [which is analogous to y_p in Eq. (10)], is given by

$$P_n = e^{-I[1 - (\cos \theta \cos \alpha_n + \sin \theta \sin \phi \cos \beta_n + \sin \theta \cos \phi \cos \gamma_n)^2]} \qquad (12)$$

where $\cos \alpha_n$, $\cos \beta_n$, and $\cos \gamma_n$ are the direction cosines of the normal to the nth heme plane relative to an arbitrary axis system. We have used the axes of the human oxyhemoglobin crystal to calculate the direction cosines.[24]

The two distributions are shown in Fig. 7a for the liganded $\beta\beta$ dimer in $\alpha_2(Co)\beta_2(FeCO)$ and in Fig. 7b for the oxyhemoglobin tetramer. The two distributions for the $\beta\beta$ dimer are similar because the angle between the planes of the two β hemes is 62.4°, not very different from the so-called magic angle of 54.7°. If the angle between the normals were equal to this value, the probability of absorbing a polarized photon by the second heme would be independent of whether a photon had been absorbed by the first and the photoselected distribution would coincide with the random distribution. The distributions that would result in cases in which limited rotation occurs during the photolysis pulse are difficult to calculate analytically, but they can be approximated by a weighted average of the two limiting distributions that reproduces the correct anisotropy. The fraction of the photoselected distribution is then given by the ratio of the limiting anisotropy at 10 nsec to 0.1, which is the limiting value for a circular absorber. We have used this approximation to calculate the populations of photoproducts that are generated by partial photolysis of the $\alpha_2(Co)\beta_2$ (FeCO) hybrid hemoglobin.

Given these populations, it is straightforward to obtain the time-dependent amplitudes associated with each of the photoproduct species. We

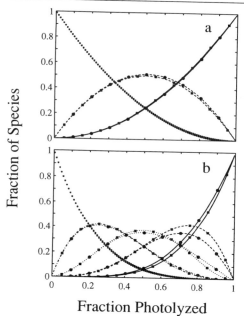

Fraction of Species

Fraction Photolyzed

FIG. 7. Populations of photoproduct species produced by random and photoselected photolysis. (a) Populations for hemoglobin in which dissociation takes place only on the two β chains corresponding to the liganded chains of $\alpha_2(Co)\beta_2(FeCO)$ hybrid hemoglobin ($\beta\beta$ dimer). (●) Undissociated (two-liganded) tetramers produced by polarized photolysis in the absence of molecular motion. (■) Undissociated (two-liganded) tetramers produced by random photolysis. (———) Single-liganded tetramers: the upper curve is for photoselected photolysis and the lower curve for random photolysis. (——) Unliganded tetramers: the two curves are plotted but are indistinguishable at the resolution of the plot. (b) Populations for the $\alpha_2\beta_2$ tetramer in which all four hemes are photodissociated with the same quantum yield. (●) Undissociated (four-liganded) tetramers produced by random photolysis. (■) Undissociated tetramers produced by photoselected photolysis. (—·—) Three-liganded tetramers: the upper curve is for random photolysis and the lower curve for photoselected photolysis. (···) Two-liganded tetramers: the upper curve is for random photolysis and the lower curve for photoselected photolysis. (———) Single-liganded tetramers: the upper curve is for random photolysis and the lower curve for photoselected photolysis. (——) Nonliganded tetramers: the upper curve is for random photolysis and the lower curve for photoselected photolysis. Note that, because the angles between heme normals are mostly smaller than 54.7°, photoselection depletes the partially photodissociated states in favor of the nonliganded photoproduct.

simply ask, for each of the basis spectra of $\mathbf{U}_D\mathbf{S}_D$, what set of amplitudes for the two species that contain deoxy hemes provides the best fit to the data. The answer is obtained from a simple, linear least-squares fit,

$$V_{Dx}(t) = \sum_i f_{xi} n_i V_{Di}(t) \tag{13}$$

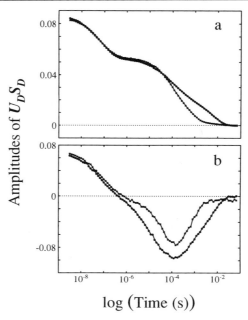

$$\log \left(\text{Time (s)} \right)$$

FIG. 8. Time-dependent amplitudes for the two photoproduct species of $\alpha_2(Co)\beta_2(FeCO)$ hybrid hemoglobin. (●) Unliganded photoproduct; (□) single-liganded photoproduct. (a) Amplitudes of U_DS_D column 1, V_{D0} and V_{D1}, obtained from a fit to the isotropic data at all levels of photolysis using Eq. (13). (b) Amplitudes of U_DS_D column 2, V_{D0} and V_{D1}.

where $V_{Dx}(t)$ is the set of time-dependent amplitudes associated with a specific column of U_DS_D, at photolysis level x, f_{xi} is the fraction of species i at photolysis level x, n_i is the number of deoxy hemes in species i, and the $V_{Di}(t)$ values are the amplitudes associated with species i. The results of this procedure are presented in Fig. 8. The two species exhibit essentially identical kinetics over the time interval from 10 nsec to about 1 μsec, showing that the geminate kinetics are independent of the number of liganded hemes. For hemoglobin, this result has been established by detailed analysis of the dependence of the kinetics in this time regime on photolysis.[31]

After about 1 μsec the kinetics diverge. The bimolecular ligand-binding kinetics for the unliganded molecule exhibit a fast phase and a slow phase, whereas ligand rebinding to the singly liganded molecule exhibits only the fast rebinding phase. The spectral change observed for the unliganded photoproduct is both faster and larger in amplitude than that observed for the singly liganded species. Geminate ligand rebinding to the singly liganded photoproduct during the first microsecond produces only the fully liganded parent molecule, $\alpha_2(Co)\beta_2(FeCO)$, therefore the kinetics observed throughout the experiment describe the time evolution of the singly liganded species. It is important to note, however, that geminate rebinding to the unliganded photoproduct produces a mixture of unliganded, singly liganded, and two-liganded species, therefore the subsequent kinetics describe the average response of a mixture of the unliganded and singly liganded species. The major features of the kinetics are well described by the allosteric model of Monod, Wyman, and Changeux[47]; the early spectral change in the unliganded species can be assigned as the quaternary conformational change from R_0 to T_0. The T_0 state has a lower affinity for CO, reflected by a slower binding rate, and so ligand rebinding to the T_0 subpopulation appears as the slow phase in the kinetics of ligand rebinding to this photoproduct species. The bimolecular phase observed for the singly liganded species is well described by two exponential relaxations, which can be assigned to the switch from R_1 to T_1, accompanied by rapid ligand rebinding to the R_1 molecules, and ligand rebinding to the resulting mixture of R_1 and T_1. Because the spectral change observed for the mixture of species produced by geminate rebinding to the unliganded photoproduct is significantly faster, the rate of the quaternary switch from R_0 to T_0 must be at least three times greater than that for the switch from R_1 to T_1.[24,48] The larger amplitude observed after the switch is consistent with the interpretation that the unliganded molecule converts completely to T_0, whereas the singly liganded molecule exists as a mixture of R_1 and T_1 at equilibrium.

These qualitative features of the photoproduct progress curves provide a sampling of the power and potential of this procedure for analysis of partial photolysis data. The curves produced by such analyses can be described much more quantitatively.[49]

Acknowledgments

We thank William A. Eaton, whose interest, challenges, and encouragement have been important ingredients in the development of the techniques described here. The use of dye fluorescence as a probe source originated from discussions with Michael Topp, and we gratefully acknowledge his suggestion. We also thank Attila Szabo, whose many contributions to describing and quantitatively understanding our measurements and the dynamics of molecules have been of immeasurable help in the interpretation of the data produced by this instrument. We also thank T. Yonetani for kindly providing the iron–cobalt hybrid hemoglobin used to prepare the sample for which data are presented.

[47] J. Monod, J. Wyman, and J.-P. Changeux, *J. Mol. Biol.* **12,** 88 (1965).
[48] W. A. Eaton, E. R. Henry, and J. Hofrichter, *Proc. Natl. Acad. Sci. U.S.A.* **88,** 4472 (1991).
[49] E. R. Henry *et al.* (in preparation).

[19] Femtosecond Measurements of Geminate Recombination in Heme Proteins

By JEAN-LOUIS MARTIN and MARTEN H. VOS

The dissociation of the ligand in heme proteins generates a nonequilibrium molecular system that evolves toward the deoxy conformation over periods of time extending from femtoseconds to milliseconds. Subsequent to photodissociation of a ligand, geminate rebinding with the heme protein is observed in the picosecond to nanosecond time scale. During this geminate recombination process, it is likely that protein relaxation plays a role in the ligand dynamics.

The quantum yield of photodissociation has been estimated to be less than 1 for many liganded heme proteins,[1] based on measurements in the microsecond time range. However, the observation of picosecond excited state dynamics and subsequent recombination of ligands in the picosecond and nanosecond time scales have led to the conclusion that the photodissociation yield is essentially 1 for any ligand.

In contrast, both the kinetics and the yield of geminate recombination are sensitive to the nature of the ligand. When the ligand is CO, the first phase of recombination is observed 100 nsec after dissociation, whereas a significant fraction of oxygen molecules and most NO molecules recombine on the picosecond time scale. These fast kinetics of rebinding in the case of NO are the consequence of a much smaller electronic barrier as compared with the other ligands. Both electronic and steric factors contribute to the inner energy barrier to ligand rebinding. The relatively small electronic contribution to the barrier when the ligand is NO makes the geminate recombination kinetics sensitive, at least in the heme vicinity, to factors altering the dynamics of relaxation of the protein. Under these conditions, the concerted use of genetic engineering, allowing single-amino acid residue replacement, and ultrafast spectroscopy techniques provide a powerful tool to identify the nature of the local protein motions that govern the binding reaction.

Early Events in Photodissociation Process

The extension of the flash-photolysis technique into the femtosecond time scale has made possible the characterization of the intermediates in

---

[1] G. Amiconi and B. Giardina, this series, Vol. 76, p. 533.

the photodissociation process. Two short-lived species are formed on the absorption of a visible photon by the heme.[2-4] We have assigned these species (Hb I* and Hb II*) to electronic excited states of the deliganded heme. The lifetime of Hb II* is 2.5 psec, as measured at its maximum absorption peak around 450 nm. This species is significantly populated only when the ligand is NO or O_2. Its short lifetime suggests that Hb II* is highly reactive toward these two different ligands. Thus the apparent low quantum yield of photodissociation for these two ligands, as estimated in the microsecond time scale, could be due in part to the fast recombination of the ligand to this highly reactive species.

The Hb I* state corresponds to an intermediate in the pathway to the ground-state unliganded (Hb†) hemeprotein. The (Hb†) species represents the first physiological state of hemoglobin in its relaxation pathway toward the deoxy conformation. In this state, the heme is five-coordinated and either in its liganded conformation or partly relaxed, but the protein is mainly in its initial liganded structural conformation. The deoxyheme spectrum corresponding to the (Hb†) species develops with a time constant of ~250 fsec after photolysis (cf. Fig. 5 for the corresponding kinetics in myoglobin). This spectrum remains essentially unchanged out to 100 nsec. The photodissociated ligand, CO excepted, recombines to the unrelaxed conformer (Hb†) on a picosecond time scale.

Measurement of Geminate Recombination Kinetics in Picosecond Time Scale

The disappearance of the unliganded (Hb†) species, $N(t)$, is monitored by probing the amplitude of the induced transient absorption in the 438-nm region. Because the maximum achievable yield of dissociation with femtosecond light pulses is only 15 to 30% in order to avoid nonlinear processes (such as stimulated Raman or continuum generation within the solvent), the population of the deoxy species is evaluated by measuring the transient difference spectra $\Delta A(t)$ at different time delays. The underlying assumption is that the observed spectral changes in the investigated spectral and time windows are due only to ligand rebinding. Two sources of distortion could a priori make this assumption invalid, particularly in the case of the extremely fast NO rebinding; a thermal relaxation of a hot heme and a contribution from the 2.5-psec lifetime-excited state (Hb II*). In the case of HbNO, probing at 438 nm (where both contributions are

[2] J. W. Petrich, C. Poyart, and J.-L. Martin, *Biochemistry* **27**, 4049 (1988).
[3] J. W. Petrich and J.-L. Martin, *Chem. Phys.* **131**, 31 (1989).
[4] J.-L. Martin and M. H. Vos, *Annu. Rev. Biophys. Biomol. Struct.* **21**, 199 (1992).

negligible) prevents any significant distortion of the observed spectral changes from vibrational or electronic relaxations.

Sample Preparation

The most extensive studies on geminate recombination in the picosecond time scale have been performed on HbNO.[5] Detailed descriptions of the preparation of human hemoglobin (Hb A) are found elsewhere[6] and are described here only briefly. A stock oxyHb A solution is diluted in 0.1 M potassium phosphate or [bis(2-hydroxyethyl)amino]tris(hydroxymethyl)methane (Bis–Tris) buffer at pH 7.0 at room temperature. Oxygen-liganded samples are obtained by equilibration under 1 atm of pure oxygen. For preparing other liganded states, the samples are first deoxygenated by equilibration under a stream of pure humidified argon and checked spectrophotometrically on the appearance of a pure deoxy spectrum. CO-liganded samples are obtained by subsequent equilibration under 1 or 0.1 atm of CO in a tonometer; NO-liganded samples are obtained by subsequent equilibration under 1 atm of pure NO. In the latter case, in which oxidation occurs at a faster rate, sodium dithionite (Merck, Darmstadt, Germany) is added.

Myoglobin can be prepared from metMb (HH III; Sigma, St. Louis, MO) by converting it into the ferrous form by addition of a 5 M excess of fresh sodium dithionite under strictly anaerobic conditions and subsequent removal of excess sodium dithionite on a Sephadex G-25 column. Samples of myoglobin with different ligands are then prepared in the same way as was described for hemoglobin.

The cuvettes (model 170-QS; Hellma, Jamaica, NY) are flushed with the gas phase used to equilibrate the hemoglobin solution. The sample is transferred in a gas-tight Hamilton syringe to the sealed cuvette.

Femtosecond Pump–Probe Spectroscopy

Any optoelectronic device is too slow to record directly spectroscopic changes occurring on the femtosecond or picosecond time scales; the required time resolution therefore is obtained by utilizing two separate femtosecond pulses. A first intense pulse, the "pump" pulse, is used to initiate the biochemical reaction, in our case ligand dissociation. The pump-induced change in the absorbance of the hemoglobin sample is

[5] J. W. Petrich, J.-C. Lambry, K. Kuczera, M. Karplus, C. Poyart, and J.-L. Martin, *Biochemistry* **30,** 3975 (1991).
[6] A. Riggs, this series, Vol. 76, p. 5.

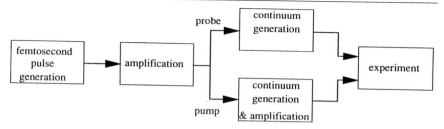

FIG. 1. General scheme of femtosecond pump–probe experiments. See text for details.

monitored by a delayed weak "probe" pulse at the appropriate wavelength. A kinetic trace is obtained by varying step by step the time delay between pump and probe pulses.

Figure 1 depicts the general scheme employed for the pump–probe experiments. High repetition rate (100-MHz) pulses with a duration of about 50 fsec and energies in the picojoule (pJ) range are generated in a ring cavity laser and further amplified at a low repetition rate (30 Hz in our case) to the hundreds of microjoules regime. The amplified pulse is split into two parts. Each of the two resulting beams will be focused in an 8-mm water cell to generate a spectral continuum (i.e., "white" light). To generate the pump beam, one of the continua is filtered spectrally and reamplified to the microjoule regime; the probe continuum is used directly as a spectroscopic source for detecting pump-induced transmission changes. The experimental arrangement employed in our laboratory is described in more detail below.

Femtosecond Laser Source

The most stable and convenient source of ultrashort pulses is provided by the colliding pulse mode-locked (CPM) technique (Fig. 2), first de-

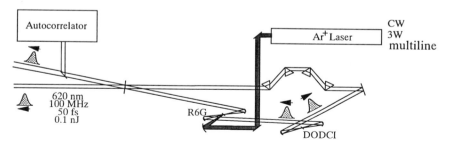

FIG. 2. Scheme of a colliding pulse mode-locked oscillator for generation of 620-nm femtosecond pulses. See text for details.

scribed by Fork et al.[7] In this bidirectional ring cavity configuration the gain medium generally employed is an ~100-μm thick jet of rhodamine 590 (R6G) (product 5901; Exciton, Dayton, OH), dissolved in ethylene glycol and adjusted to an optical density (OD) of ~0.3 at 514.5 nm. This gain medium is optically pumped by a continuous-wave (cw) argon ion laser (Innova 304; Coherent, Palo Alto, CA), operating at ~3-W output power in the multiline mode, that is, without any wavelength selection. Selecting the 514.5-nm line results in a lower rate of photodegradation of the rhodamine dye, but at the cost of a larger-frame argon laser. The CPM laser is passively mode locked by means of a saturable absorber, which is provided by a second ~30-μm ethylene glycol jet with 3,3'-diethyloxadicarbocyanide iodide (DODCI; product 6480, Exciton), adjusted to an OD of ~0.3 at 580 nm. The dyes are circulated by dye pumps (model 3760; Spectra Physics, Mountain View, CA) running at a flow adjusted to generate pressure values of 42 psi (gain jet) and 20 psi (saturable absorber jet). A shock accumulator (model 372B; Spectra Physics) is introduced in the jet circuits to minimize pressure fluctuations at the level of the nozzle. At each round trip, the counterpropagating pulses simultaneously reach (they "collide") the saturable absorber jet, which is at a distance of one-quarter of the total cavity length from the gain jet. This configuration ensures optimal recovery of the gain medium.

Pulse broadening due to group velocity dispersion is minimized by using spherical mirrors in a confocal configuration rather than lenses to focus the beams on the jets. To correct the unavoidable group velocity dispersion induced by the two jets, four intracavity glass prisms are introduced in an arrangement that yields a net negative dispersion[8] (prism angle 69°, distance within the pairs of prisms, ~26 cm). The pulse length is tuned by adjusting the optical path within the prisms by tranverse translation of one of the prisms. To minimize intracavity losses, the jets are placed at the Brewster angle and all mirrors are optimized for reflection at 620 nm; the output mirror transmits about 5%. The ~100-pJ output pulses are about 50-fsec full width at half-maximum (fwhm), a duration that is limited by the gain bandwidth of the system.

A low percentage of the energy of one of the output beams is used to track the pulse duration and its energy in real time on an oscilloscope by means of an interferometric autocorrelator equipped with a vibrating mirror.

As an alternative femtosecond source, femtosecond oscillators based on a solid state gain medium (titanium–sapphire) are now becoming com-

[7] R. L. Fork, B. I. Greene, and C. V. Shank, *Appl. Phys. Lett.* **38**, 671 (1981).
[8] R. L. Fork, O. E. Martinez, and J. P. Gordon, *Opt. Lett.* **9**, 150 (1984).

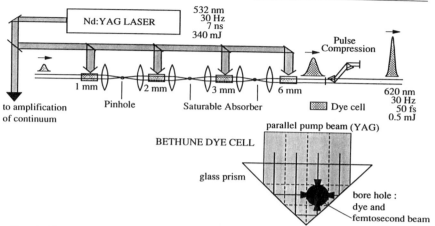

FIG. 3. Scheme of a four-stage dye amplifier and cross-section of the dye cell, which was first described by Bethune.[10] The diameter of each cell is indicated. See text for a more detailed description.

mercially available.[9] The application for the present purpose awaits the development of proper amplification schemes.

Amplification

The pulses are amplified in energy by over six orders of magnitude in a four-stage amplication system (Fig. 3), in which an energy per pulse of up to ~500 μJ is achieved at 30 Hz. The four-stage amplifier is driven by pulses of ~350 mJ at 532 nm with 7-nsec fwhm provided by a frequency-doubled Nd:YAG (neodymium:yttrium–aluminum–garnet) Laser (Quanta Ray, model GCR-4; Spectra Physics). The dye cells used at each stage are of the Bethune type,[10] that is, they consist of a hole drilled in a prism, allowing four-sided pumping. Such cells provide optimal homogeneous pumping and therewith conservation of beam quality. The cell windows are mounted at the Brewster angle with respect to the femtosecond pulse. To stay in a linear regime of amplification, the optical path length of the femtosecond pulses through the prisms is increased in the consecutive cells; they are, respectively, 2, 3, 3, and 6 cm. The pump pulses are adapted to the rectangular beam shape, required for transverse pumping, by cylindrical optics. The diameter of the beam (and the cells)

[9] J. Goodberlet, J. Wang, J. G. Fujimoto, and P. A. Schulz, Opt. Lett. **14,** 1125 (1989).
[10] D. S. Bethune, Appl. Opt. **20,** 1897 (1981).

is increased by telescope-like beam expanders located in between the consecutive stages, providing an optimized extraction of the stored energy.[11] The dyes are flowed continuously through the cells in a closed-loop circuit containing a fluid pump (micropump; Concorde, CA) and three dye reservoirs acting as microbubble traps. The dyes used are Kiton Red 620 (product 6200; Exciton) in water–2% (v/v) Ammonyx LO detergent (first stage) and sulforhodamine 640 (product 6500; Exciton) in methanol (other stages), and their concentrations are adjusted to an OD of 0.3/cell radius. One part of the pump energy from the nanosecond YAG laser is divided over the four Bethune cells according to their hole diameter (13, 24, 36, and 74 mJ for the cells 1, 2, 3, and 6 mm diameter, respectively), whereas the remaining part is used to further amplify part of the continuum (see below). The Q-switch of the pump laser is triggered by a nonamplified femtosecond pulse detected by a fast photodiode [model 4207 (Hewlett-Packard, Cupertino, CA); rise time, 0.8 nsec] in order to synchronize, within ±0.5 nsec at the level of each dye cell, the peak of the intense nanosecond YAG pulse with the incoming femtosecond pulse.

Spatial filtering (100-μm pinhole) or saturable absorber jets (malachite green, product 1264; Kodak, Rochester, NY) in between the different amplification stages prevent the *en-cascade* amplification of spontaneous emission (ASE) (see Fig. 3). The group velocity dispersion in the amplification chain is corrected by a two-prism configuration (60°/SF10 prisms; distance between prisms, 60 cm), based on the same principle as the compensation in the oscillator, yielding pulses of the same duration as at the input of the amplification chain (50 fsec).

Continuum Generation

After compression, the output beam from the amplifier is split into two portions. The resulting two beams are focused into an 8-mm water cell to generate two white light continua. The most stable continuum is obtained when the fundamental (620-nm) beam is adjusted in intensity so that it is just above the threshold for continuum generation ($\sim$3 μJ at 50 fsec, i.e., $\sim$10^{12} W/cm^2). A pump beam of the proper spectral properties and intensity is obtained by amplification of a spectrally filtered continuum.[12] Here, a spectral range of about 10 nm centered at 580 nm is selected by means of an interference filter, which is subsequently amplified as

[11] A.Migus, C. V. Shank, E. P. Ippen, and R. L. Fork, *IEEE J. Quantum Electron.* **18,** 101 (1982).

[12] J.-L. Martin, A. Migus, C. Poyart, Y. Lecarpentier, A. Antonetti, and A. Orszag, *in* "Picosecond Phenomena" (K. B. Eisenthal, R. M. Hochstrasser, W. Kaiser, and A. Lauberau, eds.), Vol. 3, p. 294. Springer-Verlag, Berlin, 1982.

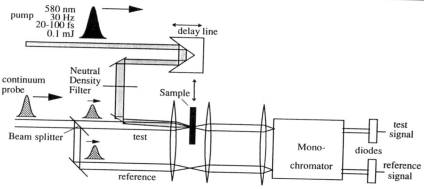

Fig. 4. Arrangement for relative timing of the pulses and detection of transmission changes. See text for details.

above to the 100-μJ regime in two stages with rhodamine 6G in methanol, applying spatial filtering in between, as described above. If very short pulses (less than 50 fsec) are required (which is generally not the case for studying geminate recombination), both pump and probe pulses can be compressed once more with prisms. In this case the spectral bandwidth of the pump pulse can also be broadened by mixing different dyes and omitting the interference filter. In both situations the pulse durations are determined by the Fourier transform of their spectral bandwidth. By tuning the spectral bandwidth of the amplified continuum and compensating for the group velocity dispersion, the duration of the actinic pulse can be adjusted from 20 to 500 fsec. This technique of continuum amplification is an easy way of obtaining femtosecond amplified pulses that are tunable both in duration and wavelength.

Dissociation of the ligand may alternatively be obtained with a near-ultraviolet (UV) pulse produced by frequency doubling the 620-nm pulses in a potassium dihydrogen phosphate (KDP) crystal. The doubling crystal should be thin to avoid excessive pulse broadening due to the dispersion between fundamental and doubled light in the crystal. A 1.5-mm crystal typically yields a pulse of 150-fsec fwhm. After filtering the remaining visible light the obtained pulse centered at ~310 nm may be used for pumping without further amplification.

Experiment

The variable in each experiment is the time after photoexcitation. This parameter is varied by changing the optical path of the pump beam relative to that of the probe beam by means of an optical delay line (Fig. 4). A pair of perpendicular mirrors is mounted on a computer-controlled stepping motor (model UT 100.100 PP; Microcontrol, Evry, France) with a step size of 1 μm (1 psec corresponds to 150-μm displacement of the mirrors). By means of neutral-density films the intensity of the pump beam is adapted so that about 20% of the monitored volume is excited on each flash.

The probe beam is split into a test beam and a reference beam. They are both focused to 60 μm on the sample (the reference may bypass the sample in an alternative configuration). Then, following strictly parallel pathways, they are focused on the entrance slit of a monochromator (model H 25; Jobin Yvon, Longjumeau, France) set at a spectral width of ~1 nm, and their intensities are detected by a pair of photodiodes. The two signals from the diodes (model UV 444B; EG&G, Vaudreuil, Quebec, Canada) are preamplified (model 142A; EG&G), amplified [model 571 (EG&G); rise time, 6 μsec], digitized, and routed to a computer where their ratio is computed on a shot-by-shot basis, allowing efficient rejection of the noise. Routinely, noise in the transmission of $<10^{-2}$ on a single shot basis is obtained. Induced absorbance changes can be detected with a precision better than 2×10^{-3} after 1 sec of signal averaging.

The pump beam is focused to the same spot size as the test beam; the overlap is maximized by monitoring the signal in real time. To obtain a fresh sample volume at each laser shot, the sample is placed in a small sealed cuvette (volume, 200 μl; optical path length, 1 mm), which is translated perpendicular to the beam direction with an average velocity of ~12 mm/sec. This procedure avoids sample degradation over the periods of data collection, that is, more than 1 or 2 days.

As previously discussed, to monitor ligand recombination only, a probe wavelength should be selected at which the transient kinetics are sensitive to the ligation state only on the time scale of measurement. For the picosecond time scale, transient spectra of heme proteins with CO ligands can be used to establish such a wavelength, because CO does not rebind to the heme in this time scale.[5] The optimal wavelength in the Soret region, where the largest signals are expected, is at the isosbestic point for the ~2.5-psec phase, which reflects partial ligand recombination from the (Hb II*) excited state.[2] For hemoglobin and myoglobin this is at 438 nm, and for protoheme it is at 425 nm.[5] Here one monitors essentially the absorption of the unliganded ground-state species (Hb†), which is stationary in the subnanosecond time scale.[13] The only remaining kinetic component that does not reflect geminate recombination is the ~300-fsec forma-

[13] S. M. Janes, G. A. Dalickas, W. A. Eaton, and R. M. Hochstrasser, *Biophys. J.* **54**, 545 (1988).

tion time of this species.[2] This time is short compared to the fastest geminate rebinding times (~ 10 psec) and therefore does not interfere with the data analysis.

Data Analysis

The raw data thus constitute a series of data points of transmitted light $I_{test}(t)$ and $I_{reference}(t)$. The quotients of these points are normalized to the baseline, yielding relative transmission $\Delta T(t)$ points and are subsequently transformed to absorption changes $\Delta A(t)$, according to the Lambert–Beer law:

$$\Delta A(t) = -\log[\Delta T(t)] \tag{1}$$

The obtained measured curves $\Delta A_m(t)$ are convolutions of the sample response function $\Delta A(t)$ with the instrument response function. The latter is itself the convolution of the pump $I_{pump}(t)$ and probe $I_{probe}(t)$ kinetic profiles and can be obtained from a cross-correlation experiment or, alternatively, from the rise time of an "instantaneously" rising signal (Fig. 5). This yields for the measured curve

$$\Delta A_m(\tau) = \int_{-\infty}^{\infty} \int_{-\infty}^{\infty} \Delta A(\tau - \tau') I_{pump}(t) I_{probe}(t + \tau') \, dt \, d\tau' \tag{2}$$

in which τ is the delay time between the pump and the probe pulse (see Fig. 5).

Experimentally obtained traces of geminate recombination may be compared with theoretical models by fitting the data. Commonly used fit functions are as follows.

1. Multiexponential decay[5] reflecting distinctly different decaying populations: For a biexponential decay this function has the form

$$\Delta A(t) = \Delta A(0)[Ae^{-t/\tau_1} + (1 - A)e^{-t/\tau_2}] \tag{3}$$

in which A is the fraction of the population decaying with time constant τ_1.
2. Power law, reflecting a distribution of decay rates[5,14]:

$$\Delta A(t) = \Delta A(0) \left(1 + \frac{t}{\tau}\right)^{-n} \tag{4}$$

Here τ and n characterize a decay rate distribution function.[5,14]

[14] R.H. Austin, K. W. Beeson, L. Eisenstein, H. Frauenfelder, and C. Gunsalus, *Biochemistry* **14**, 5355 (1975).

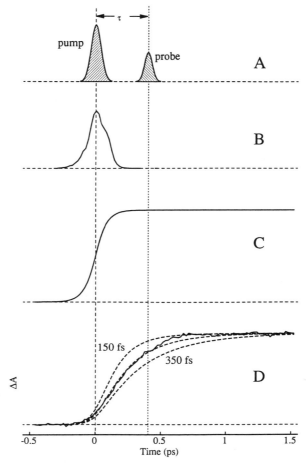

FIG. 5. (A) Principle of pump–probe spectroscopy. The probe is delayed by a variable time (τ) with respect to the pump. The autocorrelation function (B) is the convolution of the pump and the probe temporal profile. The function in B is measured by detecting the sum frequency at 250 nm of the pump wavelength (at 580 nm) and the probe wavelength (440 nm) generated in a nonlinear crystal placed at the place of the sample. Integration of the autocorrelation function (C) yields the response function for an instantaneous (relative to the pulse lengths) sample response. The experimental response curve of induced absorption of MbCO at 440 nm (D) clearly rises slower than (C), allowing the formation kinetics of the unliganded ground state species to be monitored. The dashed lines correspond to calculated responses for rise times of 150, 250 and 350 fsec.

3. Time-dependent evolution of the decay rate,[5] reflecting a progressive rise of the energy barrier for geminate recombination:

$$\Delta A(t) = \Delta A(0)e^{-\int_0^t k(t')dt'} \tag{5}$$

with

$$k(t) = k(\infty)\left[\frac{k(0)}{k(\infty)}\right]^{e^{-k_{br}t}} \tag{6}$$

Here the energy barrier evolves with a rate constant k_{br} according to

$$E(t) = (E_0 - E_{eq})e^{-k_{br}t} + E_{eq} \tag{7}$$

where E_0 and E_{eq} are the initial ($t = 0$) and final ($t = \infty$) energy barriers, respectively. The limiting rate constants are related to those limiting energy barriers for recombination as

$$k(0) = Ae^{-E_0/RT} \tag{8}$$
$$k(\infty) = Ae^{-E_{eq}/RT} \tag{9}$$

where A is an Arrhenius prefactor. Two of the three parameters, A, E_0, and E_{eq}, can be obtained from $k(0)$ and $k(\infty)$, provided one of them is known.

Examples of fits to the data using these models are shown in Figs. 6 and 7. Several algorithms are available for curve fitting; those in Figs. 6 and 7 are obtained with a Simplex procedure.[15] The quality of the fits is judged by the residuals.

Relevant Results

Geminate rebinding in the nanosecond time range, that is, with CO, follows single exponential kinetics at room temperature.[16] This is not the case when geminate rebinding occurs in the picosecond time scale, that is, with NO or O_2[5] (Figs. 6 and 7). As can be seen from Fig. 6, the data for HbNO can be fit equally well with any of the three models introduced above [see Eqs. (3)–(9)]. On the basis of molecular dynamics simulations it seems likely that both a distribution of barriers [power law, Eq. (4)] and the time dependence of the barriers [Eqs. (5)–(9)] give rise to the observed nonexponentiality of geminate recombination in the picosecond

[15] W. H. Press, B. P. Flannery, S. A. Teukolsky, and W. T. Vetterling, "Numerical Recipes." Cambridge Univ. Press, Cambridge, 1986.
[16] E. R. Henry, J. H. Sommer, J. Hofrichter, and W. A. Eaton, J. Mol. Biol. 166, 443 (1983).

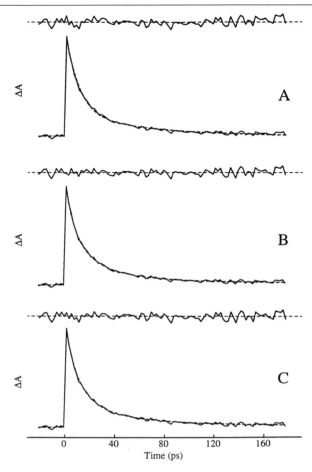

FIG. 6. Experimental curve of geminate recombination of HbNO, measured at 438 nm (data taken from Petrich *et al.*[5]). The maximal absorption change is 0.18. The signal is fit with (A) a double exponential [see Eq. (3); $\tau_1 = 10.6$ psec, $\tau_2 = 62$ psec, $A = 0.78$], (B) a power law [see Eq. (4); $\tau = 16.5$ psec, $n = 1.62$], and (C) a time-dependent rate constant [see Eqs. (5)–(9); $\tau(0) = 11.4$ psec, $\tau(\infty) = 152$ psec, $\tau_{br} = 72$ psec]. The quality of the fits can be judged from the residuals (upper traces, ×2.5).

time scale.[5] Nonexponential rebinding of CO and heme at low temperature, which takes place on the nanosecond time scale and slower, was interpreted also in terms of time-dependent distributions of barriers.[17] In this view

[17] P. J. Steinbach, A. Ansari, J. Berendzen, D. Braunstein, K. Chu, B. R. Cowen, D. Ehrenstein, H. Frauenfelder, J. B. Johnson, D. C. Lamb, S. Luck, J. R. Mourant, G. U. Nienhaus, P. Ormos, R. Philipp, A. Xie, and R. D. Young, *Biochemistry* **30**, 3988 (1991).

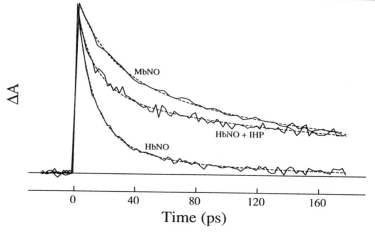

FIG. 7. Experimental curves of geminate recombination in MbNO, HbNO, and HbNO plus inositol hexaphosphate (IHP) measured at 438 nm (data taken from Petrich et al.[5]). The data are fit with a time-dependent barrier model [see Eqs. (5)–(9); the fit parameters are given in Table I. The fit of the data to MbNO includes data up to 350 psec.

the CO-rebinding kinetics at elevated temperature are monoexponential because protein relaxation and equilibration of substates is fast compared with the rebinding.

Figure 7 compares recombination kinetics of NO with myoglobin, hemoglobin, and hemoglobin with inositol hexaphosphate (IHP). Inositol hexaphosphate is a strong allosteric effector that imposes a T-state quaternary structure and weakens or cleaves the Fe–His bond in the α subunit

TABLE I
TIME CONSTANTS OBTAINED FROM FITS TO TIME-DEPENDENT BARRIER MODEL[a]

Heme protein	Time (psec)			Parameter set		
	$\tau(0)$	$\tau(\infty)$	τ_{br}	A	E_0	E_{eq}
MbNO	40.1	315	73	2.6	0.03	1.43
HbNO	11.4	152	72	8.8	0.00	1.76
HbNO + IHP	17.0	343	41	6.2	0.04	2.08

[a] See Eqs. (5)–(9), Fig. 7, and parameter sets[b] consistent with these fits. The times are the inverse of the corresponding rate constants introduced in Eq. (6).
[b] A (in 10^{10} sec^{-1}), E_0, and E_{eq} (in kilocalories per mole) are introduced in Eqs. (7)–(9).

of HbNO.[18] The decrease of the initial rate of geminate rebinding in the presence of IHP is indicative of a fast (<300 fsec) strengthening or rebinding of the Fe–His bond after photodissociation of NO.[5] Table I lists the parameters used to fit the data of Fig. 7 to the time-dependent barrier model given in Eqs. (5)–(9) and gives values for A, E_0, and E_{eq} consistent with these parameters.

Ultrafast spectroscopy of geminate rebinding proves to be successful in obtaining information on physiologically relevant protein motion near the heme pocket. Its application to studies on genetically engineered heme proteins will allow yet more detailed information to be obtained.

[18] M. F. Perutz, J. V. Kilmartin, K. Nagai, A. Szabo, and S. R. Simon, *Biochemistry* **15**, 378 (1976).

[20] Double Mixing Methods for Kinetic Studies of Ligand Binding in Partially Liganded Intermediates of Hemoglobin

By VIJAY S. SHARMA

In many respects, the double mixing stopped-flow method is similar to the single mixing method that has been described elsewhere.[1] Here we concentrate on the differences between the two methods, the type of reactions that can be studied by the double mixing method, experimental design, data treatment, and a brief discussion of the results obtained thus far. Unless otherwise mentioned, all examples discussed are for normal human hemoglobin (Hb).

Figure 1 is a schematic representation of the double mixing method. A typical double mixing experiment consists of three steps: (1) first mixing, in which equal volumes of solutions from syringes A and B are mixed in mixing chamber 1; (2) aging or incubation of the reaction mixture obtained in the first step for a certain time x, which can vary from 0 to 1 min depending on the type of reaction under investigation; (3) second mixing, in which after the desired aging time A_t, the solution from mixing chamber 1 is mixed in mixing chamber 2 (an optical cell) with an equal volume of solution from syringe C, and absorbance changes are followed as a function of time. These are all push-button operations controlled by an electronic console in a manner similar to single mixing experiments. By keeping syringe C empty and closed to mixing chamber 2 and $A_t = 0$, a double

[1] J. S. Olson, this series, Vol. 76, p. 631.

METHODS IN ENZYMOLOGY, VOL. 232

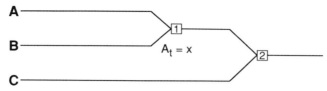

FIG. 1. General schematic of double mixing experiments. A_t, aging time; 1, first mixing; 2, second mixing.

mixing instrument is readily used as a single mixing stopped-flow spectro-photometer.

From the point of view of instrument characteristics, we must consider three requirements: (1) the volume of solutions used in each kinetic shot, (2) the range of aging time, and (3) the dead time of the instrument. Commercially available instruments can use from 2.5 to 0.065 ml of solution in each kinetic shot. In some cases, aging times of 1 min or more may be needed. During such long aging times in some instruments, there may be a contamination of the leading edge of the solution from the contents of the previous kinetic shot. Therefore, users should ask vendors to provide data showing constancy of absorbance in double mixing experiments over the range of $A_t = 0$ to $A_t = 1$ min or more. Reaction of myoglobin (Mb) with CO can be used to obtain this information. This is an important consideration and, therefore, is explained schematically in Fig. 2.

In the experiment described in Fig. 2, the absorbance at $A_t = 0$ should be the same as at $A_t = 1$ min; or the changes should be within the range acceptable to the investigator. Finally, the dead time of the instrument should be as short as possible (~ 1 msec or less) and precisely determined; otherwise it may be difficult to study faster reactions, particularly when analysis of zero-time amplitudes is required in order to draw meaningful conclusions.

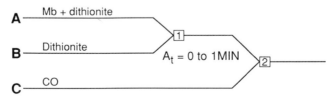

FIG. 2. Schematic of the reaction of CO with myoglobin (Mb) to check for the absence of leading-edge contamination in double mixing experiments; A_t, 0 to 1 min.

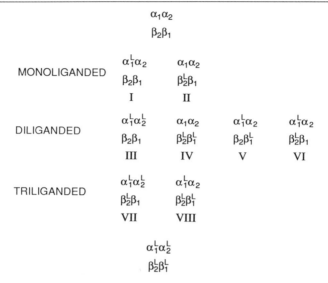

FIG. 3. Partially liganded isomers of hemoglobin. Numbering of α and β subunits follows the convention of Perutz.[9]

Types of Reactions that Can Be Studied with Double Mixing Method

Some of the reactions that can be studied using this method include the following.

1. Overall stepwise CO combination and dissociation rate constants; that is, $l'_1, \ldots, l'_4$ and $l_1, \ldots, l_4$, respectively[2-4]

2. CO combination and dissociation rate constants for various isomers of mono-, di-, and triliganded species of hemoglobin. Figure 3 shows various isomers (**I–VIII**) of partially liganded intermediates[5-9]

3. The dimer–tetramer equilibrium constants of oxyhemoglobin and the kinetics of dimer association; that is, k_1 and k_1/k_2 in Eq. (1)[10]:

[2] V. S. Sharma, M. R. Schmidt, and H. M. Ranney, *J. Biol. Chem.* **251**, 4267 (1976).

[3] V. S. Sharma, *J. Mol. Biol.* **166**, 677 (1983).

[4] V. S. Sharma and H. M. Ranney, *J. Mol. Biol.* **158**, 551 (1982).

[5] V. S. Sharma, *J. Biol. Chem.* **263**, 2292 (1988).

[6] V. S. Sharma, *J. Biol. Chem.* **264**, 10582 (1989).

[7] M. Berjis, D. Bandyopadhyay, and V. S. Sharma, *Biochemistry* **29**, 10106 (1990).

[8] V. S. Sharma, D. Bandyopadhyay, M. Berjis, J. Rifkind, and G. R. Boss, *J. Biol. Chem.* **266**, 24492 (1991).

[9] M. F. Perutz, *Nature (London)* **228**, 726 (1970).

[10] M. Berjis and V. S. Sharma, *Anal. Biochem.* **196**, 223 (1991).

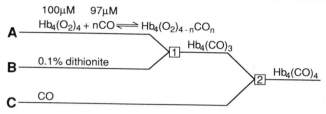

FIG. 4. Schematic representation of the double mixing experiment for the study of CO combination with $Hb_4(CO)_3$.

$$2\alpha^{O_2}\beta^{O_2} \underset{k_2}{\overset{k_1}{\rightleftharpoons}} \alpha_2^{O_2}\beta_2^{O_2} \tag{1}$$

4. The dimer–tetramer and $R \rightleftharpoons T$ equilibria in a system in which the $R \rightleftharpoons T$ equilibrium is slow [Eq. (2)]:

$$2\alpha\beta \underset{k_2}{\overset{k_1}{\rightleftharpoons}} \alpha_2\beta_2^R \underset{k_{T/R}}{\overset{k_{R/T}}{\rightleftharpoons}} \alpha_2\beta_2^T \qquad K_{R/T} = \frac{k_{R/T}}{k_{T/R}} \tag{2}$$

No unambiguously defined system corresponding to Eq. 2 has been studied so far. But in the study of valency hybrids, an equilibrium of this type was assumed and both k_1/k_2 and $K_{R/T}$ were determined.[5]

Determination of Overall Stepwise CO Combination and Dissociation Rate Constants

These experiments should be made with stripped hemoglobin solutions, so that as an approximation, differences in the reactivities of α and β subunits can be ignored, and a statistical distribution of CO can be assumed in the reaction

$$Hb_4(O_2)_4 + nCO \rightleftharpoons Hb_4(O_2)_{4-n}CO_n \qquad (n \leq 4) \tag{3}$$

Determination of l_4'

In a typical experiment,[4] one starts with deoxygenated buffer in a gas-tight syringe (0.1 M Bis–Tris, pH 7.0) and adds to it a calculated volume of oxyHb to yield about a 100 μM (heme basis) solution of hemoglobin. A calculated volume of saturated CO solution (929.64 μM at 1 atm and 25°) is added to this solution to yield 95 to 97 μM CO in the total volume. This syringe (A) is placed in port A of the double mixing stopped-flow spectrophotometer (Fig. 4). Another syringe (B), containing ~0.1% dithionite in deoxygenated buffer, is placed in port B. A CO solution in deoxygenated buffer, containing a few crystals of dithionite, is placed in the third syringe (C). The concentration of CO in this syringe (C) is approximately 6 to 10 times the concentration of free hemes generated at the first mixing. At 100 μM oxyHb and 97 μM CO, the concentration of free hemes generated at the first mixing is 1.5 μM.

Selection of Aging Time

At the first mixing, the reaction that takes place is

$$Hb_4(O_2)_{4-n}CO_n + \text{dithionite} \rightarrow Hb_4(CO)_n \qquad (4)$$

This reaction should be complete before the second mixing. The time taken for the completion of reaction (4) can be determined experimentally by operating the double mixing instrument in the single mixing mode with $A_t = 0$. Aging time for the double mixing experiment should be 5–10% more than the time taken for the completion of reaction (4).

Selection of Wavelength

Any wavelength can be selected for these experiments. However, because the isosbestic point for $T \rightarrow R$ conversion is 437.8 nm, this is a particularly suitable wavelength to use. At higher hemoglobin concentrations, it may be necessary to use an optical cell of shorter path length.

Data Treatment

Each kinetic experiment yields an analogal or digital reaction time course. Assuming random distribution of CO on four equivalent sites of $Hb_4(O_2)_4$, statistical calculations show that at 97% saturation of $Hb_4(O_2)_4$ with CO, the dominant species with unliganded hemes (after the first mixing) is $Hb_4(CO)_3$ (11%); all other species with unliganded hemes are ~0.5%.[11] The observed reaction time course is monophasic. A least-squares fit to a single exponential equation yields the value l'_4. Because the reaction is studied under pseudo-first-order conditions, the concentration of partially liganded species $Hb_4(CO)_3$ does not enter into the calculation.

Similar experiments can be made at 90% saturation of $Hb_4(O_2)_4$ with CO. The dominant species with unliganded hemes under these conditions are $Hb_4(CO)_3$, 29.2%; $Hb_4(CO)_2$, 4.9%; and all others less than 0.5%. In terms of the free hemes, which will determine the zero-time amplitudes in the reaction, 73% of the amplitude will be due to the reaction of

[11] R. Hille, J. S. Olson, and G. Palmer, *J. Biol. Chem.* **254**, 12110 (1979).

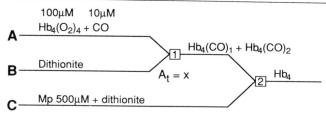

FIG. 5. Schematic representation of the double mixing experiment for the study of CO dissociation from $Hb_4(CO)_1$ and $Hb_4(CO)_2$.

$Hb_4(CO)_3$, 24% will be due to $Hb_4(CO)_2$, and 3% will be due to all other species. Ignoring the 3% due to all other species, a least-squares fit will yield the values of l_4' and l_3'. In these calculations l_4' can be kept fixed at the value obtained from experiments at 97% saturation of $Hb_4(O_2)_4$ with CO.[3]

A similar approach at 80 to 75% saturation of $Hb_4(O_2)_4$ with CO is potentially capable of yielding l_4', l_3', and l_2'. At 75% saturation of $Hb_4(O_2)_4$ with CO, $Hb_4(CO)_3$ and $Hb_4(CO)_2$ each would contribute 42% to the total zero-time amplitude, and $Hb_4(CO)_1$ about 14%; all other species would contribute ~1.5%.

Determination of the overall stepwise CO combination rate constants following the approach just described ignores differences in α- and β-subunit reactivities and assumes that the R ⇌ T equilibria for partially liganded intermediates are fast. Both of these assumptions are controversial.

In principle a similar approach using the double mixing method and the reaction of microperoxidase with carboxyHb species can yield overall stepwise CO dissociation rate constants of partially liganded species. Because the differences between various CO dissociation rate constants and absorbance changes in the reaction of carboxyHb species with microperoxidase are small, only the values of the first and last (l_1 and l_4) rate constants are determined with confidence.

Determination of l_1 and l_2

$Hb_4(O_2)_4$ (100 μM) is saturated at 10% with CO and reacted with 0.1% dithionite in the first mixing and with 5- to 10-fold excess of reduced microperoxidase (Mp) in the second mixing (see Fig. 5).[2] All solutions are prepared in deoxygenated buffers in gas-tight syringes, and anaerobic conditions are maintained by washing the stopped-flow system several times with deoxygenated buffer. The ideal wavelength for these measurements is 590 nm.

The aging time (x) should equal six times the half-time of the O_2 dissociation reaction, ~0.4 sec. At 10% saturation of oxyHb with CO after the first mixing, the dominant liganded species are $Hb_4(CO)_1$ at 73% and $Hb_4(CO)_2$ at 24%. The reaction time course is biphasic, and the values of l_1 and l_2 are obtained by least-squares analysis, using a biexponential equation.

Determination of l_4

This value[2] is obtained by reacting carboxyHb with 10-fold excess of microperoxidase or by replacing CO with NO. The double mixing method need not be used in this case. Precise measurement of initial rates of CO dissociation at 90% saturation of $Hb_4(O_2)_4$ with CO can yield the value of l_4 as well as of l_3. Because of the difficulties mentioned above, there is considerable uncertainty about the value of l_3 obtained by this method.

Preparation of Microperoxidase Solutions

A weighed quantity of solid microperoxidase (MP-11; Sigma, St. Louis, MO) is dissolved in a few drops of dilute sodium hydroxide, which is then diluted to the desired volume with an appropriate buffer. This solution is taken into a gas-tight syringe and deoxygenated by bubbling nitrogen at a slow rate for 10 min. During this procedure, enough void volume should be left over the solution in the syringe to avoid the loss of Mp solution due to slight frothing. Within 5–10 min of stopping N_2 bubbling, the Mp solution becomes free of froth and air bubbles. A few crystals of sodium dithionite are added to the solution to reduce ferric Mp to ferrous Mp. Excess dithionite should be avoided, as it can reduce thioether linkages, and the heme moiety will separate from the peptide, causing turbidity in the solution. Microperoxidase samples should be used immediately after the addition of dithionite.

Ligand Binding with Isomers of Partially Liganded Intermediates of Hemoglobin

The starting species for these studies are carboxyHb, ferric Hb, and valency hybrids $\alpha_2^{CO}\beta_2^+$ and $\alpha_2^+\beta_2^{CO}$. Ferric Hb and carboxyHb are prepared in the usual manner. Stringent precautions should be taken to avoid the presence of ferric hemes in carboxyHb solutions. The remaining two species ($\alpha_2^{CO}\beta_2^+$ and $\alpha_2^+\beta_2^{CO}$) are prepared by high-performance liquid chromatography (HPLC) using a Synchropak column CM 300 (250 × 22 mm i.d.) (Synchrom, Inc., Lafayette, IN). The gradient is obtained from 0.03 M phosphate buffer, pH 6.8 (buffer A) and 0.015 M phosphate buffer, pH 7.5 (buffer B). Details of the gradient and other details are described by Berjis *et al.*[7] Because the elution pattern varies from column to column, the gradient and flow rates described by Berjis *et al.*[7] should form the starting point and changes can then be made as required. Using this column, it was possible to load 1.5–2 ml of hemolysate (~10 g/100 ml), 60% oxidized with ferricyanide, and obtain about 1–1.5 ml of each hybrid at concentrations of approximately 1 mM. Chromatographic separation is complete in about 2 hr. The time-consuming part is concentrating the samples obtained in the chromatographic separation from about 30 ± 10 μM to 1 mM. This is the stage at which the chances of creating sample heterogeneity are considerable. The following precautions may be found to be helpful: (1) If the aim is to prepare carboxy derivatives of the hybrids, concentration may be carried out in a CO atmosphere in the concentration cell. Concentrated samples are stored in a CO atmosphere; (2) if cyano-metHb derivatives are desired, the required amounts of KCN should be added to the solutions, and a pure O_2 atmosphere should be maintained in the concentration cell; and (3) at the end of the concentration step, percentages of ferric hemes or carboxyhemes are calculated by the method of Bannerjee and Cassoly.[12] Samples in which ferric hemes or carboxyhemes are more than 50 ± 2% should be discarded. Samples once prepared should not be used after more than 2–3 days.

Kinetic Experiments

Isomers *III* and *IV*: $\alpha_2^{CO}\beta_2^+$ and $\alpha_2^+\beta_2^{CO}$

A small volume of a concentrated sample of $\alpha_2^{CO}\beta_2^+$ or $\alpha_2^+\beta_2^{CO}$ is flushed with CO for 5 min and then with N_2 for 15 min.[5,7] A calculated volume of this sample is added to deoxygenated buffer in a gas-tight syringe. This syringe is placed in port A of the double mixing stopped-flow spectrophotometer. Syringe B contains 0.1% dithionite solution in deoxygenated buffer. The CO solution, at approximately three to five times the concentration of ferric hemes in syringe A, is placed in syringe C. This would yield a CO concentration at 6 to 10 times the concentration of free hemes after the second mixing (see Fig. 6).

The aging time after the first mixing shoud be equal to the time necessary for the reduction of ferric hemes to ferrohemes and is determined separately in a control experiment using the double mixing instrument in the single mixing mode. The reaction time course for CO dissociation is obtained in a similar manner, except that syringe C contains a 10-fold excess of microperoxidase over the liganded hemes in syringe A. Depend-

[12] R. Bannerjee and R. Cassoly, *J. Mol. Biol.* **42,** 351 (1969).

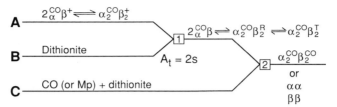

FIG. 6. Schematic representation of the double mixing experiment for the study of CO combination with or dissociation from isomer **III** or **IV**. See also Fig. 10.

ing on the sensitivity and stability of the instrument, it may be necessary to increase the protein and Mp concentrations by a factor of 10.

Isomer V: $\dfrac{\alpha_1^{CO}\alpha_2}{\beta_2\beta_1^{CO}}$

For CO combination studies,[6] an excess of carboxyHb is hybridized with ferric Hb. This is achieved by equilibrating solutions of carboxyHb with ferric Hb for 5–10 min at room temperature. As usual, all solutions are prepared in deoxygenated buffers in gas-tight syringes.

For studying the kinetics of CO dissociation, one starts with excess ferric Hb and small amounts (5–10%) of carboxyHb in syringe A. Further details of these two experiments should be clear from Fig. 7.

Isomer VI: $\dfrac{\alpha_1^{CO}\alpha_2}{\beta_2^{CO}\beta_1}$

In the case of isomer **V**, hybridization between carboxyHb and ferric Hb leads to only one major reactive species (i.e., a species that can combine with CO or from which CO can dissociate). For isomer **VI** this

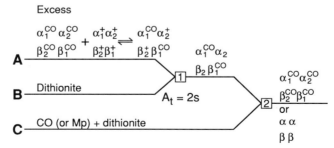

FIG. 7. Schematic representation of the double mixing experiment for the study of CO association with or dissociation from isomer **V**.

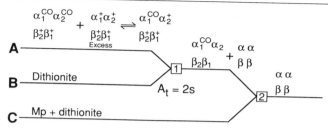

FIG. 8. Schematic representation of the double mixing experiment for studying CO dissociation from isomer **I** or **II**.

has not been possible so far. As a result, no studies of its CO combination or dissociation kinetics have been made.

*Isomers **I** and **II***: $\dfrac{\alpha_1^{CO}\alpha_2}{\beta_2\beta_1}$ and $\dfrac{\alpha_1\alpha_2}{\beta_2^{CO}\beta_1}$

Isomers **I** and **II**[8] are prepared by hybridization of isomers **III** or **IV** and a large excess of ferric Hb. There is no difficulty in studying the kinetics of CO dissociation (see Fig. 8). However, because of the presence of the large excesses of deoxyHb after the first mixing, it has not been possible to study the kinetics of CO combination with either of them.

*Isomers **VII** and **VIII***: $\dfrac{\alpha_1^{CO}\alpha_2^{CO}}{\beta_2^{CO}\beta_1}$ and $\dfrac{\alpha_1\alpha_2^{CO}}{\beta_2^{CO}\beta_1^{CO}}$

Isomers **VII** and **VIII** are prepared by hybridization of isomers **III** or **IV** and a large excess of carboxyHb. There is no difficulty in studying the kinetics of CO combination, but because of the presence of a large excess of carboxyHb it is not possible to study the CO dissociation reaction from these two isomers.

Data Treatment

Details of data treatment will vary from system to system depending on experimental conditions and detailed kinetic and mechanistic considerations. A few considerations discussed below may be helpful.

1. Initial concentrations in carboxyferric species can be calculated from the knowledge of the rate constants for the association of dimers into liganded or unliganded tetramer ($0.5-1 \times 10^6 \ M^{-1} \ \text{sec}^{-1}$) and the rate constant for the dissociation of liganded tetramer into dimers. We have provided rationale elsewhere for assuming that the association rate constants for liganded and unliganded dimers are the same.[10]

For the association of dimers from two different tetrameric species

$$e.g., \alpha^L\beta^L + \alpha^+\beta^+ \rightarrow \frac{\alpha^L\alpha^+}{\beta^+\beta^L},$$

a statistical factor of two should be applied to the association rate constant.

2. Under pseudo-first-order conditions, CO combination reactions are fast, and therefore CO combination rate constants are determined by making a least-squares fit to a single or biexponential equation. For these calculations, concentrations of individual species are not required.

3. If the reaction time course is biphasic with two rate constants close to the values expected for the R- and T-state species, it becomes necessary to reach some conclusion regarding the nature of the fast reacting species. This is generally done by varying the protein concentration and observing its effect on the zero-time amplitudes of the two phases. For CO combination studies in dilute solutions, the amplitude of the fast phase should increase with dilution of the protein solution if the fast phase is due to dimers. Unfortunately, in a complicated reaction scheme, it may require varying the protein and CO concentrations over a wide range to obtain an observable effect. This may result in reaction rates too fast to be measured by the stopped-flow method, or the absorbance changes may become too small for accurate measurements. The sudy of the CO dissociation reaction may provide some information in this regard. At present there is no easy method to resolve the fast phase of the CO combination reaction of the intermediates into contributions from dimers and tetramers. In the next section a method is discussed that may offer some possibilities in this regard.

4. Reaction schemes relevant to partially liganded intermediates become simpler if it is assumed that unliganded dimers associate into the T-state tetramer only via the formation of unliganded R-state tetramers. Arguments in favor of this assumption are discussed elsewhere.[7]

5. Dimer–tetramer association and dissociation rate constants for cyanometHb derivatives have been reported by Smith and Ackers.[13]

Study of Equilibria and Kinetics of Dimer–Tetramer Association[10]

The double mixing method is capable of studying the dimer–tetramer equilibrium of the liganded hemoglobin and the kinetics of the reaction $2\alpha\beta \rightarrow \alpha_2\beta_2$. This method has been used to study the kinetics of Hb A, Hb Rothschild, and carp Hb. The method is fast and requires only small amounts of the protein. So far this method has been used to study only

[13] F. R. Smith and G. K. Ackers, *Proc. Natl. Acad. Sci. U.S.A.* **82**, 5347 (1985).

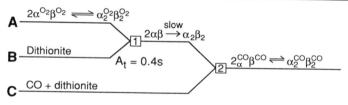

FIG. 9. Schematic of the double mixing experiment for studying the dimer–tetramer equilibria of oxyHb A.

the liganded hemoglobins, but in principle it is capable of yielding the association and dissociation rates of deoxyhemoglobins as well, provided the unliganded tetramer is significantly dissociated into dimers.

In a typical double mixing experiment for studying the dynamics of dimer–tetramer equilibria, a 3–4 μM oxyHb solution in deoxygenated buffer (syringe A) is mixed with an equal volume of 0.1% solution of dithionite (syringe B) in the first mixing. The reaction mixture is aged to remove oxygen completely from oxyHb. The products of the first mixing are then mixed with an equal volume of CO solution from syringe C (see Fig. 9).

The concentration of CO after the second mixing is 6 to 10 times the concentration of oxyHb (heme basis). The observed reaction time course for Hb A is biphasic (at $A_t = 0.4$ sec), because of the reactions of CO with deoxy dimers and tetramers.

These experiments are made at several aging times. The reaction time course at each aging time yields one data point.

Data Treatment

The zero-time amplitudes of the two phases are obtained by least-squares analysis of the reaction time courses, using a biexponential equation. The quaternary structural changes in the reaction oxyHb → deoxyHb are fast and are complete within the mixing time (3 msec) of solutions. The amplitude of the fast phase (A_f) decreases with increase in aging time, because of the association of dimers into tetramers.

From the data at various aging times, a curve of the fraction of the slow phase $[A_s/(A_s + A_f)]$ versus A_t is obtained. This curve, extrapolated to $A_t = 0$, yields initial concentrations of dimers and tetramers in the liganded Hb solution and, hence, the equilibrium constant $K_{1/2}^L$ (which is k_1/k_2) for the association of dimers into tetramers [Eq. (2) or (5)].

$$2\alpha\beta \underset{k_2}{\overset{k_1}{\rightleftharpoons}} \alpha_2\beta_2 \qquad (5)$$

$$-\frac{1}{2}\frac{[\alpha\beta]}{dt} = k_1[\alpha\beta]^2 - k_2[\alpha_2\beta_2]$$ (6)

In reaction (5), the product is deoxyHb, which in the case of Hb A shows little dissociation into dimers.

Rate equations for numerical integration are written on the basis of Eq. (6), and the variable constants k_1 or k_2 are estimated by minimizing the sum of squares of residuals of the observed and calculated values of $A_s/(A_s + A_f)$ at various aging times.

Limitations of Method

The double mixing method yields accurate values of the dimer–tetramer equilibrium constant of oxyhemoglobin and the rate constant for the association of dimers into tetramers. Accurate determination of the dissociation rate constant of unliganded tetramer with significant dissociation into dimers would require more data points at longer aging times so that the reaction can be followed to 90–95% completion. Although in the time domain of mixing experiments it can be assumed safely that the fast phase in CO combination reactions is due to the reaction of dimers, as a precaution it is advisable to confirm the dependence of relative amplitudes of the fast and slow phases on the protein concentration.

Study of Dimer ⇌ Tetramer and R ⇌ T Equilibria in System in Which R ⇌ T Equilibrium Is Slow

In the study of the kinetics of ligand association and dissociation of isomers **III** and **IV**,[5] the reaction time course is explained on the basis of the following equilibria:

$$2\alpha^{CO}\beta \xrightleftharpoons{K_{1/2}^L} \alpha_2^{CO}\beta_2^R \xrightleftharpoons{K_{R/T}} \alpha_2^{CO}\beta_2^T$$ (7)

There is some uncertainty regarding the nature of the species $\alpha_2^L\beta_2^T$ (or $\alpha_2\beta_2^{L,T}$). But this should have no effect on the method described below for resolving the fast phase into its components, due to dimers and the R-state tetramers. The schematic of the experiments carried out to achieve this aim is shown in Fig. 10.

The observed reaction time course of CO combination is biphasic and, in the absence of inositol hexaphosphate (IHP) in the CO solution in syringe C, the fast phase represents dimers as well as tetramers (see Fig. 6).

The design of the experiment described in Fig. 10 is based on the following three observations: (1) In the presence of IHP, both isomers **III** and **IV** are converted 100% to the slowly reacting species in the CO

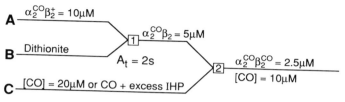

FIG. 10. Schematic representation of the double mixing experiment for the study of the kinetics of reaction (7).

combination reaction; IHP accelerates the rates of R → T conversion by a factor of more than 10; and (3) a large excess of IHP (~0.1 M at ~26 μM Hb concentration) in CO solutions does not greatly affect the rates of dimer association.[14] Therefore, in the experiment described in Fig. 10, we add excess IHP to the CO solution in syringe C. The binding of IHP to the protein and the R → T conversion should be complete within the mixing time. Therefore, the amplitude of the fast phase in this experiment corresponds only to dimers. Using this approach, the values of $K_{1/2}^{L}$ and $K_{R/T}$ for isomers **III** and **IV** can be estimated.

Results Obtained by Double Mixing Method on Partially Liganded Intermediates

The study of partially liganded intermediates, involving the separation of the intermediates, has been pursued independently in two laboratories. The contribution of the double mixing experiments has been via the study of reaction kinetics of individual isomers of partially liganded species in solutions at room temperature. Perrella et al.[15] have separated partially liganded isomers at −20° and have obtained overall stepwise rate constants using l_4' as the reference rate constant. It is worthwhile to compare the results obtained independently by the two research groups:

1. The data of Perella et al.[15] indicate that there are no large differences in the reactivities of α and β subunits at any stage of the reaction. This is in agreement with the results of double mixing studies, as well as with the results of Mathews et al.[16]

2. Both approaches indicate the presence of significant cooperativity

[14] B. L. Wiedermann and J. S. Olson, J. Biol. Chem. 250, 5273 (1975).
[15] M. Perrella, N. Davids, and L. Rossi-Bernardi, J. Biol. Chem. 267, 8744 (1992).
[16] A. J. Mathews, J. S. Olson, J.-P. Renaud, J. Tame, and K. Nagai, J. Biol. Chem. 266, 21631 (1991).

in the second step of ligation, but they differ about the magnitude of the effect.

3. Perrella et al.[15] have concluded that the major structural change takes place after the binding of the second ligand, that is, the diliganded species are structurally closer to the R state. By the double mixing method, three isomers of diliganded species have so far been studied (isomers **III**, **IV**, and **V**). About 85–70% of the reaction time courses of the three isomers were fast, with $l'_3 \approx 2$–$6 \times 10^6 M^{-1} \sec^{-1}$. The remaining isomer of diliganded species [isomer **VI**, $\alpha_1^{CO}\beta_1(\alpha_2\beta_2^{CO})$] has not been studied by the double mixing method. Perrella et al.[15] also have not studied this isomer, as it could not be resolved from isomer **V**. However, their data suggest that it might behave like the other three isomers of diliganded species (i.e., more like R-state species).

4. The four stepwise CO combination rate constants obtained in the two studies are $l'_1 = 0.14$ (0.1), $l'_2 = 0.32$ (0.7), $l'_3 = 1.6$ (1.5–6 for isomers **III**, **IV**, or **V**), $l'_4 = 6$ (6) $\mu M^{-1} \sec^{-1}$; values in parentheses were determined by double mixing. The range of error in both studies is considerable. In the study by Perrella only the overall stepwise constants have been evaluated. In double mixing studies in dilute solutions, dimers may be difficult to avoid completely. Considering these limitations and the complexity of the situation, the two studies can be considered to be in overall agreement.

5. The disagreement between the value of l'_3, $1.6 \times 10^6 M^{-1} \sec^{-1}$ (average of l'_3 for all four isomers) reported by Perrella et al.,[15] and $2 \times 10^5 M^{-1} \sec^{-1}$, obtained from double mixing experiments at 90% saturation of $Hb_4(O_2)_4$ with CO (of the type described in Fig. 4), can be reconciled if we assume that the slow component observed in these experiments represented only isomer **VI** [$\alpha_1^{CO}\beta_1(\alpha_2\beta_2^{CO})$].

6. From the point of view of the reaction model, the results of double mixing experiments have led to the suggestion that the major diliganded species is isomer **VI** [i.e., $\alpha_1^{CO}\beta_1(\alpha_2\beta_2^{CO})$. The data of Perrella et al.[15] suggest that all four diliganded isomers are formed more or less in equal amounts. But the amounts of diliganded isomers observed in that study were only a little more (~3% on the average) than the baseline noise (~5%), and isomers **V** and **VI** [i.e., $\alpha_1^{CO}\beta_1^{CO}(\alpha_2\beta_1)$ and $\alpha_1^{CO}\beta_1(\alpha_2\beta_2^{CO})$] could not be resolved from each other. The issue, therefore, should be considered undecided.

7. The results of double mixing studies are also qualitatively in agreement with the results obtained by other methods. In this regard we particularly refer to the flash-photolysis studies of Khaleque and Sawicki[17] and

[17] M. A. Klaleque and C. A. Sawicki, *Photobiochem. Photobiophys.* **13**, 155 (1986).

Zahroon and Sawicki,[18] and the single mixing studies with cyanometHb derivatives of isomers **III** and **IV** by Cassoly and Gibson.[19]

Conclusions

The double mixing method described here, along with the microperoxidase method for studying CO dissociation reactions and the HPLC method for preparation of the valency hybrids, allows us to study the kinetics of CO association with and dissociation from partially liganded intermediates of hemoglobin. All of these studies so far have been made by only two groups and, therefore, a certain degree of bias in perspective may have been unavoidable, particularly in formulating reaction schemes to fit the data. Furthermore, these studies have been made using a first-generation double mixing stopped-flow spectrophotometer with limited capabilities, such as a long dead time (3 msec or more), use of large sample volumes (2.4 ml) for each kinetic shot, and some probability of leading-edge contamination at longer aging times. More information and precision can be obtained by using state-of-the-art double mixing instruments. Of a total of 16 reactions of partially liganded intermediates, only 10 have been studied so far. The study of the remaining six reactions may be possible by using site-specific mutants of Hb or cross-linked hemoglobins.

[18] I. A. Zahroon and C. A. Sawicki, *Biophys. J.* **56**, 947 (1989).
[19] R. Cassoly and Q. H. Gibson, *J. Biol. Chem.* **247**, 7332 (1972).

[21] Hemoglobin-Liganded Intermediates

By MICHELE PERRELLA and LUIGI ROSSI-BERNARDI

The study of liganded intermediates is crucial to the clarification of the mechanisms of hemoglobin cooperativity. Information on the concentrations of the intermediates in solutions of hemoglobin at equilibrium with oxygen or carbon monoxide can be obtained by the analysis of the binding isotherms. Although a high precision has been achieved in some of the present methods for the determination of the ligand saturation,[1-3] the accuracy of these methods is questionable because they assume negligible functional heterogeneity of the subunits. In particular, methods that use

[1] S. J. Gill, this series, Vol. 76(24).
[2] K. Imai, this series, Vol. 76(25).
[3] B. W. Turner, D. W. Pettigrew, and G. K. Ackers, this series, Vol. 76(37).

METHODS IN ENZYMOLOGY, VOL. 232

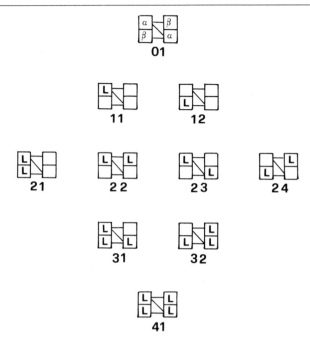

FIG. 1. Topology of ligand (L) distribution among the hemoglobin intermediates. The hybrids of species 41 and species 23 and 24 are identified here as species 31 and 32, respectively. Ackers and Smith[4] indicate the same hybrids as species 32 and 31.

absorption spectroscopy for the measurement of the ligand saturation assume that the optical properties of the intermediates can be calculated by a linear combination of the optical properties of deoxy- and fully liganded hemoglobin. However, intermediates 11, 12, 23, 24, 31, and 32 (Fig. 1)[4] may be functionally heterogeneous because of the structural differences between the α and β subunits, and intermediates 21 and 22 because of the different configurations of these subunits within the hemoglobin tetramer. The determination of the concentrations of the intermediates under dynamic conditions, such as in the course of the association reaction between hemoglobin and carbon monoxide as studied by classic optical stopped-flow techniques,[5] is subject to similar limitations. A new approach based on cryogenic techniques now allows 9 of the 10 ligation states of Fig. 1 in the reaction between hemoglobin and CO both at equilib-

[4] G. K. Ackers and F. R. Smith, *Annu. Rev. Biophys. Biophys. Chem.* **16,** 583 (1987).
[5] J. S. Olson, this series, Vol. 76(38).

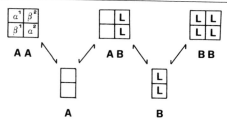

FIG. 2. Scheme of a dimer exchange reaction. Parent species AA, such as Hb^+, and parent species BB, such as HbCO, exchange dimers to yield an asymmetrical hybrid AB, such as species 21(ox), where the suffix (ox) indicates that the unliganded subunits are in the ferric state. The concentration of dimers is negligible in a concentrated solution of hemoglobin (1–5 g/dl). In aqueous media at temperatures above 0° a statistical distribution of parent species and asymmetrical oxidized hybrid is attained rapidly in a mixture of the parent species because of the high rates of the tetramer–dimer dissociation and association reactions.[4,6]

rium and under dynamic conditions to be trapped quantitatively, resolved, and identified.

General Strategy

The isolation and study of the liganded intermediates are difficult because (1) the ligand can dissociate from the heme site to rebind to a different site, and (2) the hemoglobin tetramer dissociates reversibly into dimers by cleavage along the $\alpha_1\beta_2$ and $\alpha_2\beta_1$ contacts. An asymmetrical intermediate, or hybrid, in which the ligand is not distributed symmetrically with respect to the dimer interface, dissociates into dimers that can yield on reassociation the original hybrid and its parent tetramers, as shown in Fig. 2.[6]

To stabilize the carbonmonoxy intermediates of hemoglobin, we quench thermally both the ligand and tetramer dissociation reactions by mixing rapidly the hemoglobin solution with an anaerobic cryosolvent at about −30°. The cryosolvent contains a 5- to 20-fold excess of ferricyanide to oxidize the unliganded hemes to ferric hemes. Such an oxidation provides further stabilization of the intermediates because the ligand dissociated from a ferrous heme cannot rebind to the ferric hemes. In addition, while the rate of CO dissociation from a liganded subunit in a particular intermediate depends on the functional properties of that intermediate, the rate of CO dissociation from any oxidized intermediate is likely to be

[6] M. Perrella, L. Cremonesi, L. Benazzi, and L. Rossi-Bernardi, *J. Biol. Chem.* **256,** 11098 (1981).

the same as the rate of CO dissociation from carbonmonoxyhemoglobin (HbCO) under the conditions of the quenching procedure.

The oxidation of the unliganded hemes increases the positive charge of an intermediate relative to the charge of HbCO by an amount that depends on the number of unliganded subunits of the intermediate and the pK values of the equilibrium reactions between the ferric hemes of the oxidized subunit and water. Thus the separation of the quenched species, and at the same time of the excess ferricyanide, can be carried out by isoelectric focusing at subzero temperature.

Studies of the rate of oxidation of deoxyhemoglobin (deoxyHb) under various conditions of temperature, pH, ionic strength of the cryosolvent, and concentrations of the reactants, and a similar study of the rates of tetramer dissociation and dimer association of the quenched intermediates, have not been carried out. Instead, the conditions of the thermal–chemical quenching have been established by trial and error. Controls are then carried out to test that the intermediates are quantitatively trapped and separated.

Quenching

A drawing of the thermostatted reactor used to carry out the thermal–chemical quenching is shown in Fig. 3. The quenching medium is 1 ml of 20 mM sodium phosphate, pH 7.5, at 20°, diluted in ethylene glycol–water. The proportion of organic solvent is calculated to give a 50% (v/v) mixture after the addition of the aqueous solution of hemoglobin. The pH of 10 mM phosphate in 50% (v/v) ethylene glycol is 8.5 ± 0.1 in the temperature range from −20 to −40°.[7] Anaerobicity is ensured by flowing into the reactor a stream of N_2, which is precooled in the glass coil built in the double-jacketed reactor. To remove traces of O_2, the cryosolvent is deoxygenated thoroughly in a bottle, a 1-ml aliquot is transferred to the reactor at room temperature by means of a gas-tight syringe, and dithionite (20 μl of a 0.1 M solution in 20 mM phosphate, pH 7.5, prepared in a glass vial under N_2) is added. Finally, deoxygenated ferricyanide (5–10 μl of a 0.9 M solution in water) is added to the cryosolvent before cooling.

The aqueous sample of hemoglobin (1–5 g/dl) containing the intermediates is injected into the cryosolvent at subzero temperature either through the needle of a gas-tight syringe, when the sample is at equilibrium and is collected from a tonometer, or through a needle connected to the valve of a continuous-flow apparatus placed above the reactor, when the sample

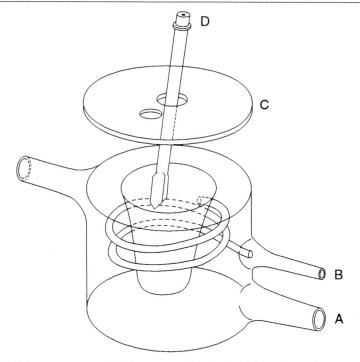

FIG. 3. Glass reactor vessel. The outer wall is insulated by a layer of neoprene foam. A, Inlet of coolant from a thermostat; B, nitrogen inlet; C, Plexiglas cover fitting the ground-glass top of the reactor; D, rotating rod of the Radiometer TTA-80 stirrer.

is under dynamic conditions,[8] for example, Hb reacting with a solution of CO. A sample of 0.5 M Tris in 50% (v/v) ethylene glycol is quenched 10 sec after quenching the protein, yielding a 20 mM concentration, to raise the pH of the cryosolvent to about 10.

The cryosolvent is mixed during each quenching by a TTA-80 stirrer (Radiometer, Copenhagen, Denmark).

The temperature of the cryosolvent is chosen so as to carry out the thermal–chemical quenching at a temperature $\leq -20°$. As an example, when a large volume of sample (0.25 ml) is quenched into the cryosolvent (1 ml) at $-30°$, the temperature rises to $-20°$, drops to $-25°$ in 10 sec, and returns to $-30°$ in 1 min under the conditions of our experiments.

---

[8] M. Perrella, L. Benazzi, L. Cremonesi, S. Vesely, G. Viggiano, and R. L. Berger, *J. Biochem. Biophys. Methods* **7,** 187 (1983).

Isoelectric Focusing of Oxidized Intermediates

The technique of isoelectric focusing at subzero temperatures is described in a previous volume in this series.[9] We describe here some of the modifications introduced since then.

We prepare the gels (9-mm long) in glass tubes of the same length (11 mm) but larger inner diameter (3.5 mm) than previously reported. This allows loading the gels with large amounts of quenched solution (up to 80 μl) in only two aliquots when dealing with dilute solutions of hemoglobin.

A recipe we find convenient for the preparation of 18–20 gels is the following: 11.33 ml of an aqueous solution containing acrylamide (5.24 g/dl) and methylenebisacrylamide (0.21 g/dl); 3.86 ml of ethylene glycol; 1.0 ml of Ampholine, pI (isoelectric pH) 6–8 [40 g/dl from Pharmacia (Uppsala, Sweden) or Sigma Chemical Co. (St. Louis, MO)]. After deoxygenation of this solution, 0.30 ml of ethyl acrylate, 2.7 ml of methanol, and 0.05 ml of tetraethylmethylenediamine are added under N_2. The mixture is swirled until all the ethyl acrylate is dissolved and 0.03 ml of ammonium persulfate (40 g/dl in water) is then added to start the polymerization at 0° for 1 hr. A layer (3 mm) of a 20% (v/v) ethylene glycol–15% (v/v) methanol–65% (v/v) water mixture on the polymerizing solutions prevents O_2 interference.

A 10% variation in the amounts of monomers and bifunctional reagent in the recipe, although yielding gels with slightly different structural properties, does not affect the sharpness and resolution of the focused intermediates or the focusing time. Gels can be prepared and stored at −30 to −20° for 1–2 weeks without loss in properties. The most frequent problem undermining the success of the separation experiment is the detachment of the gel top or of the entire gel from the wall of the glass tube. This can be prevented by soaking the tubes in detergent (Ausilab 101; Farmitalia, Milan) or chromic acid for a few hours before the preparation of the gels.

We use as the catholyte a 30% (v/v) ethylene glycol–10% (v/v) methanol–60% (v/v) water mixture (50 ml) containing 1 ml of Ampholine, pI 7–9, and 1 ml of Ampholine, pI 8–9.5 (40-g/dl solutions). The anolyte is the same hydroorganic mixture (350 ml) containing 2 ml of Ampholine, pI 6–8. The focusing equilibrium at −25° is attained in about 20 hr. In this linear pH gradient, the focusing positions of methemoglobin (Hb^+) and HbCO are 11–12 mm apart, depending on the batch of Ampholine.

Loading the quenched samples onto the gels is carried out by the use of a polyethylene tube (4-cm long) attached to the plastic tip of an automatic micropipette and precooled in the reactor for 10 sec.

[9] M. Perrella and L. Rossi-Bernardi, this series, Vol. 76(12).

Quantitation of Oxidized Intermediates

The quantitation of the separated intermediates can be carried out by two procedures: densitometry of the gels and chemical assay of the components eluted from the gels.

Densitometry

Scans can be made directly on the glass tubes containing the gels or on the extruded gels, but resolution may be impaired by diffusion during the scanning procedure carried out at room temperature. Alternatively, to minimize diffusion, scans can be made of color slides of the glass tubes containing the gels obtained immediately after removal of the tubes from the electrophoretic cell. This method requires careful standardization. Using the same film, slides are obtained of the separation of standard mixtures containing Hb^+ and HbCO in known proportions. The mixtures are prepared by injecting separately into the same quenching medium known amounts of the two proteins. Scans of the standards are carried out at varying wavelengths to select that wavelength at which the ratios of the areas of the Hb^+ and HbCO peaks are closer to the known ratios of the two quenched proteins. If no isosbestic wavelength is found an appropriate correction must be applied. Deconvolution techniques should be applied to resolve overlapping peaks. However, the improvement in the precision brought about by these techniques is limited because of the large number of peaks in the scans and, particularly, because of the irregular shape of some peaks caused by distortion of the focused protein zones.

Chemical Assay

To assay the protein, the gels are extruded by injecting water into the interface between the gels and the glass tubes. The gels are sliced and the pooled slices from three to five gels are suspended in two ml of 20 mM NaOH, 10 mM KCl for 24 hr. Pyridine (0.5 ml) is then added and after 2–4 hr in the dark the absorbance at 418 nm is read in a cuvette containing 20 mg of dithionite. An instrument was designed to slice the gels by means of a blade moved normally to the gel length by a micrometer screw. However, gel slicing carried out manually is quicker and yields reproducible results.

Table I shows a comparison of the data on the relative proportions of oxidized carbonmonoxy intermediates as determined by the chemical assay and densitometric methods. The data are the averaged values from two quenching and cryofocusing experiments carried out in duplicate, using a standard solution containing 46% Hb^+ according to the cyanomet-

TABLE I
CONCENTRATIONS OF SPECIES IN SAMPLE OF PARTIALLY OXIDIZED HbCO[a]

Species	Concentration (%)[b]		
	(a)	(b)	(c)
Hb[+]	5.35 ± 0.05	5.23 ± 0.29	5.78 ± 0.17
12(ox)	12.4 ± 0.35	14.5 ± 0.31	11.3 ± 0.57
24(ox)	20.6 ± 0.55	19.1 ± 0.61	19.0 ± 1.15
11(ox)	—	—	—
21(ox) + 22(ox)	18.4 ± 1.15	18.0 ± 0.33	18.0 ± 1.02
32(ox)	23.8 ± 0.75	22.9 ± 0.66	21.8 ± 0.06
23(ox)	—	—	—
31(ox)	8.95 ± 0.75	9.26 ± 0.34	11.5 ± 0.43
HbCO	10.2 ± 0.50	11.0 ± 0.64	12.7 ± 0.70
Ferric hemes:	42.5%	42.7%	41.1%

[a] Focused at −25° for 17 and 28 hr.
[b] (a) Chemical assay method, 17 hr of cryofocusing; (b) chemical assay method, 28 hr of cryofocusing; and (c) densitometric method, 17 hr of cryofocusing.

hemoglobin method.[10] Densitometric scans were carried out using a DU-70 Beckman (Fullerton, CA) spectrophotometer. Peak deconvolution of the scans was carried out using Jandel Scientific PeakFit (Jandel Scientific, Corte Madera, CA).

Identification of Oxidized Intermediates

The focusing position of all asymmetrical oxidized intermediates is equidistant from the focusing positions of their parent species, provided that the pH gradient is linear.[6] Thus the focusing positions of all the species in the gradient can be identified if the focusing positions of the symmetrical species are known.

The order of focusing from cathode to anode of the oxidized intermediates is the following: Hb[+], 12(ox), 24(ox), 11(ox), 21(ox) and 22(ox), 32(ox), 23(ox), 31(ox), and HbCO. The suffix (ox) indicates that the unliganded subunits of the intermediate are oxidized to the ferric state. Species 21(ox) focuses in the middle between the focusing positions of Hb[+] and HbCO. In a linear gradient the distance between Hb[+] and species 24(ox) is the same as the distance between HbCO and species 23(ox); therefore species 22(ox), the hybrid of species 24(ox) and 23(ox), must focus in the same

[10] K. A. Evelyn and H. T. Malloy, *J. Biol. Chem.* **126,** 655 (1938).

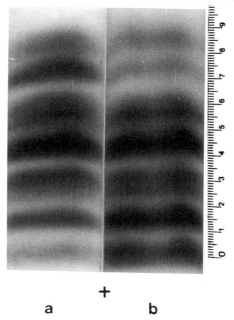

+

a   b

FIG. 4. Two glass tubes containing the gels used for the separation of samples of HbCO partially oxidized with ferricyanide. Band distortion is due to strain in the gels caused by exposure to room temperature after cryofocusing. Strain is relaxed when the gels are extruded from the tubes. Band identity from anode to cathode: HbCO, 31(ox), 23(ox) (more clearly visible on gel *b*), 32(ox) [partially overlapping 23(ox)], 21(ox) plus 22(ox), 11(ox), 24(ox) [more clearly visible on gel *a* and partially overlapping 11(ox)], 12(ox), Hb$^+$. The distances (in centimeters) from the center of the HbCO bands and the center of the bands of the symmetrical species 23(ox), 24(ox), and Hb$^+$ are 2.45/2.60, 5.90/6.00, and 8.30/8.60, respectively, on tubes *a*/*b*. The distances similarly measured for the asymmetrical species agree within ±0.05 cm with the distances calculated using the principle, described in text, of equidistance of the asymmetrical hybrids from their parents in a linear pH gradient.

position as species 21(ox). This holds true even if species 21(ox) and 22(ox) have different p*I* values. The only way to resolve these species is to make use of nonlinear gradients. However, we were not successful in resolving these species in nonlinear pH gradients. Although the p*I* values of intermediates 21 and 22 may be different, the p*I* values of 21(ox) and 22(ox) are not expected to differ significantly because these species should have similar if not identical functional and structural properties.

The focusing positions of species 24(ox) and 11(ox) and of species 23(ox) and 32(ox) are very close, as shown in Fig. 4. Partial overlapping of these components may occur if one of them is present in larger amount.

The pH gradients may sometimes deviate slightly from linearity when repeated use is made of the catholyte and anolyte solutions or when the catholyte solution is prepared using only Ampholine with a pI range of 8.5–9.[6] In these cases it is found that Hb$^+$ and species 12(ox), 24(ox), and 11(ox) are slightly closer together than the remaining species. Focusing standard solutions of partially oxidized intermediates, such as those shown in Fig. 4, helps remove ambiguities in the identification of the components.

Controls

Effect of Ferrocyanide on pI Values of Oxidized Intermediates

Ferrocyanide binds tightly to Hb$^+$ in aqueous solution and could modify its pI value at subzero temperatures. To test this possibility, a sample of Hb$^+$ containing an equimolar amount of ferrocyanide was loaded and allowed to migrate well along a gel before loading on the same gel another sample of Hb$^+$ free of ferrocyanide. The two samples yielded at equilibrium a single component. Standard solutions of partially oxidized intermediates, such as those shown in Fig. 4, containing ferrocyanide or freed of this reagent yielded identical focusing patterns. These tests indicate that ferrocyanide is removed from the protein under the conditions of cryogenic focusing. Similar tests exclude that organic phosphates, such as inositol hexaphosphate (IHP), bind tightly to hemoglobin in cryogenic focusing.

Oxidation of Unliganded Subunits

Samples of Hb are quenched in the cryosolvent containing various amounts of ferricyanide and, after focusing, the gels are inspected for the presence of unoxidized components. A 10-fold molar excess of oxidant is enough to ensure complete (>99%) oxidation when the protein concentration is in the range of 2–5 g/dl. Products of incomplete oxidation of hemoglobin are observed when quenching is not rigorously anaerobic.

Oxidation of Liganded Subunits

The time required to stop a reaction carried out in aqueous solvent at 20° by the thermal quenching procedure is about 50 msec for an organic reaction, such as the hydrolysis of 2,4-dinitrophenyl acetate, and about 100 msec for a protein reaction, such as the formation of the hybrid between hemoglobins A and S, according to the mechanism shown in Fig. 2.[8] Shorter quenching times require a mixing technique more efficient than that illustrated in Fig. 3.

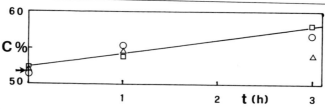

FIG. 5. Increase in the concentration of Hb^+, as determined from the concentrations of the focused intermediates, in a sample of partially oxidized HbCO incubated for various times t ($0 \leq t \leq 3$ hr) with a 20-fold molar excess of ferricyanide in: ($\bigcirc$), 10 mM phosphate in 50% (v/v) ethylene glycol, pH 8.5; ($\triangle$), same buffer plus Tris to 20 mM concentration; ($\square$), 10 mM Tris-HCl buffer in 50% (v/v) ethylene glycol. At $t = 0$ the protein was exposed to the oxidant for the time required for quenching, sample loading, and electrophoretic removal of ferricyanide. The arrow indicates the Hb^+ concentration in the sample not exposed to the oxidant. The line is drawn through the data points obtained in 10 mM Tris-HCl buffer.

Because no data at subzero temperatures are available, it can be estimated, using the data at 20° and pH 7,[11] that during 100 msec the dissociation of CO would affect <1% of the monoliganded intermediates, which are the fastest dissociating species.

At subzero temperature ($\leq -20°$) before the excess oxidant is removed by electrophoresis, the dissociation of CO exposes the liganded subunits to the oxidant. The time required to remove ferricyanide depends on the volume of sample loaded onto the gel, and it is less or equal to the stacking time of the protein onto the gel top (15–60 min). Assuming that the data on the energy of activation for the dissociation of CO from hemoglobin at pH 8.5–9 in aqueous solvent can be extrapolated to the conditions of our quenching experiments, that is −25° and pH 8.5–10,[12] and assuming also that under these conditions the oxidation of the unliganded subunits is faster than CO rebinding, it is estimated that the rate of oxidation of the liganded subunits could be as much as 7–8%/hr.

Figure 5 shows the increase in Hb^+ concentration when a quenched sample of partially oxidized hemoglobin was exposed to a 20-fold molar excess of ferricyanide at −25° under various conditions of pH. The sample was prepared by equilibrating at 20° an equimolar mixture of Hb^+ and HbCO. Because the parent species were contaminated by products of partial oxidation of hemoglobin, the mixture contained other asymmetrical hybrids in addition to species 21(ox). The total Hb^+ content was deter-

[11] M. Samaja, E. Rovida, M. Niggeler, M. Perrella, and L. Rossi-Bernardi, *J. Biol. Chem.* **262**, 4528 (1987).

[12] Q. H. Gibson and F. J. W. Roughton, *Proc. R. Soc. London, Ser. B* **147**, 44 (1957).

mined from the concentrations of the partially oxidized components measured after cryogenic isoelectric focusing. The rate of increase in Hb^+ concentration of the sample was $\leq 2\%$/hr, that is, less than estimated.

Some details of the data in Fig. 5 are shown in Table II. The concentrations of the partially oxidized components in the sample quenched in the cryosolvent free of oxidant (Table II, column a) are compared with those found when the sample was quenched in the presence of a 20-fold molar excess of oxidant in hydroorganic phosphate, pH 8.5, and in the same buffer plus Tris to 20 mM concentration. Column b in Table II shows the concentrations of the oxidized intermediates found when the sample was focused immediately after quenching in either case. Columns c and d of Table II show the concentrations found after 3 hr of incubation with the oxidant. Clearly when the time of exposure to the excess oxidant does not exceed the time required for quenching, sample loading, and electrophoretic removal of the oxidant (column b, Table II), the concentrations of the oxidized intermediates are the same as those observed in the sample not exposed to the oxidant (column a, Table II).

TABLE II

CONCENTRATIONS OF SPECIES IN SAMPLE OF PARTIALLY OXIDIZED
HbCO DETERMINED BY CHEMICAL ASSAY METHOD[a]

Species	Concentration (%)[b]			
	(a)	(b)	(c)	(d)
Hb^+	24.3 ± 1.5	24.3	29.4	24.4
11(ox) + 12(ox)	11.5 ± 0.78	11.8	18.2	15.9
21(ox) + 22(ox)	32.2 ± 1.5	31.2	19.1	27.1
31(ox) + 32(ox)	10.5 ± 0.67	10.7	16.7	16.5
HbCO	21.6 ± 0.87	22.0	16.5	16.2
Ferric hemes:	51.6%	41.5%	56.8%	54.0%

[a] After cryogenic focusing. See text for sample preparation.

[b] (a) Phosphate buffer (10 mM) in 50% (v/v) ethylene glycol, pH 8.5, at $-25°$, and same buffer plus Tris to 20 mM concentration. No added ferricyanide. (b) Phosphate (10 mM) in 50% (v/v) ethylene glycol, pH 8.5, plus 20-fold molar excess ferricyanide. Exposure to oxidant limited to the time required for quenching, sample loading onto the gels, and electrophoretic removal of ferricyanide. (c) As in (b). Three-hour incubation with ferricyanide. (d) Buffer as in (b) plus Tris to 20 mM concentration. Three-hour incubation with ferricyanide.

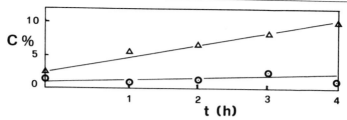

FIG. 6. Increase in the concentration of hybrid 21(ox) when an equimolar mixture of Hb$^+$ and HbCO prepared at $-25°$ was incubated for various times t in the buffers given in Fig. 5, pH 8.5 (△) and 10 (○).

Dimer Exchange Reactions

An approximate estimate of the effects of dimer exchange reactions on the concentrations of the quenched intermediates is difficult because of the limited amount of available data and the complexity of the problem. The pH and ionic strength of the quenched sample are different in the reactor and during electrophoresis. Also, they vary on the gel top during removal of salts by electrophoresis and in the gel during migration to equilibrium. In general, dimer exchange reactions are slower the higher the pH and the lower the ionic strength.[13] Thus the stability of the oxidized hemoglobin tetramer in regard to pH is expected to be maximal in the reactor and gel top (pH $\geq$ 10) and minimal at focusing equilibrium (pH 8–8.5).[6] The stability in regard to ionic strength is expected to be minimal in the reactor because of the presence in the cryosolvent of phosphate, ferricyanide, and other salts added with the quenched sample, to be varying on the gel top, and to be maximal during approach to equilibrium.

Figure 6 shows the formation of the hybrid between Hb$^+$ and HbCO when these species are quenched and incubated for various times at $-25°$ at pH 8.5 and 10. The concentration of the hybrid, species (21)ox, is 7–8% of the total after 4 hr of incubation at pH 8.5 and is not detectable at pH 10. The addition of Tris to 20 mM concentration to the hydroorganic phosphate buffer, pH 8.5, slows the rate of hybridization as observed at pH 10.

Figure 7 shows a test of the stability of a representative symmetrical hybrid under the conditions of equilibrium in focusing. A mixture of species 23(ox), HbCO, and their hybrid species 31(ox), equilibrated at 20° in phosphate buffer, pH 7, was quenched and focused at $-25°$ for times varying from 13 hr, when the focusing species have nearly reached equilibrium, to 30 hr. After 17 hr at focusing equilibrium the decrease in the

[13] M. Perrella, M. Samaja, and L. Rossi-Bernardi, *J. Biol. Chem.* **254**, 8748 (1979).

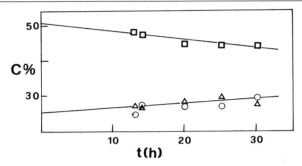

FIG. 7. Changes in the concentrations of hybrid 31(ox) (□) and its parent species 23(ox) (△) and HbCO (○) versus time t under the conditions of focusing equilibrium at $-25°$.

concentration of the hybrid was 2–3% of the total and about twice the increase in the concentration of each parent species.

The data in Table I indicate that, within the error, no significant change in the concentrations of the intermediates is observed when focusing of complex mixtures is carried out for 17 or 28 hr.

All the controls described above indicate that the thermal–chemical quenching and cryogenic focusing allow quantitative trapping of the carbonmonoxy intermediates of hemoglobin.

A "field" test of the procedure was obtained by the experiment shown in Fig. 8.[14] An equilibrium solution of oxidized carbonmonoxy intermediates at 22°, pH 7, prepared by partial ferricyanide oxidation of an aqueous sample of oxyhemoglobin and by O_2 replacement with CO, was rapidly mixed in a continuous-flow apparatus with an excess of dithionite to reduce >95% of the oxidized subunits in about 300 msec. Thus a nonequilibrium mixture was produced containing various intermediates at high and nearly statistical concentration.

Samples of the reactants were quenched at reaction times varying from 1.5 to 80 sec into a cryosolvent at $-25°$ containing enough ferricyanide to oxidize the excess dithionite and to reoxidize the reduced subunits of the intermediates. Figure 8a shows a scan of the focused components present in the mixture of oxidized intermediates before reduction. Figure 8b and c shows the modifications in the scan pattern of Fig. 8a brought about by the reactions of CO dissociation, CO rebinding, and dimer exchange in the course of approach to equilibrium, which is attained after 80 sec of reaction (Fig. 8d). The total fraction of oxidized subunits measured from the concentrations of the oxidized intermediates after ther-

[14] M. Perrella, L. Sabbioneda, M. Samaja, and L. Rossi-Bernardi, *J. Biol. Chem.* **261,** 8391 (1986).

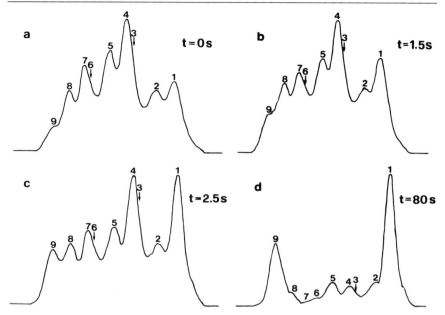

FIG. 8. Densitometer scans of gel tubes used for the separation by isoelectric focusing at $-25°$ of the products of quenching a sample of partially oxidized HbCO (4 mM heme concentration in 10 mM phosphate, 0.05 M KCl, pH 7) before reduction at 22° with dithionite (a) and after reaction with dithionite for times varying from 1.5 to 80 sec (b–d). Peak identity (from 9 to 1): Hb$^+$, 12(ox), 24(ox), 11(ox), 21(ox) plus 22(ox), 32(ox), 23(ox), 31(ox), and HbCO. The ligand saturations calculated from the scan data are as follows: (a) 0.64; (b) 0.63; (c) 0.64; (d) 0.63. [Adapted from M. Perrella *et al.*, *J. Biol. Chem.* **261**, 8391 (1986).]

mal–chemical quenching and cryogenic focusing (Fig. 8b–d) was unchanged, within error, with respect to that measured in the solution of oxidized carbonmonoxyhemoglobin before reduction (Fig. 8a), indicating quantitative trapping and recovery of the intermediates by the procedure.

The low concentration of intermediates, with respect to the concentrations of Hb and HbCO, shown in Fig. 8d, has been confirmed by the measurement of the concentrations of the intermediates in solutions of hemoglobin equilibrated with gas phases of known P_{CO} values.[15] Because the scan patterns in Fig. 8 show the dramatic change from a statistical to a nonstatistical distribution of intermediates brought about by hemoglobin

[15] M. Perrella, A. Colosimo, L. Benazzi, M. Ripamonti, and L. Rossi-Bernardi, *Biophys. Chem.* **37**, 211 (1990).

cooperativity, they can be regarded as a "photographic molecular record-ing" of cooperativity.

Acknowledgments

Previously unpublished work presented in this chapter was supported by grants from the CNR, Rome.

[22] Hemoglobin–Oxygen Equilibrium Binding: Rapid-Scanning Spectrophotometry and Singular Value Decomposition

By KIM D. VANDEGRIFF and RICHARD I. SHRAGER

Detailed knowledge of the interaction between oxygen and hemoglobin is essential both for understanding oxygen transport phenomena and for testing theories of protein structure and function. The equilibrium reaction is measured routinely simply to estimate the position and shape of the oxygen-binding curve from, respectively, the oxygen pressure at half-saturation (P_{50}) and the index of cooperativity (i.e., the Hill number, n), but thermodynamic analysis of individual binding steps is more difficult because of inherent complexities in the reaction. For example, the linearity of optical absorption with hemoglobin fractional saturation must be as-sumed in the absence of analytical tools to prove it. To test this assumption and to extract more information from a single experiment, we are develop-ing techniques using rapid-scanning spectrophotometry to measure com-plete spectra of hemoglobin during an oxygen-binding reaction and singu-lar value decomposition to resolve individual components in the transition. From these analyses, models of the equilibrium reaction are being derived using laws of mass action and matrix least squares.

Background

The reversible chemical reaction between oxygen and hemoglobin has been examined for over a century. Even though experimental methods have evolved during this time from simple gasometry to more sophisticated techniques in spectrophotometry, the reaction is still difficult to measure precisely or to interpret reliably.

The first mathematical formulation of the equilibrium was derived by Adair as a mass-action equation, with equilibrium constants describing each of the four binding steps of oxygen with the hemoglobin tetramer,[1]

$$Y = \frac{0.25a_1x + 0.5a_2x^2 + 0.75a_3x^3 + a_4x^4}{1 + a_1x + a_2x^2 + a_3x^3 + a_4x^4} \tag{1}$$

where Y is fractional saturation, a_1 through a_4 are the overall Adair constants (i.e., the product of stepwise equilibrium constants K_1 to K_4), and x is the partial pressure of oxygen (pO_2) in solution.

The hemoglobin-binding reaction was described later by allosteric theory as an equilibrium between two end-state quaternary conformations, one corresponding to the structure of deoxyhemoglobin with low oxygen affinity (i.e., the T state) and the other corresponding to the structure of oxyhemoglobin with high oxygen affinity (i.e., the R state).[2] This model has been useful particularly in the interpretation of the structural–functional mechanism of hemoglobin cooperativity,[3] but it is limited in that it predicts strictly a two-state system. Partially liganded species of hemoglobin have been identified with reaction energetics intermediate to the two end states but with quaternary structures still in either T or R conformation.[4]

Partially liganded intermediates are central to the mechanism of cooperative oxygen binding, but they are particularly difficult to measure because their low levels during the reaction[5] result in a highly sigmoid binding curve. Precise measurements have been accomplished,[6,7] but the mechanism of molecular cooperativity remains controversial. The difficulty originates from the oxygenation reaction itself, for example, the low level of partially liganded intermediates combined perhaps with differences in the reactivity of the α and β subunits[8,9] and/or spectral changes associated with the transition between low- and high-affinity forms.[10–15] In addition,

[1] G. S. Adair, J. Biol. Chem. 63, 529 (1925).
[2] J. Monod, J. Wyman, and J. P. Changeux, J. Mol. Biol. 12, 88 (1965).
[3] M. F. Perutz, Nature (London) 228, 726 (1970).
[4] G. K. Ackers, M. L. Doyle, D. Myers, and M. A. Daugherty, Science 255, 54 (1992).
[5] M. Perrella and L. Rossi-Bernardi, this volume [21].
[6] F. C. Mills, M. L. Johnson, and G. K. Ackers, Biochemistry 15, 5350 (1976).
[7] K. Imai, this series, Vol. 76, p. 438.
[8] J. S. Olson, M. E. Anderson, and Q. H. Gibson, J. Biol. Chem. 246, 5919 (1971).
[9] A. Nasuda-Kouyama, H. Tachibana, and A. Wada, J. Mol. Biol. 164, 451 (1983).
[10] Q. H. Gibson, Biochem. J. 71, 293 (1959).
[11] M. Brunori, E. Antonini, J. Wyman, and S. R. Anderson, J. Mol. Biol. 34, 357 (1968).
[12] M. L. Adams and T. M. Shuster, Biochem. Biophys. Res. Commun. 58, 525 (1974).
[13] C. A. Sawicki and Q. H. Gibson, J. Biol. Chem. 251, 1533 (1976).
[14] J. Hofrichter, J. H. Sommer, E. R. Henry, and W. A. Eaton, Proc. Natl. Acad. Sci. U.S.A. 80, 2235 (1983).
[15] A. Bellelli and M. Brunori, this volume [5].

at least two other reactions can complicate interpretation of binding data: (1) dissociation of the $\alpha_2\beta_2$ tetramer into noncooperative $\alpha\beta$ dimers[6,16] and (2) the redox reaction of the heme iron atoms, in which an oxygen-binding ferrous (Fe^{2+}) heme is oxidized to a non-oxygen-binding ferric (Fe^{3+}) heme. Experimentally, these latter effects can be limited by using a hemoglobin concentration as high as possible to minimize the effects of dimers and by performing the experiment as quickly as possible or in the presence of a methemoglobin-reducing system to minimize the effects of oxidation.[17,18]

Survey of Techniques

Gasometry provides an accurate method for obtaining single points along the oxygen-binding curve. This technique does not require a physical measurement of the oxyhemoglobin complex; rather, the volume of bound oxygen is measured directly. The disadvantages of using this technique are that experience and skill determine the precision of the experiment, a separate and precise measurement must be made of hemoglobin concentration, and the experiment is long and tedious. That is, the number of experimental points along a single curve depends on the number of gasometric readings that can be made on a single sample, with each reading taking ~15 min. Thus, for example, an experimental curve with only 10 data points takes at least 4 hr to complete: time to equilibrate 10 tonometers, each with the hemoglobin solution at a different pO_2, and a minimum of 150 min for the gasometric readings. During this interval, the hemoglobin solution will be autooxidizing, and protein degradation may occur.

Optical spectroscopy is now usually the experimental method of choice because of its speed and simplicity, and because the reaction produces a large spectral change based on hemoglobin concentration. Spectrophotometry has been combined with tonometry for expediency (i.e., a hemoglobin spectrum can be recorded faster than a gasometric reading can be made). However, like the gasometric method, time is taken for equilibration, using either separate tonometers for each data point or a single tonometer or reaction cell that is equilibrated stepwise at increasing pO_2. In either case, the number of data points that can be measured along each curve is limited, and the experiment is still time consuming. The primary advantage of using tonometry is that equilibrium is verified at each step (i.e., at each data point) along the curve.

[16] M. L. Johnson, this volume [28].
[17] A. Hayashi, T. Suzuki, and M. Shin, *Biochim. Biophys. Acta* **310**, 309 (1973).
[18] C. C. Winterbourn, B. M. McGrath, and R. W. Carrell, *Biochem. J.* **155**, 493 (1976).

In contrast to tonometry, the automatic method of Imai was developed to monitor the absorbance of a hemoglobin solution as pO_2 changed continuously.[7,19] This not only produced an equilibrium curve rapidly (e.g., typically 20–60 min for native human hemoglobin, depending on solution conditions), thus limiting the effects of methemoglobin formation, but also provided a large number of data points along each curve. Fractional saturation of hemoglobin is calculated directly from the spectral change of the hemoglobin solution, and pO_2 is measured polarographically using an oxygen electrode mounted in the reaction chamber. However, with the continuous method, unlike tonometry, equilibrium must be assumed at each measurement point along the curve, or the reverse curve must be measured to demonstrate exact reproducibility,[19] which is impossible if any oxidation has occurred. Also, because the reaction happens faster than non-rapid-scanning spectrophotometers can scan a spectral region, the transition is usually followed at a single wavelength. As with any two-state model, analysis at a single wavelength requires the further assumption that the spectral change is linearly correlated with hemoglobin fractional saturation. For the linear relation to hold, a single optical transition with a constant extinction coefficient for each binding step must take place that, in this case, would represent the transition from deoxy- to oxyhemoglobin. The actual experiment will be more complex if other reactions, such as dimer or methemoglobin formation, also occur.

There is experimental evidence both in support of and in conflict with the linear optical assumption,[20–22] with measurements showing that true isosbestic behavior is not maintained during the hemoglobin oxygenation reaction.[23,24] Nonlinear optical effects have been reported from both diode-array spectrophotometry, using the thin-layer tonometry method,[25] and a two-wavelength analysis of stopped-flow reactions by rapid mixing of oxyhemoglobin and deoxymyoglobin.[26]

In another chapter in this volume,[27] a technique is described that measures oxygen equilibrium binding of concentrated hemoglobin solutions nonoptically, thus avoiding the issue of nonlinear spectral events.[27] In this chapter, we describe a method based on rapid-scanning spectropho-

[19] K. Imai, this volume [26].

[20] F. J. W. Roughton, Biochem. J. 29, 2604 (1935).

[21] J. Rifkind and R. Lumry, Fed. Proc., Fed. Am. Soc. Exp. Biol. 26, 2325 (1967).

[22] K. Imaizumi, K. Imai, and I. Tyuma, J. Biochem. (Tokyo) 83, 1707 (1978).

[23] L. J. Parkhurst, T. M. Larsen, and H.-Y. Lee, this volume [29].

[24] M. L. Doyle, E. Di Cera, and S. J. Gill, Biochemistry 27, 820 (1988).

[25] D. W. Ownby and S. J. Gill, Biophys. Chem. 37, 395 (1990).

[26] T. M. Larsen, T. C. Mueser, and L. J. Parkhurst, Anal. Biochem. 197, 231 (1991).

[27] R. M. Winslow, A. Murray, and C. C. Gibson, this volume [23].

tometry to resolve directly unique optical transitions during continuous hemoglobin oxygenation. The rapid-scanning method has revealed small but significant spectral changes during oxygenation in addition to the large spectral difference from the primary transition of deoxy- to oxyhemoglobin.[28]

The overall utility of rapid-scanning techniques will depend on two factors: (1) Scanning rates must be fast enough to capture a complete spectrum of hemoglobin at a given pO_2 during the collection period. However, even at the fastest scanning rates time elapses during the measurement, and because of this we refer to these as "pseudoequilibrium" measurements to distinguish the possibility of kinetic events; (2) the optical spectrum must have high precision to resolve small spectral changes from the large oxy–deoxyhemoglobin spectral difference.

Spectrophotometers are becoming faster, although whether they are becoming more precise is debatable. Research has pushed this technology to the limit, and new questions are demanding the best signals that these instruments can provide. In parallel with the increasing speed of spectrophotometers, sophisticated analytical tools for data reduction are evolving, and high-speed personal computers with large memory capacities are available that can handle the huge data arrays and computational requirements of these experiments. Here we outline methods for the collection and analysis of hemoglobin spectral matrices during a pseudoequilibrium oxygenation reaction. Mathematical techniques for matrix multicomponent analysis and singular value decomposition (SVD) are described, and spectral artifacts that are resolved by these sensitive techniques are discussed.

Instrumentation

We have used an LT Quantum 1200 rapid-scanning spectrophotometer (LT Industries, Inc., Rockville, MD) to obtain complete visible spectra of hemoglobin solutions during continuous oxygenation reactions. This instrument uses a tungsten–halogen light source and a rapidly oscillating grating that scans over the 400- to 800-nm range in 200 msec. During this 200 msec, a 70-msec dark period is allowed for instrument calibration. The light is then exposed to the monochromator for 130 msec, during which time the entire range of 400–800 nm is scanned in 80 msec. From this range, we isolate the portion of the spectrum from 480 to 650 nm, which is collected in ~34 msec, for data analysis. To improve the signal-to-noise ratio, at least four scans are collected per data point to produce a single, averaged spectrum with a time resolution of 800 msec. The

[28] K. D. Vandegriff, Y. C. Le Tellier, J. R. Hess, and R. I. Shrager, *Biophys. J.* **61**, A55 (1992).

averaged scans are collected in alternating wavelength directions, with an equal number of scans in the high-to-low (800 to 400 nm) and low-to-high (400 to 800 nm) range. The instrument is calibrated such that the scans taken in either direction are indistinguishable at equilibrium. However, instrumental noise in both the vertical, or amplitude, spectral signal (i.e., due to noise in detector gain) and in the horizontal, or wavelength, position (i.e., due to slight inaccuracies in the reproducibility of wavelength position from scan to scan) produces spectral artifacts that must be accounted for in the final analysis.

A schematic diagram of the experimental apparatus is shown in Fig. 1. A temperature-controlled sample holder was designed to hold a custom-made reaction cuvette having a 1-cm light path and a fused side port for a polarographic oxygen probe (model 5331; Yellow Springs Instruments, Yellow Springs, OH). Ultrapure gas (i.e., >99.999% purity nitrogen or oxygen in a 1 : 1 O_2–N_2 mixture), humidified by bubbling in water at the same temperature as that circulated around the reaction cell, is introduced into the gas space at the top of the reaction cell through a needle inserted in a gas-tight septum. Venting takes place through a second needle in the septum. The solution is mixed continuously by a small, magnetic bar spinning just beneath the tip of the oxygen probe. (Stirring

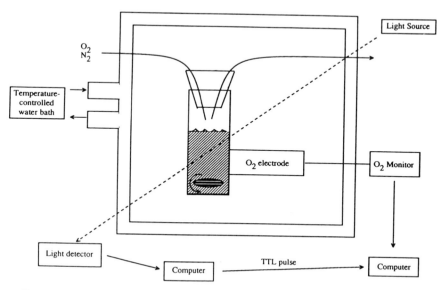

FIG. 1. Schematic diagram of the experimental system for rapid-scanning analysis of hemoglobin–oxygen equilibrium binding. Details are provided in text.

provides adequate mixing but can cause mechanical stress on the protein, and evidence of protein denaturation must be monitored during the reaction.)

The photodetector signal of the light passing through the cuvette is digitized for output by a 16-bit analog-to-digital (A/D) board in a computerized data collection system. The voltage signal from the oxygen electrode is amplified (model 5300 biological oxygen monitor; Yellow Springs Instruments), filtered digitally, and relayed to a second 16-bit A/D board in a second computerized data collection system. (Alternatively, a single computer with true multitasking capabilities could be used to collect the two signals simultaneously.) The digital filter of the oxygen voltage signal increases the signal-to-noise ratio without changing the time constant of the oxygen probe. The time constant of the oxygen probe is calculated by collecting the transient signal from a rapid change in solution pO_2 with and without the filter in line and fitting the relaxation to an exponential equation. For the binding experiments, the relaxation time of the electrode must be faster than the change in pO_2 of the solution so that readings will reflect the true equilibrium value of pO_2.

Data collection is under software control. At the initiation and end of a single averaged spectral data set, transistor–transistor logic (TTL) pulses are sent from the computer controller of the spectrophotometer to the A/D board designated for oxygen voltage readings. During the time between TTL pulses, the oxygen signal is collected at a specific sampling frequency. For example, at a frequency of 120 Hz, with 4 spectral scans averaged at 200 msec/scan, 96 oxygen voltage readings are accumulated and averaged to give a single pO_2 reading corresponding in time to the average spectrum of hemoglobin collected over the 0.8-sec period.

The voltage output from the oxygen elecrode is converted to millimeters of mercury from calibration of the output to the relative percentage of oxygen in air-saturated buffer at the barometric pressure in the laboratory with a correction for the water vapor pressure at the temperature of the experiment.

Method of Experimentation

The experiment is begun by bubbling pure N_2 gas through the reaction buffer in the sample cell. Deoxygenation of the buffer is monitored by the oxygen electrode. At the same time, a concentrated solution of hemoglobin (i.e., >2 mM in heme) is deoxygenated separately in a temperature-controlled, spinning tonometer under an N_2 atmosphere (model 237; Instrumentation Laboratory, Lexington, MA). Once deoxygenation is complete, the N_2 gas flow is directed above the surface of the fluid in the reaction cell, and deoxyhemoglobin is transferred from the tonometer to the reaction cell by a gas-tight syringe that has been flushed previously with pure N_2 gas. The hemoglobin sample is diluted in the deoxygenated buffer in the sample cell, and the oxygen voltage is monitored to assure that no oxygen was introduced during the transfer. An initial spectrum is taken to verify the concentration of deoxyhemoglobin in the reaction cell; typically, concentrations of 60–200 μM in heme are used to provide an adequate optical signal and to minimize the effect of tetramer dissociation on the measurement. Lower concentrations of hemoglobin might be used to test the consequence of tetramer dissociation on the spectral transition or to monitor the reaction in the Soret region of the spectrum. We are unable to use higher (i.e., ~1 mM) heme concentrations because the upper limit of absorbance detection is set by the fixed path length of the reaction cell.

Oxygenation is begun by switching gas valves from N_2 to 50% O_2. The flow rate is controlled by a gas-flow controller (model 4360; Matheson, Newark, CA) to provide a gentle stream of humidified gas over the surface of the hemoglobin solution. (Direct bubbling of gas into the solution or vigorous gas flow at the surface is avoided because of the potential for protein denaturation, which can cause a drift in the optical signal.) The reverse reaction can be carried out on the same sample by switching gas flow back to pure N_2. Depending on the flow rate, temperature, solution conditions, hemoglobin sample, and hemoglobin concentration, a complete oxygenation reaction with this system requires ~20–30 min, and deoxygenation requires ~40–60 min.

Analytical Procedures

The spectral data are combined in a single absorbance matrix (**A**) with a row for each wavelength and a column for each pO_2. The matrix is reduced to a range from 480 to 650 nm at 1-nm intervals for a total of 171 wavelengths, or rows (m). The final size of **A** depends on the number of pO_2 readings, or columns (n) (typically 200–400), to give matrix **A** = $m \times n$. Two procedures are used for analysis of the **A** matrix: multicomponent analysis and SVD.

Multicomponent Analysis

Multicomponent analysis determines the composition of a mixture of components when the spectrum of each of the pure parent species is known. The general procedure involves first taking a spectrum of each of the parent species. Second, the spectrum of the mixture is taken. In our case, multiple spectra are taken during the course of a single experiment (i.e., the number of spectra for multicomponent analysis in each experiment is equal to n). Third, a least-squares linear curve-fitting procedure is used to minimize the norm (sum of squares) of the residuals and to obtain the best fit combination of spectra that compose the mixture. The procedure uses the Moore–Penrose pseudoinverse of a table of the parent spectra $(\mathbf{M})$, in which each spectrum is in a separate column. The inversion returns a $p \times m$ matrix $(\mathbf{C})$, in which p is the number of parent species. Only $\mathbf{C}$ and the experimental spectra $(\mathbf{A})$ are needed for the curve fitting. The procedure is as follows.

1. Compute $\mathbf{C} = (\mathbf{M}^T\mathbf{M})^{-1}\mathbf{M}^T$, the Moore–Penrose pseudoinverse of $\mathbf{M}$.

2. Compute $\mathbf{P} = \mathbf{CA}$, where a column of $\mathbf{P}$ contains the amounts of the various parent compounds at the corresponding pO_2. These are normalized to compute percentages.

Parent spectra have been obtained on the rapid-scanning spectrophotometer for oxyhemoglobin (oxyHb), deoxyhemoglobin (deoxyHb), and methemoglobin (metHb). Using these parent spectra to create $\mathbf{M}$, multicomponent analysis is performed on each spectrum (i.e., each column) in $\mathbf{A}$ to obtain an estimate of the percentages of oxyHb, deoxyHb, and metHb in each spectrum. Fractional saturation (Y) is calculated from this analysis as

$$Y = \% \text{ oxyHb}/(\% \text{ oxyHb} + \% \text{ deoxyHb}) \tag{2}$$

in which the percentage contribution of metHb is excluded. The change in percentage metHb at each step is used to evaluate the rate of change of metHb formation ($\Delta\text{metHb}/\Delta t$) as a function of Y. It should be emphasized that these evaluations are only estimates, because to be accurate the multicomponent analysis procedure must include a parent spectrum for every optical species in the mixed spectrum. To determine the number of optical species in matrix $\mathbf{A}$, SVD is employed.

Singular Value Decomposition and Matrix Least Squares

This section describes two closely related computer-based techniques that place stringent demands on the quality of spectrophotometric data. These techniques are sensitive enough to pick up signals below the noise level of a single spectrum. The trouble is that lamps, gratings, and/or drive chains, as well as experimental designs, can deliver their own signals, which can confound the process one is trying to measure. For example, in the next section, we describe spurious signals that can arise simply because one is doing kinetics (or "pseudoequilibrium" reactions) instead of stepwise equilibrium, and in the final section, we describe an artifactual spectral component that contributes <0.05% of the total signal.

Let us keep to the context of the oxygenation of hemoglobin as measured by spectrophotometers, although these techniques apply to all manner of spectra and processes.[29,30] The data consist of a spectrum (from 480 to 650 nm in 1-nm steps to give 171 points in all) collected at each value of pO_2. As an example, if $\log(pO_2)$ ranges from -1 to 2 in steps of 0.015, which is 0.1 to 100 mmHg in increasing steps, 201 complete spectra will be recorded in all. The data are stored in matrix $\mathbf{A}$ with 171 rows and 201 columns. Each column of $\mathbf{A}$ is a spectrum of hemoglobin at a fixed pO_2, and each row of $\mathbf{A}$ is a hemoglobin oxygenation curve at a fixed wavelength.

The first question is, how many independent spectra and oxygen processes are there? With no notion of chemical mechanism, we can obtain a minimum number of independent spectra necessary to explain all the data. That is, we can find the least number of spectra needed to combine in various ratios to obtain an adequate representation of all the observed spectra. This same number serves as a minimum number of oxygenation processes to explain all the observed titration curves (rows of $\mathbf{A}$). This number is called the rank of $\mathbf{A}$, and there are many ways to estimate it. We chose a standard matrix operation called SVD, because the output from SVD will serve us in other ways. SVD decomposes the matrix $\mathbf{A}$ into three factors,

$$\mathbf{A} = \mathbf{U}\mathbf{S}\mathbf{V}^{\mathrm{T}} \tag{3}$$

such that

$$\mathbf{U}^{\mathrm{T}}\mathbf{U} = \mathbf{V}^{\mathrm{T}}\mathbf{V} = \mathbf{I} \tag{4}$$

and $\mathbf{S}$ is diagonal, $s_{1,1} \geq s_{2,2} \geq s_{3,3} \ldots$ In our case, the sizes of the factors are

$$\mathbf{U} \text{ and } \mathbf{S}: 171 \times 171; \qquad \mathbf{V}: 201 \times 171$$

The numbers on the main diagonal of $\mathbf{S}$, sorted in descending order, are the singular values of $\mathbf{A}$. The relations in Eq. (3) enable us to think of $\mathbf{A}$ as a sum of a few special components:

$$\mathbf{A} = (\mathbf{U} \text{ column } 1)s_{1,1}(\mathbf{V} \text{ column } 1)^{\mathrm{T}}$$
$$+ (\mathbf{U} \text{ column } 2)s_{2,2}(\mathbf{V} \text{ column } 2)^{\mathrm{T}} + \cdots \tag{5}$$

[29] R. I. Shrager and R. W. Hendler, *Anal. Chem.* **54**, 1147 (1982).
[30] R. I. Shrager, *Chemom. Intell. Lab Syst.* **1**, 59 (1986).

The first term of Eq. (5), that is, ($\mathbf{U}$ column 1)$s_{1,1}$($\mathbf{V}$ column 1)$^\mathrm{T}$, is a rank one matrix, the same size as $\mathbf{A}$. But being rank one, it contains only one "spectrum" ($\mathbf{U}$ column 1) in its columns, varying only in scale. Likewise, it contains only one "titration curve" ($\mathbf{V}$ column 1) in its rows, again varying only in scale. Furthermore, this rank one matrix, of all possible rank one matrices, is the best fit to $\mathbf{A}$ in the least-squares sense, and the magnitude of $s_{1,1}$ tells how much of $\mathbf{A}$ is explained by this optimal rank one term. Similar descriptions hold for the subsequent rank one terms in Eq. (5), each of which is a best rank one fit to $\mathbf{A}$ minus the previous terms. When $s_{i,i}$ is small enough to be regarded as noise, and when $\mathbf{U}$ column i and $\mathbf{V}$ column i fail to show anything that looks like signal, we can discard those parts of $\mathbf{U}$, $\mathbf{S}$, and $\mathbf{V}$ and keep only the minimal, "noiseless" representation approximating the signal in $\mathbf{A}$. It is this minimal representation property that makes SVD an appealing tool for analyzing complex mixtures. The advantages of this representation are explained in Shrager.[30]

(At this point, it may be expedient to consult a linear algebra text about matrix multiplication, transpose, diagonal, identity, inverse, and the Frobenius norm, which we refer to as norm.) The norm of $\mathbf{S}$ is the same as the norm of $\mathbf{A}$. The rank of $\mathbf{A}$ may be estimated by plotting $\log(s_{i,i})$ versus i, most of which will be a smooth curve, almost a straight line for small i, except for the first few values. These will stand out above the others as signal stands out above noise, and the number of these standouts will estimate the rank. Sometimes it is difficult to decide where the standouts end, which is appropriate. The difficulty provides a sense of doubt about concepts such as rank. One should not put too much confidence in computations that produce integer answers (decisions, if you will) from data with a continuous range of possible values. Alternate and possibly conflicting ways of estimating rank are provided in previous publications.[29,30] Rank becomes a working hypothesis, not a "hard" number.

Having chosen a rank r, we can now express our goal in matrix terms. We wish to decompose the matrix $\mathbf{A}$ into two factors,

$$\mathbf{A} = \mathbf{D}\mathbf{F}^\mathrm{T} \tag{6}$$

where $\mathbf{D}$ is a $171 \times r$ matrix, its r columns containing the spectra that are changing. The columns of $\mathbf{D}$ are plotted versus wavelength. $\mathbf{F}$ is a $201 \times r$ matrix of appearance-disappearance curves for the spectra in $\mathbf{D}$. The columns of $\mathbf{F}$ are plotted versus $\log(p\mathrm{O}_2)$.

Just as two of the three numbers must be known when solving the scalar equation $ab = c$, two of the three matrices must be known when solving $\mathbf{A} = \mathbf{D}\mathbf{F}^\mathrm{T}$. By using computer modeling in conjunction with least squares, one can often obtain a good estimate of $\mathbf{F}$ (an example of how

F is chosen is given below) and then compute **D** by the formula **D** = **A**(**F**$^{T+}$), where **F**$^{T+}$ is the Moore–Penrose pseudoinverse of **F**^{T}.[31] This process is called matrix least squares, and programs for carrying it out directly, without going through the intermediate SVD steps described below, are described in Frans and Harris.[32] The relation between **A** = **DF**^{T} (the matrix least-squares decomposition) and **A** = **USV**^{T} (the SVD) is given by

$$\mathbf{V}^{T} = \mathbf{HF}^{T} \tag{7}$$

and

$$\mathbf{D} = \mathbf{USH} \tag{8}$$

where **H** is found by making successive guesses at the Adair parameters that determine **F**, generating **F**, then applying matrix least squares,

$$\mathbf{H} = \mathbf{V}^{T}(\mathbf{F}^{T+}) \tag{9}$$

The solution parameters are those that minimize the norm of $\mathbf{S}(\mathbf{V}^{T} - \mathbf{HF}^{T})$.

This SVD-based procedure is proved in Shrager[30] to give exactly the same result as direct matrix least squares. So what are the reasons for using SVD? One reason is that the full **U**, **S**, and **V** matrices are in fact never used, because statistically indistinguishable results can be obtained by using only the first r columns of **U** and **V**, and the first r rows and columns of **S**. Thus, in terms of computing effort, $\mathbf{V}^{T} = \mathbf{HF}^{T}$ is a much smaller problem than $\mathbf{A} = \mathbf{DF}^{T}$. But equally important, SVD offers assistance in choosing a feasible model. Trying to derive a model by looking at the rows of **A** is often difficult, because the rows of **A** tend to look alike in a restricted wavelength region, and because small but independent trends tend to be swamped by larger trends and even by noise in any single row of **A**. But SVD has two advantageous properties. First, SVD collects almost all the signal in all the rows of **A** into the fewest possible columns of **V** (e.g., in our experiments, we rarely have to look past the fifth column in **V** for signal). Second, SVD tends to produce columns of **V** of contrasting shape, so that if a small but significant trend does not show up well in one column of **V**, it shows up well in another. These contrasting shapes also convey information about mechanism. If there are only two spectra in the data (e.g., deoxyHb and oxyHb with no distinction between T and R states), then the apparent rank of **A** will be $r = 2$, and only the first two columns of **V** will have significant signal. Furthermore, both columns will follow the Adair trend, a single sigmoid,

[31] G. Golub and W. Kahan, *SIAM J. Numer. Anal. B* **2**, 205 (1965).
[32] S. D. Frans and J. M. Harris, *Anal. Chem.* **56**, 466 (1984).

with contrast only in the base level and scale of the curves. (F in this case consists of two Adair curves: 0 to 1 and 1 to 0.) But if, say, there are four distinguishable species: deoxy-T, deoxy-R, oxy-T, and oxy-R, then four columns of V will contain signal, and their number of up-and-down trends versus pO_2 will absolutely preclude a simple deoxy-to-oxy mechanism.

So now the question is, for a four-species model, what can we use in place of the Adair curves in F? The Adair curves were in F because the 0-to-1 curve described the appearance of oxyHb, and the 1-to-0 curve described the disappearance of deoxyHb. Now we must describe the appearance and disappearance of four species (i.e., species = type of site: unbound T or R, or bound T or R) by postulating models and testing them. For a simple example, assume that all hemoglobin tetramers are in the T state for zero sites or any one site bound to oxygen and in the R state for any two, three, or four sites bound, with no distinction between the α and β sites. By the laws of mass action, to determine the populations of the four species (sites), let $x = pO_2$, $D = 1 + a_1 x + a_2 x^2 + a_3 x^3 + a_4 x^4$, y_i = concentration of $Hb(O_2)_i$ (i.e., $y_0 = 1/D$, and $y_i = a_i x^i / D$, $i = 1 : 4$). Then the desired populations are

$$\text{Deoxy-T} = y_0 + 0.75 y_1 \qquad \text{Deoxy-R} = 0.5 y_2 + 0.25 y_3$$
$$\text{Oxy-T} \ \ = 0.25 y_1 \qquad\qquad \text{Oxy-R} \ \ = 0.5 y_2 + 0.75 y_3 + y_4$$

So, as expected, the oxygenation model will start at $x = 0$ with all deoxy-T, finish at high x with almost all oxy-R, with oxy-T and deoxy-R as rising and falling intermediates. These are the four columns of F, the shapes of which are governed by the a values, which are adjusted by a curve-fitting program to minimize the norm of $A - DF^T$ (in full matrix least squares) or $S(V^T - HF^T)$ (in the SVD-based procedure). The choice of model depends on the combination of intermediates in the two conformational states and can be tested by allowing for the best fit to the data.

The differences between T and R spectra are subtle at best. Only a sensitive procedure can hope to detect them. And because we are using least squares rather than partial least squares, our model must include any phenomenon that our procedure can detect, not only T-to-R transitions, but also metHb formation and dimerization, unless these effects can be stabilized. (With some reformulation of the problem, SVD can ignore unchanging background.) But to return to the original point, if our instruments (e.g., spectrophotometers and electrodes), our data scrubbing (e.g., digital filters), or our experimental designs (e.g., kinetics with slow spectrophotometric scans) introduce subtle signals of their own, analysis becomes more difficult.

Kinetics and Spectrophotometer: A Study in Artifact

This section provides some corrections for errors induced by slow kinetics (i.e., pseudoequilibrium) rather than stepwise equilibrium. Time must be recorded along with pO_2, absorbance, and even wavelength in some cases. Spectra take time to gather. When the wavelength range is scanned, absorbance at each wavelength is measured at a different time. When several such spectra are averaged, the distribution of sample times is unique for each wavelength. To specify this distribution, some notation is in order:

n = number of scans combined to produce a single spectrum

Wavelength

w	wavelength in nanometers
w_{min}	minimum wavelength (start of a forward scan)
w_{max}	maximum wavelength (start of a backward scan)
w_c	central $w = (w_{min} + w_{max})/2$

Time

t	time in some standard unit
t_s	time to scan from w_{min} to w_{max}, forward or backward
t_d	dead time between successive scans
t_c	central t, midway between the start of scan 1 and the end of scan n, including dead times. For all computations below, $t_c = 0$ should be used to improve the condition of the matrices involved
c_i	central time of the ith scan relative to t_c
$f(w)$	time of sample of w relative to w_c in a forward scan
$b(w)$	time of sample of w relative to w_c in a backward scan
$t_i(w)$	time of sample of w in the ith scan relative to t_c

Absorbance

a	absorbance in optical density
$a[w, t_i(w)]$	observed a from the ith scan at w [and $t_i(w)$]
$a(w, t_c)$	estimated a at w and t_c (i.e., a deduced simultaneous spectrum at t_c, corrected for time dependence)
$a'(w, t_c)$	the first time derivative, $da(w, t_c)/dt$
$a''(w, t_c)$	the second time derivative, $d^2a(w, t_c)/dt^2$

Our focus in this context is $a(w, t_c)$, but what we obtain from the scanning procedure is a series of spectra,

$$a[w, t_i(w)] = a(w, t_c) + a'(w, t_c)t_i(w) + \tfrac{1}{2}a''(w, t_c)t_i^2(w) \qquad (10)$$

which, even when averaged, do not produce the desired spectrum. Equation (10) is a three-term Taylor series expansion of $a(w, t_c)$ with respect to time. (If the scan is so slow that three terms are not enough, chances are that the wrong experiment is being performed.) Using Eq. (10) as a model, time effects can be corrected for by fitting the curve $a[w, t_i(w)]$ versus $t_i(w)$ to a parabola at each fixed w. One needs at least 3 scans to do this because there are 3 parameters, but 10 scans or more are preferred to reduce noise. To do any of this, the functions $t_i(w)$ must be known. The linear model of $t_i(w)$ is offered here as simple to apply, but a more realistic model is to be preferred:

$$t_i(w) = t_c + c_i + f(w)$$
$$c_i = [i - (n + 1)/2](t_s + t_d) \qquad (11)$$
$$f(w) = t_i(w) - c_i = (t_s/2)(w - w_c)/(w_{max} - w_c)$$

As suggested above, $t_c = 0$ should be the convention in Eq. (11). From Eq. (11), for any wavelength, all the times at which $a(w, t)$ was sampled can be generated. Fitting a versus t to a parabola and picking off $a(w, t_c)$ is then standard procedure. Also, when all scans are forward, the $t_i(w)$ values are spaced equally, and the spacing $t_s + t_d$ is the same for all w, although the time displacement from t_c is not. Still, the common spacing allows for considerable economy of calculation, because only one matrix need be inverted instead of one for each w, that is,

$$\mathbf{G} = \left\{ \begin{bmatrix} 1 & 1 & \cdots & 1 \\ t_1(w) & t_2(w) & \cdots & t_n(w) \\ t_1^2(w) & t_2^2(w) & \cdots & t_n^2(w) \end{bmatrix}' \right\}^+ \qquad (12)$$

For each w,

$$h = \mathbf{G}[a[w, t_1(w)] \; a[w, t_2(w)] \ldots a[w, t_n(w)]]'$$
$$a(w, t_c) = h_1 - h_2 f(w) + h_3 f^2(w)$$

In the matrix in Eq. (12), the $+$ symbol once again stands for the Moore–Penrose pseudoinverse, this time of an $n \times 3$ matrix. $\mathbf{G}$ itself will be a $3 \times n$ matrix.

Some spectrophotometers allow both forward and backward scans. We will assume that the timing of a backward scan is symmetric to the timing of a forward scan [i.e., $b(w) = -f(w)$]. If the forward–backward feature is used, successive scans should alternate in direction, and the number of scans should be even to cancel out the effects of odd terms in the Taylor series in Eq. (10). When this is done, the central time, t_c, is also the midrange time $[t_1(w) + t_n(w)]/2$ for every w, but the spacings of $t_i(w)$ are different for every w. In particular, for w linear in t within each scan, we have

$$
\begin{aligned}
t_i(w) &= t_c + c_i + f(w) &&\text{for odd } i \\
t_i(w) &= t_c + c_i - f(w) &&\text{for even } i
\end{aligned}
\tag{13}
$$

t_c, c_i, and $f(w)$ are as defined in Eq. (11). This staggered spacing of $t_i(w)$ requires a different matrix inverse for every w, but because of symmetric properties, rows 2 and 3 of $\mathbf{G}$ become superfluous. The procedure for interpolating $a(w, t_c)$, including computing $\mathbf{G}$, is repeated for every w:

$$
\begin{aligned}
\mathbf{G} &= \left\{ \begin{bmatrix} 1 & 1 & \cdots & 1 \\ t_1^2(w) & t_2^2(w) & \cdots & t_n^2(w) \end{bmatrix}' \right\}^+ \\
\mathbf{H} &= [\,a[w, t_1(w)] \quad a[w, t_2(w)] \quad \cdots \quad a[w, t_n(w)]\,]' \\
a(w, t_c) &= [\mathbf{G} \text{ row } 1]\, \mathbf{H}
\end{aligned}
\tag{14}
$$

Of course, the collection of [$\mathbf{G}$ row 1] values for all w will be constant for any sampling pattern, not only for all t in the current experiment, but also for other similar experiments. So rather than recompute $\mathbf{G}$ every time it is needed, one may choose to store all the [$\mathbf{G}$ row 1] values in a file, and retrieve them, or compute them for the first t_c, and keep them in memory for subsequent t_c. The choice will depend on n, the number of w, and the memory capacity and speed of the computer.

When several "instantaneous" scans (e.g., from a photodiode array) are being combined to produce the spectrum at t, Eqs. (13) and (14) can still be used, with the following simplification: $f(w) = 0$ and, consequently, the same [$\mathbf{G}$ row 1] holds for all w.

When several scans are combined to produce a single spectrum, one may assume that errors due to the length of a single absorbance measurement are negligible compared with the discrepancies between scans. But when all measurements are taken at once, as in a photodiode array, one may decide on a single long scan to reduce noise, rather than combine several short scans. The errors now take on a new disguise: an integral. Taking t_c to be zero, a scan runs from $-t_s/2$ to $t_s/2$, common to all w. The result is an average value over that interval, namely,

$$
\bar{a}(w, 0) = 1/t_s \int_{-t_s/2}^{t_s/2} [a(w, 0) + a'(w, 0)t_s + \frac{1}{2} a''(w, 0)t_s^2]\, dt
\tag{15}
$$

again limiting the Taylor expansion in the integrand to three terms. From Eq. (15), we can deduce the following correction:

$$
a(w, t_c) = \bar{a}(w, t_c) - \tfrac{1}{24} a''(w, t_c)t_s^2
\tag{16}
$$

Unlike the multiple-scan situation, in which several samples are available for least-squares approximation of $a(w, t)$, in Eq. (16) there is no local information except $\bar{a}(w, t_c)$. This single datum cannot be used to correct itself. A less local approach must be used to approximate a'' for the correction in Eq. (16). One method is to fit a local parabola through $\bar{a}(w, t_c)$ and its neighbors in time, and use that second derivative as a''. Another method is to use a global model (e.g., the Adair curve) for all t_c, and use the second derivative of that model fitted to data at each wavelength.

Do we accomplish anything by these corrections? Remember, the corrections are being made because the data will be subjected to highly sensitive analysis methods. If, for example, the SVD analysis reveals unexplained components in the raw data $\bar{a}(w, t_c)$, and if these components disappear when the data are corrected, one can be fairly certain that they are artifacts of time dependence.

In our case, the need for a derivative correction was eliminated by doing alternate forward and backward scans. By numerically approximating second derivatives, it was determined that the corrections would be well below any effect we could measure, including noise, and so the correction procedure was not used in the experiments that we describe in the next section.

Example of Experiment

A representation of an experimental matrix $\mathbf{A}$ is shown in Fig. 2. In this experiment, human deoxyhemoglobin in 50 mM Bis–Tris, 0.1 M Cl^-, at pH 7.4 is oxygenated at 25°. Matrix $\mathbf{A}$ is evaluated by SVD to determine the rank r of the matrix and to construct the matrices $\mathbf{U}$, $\mathbf{S}$, and $\mathbf{V}$, where the columns in $\mathbf{U}$ are linear combinations of the independent spectra of each component in $\mathbf{A}$, the columns of $\mathbf{V}$ are linear combinations of the transitions of each component in $\mathbf{A}$, and $\mathbf{S}$ is used to evaluate r, or the number of components necessary to generate $\mathbf{A}$. For ideal data (i.e., the absence of noise), r is given by the number of nonzero diagonal elements in $\mathbf{S}$. For real data, all the diagonal elements of $\mathbf{S}$ are positive, but the values that are derived from noise are small. The chemical transitions that are resolved by SVD are above the noise level. For columns i where $i > r$, $\mathbf{U}$ column i and $\mathbf{V}$ column i contribute a negligible amount to the signal in $\mathbf{A}$.

Analysis of the rank of $\mathbf{A}$ is shown in Fig. 3. In this experiment, a maximum of five spectral components is attributed to $\mathbf{A}$, and the first five columns of matrices $\mathbf{U}$ and $\mathbf{V}$ from this example are represented in Fig. 4A–E. The increasing contribution of noise is evident as we move from

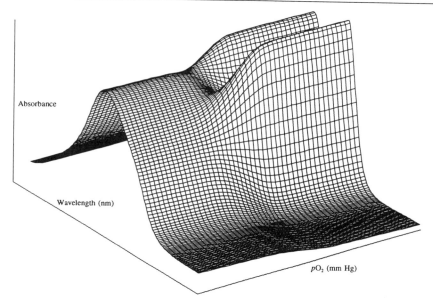

FIG. 2. Matrix **A**. Wavelength is shown along the x axis from 480 to 650 nm. Absorbance units (Au) are given on the y axis (maximum absorbance $\approx$ 1 Au), and pO_2 in millimeters of mercury is shown increasing along the z axis from 0 to $\sim$100 mmHg. For clarity, only every fifth data point is represented along the z axis.

the lower to higher numbered columns in **U** and **V**. As shown by the singular values in Fig. 3, the first two components comprise the large majority of the signal. Inspection of **U** columns 1 and 2 shows that the first component reflects primarily the linear average of the deoxyHb and oxyHb spectra in the matrix, and the second component reflects primarily the oxyHb–deoxyHb difference spectrum. The transitions of both of the major components, as shown in **V** columns 1 and 2, are sigmoid curves and reflect the oxygen-binding equilibrium. The other three components make much smaller contributions to the overall transition and, as expressed by the ratio of their singular value to the sum of the first and second component singular values, the fractions of the signal contributed by the third, fourth, and fifth components are $\sim$0.084, $\sim$0.040, and $\sim$0.032%, respectively.

A first approximation of the identity of the third component, if we assume that some of the heme–iron atoms oxidized during the experiment, is made by multicomponent analysis of matrix **A** based on three species: deoxyHb, oxyHb, and metHb. This analysis is only a first-order approximation because its accuracy depends on the completeness of matrix **M**,

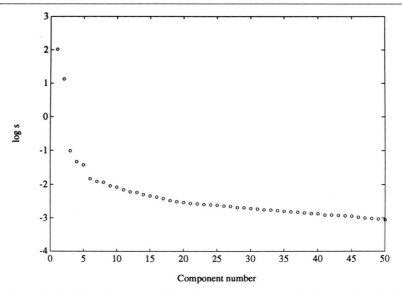

FIG. 3. Evaluation of the rank of **A**: singular values from the diagonal matrix **S** as determined by SVD of matrix **A**. Only the first 50 values are represented. The rank of matrix **A** is estimated as five by identification of five component singular values above the systematic decline in $\log(s)$.

containing a parent spectrum for every species in the experimental data, but by SVD, the rank of **A** is probably greater than three. Even so, the minor 2 components are near the level of random noise and comprise a total of less than 1 part in 1000 of the entire spectral information in **A**. In this case, multicomponent analysis provides a reasonable evaluation of the percentage of deoxyHb, oxyHb, and, to a lesser extent, metHb in each spectrum of **A**. Fractional saturation, Y, is calculated from Eq. (2), and the rate of change of metHb (ΔmetHb/Δt) can be examined according to this analysis in relation to Y (Fig. 5). It can be seen that the position of the ΔmetHb/Δt maximum is similar to that in the transition of **V** column 3.

The fourth SVD component shows a transition having some pO_2 correlation resolved just above random noise. It is difficult to interpret this signal, however, because it is so small. For the type of experiment described here, instrument noise must be decreased or the signal must be enhanced before matrix least-squares analysis can be used to model this transition successfully.

The fifth SVD component is wholly or partially an instrumental artifact that has primarily random correlation with pO_2. The signal is revealed as a derivative of the parent spectra that arises due to slight fluctuations in

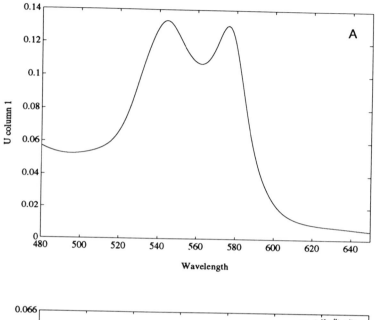

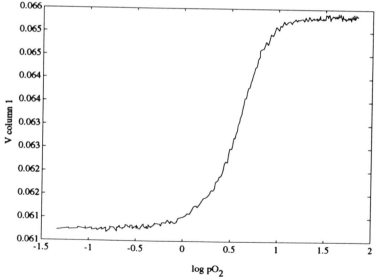

FIG. 4. The first five columns of matrices **U** and **V** by SVD of matrix **A**. Matrix **U** (top) and **V** (bottom) columns 1 (A), 2 (B), 3 (C), 4 (D), and 5 (E).

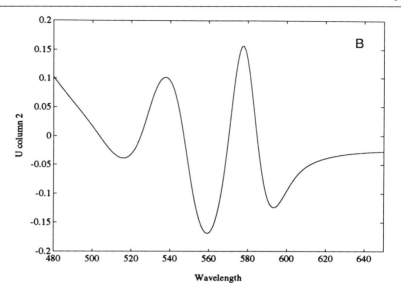

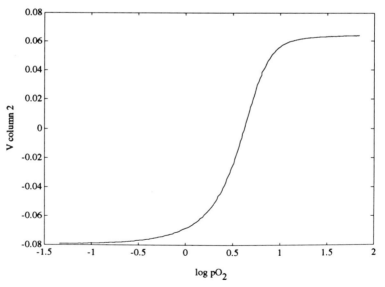

Fig. 4. (continued)

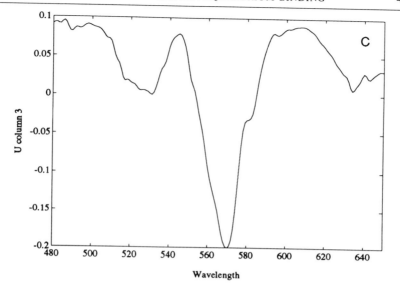

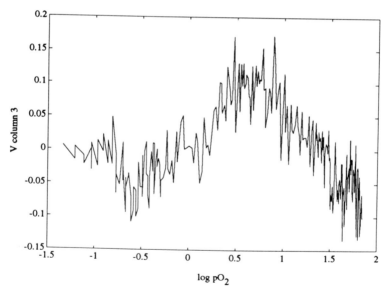

FIG. 4. (*continued*)

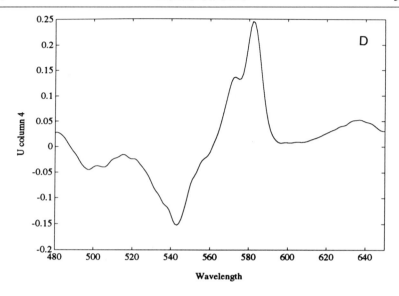

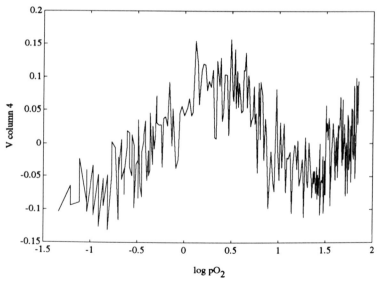

FIG. 4. (*continued*)

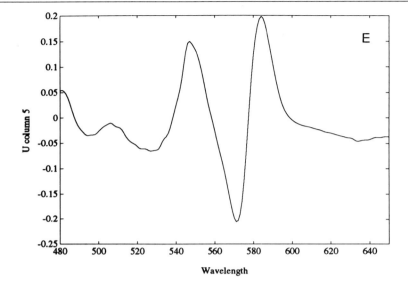

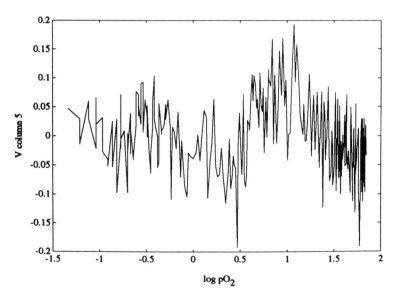

FIG. 4. (*continued*)

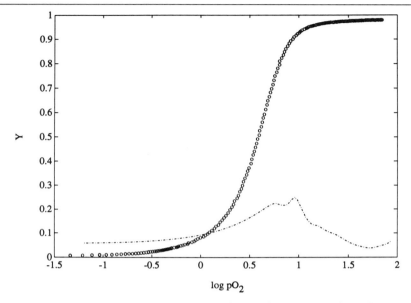

FIG. 5. Oxygen equilibrium curve (○) and rate of change in percentage of metHb (ΔmetHb/Δt) (–·–) from the binding reaction of human hemoglobin. Conditions: 60 μM (in heme) human hemoglobin A_0 in 50 mM Bis–Tris, 0.1 M Cl$^-$, pH 7.4, at 25°. Symbols give fractional saturation (Y) as a function of pO_2 (mmHg). The line (–·–) represents a smoothed, fivefold magnification in amplitude of ΔmetHb/Δt as a function of pO_2. Y was calculated from Eq. (2), using the multicomponent analysis of matrix $\mathbf{A}$ from three parent species: oxyHb, deoxyHb, and metHb. ΔMetHb was calculated as the percentage change in metHb from spectrum to spectrum as determined by multicomponent analysis of matrix $\mathbf{A}$.

wavelength position from scan to scan. The error is too small to be seen visually, but because it results from a nonrandom signal, it is resolved by SVD.

Wavelength-position error can be detected by repeated scanning of a static sample having a sharp peak, such as oxyhemoglobin, carbonmonoxyhemoglobin, or reduced cytochrome c. For the highest reproducibility of repeated scans, the sample should be scanned as fast as possible to avoid problems of protein deterioration or changes in ligand or oxidative state. Theoretically, in the absence of horizontal wavelength "jitter," SVD analysis of repeated scans of a static sample will produce a single component above the random noise of the system. Alternatively, a second SVD component in this analysis will provide the contribution of the error as $s_{2,2}$ and the derivative spectrum as $\mathbf{U}$ column 2.

The primary obstacle in the modeling effort now is in the derivation of an appropriate model for the transition to methemoglobin. Multicomponent analysis gives us an approximation of methemoglobin formation in relation to hemoglobin saturation (i.e., see Fig. 5), but it does not reveal the mechanism of oxidation in the tetramer, which may occur at any one of four binding sites, to any partially liganded species, and, perhaps, preferentially at one type of subunit. The reaction becomes notably more complex with these considerations, and modeling it is nontrivial. Information about the process of hemoglobin oxidation is required for a realistic model. As a first step, valency hybrids can be used to evaluate the effects of oxidation on the equilibrium curve.[33]

The hemoglobin conformational transition, which can be resolved as a singular spectral component if the signal measurement is sensitive and accurate enough, is another area under study. Strong allosteric effectors, such as inositol hexaphosphate[28] and bezafibrate, are being used to test this by possibly enhancing the signal. Alternatively, because the extinction coefficient for the quaternary change is much higher in the Soret region,[15] there is a better probability of detecting this signal between 410 and 450 nm by using a lower heme concentration in our reaction cell or by using the shorter path length of the thin-layer apparatus in a rapid-scanning spectrophotometric system.[25]

In the future, mathematical models will include calculations of tetramer dissociation and, perhaps, subunit differences in ligand binding. As instruments and theory improve, the critical components of the reaction will be resolved better and, once limiting assumptions are no longer required for experimental analyses, the evaluation of hemoglobin–oxygen equilibrium binding will become a simpler and more reliable process.

Acknowledgment

The opinions or assertions contained herein are the private views of the authors and are not to be construed as official or as reflecting the views of the U.S. Department of the Army or the Department of Defense.

[33] M. C. Marden, J. Kister, and C. Poyart, this volume [6].

[23] Oxygen Equilibrium Curve of
Concentrated Hemoglobin

By ROBERT M. WINSLOW, ALISON MURRAY, and CARTER C. GIBSON

Introduction

Controversy still surrounds the measurement of the hemoglobin–oxygen equilibrium curve (OEC), even though various methods have been used for many years that are able to produce continuous measurements over a wide range of P_{O_2}. Thin-layer methods that can be used for erythrocyte suspensions and hemoglobin solutions have been reviewed by Lapennas.[1] In these systems, the samples are susceptible to uncontrolled oxidation, and uncertainty is introduced by variable optical path lengths that might change as the sample dries. Moreover, optical measurements of hemoglobin oxygenation have come under suspicion with the demonstration that the optical absorbance of hemoglobin may not be linear with oxygenation.[2,3]

The methods described by Imai[4] have provided valuable data on dilute solutions of hemoglobin, but also rest on optical measurements, and can use only relatively dilute protein solutions, in which tetramer–dimer dissociation is significant.[5] Our group has described a rapid-scanning technique that has the potential to overcome many of these problems and, in addition, can give information about the oxidation state and presence of any unusual spectral components during the oxygenation measurement.[6]

Only one instrument is available commercially to measure the oxygen equilibrium curve of erythrocytes, blood, and concentrated hemoglobin solutions, the Hemox analyzer (TCS, Philadelphia, PA). This is a serviceable instrument, useful for routine laboratory and clinical purposes, but it has limitations as a research laboratory instrument. First, it uses optical analysis of the oxygen saturation, with the problems discussed above. Second, because the change in absorbance as a function of P_{O_2} is followed, estimation of the degree of hemoglobin saturation with O_2 is relative, not absolute.

[1] G. N. Lapennas, J. M. Colacino, and J. Bonaventura, this series, Vol. 76, p. 449.

[2] L. J. Parkhurst, T. M. Larsen, and H.-Y. Lee, this volume [29].

[3] M. L. Doyle, E. DiCera, and S. J. Gill, *Biochemistry* **27,** 820 (1988).

[4] K. Imai, this series, Vol. 76, p. 438.

[5] F. C. Mills, D. Gingrich, B. M. Hoffman, and G. K. Ackers, *Biochem. J.* **5,** 5350 (1976).

[6] K. D. Vandegriff and R. I. Shrager, this volume [22].

METHODS IN ENZYMOLOGY, VOL. 232

Clearly, a method is needed that does not rely on optical measurements, and that can be used with concentrated hemoglobin solutions. We have previously described such a method, designed for use with whole blood samples or erythrocyte suspensions.[7–9] The technique consists of the slow, measured addition of H_2O_2 to a sample of blood or erythrocyte suspension in a closed cuvette, with fixed volume. The measured P_{O_2} gives dissolved O_2 if the solubility coefficient is known, and the rate of addition of H_2O_2 gives the total O_2 content if the dismutation of H_2O_2 to O_2 and water is quantitative.

Preliminary unpublished experiments (1993) with chromatographically pure hemoglobin suggested that cell-free hemoglobin is too susceptible to oxidation by H_2O_2 for this method to be of value. Because of the current pressing need for a convenient method to measure the OEC in our work with hemoglobin-based blood substitutes, we have introduced improvements into the previously reported technique that permit measurements with cell-free concentrated hemoglobin solutions. The method currently used in our laboratory is described in this chapter.

Methods

Hydrogen peroxide is obtained from Sigma Chemical Corp. (St. Louis, MO), as a 30% (w/v) solution (8.8 M). It is diluted in distilled water to a final concentration of 0.44 M for use in the oxygenation experiments. Catalase is obtained from Sigma, either as a pure enzyme (C40) or as a crude extract of bovine liver (C10). Lactoperoxidase (LPO) is also obtained from Sigma. KI is reagent grade.

Apparatus

The apparatus (Fig. 1) is modified from that described previously.[9] The reaction cuvette is constructed of Lucite, and is fitted with a thermister (T) mounted in the removable portion. The temperature of the entire reaction cuvette is maintained by a circulating water bath that also is used to control the temperature of the tonometer (IL 237; Instrumentation Laboratories). The most significant change is the addition of a stirring motor (M1) mounted above the reaction cuvette (1.85-ml volume), connected to a shaft with an elliptical stirrer. The O_2 and CO_2 electrodes and

[7] R. M. Winslow, M. Swenberg, R. L. Berger, R. I. Shrager, M. Luzzana, M. Samaja, and L. Rossi-Bernardi, *J. Biol. Chem.* **252**, 2331 (1977).

[8] R. M. Winslow, J. M. Morrisey, R. L. Berger, P. D. Smith, and C. C. Gibson, *J. Appl. Physiol.: Respir., Environ. Exercise Physiol.* **45**, 289 (1978).

[9] R. M. Winslow, N. Statham, and L. Rossi-Bernardi, this series, Vol. 76, p. 511.

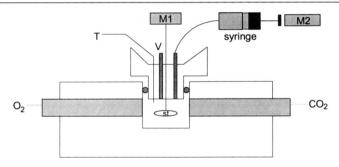

FIG. 1. Diagram of the experimental apparatus. The main body of the chamber is Lucite, and water circulates through it for temperature control. Two electrodes are mounted through the chamber and are in contact with the blood in the mixing chamber. The mixing chamber is capped by a Lucite plug sealed by an O ring. The Lucite plug contains three ports: one is for a thermister (T), one is a vent (V) to permit the exit of solution as the volume increases, and the third is an inlet for addition of H_2O_2. The stirring motor (M1) is mounted above the cuvette and connects to an elliptical stirrer (st) by a rigid shaft. A second motor (M2) drives a syringe for the delivery of H_2O_2.

amplifiers are supplied by Instrumentation Laboratories. The output from the electrodes is connected to an analog-to-digital (A/D) converter (16 bits) that is part of a PDP-11/73 computer (Digital Equipment Corporation, Maynard, MA). The H_2O_2 syringe is driven by a pump (M2) that is controlled manually, rather than under computer control as previously. In the present experiments, samples are not equilibrated with CO_2, thus titration using the P_{CO_2} electrode as previously reported is not necessary.

Data Collection and Analysis

Data collection consists of sampling time and the P_{O_2} voltage at the time intervals specified by the operator at the start of the run. Analysis of the data is carried out on a MicroVaxII computer (Digital Experiment Corporation), using a program written in Fortran. Fitting of P_{O_2} saturation data to the Adair model is done using a Fortran program version of the methods described previously.[7] Graphic display is accomplished using RS/1 (Bolt, Beranek, and Newman, Cambridge, MA).

Hemoglobin Samples

All hemoglobin samples are prepared from outdated human blood according to the procedures described elsewhere in this series.[10] "Stroma-

---

[10] R. M. Winslow and K. W. Chapman, this series, Vol. 231 [1].

free'' hemoglobin (SFH) is prepared by gentle lysis of washed erythrocytes followed by a series of filtration steps. This preparation contains many erythrocyte enzymes.[11] To prepare hemoglobin A_0, SFH is chromatographically purified in 50 to 100-g batches. All hemoglobin solutions, whether SFH or A_0, are deionized on a column of mixed-bed resin (Bio-Rad, Richmond, CA) before use. The cross-linked hemoglobin ($\alpha\alpha$Hb) is specifically modified to cross-link between the α chains at Lys-99 residues.[12]

Oxidation

In preliminary experiments, we observed that addition of H_2O_2 to purified hemoglobin solutions led rapidly to the formation of methemoglobin as iron oxidized from Fe^{2+} to Fe^{3+}. Accordingly, a systematic set of measurements was made on the effect of the type of hemoglobin and the presence or absence of catalase.

After deoxygenation of a sample in the tonometer, it is transferred to the reaction cuvette, which has been previously flushed with N_2. A small (20-μl) sample is withdrawn and diluted to 1 ml (in 0.1 M phosphate, 0.1 M KCl, or 0.1 M NaCl, depending on the experiment). The optical spectrum from 500 to 650 nm is recorded on a Cary-14 spectrophotometer (Varian Instruments Division, Palo Alto, CA), and the amount of methemoglobin is calculated by a least-squares multicomponent method.[6] After the addition of an amount of H_2O_2 sufficient to oxygenate all the hemoglobin, a second sample is withdrawn and a repeat measurement is made.

Results

Solubility Coefficient for O_2 (α)

The oxygenation data can be used to calculate the solubility coefficient for O_2 (α). For each increment in O_2 added, total O_{2T} is calculated,

$$O_{2T} = (YC) + (\alpha Pcv) \tag{1}$$

where Y is fractional hemoglobin saturation, C is total O_2 capacity for the reaction cuvette (μmol), P is P_{O_2} (torr), and cv is the cuvette volume (μl). The amount of O_2 added in each time increment is

$$\Delta O_{2T} = [(Y_iC) + (\alpha P_i cv)] - [(Y_{i-1}C) + (\alpha P_{i-1}cv)] \tag{2}$$

[11] S. M. Christensen, F. Medina, R. M. Winslow, S. M. Snell, A. Zegna, and M. Marini, J. Biochem. Biophys. Methods **17**, 143 (1988).
[12] S. R. Snyder, E. V. Welty, R. Y. Walder, L. A. Williams, and J. A. Walder, Proc. Natl. Acad. Sci. U.S.A. **84**, 7280 (1987).

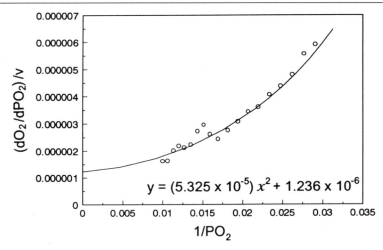

FIG. 2. Determination of the solubility coefficient (α) for oxygen. The data points are taken from the extreme high P_{O_2} range of an oxygenation experiment. The slope of the total O_2 vs P_{O_2} line is plotted against reciprocal P_{O_2} ($1/P_{O_2}$) to estimate the solubility coefficient at infinite hemoglobin saturation. At this point, hemoglobin is completely saturated with O_2, and any increase in P_{O_2} measured with the O_2 electrode is a direct measure of α.

When hemoglobin is fully saturated with O_2, $Y_i = Y_{i-1}$, and Eq. (2) reduces to

$$\alpha = (\Delta O_{2T})/(\Delta P \text{cv}) \tag{3}$$

To estimate the value of the solubility coefficient of O_2 in the solution, plots of the slope of the total O_2 as a function of P_{O_2} (dO_2/dP_{O_2}) are made as a function of reciprocal P_{O_2} (Fig. 2). In this way, it is convenient to extrapolate the curve to infinite P_{O_2} ($1/P_{O_2} = 0$) and to use the simple polynomial regression ($Y = ax^2 + b$) to estimate the limiting value of the slope. This limit is used as the value of α for a given experiment.

To determine the final solubility coefficient, it is necessary to know the exact volume of the cuvette. We estimated this value by adding hemoglobin solution to the cuvette from a glass microsyringe and recording the volume injected. The procedure resulted in a value of 1.850 ml for the current cuvette used in our laboratory. The final value of the solubility coefficient can be obtained as the limiting value of Eq. (3).

Results of determination of α for a typical experiment using a preparation of SFH are shown in Fig. 2. Using data from four separate experiments on the same sample of SFH, the extrapolated intercepts give an average value of 1.236×10^{-6} μmol/(torr·μl). Changes in protein concentration,

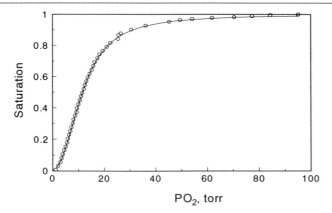

FIG. 3. An example of an experiment carried out with concentrated hemoglobin (about 3 mM heme). The solution is isoionic hemoglobin, with no buffer present. Circles represent data points, and the solid line is the Adair fit line.

solution conditions, or temperature can have significant effects on the actual value of α.[13] The classic figure for the solubility coefficient in plasma at 38° [14] is 0.00316 ml/(100 ml·torr), or 1.241×10^{-6} μmol/(torr·μl), a value in close agreement with ours. A typical final OEC is shown in Fig. 3.

Effect of Errors in α on Oxygen Equilibrium Curve

To estimate the effect of errors in α on the parameters of the OEC, a series of analyses of one data set was carried out using different values of α (Table I). Variation of α from 0.4×10^{-6} to 2.8×10^{-6} resulted in final calculated saturations of 1.024 to 0.9454. The P_{50} changed only slightly, from 2.98 to 3.10, and the Hill parameter, n, changed from 1.98 to 2.02. Interestingly, the final sum of squared residuals reached a minimum between α values of 1.0 and 1.6×10^{-6}, in agreement with our calculated value of 1.201×10^{-6} (at 37°). More significant errors occur in the values of the Adair parameters, especially a_2 and a_3, with a_2 being especially sensitive.

Oxidation of Hemoglobin

To study the oxidation of hemoglobin during the oxygenation run, a number of experiments were done using different hemoglobin prepara-

[13] C. Christophorides and J. Hedley-White, *J. Appl. Physiol.: Respir., Environ. Exercise Physiol.* **27,** 592 (1969).

[14] A. White, P. Hendler, E. L. Smith, and D. Stetten, "Principles of Biochemistry," p. 662. McGraw-Hill, New York, 1959.

TABLE I

EFFECT OF ERRORS IN α ON OXYGENATION PARAMETERS FOR ONE DATA SET[a]

Parameter	$\alpha \times 10^{-6}$ (μmol/torr·μl)				
	0.4	1.0	1.6	2.2	2.8
Final Y	1.024	1.004	0.9845	0.9651	0.9454
$a_1 \times 10$	3.314	3.739	4.195	4.681	5.204
$a_2 \times 10^2$	11.69	9.142	6.481	3.637	0.5861
$a_3 \times 10^2$	4.034	4.782	5.578	6.426	7.333
$a_4 \times 10^2$	1.273	1.229	1.181	1.132	1.080
ssr $\times 10^2$	2.022	1.705	1.933	2.680	3.932
P_{50} (torr)	2.98	3.00	3.03	3.07	3.10
n	1.98	1.99	2.00	2.01	2.02

[a] At 25°. α, O_2 solubility coefficient; Y, hemoglobin saturation; a_i, Adair parameters; ssr, sum of squared residuals for the Adair fit; n, Hill parameter.

tions, H_2O_2 concentrations, and catalase preparations. These are summarized in Table II.

Stroma-free hemoglobin (SFH) can be studied by the present method without excessive oxidation, even without the addition of catalase. Presumably this is because sufficiently active enzyme systems are copurified with hemoglobin that can dismutate H_2O_2 to O_2 and H_2O.

Chromatographically purified hemoglobin A_0 and cross-linked hemo-

TABLE II

OXIDATION OF HEMOGLOBIN[a]

Hemoglobin	n	H_2O_2 (M)	Addition	Percentage methemoglobin increase
SFH	9	0.44		1
SFH	2	0.22		4
A_0	1	0.22		67
A_0	1	0.44		82
A_0	2	0.44	C10 (crude) catalase	3
A_0	3	0.44	C40 (pure) catalase	62
A_0	1	0.44	C40 catalase + LPO + KI	4
$\alpha\alpha$Hb	1	0.44		38
$\alpha\alpha$Hb (heated)	1	0.44		38
$\alpha\alpha$Hb	1	0.44	C10 (crude) catalase	5

[a] n, Number of experiments; SFH, stroma-free hemoglobin; LPO, lactoperoxidase; $\alpha\alpha$Hb, hemoglobin cross-linked between α chains.

globin ($\alpha\alpha$Hb), in contrast, are strikingly susceptible to oxidation, presumably because the same enzymes that protect SFH are not present in the A_0 preparations.[11] Addition of a crude catalase preparation (C10; Sigma) prevents excessive oxidation, but the presence of a more purified preparation (C40; Sigma) does not. When lactoperoxidase and KI are added to the C40 catalase, oxidation is prevented. Thus, the exact mechanism of H_2O_2 dismutation is unclear.

The susceptibility of $\alpha\alpha$Hb to oxidation is interesting because it was prepared from SFH, not from chromatographically pure hemoglobin A_0. One possible explanation is that the pasteurization process to which $\alpha\alpha$Hb is exposed might inactivate enzymes that copurify with it. However, a comparison of samples that had not been heat treated showed no difference in oxidation by H_2O_2. Other experiments included addition of nonhemoglobin column fractions from the preparative high-performance liquid chromatography (HPLC) fractionations and addition of desferrioxamine. The latter was attempted because free iron and H_2O_2 are known to produce oxygen radicals that might oxidize hemoglobin in our experiments.

Reproducibility of Oxygen Equilibrium Curve

To test the reproducibility of the OEC measurements, several experiments were carried out with separate aliquots of the same sample of SFH (Table III). Examination of Table III shows that there is some variation between individual experiments. The sum of squared residual (ssr) values

TABLE III
REPLICATE MEASUREMENTS OF OXYGEN EQUILIBRIUM CURVE FOR SINGLE
HEMOGLOBIN PREPARATION[a]

Parameter	Experiment No.				Mean
	29	31	32	33	
Number of points	288	300	308	333	307
ssr $\times 10^6$	0.1625	1.017	4.355	1.245	2.060
$a_1 \times 10^1$	3.495	7.244	5.774	6.532	5.761
$a_2 \times 10^1$	1.163	1.071	1.195	3.584	1.753
$a_3 \times 10^2$	3.948	9.725	1.028	6.568	5.317
$a_4 \times 10^2$	1.414	1.336	2.042	1.381	1.543
P_{50} (torr)	2.91	2.86	2.57	2.99	2.83
Pm	2.90	2.94	2.65	2.92	2.85
n	2.01	1.91	1.82	1.98	1.93

[a] SFH34 at 25°. n, Hill parameter; ssr, sum of squared residuals; a_i, Adair parameters; Pm, median ligand concentration.

from the Adair fit show that experiment 32 had the greatest deviation from the Adair fit, and experiment 29 fit the model the best. If run 32 is eliminated, the P_{50} values for the other three curves range from 2.86 to 2.99, a very narrow interval. Likewise, the values of the Hill parameter, n, for these three curves varies only from 1.91 to 2.01. The values of Adair parameters are somewhat more variable, but again, run 32 seems to be an outlier, with values of a_3, especially, quite different from those of the other three curves.

Discussion

In our laboratory, we have attempted to understand the functional properties of whole human blood by methods that parallel those used in studies of purified hemoglobin solution. However, inside the erythrocyte the hemoglobin concentration is extremely high, 32–34 g/dl. Thus, within the erythrocyte the proportion of molecules that exist as dimers ($\alpha\beta$) as opposed to tetramers ($\alpha_2\beta_2$) is small. In addition, inside the erythrocyte many allosteric effectors are present, including H^+, CO_2, Cl^-, 2,3-diphosphoglycerate (2,3-DPG), ATP, and other salts. Furthermore, exact knowledge of intracellular pH is not possible because the Donnan equilibrium makes precise measurements of intracellular conditions difficult.[15] We have previously described that the determination of the four Adair parameters for whole blood is necessarily imprecise because of the low oxygen affinity and large standard errors involved in the curve-fitting procedures that are required for their determination.[7]

A further problem in the accurate determination of the OEC is how data are scaled, especially at high saturation. The need for scaling arises because it is difficult to measure saturation at extremely high P_{O_2}. Marden et al.[16] concluded that this scaling eliminates the usefulness of model fitting. Previous scaling procedures have been arbitrary.[4,7] Using the method described here, no scaling is needed, because the hemoglobin O_2 saturation is calculated directly.

The instrument described here represents a significant improvement over the one used previously in our laboratory. The principal improvement was the introduction of a direct stirring mechanism, with the motor mounted above the cuvette (see Fig. 1). In the previous version a magnetic stirrer was used, which was not mechanically coupled to the driving motor. Thus, when the cuvette contents were viscous or heterogeneous, the

[15] M. Samaja and R. M. Winslow, *Br. J. Haematol.* **41**, 373 (1979).
[16] M. C. Marden, E. S. Hazard, L. Leclerc, and Q. H. Gibson, *Biochemistry* **28**, 4422 (1989).

stirrer was prone to become erratic or to stop, and the resulting data were unreliable. The present version overcomes this problem.

The data presented in this chapter are not meant to be definitive analyses of the OEC of concentrated hemoglobin, but merely to illustrate the method. It is curious that the value of the Hill parameter, n, seems lower than expected in all cases. Perhaps formation of methemoglobin is responsible for this observation. Furthermore, the hemoglobin samples are unbuffered, so the pH would be expected to drop significantly as hemoglobin is oxygenated. This would have the effect of reducing oxygen affinity as the experiment progresses, thereby reducing the observed value of n. However, to our knowledge, these are the first modern data reported on concentrated hemoglobin solutions, measured by nonoptical methods, and it will be necessary to perform more experiments before fully understanding the low n values.

The second improvement in the use of the crude catalase preparation. This seems to protect hemoglobin reasonably well from oxidation by H_2O_2, but there is still some methemoglobin formed during an oxygenation experiment. The use of pure catalase with the lactoperoxidase–KI system seems to reduce methemoglobin formation somewhat, and performing the experiment at 25° rather than at 37° also helps, but prevention of some oxidation still appears an elusive goal.

The values of the Adair parameters are sensitive to the values used for the O_2 solubility coefficient (α). This is interesting because the conventional parameters of the OEC, P_{50} and n, are much less sensitive. Thus, great care must be exercised in using nonoptical methods to be sure of the actual value of α under the conditions of the experiment.

The method described here appears to be satisfactory in work with hemoglobin-based blood substitutes. The fact that hemoglobin saturation is calculated from gasometric data is distinctly advantageous compared with optical methods because in the latter method, no absolute measure of O_2 capacity is obtained. This would be important, for example, in samples with large amounts of methemoglobin.

[24] Bezafibrate Derivatives as Potent Effectors of Hemoglobin

By CLAUDE POYART, MICHAEL C. MARDEN, and JEAN KISTER

Introduction

Since the discovery of 2,3-diphosphoglycerate (DPG)[1,2] and its functional properties, the search for potent allosteric effectors of hemoglobin (Hb) has been an important field of research. Two major aspects have been studied: (1) the search for compounds that may react with Hb in erythrocytes and shift the allosteric equilibrium (either to the low- or high-affinity states) in order to use them as therapeutic agents in hypoxia or sickle cell disease and (2) the use of these potent effectors as tools for structure–function relationship studies. Certain aromatic benzaldehydes cause a shift to the left of the oxygen equilibrium curve in sickle cell suspensions.[3–6] To our knowledge no potent compound that may favor oxygen unloading in erythrocyte suspensions has been described.

The availability of allosteric effectors that could be used in erythrocyte suspensions would be advantageous in studying dilute hemoglobin solutions for the following reasons: (1) inside the erythrocytes the high concentration of Hb reduces the presence of dimers to negligible amounts; (2) the presence of the cellular enzymatic reducing system continuously keeps the formation of methemoglobin to insignificant levels; (3) erythrocyte DPG can easily be either depleted or increased in the cellular milieu, allowing one to study the function of Hb complexes both in the presence and absence of DPG; and (4) such molecules, if not toxic at the concentration necessary to affect the huge Hb content of the erythrocytes, could be developed for clinical use under various pathological conditions.

[1] R. Benesch and R. E. Benesch, *Biochem. Biophys. Res. Commun.* **26**, 162 (1967).
[2] A. Chanutin and R. R. Curnish, *Arch. Biochem. Biophys.* **121**, 96 (1967).
[3] C. R. Beddell, P. J. Goodford, G. Kneen, R. D. White, S. Wilkinson, and R. Wootton, *Br. J. Pharmacol.* **82**, 397 (1984).
[4] A. J. Keidan, I. M. Franklin, R. D. White, M. Joy, E. R. Huehns, and J. Stuart, *Lancet* **1**, 831 (1986).
[5] A. J. Keidan, M. C. Sowter, C. S. Johnson, S. S. Marwah, and J. Stuart, *Clin. Sci.* **76**, 357 (1989).
[6] F. C. Wireko and D. J. Abraham, *Proc. Natl. Acad. Sci. U.S.A.* **88**, 2209 (1991).

Clofibric acid was one of the first nonnaturally occurring molecules reported that decreased the oxygen affinity of hemoglobin.[7] In 1983 Perutz and Poyart[8] reported that the antilipidemic drug bezafibrate (Bzf), a derivative of clofibric acid, lowered the oxygen affinity of freshly drawn erythrocyte suspensions. Washing the erythrocytes in a drug-free buffer restored the oxygen affinity to control levels. No cell damage was observed. This observation was of great interest because it demonstrated that one could manipulate reversibly the oxygen affinity of Hb in its native intracellular environment; this could be achieved in the presence of a normal concentration of DPG, thus indicating that the drug and DPG have synergistic effects and different binding sites to Hb. Our hope to lower the oxygen affinity for clinical use was tempered, however, because of the low association constant of the Bzf–hemoglobin complex. More recent efforts have been directed toward the preparation of new bezafibrate derivatives in order to increase their allosteric effects.

The aim of this chapter is to report on various methods concerned with the studies of these new potent allosteric effectors in erythrocyte suspensions and to describe examples of solution studies in the presence of these effectors, using equilibrium and kinetic methods to investigate new aspects of the allosteric equilibrium. The results cited in this chapter have appeared in several articles.[9–19] We shall not describe the experimental procedures for the synthesis of these compounds.

[7] D. J. Abraham, M. F. Perutz, and S. E. V. Phillips, *Proc. Natl. Acad. Sci. U.S.A.* **80**, 324 (1983).

[8] M. F. Perutz and C. Poyart, *Lancet* **2**, 881 (1983).

[9] D. J. Abraham, P. E. Kennedy, G. S. Mehana, and F. L. Williams, *J. Med. Chem.* **27**, 867 (1984).

[10] M. F. Perutz, G. Fermi, D. J. Abraham, C. Poyart, and E. Bursaux, *J. Am. Chem. Soc.* **108**, 1064 (1986).

[11] I. Lalezari, S. Rahbar, P. Lalezari, G. Fermi, and M. F. Perutz, *Proc. Natl. Acad. Sci. U.S.A.* **85**, 6117 (1988).

[12] M. C. Marden, J. Kister, B. Bohn, and C. Poyart, *Biochemistry* **27**, 1659 (1988).

[13] I. Lalezari and P. Lalezari, *J. Med. Chem.* **32**, 2352 (1989).

[14] M. C. Marden, B. Bohn, J. Kister, and C. Poyart, *Biophys. J.* **57**, 397 (1990).

[15] I. Lalezari, P. Lalezari, C. Poyart, M. C. Marden, J. Kister, B. Bohn, G. Fermi, and M. F. Perutz, *Biochemistry* **29**, 1515 (1990).

[16] G. S. Mehana and D. J. Abraham, *Biochemistry* **29**, 3944 (1990).

[17] S. R. Randad, M. A. Mahran, A. S. Mehanna, and D. J. Abraham, *J. Med. Chem.* **34**, 752 (1991).

[18] F. C. Wireko, G. E. Kellog, and D. J. Abraham, *J. Med. Chem.* **34**, 758 (1991).

[19] D. J. Abraham, F. C. Wireko, R. S. Randad, C. Poyart, J. Kister, B. Bohn, J.-F. Liard, and M. P. Kunert, *Biochemistry* **31**, 9141 (1992).

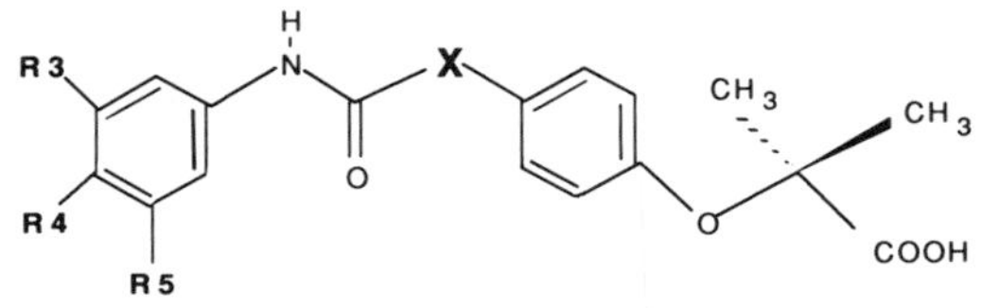

LR16	R3,R4 = Cl	X = NH
L 3,5	R3,R5 = Cl	X = NH
L 3,4,5	R3,R4,R5 = Cl	X = NH
RSR-4	R3,R5 = Cl	X = CH_2

SCHEME 1

Structures of Effectors

The bezafibrate derivatives that have been submitted to detailed structure–activity analyses in our laboratory are those synthesized by Lalezari et al.[11,13,15] In these derivatives a urea molecule has been placed between the two opposite aromatic rings of the bezafibrate molecule [2-(4-{[(aryl-amino)carbonyl]amino}phenoxy)-2-methylpropionic acid], constituting the "L" series in Scheme I; a second series of compounds was synthesized by Abraham and colleagues[16–19] and called the "R" series; in these the basic change is that one amide nitrogen of the urea is replaced by a methylene group. Various substitutions on the ring can be introduced. In the following sections, examples of some of these effectors are given.

Crystal Structure of Hemoglobin–Effector Complexes

X-Ray analyses of the hemoglobin–compound complexes have revealed that all these compounds bind at the interface of the α and β subunits deep in the water-filled central cavity at a distance from the hemes and the DPG-binding site. For simplicity this has been called the "bezafibrate-binding site."[10] One molecule of bezafibrate was found to bind to two clofibric acid sites.[9] Details concerning the Hb complexes in the crystals for these different compounds have been given in articles by Abraham et al.,[7] Perutz et al.,[10] Lalezari et al.,[15] and Wireko et al.[18] Manavalan et al.[20] have published a theoretical study for potential binding sites of various compounds to deoxyHb; they confirmed the major binding site described by Perutz et al.[10] for Bzf and its derivatives. Two main

[20] P. Manavalan, M. Prabhakaran, and M. E. Johnson, J. Mol. Biol. 223, 791 (1992).

aspects emerge from these analyses: (1) the binding region of these compounds to Hb is located mainly on α-chain residues where one molecule interacts with two α subunits and one β subunit. This binding site does not overlap with the DPG-binding site; and (2) there is no clear relationship between the structure of the compounds and the nature and stoichiometry of its binding.

Preparation of Effector Solutions

Because of their high hydrophobicity all these compounds are difficult to dissolve in aqueous buffers. We usually prepare 10 mM stock solutions of the effectors in 100 mM NaCl, 50 mM Bis–Tris buffer, pH 7.4. After addition of an excess of $NaHCO_3$, the solution is warmed to 60° and stirred for several hours. The solution should be back titrated carefully to pH 7.4 before use and prepared fresh every week. When stored in the cold a 10 mM L 3,4,5 solution becomes a gel. Many other compounds precipitate in the cold and should be redissolved by mild heating before use.

Erythrocyte Suspension Studies

Oxygen Equilibrium Binding in Erythrocyte Suspensions

Blood samples are collected in heparin from nonsmoking volunteers. The blood is centrifuged immediately for 5 min at 1500 rpm at 4° to remove the plasma and the buffy coat. The packed erythrocytes are washed three times in cold 50 mM Bis–Tris isotonic buffer, pH 7.4. An aliquot of the packed cells (30 to 50 μl) is resuspended in 4 ml of the working buffer solution containing the effector at the desired concentration. Fifteen to 20 min is allowed for equilibration of the erythrocyte suspension at 37° before recording the oxygen-binding curve. In some instances an aliquot (500 μl) of the same erythrocytes is incubated overnight at 37° in 2 ml of isotonic buffer, pH 7.4, to deplete these cells of their DPG content. Bacterial growth is avoided by the addition of chloramphenicol. The DPG-depleted cells are then washed and further processed as for the fresh erythrocytes.

The activity of the Hb–effector complexes is measured through the recording of equilibrium oxygen-binding curves with a Hemox analyzer (TCS Medical Products, Huntington Valley, PA) as described in detail in previous papers.[21] This technique consists of the simultaneous recording of the pO_2 changes on slow deoxygenation with a Clark-type oxygen

[21] J. Kister, C. Poyart, and S. J. Edelstein, *J. Biol. Chem.* **262**, 12085 (1987).

electrode and of the changes in absorbance of the Hb solution with a double-wavelength spectrophotometer. In principle this technique is similar to that described by Imai for the study of Hb.[22] At the end of the recording, the 300–400 data points are stored on tape for further analyses. These consist of the calculation of the P_{50} and n_{50} values. The former is the pO_2 at oxygen half-saturation, calculated according to the empirical Hill equation by linear regression analysis from data points determined between 40 and 60% oxygen saturation. The n_{50} value, an estimate of the heme–heme interaction, is the slope of the Hill graph at the middle portion of the curve. One major difficulty for these recordings is the incomplete saturation with oxygen of the cell suspensions in the presence of the potent allosteric effectors even under 1 atm oxygen at physiological pH (7.4) and temperature (37°). This may significantly alter the observed value of the two parameters. One may bias this difficulty by using less concentrated effector solutions. Another possibility is to test the effectors in DPG-depleted erythrocytes whose oxygen affinity is twofold higher than that of fresh erythrocytes and at a more alkaline pH. Under this condition, the observed P_{50} value can be corrected to pH 7.4 by using a Bohr factor of -0.7.[15] In all cases it is preferable to equilibrate the erythrocyte suspensions under pure oxygen for several minutes before the recordings.

Table I gives the P_{50} and n_{50} values of fresh washed erythrocyte cell suspensions recorded in the presence of the effectors listed in Scheme I, compared to those obtained with bezafibrate. The order of activity as estimated from the $\Delta \log P_{50}$ values is Bzf < LR 16 < RSR-4 ≈ L 3,5 ≤ L 3,4,5. Except for the L 3,4,5 compound, the other Hb–effector complexes maintain nearly normal cooperativity in oxygen binding.

Influence of Effectors on Erythrocyte Stability

An important point to be considered in testing these effectors in erythrocyte suspensions is the lack of deleterious effects on these cells that is manifested by the presence of even a slight degree of hemolysis. This should be done for each compound at the end of the oxygen-binding curve recording. Each recording usually takes 40 to 45 min. To test for the presence of hemolysis, the erythrocyte suspension is withdrawn from the optical cuvette of the recording system and centrifuged at 5000 rpm in the cold for 10 min. The presence of free hemoglobin in the supernatant is detected under 1 atm of carbon monoxide in the Soret region of the optical spectrum ($\varepsilon = 200$ mM^{-1} cm^{-1} at 419 nm). One can also test for the presence of methemoglobin by recording the visible spectrum at

[22] K. Imai, this series, Vol. 76, p. 438.

TABLE I

OXYGEN-BINDING MEASUREMENTS FOR ERYTHROCYTE SUSPENSIONS ON ADDITION
OF L AND R EFFECTORS[a]

Effectors	Concentration (mM)	P_{50} (mmHg)	log P_{50}	n_{50}	Δ log P_{50} ($\pm$ effector)
Fresh erythrocytes					
Control	—	25.5	1.40	2.5	—
Bzf	1.0	28.6	1.45	2.4	0.05
Bzf	5.0	37.6	1.57	2.3	0.17
LR 16	0.5	48.6	1.68	2.3	0.28
L 3,5	0.5	60.0	1.78	2.1	0.37
L 3,4,5	0.25	81.0	1.91	1.8	0.50
RSR-4	0.5	52.5	1.72	2.2	0.31
DPG-depleted erythrocytes					
Control	—	13.6	1.13	3.0	—
RSR-4	0.25	22.0	1.34	2.8	0.11
RSR-4	(0)[b]	12.4	1.09	2.7	

[a] Conditions: 140 mM NaCl (pH 7.4)–50 mM Bis–Tris Buffer, 37°. The total Hb concentration of the erythrocyte suspensions was 20 to 25 μM on a tetramer basis.

[b] Erythrocytes were first exposed to the effector and then washed in an effector-free buffer.

appropriate wavelengths in completely lysed erythrocytes at the end of the experiment.[23,24]

Reversibility of Effector–Hemoglobin Complexes

The interaction of the effectors of the bezafibrate family with intracellular hemoglobin indicates that they cross the erythrocyte membrane. That they do so without damaging the cells is proved by measuring the oxygen-binding properties of these cells after reincubating the cells previously equilibrated with the effector in an effector-free buffer for 30 min and remeasuring the P_{50} of this suspension. This is performed in DPG-depleted erythrocytes to avoid the slow decrease in DPG due to prolonged incubation at 37°. P_{50} and n_{50} values similar to the controls are observed after this procedure (Table I).

[23] R. E. Benesch, R. Benesch, and S. Young, *Anal. Biochem.* **55**, 245 (1973).
[24] O. W. van Assendelft and W. G. Zijlstra, *Anal. Biochem.* **69**, 43 (1975).

Transfer of Effectors across Erythrocyte Membrane

As is discussed below the transfer of the allosteric effectors is rapid. We have investigated whether they penetrate the erythrocytes through the membrane anion channel, also known as the protein Band 3, which is responsible for the self-exchange of chloride and bicarbonate anions. Fresh erythrocytes (10% hematocrit) are first incubated for 30 min in the dark at 37° with 20 μM (4,4'-diisothiocyanostilbene 2,2'-disulfonate (DIDS), a specific inhibitor of the erythrocyte membrane anion channel.[25] The cells are washed and equilibrated at 37° in the presence of 0.5 mM L 3,5 or RSR-4. The resulting P_{50} values for these cells are similar to the control (without DIDS), indicating that the effectors cross the erythrocyte membrane through a passive diffusion process.

Activity of Effectors in Presence of Albumin

One of the main limitations for the pharmaceutical development of these compounds as allosteric effectors of Hb in erythrocytes has been the competition between serum albumin and hemoglobin in binding these compounds.[15] Efforts have been made to obtain molecules with low affinity for albumin in order to maintain a sufficiently high Hb binding.[17] Recordings of equilibrium oxygen-binding curves have been made for erythrocyte suspensions in buffers containing human defatted serum albumin (HSA; Sigma, St. Louis, MO) at concentrations mimicking the whole blood conditions. An antifoam agent is necessary to avoid extensive foaming of the solutions containing albumin on gas admixture. We use 1520 (EU) Silicone antifoam (Dow Corning, Midland, MI), which has no deleterious effect on the erythrocytes. In whole blood the molar ratio of HSA to tetrameric Hb is approximately 0.3 (0.7 mM HSA plus plasma proteins versus 2.25 mM Hb$_4$). In the experiments described above the concentration of Hb tetramer is 20 to 25 μM. Therefore oxygen-binding curves are recorded in buffers containing 12.5, 25, 50, and 100 μM HSA at varying concentrations of the effectors. Under these conditions the molar ratio of HSA to tetrameric Hb is varied from 0.5 (at 12.5 μM HSA) to 4 (at 100 μM HSA), that is, always larger than that found in whole blood.

Table II gives the results for the two effectors L 3,5 and RSR-4. At 50 μM HSA the effect of L 3,5 is almost 80% inhibited whereas that of RSR-4 is decreased by only 30%, indicating a lower interaction of HSA with the latter effector.[19,26] In the presence of a constant concentration of HSA the order of potency of the effectors becomes RSR-13 >

[25] Z. I. Cabantchik and A. Rothstein, *J. Membr. Biol.* **10,** 311 (1972).
[26] C. Poyart, J. Kister, B. Bohn, R. S. Randad, and D. J. Abraham, *Blood* **78,** 404a (1991).

TABLE II

OXYGEN-BINDING MEASUREMENTS FOR FRESH HUMAN ERYTHROCYTES ON ADDITION OF
L AND R EFFECTORS IN PRESENCE OF HUMAN SERUM ALBUMIN[a]

Effector	Concentration (mM)	HSA (μM)	P_{50} (mmHg)	$\Delta \log P_{50}$ (± effector)	ΔA_{560}
None	—	0	27.0	—	—
L 3,5	0.5	0	60.0	0.35	0.18
		12.5	nd	nd	0.12
		50	32.0	0.07	0.06
		100	28.0	0.02	0.03
RSR-4	0.5	0	52.5	0.29	0.15
		12.5	48.5	0.25	0.11
		50	44.0	0.21	0.09
		100	36.2	0.13	0.04

[a] Conditions: 140 mM NaCl (pH 7.4)–50 mM Bis–Tris buffer, 37°. P_{50} is the partial pressure of oxygen for oxygen half-saturation; HSA, human serum albumin (defatted); hemoglobin concentration was 20–25 μM on a tetramer basis; ΔA_{560} is the maximum absorbance change at 560 nm on addition of the effector under room air. nd, Not determined.

RSR-4 > L 3,5 ≫ L 3,4,5, which is the reversed order of potency observed for these effectors in their interaction with hemoglobin in the absence of albumin.

Effector-Induced Oxygen Release

In these experiments the erythrocyte suspensions are equilibrated under room air (pO_2 = 140 torr) at 37° in the optical cuvette of the Hemox analyzer, which provides a constant stirring of the suspensions. At time 0, the effector is added, which results in a rapid increase in absorbance at 560 nm until a plateau is reached, usually within minutes. Changes in absorbance after addition of the L 3,5 and RSR-4 effectors are given in the last column of Table II and illustrated for L 3,5 in Fig. 1. An increase in absorbance at 560 nm (ε_{max} for deoxyHb in the visible part of the absorption spectrum) under room air indicates the formation of partially deoxygenated Hb in the erythrocytes and a decreased oxygen affinity of Hb, provided that the pH is held constant. The experiments are performed at fixed effector concentration in the absence and presence of HSA. Table II shows that the extent of the inhibitory effect by albumin is demonstrated by a decrease in either the P_{50} or in ΔA_{560} relative to the control values obtained in the absence of albumin. Figure 2 shows that the total change in absorbance at 560 nm (ΔA_{560}) on addition of the effectors with or without

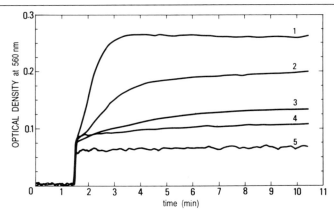

FIG. 1. Changes in absorbance at 560 nm versus time after addition of 0.5 mM L 3,5 with increased amounts of albumin (HSA) in normal fresh erythrocyte suspensions. (1) L 3,5 without HSA; (2) L 3,5 + 12.5 μM HSA; (3) L 3,5 + 50 μM HSA; (4) L 3,5 + 100 μM HSA; (5) drug-free buffer. Other experimental conditions: 0.05 M Bis–Tris buffer (pH 7.4)–0.14 M NaCl, 37°. Hemoglobin concentrations of the erythrocyte suspensions were 20 to 25 μM on a tetramer basis. The experiments were done as follows: (a) normal washed fresh erythrocyte suspensions were first equilibrated under room air in the buffer until the absorbance at 560 nm was stable; (b) the drug was then added to the suspensions, which resulted in a rapid increase in absorbance until a plateau was reached, usually within minutes; (c) when albumin was used, it was added to the buffer before the addition of erythrocyte sample; (d) the dilution effect due to the drug sample was estimated after addition of an equivalent volume of drug-free buffer (curve 5). The rationale of this method is that an increase in absorbance at 560 nm (calculated as ΔA_{560} minus the dilution effect) indicates an increase in the level of deoxyHb and is correlated to the change in oxygen affinity.

HSA is, within the limits of the experiments, linearly related to the P_{50} values of these cells.

Physiological Consequences

Figure 3 illustrates an example of the potential benefits that these effectors may provide for oxygen delivery *in vivo* at rest. The inset gives the calculated ΔY values (the oxygen delivery index) corresponding to the percentage change in oxygen saturation of the suspensions between the arterial and mixed venous blood. In going from curve 1 (Fig. 3, DPG-depleted erythrocytes) to curve 2 (Fig. 3, fresh erythrocytes) the oxygen delivery, ΔY, is increased from 3 to 23%; this is further increased to 40% on addition of 0.5 mM L 3,5. On addition of 1 mM L 3,5 the oxygen-binding curve is further right shifted but the ΔY decreases slightly. This is related to the incomplete saturation of the erythrocytes in the arterial blood due to the extremely low oxygen affinity. The inset in Fig. 3 also

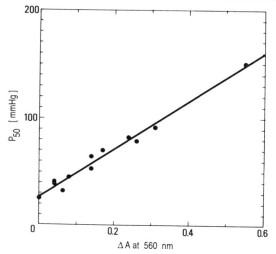

FIG. 2. Linear relationship between the oxygen affinity (oxygen equilibrium curves) and the changes in absorbance at 560 nm (ΔA_{560}) of erythrocyte suspensions on addition of the effectors. Experimental conditions were the same as those indicated in Fig. 1. The straight line was obtained by a linear regression procedure with the following equation: P_{50} (mmHg) = 220(ΔA_{560}) + 27.0 (r = 0.994).

gives the values of the oxygen saturation of the arterial blood under the condition of respiring room air (pO_2 of 150 mmHg).

Hemoglobin Solution Studies

Equilibrium Studies

Oxygen-binding studies at equilibrium on purified stripped human adult hemoglobin (Hb A) are performed in the presence of varying concentrations of the bezafibrate derivatives alone or in combination with other effectors such as DPG or inositol hexaphosphate (IHP). The technique is similar to that described for erythrocyte suspensions. The composition of the buffer system is in most instances 100 mM NaCl–50 mM Bis–Tris buffer at pH 7.2 and 25°. Catalase (20 µg/ml) and 50 µM ethylenediamine-tetraacetic acid (EDTA) are required to limit oxidation of the hemes during the 45–60 min of the recordings.[27] Figure 4 illustrates Hill plots of oxygen-binding experiments in the presence of 1 mM LR 16 (curve 2) and 0.5 mM L 3,5 (curve 3). Relative to the control (curve 1) the effectors lead

[27] J. Kister, C. Poyart, and S. J. Edelstein, *Biophys. J.* **52**, 527 (1987).

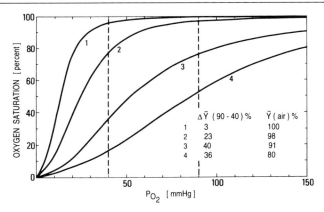

FIG. 3. Oxygen equilibrium curves for human erythrocyte suspensions without or with
L 3,5. (1) DPG-depleted erythrocytes; (2) fresh erythrocytes (DPG/Hb$_4$ $\approx$ 1); (3) fresh
erythrocytes plus 0.5 mM L 3,5; (4) fresh erythrocytes plus 1 mM L 3,5. Other experimental
conditions were the same as those indicated in Fig. 1. The vertical lines indicate the physiological pO$_2$ values in the arterial blood (90 mmHg) and in the mixed venous blood (40 mmHg).
The inset shows the change in oxygen saturation between these two lines ($\Delta Y\%$), allowing
an estimation of the capability of the erythrocytes to deliver oxygen to the tissues. The
oxygen saturation under air (pO$_2$ = 150 mmHg) is also indicated ($Y\%$).

to an important right shift of the Hill plots at the bottom portion (T state)
of the curves. The upper part of the curves with effectors is also shifted
to the right, possibly indicating an interaction of the effectors with fully
oxygenated Hb (R state). However, the upper part of the curves is poorly
defined in the presence of the potent effectors, due to incomplete saturation of Hb even under 1 atm oxygen.[15,28]

Because the Hb may not be fully oxygenated under 1 atm O$_2$, an
estimate of the percentage saturation is required at high pO$_2$ values from
the optical spectra and known ε values.

Table III gives results of the oxygen-binding parameters on addition
of the bezafibrate derivatives (Scheme 1). The order of potency for the five
compounds tested in this series is identical to that observed in erythrocyte
suspensions. Table III also confirms that the effects of these newly developed effectors and of DPG are additive, as observed in erythrocyte suspensions. This property leads to large right shifts of oxygen-binding curves
and may be used to manipulate the allosteric equilibrium in such a way
that liganded Hb tetramers in the T state may be observed. Under these
conditions, however, the oxygen affinity is so much decreased and the
unsaturation of the solution under 1 atm oxygen is so large that it precludes

[28] M. C. Marden, J. Kister, C. Poyart, and S. J. Edelstein, *J. Mol. Biol.* **208,** 341 (1989).

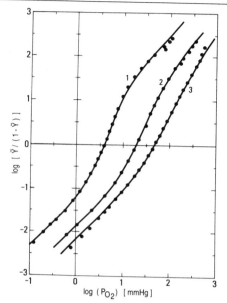

FIG. 4. Oxygen equilibrium curves for purified Hb A solutions with bezafibrate derivatives effectors. (1) 0.1 M NaCl; (2) plus 1 mM LR 16; (3) plus 0.5 mM L 3,5. Other experimental conditions were as follows: 0.05 M Bis–Tris buffer at pH 7.2, catalase (20 μg/ml), 50 μM EDTA, 25°. The hemoglobin concentration is 70 μM on a heme basis. Symbols represent one of six experimental data points. Continuous lines are the best fit of the experimental curve to the equation of the two-state allosteric model[31] with parameters given in Marden et al.[14]

analysis of the four liganded tetramers (Fig. 5). Carbon monoxide must be used to test for the functional properties of these Hb–effector complexes.[12,14]

Effector Activity in Presence of Chloride and 2,3-Diphosphoglycerate

Chloride anions are allosteric effectors of Hb A with a high-affinity site at the α_1–α_2 interface of the central cavity, a region known from X-ray analyses to be a specific binding site for bezafibrate and its derivatives. One may therefore expect that the effectors of the L or R series may compete with chloride. Figure 6 shows that the change in P_{50} on addition of 0.5 mM L 3,5 remains independent of the concentration of chloride from 7 to 600 mM, indicating the inhibition of chloride binding by the more potent L 3,5 effector.

TABLE III

OXYGEN-BINDING MEASUREMENTS FOR PURIFIED ADULT HEMOGLOBIN SOLUTIONS ON
ADDITION OF L OR R EFFECTORS[a]

Effector	Concentration (mM)	P_{50} (mmHg)	log P_{50}	n_{50}	Δ log P_{50} ($\pm$ effector)
100 mM NaCl					
Control	—	5.2	0.72	2.8	—
Bzf	0.5	7.0	0.84	2.7	0.13
Bzf	3.0	16.0	1.20	2.6	0.49
LR 16	0.5	11.2	1.50	2.6	0.33
	1.0	19.5	1.30	2.5	0.57
L 3,5	0.25	20.0	1.30	2.2	0.58
	0.5	46.0	1.66	2.1	0.95
	1.0	65.0	1.81	1.8	1.10
L 3,4,5	0.1	18.0	1.26	2.3	0.54
	0.25	32.0	1.51	1.5	0.79
	0.5	115.0	2.06	1.3	0.95
RSR-4	0.25	18.0	1.26	2.3	0.54
	0.5	42.0	1.62	2.2	0.91
100 mM NaCl + 1 mM DPG					
Control	—	15.8	1.20	2.8	—
L 3,5	0.25	43.0	1.63	2.3	0.43
L 3,4,5	0.1	32.0	1.51	1.9	0.31
RSR-4	0.25	40.2	1.60	2.3	0.40

[a] Conditions: 25°, 50 mM Bis–Tris buffer (pH 7.2), 60–80 μM heme.

The use of these effectors may also help to localize the site of a genetic mutation at the α chains. For example, Hb Suresnes [α141 (HC3) Arg $\rightarrow$ His][29] or Hb Thionville [α1 (NA1) Val $\rightarrow$ Glu, $-$1Met-Ac][30] exhibits no or little interaction with Bzf, confirming the participation of the native N–C-terminal region of the α chains to the binding of Bzf.

Determination of K_R in Presence of Strong Effectors

In the presence of strong effectors the transition from T to R state[31] may occur only after three ligands are bound. Under these conditions the triply liganded tetramers will exist with a substantial amount of T state.

[29] C. Poyart, E. Bursaux, A. Arnone, J. Bonaventura, and C. Bonaventura, *J. Biol. Chem.* **255**, 9465 (1980).

[30] C. Vasseur, Y. Blouquit, J. Kister, D. Promé, J. S. Kavanaugh, P. H. Rogers, C. Guillemin, A. Arnone, F. Galacteros, C. Poyart, and H. Wajcman, *J. Biol. Chem.* **267**, 12682 (1992).

[31] J. Monod, J. Wyman, and J. P. Changeux, *J. Mol. Biol.* **12**, 88 (1965).

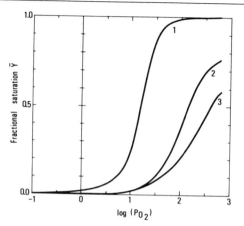

FIG. 5. Oxygen equilibrium curves of Hb A solutions with L 3,5 and IHP at pH 6. (1) 0.1 M NaCl; (2) plus 0.5 mM L 3,5; (3) plus 0.5 mM L 3,5 + 1 mM IHP. Other experimental conditions were the same as those indicated in Fig. 4.

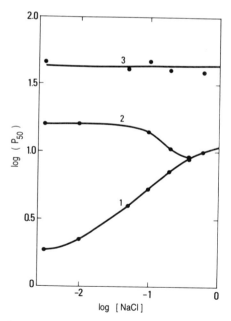

FIG. 6. Variations of oxygen affinity (log P_{50}) of Hb A solutions with increasing concentrations of chloride. (1) Chloride; (2) chloride plus 1 mM DPG; (3) chloride plus 0.5 mM L 3,5. Other experimental conditions were the same as those indicated in Fig. 4.

The binding of the fourth ligand will therefore not represent a simple binding to the R state (R_3 to R_4), but rather an average value intermediate to the pure R and T states. Even under normal conditions, there is some compensation between the R-state affinity and the allosteric parameter $L = T_0/R_0$.[31] With strong effectors, the T-state "contamination" of the upper asymptote may preclude the determination of K_R (oxygen equilibrium dissociation constant for the R state), even to within a factor of 10.[14]

The lower oxygen affinity and shift toward the T' state may lead to conditions in which the sample is less than 90% saturated under 1 atm oxygen. In these cases, there is no upper asymptote to fit, and the normalization factor for the total absorption change (0 to 100% saturation) must be estimated. Even if the saturation curve were complete, as could be measured for the higher affinity ligand CO, the T-state contribution (as mentioned above) to the binding of the fourth ligand would result in a poorly determined K_R value.[14,27]

An alternate method is to determine the on and off rates for ligand binding to the R state and calculate the affinity from the kinetic parameters: $K_{assoc} = k_{on}/k_{off}$. This is difficult for oxygen because the sample is not fully saturated. Because CO has a higher affinity ($M = K_{CO}/K_{O_2} = 250$), a tetramer with four CO molecules bound can be prepared. The dissociation (or off) rate from R_4 can be measured by ligand replacement reactions,[32] for example, CO to NO, or by addition of ferricyanide and observation of the rate of formation of metHb.[33]

The on rate can be measured by flash photolysis methods. In this case the conditions favoring the R-state rate should be used: (1) a low level of photodissociation ($<10\%$) to produce mainly R_3, and (2) a high CO concentration so that ligand recombination (R_3 to R_4) competes favorably with the allosteric transition (R_3 to T_3). Even if some T_3 is formed, the R- and T-state kinetics are clearly resolved, because the rates differ by typically a factor of 30 to 100; this is not the case in the equilibrium measurements, where one is faced with the problem of separating K_R and L from a single observed (and often poorly resolved) affinity for the upper asymptote. Ideally the R-state on rate should be faster than the R-to-T transition, so that the rebinding of R_3 to R_4 will involve tetramers still unrelaxed from the R_4 conformation.

If the oxygen- and CO-binding parameters are changed by the effectors in a similar way, then the kinetically determined shift in K_R for CO can be used to estimate the analogous shift for oxygen. One way to test this effect is by a competition experiment. A sample can be prepared with

[32] R. D. Gray and Q. H. Gibson, *J. Biol. Chem.* **246**, 7168 (1971).
[33] A. Lanir, W. S. Caughey, and S. Charache, *Eur. J. Biochem.* **128**, 521 (1982).

equal amounts of oxygen and CO bound, near 50% of each, but the exact values are not important. The effector(s) can then be added. Because the spectra for the oxy and CO forms are different, a shift favoring either ligand can be observed. With IHP and the series of bezafibrate derivatives, there is little change in the relative affinities.[14]

Separation of Allosteric and Affinity Factors

An increase in P_{50} does not necessarily imply an increase in the amount of T state (an allosteric effect). If two curves on a Hill plot are parallel, then the difference can be explained by a shift in both K_R and K_T (oxygen equilibrium dissociation constant for the T state) by the same factor, with no change in the allosteric equilibrium. There are thus two ways to explain the decreased ligand affinity on addition of effectors. For DPG, IHP, and the bezafibrate derivatives, the equilibrium and kinetic data indicate that both the affinities and the allosteric equilibrium change.[14] If we consider the shift in K_R and K_T alone, the affinity contribution can be estimated; the remainder being attributed to the allosteric factor, which should correspond to the change due to a shift in L. Generally there is a roughly equal contribution, affinity and allostery, for these effectors. It is therefore not correct to refer to these compounds only as allosteric effectors.

Triply Liganded Tetramers

A primary goal of Hb research is to determine the different possible conformations for an Hb tetramer. Even the simple 2-state model predicts 10 substates: R and T, each with 0 to 4 ligands.[31] But how do we know that they all exist? Equilibrium favors T_0 relative to R_0 by the factor $L = 10^6$, so discarding the R_0 state would not affect equilibrium simulations. Yet this state exists, because it can be produced by photodissociation of four ligands from R_4; it then relaxes to T_0 within about 100 μsec.[12]

Similarly it can be questioned whether T_3 or T_4 exist, because they also have low equilibrium populations. The use of strong effectors has shown that T_3 can be populated: at low photodissociation levels producing mainly R_3, a substantial amount of rebinding with a slow rate characteristic of T-state tetramers is observed in the presence of two strong effectors such as L 3,5 and IHP (Fig. 7)[12,34] (also see Martin et al.[35,36]). This is apparently a rebinding T_3 to T_4, which then reconverts to R_4. The slow

[34] L. P. Murray, J. Hofrichter, E. R. Henry, M. Ikeda-Saito, T. Kitagishi, T. Yonetani, and W. A. Eaton, Proc. Natl. Acad. Sci. U.S.A. 85, 2125 (1988).

[35] J.-L. Martin and M. H. Vos, this volume [19].

[36] J. W. Petrich, J. C. Lambry, K. Kuczera, M. Karplus, C. Poyart, and J.-L. Martin, Biochemistry 30, 3975 (1991).

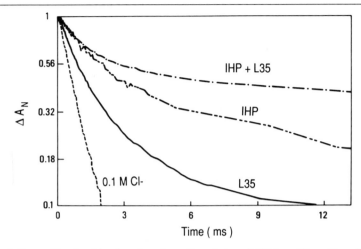

FIG. 7. Recombination kinetics after photodissociation of CO from Hb, with and without the effectors L 3,5 (0.25 mM) and IHP (1 mM). Plotted on a logarithmic scale is the normalized change in absorption versus (linear) time. Sample conditions were 20 mM NaCl (pH 7.2), 50 mM Bis–Tris buffer at 25°, equilibrated with 0.1 atm CO. The Hb concentration was 100 μM on a heme basis. The slow fraction, characteristic of T-state Hb, increases in an additive fashion for these two effectors. At the low photolysis levels used (10%), the dominant form produced is triply liganded tetramers; use of both effectors can form an appreciable amount of this species in the T state. Similar additive effects of L 3,5 and IHP have been observed for the kinetics of NO geminate recombination in the picosecond time regime.[35,36]

fraction, characteristic of T-state Hb, increases in an additive fashion for these two effectors. These kinetic results confirm equilibrium data on the stabilization of the T state of hemoglobin in the presence of bezafibrate derivatives in combination with IHP.[14,15,37]

Partially oxidized Hb samples can also be used to study triply liganded tetramers.[38–40] At high oxidation levels, the main form that binds oxygen is triply metHb. This substate can then be studied with various effectors and different ferric ligands. Although triply oxidized tetramers with high-spin ligands such as water or fluoride are easily converted to the T state by IHP alone, the use of low-spin ligands such as CN show mainly R-state properties (either high oxygen affinity or rapid CO-rebinding kinetics).

[37] S. J. Gill, M. L. Doyle, and J. H. Simmons, *Biochem. Biophys. Res. Commun.* **165,** 226 (1989).
[38] M. C. Marden, J. Kister, B. Bohn, and C. Poyart, *J. Mol. Biol.* **217,** 383 (1991).
[39] M. C. Marden, L. Kiger, J. Kister, B. Bohn, and C. Poyart, *Biophys. J.* **60,** 770 (1991).
[40] M. C. Marden, J. Kister, and C. Poyart, this volume [6].

The addition of two strong effectors can change even the low-spin species toward the T state.[38–42]

Acknowledgments

The authors are indebted to Professor D. J. Abraham for support and critical reading of this manuscript. We would also like to thank G. Delpech, B. Bohn, and G. Caron for expert editorial and technical assistance. These studies were closely coordinated with the laboratories of M. F. Perutz and G. Fermi (MRC, Cambridge, England), of I. Lalezari (Montefiore Hospital, New York, NY), and of D. J. Abraham (Richmond, VA).

[41] R. W. Noble, A. De Young, S. Vitale, M. Cerdonio, and E. E. DiIoro, *Biochemistry* **28,** 5288 (1989).

[42] R. W. Noble, A. De Young, and D. L. Rousseau, *Biochemistry* **28,** 5293 (1989).

Section IV

Mathematical Analysis and Modeling

[25] Simulation of Hemoglobin Kinetics Using Finite Element Numerical Methods

By ROBERT L. BERGER, NORMAN DAVIDS, and MICHELE PERRELLA

Introduction

Ordinary differential equations (ODEs) have been used for centuries as mathematical models of physical and chemical processes. The laws of mass action, for example, can be used to derive an ODE or system of ODEs describing a kinetic process. The advantage of using ODEs to express complex models arising from the laws of mass action is that they are easy to derive, even if they are difficult or impossible to solve analytically. Numerical solutions to such equations have been one of the great advantages of the computer age. The numerical analysis literature is full of solution methods, and a survey of that literature is beyond the scope of this chapter. The interested reader should see such topics as ordinary differential equations, stiff differential equations, initial value problems, and boundary value problems. In this chapter, we discuss one elementary method of analyzing biochemical kinetic problems to show what some of the issues are, including a discussion of how such methods appear to a user, in terms of ease of use. The general solution of a complex reaction, in this case hemoglobin (Hb) binding to carbon monoxide (CO), is then discussed in detail, including a global fit using a curve fitter.

Finite Element Approach

In simulating a rate process, we start from the law of mass action, which states that the rate of the forward reaction is proportional to the concentrations of the reactants, say A and B. If these are written as [A] and [B], $v_f = k_f[A][B]$ and in a similar manner, $v_b = k_b[C][D]$ if C and D are the products.[1] In the simplest case we might have reactant A going to B, with the back reaction going so slowly that it can be neglected. Then we have

$$dA/dt = -k_f[A] \tag{1}$$

where k_f is the rate constant for the reaction and [A] is the concentration of the particular component.

[1] W. J. Moore, "Physical Chemistry," 3rd ed., p. 169. Prentice-Hall, Englewood Cliffs, NJ, 1962.

Equation (1) may be interpreted in two ways: (1) that it is an ODE, and (2) that it is a statement about finite steps dA and dt. In the first case the ODE must be solved, which in this elementary case is easy, the solution in closed form being $A = A_0 \exp(-k_f t)$. We can, however, ignore this solution completely by refraining from having dA, dt pass to the limit, which is zero. In this way we bypass having to solve the differential equation either in closed form or numerically, because it is not generated in the first place. Thus Eq. (1) can be rewritten as $dA = -k_f[A]\, dt$ and used repetitively to provide an accounting of the increments of reaction products being created during a succession of time intervals dt.

Following this step, we adjust or update the value of A by the relation

$$[A]' = [A] + d[A] \tag{2}$$

These operations develop into a computer program when these steps are carried out for each reaction and each set of reactants, and include the proper input–output statements and control statements. The particular scheme for doing this was developed by Davids and Berger[2,3] and is called the finite element or FE method.

The above steps are inherent in the use of any digital procedure for simulating continuous processes: (1) The time must be divided into a discrete set of intervals so that the state of the system, stored as numerical values, can be updated repetitively, as described above; (2) the value of the time step dt and other numerical parameters must be chosen with care so that the computations are not only accurate but, even more fundamentally, remain stable; that is, do not cause a rapidly increasing error to build up. This point is discussed in more detail in the next section.

Stability Problem–Multiple Reaction Systems

An inherent problem with any procedure that describes a physical system with finite sequential steps is that instability occurs when initially small errors grow, tending to build up large fluctuations in a system. Instability problems may arise when multiple reactions exist in a system having widely varying orders of magnitude in rate.

As an illustration consider the following system:

$$A + B \xrightarrow{k_1} C \tag{3}$$

$$C + D \xrightarrow{k_2} A + E \tag{4}$$

---

[2] N. Davids and R. L. Berger, *Commun. ACM* **7**, 547 (1964).
[3] R. L. Berger, *Discuss. Biophys. Soc., Biophys. J.* **24**, 2 (1978).

TABLE I
COMPARISON OF EXACT SOLUTION VERSUS
FINITE ELEMENT SOLUTION[a]

	[A]/[A_0]		
$k_1 t$	Analytical	Simulation	[B]/[B_0]
0	1.000	1.000	0.000
0.2	0.819	0.818	0.164
0.4	0.670	0.669	0.268
0.6	0.549	0.547	0.331
0.8	0.449	0.448	0.363
1.0	0.368	0.366	0.374
2.0	0.135	0.134	0.302
3.0	0.050	0.049	0.204
4.0	0.018	0.018	0.144
5.0	0.007	0.007	0.114

[a] Comparison of the solutions of the kinetics of Eq. (5) according to the exact mathematical solution of Eqs. (6)–(8) and the finite element method simulation. Conditions: $[A]_0 = 1$, $k_1 = 1$, $k_3 = 1$, $k_2 = 0.1$, and $dt = 0.01$ sec.

in which k_1 is several orders of magnitude faster than k_2. In such a case the stability of the numerical model is maintained by choosing dt_d for the fast reaction, say $dt_d = (1/100)\ dt$, and then taking 100 steps for this reaction every time a single step of the slow reaction, which uses the time step dt, is calculated. We can think of this procedure as "synchronizing the clocks" associated with each of the reactions.

Several kinetic cases have been taken from the literature to validate the procedure. As an example, consider the following case of two consecutive reactions:

$$A \xrightarrow{k_1} B \underset{k_2}{\overset{k_3}{\rightleftarrows}} C \tag{5}$$

A first, irreversible unimolecular reaction is followed by a reversible one. A general treatment of such reactions as given by Benson[4] and Johnson has given a two-exponential solution[5] (extensive tables for this reaction

[4] S. W. Benson, "The Foundations of Chemical Kinetics." McGraw-Hill, New York, 1960 (see esp. Chapter III).
[5] K. A. Johnson, this series, Vol. 134, p. 677.

based on exact mathematical formulas are available[6]). The rate equations for this reaction scheme are

$$dA/dt = -k_1[A] \tag{6}$$
$$dB/dt = k_1[A] + k_2[C] - k_3[B] \tag{7}$$
$$dC/dt = -k_2[C] + k_3[B] \tag{8}$$

Table I gives a comparison with the result computed from the closed form solution, taken from Kuba and Konowalow[6] and from the FE method, assuming $[A]_0 = 1$, $k_1 = 1$, $k_3 = 1$, and $k_2 = 0.1$. The following is the FE program used, written in BASIC.

The reaction of bicarbonate and hydrochloric acid is a well-known example of this type of reaction scheme.[7] On mixing, the reactants combine quickly ($10^8 \, M^{-1} \, sec^{-1}$) to form carbonic acid. The breakdown to carbon dioxide and water is also fast, 25 per second at 25°, and the reverse reaction is slow, even at pH 7.4. Appendix A contains the Fortran program for solving this case by the finite element method, but without the curve fitter. This requires making a first estimate of the values of the rate constants. The calculation is then made and a visual appraisal is made, comparing the FE output with the experimental data. From the rates calculated, a new estimate can be made and the process repeated until satisfactory agreement is reached. Appendix B shows a more complicated case with the automatic curve fitter added. This point is discussed in greater detail later on page 523.

Association Reaction between Hemoglobin and Carbon Monoxide

Finite element analysis can be used to analyze the interaction between hemoglobin and CO in the general case, in which both the CO association and dissociation reactions occur, along with structural transition equilibria, during the course of the reaction. We consider here a special case in which substoichiometric amounts of CO are quantitatively bound by hemoglobin in a period of time during which the dissociation of the ligand is negligible. At the end of this preset reaction time, the liganded intermediates are quenched and separated by cryogenic techniques.[8,9] The experimental data are in the form of concentrations of the intermediates as a

[6] D. W. Kuba and D. D. Konowalow, "Report ARL 63-44 (ASTIA)." University of Wisconsin, Madison, 1963.

[7] L. Rossi-Bernardi and R. L. Berger, *J. Biol. Chem.* **243,** 1297 (1968).

[8] M. Perrella, L. Benazzi, L. Cremonesi, S. Vesely, G. Viggiano, and R. L. Berger, *J. Biochem. Biophys. Methods* **7,** 187 (1983).

[9] M. Perrella and L. Rossi-Bernardi, this volume [21].

```
30    'INITIALIZE:

40    A=1                          'Initial conc of A

50    B=0                          '    "    "    B

60    C=0                          '    "    "    C

70    K1=1                         'Rate const for A--->B

80    K3=1                         'Rate const for B--->C

90    K2=.1                        'Rate const for back reac B<---C

100   T=0

110   DT=.01                       'Set time step

120   KM=101                       'Set number of intervals

130   '

140   'START REACTION

150   FOR K=1 TO KM

160   LPRINT USING "#.## ###.### #.### #.###";;T,A,B,C

170   D=K1*A*DT                    'Law of Mass Action for A--->B

180   A=A-D                        'Update A

190   B=B+D                        'Update B

200   D=K3*B*DT-K2*C*DT            'Law of Mass Action for B<--->C

210   B=B-D                        'Update of B again

220   C=C+D                        'Update of C

230   T=T+DT                       'Update of time

240   NEXT K                       'End loop

250   STOP
```

SCHEME I

function of fractional CO bound. Time information is not provided because at the high values of the hemoglobin concentration required by the cryogenic technique, the CO-binding reaction is so fast that sophisticated quench-flow methods are required to study the reaction on a time-dependent basis.

Program

Figure 1a is a general schematic showing the 10 ligation states and the pathways (16 in all) that can be taken from deoxyhemoglobin (Hb) to

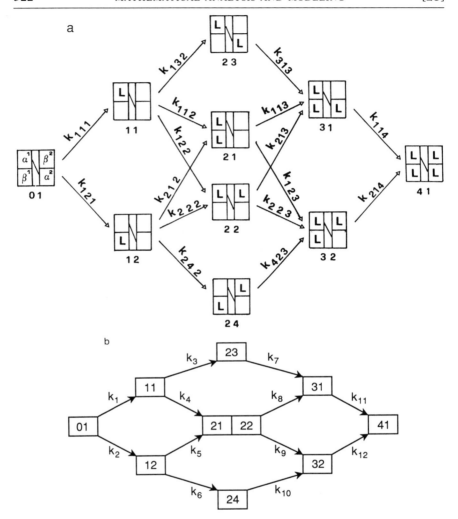

FIG. 1. (a) General schematic showing the 10 ligation states of hemoglobin and the 16 forward pathways leading from deoxyhemoglobin to fully liganded hemoglobin. Each ligation state is assigned indexes ij, where j denotes the code of the isomer in ligation state i. Each rate constant k is assigned subscripts ijk: i, matrix from which transition originates; j, matrix element where transition terminates; k, number of sites liganded. (b) Simplified 12-rate constant path. Species 21 and 22 are grouped together because they cannot be resolved by the cryogenic technique for the isolation of the intermediates.

carbonmonoxyhemoglobin (HbCO). Each ligation state is located by the number of liganded sites (horizontal position). In a previous notation each state was also located by its position from the top of the diagram (columnwide).[3] We have changed this notation for the double-liganded species to conform to another such scheme now commonly used.[10] A variable defining the rate constants is now introduced as $(KON)_{ijk}$, where i identifies the matrix element from which transition originates; j identifies the matrix element where transition terminates; and k is the number of sites liganded. Care must be taken in assigning statistical factors, 1 or 2, for the type of heme available for any given path. A value of zero for the variable corresponding to a particular transition in Fig. 1a prevents that transition from taking place. By putting in enough zeros, one can narrow the process to a single pathway or group of pathways. The general scheme for doing the simulation and optimizing the rate constants by global curve fitting is presented below, with applications to a simple four-step model of CO hemoglobin binding and then using a 12-rate constant model (Fig. 1b).

Outline of Kinetics Curve-Fit Program

Standard methods exist for determining the best fit by least squares to a set of data points, using a particular type of function, such as a polynomial or a series of exponentials.[11] The method of finite elements starts with the differential equations for the system in finite form. One then supplies the kinetics program with a set of trial parameters, that is, starting values or values determined from a previous iteration. We then go through the kinetics program and determine the error between the predicted curve and the data. These errors are then used to calculate the next mean-square error, which is compared with the previous one. A decision is then made as to whether to repeat the process with a disturbed set of parameters, until an optimum is reached. In this manner, it is possible to curve-fit a set of data by using kinetic models in place of numerical functions altogether. Figure 2 shows the scheme diagrammatically.

1. Read in and store the data, as identified by file name.
2. Enter the initial concentrations of the solutions, [Hb] and [CO].
3. Enter the assumed transition sequence case.
4. Enter the assumed starting values for the rate constants.
5. Enter the number of iterations desired.

[10] G. K. Ackers and F. R. Smith, *Annu. Rev. Biophys. Biophys. Chem.* **16**, 583 (1987).
[11] D. W. Marquardt, *SIAM J.* **2**, 431 (1963).

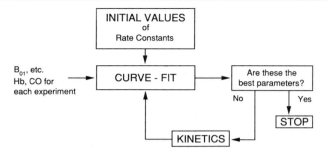

Fig. 2. Outline of the finite element solution for the hemoglobin–CO reaction.

6. The computer then begins by passing the values of the given rate constants to the subroutine CFIT, where they are stored as the vector: $parm_i$, $i = 1$, nparm.

7. These values are then passed to subroutine KINET, which works out a prediction function $y(t)$ according to the transition sequence described there. The error is then calculated from $\varepsilon(t) = y\mathrm{dat}(t) - y(t)$, where $y\mathrm{dat}(t)$ equals the observed data.

8. We then return to CFIT to obtain the sum of squares: $s = \Sigma[\varepsilon(t)]^2$.

9. Each parameter is then perturbed in turn by the relation $parm_i = parm_i(1 + 0.001)$.

10. With the above steps, the gradient vector can be set up normal to the surface of a constant sum of squares in parameter space. The Jacobean matrix of the system can then be constructed. The inversion and diagonalization of this matrix, according to the Gauss–Newton procedure, provides an updated set of parameters and a new sum of squares of error, hopefully lower in magnitude.

11. The cycle of iterations terminates when the mean-square error converges. This can be determined by specifying a preset tolerance or the number of iterations to be made. The user should review the display of the successive sum of squares and then, by typing an integer, select additional iterations as desired.

Global Fit

In general, any curve-fit scheme will yield more precise values for the parameters when more data points are available. In fact, for any kind of precision we must have the condition

$$n\mathrm{data} \gg n\mathrm{params} \tag{9}$$

In the simple four-step scheme of the reaction, in which the subunit heterogeneity is ignored, there are five values of the concentration of the intermediates, $B(i)$, available for determining four parameters, if time information is provided. If time information is not provided, there are three parameters relative to a fourth fixed parameter for making the determination. Because the sum of the concentrations of the intermediates at a certain value of the fractional CO bound equals the total hemoglobin concentration, only four values of the concentrations of the intermediates are independent. Thus for the simple four-step scheme it is possible to curve-fit the parameters by using the data from a single experiment at some fractional value of CO bound to hemoglobin. If data from experiments carried out at many different values of the fractional CO are available, a more reliable estimate of the error in the curve-fitted parameters is obtained by combining all of the data into what is called a "global" fit. This is done by simply "stacking the data" until we run out of data, as shown in Table II.

Thus, if we have $n = 10$ experiments to work with in this fashion, we obtain 90 independent data points for the adjustable parameters relative to the reference parameter and condition (9) is fulfilled.

If subunit functional heterogeneity is considered, 10 different ligation states are possible according to Fig. 1a. Because the cryogenic technique

TABLE II
STACKING METHOD

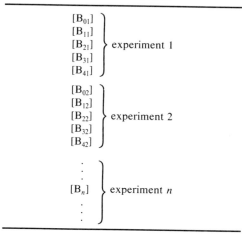

cannot determine the concentrations of intermediates B_{21} and B_{22} separately, 9 values of the concentrations of the intermediates, $B(i)$, are determined, of which only 8 values are independent, against a total of 12 parameters to curve-fit for, as shown in Fig. 1b. Thus, even if k_{11} and k_{12} are assumed to be equal reference parameters, the number of adjustable parameters (10 values of the k) is greater than the number of data points, that is, 8 concentration values, and condition (9) is not fulfilled. In such a case the global fit using the data from several experiments at different values of fractional CO bound is mandatory. The Fortran program used to do this is shown in Appendix B.

Application to Simple Four-Step Scheme

Because the functional heterogeneity of the subunits in the association reaction with CO is slight,[12] the general process in Fig. 1a can be simplified by reducing it to a simple four-step process by putting

$$k_{111} = k_{121} \tag{10}$$
$$k_{132} = k_{242} \tag{11}$$
$$k_{112} = k_{212} = k_{222} = k_{122} \tag{12}$$
$$k_{313} = k_{423} \tag{13}$$
$$k_{113} = k_{123} = k_{213} = k_{223} \tag{14}$$
$$k_{114} = k_{214} \tag{15}$$

Furthermore, when no time information is provided by the data, either the rate constants are calculated relative to a reference rate in the process or such a reference rate is assigned a value taken from the literature. Table III shows the results of the analyses of the experimental data, obtained by quench-flow and cryogenic isoelectric focusing methods[8,9] to give the intermediates, to obtain the rate constants in the association reaction[12] between Hb and CO at 20°, in 0.1 M KCl, pH 7, assuming $k_4 = 6.0 \ \mu M^{-1} \ sec^{-1}$, as obtained by Gibson using optical stopped-flow techniques[13] under similar conditions. The optimum time step dt was determined to be 0.01 msec. The analyses, starting with the data at the lowest fractional value of CO bound, were carried out by stepwise addition of the data at higher CO saturations to obtain the final global fit to all the

---

[12] M. Perrella, N. Davids, and L. Rossi-Bernardi, *J. Biol. Chem.* **267**, 8744 (1992).
[13] Q. H. Gibson, *J. Biol. Chem.* **248**, 1281 (1973).

TABLE III
RATE CONSTANTS FOR REACTION OF HEMOGLOBIN PLUS CO: FIVE-SPECIES MODEL[a]

Global fit up to experiment no.	S_{CO} (%)	Fitted parameters (μM^{-1} sec^{-1})			SSE[b] ($\times$ 10^4)
		k_1	k_2	k_3	
1	17.6	0.0777	0.199	1.081	11.403
2	18.5	0.0952	0.227	1.302	6.231
3	21.8	0.0944	0.226	1.333	6.618
4	30.1	0.0975	0.231	1.336	5.974
5	39.9	0.103	0.241	1.412	5.638
6	45.3	0.108	0.253	1.472	5.445
7	54.9	0.110	0.256	1.474	5.436
8	59.8	0.112	0.261	1.465	5.376
9	72.4	0.113	0.264	1.474	5.371
10[c]	82.7	0.113	0.265	1.472	5.368
Average values[d]:		0.1024	0.242	1.382	6.286

[a] Rate constants for the association reaction between Hb and CO at 20°, pH 7, from data on the reaction intermediates as studied by cryogenic techniques.[12] Parameters (k_1, k_2, k_3) (intrinsic rate constants, i.e., apparent rate constants divided by the statistical factors 4, 3, and 2, respectively) were fitted by the finite element method as a function of the cumulative number of experiments at different CO saturations S_{CO}, (1, 1 + 2, . . . , 1 + 2 + · · · + 10), using k_4 = 6.0 μM^{-1} sec^{-1}.[12]
[b] SSE, Sum of square errors (see text).
[c] The error in the rate constants is ±10%.
[d] Rate constants obtained by averaging the values of the parameters fitted to the data at each CO saturation value.

data. The calculated rate constants at each step were used to fit the data at all CO saturations; the sum of square errors for each fit (SSE) are shown in Table III.

Clearly, limiting values for the rate constants are reached in the step-wise process and the minimum value for SSE is attained with the parameters from the global fit.

Figure 3 shows the relative concentrations of the species in different states of ligation versus CO saturation as calculated by the finite element method, assuming negligible functional heterogeneity of the subunits and using the values of the rate constants obtained by the global fit.

If we now expand the possible pathways by utilizing the scheme of Fig. 1b, fixing the last 2 rate constants, $k_{11} = k_{12} = 3.0$ μM^{-1} sec^{-1}, and again do the global fit using all 10 experiments, we obtain the results shown in Table IV.

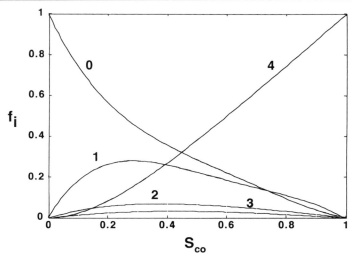

FIG. 3. Fraction f_i of liganded intermediate i ($i = 0$–4) versus CO saturation S_{CO} in the association reaction between Hb and CO at 20° in 0.1 M KCl, pH 7.[12] The curves were calculated using the rate constants obtained by the global fit to the experimental data on the concentrations of the intermediates (see Table III), assuming negligible functional heterogeneity of the subunits.

The results of this global fit to the CO-binding data, according to the 12-rate constants scheme in Fig. 1b, confirm the finding of Perrella et al.[12] that the β subunit reacts faster than the α subunit in the first CO-binding step ($k_2/k_1 = 1.91 \pm 0.28$). A similar subunit heterogeneity is observed by comparing the rate constants for the second CO-binding step along the $\alpha\alpha$–$\beta\beta$ pathway (upper pathways in Fig. 1b) and along the $\beta\beta$–$\alpha\alpha$ pathway (lower pathways in Fig. 1b), which yield the ratio $k_6/k_3 = 2.98 \pm 1.76$. Because species 21 and 22, both having one α- and one β-liganded subunit, are not equivalent and the cryogenic technique allows only the determination of the sum of their concentrations, the rate constants k_4, k_5, k_8, and k_9 are composite constants. However, within experimental error, $k_4 = k_5$ and $k_8 = k_9$. Thus the global fit confirms a previous conclusion that the CO-binding properties of species 21 and 22 are similar[12] and that all binding steps are cooperative with an acceleration after the binding of the second CO molecule.

The simple model shown above, and the more complex 12-constant model, can be solved by analytical methods and the results are in close agreement. However, when one goes to the reactions with oxygen and hemoglobin, analytical models of the 12 rate constants, including the back reactions, which cannot be neglected as in the CO-binding case, become

TABLE IV
RATE CONSTANTS FOR REACTION OF CO PLUS
HEMOGLOBIN: NINE-SPECIES MODEL[a]

Ligation of α subunits	Ligation of β subunits
$k_1 = 0.0410 \pm 0.0042$	$k_2 = 0.0782 \pm 0.0080$
$k_3 = 0.0438 \pm 0.0215$	$k_4 = 0.162 \pm 0.031$
$k_5 = 0.153 \pm 0.054$	$k_6 = 0.131 \pm 0.0428$
$k_8 = 0.454 \pm 0.117$	$k_7 = 0.133 \pm 0.220$
$k_{10} = 0.729 \pm 0.363$	$k_9 = 0.713 \pm 0.159$

[a] Rate constants, k_n ($n = 1$–12)[b], for the binding of CO to deoxyhemoglobin at 20° in 0.1 M KCl, pH 7, calculated from the concentrations of the intermediates according to the reaction scheme of Fig. 1b, assuming $k_{11} = k_{12} = 3.0$ μM^{-1} sec^{-1}, that is, one-half the value measured by optical stopped-flow kinetics for the binding of the last CO molecule.[13]

[b] Intrinsic, which is apparent constant k_n' divided by the statistical factor (in μM^{-1} sec^{-1}). The concentration is the heme concentration. $k_1 = k_1'/2$; $k_2 = k_2'/2$; $k_4 = k_4'/2$; $k_5 = k_5'/2$; $k_8 = k_8'/2$; $k_9 = k_9'/2$; in all other cases $k_n = k_n'$.

intractable in general. We thus believe that the finite element method offers an important technique for solving or simulating large kinetic problems.

General Conclusions

The FE method has the advantage that its steps are formulated easily, as they deal directly with the physics or chemistry of the problem. It simulates the problem in the sense that if the computation is stopped for any reason at an earlier time, it shows the state of the system at that time. Furthermore, it can easily deal with an additional boundary condition or the presence of additional effects, such as electric or magnetic fields, without invalidating the work up to that point. On the other hand, instabilities are difficult to deal with, just as with other numerical methods.

In a more general sense, solving kinetic equations requires a number of methods. We have found that the program MATLAB[14] serves very well. We have presented a method of solving kinetic equations using any computer and a language such as Fortran, but we have found that the

[14] MATLAB, PC or 386 version available from Mathworks, Inc., 24 Prime Park Way, Natick, MA 01760.

utilization of MATLAB allows any of the numerical methods to be written down without concern of portability, as these programs will run on any version of MATLAB. These programs are still under development.

It should be pointed out, however, that approaching a problem by numerical analysis of ODEs gives mathematical analysts the advantage of having at their disposal many established numerical and error analysis methods. They also have at their disposal such computer procedures as MATLAB, and so on, that contain within them the broad experience of many other analysts and that are universal in scope as well as convenience. Initial efforts to put the FE method into MATLAB and to utilize stiff differential equation solvers as well as curve fitters seem to be working and it is hoped these efforts will be reported in detail in the near future.[15]

Appendix A: Program without Curve Fitter

```
C   PROG: CO2.F4, CARBON DIOXIDE KINETIC MODELS

C

C      REACTIONS IN ACID SOLUTIONS

C      COMPONENTS:

C      CO2   = CARBON DIOXIDE (DISSOLVED)

C      H2CO3 = CARBONIC ACID

C      HCO3- = BICARBONATE ION(-)

C      CO3-- = CARBONATE ION (--), HCO3=H+CO3

C      TOT   = TOTAL CARB. DIOX. CONTENT OF SYSTEM (=CO2+H2CO3+HCO3-)

C      PCO2  = GASEOUS FORM OF CO2, MEASURED AS PARTIAL PRES.

C      KCO2  = HYDRATION VELOCITY CONSTANT (THERMAL)

C      KH2CO3=DEHYDRATION VELOCITY CONST.

C      HEATS:

C      DELHI = HEAT OF DE-IONIZATION CAL/MOL

C      DELH2 = HEAT OF DEHYDRATION   CAL/MOL

C      PROG. CONTROL SWITCHES

C      KS=0, TERMINATE EXECUTION

C      KS=1, RECYCLE PROGRAM

       DIMENSION Y(500),TAU(500),Q(500)
```

[15] R. I. Shrager, Division of Computer Research and Technology, National Institutes of Health, Bethesda, MD (personal communication).

```
        REAL*4 NA,NAHCO3,KCO2,KH2CO3,KCL,KH,KH2O,KHCO3,KL2CO3
        REAL*4 ISTR,MU
C       COEFF 1ST DISSOC H2CO3 (K1)
        A=3404.71
        D=14.8435
        C=0.032786
C       CONC OF H2O (MOLS/L)
        H2O=55.5
        PRINT 804
804     FORMAT (' TEMPERATURE OF EXPER (DEG-C): '$)
        READ 3,TEMP
        TABS=TEMP+273.16
C       H2CO3 DISSOC CONST
        PK1=A/TABS-D+C*TABS
        E1=10**(-PK1)
        WRITE(21,911)TEMP,PK1,E1
        IF (TEMP.EQ.37.) GO TO 10
C       ASSUMED RATE CONSTANTS (/SEC):
C       DISSOC DATA FROM LONG, BIOCHEM.HANDB. P43
C       DEHYDRATION CARBONIC ACID: H2CO3=CO2+H2O
C       VELOCITY CONST (AT 25 DEG)
        KH2CO3=25.5
C       BACK REACTION RATE (HYDRATION)
        KCO2=.0257/H2O
C       2ND DISSOC CONST HCO3 (K2), 25 DEG
        PK=10.329
C       A=2902.39
C       D=6.4980
C       C=0.02379
        GO TO 20
C
10      CONTINUE
```

```
C     VELOCITY CONST AT 37 DEG

      KH2CO3=72.00

      KCO2=.0620/H2O

20    CONTINUE

C     DISSOC CONSTS VS TEMP

C     IONIZATION  & FORMATION OF CARBONIC ACID: H+HCO3=H2CO3

C     EQUILIBR CONST H2CO3 AT 25 DEG IS 3.765

C     PK WHEN CORRECTED FOR SALT EFFECT

      PK=3.5

C     PK AT 37 DEG:

      PK=3.6

      EH2CO3=10**(-PK)

C     ASSUMED FORMATION RATE OF H2CO3 FROM H+HCO3-

      KHCO3=5E10

C     DISSOCIATION RATE (IONIZATION) OF H2CO3

      KL2CO3=EH2CO3*KHCO3

C

C     FORMATION OF H2O: H+OH=H2O

      KH=1.4E11

      KH2O=2.5E-5

      DELH1=1205.

      DELH2=1455.

C     SOLUBILITY CONSTANTS

      A=0.0301

C     COMPUTER INITIALIZATION & CONTROL CONSTS

      NFIL=23

      KM=100

      K2=0

      KPR=1

      VOL=1.0

      T=0.0

      T1=0.0
```

```
       TW=0.0

       K=0

       TW=0

       Q1=0.

       Q2=0.

       QMAX=0.
C     PREPARATION OF SOLUTION 1: HCL

       CALL ANMODE

       WRITE(NFIL,912)
912    FORMAT (' SOLUTION 1 (HCL)')

       PRINT 801
801    FORMAT (' MOLAR CONC SOLUTION 1 (HCL): )

       READ 3,HCL
3      FORMAT (2F)

       PRINT 802
802    FORMAT (' CONC SALT ADDED: )

       READ 3,KCL
C     HCL COMES TO EQUILIBRIUM (IONIZES): HCL=H+CL
C     EQUILIBRIUM CONST HCL

       H1=HCL

       CL1=HCL

       PH0=-ALOG10(H1)

       H=H1

       PH=PH0
C     CHARGE BALANCE

       CHG=H-CL

       WRITE(NFIL,901) K,K2,T,PH,HUM,OHUM,HCO3M,H2CO3M,CO2M,CHG
C     PREP. SOLUTION 2: NAHCO3

       PRINT 803
803    FORMAT (' MOLAR CONC SOLUTION 2 (NAHCO3): )

       READ 3,NAHCO3
C     NAHCO3 COMES TO EQUILIBRIUM: NAHCO3=NA+HCO3
```

```
      NA=NAHCO3

      HCO3=NAHCO3

      HCO3M=HCO3*1E3

      WRITE(NFIL,901)K,K2,T,PH,HUM
C     RESTART OF COMPUTATION HERE
1     CONTINUE
C     MIXING AFTER 1 MILLISEC, NAHCO3+HCL=H2CO3+NACL
      WRITE(NFIL,903)
903   FORMAT (1H ,'SOLUTIONS 1 & 2 BECOME MIXED')
      CL=CL1/2.0

      NA=HCO3/2.0

      HCO3=HCO3/2.0

      H=H1/2.0

      H2CO3=0.0

      H2CO3M=H2CO3*1E3

      HUM=H*1.E6*VOL

      OHUM=OH*1.E6*VOL

      CO2M=CO2*1E3

      CALL MOVABS(IX0,IY0)

      CO2=0.0
C     CARBONIC ACID FORMED: H+HCO3=H2CO3
      DT=5E-10

      DO 101 K2=1,40

      D=KHCO3*H*HCO3*DT-KL2CO3*H2CO3*DT

      H2CO3=H2CO3+D

      H=H-D

      HCO3=HCO3-D
C     EXIT AT EQUILIBRIUM
      TOL=0.1E-5

      IF (ABS(D).LT.TOL) GO TO 13

      T=T+DT
101   CONTINUE
```

```
13      CONTINUE
C       CHG BALANCE
        CHG=H+NA-CL-HCO3
        H1=H
C       CONSTANTS IN ROSSI-BERNARDI-BERGER PAPER
        A=.07
        B=.1
        C1=2.303*(1.+(EH2CO3/(B-A)))
        C2=2.303*(1.-(EH2CO3/(B-A)))
C       IONIC STRENGTH AT START OF REAC (M/L)
        ISTR=0.1
C       DEBYE HUCKEL FORMULA
        MU=ISTR
        DB=0.505*SQRT(MU)/(1.+1.6*SQRT(MU))

C       MEAN IONIC ACTIVITY COEFF
        F=10**(-DB)
        C1=2.303*(1.+(EH2CO3/F/(B-A)))
        C2=2.303*(1.-(EH2CO3/F/(B-A)))
        WRITE(21,911)CHG,DB,F,EH2CO3
911     FORMAT (1H ,4E13.3)
C       KINETICS LOOP HERE
        DT1=.001
        TMAX=KM*DT1
        DO 102 K=1,KM
        PH=-ALOG10(H)
        CHG=H-HCO3-OH-CL+NA
        HCO3M=HCO3*1E3
        H2CO3M=H2CO3*1E3
        CO2M=CO2*1E3
        IF((K-1)/KPR*KPR-(K-1))11,12,11
12      CONTINUE
```

```
C     PARTIAL PRES. CO2 (MMHG)

      PCO2=(CO2M+H2CO3M)/A

      HUM=H*1.E6*VOL

      OHUM=OH*1.E6*VOL

      WRITE (NFIL,901) K,K2,T1,PH,HUM,OHUM,HCO3M,H2CO3M,CO2M,CHG
901   FORMAT(1H ,2I3,E10.2,F6.2,2X,2E10.3,2X,3E10.3,2X,2E10.3)
11    CONTINUE
C     KINETICS PLOT
C     PLOT OF PH VS TIME

      IX=T1/TMAX*IXINT+IX0

      IY=(PH-2)/5.*IYINT+IY0

      CALL DRWABS(IX,IY)

C

C     DEHYDRATION OF CARBONIC ACID: H2CO3=H2O+CO2

      D=KH2CO3*H2CO3*DT1-KCO2*H2O*CO2*DT1

      H2CO3=H2CO3-D

      CO2=CO2+D

      H2O=H2O+D

C     INCREMENTAL HEAT PRODUCED BY DEHYDRATION

      DQ2=D*DELH2

      Q2=Q2+DQ2

C     IONIZATION OF WATER
C     H+OH=H2O
C     CALL WATER(H,OH,H2O,KH2O,KH,PH)

C

C     IONIZATION OF CARBONIC ACID (H+)+(HCO3-)=H2CO3
15    DO 103 K2=1,40

      HUM=H*1.E6*VOL

C     WRITE (NFIL,902) K2,H,HCO3,H2CO3,D
902   FORMAT(1H ,I3,4E10.3)

      D=KHCO3*H*HCO3*DT-KL2CO3*H2CO3*DT

      H2CO3=H2CO3+D
```

```
      H=H-D

      HCO3=HCO3-D

      K2M=K2


C     EXIT TEST IF EQUILIBRIUM REACHED

      IF (ABS(D).LT.TOL) GO TO 18

C     INCR HEAT PRODUCED BY DE-IONIZATION

      DQ1=D*DELH1

      Q1=Q1+DQ1

      T=T+DT

103   CONTINUE

18    CONTINUE

C     TOTAL HEAT

      Q(K)=Q1+Q2

      IF (Q(K).GE.QMAX) QMAX=Q(K)

      IF (T.GE.0.2) GO TO 2

      T1=T1+DT1

102   CONTINUE

2     CONTINUE

C

C     EQUATION ROSSI-BERGER PAPER J. Biol. Chem. 243:1297-1302,1968.

      PH=-ALOG10(H1)

      DO 104 K=1,KM

      H=10**(-PH)

      Y(K)=C1*ALOG10(H1/H)+C2*ALOG10((H1+B-A)/(H+B-A))

      TAU(K)=Y(K)/KH2CO3

      IX=TAU(K)/TMAX*IXINT+IX0

      IY=(PH-2)/5.*IYINT+IY0

      PH=PH+.1

      CALL DRWABS(IX,IY)

      IF (TAU(K).GT.TMAX) GO TO 17

104   CONTINUE
```

```
17      CONTINUE
        WRITE(5,921)TMAX
921     FORMAT(' ',F5.3)
C
        T1=0.
        DO 121 K=1,KM
        IX=T1/TMAX*IXINT+IX0
        IY=Q(K)/100.*IYINT+IY0
        T1=T1+DT1
121     CONTINUE
        WRITE (5,922)QMAX
922     FORMAT (1H ,E13.3)
        READ 501,KS
501     FORMAT (I2)
        IF (KS.NE.0) GO TO 1
        STOP
        END
C       FORMATION OF H2O
        SUBROUTINE WATER (H,OH,H2O,KH2O,KH,PH)
        KWM=200
        DO 106 KW=1,KWM
        IF (PH.LT.7) DTW=5.E-12*10**PH
        IF (PH.GE.7) DTW=5.E-2*10**(-PH)
        IF (KW.EQ.0) WRITE(NFIL,910) TW,H2O,H,OH,PH
910     FORMAT (1H ,4E12.3,F6.2)
        D=+KH*H*OH*DTW-KH2O*H2O*DTW
        H2O=H2O+D
        H=H-D
        OH=OH-D
        TW=TW+DTW
106     CONTINUE
        RETURN
        END
```

Appendix B: Program with Curve Fitter

```
C*************HGLO10.FOR*****************************************
C
C ACCEPTS DATA IN PERRELLAS FORMAT
  KINETICS OF HEMOGLOBIN-CO, ALPHA AND BETA CHAINS
C ACCEPTS CONCENTRATION OF UP TO 9 INTERMEDIATES (B) AT (QUENCHING)
C TIME
C ACCEPTS STARTING VALUES OF 10 CHOSEN ON RATE CONSTANTS
C CALCULATES BEST FIT FOR THE RATE CONSTANTS BY LEAST SQUARES
C PROCEDURE
C THE DATA ARE READ FROM A SPECIAL INPUT FILE: CO.DAT
C THE STARTING VALUES ARE READ FROM AN INPUT FILE: HEXP10.DAT
C ALL CONCS IN MOLS/LITER
C   1 M/L = 4 HEMES
C RATES IN LITERS/MOL/SEC
C FILE #4 IS THE OUTPUT FILE
C
C   THE FUNCTIONAL VALUES FOR CURVE-FIT ARE CREATED BY SIMULATING THE
C   KINETICS ITSELF FROM T=0 TO T=TC, NOT A MATH FUNCTION.
        IMPLICIT DOUBLE PRECISION (A-H,K,O-Z)
        CHARACTER*40 INFILE,OFILE
        CHARACTER*1 IX
        DIMENSION PARM(30),IRK(30,3),RK(30)
        DIMENSION X(15,15)
        COMMON NDATA,NPARM,YDAT(60),F(60),ERR(60),HB(20),CO0(20)
        COMMON PARTL(30,60),DT,TC(20),N,B(0:5,0:5,60),NC,NSW,SF,ISW1
        COMMON GRAD(30),E1(60),A(20,20),C(20,20),P(30),ALAM,IT,S0,S1
        COMMON KSW,IPARM(30,3),IYDAT(30,3,20)
        COMMON RON(12),ROFF(12),NEXP,NBDAT
C   NBDAT= COUNTER FOR THE # OF B'S ENTERED
C   NPARM= # PARMS TO BE FITTED
C   NK   = COUNTER FOR THE # OF RK'S ENTERED
```

```
C    N    = INDEX # OF EXPERIMENT

C    NEXP = # OF EXPERIMENTS

C    I    = # OF LIGANDED SITES

C    J    = # OF VALUES

C    N    = # OF EXPERIMENTS

C

   250 CONTINUE

C SET VARIABLES TO ZERO

         NPARM=0

         NBDAT=0

         NK=0

C    IN B(I,L,N) I is the # of liganded sites, N is the # of the
   experiment.

         DO 2 I=1,12

         RON(I)=0.

         ROFF(I)=0.

      2 CONTINUE

         DO 28 I=0,4

         DO 28 L=0,4

         DO 28 N=1,20

         B(I,L,N)=0.

     28 CONTINUE

         DO 11 J=1,30

         ERR(J)=0.

         F(J)=0.

         PARM(J)=0.

         YDAT(J)=0.

     11 CONTINUE

         DO 6 J=1,16

         DO 6 I=1,16

         A(J,I)=0.

         C(J,I)=0.

      6 CONTINUE
```

```
C
C ************************************************************
51      FORMAT (/' INPUT FILENAME: ')
2001    FORMAT (1A40)
2003    FORMAT (' DATA FILE: ',1A40)
502     FORMAT (3I2,2X,E15.5)
902     FORMAT (' ',3i2)
900     FORMAT (3I2)
61      FORMAT (/' # OF EXPERIMENTS, NEXP = ')
504     FORMAT (' NBDAT = ',I2)
503     FORMAT (2I2,1X,E10.4)
901     FORMAT (2F10.2)
903     FORMAT (10(F6.4,1A1))
904     FORMAT (' ',10(F6.4,1A1))
505     FORMAT (9X,I2)
83      FORMAT (F5.3)
1589    FORMAT (E15.5)
C COMPUTE MODE
C    READ-IN DATA FILE
     5 PRINT 51
       READ (*,2001) INFILE
       OPEN(UNIT=3,FILE=INFILE,STATUS='OLD')
       READ(3,2003) INFILE
       WRITE(5,2003) INFILE
C
C    ASSUMPTION: FAST CONSTS FIXED(NOT PARMS)
       NK = 12
       NPARM= 10
       DO 168 K=1,NK
       READ(3,502)IRK(K,1),IRK(K,2),IRK(K,3),RON(K)
       WRITE(5,502) IRK(K,1),IRK(K,2),IRK(K,3),RON(K)
168    CONTINUE
C    ENTER # OF B'S
```

```
      READ (3,900) NBDAT
      CALL CLOSE(3)
C
C   ENTER NO OF EXPERIMENTS
      PRINT 61
      READ(*,900) NEXP
C   CONVERT RATE CONSTS TO PARAMS
      PARM(1)=RON(1)
      PARM(2)=RON(4)
      PARM(3)=RON(8)
      PARM(4)=RON(2)
      PARM(5)=RON(5)
      PARM(6)=RON(9)
      PARM(7)=RON(3)
      PARM(8)=RON(6)
      PARM(9)=RON(7)
      PARM(10)=RON(10)
C

      PRINT 818
818   FORMAT (' OUTPUT FILE, YES(1) OR NO(0)? ')
      READ (*,900) ISW1
      IF (ISW1.EQ.0) GO TO 122
      PRINT 814
814   FORMAT (' OUTPUT FILENAME: ')
      READ (*,2002) OFILE
2002  FORMAT (1A40)
      OPEN (UNIT=4,FILE=OFILE,STATUS='NEW')
122   CONTINUE
      ISW=1
C
C  ACCEPT DATA IN PERRELLAS FORMAT
      PRINT 51
```

```
      READ (*,2001) INFILE
      OPEN (UNIT=3,FILE=INFILE,STATUS='OLD')
      DO 1622 J=1,12
      READ (3,903) (X(I,J),IX,I=1,10)
      WRITE(5,904) (X(I,J),IX,I=1,10)
1622  CONTINUE
      DO 120 N=1,10
      Hb(N)=X(N,1)
      COO(N)=X(N,2)
      TC(N)=X(N,3)
      WRITE(5,56) Hb(N),COO(N),TC(N)
      IF(ISW1.GT.0) WRITE(4,56) HB(N),COO(N),TC(N)
  56  FORMAT(/1X,'HB = ',1PE15.5,/' COO = ',E15.5/' TC = ',E15.5)
C     ASSUMPTIONS
C     NO LIGANDS
      B(0,1,N)=X(N,4)
C     ONE LIGAND
      B(1,1,N)=X(N,5)
      B(1,2,N)=X(N,6)
C     TWO LIGANDS
      B(2,1,N)=X(N,7)
      B(2,2,N)=0.
      B(2,3,N)=X(N,8)
      B(2,4,N)=X(N,9)
C     THREE LIGANDS
      B(3,1,N)=X(N,10)
      B(3,2,N)=X(N,11)
C     FOUR LIGANDS
      B(4,1,N)=X(N,12)
      I=0
      L=1
      WRITE(5,158)I,L,N,B(I,L,N)
```

```
      IF(ISW1.GT.0) WRITE(4,158) I,L,N,B(I,L,N)
      I=1
      DO 160 L=1,2
      WRITE(5,158)I,L,N,B(I,L,N)
      IF(ISW1.GT.0) WRITE(4,158) I,L,N,B(I,L,N)
  160 CONTINUE
      I=2
      DO 161 L=1,4
      WRITE(5,158)I,L,N,B(I,L,N)
      IF(ISW1.GT.0) WRITE(4,158) I,L,N,B(I,L,N)
  161 CONTINUE
      I=3
      DO 162 L=1,2
      WRITE(5,158)I,L,N,B(I,L,N)
      IF(ISW1.GT.0) WRITE(4,158) I,L,N,B(I,L,N)
  162 CONTINUE
      I=4
      L=1
      WRITE(5,158)I,L,N,B(I,L,N)
        IF(ISW1.GT.0) WRITE(4,158) I,L,N,B(I,L,N)
  120   CONTINUE
      CALL CLOSE(3)
C
      DO 59 I=1,12
      IF (RON(I).LT.1.E-15) GOTO 59
      WRITE(5,58) I,RON(I)
      WRITE (4,58) I,RON(I)
   59 CONTINUE
   58 FORMAT(' KON(',I2,') = ',1PE15.5)
C    1' KOFF(',I2,') = ',1PE15.5)
   55 CONTINUE
  158 FORMAT(' B(',I1,',',I1,',',i2,') = ',1PE15.5)
```

```
1588  FORMAT(2E15.5)
C
C Here the B's for all the experiments are lumped together
C  and stored in the array YDAT ready for curve-fit.
   57 CONTINUE
   95  format (5i2)
       N0=1
       DO 68 N=1,NEXP
       YDAT(0+N0)=B(0,1,N)
       YDAT(1+N0)=B(1,1,N)
       YDAT(2+N0)=B(1,2,N)
       YDAT(3+N0)=B(2,1,N)
       YDAT(4+N0)=B(2,3,N)
       YDAT(5+N0)=B(2,4,N)
       YDAT(6+N0)=B(3,1,N)
       YDAT(7+N0)=B(3,2,N)
       YDAT(8+N0)=B(4,1,N)
       WRITE(5,904) (YDAT(I+N0-1),IX,I=1,9)
       IF(ISW1.GT.0) WRITE(4,904) (YDAT(I+N0-1),IX,I=1,9)
       N0=N0+9
   68 CONTINUE
C
C  SCALE FACTOR FOR MILLIMOLS (KINETICS)
       SF=1.E3
       NDATA=NEXP*NBDAT
C
C   PROG RESET STARTS HERE
240     CONTINUE
       PRINT 80
80      FORMAT (' ENTER TIME STEP DT (SEC): ')
       READ (*,901) DT
       IF (ISW1.EQ.1)WRITE(4,81)DT
```

```
81     FORMAT ('/ TIME STEP DT (SEC) = ',E15.2)
C   THIS SWITCH MAY BE SET TO 1 TO PRINT THE KINETICS
       NSW=0
C DEFINE INITIAL VALUES OF PARAMETERS AND PASS TO CURVE-FIT
       CALL CFIT(PARM)
C   CLOSE OUTPUT FILE
       IF(ISW1.EQ.1) CALL CLOSE(4)
       PRINT 35
    35 FORMAT (' TYPE 0 TO TERMINATE, TYPE 1 TO RUN NEW DATA SET: ')
       READ (*,900) ISW3
       IF(ISW3.EQ.1) GO TO 250
       STOP
       END
C  *************************************************************
       SUBROUTINE KINET(PARM)
       IMPLICIT DOUBLE PRECISION (A-H,K,O-Z)
       DIMENSION PARM(30)
       COMMON NDATA,NPARM,YDAT(60),F(60),ERR(60),HB(20),COO(20)
       COMMON PARTL(30,60),DT,TC(20),N,B(0:5,0:5,60),NC,NSW,SF,ISW1
       COMMON GRAD(30),E1(60),A(20,20),C(20,20),P(30),ALAM,IT,S0,S1
       COMMON KSW,IPARM(30,3),IYDAT(30,3,20)
       COMMON RON(12),ROFF(12),NEXP,NBDAT
  904  FORMAT(' ',10(F6.4,' '))
C   RESET TIME AND ALL QUANTITIES
       T=0.
       DO 2 I=0,4
       DO 2 L=0,4
       DO 2 N=1,20
       B(I,L,N)=0.
2      CONTINUE
C   CONVERT PARAMETERS BACK TO RATE CONSTANTS
       RON(1)=PARM(1)
```

```
        RON(4)=PARM(2)

        RON(8)=PARM(3)

        RON(2)=PARM(4)

        RON(5)=PARM(5)

        RON(9)=PARM(6)

        RON(3)=PARM(7)

        RON(6)=PARM(8)

        RON(7)=PARM(9)

        RON(10)=PARM(10)

36      FORMAT(' ',I2,E15.5,3I2,E15.5)

C

        DO 500 N=1,NEXP

C    SET INITIAL VALUES FOR UNLIGANDED HEMOGLOBIN AND CO

        B(0,1,N)=HB(N)

        CO=COO(N)

        T=0.

C    KINETICS SIMULATION FROM T=0 TO T=TC

   120 CONTINUE

C

C COMPUTE

C ON REACTIONS

C FIRST LIGATION STEP    (from 0 to 1 ligand)

C B0 + CO -->  B(1,1)   ALPHA SITE LIGANDED

C B0 + CO -->  B(1,2)   BETA    '       '

C   K(1) AND K(2)

        D=RON(1)*B(0,1,N)*CO*DT

        CO=CO-D

        B(0,1,N)=B(0,1,N)-D

        B(1,1,N)=B(1,1,N)+D

        D=RON(2)*B(0,1,N)*CO*DT

        CO=CO-D

        B(0,1,N)=B(0,1,N)-D
```

```
       B(1,2,N)=B(1,2,N)+D

  111 CONTINUE

C

C SECOND LIGATION STEP   (from 1 to 2 ligands)

C B(1,J) + CO --> B(2,J)

C  K(3), K(4), AND K(5) ONLY

       D=RON(4)*B(1,1,N)*CO*DT

       B(1,1,N)=B(1,1,N)-D

       B(2,1,N)=B(2,1,N)+D

       CO=CO-D

       D=RON(5)*B(1,2,N)*CO*DT

       B(1,2,N)=B(1,2,N)-D

       B(2,1,N)=B(2,1,N)+D

       CO=CO-D

       D=RON(3)*B(1,1,N)*CO*DT

       CO=CO-D

       B(1,1,N)=B(1,1,N)-D

       B(2,3,N)=B(2,3,N)+D

       D=RON(6)*B(1,2,N)*CO*DT

       CO=CO-D

       B(1,2,N)=B(1,2,N)-D

       B(2,4,N)=B(2,4,N)+D

  113 CONTINUE

C

C THIRD LIGATION STEP   (from 2 to 3 ligands)

C B(2,J) + CO --> B(3,J)

C  K(7), K(8), AND K(9) ONLY

       D=RON(8)*B(2,1,N)*CO*DT

       CO=CO-D

       B(2,1,N)=B(2,1,N)-D

       B(3,1,N)=B(3,1,N)+D

       D=RON(9)*B(2,1,N)*CO*DT
```

```
      CO=CO-D
      B(2,1,N)=B(2,1,N)-D
      B(3,2,N)=B(3,2,N)+D
      D=RON(7)*B(2,3,N)*CO*DT
      CO=CO-D
      B(2,3,N)=B(2,3,N)-D
      B(3,1,N)=B(3,1,N)+D
      D=RON(10)*B(2,4,N)*CO*DT
      CO=CO-D
      B(2,4,N)=B(2,4,N)-D
      B(3,2,N)=B(3,2,N)+D
  114 CONTINUE
C


C FOURTH LIGATION STEP  (from 3 to 4 ligands)
C B(3,J) + CO --> B(4,J)
C K(11) AND K(12)
      D=RON(11)*B(3,1,N)*CO*DT
      B(3,1,N)=B(3,1,N)-D
      CO=CO-D
      B(4,1,N)=B(4,1,N)+D
      D=RON(12)*B(3,2,N)*CO*DT
      CO=CO-D
      B(3,2,N)=B(3,2,N)-D
      B(4,1,N)=B(4,1,N)+D
  115 CONTINUE
C
C INCREMENT TIME COUNTER
      T=T+DT
      IF(T.GE.TC(N))GOTO 500
      GO TO 120
  500 CONTINUE
```

```
C
   501 CONTINUE
C  CALCULATE VALUES PREDICTED BY KINET
        NO=1
        DO 68 N=1,NEXP
        F(0+NO)=B(0,1,N)
        F(1+NO)=B(1,1,N)
        F(2+NO)=B(1,2,N)
        F(3+NO)=B(2,1,N)
        F(4+NO)=B(2,3,N)
        F(5+NO)=B(2,4,N)
        F(6+NO)=B(3,1,N)
        F(7+NO)=B(3,2,N)
        F(8+NO)=B(4,1,N)
        NO=NO+9
901     FORMAT (f10.4)
    68 CONTINUE
C
C   Calculate errors between measured values (the B's)
C     and the predicted values)
        DO 116 I=1,NDATA
116     ERR(I)=F(I)-YDAT(I)
11      FORMAT (2F8.5)
        RETURN
        END
C*******************************************
        SUBROUTINE CFIT(PARM)
C     CURVE-FITS DATA BY SIMULATING KINETICS
C     MARQUARDT-LEVENBERG ALGORITHM
C     PARTIALS CALCULATED NUMERICALLY
C     NDATA= # OF OBSERVATIONS (DATA POINTS)
C     PARM = PARAMETERS IN MODEL
```

```
C      NPARM= # OF PARAMETERS
C      IT   = ITERATION COUNTER
C      NITER= MAX # OF ITERATIONS
C      ALAM = ITERATION PARAMETER
C      E,ERR= ERROR
       IMPLICIT DOUBLE PRECISION(A-H,K,O-Z)
       DIMENSION PARM(30),DPARM(30),PNEW(30)
       COMMON NDATA,NPARM,YDAT(60),F(60),ERR(60),HB(20),COO(20)
       COMMON PARTL(30,60),DT,TC(20),N,B(0:5,0:5,60),NC,NSW,SF,ISW1
       COMMON GRAD(30),E1(60),A(20,20),C(20,20),P(30),ALAM,IT,S0,S1
       COMMON KSW,IPARM(30,3),IYDAT(30,3,20)
       COMMON RON(12),ROFF(12),NEXP,NBDAT
       REAL LAM,LAM0,Z1,Z2,S0,E,RSQ,S1,TOL
       PRINT 810
  810  FORMAT (' MAX # OF ITERATIONS ')
       READ (*,900) NITER
  900  FORMAT (3I2)
       PRINT 811
  811  FORMAT (' INITIAL VALUE OF LAMDA: ')
       READ (*,901) ALAM
  901  FORMAT(3F)
       PRINT 816
  816  FORMAT (' LAMDA INCREMENT: ')
       READ (*,901) DLAM
C INITIALIZE ITERATION COUNTER
       IT=0.
C
C      CALC SUM OF SQUARES OF ERRORS
       CALL KINET(PARM)
       IF (NITER.EQ.0) RETURN
       S0=0.
       DO 1057 J=1,NDATA
```

```
       S0=S0+ERR(J)**2
       E1(J)=ERR(J)
1057   CONTINUE
       WRITE(5,399)
       IF (ISW1.EQ.1)WRITE(4,399)
399    FORMAT(' ITER    SSE       PARAMS ')
C      SWITCH FOR SPECIAL DISPLAYS (NOT USED)
       N1=0
       IF (N1.EQ.0) GO TO 230
       PRINT 813
813    FORMAT (' NO. OF THE PARAM TO BE SHOWN: ')
       READ (*,900) NP
C
C ********************************************
230    CONTINUE
       IT=IT+1
C    NEXT ITERATION STARTS HERE
       DO 103 I=1,NPARM
       GRAD(I)=0
103    CONTINUE
       DO 104 J=1,NPARM
       DO 104 I=1,NPARM
       A(I,J)=0.
       C(I,J)=0.
104    CONTINUE
C
C    CALCULATE NUMERICAL PARTIALS HERE DE/DPARM
C    DON'T SHOW KINETICS HERE
       KSW=NSW
       NSW=0
       DO 106 I=1,NPARM
       DPARM(I)=.001*PARM(I)
```

```
        PNEW(I)=PARM(I)+DPARM(I)

        PARM(I)=PARM(I)*1.001

        CALL KINET(PARM)

        PARM(I)=PARM(I)/1.001

        DO 1058 J=1,NDATA

        PARTL(I,J)=(E1(J)-ERR(J))/DPARM(I)

1058    CONTINUE

106     CONTINUE

11      CONTINUE

C

C   CALCULATE GRADIENT VECTOR E*DE/DP

        DO 1056 J=1,NDATA

C       CALC I-TH ROW OF JACOBIAN

C       GRADIENT VECTOR

        DO 107 I=1,NPARM

        GRAD(I)=GRAD(I)+E1(J)*PARTL(I,J)

C   MATRIX A IS COMPOSED HERE

        DO 108 I1=1,NPARM

        A(I,I1)=A(I,I1)+PARTL(I,J)*PARTL(I1,J)

108     CONTINUE

107     CONTINUE

1056    CONTINUE

C

365     CONTINUE

        DO 395 I=1,NPARM

        DO 390 J=1,NPARM

385     C(I,J)=A(I,J)

390     CONTINUE

395     CONTINUE

        ALAM1=1.+ALAM

C

C   OUTPUTS HERE
```

```
415    CONTINUE
       IF (N1.NE.0) GO TO 417
416    CONTINUE
       WRITE(5,400) IT,S0,(PARM(I),I=1,NPARM)
       IF (ISW1.EQ.1) WRITE(4,400) IT,S0,(PARM(I),I=1,NPARM)
400    FORMAT (' ',I2,E11.4,12F8.1)
417    CONTINUE
C
C    IS PRESET NO OF ITERATIONS REACHED?
       IF (IT.EQ.NITER) GO TO 687
401    CONTINUE
C
C  PERTURB C-MATRIX
       DO 111 I=1,NPARM
C******************************************************************
       DO 110 J=1,NPARM
110    C(I,J)=A(I,J)
C******************************************************************
       C(I,I)=C(I,I)*(1.+ALAM)
111    CONTINUE
C
C      MATRIX INVERSION HERE
       CALL MATIN4(C,NPARM)
C
C  CALCULATE CHANGES IN PARAMS
       DO 440 I=1,NPARM
       P(I)=0.
C      PRODUCT MATRIX (DPARM)=(C)*(GRAD)
       DO 115 J=1,NPARM
       P(I)=P(I)+C(I,J)*GRAD(J)
115    CONTINUE
440    CONTINUE
```

```
C

      IF (N1.EQ.0) GO TO 418

      WRITE(5,909)IT,PARM(NP),S0,GRAD(NP),C(NP,NP),ALAM1,P(NP)

      IF (ISW1.EQ.1) WRITE(4,909)IT,PARM(NP),S0,GRAD(NP),C(NP,NP),

     1ALAM1,P(NP)

909   FORMAT (' ',I3,F7.0,3E15.5,F8.3,F8.0)

418   CONTINUE

C

C  UPDATE PARAMS

500   DO 501 I=1,NPARM

      PARM(I)=PARM(I)+P(I)

501   CONTINUE

C

C  CALCULATE NEW ERR

C    SHOW KINETICS IF SPECIFIED

      NSW=KSW

      CALL KINET(PARM)

      SNEW=0.

      DO 5211 J=1,NDATA

      SNEW=SNEW+ERR(J)**2

      E1(J)=ERR(J)

5211  CONTINUE

C

C    TEST IF SUM OF SQUARES GETTING LARGER

C     IF (SNEW.GT.S0)GO TO 595

C  NO, UPDATE SSQ, GO AROUND ITERATION LOOP AGAIN

      ALAM=ALAM/DLAM

      S0=SNEW

      GO TO 230

C

C    SNEW GT S0

595   ALAM=ALAM*DLAM
```

```
C   RESTORE PARAMS TO PREV VALUES, TRY AGAIN
      CONTINUE
      DO 116 I=1,NPARM
      PARM(I)=PARM(I)-P(I)
116   CONTINUE
      IT=IT+1
      GO TO 365
C
C****************************************************
C     ITERATIONS DONE; COMPUTE FINAL QUANTITIES
635   CONTINUE
C     VARIANCE OF THE FIT
      V=S0/(NDATA-NPARM)
      V1=SQRT(V)
      WRITE(5,655)V1
      IF (ISW1.EQ.1) WRITE(4,655)V1
655   FORMAT(
     1' SD = ',E13.3)
      WRITE(5,675)
      IF (ISW1.EQ.1) WRITE(4,675)
675   FORMAT  (' FINAL PARAMS TETRAMER        STD ERR PARAM    COEFF VAR')
      DO 676 I=1,NPARM
      DO 676 J=1,NPARM
676   C(I,J)=A(I,J)
      CALL MATIN4(C,NPARM)
C***********************************************************************
   DO 123 I=1,NPARM
C     STAND. ERR PARAM
      D=V1*SQRT(C(I,I))
      COEF=D/PARM(I)
      TET=PARM(I)/4.
      WRITE(5,685)PARM(I),TET,D,COEF
```

```
      IF (ISW1.EQ.1) WRITE(4,685)PARM(I),TET,D,COEF
685   FORMAT (' ',F9.3,4X,F8.3,4X,2E13.3)
123   CONTINUE
C

      CALL KINET(PARM)
      WRITE(5,780)
      IF(ISW1.EQ.1) WRITE(4,780)
780   FORMAT (' OBS NO    X          EST Y    ERR')
      DO 125 J=1,NDATA
      WRITE(5,815) J,YDAT(J),F(J),ERR(J)
      IF (ISW1.EQ.1) WRITE(4,815) J,YDAT(J),F(J),ERR(J)
815   FORMAT (' ',I3,3F10.5)
  125 CONTINUE
C
C  SUBROUTINE ENDS HERE FOR NORMAL EXIT
700   RETURN
C
C  PRESET ITERATIONS REACHED
687   CONTINUE
      PRINT 919
919   FORMAT (' TYPE NO.FURTHER ITERATIONS DESIRED,OR 0 TO TERMINATE'
      READ (*,900) MORE
      IF (MORE.EQ.0) GO TO 635
      NITER=NITER+MORE
      GO TO 401
      END
C ********************************
      SUBROUTINE MATIN4(ARRAY,NPARM)
C     3 INVERSION MATRIX INVERSION BY GAUSS-JORDAN ELIMINATION
C      NO PIVOTING IS USED
C      ARRAY= MATRIX TO BE INVERTED
C      NPARM= SIZE OF MATRIX TO BE INVERTED
```

```
C       ORIGINAL MATRIX IS DESTROYED
C

        IMPLICIT DOUBLE PRECISION (A-H,K,O-Z)
        DIMENSION ARRAY(20,20)
C

        DO 604 I=1,NPARM
        STORE=ARRAY(I,I)
        ARRAY(I,I)=1.0
        DO 601 J=1,NPARM
        IF (STORE.EQ.0.0) RETURN
601     ARRAY(I,J)=ARRAY(I,J)/STORE
C

        DO 604 IK=1,NPARM
        IF(IK-I) 602,604,602
602     STORE= ARRAY(IK,I)
        ARRAY(IK,I)=0.0
        DO 603 J=1,NPARM
   603  ARRAY(IK,J)=ARRAY(IK,J)-STORE*ARRAY(I,J)
C

604     CONTINUE
        RETURN
        END
```

Acknowledgments

Acknowledgment is due to Stuart J. Davids for assistance in doing the computations in this chapter.

[26] Adair Fitting to Oxygen Equilibrium Curves of Hemoglobin

By KIYOHIRO IMAI

The oxygen equilibrium curve (OEC) of hemoglobin, that is, the curve expressing the dependence of oxygen saturation of hemoglobin (Y) on partial pressure of oxygen (p) under equilibrium, contains much information regarding various molecular properties of hemoglobin such as the binding affinities of the four oxygen molecules, the interactions among the oxygen-binding sites, the relative amounts of the high- and low-affinity states, and the free energy of the intersubunit contacts. Such information becomes more valuable when one determines a set of OECs under several experimental conditions (e.g., in the presence and absence of various allosteric effectors and at different temperatures or protein concentrations) and analyzes them collectively.

The information contained in the OEC is extracted by fitting a particular function to it. The curve-fitting procedure, which usually employs the least-squares method, yields the best-fit values of the set of the parameters involved in the function. A variety of molecular properties of hemoglobin is derived from the physical meanings of the parameters.

Several kinds of functions derived by mathematical modeling have been used for the curve fitting to the OEC. The earliest among them was the Adair equation,[1] which includes four parameters called Adair constants. Although the Adair equation was proposed in 1925, it was not applied to actual OECs until 30 years later because of the lack of experimental data that were accurate enough to yield a unique set of Adair constant values. In 1955, Roughton and collaborators[2] successfully determined the Adair constant values from accurate OECs, which were measured by their specially developed gasometric technique. The evaluation of the Adair constants was accelerated by the development of a modern automated apparatus for measuring accurate OECs[3,4] and by the construction of an on-line system of the apparatus using a microcomputer.[5]

---

[1] G. S. Adair, *J. Biol. Chem.* **63**, 529 (1925).

[2] F. J. W. Roughton, A. B. Otis, and R. L. J. Lyster, *Proc. R. Soc. London, Ser. B* **144**, 29 (1955).

[3] K. Imai, H. Morimoto, M. Kotani, H. Watari, W. Hirata, and M. Kuroda, *Biochim. Biophys. Acta* **200**, 189 (1970).

[4] K. Imai, this series, Vol. 76, p. 438.

[5] K. Imai and T. Yonetani, *Biochim. Biophys. Acta* **490**, 164 (1977).

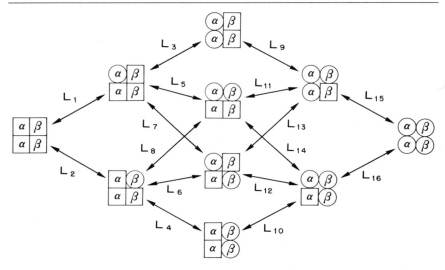

FIG. 1. Generalized scheme of oxygen-binding equilibria of tetrameric hemoglobin. The square and round symbols indicate deoxy and oxy subunits, respectively. L_1 to L_{16} are microscopic intrinsic equilibrium constants for the individual oxygen-binding steps.

The extensive studies on the evaluation of the Adair constants by the author and collaborators[6–8] indicate that the dominant factors that influence the reliability of the Adair constant evaluation reside in both the accuracy of the experimental data and the use of the least-squares method. The latter is the principal subject of this chapter.

Adair Scheme

Figure 1 shows a generalized scheme of oxygen binding to tetrameric hemoglobin with two types of subunits, α and β. This scheme presents 16 equilibria among the ten oxygen-binding states. The 16 microscopic association constants, L_1 to L_{16}, corresponding to the individual equilibria, can be distinguished as long as the α and β subunits are not equivalent with respect to their affinity for oxygen and the interactions with the neighboring subunits. Because the generalized scheme includes closed thermodynamic cycles and because of the restriction of the law of energy conservation, only 9 of the 16 equilibrium constants are independent. Here, we

[6] K. Imai, this series, Vol. 76, p. 470.
[7] K. Imai, "Allosteric Effects in Haemoglobin." Cambridge Univ. Press, Cambridge, 1982.
[8] K. Imai, *Biophys. Chem.* **37**, 197 (1990).

choose L_1, L_2, L_3, L_4, L_5, L_7, L_9, L_{10}, and L_{15} as the independent constants. The concentrations of the molecular species combining with oxygen molecules of different numbers are then given by

$$[HbO_2] = 2(L_1 + L_2)p[Hb]$$
$$[Hb(O_2)_2] = (L_1L_3 + 2L_1L_5 + 2L_1L_7 + L_2L_4)p^2[Hb]$$
$$[Hb(O_2)_3] = 2(L_1L_3L_9 + L_2L_4L_{10})p^3[Hb]$$
$$[Hb(O_2)_4] = L_1L_3L_9L_{15}p^4[Hb]$$

(1)

Here, all the L terms are "intrinsic" equilibrium constants in the sense that they are free of statistical factors, that is, the mass-action law is applied to statistically weighted concentrations of the intermediate species of hemoglobin. The fractional saturation of hemoglobin with oxygen is then given by

$$Y = \frac{[HbO_2] + 2[Hb(O_2)_2] + 3[Hb(O_2)_3] + 4[Hb(O_2)_4]}{4([Hb] + [HbO_2] + [Hb(O_2)_2] + [Hb(O_2)_3] + [Hb(O_2)_4])}$$

(2)

which yields

$$Y = \frac{A_1p + 2A_2p^2 + 3A_3p^3 + 4A_4p^4}{4(1 + A_1p + A_2p^2 + A_3p^3 + A_4p^4)}$$

(3)

where

$$A_1 = 2(L_1 + L_2)$$
$$A_2 = L_1L_3 + 2L_1L_5 + 2L_1L_7 + L_2L_4$$
$$A_3 = 2(L_1L_3L_9 + L_2L_4L_{10})$$
$$A_4 = L_1L_3L_9L_{15}$$

(4)

If the $\alpha_1\beta_1$ interface is equivalent to the $\alpha_1\beta_2$ interface, the 2 intermediate species in the middle of the scheme are merged into 1, and the number of the microscopic equilibria is reduced to 12. The number of independent constants becomes eight, because $L_5 = L_7$, $L_6 = L_8$, $L_{11} = L_{13}$, and $L_{12} = L_{14}$.

If the α and β subunits are entirely equivalent, all the molecular species combining with the same number of oxygens are merged into one and, as a consequence, the oxygen binding is described by only four equilibria:

$$Hb + 4O_2 \rightleftharpoons HbO_2 + 3O_2 \rightleftharpoons Hb(O_2)_2 + 2O_2 \rightleftharpoons Hb(O_2)_3 + O_2 \rightleftharpoons Hb(O_2)_4$$

(5)

In this case,

$$[Hb(O_2)_i] = A_ip^i[Hb]$$

(6)

and the A terms in Eq. (3) are given by

$$
\begin{aligned}
A_1 &= K_1' \\
A_2 &= K_1' K_2' \\
A_3 &= K_1' K_2' K_3' \\
A_4 &= K_1' K_2' K_3' K_4'
\end{aligned}
\tag{7}
$$

Here, K_i' ($i = 1$ to 4) is the equilibrium constant for the ith binding step, and A_i ($i = 1$ to 4) is the overall equilibrium constant for the binding of the first i oxygen molecules in Eq. (5). Because the binding of O_2 to $Hb(O_2)_{i-1}$ is enhanced by a factor of the number of empty sites, $4 - (i - 1)$, and the dissociation of O_2 from $Hb(O_2)_i$ is enhanced by a factor of the filled sites, i, each K_i' can be corrected for these statistical factors to obtain an "intrinsic" equilibrium constant, K_i, as follows:

$$
K_i = \{i/[4 - (i - 1)]\} K_i'
\tag{8}
$$

Using the stepwise intrinsic equilibrium constants, K_i, Eq. (3) is written as

$$
Y = \frac{K_1 p + 3 K_1 K_2 p^2 + 3 K_1 K_2 K_3 p^3 + 3 K_1 K_2 K_3 K_4 p^4}{1 + 4 K_1 p + 6 K_1 K_2 p^2 + 4 K_1 K_2 K_3 p^3 + K_1 K_2 K_3 K_4 p^4}
\tag{9}
$$

Equation (3) or Eq. (9) is called the Adair equation in honor of Adair's pioneer work.[1] The A parameters and K parameters are referred to as the overall Adair constants and the stepwise Adair constants, respectively. Equation (3) is the most general description of the oxygen-binding equilibria of hemoglobin, being valid as long as hemoglobin binds oxygen in the tetrameric state, no matter whether the constituent subunits are equivalent or not. The definition of the A terms depends on the nature of the subunits and the mode of the intersubunit interactions.

It should be noted here that it is impossible to evaluate the individual L terms from an OEC alone, because any generalized scheme is reduced to Eq. (3), for which only the four A terms can be evaluated separately. Data from other kinds of experiments[9-13] are needed.

Usually the subunit inequivalence and intersubunit interaction inequivalence are not conspicuous, and the oxygen binding is described well in terms of the simplest set of the four K terms. Even if these conditions are

[9] N. Makino and Y. Sugita, *J. Biol. Chem.* **253**, 1174 (1978).
[10] K. Imai, M. Ikeda-Saito, and T. Yonetani, *J. Mol. Biol.* **138**, 635 (1980).
[11] M. Perrella and L. Rossi-Bernardi, this series, Vol. 76, p. 133.
[12] N. Shibayama, H. Morimoto, and T. Kitagawa, *J. Mol. Biol.* **192**, 331 (1986).
[13] M. Perrella, A. Colosimo, L. Benazzi, M. Ripamonti, and L. Rossi-Bernardi, *Biophys. Chem.* **37**, 211 (1990).

not satisfied, the K values impose definite constraints on the description of the hemoglobin–oxygen equilibria, giving their evaluation a certain significance.

On-Line System for Oxygen Equilibrium Data Acquisition and Analysis

The OEC, which is suitable for least-squares curve fitting, is composed of as many as 30 to 100 data points. To analyze a number of OECs it is useful to construct an on-line system that allows acquisition of the oxygen equilibrium data in real time, that is, simultaneously with the measurement and analysis immediately after the measurement.

Figure 2 shows such a system.[8] This system consists of an automatic oxygenation apparatus described previously[4,7] and a microcomputer (model PC-98XA; Nippon Electric Co., Tokyo, Japan) that is interfaced to the apparatus via an 8-channel 12-bit analog-to-digital (A/D) converter. The apparatus generates two voltage signals corresponding to the p value and the absorbance value, which is proportional to Y, of the hemoglobin sample. The absorbance data acquisition uses one channel (CH1) of the converter and the resolution is 0.00024 absorbance unit (= 1 absorbance unit/2^{12}). The dynamic range of oxygen pressure data acquisition is expanded by using two channels, one of which (CH3) covers the range from the maximum pressure down to 10% thereof, while the other (CH2) covers the range from 10% maximum pressure down to zero. The practical dynamic range of oxygen pressure measurement is as wide as 0.01 to 760 mmHg. The sampling of the data points on the deoxygenation curve is made in such a way that each successive point is taken when its oxygen pressure value decreases to α times the last sampled value (α is a selectable factor smaller than unity, typically being 0.9). The sampling of the data points on the reoxygenation curve is made in the same way except that the factor is replaced by $1/\alpha$. By sampling in this way, the data points are collected at constant intervals on a $\log p$ scale. The range of saturation measurement with reasonable accuracy is 1 to 99.9%. The range of the suitable concentration of hemoglobin samples is 1 μM to 2 mM on a heme basis. The measurement time for the deoxygenation curve is approximately 30 min (60 to 180 μM hemoglobin, 25°, pH 7.4) and that for the reoxygenation curve is shorter. Usually, the number of the collected data points per deoxygenation or reoxygenation curve is 60 to 70.

Figure 3 shows a flow chart for data processing. The oxygen equilibrium data acquired are stored on a hard disk as a (P, A) data file (P, oxygen pressure; A, absorbance). The data are extrapolated toward both ends on the color CRT display to determine the exact 0 and 100% oxygen saturation points and to convert the A values to saturation values, generat-

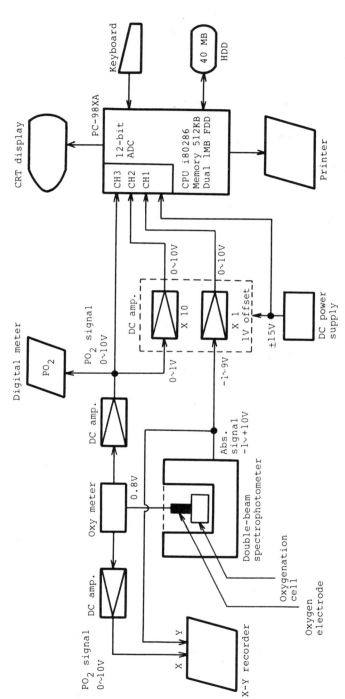

FIG. 2. An on-line system for oxygen equilibrium data acquisition and analysis used by the author. The automatic oxygenation apparatus, described previously,[4,7] is interfaced to a microcomputer (model PC-98XA; Nippon Electric Co.), via an 8-channel 12-bit A/D converter. The voltage signals proportional to absorbance and oxygen pressure changes are fed to the converter after amplification and offset processing.

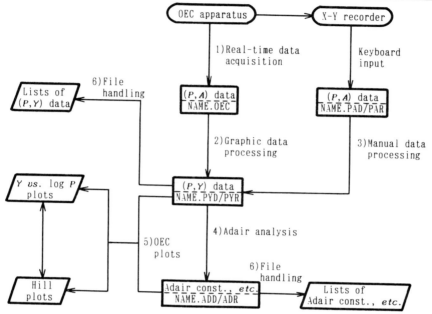

FIG. 3. Flow chart of oxygen equilibrium data processing with the on-line system (Fig. 2). P, Oxygen pressure; A, absorbance of hemoglobin sample; Y, oxygen saturation of hemoglobin. The words given underneath the broken line within the blocks indicate typical forms of the file name under which the data are stored on the hard disk. PYD and PYR indicate deoxy (P, Y) data and reoxy (P, Y) data, respectively, and so forth.

ing a (P, Y) data file (Y, oxygen saturation). Alternatively, one can make the (P, Y) data file by typing the (P, A) data, which are read by eye from the OEC recorded on X–Y recorder chart paper, together with the manually extrapolated A values. Then the (P, Y) data are used for the Adair fitting, generating a data file of the best-fit values of the Adair constants and other oxygenation parameters derived from the Adair constants. The analyzed data are printed out in graphic forms or as numerical tables. The graphic resolution of the display is 750×1120 dots. The computer programs are written in BASIC and run on an MS-DOS (Microsoft Corp., Redmond, WA) operating system.

Adair Fitting to Oxygen Equilibrium Curves

Nonlinear Least-Squares Method

When the fitting model function is nonlinear with respect to the parameters that are to be evaluated, as is the case for the Adair equation, it must be linearized by expanding to the first order in a Taylor expansion so that the problem is reduced to the least-squares method for linear functions. Because of this approximation, the minimization proceeds iteratively, starting with approximate values of the parameters. There are several versions of this method (called the Gauss–Newton method), depending on the way of efficient or smooth approach to the minimum. Most commonly used is the Levenberg–Marquardt method.[14] These nonlinear methods are described in detail with practical computer programs in several books.[15-17]

The method used by the author for Adair fitting is based on Rubin[18] and was modified by the author so that weights can be applied to the observed points and a routine for estimating the errors of the best-fit parameter values can be included.[19] Because the whole procedure has been described previously in full detail,[6-7] it is not repeated here. The source program list written in BASIC or Fortran is available on request.

Conversion of (P, A) Data to (P, Y) Data: Extrapolation Procedure

The absorbance values are converted to oxygen saturation values using the equation

$$Y_j = \frac{\text{Abs}_j - \text{Abs}_0}{\text{Abs}_\infty - \text{Abs}_0} = \frac{\text{Abs}_j - \text{Abs}_0}{\Delta\text{Abs}_T} \qquad (j = 1 \text{ to } N) \tag{10}$$

where Abs_0, Abs_∞, and Abs_j represent the absorbance values at $p = 0$, $p = \infty$, and p for the jth point (p_j), respectively. ΔAbs_T is the total absorbance change on full deoxygenation or reoxygenation, and N is the total number of data points collected. Abs_0 is determined by extrapolating the bottom portion of the curve toward $p = 0$ along an Abs_j vs p_j plot. Abs_∞ is determined by extrapolating the top portion toward $p = \infty$ along an Abs_j vs $1/p_j$ plot. These graphical extrapolations are performed on the CRT display by fitting the quadratic function

$$\text{Abs} = ax^2 + bx + c \tag{11}$$

[14] D. W. Marquardt, *J. Soc. Ind. Appl. Math.* **11**, 431 (1963).
[15] P. R. Bevington, "Data Reduction and Error Analysis for the Physical Sciences." McGraw-Hill, New York, 1969.
[16] M. E. Magar, "Data Analysis in Biochemistry and Biophysics." Academic Press, New York, 1972.
[17] W. H. Press, B. P. Flannery, S. A. Teukolsky, and W. T. Vetterling, "Numerical Recipes. The Art of Scientific Computing." Cambridge Univ. Press, Cambridge and New York, 1986.
[18] D. I. Rubin, *Chem. Eng. Prog., Symp. Ser.* **59**, 90 (1963).
[19] K. Imai, *Biochemistry* **12**, 798 (1973).

($x = p$ and $1/p$ for bottom and top extrapolations, respectively), to the endmost observed points (usually 15 to 20 points). The best-fit c value is made equal to Abs_0 or Abs_∞.

This extrapolation procedure is important to obtain accurate saturation values at the extremes of the OEC. If one sets the end points to 0 and 100% saturation values, this simplification will cause serious errors in the succeeding Adair analysis. Previously, the author examined the effect of the extrapolation on the best-fit values of the A terms for highly purified Hb A (human adult hemoglobin) samples of 1.2 mM concentration by intentionally changing the Abs_0 and Abs_∞ values by 0.1% of ΔAbs_T.[8] Even such minute changes in the extrapolated values caused marked variations in A_2 and A_3 in the opposite direction, the A_2 value even becoming negative in some cases.

One can include the extrapolation procedure into the least-squares curve-fitting procedure by defining the fitting function as a combination of Eqs. (3) and (10) and adding two unknown parameters, Abs_0 and Abs_∞ (six parameters as the total). This way is recommended when highly accurate (P, A) data are obtained, because subjective effects, which may be introduced by the quadratic graphic extrapolation, can be excluded. However, the estimated errors of the best-fit parameter values will become greater because the freedom is reduced by the increase in the number of parameters to be evaluated.

Standard Deviation of Observed Y Values and Weighting

When the OEC is measured by the automatic oxygenation apparatus, the standard deviation of Y, S_Y, depends on Y; the dependence is simulated roughly by a parabolic curve, $S_Y = 0.08Y(1 - Y)$, indicating that the Y value becomes more accurate as Y approaches either 0 or 1.[3] The bell-shaped dependence of S_Y on Y may be explained as follows: the errors of absorbance show an even distribution with respect to Y, whereas a given relative error of p, $\Delta p/p_j$, yields the largest errors of Y around $Y = 0.5$ and smaller ones on both sides of $Y = 0.5$. This is because the OEC, as expressed by means of a Y vs log p plot, is steepest around $Y = 0.5$ and becomes less steep at both ends. The parabolic dependence of S_Y is equivalent to the case in which the observed points are uniformly scattered along the ordinate of the Hill plot {log[$Y/(1 - Y)$] vs log p}.[6]

According to the experimentally determined parabolic dependence of S_Y, the jth point of the OEC is weighted by the factor

$$w_j^2 = N[Y_j(1 - Y_j)]^{-2} \bigg/ \sum_{j=1}^{N} [Y_j(1 - Y_j)]^{-2} \qquad (12)$$

where w_j^2 is normalized to give $w_1^2 + w_2^2 + \cdots + w_N^2 = N$. This weighted curve fitting is equivalent to the unweighted (or equally weighted) curve fitting on the Hill plot.

The weighted curve fitting produces an uneven distribution of the residuals of Y. Figure 4 shows the distributions of $\delta p/p$ and δY as plotted against Y, where δp and δY are the residuals (i.e., the observed value minus the value calculated from the best-fit Adair constants) of p and Y, respectively. These distribution plots were obtained from 10 sets of oxygen equilibrium data for 180 μM Hb A samples under different pH conditions. In the $\delta p/p$ vs Y plots, $\delta p/p$ is roughly independent of Y in the range of $Y = 0.2$ to 0.8, whereas it becomes larger in magnitude sharply as Y approaches 0 but gradually as Y approaches 1. The δY vs Y plots show a bell-shaped distribution. These distributions reproduce the previously observed dependences of the standard deviation of p and Y on Y.[3]

However, the plots in Fig. 4 are not symmetric with respect to the baseline, indicating that the errors of p and Y do not occur randomly and are somewhat biased. The asymmetric distributions of residuals suggest

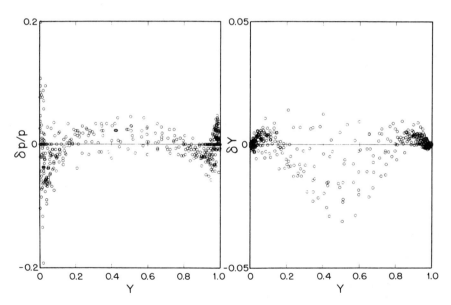

FIG. 4. Distribution of $\delta p/p$ and δY as plotted against Y. δp and δY are residuals of oxygen pressure, p, and oxygen saturation, Y, that is, the observed value minus the value calculated from the best-fit Adair constant values. The experimental data used for this analysis were taken from 10 oxygen equilibrium curves determined for 180 μM Hb A in 0.05 M Bis–Tris buffers (pH 5.8, 6.2, 6.6, 7.0, and 7.4) or 0.05 M Tris buffers (pH 7.4, 7.8, 8.2, 8.6, and 9.0) with minimum Cl$^-$ added and at 25° (K. Imai, unpublished data, 1991).

that the Adair equation does not strictly express the actual OEC. Three possible causes may be given: (1) a slow response of the oxygen electrode that follows behind the actual changes of p, (2) autoxidation of hemoglobin accompanying one-directional absorbance changes during measurement, and (3) deviation from the Adair model by the existence of oxygen-linked tetramer–dimer equilibrium. Cause 1 is not likely because both the deoxygenation and reoxygenation curves show the same pattern of distribution. Cause 2 is also unlikely because the situation remains unaltered when the OECs are determined at different wavelengths, where autoxidation exerts opposite influences on the oxygenation-induced absorbance change. Last, cause 3 is probably unlikely because the distributions of residuals show the same pattern when the OECs are determined for concentrated (1.2 mM) Hb A samples. The individual data points making up a single OEC, as determined by the automatic apparatus, are not completely independent of each other, and this partial dependence can generate a systematically biased distribution. Regardless, the asymmetric distributions must at present be attributed to some unknown cause(s).

Importance of Weighting

Generally speaking, the standard deviation of the variables in the model function must be explored experimentally, and the observed points must be weighted appropriately according to the distribution of standard deviation as long as the distribution is not even.

The author tested the effect of the weighting on the best-fit values of the A terms for 1.2 mM pure Hb A samples.[8] When no weight was applied to the data points (or when equal weights were applied) the best-fit values of A_1, A_2, and A_3 showed significant differences from those obtained by the weighted fitting. When Abs_∞ was floated in the least-squares minimization, the A_3 value even became negative. Thus, it is important to use appropriate weighting for obtaining physically meaningful parameter values.

Importance of Bottom and Top Data

With the early version of our automatic oxygenation apparatus[3] it was not easy to obtain sufficiently accurate oxygen equilibrium data so that the Adair constants could be evaluated with reasonable accuracy. The subsequent improvement of the apparatus[4,5,7] enabled us to obtain data accurate enough for the Adair fitting and facilitated a series of oxygen equilibrium studies based on the Adair scheme. The improvement in accuracy that made the Adair fitting easier was due principally to being able to determine oxygen saturation over a wide range, from about 1 to about

TABLE I

EFFECT OF PARTIAL DELETION OF DATA POINTS ON STEPWISE ADAIR CONSTANTS AND
STANDARD ERRORS

Deletion of data points	$K_1{}^a$ (%SE)[b]	K_2 (%SE)	K_3 (%SE)	K_4 (%SE)	RMS[c]
PHOS data					
No deletion	0.0151	0.0152	0.347	3.20	3.7×10^{-4}
	(1.9)	(22.4)	(22.6)	(3.8)	
Bottom[d]	0.0161	0.00901	0.570	3.09	3.4×10^{-4}
	(2.4)	(36.5)	(36.6)	(3.5)	
Middle[e]	0.0144	0.0277	0.231	3.26	2.4×10^{-4}
	(1.8)	(13.3)	(13.6)	(4.3)	
Top[f]	0.0145	0.0318	0.107	5.17	6.6×10^{-4}
	(2.0)	(15.9)	(21.5)	(14.6)	
Bottom and top[g]	0.0154	0.0207	0.187	4.26	9.7×10^{-4}
	(2.7)	(25.4)	(28.1)	(12.4)	
BISTRIS data					
No deletion	0.0188	0.0566	0.407	4.28	4.9×10^{-4}
	(4.5)	(22.1)	(22.1)	(5.5)	
Bottom[d]	0.0225	0.0249	0.810	4.06	3.7×10^{-4}
	(4.0)	(34.9)	(34.8)	(4.1)	
Middle[e]	0.0165	0.102	0.332	4.26	3.2×10^{-4}
	(4.6)	(14.7)	(14.9)	(6.7)	
Top[f]	0.0155	0.172	0.0527	13.2	1.3×10^{-3}
	(5.5)	(13.4)	(37.1)	(35.1)	
Bottom and top[g]	0.0192	0.0965	0.126	7.90	1.3×10^{-3}
	(5.0)	(16.4)	(23.6)	(17.8)	

[a] Given in $(mmHg)^{-1}$.
[b] %SE, Percentage standard error.
[c] RMS, Root mean square of residuals of Y.
[d] The data points in the range $Y < 0.01$ were deleted.
[e] The data points in the range $0.1 < Y < 0.9$ were deleted.
[f] The data points in the range $Y > 0.99$ were deleted.
[g] The data points in the range $Y < 0.01$ and $Y > 0.99$ were deleted.

99.9%. The importance of the determination of the extremes of OEC over this wide range is equivalent to saying that oxygen saturation must be determined over a wide range so that both the lower and upper asymptotes of the Hill plot can be defined definitely. Owing to the so-called asymmetric nature of the OEC, oxygen saturation must be determined up to 99.9% but only down to 1% to define the upper and lower asymptotes, respectively.

The importance of the bottom and top data is demonstrated by the following analysis. Table I lists the best-fit values of the stepwise Adair constants for 1.2 mM Hb A in either 0.1 M phosphate buffer (designated as PHOS data) or 0.1 M Bis–Tris buffer, containing 0.1 M Cl$^-$ (designated as BISTRIS data) at pH 7.4.[8] The Adair fitting was performed further for four other sets of data derived from the original data: (1) bottom truncated by deleting the data points at $Y < 0.01$, (2) top truncated by deleting the points at $Y > 0.99$, (3) top and bottom truncated, and (4) middle portion ($0.1 < Y < 0.9$) deleted. The results are listed in Table I and illustrated in Fig. 5, in which the data points are compared with the lines calculated from the best-fit K values. The bottom truncation causes significant changes in value and increases in estimated standard error (%SE) for K_2 and K_3 rather than for K_1 or K_4. In contrast, the influences of the top truncation are more global: the values of K_2, K_3, and K_4 are affected significantly and %SE for K_4 shows a striking increase. The double truncation at the bottom and top causes increases in %SE for all the K terms

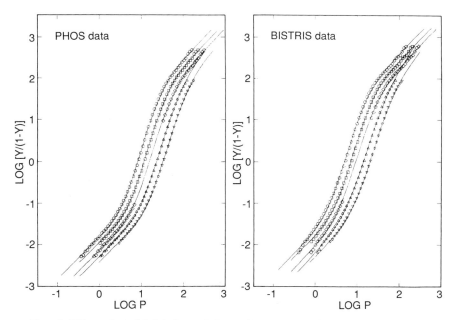

FIG. 5. Effect of partial deletions of data points on Adair fitting as presented by Hill plots. The symbols indicate observed points and the lines were calculated from the best-fit values of the Adair constants listed in Table I. The oxygen equilibrium data used are for 1.2 mM Hb A in 0.1 M phosphate or 0.1 M Bis–Tris at pH 7.4 and 25°.[8] *Left:* Deoxygenation data obtained using phosphate buffer; *Right:* Deoxygenation data obtained using Bis–Tris buffer. In each graph, the leftmost plot represents the original, intact data and the plots for partially deleted data sets are shifted gradually for comparison toward the right at constant intervals, $\Delta \log p = 0.15$. From left to right: no deletion, bottom portion ($Y < 0.01$) truncated, middle portion ($0.1 < Y < 0.9$) deleted, top portion ($0.99 < Y$) truncated, and both bottom and top truncated.

except for K_2 (see Table I) and also an increase in the root mean square (RMS) of residuals of Y, which means that the fitting becomes worse. It is surprising that even when the points of the major portion of the OEC ($0.1 < Y < 0.9$) are deleted, the changes of the K values are not striking, and %SE for K_2 and K_3 and RMS become smaller. This unexpected result may be related to the asymmetric distribution of the residuals shown in Fig. 4. The present analysis certainly indicates that precision determination of the top portion of OEC is important for the evaluation of K_2, K_3, and K_4, whereas precision determination of the bottom portion is important for the evaluation of K_2 and K_3. K_1 can be evaluated more easily than the other K terms.

The importance of the bottom and top data was first pointed out by Roughton et al.[2] By a special gasometric technique, they were able to determine oxygen saturation to an accuracy of 0.05% saturation at the bottom ($Y = 0–2\%$) and top ($Y = 98–99.5\%$) ends and to an accuracy to 0.5% saturation in the middle range ($Y = 5–95\%$). The data, 10-fold more accurate at both ends, and least-squares fitting using 100-fold heavy weights for the end points, enabled them to obtain a unique set of Adair constant values.

The behavior of the K terms on partial deletions of the data points shown in Table I implies the existence of a compensatory effect between K_1 and K_2 and between K_3 and K_4. Previously, Gill et al.[20] showed that there are close correlations between K_1 and K_2 and between K_3 and K_4. In this sense the determination of the bottom and top of the OEC is essential to resolve these K terms. Adair fitting, applying heavy weights to the bottom and top portions of OEC, is suited for their resolution.

Errors of Best-Fit Adair Constant Values

The standard errors of the A or K terms estimated by the least-squares procedure do not exactly reflect the true reliability of their best fit values, because the data points making up the given OEC are not completely independent, as described above.

The author calculated the standard deviation for the best-fit values of the K terms, which had been obtained with control Hb A samples prepared in a series of oxygen equilibrium studies of abnormal hemoglobins. Twenty-two OECs determined for 60 μM Hb A were Adair fitted, and the mean and percentage standard deviation values for the K terms together with those for oxygen pressure at half-saturation, P_{50}, maximal slope of the

[20] S. J. Gill, E. Di Cera, M. L. Doyle, G. A. Bishop, and C. H. Robert, *Biochemistry* **26,** 3995 (1987).

TABLE II

MEAN AND PERCENTAGE STANDARD DEVIATION
VALUES OF OXYGEN EQUILIBRIUM PARAMETERS
AND METHEMOGLOBIN CONTENT FOR
ADULT HEMOGLOBIN[a]

Parameter	Mean	SD (%)
$K_1{}^b$	0.0517	37.9
$K_2{}^b$	0.0398	41.5
$K_3{}^b$	0.453	55.8
$K_4{}^b$	6.90	29.4
$P_{50}{}^c$	4.14	6.7
$n_{max}{}^d$	2.99	3.3
metHb^e	5.5	42.0

[a] Twenty-two OECs determined at 25° for 60 μM Hb A in 0.05 M Bis–Tris buffer (pH 7.4) containing 0.1 M Cl$^-$ were used for the statistical analysis.
[b] In (mmHg)$^{-1}$.
[c] Oxygen pressure at half saturation (in mmHg).
[d] Maximal slope of the Hill plot.
[e] Percentage content of metHb in total hemoglobin.

Hill plot, n_{max}, and percentage metHb content were obtained (Table II). These standard deviation values include the errors originating from not only the scattering of data points in the single OECs, but the sample-to-sample differences in actual pH values, impurities, metHb content, and so on. It should be noted that the experimental conditions used for the series of abnormal hemoglobin studies are not suitable for precision determinations of OECs in the sense that the dilute hemoglobin samples, which were separated from the abnormal fractions by isoelectric focusing, were not stable as judged from the high levels of metHb content. If one works at higher protein concentrations, parameter values of smaller error will be obtained.

In the early stages of Adair-fitting studies by the author and collaborators,[21-24] some inconsistency in the reported best-fit Adair constant values was noted. It is now ascribed mainly to disagreement of the nominal and actual pH values of hemoglobin samples: the temperatures at which pH

[21] I. Tyuma, K. Shimizu, and K. Imai, *Biochem. Biophys. Res. Commun.* **43**, 423 (1971).
[22] I. Tyuma, K. Imai, and K. Shimizu, *Biochem. Biophys. Res. Commun.* **44**, 682 (1971).
[23] K. Imai and I. Tyuma, *Biochim. Biophys. Acta* **293**, 290 (1973).
[24] I. Tyuma, K. Imai, and K. Shimizu, *Biochemistry* **12**, 1491 (1973).

was measured differed from that at which OECs were measured (i.e., 25°), and those temperature differences caused the disagreement of pH values, especially because Tris and Bis–Tris buffers, which have large temperature coefficients, were used.

Magnitude of A_3 Value

By using the thin-layer optical technique Gill and co-workers[25,26] determined OECs of concentrated (2–12 mM) Hb A solutions and evaluated the Adair constants by a least-squares method. They obtained an unanticipated result that the best-fit A_3 value (β_3 according to their designation) was too small to be determined, indicating that the population of the triply liganded species was immeasurably small.[20] When all the A terms were constrained to be nonnegative in the least-squares procedure, the best-fit value of A_3 was always zero, making the K_4 value undefinable (note that $K_4 = 4A_4/A_3$). The author of this chapter and collaborators had not encountered such a situation in earlier work. Gill *et al.* suggested that the protein concentrations (60–600 μM) used by the author's group were so low that partial tetramer-to-dimer dissociation of hemoglobin could exert an influence on the observed Adair constants. Ackers and co-workers[27–29] stressed that the effect of dimeric species must be taken into account in the oxygen equilibrium analysis unless the hemoglobin concentration is higher than 1 mM.

The author then determined highly accurate OECs for pure Hb A of high concentration (1.2 mM) and examined whether the experimental data were able to be accommodated within the Adair scheme with $A_3 = 0$.[8] It was a key diagnosis to test the slope of the upper asymptote of the Hill plot because, if A_3 was actually zero, then the expression

$$\log \frac{Y}{1 - Y} = 2 \log p + \log \frac{A_4}{3A_2} \qquad \text{as } p \to \infty \qquad (13)$$

indicated that the slope was 2 rather than unity. It was concluded that the major cause of the occurrence of zero or negative A_3 values in the Adair fitting by Gill *et al.*[20] was the use of equal weighting (i.e., no weighting) in spite of an uneven distribution of the residuals along the log

[25] D. Dolman and S. J. Gill, *Anal. Biochem.* **87**, 127 (1978).
[26] S. J. Gill, this series, Vol. 76, p. 427.
[27] G. K. Ackers and H. R. Halvorson, *Proc. Natl. Acad. Sci. U.S.A.* **71**, 4312 (1974).
[28] G. K. Ackers, M. L. Johnson, F. C. Mills, H. R. Halvorson, and S. Shapiro, *Biochemistry* **14**, 5128 (1975).
[29] M. L. Johnson and G. K. Ackers, *Biophys. Chem.* **7**, 77 (1977).

p scale. In fact, the Adair fitting to the oxygen equilibrium data of the author for 1.2 mM Hb A with equal weighting and floated Abs_∞ gave negative A_3 values, as already described above. Myers et al.[30] also reported that equally weighted fitting results in an unusually low value for A_3. Another conclusion by the author[8] was that the zero A_3 value cannot be accommodated within accurate equilibrium data because those data are not consistent with an asymptote slope of 2 for the upper end of the Hill plot. The Adair fitting to the 1.2 mM Hb A data with weighting consistently gives positive A_3 values that are comparable to A_4 values. These values yield K_4 values ranging from 3 to 5 mmHg^{-1}. On the contrary, the virtually zero A_3 value makes the K_R vlue as large as 21 mmHg^{-1} (Di Cera et al.[31]), which is inconsistent with the generally accepted affinity constant for the R state.[32]

In work by Gill et al., the reversibility of the OEC, that is, agreement of the OEC measured with stepwise increases in p and the OEC measured with stepwise decreases in p for a single hemoglobin sample, has not yet been tested at a high saturation range: in their reversibility test,[20] the reoxygenation was achieved only up to $p = 30$ mmHg ($Y = 0.986$), whereas the deoxygenation started at $p = 450$ mmHg ($Y = 0.999$). Such a test is particularly important for their thin-layer technique, which requires a definite period of time to establish the equilibrium at each value of p.

Criteria for Correct Evaluation of Adair Constants

When an automatic oxygenation apparatus that yields a continuous OEC is used, the agreement of the deoxygenation and reoxygenation curves for a single hemoglobin sample must be tested so that one can be confident of observing the true equilibrium.

In view of the importance of bottom and top data in Adair fitting, it is also important to express the experimental and analyzed data by means of a Hill plot, in which both ends are enlarged compared with the conventional Y vs log p plot.

From the discussions given above and described previously,[8] the criteria, from both experimental and analytical standpoints, for a correct evaluation of the Adair constants are as follows.

1. Deoxygenation and reoxygenation curves must agree throughout the entire range of saturation.

[30] D. Myers, K. Imai, and T. Yonetani, Biophys. Chem. **37**, 323 (1990).
[31] E. Di Cera, C. H. Robert, and S. J. Gill, Biochemistry **26**, 4003 (1987).
[32] M. C. Marden, J. Kister, C. Poyart, and S. J. Edelstein, J. Mol. Biol. **208**, 341 (1989).

2. The standard deviation of the observed variables must be determined experimentally, and appropriate weights based on the standard deviation data must be used in the least-squares fitting.

3. Positive values for all the Adair constants must be obtained without any constraint on them.

4. The best-fit values of the Adair constants must be consistent with their physical meanings.

5. Agreement between the experimental and analyzed data must be tested by Hill plot, not by the conventional Y vs log p plot.

[27] Weighted Nonlinear Regression Analysis of Highly Cooperative Oxygen Equilibrium Curves

By MICHAEL L. DOYLE, DAVID W. MYERS, GARY K. ACKERS, and RICHARD I. SHRAGER

Introduction

Purpose and Scope

The highly cooperative nature of hemoglobin (Hb) oxygenation makes resolution of the oxygen-binding equilibrium constants of the intermediates a challenging problem, both experimentally and in terms of least-squares regression analysis. To meet this challenge experimentally, high-precision oxygen equilibrium methods were devised,[1,2] whereby fractional oxygen saturation is measured spectrophotometrically and oxgyen activity is measured with a Clark polarographic electrode.[3] This widely used method enables a high degree of precision in measuring oxygen equilibrium curves (tenths of a percent of fractional saturation). Even so, resolution of equilibrium constants of the intermediate oxygenated species requires careful attention at the stage of regression analysis.

An important strategy for improving parameter resolvability during regression analysis involves weighting data points according to their relative precision. This chapter outlines a general method for direct determination of the precision of data points obtained by the Clark electrode/spectrophotometric oxygen equilibrium technique, that is, the instrumental

[1] K. Imai, "Allosteric Effects in Haemoglobin." Cambridge Univ. Press, Cambridge, 1982.
[2] A. H. Chu, B. W. Turner, and G. K. Ackers, *Biochemistry* **23**, 604 (1984).
[3] I. Fatt, "Polarographic Oxygen Sensor: Its Theory of Operation and Its Application in Biology, Medicine, and Technology." CRC Press, Cleveland, OH, 1976.

uncertainty of the spectrophotometer and the oxygen electrode, and describes a procedure for incorporating the measured instrumental uncertainty into a continuous "weighting function" of the absorbance-versus-oxygen activity equilibrium curves.

Special treatment is required for cases in which error is present in the independent variable (oxygen activity here) because standard least-squares regression algorithms assume all the experimental noise originates from measurements of the dependent variable. Of particular concern is the possibility that uncertainty in the independent variable may lead to bias error (precise but inaccurate parameter estimates). Such bias error is well known in the case of regression to a line[4] and cannot be corrected by standard weighted regression. The means of incorporating noise in the independent variable, called total least-squares regression, is reviewed below by way of application to Hb–oxygen equilibrium curves.

Finally, a general Monte Carlo method is described for evaluating the performance of various weighting functions in terms of parameter resolvability and bias. The Monte Carlo method involves simulating many oxygen equilibrium data sets with noise properties that mimic actual instrumental noise. The synthetic data sets are then analyzed by weighted nonlinear regression and the statistics of the resulting distributions of best-estimate parameters are evaluated. An important benefit of the method is that the true parameters are known in advance.

Weighted Nonlinear Least-Squares Regression

The goal of the regression analysis that follows is to estimate physical parameters of a fitting function (model) governing the response of the dependent variable (absorbance) on the independent variable (oxygen activity). The fitting function for Hb–oxygen equilibrium curves includes the four oxygen-binding equilibrium constants of tetrameric Hb, and almost always includes oxygenation-linked dimer–tetramer assembly equilibrium constants due to dissociation into dimers. Least-squares regression algorithms serve to find the best parameter estimates for a given model by minimizing the sum-of-squared distances between the dependent variables, Y_i, of the data set and the fitting function, $f(X_i)$, calculated at the corresponding values of the independent variable [numerator, Eq. (1)]. Because the fitting parameters are expressed nonlinearly in the fitting function, that is, the second derivatives of the function with respect to

---

[4] N. R. Draper and H. Smith, "Applied Regression Analysis," 2nd ed., Wiley, New York, 1981.

each parameter are not identically equal to zero, a nonlinear least-squares regression algorithm is required for analysis of the data.

The method of weighted least-squares analysis takes into account the fact that some data points are known to higher precision than others and therefore "weights" them more heavily. The fundamental operation of nonlinear regression algorithms is to search iteratively for a best estimate vector of model parameters that minimizes the familiar χ^2 merit function.

$$\chi^2 \equiv \sum_{i=1}^{N} \left[\frac{Y_i - f(X_i)}{\sigma_i} \right]^2 = \sum_{i=1}^{N} w_i [Y_i - f(X_i)]^2 \tag{1}$$

Here σ_i is the standard error in measuring the ith-dependent variable, Y_i. The weight, w_i, assigned to each data point in Eq. (1) is thus inversely proportional to the square of the standard error in measuring Y_i ($w_i = 1/\sigma_i^2$). χ^2 may be minimized by a variety of iterative procedures.[4,5]

Form of Fitting Equation

For the spectrophotometric/Clark electrode oxygen equilibrium technique, the fitting function is cast in terms of absorbance measurements as

$$A(X_i) = A_0 + (A_\infty - A_0)\overline{Y}(X_i) \tag{2}$$

Here $\overline{Y}(X)$ is the fitting model describing the equilibrium oxygen-binding properties of Hb. The parameters A_0 and A_∞ are the asymptotic absorbance values for the deoxygenated and fully oxygenated Hb sample.

A popular functional transformation of oxygen equilibrium data, but one that is risky when applied to highly cooperative isotherms, is the fractional saturation function [Eq. (3)]. Here absorbance measurements are transformed into the fractional saturation domain according to

$$\overline{Y}(X_i) = \frac{A(X_i) - A_0}{A_\infty - A_0} \tag{3}$$

The practical benefit of this transformation is that, by assuming A_0 and A_∞ are known accurately, the fitting problem is reduced by two parameters. Thus, by constraining these two asymptotic parameters, resolvability of the remaining parameters should in principle be improved.

The fractional saturation transformation frequently seems straightforward, based on the precision with which the parameters A_0 and A_∞ usually can be estimated by extrapolation procedures (e.g., $\pm 0.1\%$ of the total change $A_\infty - A_0$). However, the apparent resolvability benefit of this

[5] M. L. Johnson and S. G. Frasier, this series, Vol. 117, p. 301.

transformation is offset by the following factors, which can be serious in the case of highly cooperative systems. First, the precision of A_0 and A_∞ is not always sufficient to justify the transformation. For example, the oxygen affinity of cobaltous Hb is so low that the upper asymptotic parameter, A_∞, is not experimentally determined even under an atmosphere of pure oxygen.[6] Thus, extrapolated values for A_∞ depend in large part on the mathematical model chosen for the extrapolation so that neither the precision nor the accuracy of A_∞ could be known *a priori*. In practice, analysis of cobaltous Hb–oxygen equilibrium curves required incorporation of A_∞ as a fitting parameter.

Second, the accuracies of determining A_0 and A_∞ are considerably poorer than their precision whenever sample decomposition occurs. Unfortunately, sample decomposition (e.g., oxidation of heme irons from ferrous to ferric) is the rule rather than the exception in the course of oxygen equilibrium measurements with Hb. Typically, one is faced with approximately 0.5 to 1% oxidation of the heme irons: an amount that is insignificant in the case of noncooperative isotherms but, nevertheless, could lead to significant bias error in parameter estimation of highly cooperative isotherms, as indicated by the work of Marden *et al.*[7]

Determination of Instrumental Noise

Measurement of Instrumental Noise for Dependent Variable

Figure 1 shows the instrumental noise associated with a spectrophotometer in the visible region. The functional dependence of the noise on absorbance is typical for a high-precision double-beam spectrophotometer,[8] although the magnitude of the noise will vary somewhat between different instruments. The data in Fig. 1 were fitted to the arbitrary second-order linear response function [Eq. (4)].

$$\sigma_{A_i} = a + bX + cX^2 \qquad (4)$$

The best-fit constants are found to be $a = 1.68 \times 10^{-4}$, $b = 8.63 \times 10^{-5}$, and $c = 1.07 \times 10^{-4}$. Equation (4) is used as a predictor of spectrophotometric noise as a continuous function of the absorbance. This noise model is incorporated in computer simulations of data and then used to calculate

[6] M. L. Doyle, P. C. Speros, V. J. LiCata, D. Gingrich, B. M. Hoffman, and G. K. Ackers, *Biochemistry* **30**, 7263 (1991).

[7] M. C. Marden, J. Kister, C. Poyart, and S. J. Edelstein, *J. Mol. Biol.* **208**, 341 (1989).

[8] H. A. Strobel and W. R. Heineman, "Chemical Instrumentation: A Systematic Approach," 3rd ed. Wiley, New York, 1989.

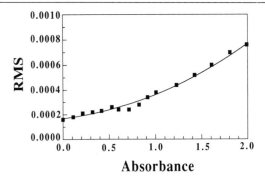

FIG. 1. Precision plot for absorbance measurements as a function of absorbance for a high precision, double-beam spectrophotometer (Cary 219; Varian Analytical Instruments). RMS, Root-mean-squared deviation about the mean absorbance value. The detection wavelength was 577 nm, spectral bandwidth was 1 nm, and time constant to reach 98% of signal was 1 sec. Absorbance values were obtained by use of a steel screw masking cuvette. The curve represents the best fit to the data according to the empirical response function [Eq. (4)].

weighting functions for regression analysis of the simulated data (below). On the basis of Eq. (2) and Fig. 1, the relative noise on each data point will depend on both the absolute absorbance and the total range $A_\infty - A_0$. The example used below assumes a total change in absorbance of 0.25 au ($A_0 = 0.5$ au and $A_\infty = 0.75$ au) for complete oxygenation, corresponding to an oxygen equilibrium curve measured at 577 nm with a total Hb concentration of 50 μM heme.

Additional performance characteristics of the spectrophotometer, such as long-term stability and linearity of response with concentration, should also be checked. Errors of this type are systematic in nature and therefore incapable of being corrected for by weighting factors. They may be corrected only by repairing the instrument itself, or perhaps by modifying the model for the fitting equation.

Measurement of Instrumental Noise for Independent Variable

The standard deviation for measuring oxygen activity with a Beckman (Fullerton, CA) model 39065 polarographic electrode[3] has been reported by Doyle and Ackers.[9] Here we summarize their results. The activity of oxygen in solution, X_i, as monitored with an oxygen electrode is based on the following relationship:

[9] M. L. Doyle and G. K. Ackers, *Biophys. Chem.* **42,** 271 (1992).

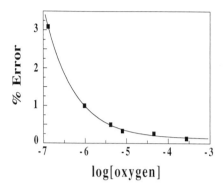

FIG. 2. Percentage random error of the Beckman model 39065 polarographic oxygen electrode versus the logarithm of the oxygen molarity. Solid curve is the fit to Eq. (6) with parameter best estimates given in text. (Taken from Ref. 9.)

$$X_i = X_T \left(\frac{mv_i - mv_0}{mv_T - mv_0} \right) \tag{5}$$

Here X_T is the total oxygen activity, calculated from the barometric pressure, water vapor pressure, percentage of oxygen in the gas mix, and solubility data.[10] The millivolt terms mv_i, mv_0, and mv_T correspond to the observed oxygen electrode output voltages at oxygen activity values X_i, zero, and X_T, respectively.

The voltage signal of the oxygen electrode is amplified with a Keithley model 150B microvolt ammeter (using three orders of magnitude of input sensitivity). The ±1-V output of model 150B, equipped with a series RC circuit as a high frequency-reject filter (halftime of 0.15 sec), is then monitored with a Nicolet model 3091 oscilloscope at a sampling rate of 1 point/sec. Temperature is regulated with a Lauda K-2/R water bath to a precision of ±0.02°. Solution conditions are those used in oxygen equilibrium experiments[2]: 0.1 M Tris–base, 0.1 M NaCl (0.18 M total chloride), 1 mM Na$_2$EDTA, pH 7.40, and 21.5°. Matheson-certified standard gas mixtures of oxygen in nitrogen (certified to an accuracy of ±2% of the oxygen partial pressure) are employed to attain various oxygen partial pressures.

The standard error in the oxygen electrode is measured by equilibrating a buffer sample in the oxygenation apparatus with known partial pressures of oxygen. Figure 2 shows the random error in the oxygen electrode measured over a range of oxygen activity. The standard error at each

[10] E. Wilhelm, R. Battino, and R. J. Wilcock, *Chem. Rev.* **77**, 219 (1977).

equilibrated oxygen activity value is calculated as the square root of the variance for fitting millivolt-versus-time data to an arbitrary exponential. The entire set of standard errors (Fig. 2) is then adequately fitted to Eq. (6) as a means of obtaining the oxygen electrode noise as a continuous function of the oxygen activity:

$$\%\sigma_{mv_i} = a[O_2]^b + c \tag{6}$$

The constant values are determined as $a = 1.79 \times 10^{-4}$, $b = -0.612$, and $c = 0.129$. Thus, the absolute value of the electrode noise is $\sigma_{mv_i} = (mv_i \times \%\sigma_{mv_i})/100$.

In addition to the random error of the measurements in Fig. 2, there are also systematic error sources. For example, Eq. (5) contains four experimental variables and only one of them, mv_i, is independent for all data points of a given isotherm. The other three measurements (X_T, mv_T, and mv_0) are made only once per isotherm and therefore represent systematic error sources. The impact that these systematic errors have on parameter resolvability and bias in oxygen equilibrium curves of Hb has been reported elsewhere.[9] We do not discuss these errors in the present chapter because there are no general data analysis procedures, other than changing the fitting equation, that can correct systematic error sources.

Implicated Weighting Functions

The existence of instrumental noise in both the dependent and independent variable measurements for oxygen equilibrium data suggests several possible weighting functions. The benefit of using one weighting function rather than another will depend on the specific noise properties of the data acquisition system. Below we describe these weighting functions in terms of the variance, σ_i^2, which is a measure of the noise inherent to each data point. The weight assigned to each data point may then be calculated as the inverse of the variance.

Uniform Weighting

The simplest weighting function we consider is uniform weighting, whereby the variance is assumed constant for all data points.

Spectrophotometric Noise

The variance of each data point is expressed as a function of absorbance according to the following spectrophotometric noise (SPEC) model, which was determined empirically from the data in Fig. 1:

$$\sigma_i^2 = [a + bX + cX^2]^2 \tag{7}$$

The constant values a, b, and c are given above, following Eq. (4). As indicated by Fig. 3, the minor differences in spectrophotometric noise between data points in a typical oxygen equilibrium curve lead to a weighting function, which is for all practical purposes equivalent to a uniform weighting function.

Spectrophotometric and Propagated Oxygen Electrode Noise

The third weighting procedure (SPEC/PROPX) is an empirical attempt to compensate both for spectrophotometer and oxygen electrode noise. The strategy is to propagate error from the independent variable domain into the dependent variable, and to add it to the already-present error in the dependent variable. Thus the variance at each point is expressed as a linear combination of instrumental uncertainties in both the spectrophotometer and oxygen electrode as

$$\sigma_i^2 = \left(\frac{\partial A_i}{\partial X}\right)^2 (\sigma_{X_i})^2 + (\sigma_{A_i})^2 \tag{8}$$

The noise terms, σ_{X_i} and σ_{A_i}, correspond to the noise model functions given above [Eqs. (4) and (6)].

Propagating error from the independent variable domain to the dependent variable and then weighting accordingly has been suggested for the linear case.[11] However, there is no theory for this procedure in the nonlinear case, and it is unclear whether such a procedure could in principle improve parameter resolvability for general nonlinear problems.

Hill Transform

The fourth weighting procedure is identical to transforming fractional saturations into Hill plot data $\{\log[\bar{Y}_i/(1 - \bar{Y}_i)]\}$ and then weighting uniformly:

$$\sigma_i^2 = [\bar{Y}_i(1 - \bar{Y}_i)]^2 \tag{9}$$

The Hill transformation weighting function demands attention in the present chapter as a result of its broad usage in the analysis of oxygen equilibrium data. The broad usage of the Hill transform derives in part from its theoretical value as a graphical tool for estimating certain parameters such as the first and last oxygen-binding constants, P_{50}, and the familiar Hill

[11] H. S. Tan and W. E. Jones, *J. Chem. Educ.* **66**, 650 (1989).

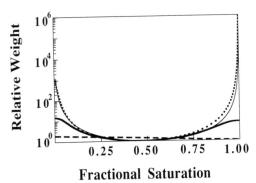

Fractional Saturation

FIG. 3. Normalized weighting functions for regression analysis of oxygen equilibrium curves versus fractional oxygen saturation of Hb. Weighting values are equal to the inverse of the variance about each experimental observation. The curves correspond to the following weighting functions: SPEC (– – –), PROPX (⋯), SPEC/PROPX (—), and the Hill plot transform (—).

slope.[12] However, aside from mere coincidence, it seems unlikely that this graphical procedure would best represent the noise properties of a wide range of oxygen equilibrium data acquisition systems. (But see Imai[1,13,14] for evidence supporting use of the Hill transform under certain conditions.)

The weighting procedures described above are summarized in Fig. 3 as a function of the fractional oxygen saturation of tetrameric Hb. The SPEC weighting function is seen to be essentially flat, whereas the SPEC/PROPX function places slightly more weight at low and high saturations. In contrast, the Hill transform assigns extremely large relative weight to data points at low and high fractional saturations, so that differences in relative weight of 10,000- to 100,000-fold would be used for typical Hb A–oxygen equilibrium curves. Also shown is the predicted weighting function where experimental noise originates solely from measurements of oxygen activity, that is, only the first term of Eq. (8) (called PROPX in Fig. 3). As seen, a weighting function reflecting noise exclusively in the independent variable closely resembles the Hill transform weighting function.

[12] J. Wyman and S. J. Gill, "Binding and Linkage: Functional Chemistry of Biological Macromolecules." University Science Books, Mill Valley, CA, 1990.

[13] K. Imai, *Biophys. Chem.* **37**, 197 (1990).

[14] K. Imai, this volume [26].

Total Least Squares

The fitting of both independent and dependent variables is called total least squares (TLS). Total least squares is more difficult to implement than ordinary least squares, but it offers the possibility of producing parameter estimates with noticeably less variance and bias. The idea has been discussed since 1878,[15] but algorithm development has blossomed only recently, spurred by the seminal paper of Golub and Van Loan.[16] A book by Van Huffel and Vandewalle[17] gives a survey of the field, concentrating on the general linear problem. Fitting oxygen equilibrium curves is a nonlinear problem, but there is nothing precluding its treatment by TLS. There is a much smaller but significant literature on the nonlinear problem[18,19] (see also Doyle and Ackers,[9] and references therein). Here we present our own techniques that take advantage of some properties of the fractional saturation equation for equilibrium binding curves. Our goal is not to improve on other methods, but rather to demonstrate that in the present case, one does practically as well with less elaborate least-squares methods. Still, it is possible that future problems may require TLS analysis, so we shall describe how it was done in our case.

Total least squares minimizes weighted distance vectors, $S(Z_i)$, between the data points and the fitting function according to the following χ^2 function:

$$\chi^2_{TLS} \equiv \sum_{i=1}^{N} S(Z_i) \tag{10}$$

where $S(Z_i)$ is given by

$$S(Z_i) = \left[\frac{A_i - A(Z_i)}{\sigma_{A_i}}\right]^2 + \left(\frac{X_i - Z_i}{\sigma_{X_i}}\right)^2 \tag{11}$$

Here σ_{A_i} and σ_{X_i} are the standard deviations in measuring the dependent and independent variables, and Z_i is the optimal value of the independent variable, which must be calculated for each data point and iteration. Because A_i is measured by spectrophotometer, and X_i by electrode, the errors of the two measurements are regarded as uncorrelated {otherwise, a cross-correlation term involving $[A_i - A(Z_i)](X_i - Z_i)$ would be required in Eq. (11)}.

[15] R. J. Adcock, *Analyst* **5**, 53 (1878).
[16] G. Golub and C. Van Loan, *SIAM J. Numer. Anal.* **17**, 883 (1980).
[17] S. Van Huffel and J. Vandewalle, "The Total Least Squares Problem: Computational Aspects and Analysis." SIAM, Philadelphia, 1991.
[18] H. Schwetlick and V. Tiller, *Technometrics* **27**, 17 (1985).
[19] P. T. Boggs, R. H. Byrd, and R. B. Schnabel, *SIAM J. Sci. Stat. Comput.* **8**, 1052 (1987).

Ordinary least squares assumes $Z_i = X_i$, so that only one squared error is required for each observation. In contrast, Eq. (11) involves two squared error terms. One way to approach this problem is to find a zero (i.e., a root) of the equation,

$$S'(Z_i) = 2\left[\frac{A_i - A(Z_i)}{\sigma_{A_i}}\right] A'(Z_i) + 2\left(\frac{X_i - Z_i}{\sigma_{X_i}}\right) = 0 \qquad (12)$$

where $S'(Z_i) = dS(Z_i)/dZ$ and $A'(Z_i) = dA(Z_i)/dZ$. Because $A(Z)$, the oxygen equilibrium absorbance function, is monotonically increasing, it is easily proved that all roots Z of Eq. (12) satisfy

$$Z > X, \text{ if } A > A(X) \qquad \text{and} \qquad Z < X, \text{ if } A < A(X) \qquad (13)$$

that is, $Z = X$ is never a root of Eq. (12), but it does serve as a reasonable first estimate in an iterative root-finding algorithm. Using $Z = X$ as one end of a search interval, we chose Z equal to twice the maximum X value or Z equal to 0 as upper or lower ends of the search interval, respectively, according to Eq. (13). If the search interval is a bracket, that is, if $S'(Z)$ has opposite signs at the ends of the interval, we proceed with the search. Otherwise we accept $Z = X$ as the solution. There are several reliable methods for finding a root when starting with a guaranteed bracket.[20,21] In our work we used a slightly slower scheme that was easier to code, but the particular method is irrelevant here, as long as the roots are found to high precision. We required that Z be precise to eight decimal digits to prevent any numerical error from distorting our statistics.

Monte Carlo Methods

Data Synthesis

Noise-free data sets were simulated for the cooperative oxygen-binding equilibrium curve of tetrameric human Hb according to Eq. (14):

$$A(X) = A_0 + [A_\infty - A_0]\bar{Y}(X) \qquad (14)$$

where, for simplicity and computational cost effectiveness, we use the idealized fractional saturation function [Eq. (15)], which assumes the absence of dissociation into dimers.

[20] R. P. Brent, "Algorithms for Minimization without Derivatives." Prentice-Hall, Englewood Cliffs, NJ, 1973.
[21] R. I. Shrager, *Math. Comput.* **44**, 151 (1985).

$$\overline{Y}(X) = \frac{\displaystyle\sum_{j=1}^{4} j e^{-\Delta G_j/RT} X^j}{4\left(1 + \displaystyle\sum_{j=1}^{4} e^{-\Delta G_j/RT} X^j\right)} \qquad (15)$$

Here $\overline{Y}(X)$ is the fractional oxygen saturation at oxygen activity X, R is the gas constant, T is temperature, and the Adair constant free energies (which represent equilibria between deoxygenated Hb and the ith ligation state) are denoted by ΔG_j. The true values of the four Adair constant free energies were taken from Chu et al.[2] Noise-free data were generated as 60 points equally spaced in the logarithm of oxygen molarity from approximately -3.5 to -6.2.

We note that in the analysis of real experimental data, the assumption that dimers make a negligible contribution to the shape of the isotherm is almost never justified for highly cooperative systems, because the population of dimer species usually surpasses the populations of some of the intermediate oxygenated tetrameric species. Whether the fitting model must also include dimers depends on several variables, such as the concentration of Hb studied, the dimer–tetramer assembly energetics, and the accuracy of the parameters required from the data analysis.[22–24]

Pseudorandom noise was superimposed on the perfect data by means of the Box–Muller transformation of uniform deviates.[25] Uniform deviates were obtained from Hewlett-Packard (Palo Alto, CA) 9000/835 system calls in single precision and further randomized with the shuffling algorithm RAN0 as described by Press et al.[26] The sequence of deviates was identical for all four weighting functions studied, that is, they were all initiated with the same seed. The mean and root-mean-square deviation about the mean of the distribution of 60,000 pseudorandom normal deviates were -0.0003 and 0.9955, respectively. The serial correlation statistic[27] of the sequence of these pseudonormal deviates was equal to the expected statistic for a normal distribution at the 95% confidence level for sequence lengths of six and longer.

[22] M. L. Johnson and G. K. Ackers, Biophys. Chem. 7, 77 (1977).
[23] M. L. Johnson and A. E. Lassiter, Biophys. Chem. 37, 231 (1990).
[24] M. L. Johnson, this volume [28].
[25] G. E. P. Box and M. E. Muller, Ann. Math. Stat. 29, 610 (1958).
[26] W. H. Press, B. P. Flannery, S. A. Teukolsky, and W. T. Vetterling, "Numerical Recipes: The Art of Scientific Computing." Cambridge Univ. Press, Cambridge, 1986.
[27] D. E. Knuth, "The Art of Computer Programming," Vol. 2. Addison-Wesley, Reading, MA, 1969.

Pseudorandom noise was first superimposed on the dependent variable of each data point according to Eq. (16)

$$A_i = A(X_i) + \Phi_1 \sigma_{A_i} \qquad (16)$$

and then on the independent variable:

$$X_i = X_T \left[\frac{(mv_i + \Phi_2 \sigma_{mv_i}) - mv_0}{mv_T - mv_0} \right] \qquad (17)$$

Here the Φ terms are independent, Gaussian noise generators with mean values of zero and standard deviations of unity. The standard deviation functions σ_{A_i} and σ_{mv_i} were calculated from direct measurements of instrumental uncertainty as described by Eqs. (4) and (6), respectively (see also Figs. 1 and 2).

Monte Carlo Nonlinear Regression Algorithm

The synthetic data sets were fitted to Eq. (2) by nonlinear regression for the four Adair constant free energies and asymptotic parameters A_0 and A_∞. Initial parameter estimates were supplied to the nonlinear regression algorithm by a linear least-squares calculation of the parameters, based on the linearized form of the fractional saturation function as described in the appendix. The linear least-squares solver LFIT[26] was used with uniform weighting for all simulations described in this chapter, thus providing the same set of initial guess vectors to all weighting function cases studied.

After obtaining initial parameter estimates for each equilibrium curve, nonlinear least-squares regression was carried out in double precision using the simplex algorithm[28] to minimize the generalized χ^2 function [Eq. (1)]. The simplex algorithm was chosen for its suitability for use with the TLS procedure, because we wished to avoid programming the partial derivatives of Z and $A(Z)$ with respect to the fitting parameters. Convergence of the algorithm was established as a change of 10^{-9} or less in χ^2. After convergence with each equilibrium curve, the best estimates of the parameters were stored. The statistics of the distributions of best-estimate parameters were then evaluated after analysis of 1000 equilibrium curves.

To assess whether the simplex algorithm was finding the global minimum for each isotherm, it was particularly important to vary the size of the starting simplex (i.e., the matrix of initial-estimate vectors used to start the simplex algorithm[26]) and the magnitude of the convergence crite-

[28] J. A. Nelder and R. Mead, *Comput. J.* **7**, 308 (1965).

rion. Small simplex sizes (e.g., vectors of initial parameter estimates nearly identical to the true parameter values) when combined with less stringent convergence criteria (e.g., 10^{-6} change in χ^2 or greater) led to parameter distributions corresponding to local minima. Other conditions such as the seed of the random number generator and number of simulations (several hundred trials sufficiently defines the distributions for the present purposes) were also explored.

Results and Discussion

Evaluation of Parameter Distributions

One-thousand synthetic data sets, generated according to Eqs. (14)–(17), were analyzed by weighted nonlinear regression, and the resulting distributions of best-estimate parameters are shown in Fig. 4 (see also Table I). The free energies for binding one and four oxygens, ΔG_1 and ΔG_4, and the absorbance asymptotic parameters, A_0 and A_∞, are in all cases well resolved. In contrast, the free energies for binding two or three oxygens, ΔG_2 and ΔG_3, are more difficult to resolve because of the low populations of intermediate species with a total of two or three oxygens bound. It is with these later parameters that an efficient weighting function is most helpful.

For practical purposes, the uniform, SPEC, SPEC/PROPX, and TLS weighting methods are equally proficient in resolving the six fitting parameters. Thus, any of these methods may be regarded as optimal weighting functions for the present set of experimental errors. Bias error (Table I) for all weighting functions is insignificant when compared to upper and lower 68% confidence limits. On the other hand, the Hill transform was considerably worse in terms of parameter resolution. This is especially evident for the parameters ΔG_2 and ΔG_3, for which the distributions about the median values are significantly broader and are also noticeably bimodal in character.

Another indicator of the appropriateness of a weighting function is the distribution of variances of fit (Fig. 4G). The similarity in the variance distributions of the uniform, SPEC, and SPEC/PROPX weighting functions indicates the equivalence of these functions in their description of the equilibrium curve data. However, the Hill transform is seen to be a poor weighting function because of the significantly larger variances of fit and broader distribution of variances. The larger variance of fit seen with the Hill transform reflects the deemphasis on data points in the middle region of the isotherm, where the intermediate oxygenated species contribute most to the data.

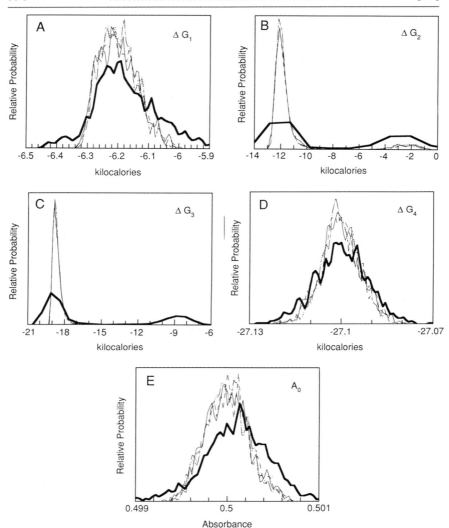

FIG. 4. Probability distributions for Monte Carlo nonlinear regression estimation of parameters in Eqs. (14) and (15) for various weighting functions [(A) ΔG_1, (B) ΔG_2, (C) ΔG_3, (D) ΔG_4, (E) A_0, (F) A_∞, (G) variance] as described in text (curves overlay in some instances): uniform (—), SPEC (– – –), SPEC/PROPX (–·–·–), TLS (···), and the Hill transform (—). Random error was present on both dependent and independent observables according to Eqs. (16) and (17). True values of the Adair constant free energies were taken from Chu *et al.*[2] as −6.2 (A), −12 (B), −18.7 (C), and −27.1 (D) in units of kilocalories, and asymptotic absorbance parameters for (E) deoxyHb and (F) oxyHb were 0.5 and 0.75 au, respectively. Distributions of the variances of fit are shown in (G). The variances were calculated on convergence as the sum-of-squares [numerator Eq. (1)] divided by the degrees of freedom.

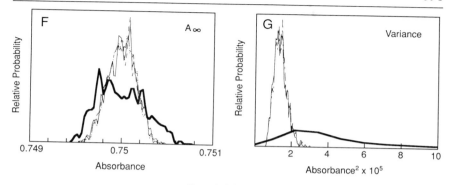

FIG. 4. (*Continued*)

It should be pointed out that the validity of the SPEC/PROPX, PROPX and Hill transform weighting functions may not hold in general. There is certainly no exact theory for their use. In fact, a question that arises from the existence of noise in the independent variable is whether it may lead to bias error, because it is well known to do so in the case of regression to a line.[4] Inspection of the parameter biases given in Table I indicates that this concern is not problematic for the specific instrumental noise properties and biophysical parameters used in this study. The generality of this conclusion, however, is unknown.

Weighting Functions Based Solely on Residuals

Examination of residuals [the deviations between data points and fitting function, $Y_i - f(X_i)$] is an important part of the data analysis process. The magnitudes and trends in the residuals originate from two sources: (1) random experimental error and (2) systematic experimental error (an inaccurate fitting model). For cases in which the fitting model is exact, the residuals should exhibit randomly distributed positive and negative deviations corresponding in magnitude to instrumental noise. Thus, in principle a weighting function could be deduced indirectly on the basis of examination of residuals. To obtain a large enough sample size in the case of Hb–oxygen equilibrium curves it would be desirable to measure multiple isotherms under strictly identical solution conditions. Unfortunately, in view of the hypersensitive allosteric nature of hemoglobin to solution conditions (especially pH), the general problems of sample decomposition, and the steepness of the isotherm, the direct method of determining instrumental uncertainties, as outlined in this chapter, is more reliable. For example, in the context of the high precision of typical

TABLE I
PARAMETER RESOLVABILITY FOR VARIOUS WEIGHTING FUNCTIONS[a]

True parameters	Uniform	SPEC	SPEC/PROPX	Hill transform	TLS
$\Delta G_1 = -6.2$	0.0050	0.0053	0.0044	-0.0134	0.0048
	(0.061, 0.072)	(0.061, 0.072)	(0.056, 0.065)	(0.079, 0.123)	(0.056, 0.066)
$\Delta G_2 = -12.0$	-0.037	-0.033	-0.023	-0.314	-0.021
	(0.36, 0.93)	(0.36, 0.92)	(0.31, 0.73)	(0.89, 9.55)	(0.31, 0.72)
$\Delta G_3 = -18.7$	0.016	0.014	0.022	-0.050	0.021
	(0.24, 0.43)	(0.24, 0.43)	(0.22, 0.37)	(0.52, 9.89)	(0.22, 0.37)
$\Delta G_4 = -27.1$	4.2×10^{-4}	4.0×10^{-4}	4.2×10^{-4}	-1.8×10^{-5}	4.4×10^{-4}
	(0.0056, 0.0070)	(0.0056, 0.0070)	(0.0055, 0.0062)	(0.0085, 0.0085)	(0.0055, 0.0062)
$A_0 = 0.500$	-1.8×10^{-5}	-1.2×10^{-5}	-1.9×10^{-5}	-7.9×10^{-5}	-1.8×10^{-5}
	$(2.3 \times 10^{-4}, 2.4 \times 10^{-4})$	$(2.4 \times 10^{-4}, 2.3 \times 10^{-4})$	$(2.3 \times 10^{-4}, 2.3 \times 10^{-4})$	$(3.4 \times 10^{-4}, 3.3 \times 10^{-4})$	$(2.3 \times 10^{-4}, 2.3 \times 10^{-4})$
$A_\infty = 0.750$	7.7×10^6	9.8×10^6	13.0×10^6	-5.9×10^6	11.9×10^6
	$(1.6 \times 10^{-4}, 1.5 \times 10^{-4})$	$(1.7 \times 10^{-4}, 1.5 \times 10^{-4})$	$(1.6 \times 10^{-4}, 1.5 \times 10^{-4})$	$(2.4 \times 10^{-4}, 2.9 \times 10^{-4})$	$(1.6 \times 10^{-4}, 1.5 \times 10^{-4})$
Variance	1.37×10^{-7}	1.37×10^{-7}	1.40×10^{-7}	9.69×10^{-7}	NA

[a] Free energies are in kilocalories and the asymptotic parameters A_0 and A_∞ are in absorbance units. Parameter errors are in the top row for each parameter and 68% error limits about the median of each distribution are in parentheses (lower limit, upper limit). Parameter error is equal to the observed median of the distribution minus the true parameter value.

oxygen equilibrium measurements (0.15% of the total absorbance change predicted from the variances of fit in Table I), small amounts of systematic error may contribute to the residuals, most likely as "runs" of residuals in the middle of the transition curve. Common sources of systematic error during oxygenation of Hb include oxidation of the heme irons, general denaturation problems, possible differences in extinction coefficients of the intermediates, and dissociation into dimers (unless explicitly accounted for in the binding partition function).

Relation between Monte Carlo Simulations and Real Data

The advantage of Monte Carlo simulations with synthetic data is that the true fitting model and parameter values are known in advance. Therefore the effects of instrumental noise models and weighting functions can be evaluated unambiguously. Interestingly, the parameter bias error in the present study (Table I) was small and independent of the weighting function used, whereas analysis of experimentally derived data with different weighting functions (uniform versus Hill transform) showed a dependence of the best-estimate parameters on weighting function.[29] The implication of this discrepancy is that the real data contain systematic error. if so, the fitting equation should be modified to include the functional dependence of the systematic error source. The existence of dimers, for example, can be accommodated in a straightforward manner as discussed above. On the other hand, oxidation is difficult to incorporate into the binding partition function because the thermodynamic linkage between oxidation and oxygen binding would have to be determined; multiple-wavelength studies, however, may afford a first-order correction in this regard.[30,31]

Oxygen-Linked Dissociation into Dimers as Major Source of Systematic Error

Ten percent of oxygenated human Hb is dissociated into dimers under solution conditions of 0.1 M Tris–base, 0.1 M NaCl, pH 7.4, 21.5°, and 100 μM heme. Even at as high as 1 mM heme, dimers are present at 3%, an amount that is considerable in the context of the precision of typical oxygen-binding data. Thus, the oversimplified fractional saturation model given by Eq. (15) is clearly invalidated for highly cooperative equilibrium curves. The underlying problem is that the shape of the equilibrium curve

[29] D. Myers, K. Imai, and T. Yonetani, *Biophys. Chem.* **37**, 323 (1990).
[30] K. D. Vandegriff, Y. C. Le Tellier, J. R. Hess, and R. I. Shrager, *Biophys. J.* **61**, A55 (1992).
[31] K. D. Vandegriff and R. I. Shrager, this volume [22].

is influenced to a greater extent by binding properties of the dimers than by the intermediate oxygenated tetrameric species of interest, such as the doubly and triply oxygenated species that are present at only 2–3% maximum.[2]

The degree to which dimers cause bias error when not explicitly accounted for in the binding partition function can be determined by computer simulations. The reader is referred to previous studies for more details.[22-24] Fortunately, in practice dimers may be readily incorporated into the fitting equation, although global regression of multiple isotherms measured over a range of Hb concentration is usually required.[2,6]

Multiple Experimental Probes to Improve Reliability of Parameter Estimates

If systematic error cannot be corrected by modification of the fitting function, reliability of the parameter estimates is improved by using multiple types of experimental methods. Thus if two distinct experimental methods, each potentially susceptible to different types of systematic error, yield the same estimate for a given parameter, the confidence in that parameter estimate is much greater. An example is direct subunit-assembly equilibrium measurements versus the combined database of oxygen equilibrium curves as a function of Hb concentration (which probes dimer–tetramer, subunit-assembly equilibria).[2,32] In those studies, agreement was found between oxygen equilibrium data versus (1) analytical gel chromatography data for estimating the oxygenated Hb dimer–tetramer assembly free energy and (2) kinetically determined estimates of deoxygenated Hb dimer–tetramer assembly free energies and oxygen-binding free energies of dimers for several recombinant Hbs.

Concluding Remarks

A general method is presented for deducing an appropriate weighting function for nonlinear least-squares regression. The influence of instrumental uncertainties in both dependent and independent variables on parameter resolution and bias is also outlined. Improved resolution is seen when a weighting function is used that most accurately describes the random instrumental uncertainties. Bias error is insignificant for the weighting functions studied when systematic error is absent.

For the case of Hb–oxygen equilibrium curves measured with a high-precision spectrophotometric/Clark electrode apparatus, the noise in the

[32] M. L. Doyle, G. Lew, A. De Young, L. Kwiatkowski, A. Wierzba, R. W. Noble, and G. K. Ackers, *Biochemistry* **31**, 8629 (1992).

oxygen activity measurement was found to be insignificant compared with the uncertainty in measuring fractional saturation with a high-precision spectrophotometer. Consequently, a weighting function that corresponds predominantly to noise in the spectrophotometer was found to offer optimal parameter resolvability. Moreover, because the spectrophotometric noise between 0 and 1 absorbance unit was nearly constant (Fig. 1), a uniform weighting function was found to be equally effective.

However, other spectrophotometers and/or oxygen electrode systems may exhibit unique noise properties, and different experimental methods, such as gasometric techniques, almost certainly involve a different set of noise properties. It is therefore recommended that the noise properties of each experimental system be characterized individually, and that the most appropriate weighting function be deduced according to Monte Carlo simulations with synthetic data.

Appendix: Obtaining Initial Estimates of Equilibrium Constants

Fitting ligand-binding equilibrium curves to data can be time consuming when the initial estimates of the equilibrium constants are poor. Additionally, for Monte Carlo nonlinear regression studies it is desirable to obtain objective initial parameter estimates for each synthetic equilibrium curve analyzed. We present here a method for initializing parameter estimates of the stoichiometric ligand-binding equilibrium constants of nondissociating macromolecule systems. The example given below pertains to a macromolecule that bind four ligands, but the method can be extended readily to any number of ligand-binding sites. It may also be extended to include absorbance asymptotic parameters [see Eq. (14)].

In this method the fractional saturation function is recast as a linear problem, that is, one in which all parameters (a values) are first-power multipliers of terms that do not contain other parameters. Using x for ligand activity and y for observed saturation, the model may be stated as follows:

$$y = \frac{0.25a_1x + 0.5a_2x^2 + 0.75a_3x^3 + a_4x^4}{1 + a_1x + a_2x^2 + a_3x^3 + a_4x^4} + \text{error} \qquad (A.1)$$

Equation (A.1) is called nonlinear in the a values because there are a values in the denominator. Multiplying Eq. (A.1) by its denominator and collecting terms with a values on the right-hand side, we obtain a linear form,

$$y = a_1x(0.25 - y) + a_2x^2(0.5 - y) + a_3x^3(0.75 - y) + a_4x^4 + \text{error}'$$
$$(A.2)$$

which can be presented to any of dozens of linear least-squares solvers on the market; in other words, fit the data y to the expression on the right-hand side of Eq. (A.2) by adjusting the a values, which now can be found in one step because they are linear. True, data y also appear in the fitting function, but that is not against the rules.

This would be the end of the procedure except for the problem of weights. Whether we use Eq. (A.1) or Eq. (A.2) for fitting, we are trying to minimize the sum

$$\sum_{i=1}^{N} w(i)[y(i) - f(i)]^2 \tag{A.3}$$

where $f(i)$ is the computed expression that approximates $y(i)$, and $w(i)$ is the weight, a measure of the importance of the observation. As indicated throughout this chapter, weights must be chosen carefully to give optimal results. The trouble with multiplying by the denominator of Eq. (A.1) to obtain Eq. (A.2) is that the errors in Eq. (A.1) are also multiplied by the same denominator [hence the prime superscript on the error term in Eq. (A.2)]. Furthermore, we do not know what that denominator is, becuase we have not solved for the a values yet. In Winslow et al.,[33] there is a derivation of a method for approximating the denominator of Eq. (A.1) directly from the data, provided the data are reasonably dense from $x = 0$ to the maximum x. This leads to the following algorithm for generating first estimates of the a values:

```
        dimension x(100), y(100), w1(100), w2(100), a(4), c(100,4), d(100)
        integer nobs
c       x = ligand pressure, > 0, monotonically increasing.
c       y = observed saturation, estimated from absorbance data.
c           NOTE: as this program is now written, the maximum value
c           of theoretical saturation should be 1, not 4.
c       w1 = weights for use with the nonlinear Adair model (A1).
c       w2 = square roots of weights for use with linear model (A2).
c       a = Adair equilibrium constants, to be initialized here.
c       c = matrix of coefficients for the linear model.
c       d = vector of right hand sides for the linear model.
c       nobs = number of observations.
c-------Integrate 4y/x from 0 to each x(i) by trapezoidal rule.
c-------Set w2 to sqrt(w1) times the exponential of the integral.
c-------The w2's are actually square roots of the weights for (A2).
        trap=2.*y(1)
        w2(1)=sqrt(w1(1))*exp(-trap)
        do i=2,nobs
            trap=trap + 2.*(y(i-1)/x(i-1) + y(i)/x(i))*(x(i)-x(i-1))
            w2(i)=sqrt(w1(i))*exp(-trap)
```

[33] R. M. Winslow, M. Swenberg, R. L. Berger, R. I. Shrager, M. Luzzana, M. Samaja, and L. Rossi-Bernardi, J. Biol. Chem. 252, 2331 (1977).

```
      enddo
c-------Prepare the linear least-squares problem.
c-------Note that we will multiply the w2's directly into
c-------the c's and d's, thus enabling a solution by a solver
c-------that does not handle weights explicitly.  If, however,
c-------your solver requires weights, give it a vector of 1's.
      do i=1,nobs
        d(i)=w2(i)*y(i)
        do j=1,4
          c(i,j)=w2(i)*x(i)**j*(.25*j-y(i))
        enddo
      enddo
c-------Solve for the a's that minimize the 2-norm
c-------of the matrix expression (C*a-d), using
c-------your favorite linear least-squares routine, e.g.:
      call leastsq(c,d,a,nobs,4,100)
c-------Use the a's generated by the above step as first
c-------estimates in fitting (A1) to y with weights w1.
```

[28] Dimer–Tetramer Equilibrium in Adair Fitting

By Michael L. Johnson

The analysis of ligand (oxygen)-binding data for normal and mutant hemoglobins usually involves the estimation of the Adair binding constants by "fitting" the experimental data to a binding equation.[1,2] The fitting procedure is normally one of several different nonlinear least-squares algorithms.[3] The statistically valid application of any linear or nonlinear least-squares technique requires that several assumptions are made about the nature of the data, the experimental uncertainties of the data, and the fitting function.[3] These assumptions are discussed in detail elsewhere.[3,4] Consequently, a general discussion of these assumptions is not presented here. This chapter discusses the consequences of one specific least-squares assumption on the analysis of Adair constants, that assumption being that the fitting equation is the correct mathematical description of the molecular interactions being considered.

Because human hemoglobin A_0 exists predominantly as a tetramer with one binding site per monomeric unit, the most common choice for a fitting equation is the standard four-binding site Adair equation,

[1] G. S. Adair, *Proc. R. Soc. Ser. London, A* **109**, 292 (1925).
[2] G. S. Adair, *J. Biol. Chem.* **63**, 529 (1925).
[3] M. L. Johnson and S. G. Frasier, this series, Vol. 117, p. 301.
[4] M. L. Johnson and L. M. Faunt, this series, Vol. 210, p. 1.

$$\overline{Y}_4 = \frac{1}{4}\frac{\partial \ln Z_4}{\partial \ln X} = \frac{1}{4}\frac{K_{41}X + 2K_{42}X^2 + 3K_{43}X^3 + 4K_{44}X^4}{1 + K_{41}X + K_{42}X^2 + K_{43}X^3 + K_{44}X^4} \tag{1}$$

where X is the free ligand (oxygen) concentration and Z_4 is the macroscopic analog of the partition function, sometimes called a binding polynomial.

$$Z_4 = 1 + K_{41}X + K_{42}X^2 + K_{43}X^3 + K_{44}X^4 \tag{2}$$

The K_{ij} values in Eq. (2) are the product Adair constants for the binding of j ligands (oxygen molecules) to a macroscopic species with i binding sites. Sometimes the Adair constants are written as stepwise Adair constants, denoted as k_{ij}. A stepwise Adair constant refers to the equilibrium constant to bind the jth ligand (oxygen) to a species that already contains $j - 1$ ligands (oxygens). The relationship between the product and stepwise Adair constants is given in Eqs. (3)–(6).

$$K_{41} = k_{41} \tag{3}$$
$$K_{42} = k_{41}k_{42} \tag{4}$$
$$K_{43} = k_{41}k_{42}k_{43} \tag{5}$$
$$K_{44} = k_{41}k_{42}k_{43}k_{44} \tag{6}$$

Are Eqs. (1) and (2) the correct mathematical formulation for estimating the Adair binding constants?[5,6] It has been shown that hemoglobin A_0 dissociates into $\alpha\beta$ dimers at the dilute hemoglobin concentrations where ligand (oxygen)-binding experiments are usually performed.[7] This reaction scheme is presented in Fig. 1. Furthermore, it has been shown for hemoglobin Kansas[8] that a significant amount of dimers exist at the concentrations found within the erythrocyte. The dimeric species also bind ligands (oxygens) according to a binding isotherm (saturation function) analogous to Eqs. (1) and (2):

$$\overline{Y}_2 = \frac{1}{2}\frac{\partial \ln Z_2}{\partial \ln X} = \frac{1}{2}\frac{K_{21}X + 2K_{22}X^2}{1 + K_{21}X + K_{22}X^2} \tag{7}$$

$$Z_2 = 1 + K_{21}X + K_{22}X^2 \tag{8}$$

The dimer–tetramer association of human hemoglobin is a reversible equilibrium, with the dimers having a higher affinity for oxygen than the tetramers.[9] This implies that the fraction of dimers present will be a func-

[5] M. L. Johnson and G. K. Ackers, *Biophys. Chem.* **7**, 77 (1977).
[6] M. L. Johnson and A. E. Lassiter, *Biophys. Chem.* **37**, 231 (1990).
[7] G. K. Ackers and H. R. Halvorson, *Proc. Natl. Acad. Sci. U.S.A.* **71**, 4312 (1974).
[8] D. H. Atha, M. L. Johnson, and A. F. Riggs, *J. Biol. Chem.* **254**, 12390 (1979).
[9] F. C. Mills, M. L. Johnson, and G. K. Ackers, *Biochemistry* **15**, 1093 (1976).

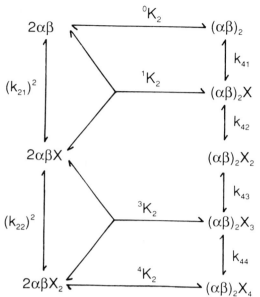

FIG. 1. The seven-parameter oxygenation-linked dimer $\alpha\beta$ to tetramer, $(\alpha\beta)_2$, reaction scheme as defined by Ackers and Halvorson.[7] The ${}^iK_2$ values are the subunit assembly constants to form a tetramer with i ligands bound. The k_{ij} values denote the stepwise Adair constants for binding oxygen to hemoglobin. Only 7 of the 10 equilibrium constants shown here are independent. The other three can be calculated from the seven independent parameters by following closed reaction paths. For example, ${}^1K_2 = {}^0K_2(k_{41}/k_{21})$. [Reprinted with permission from M. L. Johnson and A. E. Lassiter, *Biophys. Chem.* **37**, 231 (1990).]

tion of the fractional saturation with oxygen. The converse is also true. Therefore, the binding isotherms for tetramers [Eq. (1)] and dimers [Eq. (7)] must be combined with a dimer–tetramer self-association reaction to form the complete saturation function for the linked system[7] [Eq. (9)],

$$\overline{Y_{2,4}} = \frac{Z_2' + Z_4'[\sqrt{(Z_2)^2 + 4{}^0K_2Z_4(P_T)} - Z_2]/(4Z_4)}{Z_2 + \sqrt{(Z_2)^2 + 4{}^0K_2Z_4(P_T)}} \tag{9}$$

$$Z_2' = K_{21}X + 2K_{22}X^2 \tag{10}$$

$$Z_4' = K_{41}X + 2K_{42}X^2 + 3K_{43}X^3 + 4K_{44}X^4 \tag{11}$$

where the primed Z values are the derivatives of the unprimed Z values with respect to the natural logarithm of X, ${}^0K_2$ is the dimer-to-tetramer association constant for the unliganded species, and P_T is the molar concentration of heme (hemoglobin monomers). In the limit of zero concentration of hemoglobin, the linked saturation function [Eq. (9)], will approximate the saturation function for the dimers [Eq. (7)]. In the limit of infinite hemoglobin concentration, the linked saturation function [Eq. (9)], will approximate the saturation function for the tetramers [Eq. (1)].

Here is the paradox. We know that nonlinear least squares will not provide accurate estimates of the Adair binding constants if the fitting equation is incorrect. We know that the standard four-binding site Adair formulation [Eq. (1)], is incorrect except in the limit of high concentrations of hemoglobin. We know that for some mutants, like hemoglobin Kansas, there is a significant fraction of dimers present at the concentrations found within the erythrocyte. What we do not know is how high the hemoglobin concentration must be for the four-site Adair binding isotherm [Eq. (1)] to be a sufficiently close approximation to the complete linked binding isotherm [Eq. (9)] to provide accurate estimates of the Adair constants. One definition of sufficiently close is when the least-squares parameter estimation procedure provides Adair constants without significant systematic uncertainties. How high the hemoglobin concentration must be depends on the actual values of the Adair binding constants for dimers, the Adair binding constants for tetramers, and the dimer–tetramer association constant. However, without having performed the analysis in terms of the complete linked binding isotherm [Eq. (9)], we cannot be sure that the Adair constants that have been evaluated are without significant systematic uncertainties. Therefore, to be sure that the use of Eq. (1) is acceptable the analysis must first be performed with Eq. (9). This chapter outlines a procedure that can answer the question of how high a hemoglobin concentration is needed to provide reasonable estimates of the Adair constants. The procedure is then used to provide an answer for the only two hemoglobins for which the complete linkage scheme has been resolved, that is, hemoglobins A_0[9,10] and Kansas.[8]

Methods

The general method is simple. Synthetic noise-free data are calculated according to Eq. (9) (*i.e.*, the complete linked saturation function for ligand binding and self-association). For the calculations presented here the values of the various equilibrium constants have been taken from the literature.[8-10] A different synthetic data set is generated for each of a series of different hemoglobin concentrations. These synthetic data are then least-squares fit[3,4] with the tetramer-only binding isotherm [Eq. (1)] to estimate the apparent values for the tetramer Adair binding constants.

[10] A. H. Chu, B. W. Turner, and G. K. Ackers, *J. Biol. Chem.* **256,** 1199 (1981).

TABLE I
LINKAGE EQUILIBRIUM CONSTANTS[a]

Constant	Hb A_0	Hb Kansas	Units
K_{21}	3.09×10^6	1.93×10^6	M^{-1}
K_{22}	2.39×10^{12}	6.00×10^{11}	M^{-2}
K_{41}	4.26×10^4	8.44×10^4	M^{-1}
K_{42}	8.22×10^8	1.50×10^9	M^{-2}
K_{43}	7.96×10^{13}	1.43×10^{13}	M^{-3}
K_{44}	1.24×10^{20}	2.07×10^{17}	M^{-4}
$^0K_2$	4.40×10^{10}	1.59×10^{10}	M^{-1}

[a] From the literature: Hb A_0 from Ackers et al.[9,10]; Hb Kansas from Atha et al.[8]

The differences between the apparent values and the actual values used to generate the data are thus a direct measure of the amount of systematic error in the Adair constants introduced by using the tetramer-only saturation function.

Each simulated data set encompasses a range of fractional saturations from 0.00 to 0.99 and contains 50 data points equally spaced in pO_2. Normally, simulated experiments should include a realistic amount of simulated pseudorandom noise. However, to address the question of systematic uncertainties introduced by the form of the fitting equations pseudorandom noise is neither needed nor desired. These simulations approximate the data from an Imai apparatus,[8–14] but the results apply to any method for the evaluation of Adair constants from binding data.

The "induced systematic errors" are calculated as a free-energy change of the apparent values from the true values of each of the tetramer Adair constants, as in Eqs. (12) and (13),

$$\delta\Delta G_{4i} = RT \ln\left(\frac{K_{4i,\text{app}}}{K_{4i}}\right) \tag{12}$$

$$\delta\Delta g_{4i} = RT \ln\left(\frac{k_{4i,\text{app}}}{k_{4i}}\right) \tag{13}$$

where T is the absolute temperature and R is the gas constant.

Two different sets of equilibrium constants have been used for these simulations, one each for hemoglobins A_0[9,10] and Kansas.[8] These equilibrium constants are shown in Table I. The hemoglobin A_0 constants pre-

[11] K. Imai, *Biochemistry* **12**, 798 (1973).
[12] K. Imai, *J. Biol. Chem.* **249**, 7607 (1974).
[13] K. Imai and T. Yonetani, *J. Biol. Chem.* **250**, 2227 (1975).
[14] K. Imai and T. Yonetani, *J. Biol. Chem.* **250**, 7903 (1975).

TABLE II
INDUCED SYSTEMATIC ERRORS FOR
HEMOGLOBIN A_0

Term	Error (kcal/mol)	
	Heme (1.0 mM)	Heme (3.0 mM)
$\delta\Delta G_{41}$	−0.07	−0.04
$\delta\Delta G_{42}$	−0.26	−0.16
$\delta\Delta G_{43}$	0.14	0.07
$\delta\Delta G_{44}$	−0.06	−0.03
$\delta\Delta g_{42}$	−0.19	−0.12
$\delta\Delta g_{43}$	0.40	0.23
$\delta\Delta g_{44}$	−0.20	−0.10

sented in Table I are a composite of five data sets from the literature.[9,10] The experimental conditions for the hemoglobin A_0 data[9,10] are 0.1 M Tris, 0.1 M NaCl, 1.0 mM disodium ethylenediaminetetraacetic acid (Na_2EDTA), pH 7.4, at 21.5°. Hemoglobin Kansas is a low-affinity mutant of human hemoglobin at β_{102} (asparagine → threonine). Experimental conditions for the hemoglobin Kansas data[8] are 0.05 M Tris, 0.1 M NaCl, 1.0 mM EDTA, pH 7.5, at 20°. The thermodynamic reference state for these data sets, and this chapter, is 1M O_2 and 1M heme (hemoglobin monomer) concentration. Thus, all equilibrium constants are expressed in units of molar oxygen and heme.

Results

This section investigates the validity of the use of the tetramer binding isotherm [Eq. (1)] as a high hemoglobin concentration approximation of the complete dimer–tetramer-linked binding isotherm [Eq. (9)]. Unfortunately, the validity of the approximation depends on the actual values of the equilibrium (Adair) constants, the hemoglobin concentration, and the level of acceptable induced systematic error in the Adair constants. Because it is impossible to test the validity at all possible values of the Adair constants only two specific cases will be investigated: hemoglobins A_0[9,10] and Kansas.[8]

Tables II and III present the induced systematic errors for hemoglobins A_0 and Kansas when the data are simulated for 1.0 and 3.0 mM heme with Eq. (9) and then fit with Eq. (1). The induced systematic errors, $\delta\Delta G_{4i}$, as a function of hemoglobin monomer concentration for hemoglobin

TABLE III
INDUCED SYSTEMATIC ERRORS FOR
HEMOGLOBIN KANSAS

Term	Heme (1.0 mM)	Heme (3.0 mM)
$\delta\Delta G_{41}$	0.02	0.01
$\delta\Delta G_{42}$	−0.24	−0.15
$\delta\Delta G_{43}$	−0.22	−0.14
$\delta\Delta G_{44}$	−0.48	−0.29
$\delta\Delta g_{42}$	0.26	0.16
$\delta\Delta g_{43}$	−0.02	−0.01
$\delta\Delta g_{44}$	0.26	0.15

A_0 and Kansas, respectively, are presented in Figs. 2 and 3. The values of the Adair constants used for these simulations are presented in Table I. From Fig. 2 it is evident that the induced systematic errors in the second and third Adair constants of approximately 0.5 kcal/mol are introduced at heme concentrations of 100 μM for hemoglobin A_0. For hemoglobin Kansas induced systematic errors of 0.5 kcal/mol, or more, occur in the second, third, and fourth Adair constants at a concentration of 100 μM.

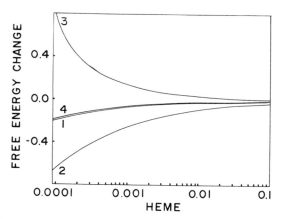

FIG. 2. The free-energy change of the induced systematic errors, kilocalories per mole, for hemoglobin $A_0{}^{9,10}$ as a function of molar heme (hemoglobin monomer) concentration. The index refers to individual $\delta\Delta G_{4i}$ values. Some of these values are also listed in Table II. [Reprinted with permission from M. L. Johnson and A. E. Lassiter, *Biophys. Chem.* **37**, 231 (1990).]

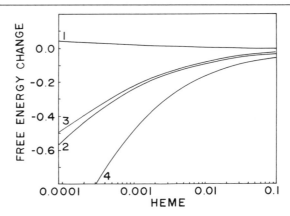

FIG. 3. The free-energy change of the induced systematic errors, in kilocalories per mole, for hemoglobin Kansas[8] as a function of molar heme (hemoglobin monomer) concentration. The index refers to individual $\delta\Delta G_{4i}$ values. Some of these values are also listed in Table III. [Reprinted with permission from M. L. Johnson and A. E. Lassiter, *Biophys. Chem.* **37**, 231 (1990).]

The acceptable level of induced systematic error depends on the reason that the Adair binding constants are being measured. If the object is to find a smooth empirical curve that describes the experimental data with reasonable precision then the actual values of the Adair constants, and the form of the fitting equation, are not particularly critical. If, however, the Adair constants are to be used to distinguish between thermodynamic and/or mechanistic models of cooperativity then it has been shown in many studies[5,15–21] that they need to be determined to within a few tenths of 1 kcal/mol. The current limit of precision for the determination of Adair constants is usually claimed to be a few tenths of 1 kcal/mol. Thus, the level of induced systematic errors that can be tolerated is less than a few tenths of 1 kcal/mol. From Tables II and III it is obvious that even at 3.0 mM heme significant induced systematic errors exist for both hemoglobins A_0 and Kansas.

[15] J. Monod, J. Wyman, and J.-P. Changeux, *J. Biol. Chem.* **12**, 88 (1965).

[16] A. Szabo and M. F. Karplus, *J. Mol. Biol.* **72**, 163 (1972).

[17] G. K. Ackers and M. L. Johnson, *J. Mol. Biol.* **147**, 559 (1981).

[18] M. L. Johnson and G. K. Ackers, *Biochemistry* **21**, 201 (1982).

[19] M. L. Johnson, B. W. Turner, and G. K. Ackers, *Proc. Natl. Acad. Sci. U.S.A.* **81**, 1093 (1984).

[20] M. L. Johnson, *Biochemistry* **27**, 833 (1988).

[21] M. Straume and M. L. Johnson, *Biophys. Chem.* **56**, 15 (1989).

Discussion

This chapter has outlined, and applied, a method for the evaluation of the minimum hemoglobin concentration that can be employed if the effects of dimeric species are to be ignored. For the two examples shown at 3.0 mM heme, systematic errors as great as 0.2 to 0.3 kcal/mol are expected to be induced by neglecting dimers. However, the answer is more complex than this suggests.

It is important to realize that these induced errors are systematic. They will not be cancelled out by measuring more binding isotherms. For a given hemoglobin system they will occur in the same direction and magnitude for every experiment. Consequently, they are expected to bias any conclusions based on the values of the Adair binding constants.

Table II shows that the induced systematic error for the evaluation of the third tetramer stepwise Adair constant, $\delta\Delta g_{43}$, is approximately 0.40 kcal/mol at 1.0 mM heme. This induced systematic error will lead to the conclusion that the concentration of the triply liganded species is significantly lower than reality.

Is the concentration of the triply liganded species always underestimated by neglecting the presence of the dimers? That is, is the free-energy change of this induced systematic error always positive? The triply liganded values for hemoglobins A_0 and Kansas (see Tables II and III) have the opposite signs. A comparison of Figs. 2 and 3 suggests no consistent trends between the two hemoglobins. Consequently, the direction and magnitude of the induced systematic uncertainties for the triply liganded species, and other species, cannot be predicted without prior knowledge of the particular values of the Adair constants.

Can measures of the overall oxygen affinity, like the median ligand concentration, be used to predict the minimum heme concentration that can be used without introducing systematic uncertainties in the Adair constants? The median ligand concentration depends only on K_{44} for the tetramer case [Eq. (1)], and is a function of K_{22}, K_{44}, and $^0K_2$ for the complete ligand-linked subunit assembly case [Eq. (9)].[7] Therefore, the median ligand concentration will not be affected by changes in K_{41}, K_{42}, and K_{43}. Hemoglobin A_0 does not have a large induced systematic error for the last oxygenation step of the example in Fig. 2, $\delta\Delta G_{44}$. Therefore, it is expected, and experimentally observed,[6] that the median ligand concentration for hemoglobin A_0 will not change at high concentration whereas the values of K_{42} and K_{43} will change with concentration. However, hemoglobin Kansas does show a large shift in median ligand concentration due to an induced systematic error in the last step of oxygenation.[6] Again, the analyses of the two hemoglobins are not consistent. It also has been shown that the induced systematic errors are not concomitant with an increase in the variance of fit.[6]

It is evident that neglecting the dimeric species will induce systematic errors in the evaluation of the Adair binding constants even at high hemoglobin concentrations. It is also important to note that the induced systematic errors expected for hemoglobin A_0 have different magnitudes and directions from the induced systematic errors expected for hemoglobin Kansas. Thus, these results cannot be used to predict the expected uncertainties for other hemoglobins, mutants, or buffer conditions. The direction and magnitude of these induced systematic errors cannot be predicted *a priori*; their prediction requires the unique values of the various Adair constants obtained by an analysis in terms of the complete oxygenation-linked dimer–tetramer reaction scheme shown in Fig. 1.

Acknowledgments

The author acknowledges the support of the NSF Science and Technology Center for Biological Timing at the University of Virginia, the Diabetes Endocrinology Research Center at the University of Virginia, and National Institutes of Health Grants GM-28928 and DK-38942.

[29] Effects of Wavelength on Fitting Adair Constants for Binding of Oxygen to Human Hemoglobin

By Lawrence J. Parkhurst, Todd M. Larsen, and Horng-Yuh Lee

Since the end of the last century measurements of the binding of oxygen to blood and to hemoglobin (Hb)[1] have been important in physiology and biochemistry. The first crude measurements of oxygen-binding capacity appear to be those of Bert in 1878,[1a] who reported a few isotherms and showed that the binding of oxygen in blood was exothermic. The first quantitative measurements on hemoglobin were those of Hüfner,[2] but it was not until 1904 that Bohr[3] established the sigmoidal character of the binding curve. The development of the gasometric method by Haldane and

---

[1] Standard abbreviations of Hb and HbO_2 denote deoxy and oxy forms of hemoglobin, but Hb is also used generically for hemoglobin. The distinction will be clear from the context.

[1a] P. Bert, "La Pression Barometrique. Recherches de Physiologie Expérimentale," p. 630. Masson, Paris, 1878.

[2] G. Hüfner, *Arch. Anat., Physiol. Wiss. Med.* **1**, 1890.

[3] C. Bohr, *Zentralbl. Physiol.* **17**, 688 (1904).

Douglas[4] and Barcroft and Peters[5] was shortly followed by spectroscopic measurements of oxygen binding.[6] Refined gasometric measurements by Van Slyke and Neill,[7] Adair,[8] and Roughton et al.[9] allowed a determination of the four sequential binding constants (Adair constant[9a]) for mammalian hemoglobins. Those methods required large amounts of protein, did not allow multiple measurements on the same sample, and required considerable experimental expertise. In principle the gasometric methods were free of any wavelength effects, but it was not trivial to correct for progressive oxidation of the protein (which is now usually assayed spectrophotometrically) during the runs or to assure that all samples had equivalent or known and negligible amounts of methemoglobin (metHb). Following the development of dual-beam spectrophotometers, the tonometric methods[10] and thin-layer[11,12] and automated methods were developed, in which the oxygen concentration was determined by dilution factors, calibrated flow methods,[13] or by oxygen electrodes.[14]

That there might be differences between the fractional change in absorbance and fractional saturation of the hemoglobin with oxygen was appreciated as early as 1935 by Roughton,[15] who found, using a crude spectrophotometric measurement, that there was no essential difference between fractional saturation measured gasometrically and spectrophotometrically. In 1967, Rifkind and Lumry[16] compared gasometric and spectrophotometric determinations of oxygen binding to Hb and reported that the two measurements did not exactly agree. That work was questioned by Anderson and Antonini,[17] who used CO rather than oxygen as the

[4] J. S. Haldane and C. G. Douglas, "Respiration." Yale Univ. Press, New Haven, CT, 1922.

[5] J. Barcroft and F. Peters, J. Physiol. (London) 39, 143 (1909–1910).

[6] J. Barcroft, Proc. R. Soc. London, Ser. B 118, 242 (1935).

[7] D. D. Van Slyke and J. M. Neill, J. Biol. Chem. 61, 523 (1924).

[8] G. S. Adair, J. Biol. Chem. 63, 529 (1925).

[9] F. J. W. Roughton, R. L. Lyster, and A. B. Otis, Proc. R. Soc. London, Ser. B. 144, 29 (1955).

[9a] Various association constants have been termed Adair constants in the literature. Our K_i constants are the same as k_i of Edsall and Wyman. [J. T. Edsall and J. Wyman, Biophys. Chem. 1, 624 (1958).]

[10] D. W. Allen, K. F. Guthe, and F. J. W. Roughton, J. Biol. Chem. 187, 393 (1951).

[11] H. Sick and K. Gersonde, Anal. Biochem. 47, 46 (1972).

[12] D. Dolman and S. J. Gill, Anal. Biochem. 87, 124 (1978).

[13] F. C. Knowles and Q. H. Gibson, Anal. Biochem. 76, 458 (1976).

[14] K. Imai, H. Morimota, M. Kotani, H. Watari, W. Hirata, and M. Kuroda, Biochim. Biophys. Acta 200, 189 (1970).

[15] F. J. W. Roughton, Biochem. J. 29, 2604 (1935).

[16] J. Rifkind and R. Lumry, Fed. Proc., Fed. Am. Soc. Exp. Biol. 26, 2325 (1967).

[17] S. R. Anderson and E. Antonini, J. Biol. Chem. 243, 2918 (1968).

ligand, but showed that fractional saturation and normalized fractional absorbance changes at 430 and 555 nm were the same, within their experimental error. In 1983 Nasuda-Koyama et al.[18] reported that the apparent P_{50} depended on wavelength, particularly in the region of 585 nm. The authors, in agreement with earlier work of Enoki and Tyuma,[19] reported that isosbesty was maintained, however, throughout the oxygenation process. It has become common[20-25] to assume that fractional saturation is directly proportional to fractional absorbance change in fitting for the binding constants or to assume that any differences are of little consequence.

Ackers and co-workers[22-26] have addressed the problem of dimers in equilibrium with tetramers and have developed an expression for the fractional saturation of the total hemoglobin in solution as a function of heme concentration and oxygen activity. This thermodynamic model requires two binding constants to describe ligation to the dimer, four constants for the tetramer, and one additional constant (which can be determined independently) that describes the equilibrium of dimers and tetramers. In an actual equilibrium measurement, however, the absorbances corresponding to infinite and zero oxygen must also be fitted, thus raising the number of fitting parameters to eight. If there are significant wavelength effects, then there may be as many as 6 additional spectrophotometric parameters, raising the total fitting parameters to 14 at a given wavelength. Clearly some simplifications are required, and in principle the easiest involves working at sufficiently high concentrations that dimers are indeed negligible.[21] In general, compromises have always been made in fitting oxygenation data in that dimers have been neglected, metHb formation has been assumed to be negligible, absorbance end points have been assumed, or wavelength effects have been neglected. These matters are discussed below.

Formation of metHb during oxygen equilibria will clearly generate extraneous wavelength effects. Not only may the two chains of human Hb oxidize at different rates, but the metHb chains appear to be nonrandomly

[18] A. Nasuda-Kouyama, H. Tachibana, and A. Wada, J. Mol. Biol. 164, 451 (1983).
[19] Y. Enoki and I. Tyuma, Jpn. J. Physiol. 14, 280 (1964).
[20] K. Imai, "Allosteric Effects in Hemoglobin." Cambridge Univ. Press, Cambridge, 1982.
[21] S. J. Gill, E. Di Cera, M. L. Doyle, G. A. Bishop, and C. H. Robert, Biochemistry 26, 3995 (1987).
[22] F. C. Mills and G. K. Ackers, Proc. Natl. Acad. Sci. U.S.A. 76, 273 (1979).
[23] F. C. Mills, G. K. Ackers, H. T. Gaud, and S. J. Gill, J. Biol. Chem. 254, 2875 (1979).
[24] A. H. Chu, B. W. Turner, and G. K. Ackers, Biochemistry 23, 604 (1984).
[25] M. L. Johnson, B. W. Turner, and G. K. Ackers, Proc. Natl. Acad. Sci. U.S.A. 81, 1093 (1984).
[26] G. K. Ackers and H. R. Halvorson, Proc. Natl. Acad. Sci. U.S.A. 71, 4312 (1974).

distributed between dimers and tetramers,[27] thus complicating any model designed to correct for their presence in order to obtain oxygen-binding constants. Various hemoglobins, particularly those of fish in the T state, are susceptible to autoxidation,[28] some even in the HbCO form, and cannot tolerate even gentle stirring, such as is required for measurement of oxygen activity with an oxygen electrode.

We have developed enzymatic depletion of oxygen in the solution as an alternative to equilibration with a gas phase and have also introduced the use of myoglobin as an oxygen sensor to eliminate stirring altogether.[29] Even if an oxygen electrode is used, enzymatic depletion greatly reduces the vigor with which the solution must be stirred. If metHb formation is still suspected, we have reported tests for such occurrence[29] and have found that the presence of dithiothreitol (DTT) essentially eliminates autoxidation of even myoglobin during the time required for oxygen equilibrium determinations. These gentle enzymatic procedures have allowed us to determine[29,30] that isosbesty is not maintained throughout the deoxygenation process and that the apparent Adair parameters are wavelength dependent in the 500- to 600-nm region of the spectrum. It was reported that, for Hb in concentrated solutions and with observation in the Soret band, that although the isosbestic condition was not maintained, the effect on binding constants was insignificant;[31] however, a more recent study[32] in the same wavelength region concluded that the apparent binding constants were indeed wavelength dependent.

The tracking of absorbance changes over time at the static isosbestic point is best carried out in a dual-beam instrument such as a Cary 218 (Varian Instruments Division, Palo Alto, CA) or equivalent. The wavelength dependence of apparent fractional saturation is easily shown in plots of $\Delta"Y,"$ differences in apparent fractional saturation, generated from diode-array or rapid-scanning spectrophotometry, as described below. Errors in oxygen activity that can distort the equilibrium curves should also be considered. Propagation of random error has been treated elsewhere.[29] In using an oxygen electrode, a procedure to smooth the data using exponential splines[33] has been developed that is appropriate when the deoxygenation process, as carried out enzymatically, can be regarded

[27] L. Cordone, A. Cupane, M. Leone, V. Militello, and E. Vitrano, *Biophys. Chem.* **37**, 171 (1990).

[28] M. Astatke, W. McGee, and L. J. Parkhurst, *Comp. Biochem. Physiol. B* **101B**, 683 (1992).

[29] T. M. Larsen, T. C. Mueser, and L. J. Parkhurst, *Anal. Biochem.* **197**, 231 (1991).

[30] T. M. Larsen, Ph.D. Dissertation, University of Nebraska, Lincoln (1991).

[31] M. L. Doyle, E. DiCera, and S. J. Gill, *Biochemistry* **27**, 820 (1988).

[32] D. W. Ownby and S. J. Gill, *Biophys. Chem.* **37**, 395 (1990).

[33] L. J. Parkhurst and T. M. Larsen, *Comput. Chem.* (to be published).

as piecewise exponential. Such smoothed data can then be incorporated into deconvolution procedures to recover the distortions from the electrode time constant.

Methodology

Hemoglobin is prepared according to the procedure of Geraci et al.[34] At 4° the Hb (2 ml, 2 mM in heme) solution is passed over a 20 × 2 cm diameter G-25 Sephadex (G-25-150; Sigma Chemical Co., St. Louis, MO) column equilibrated with 0.1 M Tris-HCl (T-3253; Sigma)–0.1 M NaCl (pH 7.4) buffer (total Cl⁻, 200 mM) to remove the phosphate from the Hb solution.[35] The pH is measured at 21°. To allow complete removal of phosphate, the flow rate is adjusted for a total elution time of 3 hr.

In these studies, oxygen activities are determined in a 1-cm path length cuvette, using an oxygen electrode, with the heme concentration approximately 100 μM. In some studies to elucidate the wavelength effect, only the time dependence of the oxygen depletion is of interest, and no electrode is employed. If needed, oxygen activity can be estimated from the absorbance changes by reference to solutions in which the activity is directly monitored. In other studies, a thermostatted 1-mm path length cuvette is employed, and the heme concentration is approximately 0.9 mM.

The reactions in the 1-cm path length cuvette are monitored using a modified thermostatted cuvette holder as described elsewhere.[36] The oxygen activity is monitored by a YSI 5331 Clark oxygen probe (Yellow Springs Instruments Co., Yellow Springs, OH) with sensitive membranes (model 5776). The probe is connected to a YSI model 53 oxygen monitor, modified for 10-V output, which is the approximate setting for the initial oxygen readings. The oxygen monitor is interfaced to a microcomputer using an IBM XT/AT-based data acquisition and control adapter (DACA) (Mendelson Elec. Co., Dayton, OH), with the acquisition software written in QuickBasic 4.0 (Microsoft Corp., Redmond, WA). During a deoxygenation run, data collection is continued until the O_2 electrode voltage is constant to within 0.015 V for at least 2 min. The distortion introduced by the time constant of the oxygen electrode is corrected by using linear response theory (see section "Treatment of Oxygen Electrode Data"). The response of the system to a step function is exponential decay, with a value for R of 0.7 sec⁻¹.

The Hb absorbances are monitored by a model 8452A Hewlett-Packard diode array spectrophotometer (Hewlett-Packard Co., Palo Alto, CA) and

[34] G. Geraci, L. J. Parkhurst, and Q. H. Gibson, J. Biol. Chem. **244**, 4664 (1969).
[35] M. Berman, R. Benesch, and R. E. Benesch, Arch. Biochem. Biophys. **145**, 236 (1971).
[36] T. M. Larsen and L. J. Parkhurst, Anal. Biochem. (submitted for publication).

a Cary 210 spectrophotometer (Varian Instruments Division). The diode array spectrophotometer is interfaced to a model A000 286 CompuAdd PC (CompuAdd Corporation, Austin, TX) and run with the HP 89531A MS-DOS–UV/Vis operating software revision A.02.00 (Hewlett-Packard). For a typical 30-min run with the diode array spectrophotometer, 450 O_2 electrode data points are collected with the corresponding 450 absorbances at each even-numbered wavelength from 530 to 590 nm, inclusive. Data collection is continued until the Hb absorbance is constant within 0.00045 (standard deviation) in absorbance. The Cary 210 is interfaced to a microcomputer using an IBM XT/AT-based data acquisition and control adapter with the acquisition software written in QuickBasic 4.0. Typically three wavelengths are monitored for each deoxygenation run. For each wavelength, 5000 absorbances over the 30-min duration of Hb deoxygenation are collected and averaged in blocks of 20 to a final 250 points.

The concentrations of materials present in the Hb solution are as follows: 100 μM Hb; 7.8 mM DTT (Cat. No. 233153; Calbiochem, La Jolla, CA); 1 mg of BSA per milliliter (Cat. No. 126609; Calbiochem); 0.06 μM protocatechuate 3,4-dioxygenase (EC 1.13.11.3) (P-8279; Sigma); 2700 μM protocatechuate (P-5630; Sigma). The buffer is 0.1 M Tris-HCl (200 mM Cl$^-$), pH 7.4. The temperature is 21°. (For the wavelength studies at low concentration in Fig. 4A and B, the heme concentration is 56 μM.) The deoxygenation is initiated by the rapid addition and mixing of 50 μl of 124 mM protocatechuate. For the wavelength studies at 0.9 mM in heme, the concentration of protocatechuate is 9.5 mM. (Although DTT was needed for the earlier work with Mb and the glucose oxidase system,[29] it is not essential for studies without Mb and is not included in all runs.)

Treatment of Oxygen Electrode Data

The oxygen electrode data are smoothed by exponential splines and then deconvoluted. Least-squares exponential splines (LSES) were developed to fit the oxygen electrode time-dependent decay curves. The splines are sums of three exponentials or two exponentials plus a constant. The constraints that the splines and their first and second derivatives match at the joining points ("knots") result in piecewise smooth fits to the data. A Simplex algorithm is used to find the best fitting two or three decay constants over each segment, subject to the constraints on the linear parameters. The length of each fitting segment is increased until statistical criteria for goodness of fit fail. These criteria involve both an F test and a runs analysis. The LSES fits are superior to those obtained from polynomial filters,[37] especially for noisy data. This superiority is clearly evident in derivative curves, which lack the characteristic oscillations generated by polynomial filters. The Adair constants derived from the LSES-smoothed simulated noisy data are significantly better than those obtained from polynomial filtering of the same data.[33] Details of the LSES equations can be found elsewhere.[33] The number of exponential splines used in smoothing a data set of 450 points with 4-sec intervals is approximately 30. The exponential spline-smoothed data are then deconvoluted as follows.

The distortion introduced by the time constant of the oxygen electrode is corrected by using linear response theory. The oxygen electrode is subjected to a sudden and complete removal of oxygen by the rapid injection and mixing of 50 μl of a 5% (w/w) sodium dithionite solution (Manox brand; Holdman and Hardman, Miles Platting, Manchester, England) into 2.3 ml of air-equilibrated buffer monitored by the electrode. By injecting dithionite we simulate subjecting the system to a scaled complement of the Heaviside unit step function (USF),[38] that is, to a function that has a constant value for times less than time zero, and is zero for times after time zero. The response of the system was exponential decay, with a value for R (see below) of 0.7 sec^{-1} for the YSI-sensitive membrane. [For the standard membrane (model 5775), a typical value is 0.1 sec^{-1}]. Although it is possible to carry through various limiting operations for dealing with the complement of the USF, it is more convenient to regard the actual deoxygenation as corresponding to the USF, in which case the system response is an exponential complement. One then transforms back when required to recover the actual decaying voltage, $V_{(t)}$. Let $g(s)$ be the Laplace transform of the time domain signal $W(t)$, $g(s) = \mathcal{L}[W(t)]$. $W(t)$ is a normalized voltage output:

$$W(t) = \frac{V_1 - V_{(t)}}{V_1 - V_\infty} \tag{1}$$

where V_1 is the initial voltage, and V_∞ is the voltage following injection of 10 μl of 1% dithionite to obtain the voltage offset for zero oxygen, following a deoxygenation run.

The transfer function $G(s)$ of the instrument is then given by Eq. (2):

$$G(s) = \frac{g_{\text{output}}(s)}{g_{\text{input}}(s)} = \frac{\mathcal{L}(1 - e^{-RT})}{\mathcal{L}(\text{USF})} = \frac{R}{(s + R)} \tag{2}$$

[37] A. Savitzky and M. J. E. Golay, *Anal. Chem.* **36,** 1627 (1964).
[38] E. J. Berg, "Heaviside's Operational Calculus," p. 1. McGraw-Hill, New York, 1936.

where R is the reciprocal time constant for the oxygen electrode. In the time domain, $G(t)$ is merely the time derivative of the instrument response to the USF. Inversion of the Laplace transform gives the familiar expression

$$W(t)_{output} = \int_0^t [Re^{-R(t-\tau)}]W(\tau)_{input} \, d\tau \tag{3}$$

a Volterra integral equation of the first kind. The kernel of this equation is $G(t, \tau)$, more commonly denoted by $K(t, \tau)$, which, for our system, is $R \exp[-R(t - \tau)]$. Symbolically, we can write the above equation as

$$W(t)_{output} = K * W(t)_{input} \tag{4}$$

where the asterisk ($*$) denotes a convolution. Following Volterra,[39] we can integrate this equation by parts, obtaining

$$W(t)_{output} = K(t, \tau)\theta(t) - \int_0^t \frac{\partial K(t, \tau)}{\partial t} \theta(\tau) \, d\tau \tag{5}$$

where

$$\theta(t) = \int_0^t W(t)_{input} \, dt \tag{6}$$

This leads to the following Volterra equation of the "second" kind:

$$\frac{W(t)_{output}}{K(t, \tau)} = \theta(t) - \int_0^t \frac{\partial K(t, \tau)/\partial \tau}{K(t, \tau)} \theta(\tau) \, d\tau \tag{7}$$

which in our case is just Eq. (8),

$$\frac{W(t)_{output}}{R} = \theta(t) - K * \theta \tag{8}$$

which can be rearranged to Eq. (9):

$$\theta = \frac{W_{output}}{R} + K * \theta \tag{9}$$

This equation can be solved by the method of successive substitutions, giving for the nth iteration:

$$\theta_n = \frac{W_{output}}{R} + \frac{1}{R}\sum_{j=1}^n K^j * W_{output} \tag{10}$$

---

[39] V. Volterra, "Theory of Functionals and of Integral and Integro-Differential Equations," p. 53. Dover, New York, 1959.

a series that can be shown[40] to converge absolutely and uniformly to the continuous function θ(input). The expression $K^j * W$ is understood to denote application of the convolution operator j times. One stops the iteration when $\theta_n - \theta_{n-1}$ is within the required error bounds for all required times. One then obtains $W(t)_{\text{input}}$ from

$$W(t)_{\text{input}} = \frac{d\theta_n(t)}{dt} \tag{11}$$

In practice, this procedure appeared, in terms of rate of convergence, equivalent to but somewhat less convenient than the following, for which convergence may not generally be obtained. P. van Cittert[41] suggested a two-cycle procedure for deconvoluting the effect of a finite slit width on the intensities of spectral lines, a problem that leads to a Fredholm integral equation of the first kind, from minus infinity to plus infinity. This can be made equivalent to our Volterra integral equation by choosing the kernel $K(t, \tau)$ to be 0 for both $\tau > t$ and $t < 0$. The suggestion of van Cittert can be generalized to obtain an iterative procedure as follows:

$$F_1 = W_{\text{output}}, \quad F_{n+1} = F_1 - K * F_n + F_n \tag{12}$$

where iteration continues until the difference between F_{n+1} and F_n is negligible. This deconvolution procedure can be written in compact form in terms of F_1, the observed output, as

$$F_n \sum_{j=1}^{n} \binom{n}{j} (-1)^{j-1} K^{j-1} * F_1 \tag{13}$$

where $\binom{n}{j}$ is the binomial coefficient.

In practice, Eq. (12) was implemented, and for a value of $R = 0.7$ sec^{-1} and our operating conditions, convergence was achieved at $n = 8$.

Simpson's rule is used for numerical integration in the deconvolution procedure. After each iteration, every even-numbered data point and its corresponding time point is lost in the integration. A five-point (fourth-order) shifting Lagrange interpolation routine is used to regain those data points. For integration from t to 0, a cut-off requiring $R(t - \tau) < 30$ is implemented, because for values of τ leading to $\exp[-(t - \tau)] < \exp(-30)$, the kernel is so small that the contribution to the integral of Eq. (3) becomes negligible.

[40] W. V. Lovitt, "Linear Integral Equations," p. 13. Dover, New York, 1950.
[41] P. H. van Cittert, Z. Phys. **69**, 298 (1931).

After the deconvolution procedure is completed, the oxygen activity at each of the original 450 time points was

$$X_i(t) = X_1[1 - F_n(t)] \tag{14}$$

where $F_n(t)$ denotes the deconvoluted normalized data vector W, and X_1 denotes initial oxygen activity under ambient conditions. The initial oxygen activity is calculated from solubility tables[42] using Henry's law, correcting for the barometric pressure and the pressure of the water vapor by using readings of the relative humidity from a wet bulb–dry bulb thermometer and a table of the vapor pressure of water.

Equations for Apparent Fractional Saturation

From Beer's law one has

$$A = l\Sigma(\text{Hb}X_i)\varepsilon_i \tag{15}$$

where A is the absorbance of the Hb solution at oxygen activity X, ε_i is the molar absorptivity of the species with i bound oxygen molecules, and l is the path length, set equal to 1 in what follows. We assume the effects of dimers are negligible and that the partition function can be represented by the binding polynomial D:

$$D = 1 + \beta_1 X + \beta_2 X^2 + \beta_3 X^3 + \beta_4 X^4, \qquad \beta_j = \prod_{i=1}^{j} K_i \tag{16}$$

where the K values are Adair constants.

When the O_2 activity is zero, the equation for absorbance, A_0, is

$$A_0 = P_T\varepsilon_0 \tag{17}$$

where P_T is total protein and is

$$P_T = [\text{Hb}]D \tag{18}$$

When the O_2 activity is infinite (complete saturation of the Hb), the equation for absorbance, A_∞, is

$$A_\infty = P_T\varepsilon_4 \tag{19}$$

At any X:

$$A = [\text{Hb}] \sum_{i=0}^{4} \varepsilon_i\beta_i X^i = \left(\frac{P_T}{D}\right) \sum_{i=0}^{4} \varepsilon_i\beta_i X^i \tag{20}$$

[42] C. D. Hodgman, ed., "Handbook of Chemistry and Physics," 41st ed., p. 1706. Chem. Rubber Publ. Co., Cleveland, OH, 1960.

where $\beta_0 = 1$, and the *apparent* fractional saturation, "Y," at ligand activity X, is

$$"Y" = \frac{A - A_0}{A_\infty - A_0} = \left(\frac{1}{D}\right) \sum_{i=1}^{4} \left(\frac{\varepsilon_i - \varepsilon_0}{\varepsilon_4 - \varepsilon_0}\right) \beta_i X^i \tag{21}$$

If the spectrophotometric assumption were correct, the summation term in parentheses would equal $i/4$. If the spectrophotometric assumption does not hold, then a new dimensionless parameter, E_i, can be defined as the ratio of actual fractional absorptivity change after the binding of i ligands to that in the ideal case:

$$E_i = \frac{4}{i} \left(\frac{\varepsilon_i - \varepsilon_0}{\varepsilon_4 - \varepsilon_0}\right) \tag{22}$$

and "Y" can be written as Eq. (23):

$$"Y" = \frac{E_1\beta_1 X + 2E_2\beta_2 X^2 + 3E_3\beta_3 X^3 + 4E_4\beta_4 X^4}{4D} \tag{23}$$

where $E_4 = 1$. If all E values $= 1$, then "Y" is equal to the actual fractional saturation. The apparent fractional saturation in Eq. (21) measured at wavelength j will be denoted by "Y_j" with the corresponding E_{ij} coefficients. If the difference in fractional absorbance change is taken for two wavelengths j and k, then we can define a quantity $\Delta"Y"$ as

$$\Delta"Y" = \frac{(E_{1j} - E_{1k})\beta_1 X + 2(E_{2j} - E_{2k})\beta_2 X^2 + 3(E_{3j} - E_{3k})\beta_3 X^3}{4D} \tag{24}$$

which must equal zero at infinite and zero ligand activity. There may be one or two additional real and positive zeroes that derive from the numerator of Eq. (24).

Data Fitting

For fitting data at single wavelengths Eq. (25) (nine parameters) was used in a nonlinear least-squares Simplex[43] program to minimize the variance. The absorbance is

$$A = A_0 + \frac{(A_\infty - A_0)N'}{4D} \tag{25}$$

---

[43] J. H. Noggle, "Physical Chemistry on a Microcomputer," p. 148. Little, Brown, Boston, 1985.

where

$$N' = \sum_{i=1}^{4} iE_i\beta_i X^i \qquad (26)$$

with $E_4 = 1$. Typically 2000 iterations were sufficient for convergence. For each data set a number of different initial points in parameter space were selected. For global fitting over 5 wavelengths, 29 variables were included in the fitting algorithm; however, the 4 Adair constants (K_1, K_2, K_3, K_4) were required to be invariant with wavelength and the wavelength dependence was determined by the E values. If all E values are set equal to 1 in Eqs. (21)–(23), the familiar equation that assumes fractional saturation is equal to the normalized fractional absorbance change is obtained. That equation has six adjustable parameters. A global fitting over 5 wavelengths with this model, requiring the Adair constants to be invariant, has 14 variables. For the global fittings, typically 10,000 iterations were required for convergence for each initial starting parameter set.

Equation (25) has six fitting parameters if all E values are set equal to one. By fitting absorbance data at the extremes of oxygen (X), one can obtain four of these parameters $(K_1, K_4, A_0, A_\infty)$. The Hill numbers (n^*) at $\underline{X}$ and $\overline{X}$ provide two additional quantities from which to extract K_2 and K_3. The quantity $\overline{X}$ is oxygen activity corresponding to half-saturation and is used here in preference to P_{50}. Gill et al.[44] have shown how to obtain K_2 and K_3 from $\overline{X}$, n^*, K_1, and K_4. These equations assume that one has true values for $\overline{X}$ and n^* and not apparent values from "Y". The Gill equations, in terms of our K values, are

$$K_2 = \frac{J_1 - J_2}{J_3} \qquad (27)$$

$$K_3 = \frac{J_4}{\overline{X}(J_1 - J_2)} \qquad (28)$$

where

$$J_1 = 4 + \frac{12}{K_1\overline{X}} + 12K_4\overline{X} + \frac{32K_4}{K_1} \qquad (29)$$

$$J_2 = \frac{n^*}{2}\left(8 + \frac{12}{K_1\overline{X}} + 12K_4\overline{X} + \frac{16K_4}{K_1}\right) \qquad (30)$$

$$J_3 = (2 + 4K_4\overline{X})n^*\overline{X} \qquad (31)$$

[44] S. J. Gill, H. T. Guad, J. Wyman, and B. G. Barisas, Biophys. Chem. 8, 53 (1978).

$$J_4 = \left(2 + \frac{4}{K_1\bar{X}}\right) n^*$$ (32)

The correct limiting equations (including E values) are, for low and high oxygen activities, respectively:

$$A = A_0 + [A_\infty - A_0](S_1X + C_1X^2 + \cdots)$$ (33)

where $S_1 = E_1K_1/4$, and $C_1 = (2E_2K_1K_2 - E_1K_1^2)/4$, and

$$\frac{1}{(A - A_0)} = \left(\frac{1}{A_\infty - A_0}\right)\left[1 + \frac{S_2}{X} + \left(\frac{C_2}{X}\right)^2\right]$$ (34)

where $S_2 = (4 - 3E_3)/(4K_4)$, and $C_2 = [2\beta_2(2 - E_2) - 3E_3\beta_3(4 - 3E_3)]/(4K_4)$.

Because there are actually nine parameters to be determined, one needs nine independent descriptions of the apparent binding curve, and these might be the limiting intercepts, slopes, and curvatures, given above, as well as the midpoint oxygen activity, and the slope and second derivative of "Y" at that point. There appears to be little advantage in fitting the data for human Hb in this manner. We do make use of some of these equations below in assessing wavelength effects on Adair constants, however.

Results and Discussion

Not all hemoglobins show wavelength effects such as those discussed here. For instance, the cyanomet ferrous hybrids[30,45] show isosbestic points in the 580-nm region that remain constant during deoxygenation. The tetrameric hemoglobins of the turtle *Pseudemys* show large changes in oxygen affinity with pH and also a strong tendency toward autoxidation. Once the latter effects are eliminated, however, the wavelength dependencies of fractional saturation appear to be minor.[46]

The extent to which dimers in equilibrium with tetramers complicate the analysis of oxygen binding depends on the ratio of heme concentration (H) to the tetramer–dimer dissociation constant (K_{TD}), on the existence of cooperativity in dimers, and on the difference between the intrinsic affinities of tetramer and dimer for the last oxygen to be bound. If we let Z_4 denote the binding polynomial for tetramers [D in Eq. (16)], and Z_2

[45] L. J. Parkhurst, K. M. Parkhurst, and T. M. Larsen, *J. Biol. Chem.* (submitted for publication).

[46] R. C. Steinmeier, T. M. Larsen, and L. J. Parkhurst, in preparation.

the corresponding polynomial for dimers (with β_{1D} and β_{2D} as the corresponding coefficients) and $Z' = dZ/dX$, (X is ligand activity), an expression can be written for fractional saturation Y in terms of heme concentration (H), and the tetramer–dimer equilibrium constant, K_{TD}, by defining J as

$$J = [(1 + 4\beta_{2D}^2 Z_4 H)/(K_{TD}\beta_4 Z_2^2)]^{1/2} \tag{35}$$

and then,

$$Y = \{[4Z_2 Z_4 + Z_4 Z_2(J - 1)](X)\}/[4Z_2 Z_4(J + 1)] \tag{36}$$

an expression equivalent to that of Ackers and Halvorson.[26] In one study[47] the kinetics of HbCO were followed to low concentrations of CO and hemeprotein. The assumption was that dimers were always quickly reacting toward CO. Any slowly reacting dimers would have distorted the fitting to give a lower value for K_{TD}. The value for K_{TD}, however, was in remarkable agreement with that found for HbO_2 by gel filtration.[48] Additional slowly reacting dimers (implying cooperativity) would have been detected unless the R → T conformational change was *very* much slower than that for the tetramers. Earlier flow-flash results[49] provide further evidence that little cooperativity occurs in the $\alpha_1\beta_1$ dimers, allowing us to set, for the dimers, $K_1 = 4K_2$, or $\beta_{1D} = 2(\beta_{2D})^{1/2}$. If there is no tetramer enhancement, then K_2 for the dimer $= 2K_4$ for the tetramer. A reinvestigation of the equilibrium data by Straume and Johnson[50] concluded that tetramer enhancement may be as low as 0.18 kcal/mol. Kinetic studies analyzed by Gibson and Edelstein[51] called into question the occurrence of tetramer enhancement. Kinetic studies by Lary and Philo[52] on hemoglobin and in our laboratory on cyanomet hybrids of human hemoglobin[45,53] show that tetramer enhancement is most likely less than 0.16 kcal/mol and may be no more than about one-third that value, or well within experimental error. It must be emphasized that the original oxygen equilibrium data[22,24,54] were not analyzed to include wavelength effects. Assuming that tetramer enhancement is negligible, the consequence of studying Hb equilibria at 100 μM in heme compared to an "infinite" concentration of

[47] K. D. Martin and L. J. Parkhurst, *Anal. Biochem.* **186**, 288 (1990).
[48] G. K. Ackers, M. L. Johnson, F. C. Mills, and S. H. C. Ip, *Biochem. Biophys. Res. Commun.* **69**, 135 (1976).
[49] Q. H. Gibson and L. J. Parkhurst, *J. Biol. Chem.* **243**, 5521 (1968).
[50] M. Straume and M. L. Johnson, *Biophys. J.* **56**, 15 (1989).
[51] Q. H. Gibson and S. J. Edelstein, *J. Biol. Chem.* **262**, 516 (1987).
[52] J. W. Lary and J. S. Philo, *J. Biol. Chem.* **265**, 139 (1990).
[53] K. M. Parkhurst, Ph.D. Dissertation, University of Nebraska, Lincoln (1991).
[54] F. C. Mills, M. L. Johnson, and G. K. Ackers, *Biochemistry* **15**, 5350 (1976).

heme was investigated computationally as follows. We used a value of 0.995 μM for K_{TD} and values of 0.0263, 0.0867, 0.044, and 0.711 for K_1 to K_4, respectively, in reciprocal micromolar units in Eq. (36) (values were from a data set obtained using an oxygen electrode[55]). We then analyzed the synthetic data according to a simple binding polynomial as if there were no dimers present. The data could be analyzed in a number of ways. First, consider the use of the Gill equations [Eqs. (27)–(32)] and measurements at the extremes and at the center of the binding isotherm. At infinite oxygen levels, the limiting slope gave a value for K_4 of 0.711, in exact agreement with the input data. For oxygen-saturated solution, the values were 0.663 and 0.667 for no dimer present and for the 100 μM heme solution, respectively, a difference of less than 1%. At 0.1 μM in oxygen, the limiting slopes gave values of 0.0302 and 0.0346, for the same respective heme concentrations, a difference of about 15% from each other, and both larger than the true value owing to the slight curvature in the Y vs X plots even at 0.1 μM in oxygen. These values for K_1 and K_4 were then used, together with calculated values for $\bar{X}$ and the Hill number, to obtain estimates for K_2 and K_3. For infinite heme, the values for K_2 and K_3 were, respectively, 0.0665 and 0.0544, and for 100 μM in heme, they were 0.0770 and 0.0483, respectively, differences of 15 and 12%. The synthetic 100 μM curve was also analyzed by nonlinear least squares for the four Adair constants. Y (100 points) varied from 0.9997 ($X = 1250$ μM) to 0.012 ($X = 1.3$ μM), with the X values decreasing exponentially between the two values. There were five runs of residuals in the fits, owing to the difference between the simulated Y and an incorrect theoretical Y based on only four Adair constants and a binding polynomial representation of the binding. The root mean square (rms) residual was 0.00062 in Y, the maximum was 0.0017 ($Y = 0.14$) and that at $Y = 0.5$ was 0.0011. [This difference is to be compared with those for $\Delta``Y"$ (Figs. 3 and 4) for the wavelength effect.] For a total change in absorbance of 0.4, the largest residuals in absorbance would be about 0.0004, and we conclude, subject to the assumptions above regarding dimers, that differences between the actual curve and that represented by a binding polynomial would be lost in the noise of current spectrophotometric measurements. For the above curve, the values extracted for the four Adair constants were 0.0335, 0.0920, 0.0292, and 0.973, for K_1 to K_4, respectively, values that differed from the values for the tetramer alone (infinite heme case) by factors of 1.3, 1.06, 0.66, and 1.4. In this instance, a better value for K_4 would have been obtained from the limiting slope of a double-

[55] T. M. Larsen, Ph.D. Dissertation, p. 96. University of Nebraska, Lincoln (1991).

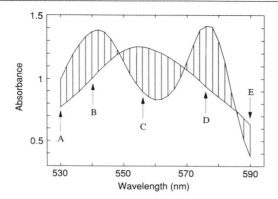

FIG. 1. Absorbances as a function of wavelength for each of 450 times at 31 wavelengths from 530 to 590 nm during the deoxygenation of HbO_2 at 21°, pH 7.4, 0.1 M Tris-HCl (200 mM Cl$^-$). The wavelengths marked by arrows are those used in the fitting procedures to obtain binding constants. Because the nominal wavelengths do not correspond to the maximum of the α peak, the α/β peak ratio appears lower than for a scan of the same HbO_2 in a Cary spectrophotometer.

reciprocal plot than from fitting the entire binding curve. Imai[56] has pointed out that except for the apparent K_1 (which Imai found to differ from that for infinite heme concentration by 38%), the concentration dependencies of the Adair constants are *less* than their experimental errors for hemoglobin concentrations > 60 μM in heme. The value of any of the constants depends on the actual distribution of the data points with respect to the sensitivity coefficient for that parameter. We conclude from these simulated results as well as from comparisons of results between different laboratories, that the effects of variations in apparent fractional saturation with wavelength are of greater consequence than are effects arising from the presence of dimers for human hemoglobin at pH 7.4 and for heme concentrations equal to or greater than 100 μM. The matter is made even more complex, however, because the wavelength effects appear to have a concentration dependence (Figs. 3 and 4), which, if the spectrophotometric parameters are neglected, can lead one to overestimate the importance of concentration effects on Adair constants.

Figure 1 shows absorbance as a function of time at the 31 wavelengths monitored during the deoxygenation process. Absorbances for initial

[56] K. Imai, "Allosteric Effects in Hemoglobin," p. 81. Cambridge Univ. Press, Cambridge, 1982.

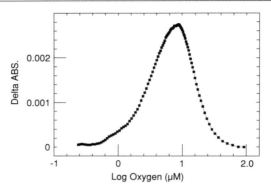

FIG. 2. The absorbance change for an identical sample of Hb obtained from measurements in an automated Cary 210 spectrophotometer at the overall isosbestic point for HbO_2 − Hb. Data were collected at 585.9 and 586.0 nm and were interpolated to obtain the data shown. The calculated isosbestic point was 585.93 nm.

and final times are connected by straight line segments and show, respectively, absorption spectra for very nearly 100% HbO_2 and Hb. The 450 data points collected at each wavelength are represented by single points, but at the resolution of Fig. 1 they appear as vertical lines. The five wavelengths used for the global fitting are indicated in Fig. 1 as points A–E.

Figure 2 shows absorbance changes at the overall isosbestic point determined in the same cell but using a Cary 210 spectrophotometer for data collection. The measurements were made at 585.9 and 586.0 nm and were interpolated to give the curve for 585.93 nm, that wavelength at which the initial and final absorbances were identical.

Figure 3A and B shows the differential apparent fractional saturations at various pairs of wavelengths from 530 to 590 nm for the 0.9 mM heme solution, and Fig. 4A and B shows corresponding figures for 56 μM heme. The patterns are consistent from run to run, but, owing to the sensitivity to end-point absorbances, there is some variation in the y-axis values. Time is plotted rather than X. For Fig. 3A and B, half-saturation is at 140 sec; for Fig. 4A and B, half-saturation is at 270 sec. The curves labeled K are for spectra that differ by only 2 nm and are expected to be nearly zero. They are included only to give an indication of the intrinsic noise in such plots. For similar curves, plotted vs X, for the five wavelengths used in the global fitting, the data can be fit well[57] according to Eq. (24) above using the globally fitted Adair constants and the "local" E values.

[57] T. M. Larsen, Ph.D. Dissertation, p. 116. University of Nebraska, Lincoln (1991).

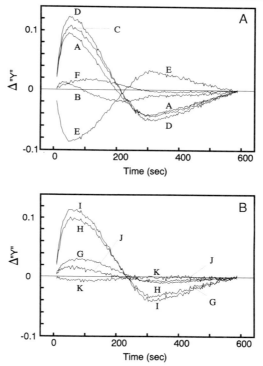

FIG. 3. (A and B) Plots of Δ"Y," differences of apparent fractional saturation, for the deoxygenation of hemoglobin, measured on a sample 0.9 mM in heme. The wavelength differences (nm) are as follows: (A) A, 590–576; B, 590–556; C, 590–540; D, 590–530; E, 576–556; F, 576–540; (B) G, 576–530; H, 556–540; I, 556–530; J, 540–530; K, 558–556.

Figures 3 and 4 show clearly that the spectrophotometric assumption (all $E = 1$) is untenable. As we have discussed in detail elsewhere[29,30] these spectroscopic effects cannot derive from formation of Hb$^+$ because there is no detectable Hb$^+$ generated during the deoxygenation process and, furthermore, such changes for several of the wavelengths would be in a direction opposite to that found.

Wavelength effects in Hb that may relate to the above effects are well known. In 1959 Gibson[58] reported a difference in the Soret region between deoxy forms of the R and T states. In other work[59,60] spectroscopic differences between the hemes of the α and β chains were used to elucidate

[58] Q. H. Gibson, *Biochem. J.* **71**, 293 (1959).
[59] Q. H. Gibson, L. J. Parkhurst, and G. Geraci, *J. Biol. Chem.* **244**, 4668 (1969).
[60] J. S. Olson, M. E. Anderson, and Q. H. Gibson, *J. Biol. Chem.* **246**, 5919 (1971).

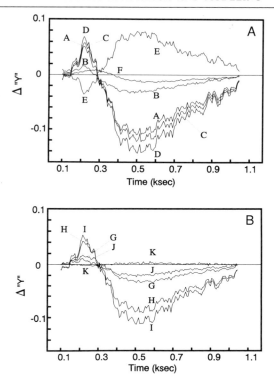

FIG. 4. (A and B) Plots of Δ"Y," differences of apparent fractional saturation, for the deoxygenation of hemoglobin, measured on a sample 56 μM in heme. The wavelength differences (nm) are as follows: (A) A, 590–576; B, 590–556; C, 590–540; D, 590–530; E, 576–556; F, 576–540; (B) G, 576–530; H, 556–540; I, 556–530; J, 540–530; K, 558–556.

details of ligand kinetics in Hb. Adams and Schuster[61] have reported that T-state and R-state oxyHb differ with a maximum $\Delta\varepsilon$ of 6.9 mM^{-1} cm^{-1} at 584.5 nm. This effect would be expected to be a major contributor to E_1, because E_1 should reflect the presence of a significant population of T-state oxygenated molecules. On the other hand, Adams and Schuster[61] reported an isosbestic point for their effect at 546 nm, where we find that E_1 is significantly different from unity. Work in our laboratory on α and β cyanomet valency hybrids has shown that the isosbestic points in the region of 585 nm differ by 3 nm. The macroscopic fitting E values can be expected to reflect both R and T states as well as chain differences associated with ligation.

[61] M. L. Adams and T. M. Schuster, *Biochem. Biophys. Res. Commun.* **58,** 525 (1974).

TABLE I
OXYGEN-BINDING PARAMETERS[a]

Parameter	Global (nine parameters)	Nine parameters	Global (six parameters)	Six parameters
K_1 (μM^{-1})	0.028	0.025 ± 0.002 (0.023–0.027)	0.079	0.091 ± 0.007 (0.080–0.104)
K_2 (μM^{-1})	0.095	0.097 ± 0.013 (0.078–0.117)	0.038	0.022 ± 0.006 (0.013–0.030)
K_3 (μM^{-1})	0.036	0.042 ± 0.004 (0.034–0.049)	0.016	0.043 ± 0.028 (0.022–0.107)
K_4 (μM^{-1})	0.752	0.718 ± 0.018 (0.70–0.74)	2.86	2.17 ± 0.49 (1.06–2.64)
σ_{fit} (Abs)	0.0009	0.0008	0.0019	0.0011

[a] Oxygen-binding parameters and the σ fit (RMS deviation in absorbance) values are reported for four types of fitting. These procedures include global fitting according to the 9- and 6-parameter models at 5 wavelengths, a total of 29 and 14 parameters, respectively, and the average values of the Adair constants for fitting individual wavelengths to the same two models. For the global fitting, convergence from various starting points was to the same point in parameter space within the number of significant figures shown. The numbers in parentheses give the range of the parameter. The standard error of the mean is also given for the single-wavelength fits. These data pertain to the oxygenation of human Hb at 21°, in 0.1 M Tris-HCl–0.1 M NaCl, pH 7.4 (200 mM total Cl$^-$). The five wavelengths employed for the fittings were 590, 576, 556, 540, and 530 nm.

Brand and colleagues[62–65] have developed "global fitting" to extract fluorescence lifetimes from complex fluorescence decay processes. In our application of the procedure we have also found rapid convergence of the variance to a minimum and have found that the simultaneous fitting of data at several wavelengths, when spectrophotometric coefficients are included, gives rapid convergence for the global parameters, the Adair constants. The global fits for the nine-parameter model were in excellent agreement with the individual fits for the same model and K_4 was always less than the kinetic limit (Table I). The Δ"Y" values shown in Fig. 4 were in general less than 2% at the half-saturation point. Nevertheless this small difference can have major effects on the values obtained for the fitted Adair constants (Table I). It is not sufficient, however, merely to include data at a number of wavelengths in a global fitting routine, because for the minimization to be effective, the E values at the various

[62] J. R. Knutson, J. M. Beechem, and L. Brand, *Chem. Phys. Lett.* **102**, 501 (1983).
[63] J. M. Beechem, J. R. Knutson, J. B. Alexander Ross, B. W. Turner, and L. Brand, *Biochemistry* **22**, 6054 (1983).
[64] D. G. Walbridge, J. R. Knutson, and L. Brand, *Anal. Biochem.* **161**, 467 (1987).
[65] M. K. Han, J. R. Walbridge, J. R. Knutson, L. Brand, and S. Roseman, *Anal. Biochem.* **161**, 479 (1987).

wavelengths must differ by other than scale factors. For this purpose, the absorbance changes in the 530- to 590-nm region are ideal, but compromises must be made in optical path length and heme concentration.

The E values in Eq. (23), being the ratio of two terms [Eq. (22)], will have singularities at the HbO_2–Hb isosbestic wavelengths because it is at those wavelengths that the denominator term in the E values is zero. The term in the numerator is the difference of two molar absorptivities, which can be decomposed further into two differences. One difference is that of the i ligated hemes in HbX_i and i unligated hemes in Hb, and the other is that of the $n - i$ unligated hemes in HbX_i and $n - i$ unligated hemes of Hb. The molar absorptivity of the i ligated hemes (and $n - i$ unligated hemes) in HbX_i is in turn a weighted average of molar absorptivities for the various microstates having i ligated and $n - i$ unligated hemes, the weighting factor being the mole fraction of that microstate. Simulations based on the Adams–Schuster effect,[61] on having oxy–deoxy difference spectra for the α- and β-hemes differing by up to 3 nm, and on a 15% difference of absorptivities for Hb in T and R conformations, gave values for E values that lay between 1.5 and 0.5 for wavelengths no closer than 3 nm to the overall HbO_2–Hb isosbestic wavelength.[30] The values for E values from the global fitting are all between 0.8 and 2.5. At this stage of development, the error in the E values (ca. 20%) is such that they should be regarded as necessary global fitting parameters and not yet as useful for extracting more detailed information on populations of the conformers that comprise a given ligated form.[30]

The fitting is summarized in Table I, where the parameter means, the standard deviations of the means, and the ranges are shown for the five wavelengths used in the global fitting. For the nine-parameter fits the values for the four Adair constants are within one standard deviation of those determined using Mb as an O_2 sensor.[29] There is little difference between σ_{fit} for the global nine-parameter fitting and the average σ_{fit} for the individual nine-parameter fits. There is nearly a twofold difference, however, between the global and the average individual σ_{fit} values for six parameters, which also points to the need to include additional spectroscopic parameters in the absorbance fitting. In general the Adair constants determined by fitting individual wavelengths showed the greatest variations with wavelength for the six-parameter model, with the largest variations found in K_4 and K_3. A value for K_1 ($0.04 \mu M^{-1}$) can be calculated from four rate constants[66] determined in phosphate buffer. We have obtained a corresponding equilibrium value of $0.025 \mu M^{-1}$ for Hb in $0.1 M$ phosphate, pH 7.4, by nine-parameter fitting, which suggests that the values of K_1

[66] C. A. Sawicki and Q. H. Gibson, *J. Biol. Chem.* **252**, 7538 (1977).

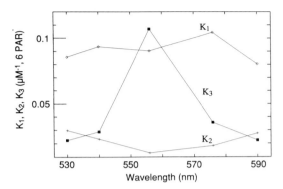

FIG. 5. A plot of the Adair constants K_1, K_2, and K_3 (μM^{-1}) vs wavelength, obtained from fitting the data at the five individual wavelengths by the six-parameter model. $(\diamond)\ K_1$; $(+)\ K_2$; $(\blacksquare)\ K_3$.

shown in Fig. 5 are too high. The Adair constant K_3 is strongly correlated with the constant K_4 and is expected to show large variations that reflect variation in K_4, and that is evident in a comparison of Figs. 5 and 7. Figure 6 shows the same Adair constants, K_1 through K_3, obtained from fitting the five wavelengths with the nine-parameter model. As can be seen, these constants have small variation and their order does not change with wavelength.

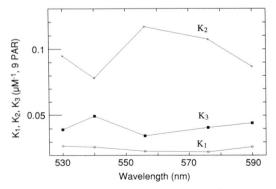

FIG. 6. A plot of the Adair constants K_1, K_2, and K_3 (μM^{-1}) vs wavelength, from fitting according to the nine-parameter model. $(\diamond)\ K_1$; $(+)\ K_2$; $(\blacksquare)\ K_3$.

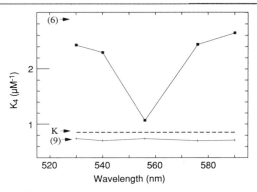

FIG. 7. Values for K_4 obtained for each wavelength individually according to the six-parameter (■) and nine-parameter (+) models. The arrows designated (6) and (9) denote values for K_4 obtained from global fittings according to the six- and nine-parameter models, respectively. The dashed line, denoted "K", is the kinetically determined upper limit for K_4.

Figure 7 shows values of K_4 obtained by different fitting procedures. The solid squares depict values obtained at each of the five wavelengths with a six-parameter fitting and the plus signs depict the values at the same wavelengths obtained with a nine-parameter fitting. By measuring the rate of O_2 release in a relaxation experiment, a rate constant can be obtained for O_2 dissociation that should be appropriate for the O_2 dissociation rate constant that can be associated with K_4. An upper limit for the association constant for triliganded hemoglobin is obtained by photolyzing HbO_2 at low percent photolysis. The ratio of these two rate constants must provide an upper bound for the Adair constant K_4.[67] Measurements made in our laboratory and elsewhere[45,52,67] for Hb in Tris buffer yield an upper limit of 0.85 μM^{-1} and that is shown by the heavy dashed line marked by the arrow designated "K" in Fig. 7. All of the values for K_4 obtained from the six-parameter model lie above this limit and must be incorrect. Values below this kinetic limit reflect, in a simple two-state allosteric model, the extent to which the T state contributes to triliganded Hb and affects the association constant for that species.

For data collected over the wavelength range 530 to 590 nm, the wavelength effects can be assessed as follows. From the global analysis of the equilibrium data over five wavelengths, the best fitting K values were used to fit the remaining wavelength data in order to obtain E values. These parameters were then used to generate limiting slopes and apparent values for $\overline{X}$ and the Hill number for use in the Gill relations [Eqs. (27)

[67] K. M. Parkhurst, Ph.D. Dissertation, p. 143. University of Nebraska, Lincoln (1991).

to (32)]. The ratios of "true" Adair constants to the best-fitting apparent constants ranged as follows (we discarded wavelengths where the apparent K_4 was less than 0); K_1, 0.26–0.53; K_2, 0.32–0.61; K_3, 1.2–3.5; K_4, 0.6–2.6. Over the wavelength range 530–590 nm, the first two Adair constants will be overestimated and K_3 will be underestimated if wavelength effects are excluded in this type of fitting. More widely ranging differences were generally found, however, when the entire curves were fit by six parameters at the five selected wavelengths. In this case (Table I) the ranges for the ratios of globally fitted constants (nine-parameter fits) to those at individual wavelengths for the six-parameter fits were as follows: K_1, 0.35–0.37; K_2, 3.2–7.2; K_3, 0.34–1.6; K_4, 0.71–0.29. For the six-parameter model global fit, K_1 and K_4 in particular were larger than the nine-parameter global fits by factors of 2.8 and 3.8, respectively. In the case of K_4, the values obtained neglecting the spectrophotometric factors exceed values estimated from kinetics (see Fig. 7). In summary, for the wavelength range 530–590 nm, exclusion of E values in fitting Adair constants can lead to large differences with respect to the Adair constants obtained by including the E values. The differences can amount to 4-, 8-, 3-, and 3.5-fold for K_1 to K_4, respectively.

Figure 8 shows how "Y" is inadequately fit by the six-parameter model even at 590 nm, where the misfit derives mainly from E_1 (2.3) and E_2 (1.7). In Fig. 8, the limiting slopes and intercepts were fit individually, then the apparent $\overline{X}$ and n^* values were used to generate all four K values, assuming all E values were equal to 1. The theoretical curve, generated from the six-parameter model, misses the actual curve at the lower turning point.

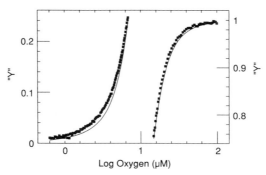

FIG. 8. Apparent fractional saturation of hemoglobin, 590 nm, vs the log of oxygen activity for the extremes of the binding isotherms. The theoretical curve depicts a six-parameter fit to the data where the limiting intercepts and slopes and apparent $\overline{X}$ and n^* values were used to obtain the four apparent Adair constants (see text for details). The left y axis is for the left half of the plot, and the right y axis is for the right half of the plot.

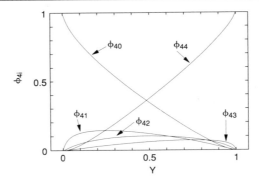

FIG. 9. Mole fractions (Φ) of various species for the tetramer (denoted by the first subscript, 4) with i (second subscript) bound ligands as a function of overall fractional saturation. The Adair constants from the global (nine-parameter) column of Table I were used to obtain the curves.

Note that at high X, for $(A_\infty - A_0) > 0$, K_4 obtained from the limiting slope [Eq. (34) with all $E = 1$] will appear to be negative if E_3 is > 1.33. For the data set depicted in Fig. 1, we found that the apparent K_4 (for data at 550 nm) was negative because E_3 was 1.48. At 590 nm, however, E_3 (0.96) was sufficiently close to 1 that K_4 was within 7% of that calculated from the nine-parameter global fits.

Figure 9 shows a plot of mole fractions of the various tetrameric species vs fractional saturation from the global nine-parameter K values of Table I. It can be seen that significant concentrations of all intermediates occur, a conclusion at variance with findings that β_3 is essentially zero.[21] Our conclusion is also consistent with evidence from kinetic data that the triliganded species occurs at significant concentrations at equilibrium.[68]

Further evidence supporting the importance of wavelength effects and the relative unimportance of dimers for our conditions (heme concentration $\geq 100 \ \mu M$) is the following. Our results are in agreement with other studies[21,24] with regard to the values of the half-saturation activities and the Hill number, and our value of K_4 for the six-parameter model fit is in agreement with that reported by Chu et al.,[24] 1.56 μM^{-1}, corrected to our conditions, with a range at the 68% confidence level of 0.88 to 2.91 μM^{-1}. Gill et al.[21] fit data for a solution 4 mM in heme according to a six-parameter model, using β values rather than K values. The experimental conditions for Gill et al.[21] were 180 mM Cl$^-$ at 21.5°. Our conditions were 200 mM Cl$^-$ at 21°. After making these

[68] M. Zhang, F. A. Ferrone, and A. J. Martino, Biophys. J. 58, 333 (1990).

small corrections for temperature and $[Cl^-]$,[69] and converting β to K values, their values for K_1, K_2, and K_3 were in agreement with our K values obtained as the mean of the five wavelength fits for six parameters, within our standard error of the mean. For K_4, our value agreed with their value within their 68% confidence region. The mean values we calculate from the data of Gill *et al.*[21] for 576 nm are (μM^{-1}) as follow: K_1, 0.0916; K_2, 0.0296; K_3, 0.040; K_4, 1.210. This agreement shows clearly that the observed absorbance changes, in the same wavelength region, are in agreement between the two laboratories (although the heme concentrations differed by a factor of 40), justifying our neglecting the effects of dimers for our conditions. The differences between the above K values and those in Table I for the global nine-parameter fits must reflect primarily wavelength effects.

The analysis described here is based on global fitting over a number of wavelengths to extract both the global Adair constants and the three local spectroscopic constants. Alternative approaches or extensions include the following. The Adair constants might be assumed to follow a two-parameter Van't Hoff relationship over a narrow temperature range, reducing the number of global parameters, but the E values might well vary with temperature if preferential and temperature-dependent $\alpha-\beta$ ligation occurred. Various models based on microstates could be constructed and the E values written in terms of those states,[18] provided one has some set of basis spectra. The numerators of the E values might be required, as a function of wavelength, to be simple scale factors of $Hb-HbO_2$ difference spectra with variable wavelength shifts, so as to reduce the number of fitting parameters by making some part or all of the E values also global parameters. Singular value decomposition (SVD) techniques provide another means for analyzing equilibrium data and provide a means for correcting for progressive metHb formation. For observations in the Soret region of Hb A_0 with 30 mM inositol hexaphosphate SVD analysis has shown[32] that β_3 is not equal to zero and that fractional saturation is not proportional to fractional changes in absorbance.

Higher precision spectrophotometers with increasingly sophisticated computerized data acquisition and noise reduction techniques may allow the E values as well as the Adair constants to be better defined. Work at high concentrations will require special cells for multiple path lengths with oxygen electrode or optical sensing of oxygen concentration. These instrumental and computational advances will be required to accommodate

---

[69] K. Imai, "Allosteric Effects in Hemoglobin," p. 116. Cambridge Univ. Press, Cambridge, 1982.

the large number of modified[70,71] and genetically engineered hemoglobins now being produced.

Acknowledgment

Grant Support NIH DK36288, American Heart Association Grant-in-Aid, Nebraska Affiliate.

[70] K. D. Vandegriff, R. J. Rohlfs, and R. M. Winslow, in "Blood Substitutes" (T. M. S. Change and R. P. Geyer, eds.), p. 647. Dekker, New York, 1989.
[71] K. D. Vandegriff, F. Medina, M. A. Marini, and R. M. Winslow, J. Biol. Chem. 264, 17824 (1989).

[30] Adair Equation: Rederiving Oxygenation Parameters

By EDWARD C. DELAND

A central problem in hemoglobin (Hb) research over the years has been the accurate determination of the oxygen-binding parameters at the four sites in the complex protein. This apparently simple task of fitting a function to the observed saturation data curve has, however, become increasingly complicated. Because of cooperativity (or heme–heme interaction, as first designated by Bohr in 1903[1]), the very definition of the observed binding parameters comes into question and has led to the study of myriad ancillary structural and functional properties of the protein.

Mathematical models of hemoglobin structure and function have been devised for many years to aid this study; however, these models generally have been limited in scope because the conceptual complexity of the protein does not readily admit explicit modeling. Here we describe a mathematical method that allows biochemical systems of arbitrary complexity to be modeled, and we apply it to problems from work on isoionic and cross-linked human hemoglobin.[2]

It is possible to calculate the equilibrium distribution of reaction products of a complex system by finding the minimum of the Gibbs free-energy function for the set of reactions at the given temperature and pressure. Further, if some of the reaction parameters in the model are unknown,

[1] C. Bohr, Zentralbl. Physiol. 17, 682 (1903).
[2] This work was supported by the Blood Research Group, Letterman Army Institute of Research, U.S. Army Medical Research and Development Command, Presidio of San Francisco, Col. R. M. Winslow, Director.

they may be approximated by fitting the mathematical model to observed data. Such a method for biochemical research was proposed by Dantzig *et al.*[3] in 1961, based on previous work by White *et al.*,[4] and this work evolved into the current computer programs. Model constructs of biochemical detail using this method are called "isomorphic" because they profess to be one-to-one copies of the hypothetical system under study. A theoretical model would essentially consist of a list of all pertinent biochemical reactions in the system and, because computation is computerized, the method is limited primarily only by our detailed knowledge of that list and availability of critical data for validation.

With the benefit of an explicit complex model, multivariate relationships can arise to be examined that are ordinarily fairly difficult to perceive. One of these is the known variability of the Adair "constants" under a variety of different experimental conditions. Here, we begin a mathematical analysis and review of the problem with the goal of clarifying certain basic concepts relating to the oxygenation parameters.

Statement of Problem

The simple sigmoidal shape of the hemoglobin saturation curve can, of course, be fit arbitrarily well by any number of functions, a sum of exponentials, for example. However, such a function, not being derived from fundamental biochemistry, will not help to explain why the protein produces a curve. Researchers studying hemoglobin, therefore, have tried to argue from fundamental principles. If we suppose, for example, that n oxygen molecules can bind to the hemoglobin each with the identical binding constant K as in the equation

$$Hb + nO_2 \rightleftharpoons Hb(O_2)_n \tag{1}$$

then the oxygen saturation of Hb, y, can be written as the ratio of concentrations of liganded sites to the total number of sites,

$$y = \frac{Hb(O_2)_n}{Hb + Hb(O_2)_n} \tag{2}$$

or

$$y = \frac{Kp^n}{1 + Kp^n} \tag{3}$$

[3] G. B. Dantzig, J. C. DeHaven, I. Cooper, S. H. Johnson, E. C. DeLand, H. E. Kanter, and C. F. Sams, *Perspect. Biol. Med.* **4**(3), 324 (1961).
[4] W. B. White, S. M. Johnson, and G. B. Dantzig, *J. Chem. Phys.* **28**(5), 751 (1958).

where the chemical symbols indicate concentration, a convention we adopt for most of this chapter, and p is oxygen pressure.

This equation, first proposed by Hill[5] in 1910, introduced two important concepts (in addition to the very idea of an abstract mathematical model based on chemical principles): the first, suggested by Hill, that the equilibrium constant, K, may be regarded as proportional to the probability of finding the protein occupied by oxygen, and the second, that by adjusting just two fixed parameters it may be possible to "model" the observed saturation curve.

We now know that two parameters are not sufficient to describe the complexity of the saturation curve. In fact, we show here that perhaps several dozen parameters may be necessary, and can be derived.

Four-Parameter Model

After the historical determination that Hb has exactly four oxygen-binding sites, Adair[6] developed the famous four-parameter formula for calculating the saturation curve,

$$y = \frac{a_1 p + 2a_2 p^2 + 3a_3 p^3 + 4a_4 p^4}{4(1 + a_1 p + a_2 p^2 + a_3 p^3 + a_4 p^4)} \tag{4}$$

where $a_i = \Pi_1^i k_j$, $i = 1, \ldots, 4$, the k_j values are the sequential oxygen equilibrium constants at the four sites, and p is the pressure of oxygen (in mmHg). The Adair equation, the basis of considerable subsequent work, produces a saturation curve having certain necessary qualitative aspects as well as a rational, that is a chemical, theoretical base. However, it apparently has two shortcomings: first, it still gives puzzling characteristic errors with respect to an observed saturation curve,[7] and, second, it does not in itself explain the observed shifts of the curve with changes of pH, P_{CO_2}, or temperature.

From the theoretical equation, Eq. (4), we can formally derive a corresponding equation showing the ratio of saturated sites to total sites on the hemoglobin, that is, the fractional saturation observed in the laboratory. Because

$$Hb + jO_2 \rightleftharpoons Hb(O_2)_j; \qquad a_j, j = 1, \ldots, 4 \tag{5}$$

[5] A. V. Hill, *J. Physiol. (London)* **40**, 4 (1910).

[6] G. S. Adair, *J. Biol. Chem.* **36**, 529 (1925).

[7] R. M. Winslow, M. Samaja, N. J. Winslow, L. Rossi-Bernardi, and R. I. Schrager, *J. Appl. Physiol.: Respir., Environ. Exercise Physiol.* **54**(2), 524 (1983).

or

$$Hb(O_2)_j = a_j p^j / Hb \tag{6}$$

where we have changed the units of oxygen concentration to millimeters of mercury, we can substitute in Eq. (4) to obtain

$$y(p) = \frac{HbO_2 + 2Hb(O_2)_2 + 3Hb(O_2)_3 k + 4Hb(O_2)_4}{4[1 + HbO_2 + Hb(O_2)_2 + Hb(O_2)_3 + Hb(O_2)_4]} \tag{7}$$

which does not involve the a_j parameters, but just the ratio of oxygenated sites to all sites on the Hb at a given pressure p.

Laboratory data, whether obtained by gasometric, optical, or some other means, measures and reports as observed data the saturation ratio of Eq. (7). In this chapter we ask why the data calculated by Eq. (4) does not yield the same result as the observed protein data provided by Eq. (7)? A simple answer is that Eq. (4) is a simplistic mathematical model of the protein behavior involving the a_j values explicitly, whereas Eq. (7) is implicitly symbolic of the complex protein itself. The subtlety and complexity of the protein function is not represented in Eq. (4), but it is in Eq. (7). As is discussed briefly below, many attempts have been made to modify Eq. (4) so that it would better predict the observed data. We show, first, that, for precision, it is necessary for Eq. (4) to include explicitly the effector reactions (as well as the oxygenation reactions), but also that these effects are manifested only in the variability of the A_j coefficients; these parameters are not simple constants. Specifically, then, if the a_j values are not constant under a variety of experimental conditions, exactly (mathematically) why and how do the a_j values change with pH, P_{CO_2}, or the other effectors?

Intrinsic vs Observed Constants

With Eq. (4), it is necessary to refit the equation to the observed data after, say, a change in ambient pH, which results in a new set of binding parameters, a_j.[7,8] Is it possible, however, for the parameters to be adjusted automatically in the theoretical model just as they are in the protein as a consequence of a change of conditions? The principles on which common effectors [pH, P_{CO_2}, 2,3-diphosphoglycerate (2,3-DPG), Cl^-, stereochemistry, and temperature] might work are reasonably well-known chemical concepts, such as competition for the same binding site or altering a local

[8] R. M. Winslow, J. M. Morrissey, R. L. Berger, P. D. Smith, and C. C. Gibson, *J. Appl. Physiol.: Respir., Environ. Exercise Physiol.* **45**(2), 289 (1978).

charge field,[9-13] although explanatory details for a particular protein such as Hb may still be conjectural. Still, we wish to incorporate the effectors into Eq. (4) using basic chemical principles in such a way that the "observed" or effective a_j values are modified as in the real protein under varying experimental conditions. We do this by simulating theoretical hypotheses for the effectors and validating against observed data.

Early hypotheses in the literature invoked cooperativity of the protein,[14] which essentially suggests that as the sequential oxygen molecules are bound, each binding affects the local environment to alter the conditions under which the next will be bound. Thus, although the intrinsic k_j values may be nearly equal on the deoxyhemoglobin molecule, the effective a_j values are quite different as oxygenation proceeds. As to exactly how the intrinsic constants are modified: "It must be assumed that these interactions are mediated by some kind of molecular transition (allosteric transition) which is induced or stabilized in the protein when it binds an 'allosteric ligand.' "[15]

The emergent concept is that the individual monomers have an intrinsic binding constant for oxygen and, whether α or β, they may be nearly equal or differ by a small factor (see Baldwin[16] and Baldwin and Chothia),[17] but bound into the tetramer the observed binding constants may be different and variable depending on experimental circumstances. In particular, the different values of the observed a_j are a consequence of the effectors operating on (modifying) the intrinsic constant.

The mathematical goal, here, is to detail explicitly these phenomena by incorporating additional molecular hypotheses into the fundamental Adair equation. Pauling was the first to show such an explicit relationship of an effector (pH) in a mathematical model. Thus, Pauling,[18] in 1935, developed a modified Adair equation with the following form:

$$y = \frac{Kp + (2\alpha + 1)K^2p^2 + 3\alpha^2 K^3 p^3 + \alpha^4 K^4 p^4}{1 + 4Kp + (4\alpha + 2)K^2 p^2 + 4\alpha^2 K^3 p^3 + \alpha^4 K^4 p^4} \tag{8}$$

[9] J. V. Kilmartin, *Br. Med. Bull.* **32**(3), 209 (1976).

[10] M. F. Perutz, *Ann. Rev. Biochem.* **48**, 327 (1979).

[11] J. V. Kilmartin, *FEBS Lett.* **38**(2), 147 (1974).

[12] L. Rossi-Bernardi and F. J. W. Roughton, *J. Physiol. (London)* **189**, 1 (1967).

[13] K. Imaizumi, K. Imai, and I. Tyuma, *J. Biochem. (Tokyo)* **86**, 1829 (1979).

[14] M. F. Perutz, M. G. Rossmann, A. F. Culls, H. Muirhead, G. Will, and A. C. T. North, *Nature (London)* **185**, 416 (1960).

[15] J. Monod, J. Wyman, and J. Changeux, *J. Mol. Biol.* **12**, 88 (1965).

[16] J. M. Baldwin, *Prog. Biophys. Mol. Biol.* **29**, 225 (1975).

[17] J. M. Baldwin and C. Chothia, *J. Mol. Biol.* **129**, 175 (1979).

[18] L. Pauling, *Proc. Natl. Acad. Sci. U.S.A.* **21**, 186 (1935).

where α is a stabilizing ($\alpha > 1$) interaction coefficient between adjacent hemes, and the single oxygenation constant is a parameter. Pauling, further recognizing that the Adair "constants" cannot be constant over the entire curve, introduced a function of pH to relate the intrinsic and observed binding constant,

$$K' = K \frac{[1 + (\beta A/H^+)]^2}{[1 + (A/H^+)]^2} \tag{9}$$

where β is an interaction constant, and A is the proton ionization constant. Although this model does not yet explain or detail just how the α and β coefficients work to modify the binding constants at the molecular level, it has an explicit awareness of the variability of a_j values with P_{O_2} because, for example, the pH changes.

Several subsequent models of the saturation curve have been offered; one in particular specifically incorporates the stereochemical effector that will be important in the present discussion, the Monod–Wyman–Changeux (MWC) model.[15]

$$y = \frac{LK_Tp(1 + K_Tp)^3 + K_Rp(1 + K_Rp)^3}{L(1 + K_Tp)^4 + (1 + K_Rp)^4} \tag{10}$$

The two probable states of the molecule, designated R and T, are reversibly accessible and they differ by the distribution of energy throughout the interdimeric bonds and, indeed, throughout the entire tetramer. As a result, the probability for oxygen binding at the stereolabile sites (the "allosteric effect") is altered by the factor L as the transition occurs. This mathematical idea is similar to Pauling's in that an empirical constant is introduced to account for and correct persistent, characteristic errors of previous theoretical models. The new parameters are also based on or derived from a novel molecular hypothesis. A drawback is that the parameter is empirical rather than arising from first principles, say, for example, the calculation of charge distribution. Consequently, these models run the same risk of lack of generality as the Adair theory, and, indeed, this has proved to be the case (discussed below).

Subsequently, Ackers, in a series of papers, devised chemical and thermodynamic formulations of these complex phenomena and proposed distinct theoretical hypotheses to explain observational data. In particular (e.g., in Ackers[19] and Smith and Ackers),[20] the authors propose specific empirical relationships among the effectors and the observed binding coefficients for oxygen, and they offer considerable experimental verification.

[19] G. K. Ackers, *Biochemistry* **18**(5), 3372 (1979).
[20] F. R. Smith and G. K. Ackers, *Proc. Natl. Acad. Sci. U.S.A.* **82**, 5347 (1985).

Then, in 1987 Ackers and Smith[21] showed that the earlier model of Eq. (10) is not consistent with observations and proposed a three-state molecular hypothesis, again with arbitrary constants.

Trial theoretical models of these several molecular hypotheses have been devised, and we have proposed critical tests to aid distinguishing among them.

Adair Mathematical Model

By a "mathematical model" we mean, here, a list of biochemical reactions hypothesized to be relevant and significant for determining the observed function of the protein, plus relationships or conditions among the products, such as stereochemical constraints, that may preclude or alter the probability of an event. Such a model could be subjected to the exact same protocol, say, titration, as in the laboratory and the calculated reaction products displayed for comparison and analysis.

For example, a consequence of the previous work has been the considerable attention focused on the intermediate states of hemoglobin during oxygenation. Such information would aid theoretical design, but direct observation of these compounds and the transition states of the tetramer, as in the MWC or Ackers models, has proved difficult (except for carbonmonoxyhemoglobin[22]) so that validation of a particular molecular hypothesis generally awaits critical data. A mathematical simulation of the protein function, however, will necessarily calculate the quantity and variety of intermediate compounds resulting from a particular molecular hypothesis, however complicated.

In the following, we show detailed models of distinct molecular hypotheses, beginning with the early Adair[6] and Roughton[23] models. Both the Adair and Roughton models were devised before the remarkable work of Chanutin and Churnish[24] (and simultaneously Benesch and Benesch[25]) demonstrated the importance of 2,3-DPG, so these models do not include an effect of the phosphate binding.

The Adair molecular hypothesis of 1925 was simply that four oxygens

[21] G. K. Ackers and F. R. Smith, *Annu. Rev. Biophys. Biophys. Chem.* **16**, 583 (1987).

[22] M. Perrella and L. Rossi-Bernardi, this volume [21].

[23] F. J. W. Roughton, F. R. S., E. C. DeLand, J. C. Kernohan, and J. W. Severinghaus, *in* "Oxygen Affinity Hemoglobin and Red Cell Acid-Base Status" (P. Astrup and M. Rorth, eds.), p. 131. Academic Press, New York, 1972.

[24] A. Chanutin and R. R. Churnish, *Arch. Biochem. Biophys.* **121**, 96 (1967).

[25] R. Benesch and R. E. Benesch, *Biochem. Biophys. Res. Commun.* **26**, 162 (1967).

TABLE I
ADAIR MODEL FOR HEMOGLOBIN OXYGENATION[a]

Input reactants	Input (mol)	Gas phase (mole fraction)
O_2		3.500×10^{-2}
CO_2		1.000×10^{-8}
N_2		9.039×10^{-1}
H_2O	5.513×10^1	6.107×10^{-2}
H^+	-2.286×10^{-2}	
Hb	4.901×10^{-3}	

[a] A computer-based model of the Adair molecular hypothesis: four oxygen molecules per Hb tetramer (P_{CO_2}, 2,3-DPG, $Cl^- = 0$, 37°). The input quantities of reactants are for a 4.9-mmol/kg Hb solution.

bind to one Hb, and we therefore have the simple isomorphic structure of Eq. (7) and Tables I–III.

The Adair model is a two-phase system in which, as Table I shows, 4.9 mmol of Hb tetramer is dissolved in 1 liter of water, about the concentration in erythrocytes (1 liter $H_2O = 55.137$ mol, 37°). Because the Adair model was originally devised for blood, this demonstration will use a normal P_{50} of 26.6 mmHg O_2. The remainder of the gas phase is saturated with water (water vapor = 46.4 mmHg at 37°), has essentially no CO_2, but enough P_{N_2} to make 1 atm pressure. The gas solubility constants at 37° are shown in the gas phase entries of Table II. The H^+ content is adjusted to make the resulting protein solution isoionic, pH 7.32 at 50% saturation. HCO_3^- is included for a later experiment, but here, because the P_{CO_2} is small in the gas phase, bicarbonate and the carbamino reactions will be negligible.

A typical calculation result is given in Table III, in which output species are listed in moles, but because the amount of solvent water is given they could be calculated in any concentration unit. Here, the input was 1 liter of water plus the volume of the protein (about 50 ml/mM Hb) so that the output values are equivalent to molal units. Elsewhere in this discussion, we deduct from the input an amount of water equal to the protein volume, so that the output species are in molar units. The amount of water required to make HCO_3^- is added, but the water for initial solvation of the protein, perhaps 1000 molecules per tetramer, is presumed to be included with the input protein.

A simulation of blood might consist of varying the P_{O_2} (and inversely the P_{N_2} to maintain 1 atm pressure) to calculate a saturation curve, follow-

TABLE II
REACTION EQUATIONS FOR ADAIR MODEL[a]

Reaction products	Reaction constant	Reactants (with reaction coefficients)		
Gas phase	Solubility (cm^3/cm^3)			
O_2	0.026	(1) O_2		
CO_2	0.0440	(1) CO_2		
N_2	0.0146	(1) N_2		
H_2O	2.79569	(1) H_2O		
Hb solution	pK (37°)			
O_2	0.0	(1) O_2		
CO_2	0.0	(1) CO_2		
N_2	0.0	(1) N_2		
H_2O	0.0	(1) H_2O		
H^+	0.0	(1) H^+		
OH^-	13.0958	(1) H_2O	(-1) H^+	
HCO_3^-	6.13	(1) CO_2	(1) H_2O	(-1) H^+
	log a_j			
Hb	0	(1) Hb		
HbO_2	-1.222	(1) Hb	(1) O_2	
$Hb(O_2)_2$	-2.428	(1) Hb	(2) O_2	
$Hb(O_2)_3$	-3.617	(1) Hb	(3) O_2	
$Hb(O_2)_4$	-5.184	(1) Hb	(4) O_2	

[a] Expected chemical reaction products with their pK values are shown. The products are shown in the left-hand column, the equilibrium constants (pK values or solubilities) in the middle column, and the reactants in the columns to the right (integers are reaction coefficients).

ing the protocol in the laboratory. The a_j values can be determined by using this entire mathematical model as an arbitrary function in standard regression programs. When this is done, using the Severinghaus[26] data for blood, the best fit a_j values are shown in Table II. The k_j values calculated from these a_j values are 0.060, 0.062, 0.065, and 0.027, which compare fairly well with literature values (see Table VII, below). Fitting the model will give the same results as fitting the Adair equation itself because this is an isomorphic, that is, a one-to-one model of the Adair hypothesis. Figure 1a and b is calculated from Eq. (4), but could have been generated by using this model with the same a_j values.

The entire curve (Fig. 1a) does not show clearly the characteristic errors of the Adair equation. Typically, the calculated curve compared with the observed curve for blood, (i.e., with the full complement of effectors for blood), shows too great an affinity for oxygen below about

[26] J. W. Severinghaus, *J. Appl. Physiol.* **21**, 1108 (1966).

TABLE III
ADAIR HEMOGLOBIN MODEL[a]

Output species	Gas phase (mole fraction)	Hb solution[b] (mol)
O_2	3.500×10^{-2}	3.923×10^{-6}
CO_2	1.315×10^{-8}	1.948×10^{-10}
N_2	9.037×10^{-1}	8.066×10^{-4}
H_2O	6.107×10^{-2}	5.514×10^{1}
H^+		4.186×10^{-8}
OH^-		3.090×10^{-7}
HCO_3^-		1.303×10^{-9}
Hb		1.589×10^{-3}
HbO_2		8.500×10^{-4}
$Hb(O_2)_2$		4.470×10^{-4}
$Hb(O_2)_3$		3.352×10^{-7}
$Hb(O_2)_4$		2.014×10^{-3}

[a] Calculated equilibrium distribution of species for the Adair model in Tables I and II. $P_{O_2} = 26.6$ mmHg, 37°.
[b] pH 7.321.

10 mmHg P_{O_2}, and, after crossing the observed curve at about 45 and 80 mmHg, too low an affinity above 80 mmHg. In Fig. 1b, two expanded portions of a calculated curve show these "characteristic" errors, using the 1966 blood saturation data of Severinghaus.[26]

Subsequently, improved instrumentation and technique have reduced but not eliminated these errors. That is because the errors are inherent in the Adair assumption that the a_j values are constant over the whole curve, even though the effectors may be changing.

One approach to reconcile the observed data rests on the 2,3-DPG effect addressed indirectly by the MWC model. Calculations using the Adair model do not show the delayed first-stage affinity (when 2,3-DPG is bound to Hb) or the increased fourth-stage affinity (when the probability of 2,3-DPG binding because of stereochemical rotation is small). With respect to the model, this means that 2,3-DPG binding should decrease the effective value of a_1 and increase the effective values of a_3 and a_4. However, that linkage is not reflected explicitly in the Adair equation, where the a_j values remain constant. If this effect could be built into the Adair model, the calculated curve would be delayed at the bottom and pushed to the left at the top, a translation of the curve to the right plus a counterclockwise rotation. This is an observed result calculated by the MWC model.

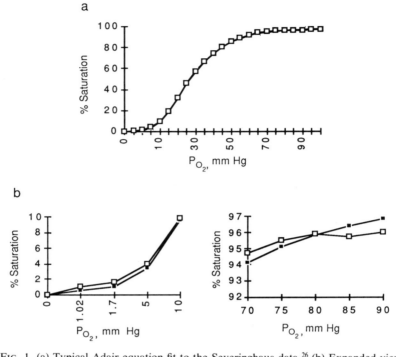

FIG. 1. (a) Typical Adair equation fit to the Severinghaus data.[26] (b) Expanded view of typical curve fit by the Adair equation to experimental human Hb saturation data. The open squares are calculated by the model of Table I (see text for details). Filled squares: Severinghaus data.

Because of this and similar shortcomings, the Adair model is expected to fail as a general model; four parameters are not enough to incorporate the complicated Hb functions. Subsequent literature models have attempted to cure this problem, but a successful theoretical model will necessarily incorporate the effectors explicitly, as in the Roughton model described in the next section.

Roughton Isomorphic Model

A more fundamental approach, taken in the Roughton model,[23] explicitly includes the Bohr and CO_2 effects biochemically linked to oxygenation so that, if the Roughton molecular hypotheses are correct, the a_j values will be internally modified with changing experimental pH or P_{CO_2} conditions. Specifically, Roughton proposed that CO_2 was bound at the N

TABLE IV
INPUT REACTANTS FOR ROUGHTON HEMOGLOBIN MODEL[a]

Input reactants and sites	Input (mol)	Gas phase (mole fraction)
O_2	9.843×10^{-3}	3.500×10^{-2}
CO_2	1.787×10^{-2}	5.263×10^{-2}
N_2	5.402×10^{-4}	8.505×10^{-1}
H_2O	3.859×10^{1}	6.185×10^{-2}
H^+	-2.603×10^{-1}	
Na^+	9.667×10^{-3}	
K^+	1.019×10^{-1}	
Ca^{2+}	6.608×10^{-5}	
Mg^{2+}	1.291×10^{-3}	
Cl^-	5.368×10^{-2}	
HPO_4^{2-}	1.404×10^{-2}	
Hb	4.901×10^{-3}	

Oxylabile sites	No. of sites (per Hb)
Asp	4.0
αVal	2.0
βHis	2.0
βVal	2.0
Oxystable sites	
εHis	2.0
Asp/Glu	54.0
–COOH sites	8.0
His	16.0
Tyr	12.0
Lys	40.0
Arg	12.0

[a] Theoretical model of human Hb (4.9 mM in tetramer) devised by Roughton.[23]

terminals of each α and β chain, that this terminal was also a Bohr group, that there were 3 additional Bohr groups per heme [an aspartic group, a valine (which also binds CO_2), and a histidine], and that there were about 144 more oxystable H^+ buffering sites on the tetramer. Tables IV–VI are a listing of the Roughton model.

Although 2,3-DPG is not present (indeed, was not known at the time), Roughton did include an inorganic phosphate as a buffer, and he put in Na^+, K^+, Ca^{2+}, and Mg^{2+} for use in subsequent protocols. In Table IV the input reactants are listed plus the names of the oxylabile and oxystabile sites. The latter species also are shown in Table IV with their frequencies on the Hb molecule.

TABLE V
REACTION EQUATIONS FOR ROUGHTON HEMOGLOBIN MODEL[a]

Reaction products and sites	Reaction constant	Reactants (with reaction coefficients)		
Gas phase	Solubility (cm^3/cm^3)			
O_2	0.026	(1) O_2		
CO_2	0.440	(1) CO_2		
N_2	0.0146	(1) N_2		
H_2O	2.79569	(1) H_2O		
Hb solution	pK (37°)			
O_2	0.0	(1) O_2		
CO_2	0.0	(1) CO_2		
N_2	0.0	(1) N_2		
H_2O	0.0	(1) H_2O		
H^+	0.0	(1) H^+		
OH^-	13.0958	(1) H_2O	(-1) H^+	
Na^+	0.0	(1) Na^+		
K^+	0.0	(1) K^+		
Ca^{2+}	0.0	(1) Ca^{2+}		
Mg^{2+}	0.0	(1) Mg^{2+}		
Cl^-	0.0	(1) Cl^-		
HCO_3^-	6.13	(1) CO_2	(1) H_2O	(-1) H^+
HPO_4^{2-}	0.0	(1) HPO_4^{2-}		
$H_2PO_4^-$	7.19	(1) HPO_4^{2-}	(1) H^+	
	$\log a_j$			
Hb	—	(1) Hb		
HbO_2	-1.850	(1) Hb	(1) O_2	
$Hb(O_2)_2$	-3.254	(1) Hb	(2) O_2	
$Hb(O_2)_3$	-4.960	(1) Hb	(3) O_2	
$Hb(O_2)_4$	-5.816	(1) Hb	(4) O_2	
Deoxy sites	pK (37°)			
DAsp	4.900	(1) $DAsp^-$	(1) H^+	
$\beta DHis^+$	8.100	(1) $\beta DHis$	(1) H^+	
$\alpha DVal^+$	7.700	(1) $\alpha DVal$	(1) H^+	
$\alpha DValCO_2$	11.9770	(1) $\alpha DVal$	(-1) H^+	(1) CO_2
$\beta DVal^+$	7.300	(1) $\beta DVal$	(1) H^+	
$\beta DValCO_2$	11.764	(1) $\beta DVal$	(-1) H^+	(1) CO_2
Oxy sites				
OAsp	5.500	(1) $OAsp^-$	(1) H^+	
$\beta OHis^+$	7.200	(1) $\beta OHis$	(1) H^+	
$\alpha OVal^+$	7.300	(1) $\alpha OVal$	(1) H^+	
$\alpha OValCO_2$	11.280	(1) $\alpha OVal$	(-1) H^+	(1) CO_2
$\beta OVal^+$	7.300	(1) $\beta OVal$	(1) H^+	
$\beta OValCO_2$	12.100	(1) $\beta OVal$	(-1) H^+	(1) CO_2
Oxystable sites				
HCOOH	4.000	(1) $HCOO^-$	(1) H^+	
Asp/Glu	4.500	(1) Asp/Glu^-	(1) H^+	
εHis^+	7.000	(1) εHis	(1) H^+	
His^+	7.000	(1) His	(1) H^+	
Tyr	9.800	(1) Tyr^-	(1) H^+	
Lys^+	10.350	(1) Lys	(1) H^+	
Arg^+	12.000	(1) Arg	(1) H^+	

[a] Theoretical model of human Hb devised by Roughton[23] (integers are reaction coefficients).

TABLE VI
CALCULATED OUTPUT FOR ROUGHTON HEMOGLOBIN MODEL[a]

Products	Gas phase	Hb solution[b]	Deoxy sites	Oxy sites	Oxystable sites
Total (mol)	1.0	3.871×10^1	2.538×10^{-2}	2.363×10^{-2}	7.058×10^{-1}
O_2	3.500	3.922×10^{-5}			
CO_2	5.263	9.447×10^{-4}			
N_2	8.505	5.390×10^{-4}			
H_2O	6.185	3.849×10^1			
H^+		4.186×10^{-8}			
OH^-		3.090×10^{-7}			
Na^+		9.674×10^{-3}			
K^+		1.019×10^{-1}			
Ca^{2+}		6.620×10^{-5}			
Mg^{2+}		1.292×10^{-3}			
Cl^-		5.358×10^{-2}			
HCO_3^-		1.303×10^{-2}			
HPO_4^{2-}		7.677×10^{-3}			
$H_2PO_4^-$		6.363×10^{-3}			
Hb		1.589×10^{-3}			
HbO_2		8.500×10^{-4}			
$Hb(O_2)_2$		4.470×10^{-4}			
$Hb(O_2)_3$		3.352×10^{-7}			
$Hb(O_2)_4$		2.014×10^{-3}			
Asp			4.169×10^{-5}	1.638×10^{-4}	
Asp^-			9.762×10^{-3}	9.640×10^{-3}	
βHis^+			5.030×10^{-3}	1.853×10^{-3}	
βHis			7.432×10^{-4}	2.175×10^{-3}	
αVal^+			2.188×10^{-3}	2.671×10^{-3}	
αVal			8.120×10^{-4}	2.490×10^{-3}	
$\alpha ValCO_2$			1.029×10^{-3}	6.128×10^{-3}	
βVal^+			2.151×10^{-3}	1.768×10^{-3}	
βVal			2.005×10^{-3}	1.648×10^{-3}	
$\beta ValCO_2$			1.616×10^{-3}	6.128×10^{-4}	
HCOOH					2.115×10^{-5}
$HCOO^-$					3.919×10^{-2}
Asp/Glu					4.509×10^{-4}
Asp/Glu^-					2.642×10^{-1}
εHis					2.751×10^{-2}
εHis^+					5.092×10^{-2}
His					6.365×10^{-3}
His^+					3.438×10^{-3}
Tyr^-					5.865×10^{-2}
Tyr					1.724×10^{-4}
Lys^+					1.959×10^{-1}
Lys					1.623×10^{-4}
Arg^+					5.882×10^{-2}
Arg					1.091×10^{-6}

[a] Typical calculated equilibrium distribution (in moles) of species for the Roughton Hb model at 50% saturation.
[b] pH 7.321.

The reaction equations and buffering reactions are shown in Table V. The buffering species, sorted under deoxygenated (D), oxygenated (O), and the oxystable sites, are included in the model via an ancillary calculation of approximately 154 simultaneous equations. An equilibrium distribution of species is calculated for the entire model by minimizing the Gibbs free-energy function.

The Roughton oxygenation parameters, a_j, determined by fitting this entire chemical model to the saturation curve, are shown for the Hb reactions in Table V. These parameters fall within the range of modern a_j values (such as Winslow[7]); the implied k_j values are 0.014, 0.040, 0.020, and 0.139.

A typical calculated equilibrium result for the species in the Roughton model is listed in Table VI, in which the Hb is 50% saturated. This model could be used to simulate an experimental H^+ or CO_2 titration protocol. For example, because the oxylabile Bohr sites have different pK values for oxy- or deoxyHb, the calculated titration curves for the deoxy molecule will necessarily be different from the oxygenated; the isoionic points are about pH 7.25 for oxyHb in solution and pH 7.41 for deoxyHb, and a similar difference will persist from pH 4 to pH 11.[27] Outside of this range, the two curves intersect. Figure 2 is a plot of the calculated titration experiment.

Intrinsic Constants Revisited

Because in this Roughton model the Bohr and carbamino reactions are linked to the oxygenation reaction and it includes the oxystable H^+ buffering, this model is more general than the Adair model in the sense that changes in pH and P_{CO_2} will be reflected automatically in the changing a_j values as experimental conditions change. Therefore, if there were no other effectors than pH and P_{CO_2}, the oxygenation parameters a_j of the model would not need to be changed to simulate changing laboratory conditions.[28] This is not true because there are other effectors such as 2,3-DPG and Cl^-, so that this simple model would not fit the data of a

---

[27] For this work, we have used the pK values and other titration data from the work of M. A. Marini and L. Forlani, found in the Proceedings of the 5th International Symposium on Blood Substitutes, San Diego, March 17–20, 1993.

[28] This discussion presumes that the Roughton hypotheses on which the model was constructed are correct. However, the Roughton hypotheses were devised some time ago, and better data are now available, along with more sophisticated conjectures concerning the complex interactions of Hb. Still, as will be shown in the discussion of modern data in Table VII, if the effectors are held constant or set to zero, all four k_j values are nearly equal, as in the Roughton model.

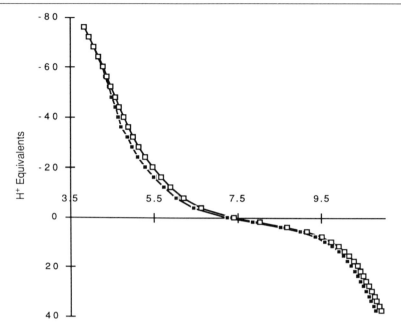

FIG. 2. Titration of the Roughton Hb model,[23] using titration data from Marini.[27] Isoionic pH: oxygenated (■), 7.25; deoxygenated (□), 7.41 (horizontal axis is pH).

laboratory experiment involving 2,3-DPG or Cl^- (except by manually changing the a_j values in a way that would reflect the 2,3-DPG and Cl^- effects).

However, supposing it were true that the Bohr and Haldane effects were the only effectors and the Roughton hypotheses were correct, then the constant a_j values in the mathematical model would be the intrinsic oxygenation parameters. By "intrinsic," we mean here the potential binding parameters for the isoionic deoxyHb A_0 tetramer, before binding with oxygen. These intrinsic parameters are then modified by internal crosslinking to the effective or experimentally observed a_j values by competitive interactions with H^+ and CO_2 during oxygenation, just as in the protein.

In fact, this concept can be made mathematically explicit and the degree of modification calculated. The Roughton model can illustrate the process. We wish to take account of the fact that the H^+ and P_{CO_2} effectors treat each individual monomer in the tetramer differently, depending on whether they are α or β subunits and whether they are oxygenated or deoxygenated.

We first symbolically represent the list of possible product species in solution (from Table V). Consider the equations relating only to the first stage of oxygenation:

$$
\begin{array}{ll}
k_1 & \mathrm{Hb} + \mathrm{O}_2 \rightleftharpoons \mathrm{HbO}_2 \\
k_{\mathrm{DA}} & \mathrm{DAsp}^- + \mathrm{H}^+ \rightleftharpoons \mathrm{DAsp} \\
k_{\beta\mathrm{DH}} & \beta\mathrm{DHis} + \mathrm{H}^+ \rightleftharpoons \beta\mathrm{DHis}^+ \\
k_{\alpha\mathrm{DV}} & \alpha\mathrm{DVal} + \mathrm{H}^+ \rightleftharpoons \alpha\mathrm{DVal}^+ \\
k_{\alpha\mathrm{DC}} & \alpha\mathrm{DVal} - \mathrm{H}^+ + \mathrm{CO}_2 \rightleftharpoons \alpha\mathrm{DValCO}_2 \\
k_{\beta\mathrm{DV}} & \beta\mathrm{DVal} + \mathrm{H}^+ \rightleftharpoons \beta\mathrm{DVal}^+ \\
k_{\beta\mathrm{DC}} & \beta\mathrm{DVal} - \mathrm{H}^+ + \mathrm{CO}_2 \rightleftharpoons \beta\mathrm{DValCO}_2 \\
k_{\mathrm{OA}} & \mathrm{OAsp}^- + \mathrm{H}^+ \rightleftharpoons \mathrm{OAsp} \\
k_{\beta\mathrm{OH}} & \beta\mathrm{OHis} + \mathrm{H}^+ \rightleftharpoons \beta\mathrm{OHis}^+ \\
k_{\alpha\mathrm{OV}} & \alpha\mathrm{OVal} + \mathrm{H}^+ \rightleftharpoons \alpha\mathrm{OVal}^+ \\
k_{\alpha\mathrm{OC}} & \alpha\mathrm{OVal} - \mathrm{H}^+ + \mathrm{CO}_2 \rightleftharpoons \alpha\mathrm{OValCO}_2 \\
k_{\beta\mathrm{OV}} & \beta\mathrm{OVal} + \mathrm{H}^+ \rightleftharpoons \beta\mathrm{OVal}^+ \\
k_{\beta\mathrm{OC}} & \beta\mathrm{OVal} - \mathrm{H}^+ \, \mathrm{CO}_2 \rightleftharpoons \beta\mathrm{OValCO}_2
\end{array}
\tag{11}
$$

where the subunits are α and β, the prefixes D and O stand for deoxygenated and oxygenated states, k_1 is the association constant for Hb plus O_2, and the remaining constants are dissociation constants. From these chemical equations, we can write the implied mass action equations and this will allow derivation of the relationship between the intrinsic and observed parameters. Again (as for the Adair equation) we start with the concentration ratio of total oxy sites to deoxy sites (in the first stage of oxygenation):

$$
\frac{\mathrm{HbO}_2}{\mathrm{Hb}} = k_1' p = \frac{\mathrm{HbO}_2 + \mathrm{OAsp} + \beta\mathrm{OHis} + \alpha\mathrm{OVal}^+ + \alpha\mathrm{OValCO}_2 + \beta\mathrm{OVal}^+ + \beta\mathrm{OValCO}_2}{\mathrm{Hb} + \mathrm{DAsp} + \beta\mathrm{DHis} + \alpha\mathrm{DVal}^+ + \alpha\mathrm{DValCO}_2 + \beta\mathrm{DVal}^+ + \beta\mathrm{DValCO}_2}
$$

$$
= \frac{k_1 p + k_1 p \dfrac{\mathrm{H}^+}{k_{\mathrm{OA}}} + k_1 p \dfrac{\mathrm{H}^+}{k_{\beta\mathrm{OH}}} + k_1 p \dfrac{\mathrm{H}^+}{k_{\alpha\mathrm{OV}}} + k_1 p \dfrac{\mathrm{H}^+}{k_{\alpha\mathrm{OC}}} + k_1 p \dfrac{\mathrm{H}^+}{k_{\beta\mathrm{OV}}} + k_1 p \dfrac{\mathrm{H}^+}{k_{\beta\mathrm{OC}}}}{1 + \dfrac{\mathrm{H}^+}{k_{\mathrm{DA}}} + \dfrac{\mathrm{H}^+}{k_{\beta\mathrm{DH}}} + \dfrac{\mathrm{H}^+}{k_{\alpha\mathrm{DV}}} + \dfrac{\mathrm{H}^+}{k_{\alpha\mathrm{DC}}} + \dfrac{\mathrm{H}^+}{k_{\beta\mathrm{DV}}} + \dfrac{\mathrm{H}^+}{k_{\beta\mathrm{DC}}}}
\tag{12}
$$

where we have divided numerator and denominator by Hb and substituted from the mass action equations. Factoring out $k_1 p$,

$$
k_1' p = k_1 p \frac{1 + \dfrac{\mathrm{H}^+}{k_{\mathrm{OA}}} + \dfrac{\mathrm{H}^+}{k_{\beta\mathrm{OH}}} + \dfrac{\mathrm{H}^+}{k_{\alpha\mathrm{OV}}} + \dfrac{\mathrm{H}^+}{k_{\alpha\mathrm{OC}}} + \dfrac{\mathrm{H}^+}{k_{\beta\mathrm{OV}}} + \dfrac{\mathrm{H}^+}{k_{\beta\mathrm{OC}}}}{1 + \dfrac{\mathrm{H}^+}{k_{\mathrm{DA}}} + \dfrac{\mathrm{H}^+}{k_{\beta\mathrm{DH}}} + \dfrac{\mathrm{H}^+}{k_{\alpha\mathrm{DV}}} + \dfrac{\mathrm{H}^+}{k_{\alpha\mathrm{DC}}} + \dfrac{\mathrm{H}^+}{k_{\beta\mathrm{DV}}} + \dfrac{\mathrm{H}^+}{k_{\beta\mathrm{DC}}}}
\tag{13}
$$

Thus k_1', the observed constant, is derived from the intrinsic k_1 by the linkage equations. These links are, of course, explicit in the model, and, in the calculation, the ratio of oxy to deoxy sites will be given by k_1'. Similar equations can be derived for subsequent stages of oxygenation, but for the jth stage, the right-hand side of Eq. (13) will be raised to the jth power,

$$k_j' p = k_j p \left(\frac{1 + \dfrac{H^+}{k_{OA}} + \dfrac{H^+}{k_{\beta OH}} + \dfrac{H^+}{k_{\alpha OV}} + \dfrac{H^+}{k_{\alpha OC}} + \dfrac{H^+}{k_{\beta OV}} + \dfrac{H^+}{k_{\beta OC}}}{1 + \dfrac{H^+}{k_{DA}} + \dfrac{H^+}{k_{\beta DH}} + \dfrac{H^+}{k_{\alpha DV}} + \dfrac{H^+}{k_{\alpha DC}} + \dfrac{H^+}{k_{\beta DV}} + \dfrac{H^+}{k_{\beta DC}}} \right)^j \quad (14)$$

In a calculation of the saturation curve, the k_j' values of Eq. (14) must replace the implicit k_j parameters in the Adair equation. Because

$$a_i' = \prod_1^i k_j', \qquad i = 1, \ldots, 4 \quad (15)$$

following Eq. (5), we write

$$Hb + jO_2 \rightleftharpoons Hb(O_2)_j, \qquad a_j', j = 1, \ldots, 4 \quad (16)$$

and simply use the a_j values in the Adair equation:

$$y = \frac{a_1' p + 2a_2' p^2 + 3a_3' p^3 + 4a_4' p^4}{4(1 + a_1' p + a_2' p^2 + a_3' p^3 + a_4' p^4)} \quad (17)$$

Now the Adair equation is clearly a function of pH and P_{CO_2} because the k_1' values and hence the a_j' values are derived from Eq. (14).

Equation (17) is the intermediate model we have been seeking: the modified Adair model with variable coefficients a_j. However, although conceptually correct, it is an ungainly procedure to calculate these algebraic equations because, for example, the H^+ ion concentration is known in the context of the buffering system and must be calculated separately. It is easier simply to use the simulation model, in which the pH and all of the linkages are simultaneously calculated. The model (a list of reactions) is merely a more complicated Adair-type equation.

In a more complicated hypothetical tetramer model, this algebraic relationship is more complicated, but conceptually similar in that each hypothetical configuration of the molecule is represented in the algebra.[29] In this model, the intrinsic parameter is modified primarily by competitive configurations, but the concepts of stereochemical "blocking" or variable local potential fields, allostery, and probability configurations can be included.

Cooperativity

This proposed structure is not yet cooperative in the classic sense. Because Roughton regarded the oxylabile Bohr and Haldane sites to be effective in the vicinity of a single heme no specific linkage was hypothesized among hemes. Roughton, of course, was aware of the changes in tertiary geometry ascribed to the breaking of salt bridges and intramolecular bonds. Roughton proposed a regulatory mechanism, but lacked the data on Cl^- and 2,3-DPG. In a more complete model, mechanisms of linkage must still be established involving Cl^-, 2,3-DPG, and stereochemistry.

This model may nevertheless appear to be cooperative, because the sequential a_j values change dramatically, even though Roughton's individual k_j values are still independent. However, there is some confusion about this point in the literature, where a_4 is commonly regarded as a measure of the energy of binding of the fourth oxygen ligand (see, e.g., Dickerson and Geis[30]), which is k_4, not a_4. Whereas it may appear that the fourth oxygen is tightly bound because a_4 may be 10^4 times a_1, in fact, the k_j values do not exhibit this property so dramatically, if at all. In Hb A_0, cross-linked $\alpha\alpha$Hb, or blood, the individual k_j values are not as widely separated; k_2 may be larger or smaller than k_1, and k_4 is generally less than five times k_1 and may even be approximately equal to k_1 under some circumstances (Table VII).

In an analysis of cooperativity and linkage leading to regulation and control of oxygen binding, it is the individual binding parameters k_j that are being operated on by effectors. Table VII shows typical sets of the k_j values for various laboratory conditions as calculated from the data in

---

[29] A fundamental problem exists in this regard: many of the reaction parameters, such as the H^+ ionization pK values, are gathered by observation of protein function in the laboratory. Hence these parameters are observed or effective parameters, as modified in context, and perhaps calculated from the heats of ionization. Whereas, just like the oxygenation parameters, all reaction constants in the model should have the intrinsic values, which are then modified in context, just as in the protein. Frequently, as in Roughton's case, observed pK values are used simply because nothing else is available.

[30] R. E. Dickerson and I. Geis, "Hemoglobin: Structure, Function, Evolution, and Pathology." Benjamin/Cummings, Menlo Park, CA, 1983.

TABLE VII
VALUES FOR $k_j{}^a$

Source	DPG/Hb	pH	P_{CO_2}	k_1	k_2	k_3	k_4
Hb A_0	0	7.4	0	0.326	0.095	k_3k_4 =	0.176^b
Hb A_0	0.4	7.4	0	0.098	0.24	k_3k_4 =	0.0158^b
$\alpha\alpha$Hb	0	7.4	0	0.124	0.0157	0.154	0.187
$\alpha\alpha$Hb	0.4	7.4	0	0.103	0.0034	0.08	0.392
Blood	1.0	7.4	40	0.022	0.057	0.069	0.023
Blood	0.4	7.4	40	0.012	0.158	0.023	0.050
Blood	0.27	7.31	70	0.036	0.008	0.24	0.062
Blood	0.27	7.39	70	0.029	0.012	0.16	0.063
Blood	0.27	7.48	70	0.029	0.024	0.095	0.098
Blood	0.27	7.59	70	0.017	0.064	0.052	0.099

^{a} Typical literature values calculated from a_j values in Winslow et al.[7] and Vandegriff et al.[31]

^{b} An a_3 value is not available; we use the product k_3k_4.

Winslow et al.[7] and Vandegriff et al.[31] Under the wide conditions of Table VII, the values of the four binding parameters k_j are remarkably similar: the value of k_4 does not exceed k_1 by more than a factor of 6, and is generally much less.

The k_j values in the tetramer are much smaller than the binding constants for the isolated monomers, which differ by about a factor of two (1.9 mmHg^{-1} for the α chain and 3.4 mm Hg^{-1} for the β chain[16]). If this factor persists in the tetramer the fraction of the cooperativity to be accounted for by the effectors is not great, probably no more than a factor of 2 to 5. This is the order of magnitude found in simple chemical competitive interaction models, such as the Roughton model.

To illustrate this point, Cl$^-$ binding was simulated using the Roughton model and the suggestions in Chiancone et al.[32] and Haire and Hedlund[33] to locate the binding at the N-terminal α valines, also used to bind CO$_2$. With the Cl$^-$ concentration very small (10^{-7} mM), the k_j values are determined to give a P_{50} of 1 mmHg. With the same k_j values and the same pH (7.32, 37°), when the Cl$^-$ is increased to 10 mM, the saturation drops to 40% at 1 mmHg. Conversely, if the pH changes, the k_j values must change significantly to maintain the same saturation. Thus, in Table VII, for the four blood samples having 2,3-DPG/Hb = 0.27 (and approximately

[31] K. D. Vandegriff, F. Medina, M. A. Marini, and R. M. Winslow, J. Biol. Chem. 264, 17824 (1989).

[32] E. Chiancone, J. E. Norne, S. Forsen, J. Bonaventura, M. Brunori, E. Antonini, and J. Wyman, Eur. J. Biochem. 55, 385 (1975).

[33] R. N. Haire and B. E. Hedlund, Proc. Natl. Acad. Sci. U.S.A. 74(10), 4135 (1977).

the same P_{50}[7]), each of the four k_j values increment in an orderly manner as the pH increases, as might be expected. Also, k_1 and k_3 are inversely proportional to k_2 and k_4.[34]

Furthermore, it appears that cooperativity may essentially be eliminated simply by holding effector conditions constant. Thus, in Table VII, the cross-linked Hb, having zero 2,3-DPG and zero P_{CO_2}, shows all four k_j values approximately equal.

Unfortunately, k_3 is not always determinable by the fitting algorithms usually employed[35] and, in these cases, k_4 is also indeterminate. However, in Vandegriff et al.,[36] the fourth oxygenation constant, k_4, was determined by alternative, kinetic considerations and then held fixed during the analysis of binding, thus forcing a_3 to acquire a value. The shortcoming of this method is that as advocated here, the observed Adair "constant" k_4 must be allowed to vary with changes in effectors along the saturation curve. If, as in Vandegriff et al.,[36] effectors are held constant along a saturation

---

[34] This argument can be made rigorously, but conceptually, consider fitting the curve through the saturation point at 10 mmHg P_{O_2}: if k_1, which is the slope of the curve at the origin, is increased, the fitted curve will pass above and miss the data point at 10. k_2, which is also effective near the origin, must be decreased to compensate. A similar but more complicated argument applies at the top of the curve where k_3 and k_4 are effective.

[35] Calculations of k_4 from measured a_j values are usually conjectural because of an artifact in the process. When fitting the Adair (A) to the observed saturation curve (S), the partial derivative of the error, which we may call $\Sigma(A - S)^2$, with respect to a_3 is essentially zero. That is, the process of fitting the error is insensitive to and hence ambivalent about the value of a_3; a_3 can be set arbitrarily without affecting the goodness of fit. The fact that a_3 is frequently found to be near zero does not necessarily imply the absence of the third stage of oxygenation from the reaction mixture. Parameter a_3 still plays an important role in the real chemistry, of course; but its value cannot be determined in the usual manner. If, in the Adair equation, a_j were replaced by products of k_j during the fitting process, the procedure would give satisfactory results for k_j (including k_3), but calculation of the sensitivity coefficients would become considerably more complicated. This result is not obvious from an examination of the partial derivative with respect to a_3, which is

$$\frac{\partial A}{\partial a_3} = (3*p^\wedge 3)/[4*(1 + a1*p + a2*p^\wedge 2 + a3*p^\wedge 3 + a4*p^\wedge 4)]$$
$$- [p^\wedge 3*(a1*p + 2*a2*p^\wedge 2 + 3*a3*p^\wedge 3 + 4*a4*p^\wedge 4)]/$$
$$[4*(1 + a1*p + a2*p^\wedge 2 + a3*p^\wedge 3 + a4*p^\wedge 4)^\wedge 2]$$

but during calculation either the analytical or numerical derivative of the error function is nearly zero with respect to a_3. This problem does not always occur, but it is not known why, in some cases, a_3 can be determined. A conjecture is that if a_3 is indeterminate, the third stage of oxygenation is, in fact, improbable in the reaction mixture, but this has not been demonstrated.

[36] K. D. Vandegriff, Y. C. Le Tellier, R. M. Winslow, R. J. Rohlfs, and J. S. Olson, *J. Biol. Chem.* **266**, 17049 (1991).

curve, the a_j values may be determined, but would have a different value for a different setting of effectors.

Adair Derivatives

The fact that the Adair equation can be differentiated has been useful because it yields a guide to the value of k_1. But an interesting innovation is that it also yields a guide to the value of k_2 for Adair-type mathematical models. The first derivative (with the a_j constant) is

$$dy/dp = \frac{a_1 + 4a_2p + 9a_3p^2 + 16a_4p^3}{4(1 + a_1p + a_2p^2 + a_3p^3 + a_4p^4)}$$
$$- \frac{(a_1 + 2a_2p + 3a_3p^2 + 4a_4p^3)(a_1p + 2a_2p^2 + 3a_3p^3 + 4a_4p^4)}{4(1 + a_1p + a_2p^2 + a_3p^3 + a_4p^4)^2}$$

(18)

and the limit of this expression as $p \to 0$ is

$$\lim_{p \to 0} dy/dp = a_1/4$$

so that near the origin, a_1, that is k_1, can be calculated from the slope of the saturation curve. This may be the intrinsic constant for the first monomer to bind in the tetramer at very small P_{O_2} because cooperativity, if any, has not yet had an effect. Also, the derivative of the Roughton chemical model gives the same result.

The second derivative of the Adair [i.e., the derivative of Eq. (18) with respect to p, not shown here], is a measure of the curvature of the saturation curve. This also gives an interesting limit near the origin where the saturation curve is turning upward:

$$\lim_{p \to 0} d^2y/dp^2 = -a_1^2/2 + a_2$$

This limit near the origin should be positive, indicating an upward curvature. Therefore, the expression must be greater than zero, that is, $a_2 > a_1^2/2$ or $k_2 > k_1/2$.

Near the origin, then, k_2 must be greater than $k_{1/2}$. This is a condition not always met by published sets of a_j values. For example, it is not true for all the sets of Table VII, but this condition must apply; the curve does not turn down at the origin. It is obviously possible for the first increments of the curve calculated by the Adair equation to have negative curvature, but it is not chemically sensible and published details of the curve near the origin have positive curvature. Detail of the function of the protein near the origin would be useful because it would help to establish the initial values of k_1 and k_2.

Unfortunately, the third derivative of the Adair (not listed here) is not as useful. This derivative is the rate of change of the curvature of the saturation curve. Near the origin its value would be nonzero if the curvature of the sigmoid curve were constant. From details of the normal blood curve, the curvature might be constant until about 5 mmHg for P_{O_2} (then the curve straightens out), and so the limit of the third derivative at the origin might be zero. The limit of the third derivative at the origin is

$$\lim_{p \to 0} d^3y/dp^3 = [(3a_1^3)/2] - [(9a_1a_2)/2] + (9a_3)/2]$$

or

$$a_1^3 + 3a_3 = 3a_1a_2$$

However, although this is a mathematically valid equality, the problems with determining the value of a_3 do not allow either accurate evaluation of the expression or whether the data curve has zero curvature at the origin.

Conclusion

At a given moment most of the tetrameric hemoglobin molecules in solution with high probability will be either completely oxygenated (oxy) or completely deoxygenated (deoxy). But, the tetramer, being a complex system of variable bonds, linkages, and critical bindings, can assume a vast number of intermediate states depending on ambient variations of pH, P_{CO_2}, 2,3-DPG, or Cl^- concentration. By this view, the probability of finding the molecule in a particular state is an aggregate function of the microscopic probabilities for each binding, linkage, or bonding that can occur locally. We speak of the global properties of the solution, such as O_2 saturation or bound 2,3-DPG concentration, because the law of large numbers allows such an overall summation or averaging operation. In particular, the probability of finding a given molecule in a certain oxygenated state can be altered by changing any of the ancillary microscopic probabilities, such as the buffering of protons, as might occur during a saturation curve experiment.

Keeping track of and linking these details of the protein can be accomplished readily by constructing a theoretical model of the type described. Under a particular molecular hypothesis, the simultaneous biochemical equations simulating the protein automatically impose linkages and constraints that are a consequence of the user's hypotheses about the system. The basic concept is that various hypothetical protein structures can be proposed and tested by incorporating their tenets into the mathematical chemistry of a model. The procedures for modeling are not new, they are basic biochemistry; but a novel aspect is that a computer-based program will support analysis of systems as complex as the Hb protein.

Encouragingly, these straightforward biochemical simulations are powerful. Even the simple Roughton model used as a demonstration is able to show the algebraic linkage between the effectors and ligand binding. It appears that this biochemical linkage alone may account for enough cooperativity to simulate protein function; and the theoretical k_j values derived from the model are in the range of k_j values expected from the literature.

[31] Linkage Thermodynamics

By Enrico Di Cera

Introduction

Biological macromolecules have the unique ability to link processes of widely different nature. Wyman was the first to suggest that the basis of such functional flexibility, or plasticity, was to be found in the existence of an equilibrium involving different macromolecular forms, or conformations, that can be stabilized preferentially by different thermodynamic driving forces.[1,2] When this elegant hypothesis was cast in terms of the powerful principles of equilibrium thermodynamics laid down by Gibbs, it culminated in the formulation of the theory of "reciprocal effects,"[3] which we now know as "linkage thermodynamics."[4,5] Wyman's idea of linked functions is a fundamental thermodynamic principle in its own right whose implications are yet to be fully exploited. The importance of this principle is best illustrated by the fact that one of its straightforward consequences, that is, the celebrated theory of allosteric transitions in biological macromolecules,[6] is perhaps more famous than the principle itself. The theory proposed by Wyman has made it possible to recognize and explore the close connection between functional properties and struc-

[1] J. Wyman, *Adv. Protein Chem.* **4**, 407 (1948).
[2] J. Wyman and D. W. Allen, *J. Polym. Sci.* **7**, 499 (1951).
[3] J. Wyman, *Adv. Protein Chem.* **19**, 223 (1964).
[4] E. Di Cera, S. J. Gill, and J. Wyman, *Proc. Natl. Acad. Sci. U.S.A.* **85**, 5077 (1988).
[5] E. Di Cera and J. Wyman, *Proc. Natl. Acad. Sci. U.S.A.* **88**, 3494 (1991).
[6] J. Monod, J. Wyman, and J. P. Changeux, *J. Mol. Biol.* **12**, 88 (1965).

tural transitions in biological macromolecules. In some cases the different energetics reflecting different functional states are linked to subtle structural changes. In other cases, these differences map into drastic structural rearrangements of macromolecular components. Hemoglobin is a beautiful example of how structural transitions can be used to accomplish an important physiological function.[7] This protein is the prototype of cooperative transitions in biological macromolecules and has long provided the ideal ground to test and develop new ideas in linkage thermodynamics. In no other macromolecular system has the interplay between theoretical and experimental investigation produced so many spectacular results and opened so many puzzling questions as in the case of hemoglobin. The questions that remain to be answered about hemoglobin cooperativity are no less important and exciting than the answers already at hand. Likewise, the new ideas to be explored in linkage thermodynamics are no less important and exciting than the basic principles with which we are familiar. Hence, the need to explore the fundamental properties of the hemoglobin system, as well the theoretical basis of ligand-binding cooperativity in biological macromolecules, continues to be a high priority in molecular biophysics. This chapter deals with linkage thermodynamics of hemoglobin. It briefly summarizes some basic concepts in ligand-binding thermodynamics and provides a rather detailed account of new theoretical developments that have cast the entire problem of cooperativity in biological macromolecules within a new framework. A detailed account of the "classic" concepts of linkage thermodynamics is presented in an excellent book[8] to which the interested reader is referred.

Partition Function

From a statistical thermodynamic point of view, a biological macromolecule can be seen as a generalized ensemble in the Gibbs sense. The macromolecule is in fact a system open to chemical and physical quantities coupled to thermodynamic driving forces such as pressure, temperature, and chemical potentials. The partition function can be constructed as the ratio,

$$Z = \frac{\text{sum over all configurations}}{\text{reference configuration}} \tag{1}$$

[7] M. F. Perutz, *Q. Rev. Biophys.* **22**, 139 (1989).
[8] J. Wyman and S. J. Gill, "Binding and Linkage." University Science Books, Mill Valley, CA, 1990.

where Z stands for the German *Zustandsumme*, or sum over states, as originally defined by Planck.[9] The connection with thermodynamics is obtained by identifying the thermodynamic potential associated with Z. Following Guggenheim's approach,[10] one can write

$$\Psi(\{\Omega\}) = -k_B T \ln Z(\{\Omega\}) \tag{2}$$

where k_B is the Boltzmann constant, T is the absolute temperature, and $\{\Omega\}$ is the set of independent thermodynamic quantities entering the definition of an ad hoc potential Ψ. It is clear from Eq. (2) that a number of different thermodynamic potentials can be constructed depending on the particular choice of independent variables.[10–12] The conditions under which biological systems operate put some restrictions on the set of independent thermodynamic quantities that can be accessed experimentally. When studying biological systems our emphasis concentrates on intensive quantities and their mutual interference. The partition function for a macromolecule M interacting with a ligand X whose activity is x, is a polynomial expression such as

$$Z(x) = \sum_{j=0}^{N} A_j x^j \tag{3}$$

where N is the total number of binding sites for ligand X, per mole of macromolecule, while A_j is the overall equilibrium constant for the reaction $M + jX = MX_j$, and hence $A_0 = 1$ by definition. It is implicit in the formulation of $Z(x)$ that the macromolecule does not change its aggregation state on ligand binding.

The mathematical form of the partition function $Z(x)$ is a direct consequence of the law of mass action and can be understood readily in terms of the existence of discrete energy levels. In a system containing N binding sites, there are $N + 1$ distinct liganded configurations. Each of these configurations can be assigned an energy level,

$$E_j = \varepsilon_j - j\mu_X \tag{4}$$

where μ_X is the chemical potential of ligand X, and ε_j is the intrinsic energy level of the j-liganded configuration. The sum over all energy states normalized by the energy level of the reference configuration gives the partition function,

[9] R. Kubo, "Statistical Mechanics." North-Holland Publ., Amsterdam, 1990.
[10] E. A. Guggenheim, *J. Chem. Phys.* **7**, 103 (1939).
[11] T. L. Hill, "Statistical Mechanics." Dover, New York, 1960.
[12] W. Stockmayer, *J. Chem. Phys.* **18**, 58 (1950).

$$Z(\mu_X) = \frac{\exp(-\beta E_0) + \exp(-\beta E_1) + \cdots + \exp(-\beta E_N)}{\exp(-\beta E_0)} \tag{5}$$

where $\beta = 1/k_B T$. Hence,

$$Z(\mu_X) = \sum_{j=0}^{N} \exp[\beta(j\mu_X - \Delta\varepsilon_j)] \tag{6}$$

where $\Delta\varepsilon_j = \varepsilon_0 - \varepsilon_j$ is the energy difference between the j-liganded configuration and the reference unliganded configuration. This energy difference is related to the standard free energy change per mole of macromolecule in going from the unliganded to the j-liganded configuration, that is,

$$\Delta G_j = \mathbb{N}\Delta\varepsilon_j = -RT \ln A_j \tag{7}$$

where $\mathbb{N}$ is the Avogadro number and A_j is the equilibrium constant in Eq. (3). The relationship above provides the connection between the discrete energy levels of the system, the Gibbs free-energy changes associated with ligand binding, and the relevant equilibrium A constants. The polynomial expansion [Eq. (3)] is obtained readily from the relationship above and the basic definition of the chemical potential $\mu_X = k_B T \ln x$.

The definition of the partition function $Z(x)$ is of paramount importance in the analysis of ligand-binding processes. All information on the energetic aspects of the system is stored in the A coefficients. The quantitative characterization of these coefficients under a variety of experimental conditions reveals the thermodynamic driving forces involved in the changes of functional significance. For example, the Gibbs–Helmholtz equation applied to ΔG_j is,

$$(\partial \ln A_j/\partial\tau)_P = -\Delta H_j/R \tag{8}$$

where $\tau = 1/T$. This equation yields the standard enthalpy change associated with the binding reaction $M + jX = MX_j$. Likewise, the partial derivative

$$(\partial \ln A_j/\partial P)_T = -\Delta V_j/RT \tag{9}$$

gives the isothermal volume change associated with the same reaction. If the volume of the macromolecule with j ligand molecules bound is smaller than that of the sum of the volumes of the individual reagents free in solution ($\Delta V_j < 0$), then the reaction is favored by an increase in pressure. Otherwise ($\Delta V_j > 0$) the reaction is favored by a decrease in pressure.

The effect of a second ligand on macromolecular equilibria is interpreted in an analogous way. The partial derivative

$$(\partial \ln A_j / \partial \ln y)_{T,P} = \Delta Y_j \qquad (10)$$

where y is the ligand activity of ligand Y, gives the change in A_j due to binding of Y. If the number of ligand molecules Y bound to M increases on binding of j molecules of X ($\Delta Y_j > 0$), then the reaction $M + jX = MX_j$ is favored by an increase in the chemical potential of Y. On the other hand, if the number of ligand molecules Y bound to M decreases on binding of j molecules of X ($\Delta Y_j < 0$), the reaction $M + jX = MX_j$ is favored by a decrease in the chemical potential of Y. The mutual interference of physical and chemical effects is an elegant consequence of the Le Chatelier principle applied to a macromolecular system. The response of the system to a given perturbation is such to oppose the effect of the perturbation itself. When a binding equilibrium is perturbed by an increase in pressure, the macromolecule reacts by assuming a conformation of smaller volume and, therefore, by opposing and minimizing the effect of the perturbation. Likewise, when the binding equilibrium is perturbed by an increase in the chemical potential of a second ligand Y, the macromolecule reacts by assuming a conformation in which more molecules of Y are bound. In either case, the response observed is subject to the fundamental principles of thermodynamic stability embodied by the second law. Hence, the basis for the existence of linked functions and reciprocal effects in ligand binding to biological macromolecules can be understood in rigorous thermodynamic terms. The linkage between proton and oxygen binding to hemoglobin (Bohr effect) is a beautiful example of how macromolecules can use physical principles encapsulated by linked functions to accomplish their physiological role.

Thermodynamic Response Functions

Given the importance of the A coefficients of the partition function $Z(x)$, much attention should be devoted to resolving these parameters from analysis of experimental binding data. Some key operational aspects arising in this connection are dealt with elsewhere.[13,14] Experimental investigation of binding processes can access the information stored in the partition function only indirectly through a set of thermodynamic response functions. These are the average number of liganded sites $X(x)$ and the

[13] E. Di Cera, this series, Vol. 210, p. 68.
[14] M. L. Johnson and L. M. Faunt, this series, vol. 210, p. 1.

binding capacity $B(x)$. The function $X(x)$ is obtained by differentiation of $Z(x)$ as follows:

$$X(x) = \frac{d \ln Z(x)}{d \ln x} = \frac{\sum\limits_{j=0}^{N} jA_j x^j}{\sum\limits_{j=0}^{N} A_j x^j} = \langle j \rangle \tag{11}$$

and is bounded from zero to N. The variance of the distribution of liganded sites is

$$B(x) = \frac{dX(x)}{d \ln x} = \frac{\sum\limits_{j=0}^{N} j^2 A_j x^j}{\sum\limits_{j=0}^{N} A_j x^j} - \frac{\left(\sum\limits_{j=0}^{N} jA_j x^j\right)^2}{\left(\sum\limits_{j=0}^{N} A_j x^j\right)^2} = \langle j^2 \rangle - \langle j \rangle^2 \tag{12}$$

and gives the binding capacity of the macromolecule. The two quantities $X(x)$ and $B(x)$ are the key response functions of the system that can be accessed experimentally. In the case of hemoglobin, $X(x)$ is measured by means of the Imai apparatus[15] and $B(x)$ is measured by means of the Gill cell.[16] The function $X(x)$ is a measure of how much *is* bound as a function of the ligand activity x, whereas $B(x)$ is roughly a measure of how much *can be* bound as a function of x. The response functions $X(x)$ and $B(x)$ for hemoglobin under nearly physiological conditions[17] are depicted in Figs. 1 and 2. The binding capacity parallels physical response functions such as the heat capacity and the compressibility and is strictly related to the stability principles embodied by the second law. The binding capacity is always positive and an increase/decrease in the chemical potential of X always leads to an increase/decrease of $X(x)$. This fundamental property of $B(x)$ can be understood using arguments developed in the previous section. The increase in $X(x)$ when $\ln x$ increases is nothing but the response of the macromolecule to a change in the chemical potential of a second ligand Y, where Y is in this case the same as X. Hence, the perturbation induced by an increase in $\ln x$ drives the macromolecule into a conformation where more molecules of X are bound, so that the property of the binding capacity is a direct consequence of the Le Chatelier principle applied to ligand binding. From a mathematical point of view, this implies

[15] K. Imai, H. Morimoto, M. Kotani, H. Watari, W. Hirata, and H. Kuroda, *Biochim. Biophys. Acta* **534**, 189 (1970).

[16] D. Dolman and S. J. Gill, *Anal. Biochem.* **87**, 127 (1978).

[17] E. Di Cera, M. L. Doyle, M. S. Morgan, R. De Cristofaro, R. Landolfi, B. Bizzi, M. Castagnola, and S. J. Gill, *Biochemistry* **28**, 2631 (1989).

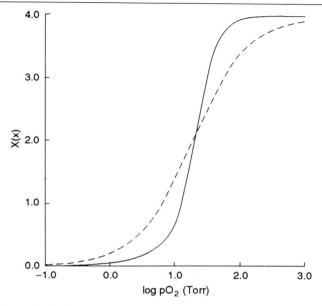

FIG. 1. Thermodynamic response function $X(x)$ for hemoglobin under nearly physiological conditions. The continuous curve plotted as a function of the logarithm of oxygen partial pressure was computed with the overall equilibrium constants reported in Di Cera *et al.*[17] The curve $X(x)$ for a noncooperative system with $N = 4$ is also shown by dashed lines for comparison.

that $X(x)$ is a monotonic function of $\ln x$ whose derivative vanishes only at $x = 0$ and $x = \infty$. Hence, there is no finite value of x for which $B(x) = 0$.[18] The derivative $dX(x)/d \ln x$ can be arbitrarily small for finite x, but not zero. We shall now consider the fundamental properties of the response functions in detail. Some of these properties are well known from previous discussions of ligand binding and linkage.[3,19,20] Other properties hinge on the results of new theoretical developments.

Properties of $X(x)$

Wyman was the first to recognize a fundamental property of the function $X(x)$ that can be exploited from analysis of experimental data. If

[18] E. Di Cera, *J. Chem. Phys.* **96,** 6515 (1992).
[19] T. L. Hill, "Cooperative Theory in Biochemistry." Springer, New York, 1984.
[20] E. Di Cera, S. J. Gill, and J. Wyman, *Proc. Natl. Acad. Sci. U.S.A.* **85,** 449 (1988).

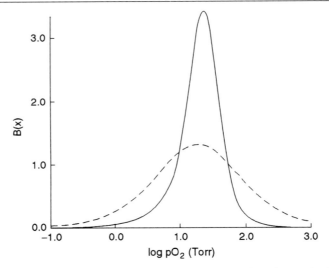

FIG. 2. Thermodynamic response function $B(x)$ for hemoglobin under nearly physiological conditions. The continuous curve plotted as a function of the logarithm of oxygen partial pressure was computed with the overall equilibrium constants reported in Di Cera et al.[17] The curve $B(x)$ for a noncooperative system with $N = 4$ is also shown by dashed lines for comparison.

measurements of $X(x)$ are available, it is always possible to obtain the coefficient A_N of the partition function. The Wyman integral equation[3]

$$\langle \ln x \rangle = N^{-1} \int_0^N \ln x \, dX(x) = -N^{-1} \ln A_N = \ln x_m \qquad (13)$$

defines the mean value of $\ln x$ associated with the mean ligand activity x_m, where the unliganded and fully liganded configurations of the macromolecule are equally populated. The mean ligand activity should not be confused with $x_{1/2}$, which gives the value of x at half-saturation where $X(x_{1/2}) = N/2$. The importance of x_m in the analysis of linkage effects is twofold. First, it can always be obtained directly from $X(x)$ by numerical integration and without nonlinear least-squares analysis. Second, the logarithm $\ln x_m$ represents the average work (in RT units) spent in the binding process per site, or else the average free energy of binding per site. Values of x_m can be derived in a straightforward way from analysis of $X(x)$ measured under a variety of conditions of interest. Measurements made as a function of temperature and the use of Eq. (8) yield

$$(\partial \ln x_m / \partial \tau)_P = \Delta H_m / R \qquad (14)$$

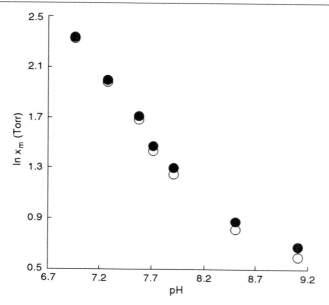

FIG. 3. Values of $\ln x_m$ for oxygen binding to human hemoglobin in the pH range 6.95–9.10 obtained from Di Cera *et al.*[21] ($\bigcirc$). The results of the analysis based on numerical integration of binding capacity data according to Eq. (28) in text ($\bullet$) agree with those obtained by nonlinear least squares ($\bigcirc$).

and hence the average enthalpy of binding per site. Likewise, measurements as a function of pressure yield

$$(\partial \ln x_m/\partial P)_T = \Delta V_m/RT \tag{15}$$

which is the average volume change per site due to binding. Specifically, ΔV_m is the value of ΔV_N in Eq. (9) divided by the number of binding sites N. Finally, measurements of $X(x)$ as a function of the chemical potential of a second ligand yield

$$(\partial \ln x_m/\partial \ln y)_{T,P} = -\Delta Y_m \tag{16}$$

and hence the average number of molecules Y exchanged per X binding site on ligation. The quantity ΔY_m is the difference in molecules Y bound to the fully liganded and unliganded configurations, ΔY_N in Eq. (10), divided by the number of X binding sites N. The values of x_m in the case of the Bohr effect of human hemoglobin[21] are reported in Fig. 3 by open circles.

[21] E. Di Cera, M. L. Doyle, and S. J. Gill, *J. Mol. Biol.* **200**, 593 (1988).

The calculation of x_m from measurements of $X(x)$ is rather simple. Wyman pointed out that Eq. (13) can be rewritten as[3]

$$N^{-1} \int_0^N (\ln x - \ln x_m) \, dX(x) = 0 \tag{17}$$

whereby integration by parts yields

$$\int_{-\infty}^{\ln x_m} X(x) \, d \ln x = \int_{\ln x_m}^{\infty} [N - X(x)] \, d \ln x \tag{18}$$

The value of $\ln x_m$ is, therefore, the value of $\ln x$ when the area under the $X(x)$ curve, Q_u, at the left-hand side of Eq. (18) equals the area Q_a above the $X(x)$ curve up to the asymptote $X(x) = N$. The Wyman equal-area rule is an elegant property of $X(x)$ that is widely used in practical applications. There is, however, a most important property of x_m that seems to have been neglected so far. From the definition of $Q_u(x)$ and $Q_a(x)$ as

$$Q_u(x) = \int_{-\infty}^{\ln x} X(x) \, d \ln x \tag{19}$$

$$Q_a(x) = \int_{\ln x}^{\infty} [N - X(x)] \, d \ln x \tag{20}$$

it follows immediately from Eq. (18) that

$$\ln x_m = \ln x + N^{-1}[Q_a(x) - Q_u(x)] \tag{21}$$

for any value of x. Therefore, $\ln x_m$ is an invariant property of the binding curve $X(x)$ because for any two points x_1 and x_2 one necessarily has

$$\ln x_1 + N^{-1}[Q_a(x_1) - Q_u(x_1)] = \ln x_2 + N^{-1}[Q_a(x_2) - Q_u(x_2)] \tag{22}$$

As a result of this invariant property, the mean ligand activity can be measured from any point along $X(x)$. Given the arbitrary point $\ln x_1$, the value of $Q_u(x_1)$ is given by the area under the curve $X(x)$ from $-\infty$ to $\ln x_1$ and likewise the value of $Q_a(x_1)$ is given by the area above the $X(x)$ curve up to the asymptote $X(x) = N$, from $\ln x_1$ to ∞. The difference $Q_a(x_1) - Q_u(x_1)$ divided by N and added to $\ln x_1$ gives $\ln x_m$, regardless of the point x_1, as illustrated in Fig. 4. The Wyman equal-area rule is embodied by the important invariant property [Eq. (21)] as a special case. The practical advantage of the invariant property [Eq. (21)] is evident, because computing $\ln x_m$ from any point along $X(x)$ is much easier than finding the exact value of $\ln x$ where $Q_a(x) = Q_u(x)$.

Another property of the binding curve $X(x)$ arises in connection with the analysis of its shape. The binding curve $X(x)$ plotted versus $\ln x$ is symmetric when it can be reproduced exactly on a rotation of $180°$ around

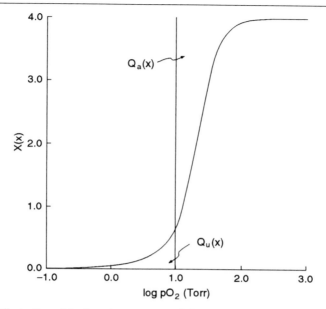

FIG. 4. Illustration of the invariant property of the mean ligand activity x_m based on Eq. (21) in text. The value of $\ln x_m$ can be obtained from any point along the $X(x)$ curve, for example, a point $\ln x_1$ as depicted by the vertical line, by computing $Q_u(x)$, the area under the curve from $-\infty$ to $\ln x_1$, and $Q_a(x)$, the area above the curve up to the asymptote $X(x) = N$ ($N = 4$ in this case) from $\ln x_1$ to ∞. The value of $\ln x_m$ is obtained from Eqs. (21) and (22) in text as $\ln x_m = \ln x_1 + [Q_a(x_1) - Q_u(x_1)]/4$. The arbitrary point depicted by the vertical line corresponds exactly to $\ln x_m$ if the two areas are the same. In general, the arbitrary point lies to the left of $\ln x_m$ if $Q_a(x_1) > Q_u(x_1)$, and to the right if $Q_a(x_1) < Q_u(x_1)$. The reader should verify that repeating the calculation using different points along the curve yields similar values of $\ln x_m$.

a point of coordinates $\ln x_s$, $X(x_s)$. The point x_s defines the center of symmetry of $X(x)$. Interest in the shape of $X(x)$ stems from the fact that symmetry introduces mathematical constraints among the otherwise independent A coefficients of the partition function $Z(x)$. From a mathematical standpoint the binding curve $X(x)$ is symmetric if, and only if, the condition[22]

$$X(x_s\lambda) + X(x_s\lambda^{-1}) = N \qquad (23)$$

holds for any $\lambda \geq 0$. At the center of symmetry, x_s, one necessarily has $X(x_s) = N/2$ by definition. Hence, x_s coincides with $x_{1/2}$, the value of x

[22] E. Di Cera, K.-P. Hopfner, and J. Wyman, *Proc. Natl. Acad. Sci. U.S.A.* **89**, 2727 (1992).

at half-saturation, whenever $X(x)$ is symmetric. It is quite instructive to prove also that $x_s = x_m$. This is because Eq. (23) necessarily implies

$$\int_{\ln x_s\lambda^{-1}}^{\ln x_s} X(x) \, d\ln x = \int_{\ln x_s}^{\ln x_s\lambda} [N - X(x)] \, d\ln x \qquad (24)$$

or

$$\int_{\ln x_s\lambda^{-1}}^{\ln x_s\lambda} [X(x) - N/2] \, d\ln x = 0 \qquad (25)$$

Integration leads to

$$\sum_{j=0}^{N} A_j x_s^j \lambda^j = \sum_{j=0}^{N} A_j x_s^j \lambda^{N-j} \qquad (26)$$

Because Eq. (26) must hold for any λ, one necessarily has $x_s = A_N^{-1/N}$, and hence $x_s = x_m$ as predicted. Hence, if $X(x)$ is symmetric, the center of symmetry is the same as $x_{1/2}$ and the mean ligand activity x_m, so that $x_m = x_{1/2}$. The proof that the reverse is not necessarily true, that is, that $x_m = x_{1/2}$ does not necessarily imply that $X(x)$ is symmetric, is left to the interested reader as an exercise. Elimination of x_s from Eq. (26) yields the condition of symmetry for the equilibrium constants as follows,

$$A_j A_N^{-j/N} = A_{N-j} A_N^{j/N-1} \qquad (27)$$

which must hold for any j. The above condition is a well-known result first obtained by Wyman[1] and implies that $X(x)$ is symmetric if the partition function $Z(x)$ is itself symmetric when x is expressed in $x_s = x_m$ units. For $N = 1$ or $N = 2$ the condition is a mere tautology. For $N = 3$ the condition is $A_3 = (A_2/A_1)$.[3] In the case of human hemoglobin, $N = 4$, and the condition of symmetry is $A_4 = (A_3/A_1)^2$ and does not depend on A_2. In general, the condition of symmetry involves all A values in the partition function for N odd and all A values but $A_{N/2}$ for N even. Therefore, the doubly liganded species makes no contribution to the symmetry or asymmetry of the binding curve of a protein such as hemoglobin, and the same applies to the half-saturated species of a protein containing an even number of binding sites. We shall consider the consequences of this fact in the hemoglobin system in Hemoglobin Cooperativity (below).

Properties of $B(x)$

In addition to its fundamental properties related to thermodynamic stability,[20] the binding capacity $B(x)$ has a number of intriguing features that have been pointed out only recently.[23] We have seen that the coefficient A_N of the partition function is related to the mean ligand activity and can always be determined from numerical integration of experimental values of $X(x)$. Information on the other coefficients of the partition function necessarily requires nonlinear least-squares analysis. There is no other set of relationships such as Eq. (13) that allows one, at least in principle, to uniquely determine the coefficients of $Z(x)$ from measurements of $X(x)$. Such a set of relationships, if it existed, would provide an important and direct connection between the partition function and experimental measurements that could be exploited in practical applications. It is here that the binding capacity comes into play. Let us rewrite the Wyman integral equation [Eq. (13)] in the equivalent form,

$$\langle \ln x \rangle = N^{-1} \int_{-\infty}^{\infty} B(x) \ln x \, d \ln x = -N^{-1} \ln A_N = \ln x_m \qquad (28)$$

which follows directly from the definition of $B(x)$. The change of variable $y = \ln x$ leads to

$$\langle y \rangle = N^{-1} \int_{-\infty}^{\infty} y B(y) \, dy \qquad (29)$$

The form of Eq. (29) is reminiscent of one of the basic equations of probability theory and suggests that $B(y)$ is the "probability density function" associated with the independent variable y. Calculation of $\ln x_m$ is thus equivalent to calculation of the first moment of $B(y)$. All higher moments

$$\langle y^k \rangle = N^{-1} \int_{-\infty}^{\infty} y^k B(y) \, dy \qquad (30)$$

are defined in an analogous way. Hence, numerical integration of measurements of $B(y)$ can be used to compute not only $\ln x_m$, but all the other moments. If one could obtain analytical expressions of the moments $\langle y^k \rangle$ in terms of the equilibrium A constants, then measurements of the binding capacity could be used to extract much more information about the partition function in a direct way. To find an analytical expression for the moments it is necessary to construct the generating function

$$G(\omega) = \int_{-\infty}^{\infty} e^{\omega y} B(y) \, dy \qquad (31)$$

Once the generating function is known, the moments associated with the binding capacity $B(y)$ are derived by differentiation as follows:

[23] E. Di Cera and Z.-Q. Chen, *Biophys. J.* **65**, 164 (1993).

$$\langle y^k \rangle = N^{-1} \left[\frac{\partial^k G(0)}{\partial \omega^k} \right] \tag{32}$$

The function $G(\omega)$ exists for $0 < \omega < 1$ because the integral

$$\int_{-\infty}^{\infty} B(y) \, dy = N \tag{33}$$

exists and is bounded, as shown in Fig. 2. The same does not apply to the binding curve $X(x)$, because the integral

$$\int_{-\infty}^{\infty} X(y) \, dy \tag{34}$$

diverges (see Fig. 1). This is a fundamental difference between the two response functions $X(x)$ and $B(x)$ that should be borne in mind when considering their statistical properties.

To solve Eq. (31) it is convenient to rewrite the partition function as

$$Z(x) = (1 + \alpha_1 x)(1 + \alpha_2 x) \cdots (1 + \alpha_N x) \tag{35}$$

where α_j is the jth coefficient of the factorization related to the jth root of $Z(x)$.[11,24] The coefficients can be real and positive or pairs of complex conjugate. The analytical expression for $B(x)$ assumes a rather compact form when cast in terms of the α values, that is,

$$B(x) = \sum_{j=1}^{N} \frac{\alpha_j x}{(1 + \alpha_j x)^2} \tag{36}$$

Hence, the solution of Eq. (31) is obtained as the combination of N similar terms after integration along a suitable path in complex plane. The analytical form of $G(\omega)$ in the general case is[23]

$$G(\omega) = \pi \omega \operatorname{cosec} \pi \omega \sum_{j=1}^{N} \alpha_j^{-\omega} \tag{37}$$

The moments $\langle y^k \rangle$ can be derived from the Taylor expansion of $G(\omega)$ split into two terms as follows:

$$G(\omega) = f(\omega) h(\omega) \tag{38}$$
$$f(\omega) = \pi \omega \operatorname{cosec} \pi \omega \tag{39}$$
$$h(\omega) = \alpha_1^{-\omega} + \alpha_2^{-\omega} + \cdots + \alpha_N^{-\omega} \tag{40}$$

[24] C. N. Yang and T. D. Lee, *Phys. Rev.* **87**, 404 (1952).

The Taylor expansion of $f(\omega)$ is

$$f(\omega) = 1 + \sum_{k=1}^{\infty} \frac{2(2^{2k-1} - 1)|B_{2k}|(\pi\omega)^{2k}}{(2k)!} \tag{41}$$

where B_{2k} is the $2k$th Bernoulli number.[25] The Taylor expansion of $h(\omega)$ is

$$h(\omega) = N - \sum_{j=1}^{N} \omega \ln \alpha_j + \sum_{j=1}^{N} \frac{\omega^2(\ln \alpha_j)^2}{2!} - \sum_{j=1}^{N} \frac{\omega^3(\ln \alpha_j)^3}{3!} + \cdots \tag{42}$$

$$= N - C_1\omega + C_2\omega^2/2! - C_3\omega^3/3! + \cdots$$

Hence,

$$\langle y \rangle = -N^{-1}C_1 \tag{43}$$
$$\langle y^2 \rangle = \pi^2/3 + N^{-1}C_2 \tag{44}$$
$$\langle y^3 \rangle = -\pi^2 C_1 - N^{-1}C_3 \tag{45}$$
$$\langle y^4 \rangle = 7\pi^4/15 + 2\pi^2 C_2 + N^{-1}C_4 \tag{46}$$

and so on. The first moment equals $\ln x_m$. The second moment includes the sum of squares of the logarithm of each coefficient α, and in general the kth moment includes the sum of $\ln \alpha$ terms up to the kth power. For a macromolecule containing N binding sites, the first N moments of the binding capacity uniquely define the N independent coefficients of the partition function. In the case of human hemoglobin, the relationships [Eqs. (43)–(46)] can in principle be used to resolve the equilibrium A constants directly from numerical integration of binding capacity measurements.

The foregoing analysis suggests a new definition for the binding capacity. As we have seen, the logarithm of x_m is the average work (in RT units) done per binding site on ligation. Because $\ln x_m$ is an average value, it is appropriate to look for the complete distribution to which it is associated. This is equivalent to finding out the probability that $\ln x_m$ has a given value $y = \ln x$. The answer is provided by the value of $B(y)/N$ for a given y. Hence, the binding capacity, divided by the number of sites N, is the distribution of the values of the work done per binding site on ligation, or else the distribution of $\ln x_m$ values. The variance of this distribution is a direct measure of cooperativity, and the standard deviation is a measure of the dispersion or "confidence interval," σ, associated with $\ln x_m$.

To understand the connection between the variance of the distribution, $\sigma^2 = \langle y^2 \rangle - \langle y \rangle^2$, and cooperativity we first explore the properties of a noncooperative "reference system," in which all sites are alike and

[25] I. S. Gradshteyn and I. M. Ryzhik, "Tables of Integrals, Series and Products." Academic Press, New York, 1980.

independent. A system containing N such independent sites has a partition function

$$Z(x) = (1 + \alpha x)^N \tag{47}$$

The first two moments for the noncooperative system are

$$\langle y \rangle = -\ln \alpha \tag{48}$$
$$\langle y^2 \rangle = \pi^2/3 + \ln^2\alpha \tag{49}$$

They are independent of N and depend solely on the coefficient α, which represents the equilibrium constant for binding a given site. The variance of y values is

$$\sigma^2 = \langle y^2 \rangle - \langle y \rangle^2 = \pi^2/3 \tag{50}$$

and is independent of N and α. Consider next the case of a highly cooperative system for which the partition function is given by

$$Z(x) = 1 + \alpha^N x^N \tag{51}$$

It is straightforward to prove that the first two moments are in this case[23]

$$\langle y \rangle = -\ln \alpha \tag{52}$$
$$\langle y^2 \rangle = N^{-2}\pi^2/3 + \ln^2\alpha \tag{53}$$

The first moment is identical with that of the (noncooperative) reference system, whereas the second moment is now a function of the number of interacting sites, N. One sees from Eqs. (52) and (53) that the first moment, and hence the mean ligand activity, contains no information on the cooperative properties of the system. On the other hand, the variance

$$\sigma^2 = N^{-2}\pi^2/3 \tag{54}$$

differs from that of a noncooperative system and decreases as $1/N^2$. Another limiting case of interest is the one in which all sites are different and independent of each other. The partition function is then

$$Z(x) = (1 + \alpha_1 x)(1 + \alpha_2 x) \cdots (1 + \alpha_N x) \tag{55}$$

where all coefficients are real and positive and reflect the equilibrium constants for the binding reaction to each site. The first two moments are

$$\langle y \rangle = -N^{-1} \ln(\alpha_1\alpha_2 \cdots \alpha_N) \tag{56}$$
$$\langle y^2 \rangle = \pi^2/3 + N^{-1}(\ln^2\alpha_1 + \ln^2\alpha_1 + \cdots + \ln^2\alpha_N) \tag{57}$$

and the variance is

$$\sigma^2 = \pi^2/3 + N^{-2}/2 \sum_{i=1}^{N} \sum_{j=1}^{N} (\ln \alpha_i - \ln \alpha_j)^2 \tag{58}$$

The variance is always larger than that of the reference system. Heterogeneity of the sites translates into a binding capacity that is more spread over the $\ln x$ axis. In general, the variance is given by[23]

$$\sigma^2 = \pi^2/3 + N^{-2}/2 \sum_{i=1}^{N} \sum_{j=1}^{N} (\ln r_i - \ln r_j)^2 - N^{-1} \sum_{j=1}^{N} \theta_j^2 \tag{59}$$

where each α_j in the partition function [Eq. (35)] has been expressed uniquely in polar coordinates r_j and θ_j. The definition of σ^2 contains three components. The first term, $\pi^2/3$, gives the variance associated with a noncooperative system for which all r values are equal and all θ values are zero. The second term is always positive and reflects the contribution arising from the differences of the norms of the α coefficients. The third component is negative and is a function solely of the angles in complex plane associated with the α coefficients of the partition function. A suitable measure of cooperativity can be defined as the ratio between the standard deviation of the binding capacity of the reference noncooperative system and that of the system under investigation, that is,

$$c = \pi/\sigma\sqrt{3} \tag{60}$$

The value of c equals 1 for a noncooperative system and is greater than 1 for cooperative systems. As expected, c cannot exceed N, the total number of sites in the system. One sees from Eq. (59) that $c > 1$ necessarily demands at least two α coefficients of the partition function to be complex conjugate. This condition is, however, not sufficient, because $\sigma^2 < \pi^2/3$ and hence $c > 1$ only if the sum of the angles in Eq. (59) exceeds the double sum of squares.

The value of c provides a macroscopic measure of cooperativity that can be obtained directly from binding capacity data without nonlinear least-squares analysis. Examples of such measurements for the Bohr effect of human hemoglobin are given in Di Cera et al.[21] Calculation of the first two moments directly from these binding capacity data using numerical integration[23] yields the results depicted in Figs. 3 and 5. The values of $\ln x_m$ as a function of pH (see Fig. 3) obtained by direct integration (filled circles) are in excellent agreement with those derived by nonlinear least-squares analysis (open circles). The values of c computed according to Eq. (60) as a function of pH are sketched in Fig. 5 (filled circles) along

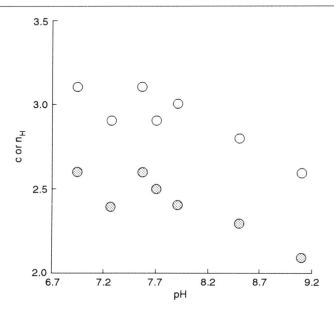

FIG. 5. Comparison between the Hill coefficient n_H (○) and the measure of cooperativity c (●) in Eq. (60), based on the first two moments derived from numerical integration of binding capacity data on the Bohr effect of hemoglobin.[21] Although the two measures of cooperativity differ by about 0.4 units, the dependence of c and n_H on pH is remarkably similar.

with the values of the Hill coefficient (open circles), which gives the classic measure of cooperativity.[3] Although the two measures of cooperativity differ by about 0.4 units, the dependence of c and n_H on pH is remarkably similar, thereby implying that c is indeed monitoring the cooperative properties of the system.

Linkage Thermodynamics of Site-Specific Binding Processes

The thermodynamic treatment described so far deals with the properties of the macromolecule as a whole. This is the situation of practical interest when experimental measurements of ligand binding are performed using probes that cannot detect the contribution arising from each separate site. The availability of new and powerful techniques able to resolve site-specific binding curves[26–30] has made the development of site-specific thermodynamics[31] particularly timely and important. These processes provide a local description of cooperative transitions, as opposed to the global description derived from the classic principles of linkage thermodynamics, which deals only with the properties of the macromolecule as a whole.

The importance of accessing information relative to local processes, occurring at the level of each constituent binding site of a biological macromolecule, can be illustrated with the simplest case of interest. Consider the case of two binding sites and the corresponding global partition function

$$Z(x) = 1 + A_1 x + A_2 x^2 \tag{61}$$

The site-specific version of $Z(x)$ is

$$Z(x) = 1 + (K_1 + K_2)x + c_{12}K_1 K_2 x^2 \tag{62}$$

where K_1 and K_2 are the site-specific equilibrium binding constants for sites 1 and 2, and c_{12} is an interaction parameter reflecting positive ($c_{12} > 1$), negative ($c_{12} < 1$), or absence of ($c_{12} = 1$) interactions between the sites. The analysis of the global properties $X(x)$ and $B(x)$ derived from $Z(x)$ provide information on the equilibrium binding constants A_1 and A_2. Are these parameters sufficient to define the cooperative properties of the system under consideration? Macroscopically, cooperativity is completely defined by the sign of the expression

$$\Delta = 4A_2 - A_1^2 \tag{63}$$

Positive and negative cooperativity imply $\Delta > 0$ and $\Delta < 0$, respectively, whereas $\Delta = 0$ implies absence of cooperative interactions. If A_1 and A_2 are expressed in terms of site-specific parameters Eq. (63) becomes

$$\Delta = 4c_{12}K_1 K_2 - (K_1 + K_2)^2 \tag{64}$$

Global and local cooperativity patterns coincide only if $K_1 = K_2$. If the two sites have different affinities, then the sign of Δ does not necessarily reflect the exact nature of the interactions involved. For example, if $K_1/$

[26] G. Viggiano and C. Ho, *Proc. Natl. Acad. Sci. U.S.A.* **76,** 3673 (1979).
[27] G. Simonneaux, A. Bondon, C. Brunel, and P. Sodano, *J. Am. Chem. Soc.* **110,** 7637 (1988).
[28] M. Perrella, L. Sabbioneda, M. Samaja, and L. Rossi-Bernardi, *J. Biol. Chem.* **261,** 8391 (1986).
[29] T. Inubushi, C. D'Ambrosio, M. Ikeda-Saito, and T. Yonetani, *J. Am. Chem. Soc.* **108,** 3799 (1986).
[30] G. K. Ackers and F. R. Smith, *Annu. Rev. Biophys. Biophys. Chem.* **16,** 583 (1987).
[31] E. Di Cera, *Biophys. Chem.* **36,** 147 (1990).

$K_2 = 10$, then $\Delta > 0$ only if $c_{12} > 3.025$ and not $c_{12} > 1$. For $K_1/K_2 = 100$, the minimum value of c_{12} for $\Delta > 0$ exceeds 25. Site heterogeneity may lead to a gross underestimation of the true cooperative properties of the interaction. Actually, site heterogeneity can be such as to overwhelm positive interactions totally and give rise to a global binding curve that shows negative cooperativity. This fact is readily understood not only from Eq. (64), but also from arguments developed in the section Properties of $B(x)$ (above) regarding the conflicting contributions of site heterogeneity and cooperative interactions to the value of σ^2. The global cooperative properties of the system, as described by the principles of classic linkage thermodynamics,[3] are not sufficient to resolve unequivocally the detailed network of interactions that leads to the effects observed experimentally. Local, site-specific effects set the rules of cooperativity in biological macromolecules.

The foregoing considerations make it clear that a new approach is necessary to resolve site-specific parameters and the energetics underlying cooperative transitions at the local level. For a macromolecule containing N binding sites there are N independent parameters in the global description, but $2^N - 1$ in the local description. Hence, there is no way one can access information at the local level from thermodynamic quantities that reflects the global behavior of the macromolecule. When binding processes can be followed at the level of each site separately, then it is possible to construct site-specific binding isotherms that reflect the properties of subsystems open to interactions with the rest of the macromolecule. Let $X_j(x)$ be the probability of site j being liganded when the ligand activity is x. To derive an analytical expression for $X_j(x)$ we proceed as follows. For a macromolecule of N sites the partition function contains 2^N possible configurations, half of which contains site j in its unliganded form and half in its liganded form. The configurations with site j unliganded sum up to give a "contracted" partition function, $^0Z_j(x)$, that encapsulates the binding properties of the remaining $N - 1$ sites constrained by the unliganded form of site j. If K_j denotes the site-specific binding constant to site j when all other sites are unliganded, then $^0Z_j(x)$ is a function of all K values but K_j. The configurations with site j liganded sum up to give another expression that encapsulates the binding properties of the remaining $N - 1$ sites constrained by the liganded form of site j. The sum is of the form $K_j x^1 Z_j(x)$ because all configurations necessarily contain the term $K_j x$, and $^1Z_j(x)$ is another contracted partition function that is a function of all K values but K_j. Hence,[31]

$$Z(x) = {}^0Z_j(x) + K_j x {}^1Z_j(x) \tag{65}$$

and

$$X_j(x) = K_j x \frac{{}^1Z_j(x)}{Z(x)} = 1 - \frac{{}^0Z_j(x)}{Z(x)} \tag{66}$$

Unlike $X(x)$, $X_j(x)$ cannot be obtained from $Z(x)$ by differentiation with respect to $\ln x$, except in the trivial case in which $X_j(x) = X(x)/N$. The connection between global and local description is provided by the conservation relationship

$$X(x) = X_1(x) + X_2(x) + \cdots + X_N(x) \tag{67}$$

which follows naturally from the definition of $X_j(x)$. The site-specific binding capacity $B_j(x) = dX_j(x)/d \ln x$ also obeys a conservation relationship,

$$B(x) = B_1(x) + B_2(x) + \cdots + B_N(x) \tag{68}$$

that connects the properties of individual sites to those of the macromolecule as a whole. As seen in the case of the global description, the response functions $X_j(x)$ and $B_j(x)$ have a number of important properties. The conservation relationships [Eqs. (67) and (68)] may lead to the conclusion that these properties are nothing but a reproduction of global properties on the local scale, but this is by no means the case.

The basic difference between the global binding isotherm $X(x)$ and the analogous local quantity $X_j(x)$ can be understood in terms of a number of considerations, most notably those involving the shape of the binding curve. When all sites are independent, then each $X_j(x)$ behaves just as $X(x)$ for $N = 1$ and is always symmetric. When all sites are identical and interact equally, then $X_j(x)$ is merely $X(x)/N$ and reproduces $X(x)$ in the local scale, so that the symmetry properties of an individual site coincide with those of the macromolecule as a whole. In general, however, the symmetry properties of $X(x)$ cannot be defined uniquely from those of individual sites and vice versa.[22] Even in the simplest case of $N = 2$, it is not easy to predict the symmetry properties of $X_1(x)$ and $X_2(x)$ if $X_1(x) \neq X_2(x)$, although the sum $X_1(x) + X_2(x)$ is always symmetric. Individual sites are subsystems open to interaction with the rest of the macromolecule, and this makes the local properties different from analogous global quantities. In fact, for $N = 2$ the site-specific binding isotherms are never symmetric, unless the two sites are identical or independent. For $N \geq 3$, $X_j(x)$ can be symmetric or asymmetric, and the conditions of symmetry are in general completely decoupled from those of the macromolecule as a whole. The importance of the shape of the binding curve in the local picture is demonstrated by the fact that when $X_j(x)$ is symmetric, then the site-specific mean ligand activity x_{jm} assumes a simple form. The quantity x_{jm} represents the average work done to ligate site j and parallels the quantity x_m in the global picture. The temperature dependence of $\ln x_{jm}$

$$(\partial \ln x_{jm}/\partial \tau)_P = \Delta H_{jm}/R \tag{69}$$

gives the average enthalpy of binding to site j. Likewise,

$$(\partial \ln x_{jm}/\partial P)_T = \Delta V_{jm}/RT \tag{70}$$

gives the average volume change on binding to site j. Finally,

$$(\partial \ln x_{jm}/\partial \ln y)_{T,P} = -\Delta Y_{jm} \tag{71}$$

gives the number of molecules of a second ligand Y exchanged on binding to site j, for given values of T and P.[5] Although x_{jm} can be measured directly from numerical integration of the binding curve, using the equal-area rule of Wyman, its connection with site-specific parameters is rather complex in general.[22] On the other hand, symmetry of $X_j(x)$ implies that

$$x_{jm} = K_j^{-1} \sqrt{\alpha_{N-1}/\beta_{N-1}} \tag{72}$$

where α_{N-1} and β_{N-1} are the coefficients of the highest power of x in the contracted partition functions ${}^0Z_j(x)$ and ${}^1Z_j(x)$, respectively.[22] The differences between $B(x)$ and $B_j(x)$ are of no less importance. Although $B(x)$ is always positive, as required by thermodynamic stability, no such restriction applies in general for the binding capacity of an individual site. In fact, Eq. (68) requires only the sum of the individual binding capacities to be positive, so that $B_j(x)$ can be positive or also negative. Negative values of $B_j(x)$ can be observed for $N \geq 3$ and play an important role in the equilibrium and nonequilibrium properties of the system.[18,31-33] A relevant example is provided by the site-specific redox properties of cytochrome aa_3, in which negative electron binding capacity values have been documented experimentally.[34]

Notwithstanding, site-specific binding curves generate a substantial amount of information on the properties of subsystems; the local picture is much more intriguing than one can actually imagine and in general even more information is needed to resolve site-specific parameters. Additional information is gathered from the contracted forms of the partition function of the system as a whole. From the point of view of parameter resolvability, each partition function can be seen as a constraint among physical parame-

[32] E. Di Cera, *J. Chem. Phys.* **92**, 3241 (1990).
[33] E. Di Cera and P. E. Phillipson, *J. Chem. Phys.* **93**, 6006 (1990).
[34] N. Kojima and G. Palmer, *J. Biol. Chem.* **258**, 14908 (1983).

ters of interest. For a macromolecule containing N sites, there are 2^N site-specific configurations and $2^N - 1$ independent site-specific constants. For each power of x, x^j, in the polynomial expansion $Z(x)$, there are as many as $n_j = N!/j!(N - j)!$ independent site-specific terms. Resolution of all these terms requires an equal number of constraints. From the partition function $Z(x)$ of the system as a whole, one can resolve N coefficients, the equilibrium A constants, each A_j being the sum of n_j site-specific terms. Hence, a total of n_j independent partition functions are necessary to resolve the n_j independent site-specific terms defining A_j. The maximum value of n_j in the polynomial expansion $Z(x)$ gives the number of partition functions that must be accessed experimentally to resolve the entire set of site-specific parameters. It is quite straightforward to prove that this number, n_Z, is equal to

$$n_Z = \binom{N}{[N/2]} = \begin{cases} N!/\{(N/2)!\}^2 & \text{for } N \text{ even} \\ N!/\{(N + 1)/2\}!\{(N - 1)/2\}! & \text{for } N \text{ odd} \end{cases} \tag{73}$$

Because measurements of $X_j(x)$ provide information on $Z(x)$ and $^0Z_j(x)$, a total of $N + 1$ partition functions can be accessed from N site-specific isotherms. This substantial amount of information is, however, only sufficient for $N \leq 3$. For $N > 3$ the value of n_Z rapidly exceeds $N + 1$. Already for $N = 4$, one has $n_Z = 6$, whereas for $N = 5$ a value of $n_Z = 10$ shows that all site-specific binding curves provide about half of the total number of constraints needed. When N describes the number of ionizable residues of a macromolecule such as hemoglobin, resolution of all site-specific parameters becomes an impossible task. The foregoing considerations point out the extraordinary complexity of cooperative transitions at the local level. Although approximate treatments are available to simplify the problem in the general case,[18] nonetheless it is extremely important to fully exploit the generality of the theory of contracted partition functions in those cases, including hemoglobin, in which all necessary constraints can be accessed reasonably by experimentation.

Hemoglobin Cooperativity

We are now in the position to cast the problem of hemoglobin cooperativity in terms of site-specific linkage thermodynamics. The global properties of hemoglobin are encapsulated by the partition function,

$$Z(x) = 1 + A_1x + A_2x^2 + A_3x^3 + A_4x^4 \tag{74}$$

where the A values have been defined already, and x is the activity of a heme ligand, say, oxygen or carbon monoxide. The partition function can alternatively be cast in terms of stepwise equilibrium constants as follows:

$$Z(x) = 1 + 4a_1x + 6a_1a_2x^2 + 4a_1a_2a_3x^3 + a_1a_2a_3a_4x^4 \qquad (75)$$

with the obvious equivalence relationships

$$A_1 = 4a_1 \qquad\qquad a_1 = A_1/4 \qquad\qquad (76)$$
$$A_2 = 6a_1a_2 \qquad\qquad a_2 = 2A_2/3A_1 \qquad\qquad (77)$$
$$A_3 = 4a_1a_2a_3 \qquad\qquad a_3 = 3A_3/2A_2 \qquad\qquad (78)$$
$$A_4 = a_1a_2a_3a_4 \qquad\qquad a_4 = 4A_4/A_3 \qquad\qquad (79)$$

The stepwise constants give a measure of the affinity of each successive binding step and are often used as a convenient measure of the change in binding affinity at each ligation step. Although they are perfectly equivalent from a thermodynamic point of view, overall and stepwise constants are *not* equivalent from an operational standpoint. The overall constants should be preferred in nonlinear least-squares analysis of experimental data, because the resulting equations are less ill conditioned than those cast in terms of stepwise constants.[35,36] Once the A values are derived by nonlinear least squares, the stepwise constants can always be arrived at from Eqs. (76)–(79). In the case of hemoglobin, there is a substantial increase in binding affinity as oxygenation proceeds. The constant a_4 is typically 100–1000 times bigger than a_1, depending on solution conditions. The increase in affinity builds up progressively with ligation, usually in an uniform fashion, and leads to the macroscopic pattern $a_1 < a_2 < a_3 < a_4$. Nonuniform macroscopic patterns such as $a_1 > a_2 < a_3 < a_4$ are also observed, especially under physiological conditions.[37] Additional relationships among the stepwise constants are provided by consideration of the shape of the oxygen-binding curve. Under physiological conditions the binding curve is strongly asymmetrical and displays higher cooperativity at high saturation.[17,37] This peculiar feature of hemoglobin nicely matches the physiological need of unloading oxygen at high saturation with high efficiency. The condition for symmetry in the case of hemoglobin is $A_4 = (A_3/A_1)^2$, or else $a_1a_4 = a_2a_3$. The asymmetry observed experimentally implies $a_1a_4 > a_2a_3$, which means that the relative increase in affinity between the third and fourth ligation steps overwhelms that observed between the first and second steps. The two fundamental features of the

[35] Y. Bard, "Nonlinear Parameter Estimation." Academic Press, New York, 1974.

[36] M. E. Magar, "Data Analysis in Biochemistry and Biophysics." Academic Press, New York, 1972.

[37] F. J. W. Roughton, A. B. Otis, and R. J. L. Lyster, *Proc. R. Soc. London, Ser. B* **144,** 29 (1955).

hemoglobin binding curve under physiological conditions, that is, its highly cooperative and asymmetric nature, have long provided the basis for the development of a number of mechanistic models of cooperativity. Some of these models correctly account for the patterns of cooperativity and asymmetry observed experimentally and also provide plausible explanations for many of the kinetic and structural properties of hemoglobin. However, the spectacular results emerging from the study of the intermediate states of ligation[38] have revealed the complexity of the problem of hemoglobin cooperativity fully and have left little ground to the simplistic assumptions of allosteric models. Some serious limitations of these models in describing the fundamental site-specific properties of hemoglobin also have been discussed in theoretical terms.[31,39] Our understanding of hemoglobin cooperativity at the local level must rely on the model-independent principles of site-specific thermodynamics.

The site-specific partition function for hemoglobin is

$$Z(x) = 1 + 2(K_\alpha + K_\beta)x + [c_{\alpha\alpha}K_\alpha^2 + c_{\beta\beta}K_\beta^2 + 2(c_{\alpha\beta} + c'_{\alpha\beta})K_\alpha K_\beta]x^2$$
$$+ 2(c_{\alpha\alpha\beta}K_\alpha + c_{\alpha\beta\beta}K_\beta)K_\alpha K_\beta x^3 + c_{\alpha\alpha\beta\beta}K_\alpha^2 K_\beta^2 x^4 \quad (80)$$

where K_α and K_β are the binding affinities of the two chains and the c values are the appropriate interaction constants. A distinction between $c_{\alpha\beta}$ and $c'_{\alpha\beta}$ must be made to take into account differential pairwise interactions between $\alpha_1\beta_2$ and $\alpha_1\beta_1$ pairs. The existence of 2 pairs of identical subunits in the hemoglobin tetramer reduces the number of independent parameters of the local description from 15 to 9. The complexity of the description based on Eq. (80) is appreciated readily by recasting the problem of the asymmetry of the oxygen-binding curve in terms of the local picture. From a model-dependent analysis, Weber and Peller have concluded that the asymmetry is a consequence of the existence of two types of chains in the hemoglobin tetramer and is due either to asymmetric pairwise interactions between $\alpha\alpha$ and $\beta\beta$ pairs[40] or to binding heterogeneity of the two subunits.[41] Symmetry or asymmetry of the global binding curve does not involve A_2, and therefore it cannot depend on any of the second-order or pairwise interaction constants $c_{\alpha\alpha}$, $c_{\beta\beta}$, $c_{\alpha\beta}$, or $c'_{\alpha\beta}$. Symmetry requires

$$c_{\alpha\alpha\beta\beta}(K_\alpha + K_\beta)^2 = (c_{\alpha\alpha\beta}K_\alpha + c_{\alpha\beta\beta}K_\beta)^2 \quad (81)$$

[38] G. K. Ackers, M. L. Doyle, D. Myers, and M. A. Daugherty, *Science* **255**, 54 (1992).
[39] G. K. Ackers, *Biophys. Chem.* **36**, 371 (1990).
[40] G. Weber, *Nature (London)* **300**, 603 (1982).
[41] L. Peller, *Nature (London)* **300**, 661 (1982).

and depends on the association constants of the two chains and third- and fourth-order interaction constants. Even if $K_\alpha = K_\beta$, asymmetry can be observed whenever $4c_{\alpha\alpha\beta\beta} \neq (c_{\alpha\alpha\beta} + c_{\alpha\beta\beta})^2$, and this condition can be satisfied even if $c_{\alpha\alpha\beta} = c_{\alpha\beta\beta}$. Therefore, symmetry or asymmetry demands neither subunit heterogeneity nor asymmetric interactions.[22] On the other hand, if $c_{\alpha\alpha\beta\beta}$, $c_{\alpha\alpha\beta}$, and $c_{\alpha\beta\beta}$ are modeled in terms of pairwise interactions so that $c_{\alpha\alpha\beta\beta} = c_{\alpha\alpha}c_{\beta\beta}c_{\alpha\beta}^2 c_{\alpha\beta}'^2$, $c_{\alpha\alpha\beta} = c_{\alpha\alpha}c_{\alpha\beta}c_{\alpha\beta}'$, and $c_{\alpha\beta\beta} = c_{\beta\beta}c_{\alpha'\beta}c_{\alpha\beta}'$ then asymmetry necessarily demands $c_{\alpha\alpha} \neq c_{\beta\beta}$ as found by Weber. Not surprisingly, the asymmetric nature of the global binding curve *cannot* provide unequivocal information on subunit interactions, because, in general, site-specific parameters cannot be resolved from analysis of global binding curves. The number of partition functions that must be accessed to resolve these parameters in the case of hemoglobin is reduced from six to four, due to the presence of two identical pairs of chains. A number of contracted partition functions can be defined for the hemoglobin system. The set obtained by keeping one chain in a particular liganded configuration is given by

$$
{}^{0}Z_\alpha(x) = 1 + (K_\alpha + 2K_\beta)x + [c_{\beta\beta}K_\beta
$$
$$
+ (c_{\alpha\beta} + c_{\alpha\beta}')K_\alpha]K_\beta x^2 + c_{\alpha\beta\beta}K_\alpha K_\beta^2 x^3 \quad (82)
$$

$$
{}^{1}Z_\alpha(x) = 1 + [c_{\alpha\alpha}K_\alpha + (c_{\alpha\beta} + c_{\alpha\beta}')K_\beta]x + (c_{\alpha\beta\beta}K_\beta
$$
$$
+ 2c_{\alpha\alpha\beta}K_\alpha)K_\beta x^2 + c_{\alpha\alpha\beta\beta}K_\alpha K_\beta^2 x^3 \quad (83)
$$

$$
{}^{0}Z_\beta(x) = 1 + (K_\beta + 2K_\alpha)x + [c_{\alpha\alpha}K_\alpha
$$
$$
+ (c_{\alpha\beta} + c_{\alpha\beta}')K_\beta]K_\alpha x^2 + c_{\alpha\alpha\beta}K_\alpha^2 K_\beta x^3 \quad (84)
$$

$$
{}^{1}Z_\beta(x) = 1 + [c_{\beta\beta}K_\beta(c_{\alpha\beta} + c_{\alpha\beta}')K_\alpha]x
$$
$$
+ (c_{\alpha\alpha\beta}K_\alpha + 2c_{\alpha\beta\beta}K_\beta)K_\alpha x^2 + c_{\alpha\alpha\beta\beta}K_\alpha^2 K_\beta x^3 \quad (85)
$$

where ${}^{1}Z_\alpha(x)$ is the partition function of hemoglobin when one α chain is kept liganded, and so forth. From the foregoing expressions, it is straightforward to verify that[31]

$$
Z(x) = K_\alpha x {}^{1}Z_\alpha(x) + {}^{0}Z_\alpha(x) = K_\beta x {}^{1}Z_\beta(x) + {}^{0}Z_\beta(x) \quad (86)
$$

The chain-specific binding curves are obtained from the contracted partition functions [Eqs. (82)–(85)] as

$$
X_\alpha(x) = K_\alpha x \frac{{}^{1}Z_\alpha(x)}{Z(x)} = 1 - \frac{{}^{0}Z_\alpha(x)}{Z(x)} \quad (87)
$$

$$
X_\beta(x) = K_\beta x \frac{{}^{1}Z_\beta(x)}{Z(x)} = 1 - \frac{{}^{0}Z_\beta(x)}{Z(x)} \quad (88)
$$

with the obvious conservation relationship

$$X(x) = 2[X_\alpha(x) + X_\beta(x)] \tag{89}$$

Because of these conservation conditions, measurements of $X(x)$ and either $X_\alpha(x)$ or $X_\beta(x)$ provide information on a total of two independent partition functions. All the coefficients of the powers of x in $Z(x)$ contain at most two independent terms, with the exception of the coefficient of x^2, which contains four terms. Consequently, measurements of $X_\alpha(x)$ or $X_\beta(x)$, along with the global binding curve $X(x)$, can be used to resolve K_α, K_β, $c_{\alpha\beta\beta}$, $c_{\alpha\alpha\beta}$, and $c_{\alpha\alpha\beta\beta}$, but not $c_{\alpha\alpha}$, $c_{\beta\beta}$, $c_{\alpha\beta}$, and $c'_{\alpha\beta}$, separately. For example, assume that $X(x)$ and $X_\alpha(x)$ are measured and define the coefficients of each power of x in $^0Z_\alpha(x)$ as B_1, B_2, and B_3. Then the B values derived from $^0Z_\alpha(x)$ and the A values derived from $Z(x)$ yield

$$K_\alpha = A_1 - B_1 \tag{90}$$
$$K_\beta = B_1 - A_1/2 \tag{91}$$
$$c_{\alpha\alpha\beta\beta} = A_4(K_\alpha K_\beta)^{-2} \tag{92}$$
$$c_{\alpha\beta\beta} = B_3(K_\alpha K_\beta^2)^{-1} \tag{93}$$
$$c_{\alpha\alpha\beta} = (A_3/2 - B_3)(K_\alpha^2 K_\beta)- \tag{94}$$
$$c_{\alpha\alpha}K_\alpha^2 - c_{\beta\beta}K_\beta^2 = A_2 - 2B_2 \tag{95}$$

Resolution of the second-order interaction constants demands measurements of the other two independent partition functions that can provide information on $c_{\alpha\alpha}$ or $c_{\beta\beta}$ and $c_{\alpha\beta}$ or $c'_{\alpha\beta}$. The set of contracted partition functions generated from Eqs. (82)–(85) by keeping a second site in a fixed liganded configuration is needed for this purpose. We first consider those configurations in which identical chains are kept in the same liganded state, that is,

$$^{00}Z_{\alpha\alpha}(x) = 1 + 2K_\beta x + c_{\beta\beta}K_\beta^2 x^2 \tag{96}$$
$$^{11}Z_{\alpha\alpha}(x) = c_{\alpha\alpha} + 2c_{\alpha\alpha\beta}K_\beta x + c_{\alpha\alpha\beta\beta}K_\beta^2 x^2 \tag{97}$$
$$^{00}Z_{\beta\beta}(x) = 1 + 2K_\alpha x + c_{\alpha\alpha}K_\alpha^2 x^2 \tag{98}$$
$$^{11}Z_{\beta\beta}(x) = c_{\beta\beta} + 2c_{\alpha\beta\beta}K_\alpha x + c_{\alpha\alpha\beta\beta}K_\alpha^2 x^2 \tag{99}$$

These partition functions apply to hemoglobin valency mutants in which either type of chain is kept in a particular ligation state.[29,30,42] For example, $^{00}Z_{\alpha\alpha}(x)$ is the partition function of a mutant in which the α chains are kept in the unliganded configuration, and so forth. Although these mutants would provide information on $c_{\alpha\alpha}$ and $c_{\beta\beta}$ separately, the constants $c_{\alpha\beta}$ and $c'_{\alpha\beta}$ can be resolved only from the partition functions,

[42] C. Ho, Adv. Protein Chem. 43, 153 (1992).

$$^{00}Z_{\alpha\beta}(x) = 1 + (K_\alpha + K_\beta)x + c_{\alpha\beta}K_\alpha K_\beta x^2 \tag{100}$$

$$^{11}Z_{\alpha\beta}(x) = c_{\alpha\beta} + (c_{\alpha\alpha\beta}K_\alpha + c_{\alpha\beta\beta}K_\beta)x + c_{\alpha\alpha\beta\beta}K_\alpha K_\beta x^2 \tag{101}$$

$$^{00}Z_{\alpha\beta'}(x) = 1 + (K_\alpha + K_\beta)x + c'_{\alpha\beta}K_\alpha K_\beta x^2 \tag{102}$$

$$^{11}Z_{\alpha\beta'}(x) = c'_{\alpha\beta} + (c_{\alpha\alpha\beta}K_\alpha + c_{\alpha\beta\beta}K_\beta)x + c_{\alpha\alpha\beta\beta}K_\alpha K_\beta x^2 \tag{103}$$

where the subscripts $\alpha\beta$ and $\alpha\beta'$ refer to the $\alpha_1\beta_1$ and $\alpha_1\beta_2$ pairs. Hence, information on the binding properties of the $\alpha_1\beta_1$ and $\alpha_1\beta_2$ pairs inside the hemoglobin tetramer is critical to fully resolve the energetics of hemoglobin cooperativity at the local level.

The foregoing analysis makes it clear that a substantial amount of information should be gathered on the behavior of subsystems, as encapsulated by the contracted partition functions, in order to solve the site-specific pattern of hemoglobin cooperativity. Knowledge of the binding properties of individual chains in the hemoglobin tetramer is not sufficient, nor is it sufficient to measure the properties of valency mutants in which identical chains are kept in the same ligation state. Resolution of the binding properties of valency mutants, in which either $\alpha\beta$ pair is kept in a particular ligation state is absolutely necessary to solve the second-order interaction constants completely. Valency mutants that can mimic either ligation state of the hemoglobin chains provide powerful tools for exploring cooperativity at the local level. It is therefore important that the binding properties of these mutants reflect those of subsystems and are not perturbed by the chemical modification. The number n_Z of partition functions necessary to solve all site-specific coefficients provides the minimum number of necessary constraints. Hence, a solution of the problem can always be found by properly choosing n_Z partition functions. The key question is whether this solution is also unique. This is equivalent to establishing whether or not the valency mutants used to probe the energetics of subsystems introduce perturbations of the relevant binding parameters. The answer to this question is found by obtaining information on more than n_Z partition functions and checking the results by means of "self-consistency" relationships. For example, if $X(x)$, $X_\alpha(x)$, and $X_\beta(x)$ are measured independently, then the A, B, and C coefficients of $Z(x)$, $^0Z_\alpha(x)$, and $Z_\beta(x)$ can be arrived at. Hence, the conditions

$$3A_1 = 2(B_1 + C_1) \tag{104}$$

$$A_2 = B_2 + C_2 \tag{105}$$

$$A_3 = 2(B_3 + C_3) \tag{106}$$

must be satisfied. Likewise, measurements of $^{00}Z_{\alpha\alpha}(x)$ and $^{00}Z_{\beta\beta}(x)$ yield the coefficients of two polynomials of second degree and the sum of the coefficients of x for the two partition functions must equal A_1, and so forth, for the other valency mutants. If the self-consistency relationships are not satisfied, then it can be concluded that the modifications used in generating mutants introduce significant perturbations of the energetics. In general, at least $n_Z + 1$ partition functions must be measured to assess the presence of any perturbation. Hence, $n_Z + 1$ is also the minimum number of independent partition functions that must be measured to guarantee that the solution of all site-specific parameters is unique and not biased by perturbations.

The complexity of the problem of hemoglobin cooperativity cannot be overlooked or cast in terms of simplistic assumptions. Consideration of the fundamental role played by site-specific effects in ligand-binding processes and hemoglobin cooperativity has opened new and exciting lines of investigation in which, again, the interplay between experimental strategies and powerful thermodynamic principles is going to play a major role.

Acknowledgments

This work was supported by National Science Foundation Grant DMB91-04963. The author is an Established Investigator of the American Heart Association and Genentech.

Author Index

Numbers in parentheses are footnote reference numbers and indicate that an author's work is referred to although the name is not cited in the text.

H

Subject Index

A

and myoglobin dynamics, comparison, 178

conformational substates, 213–215

cooperativity, 632, 656

  mathematical analysis and modeling, 650–653

  resonance Raman spectroscopy studies, 208–209

  site-specific linkage thermodynamics, 677–683

crocodilian, circular dichroism in near-UV region, 254–255

crustacean, circular dichroism in far-UV region, 251

crystallization

  low-salt conditions, 16

  from polyethylene glycol, 16–17

crystals

  cocrystallization with effectors and drugs, 19

  DPG binding, 16, 19

  high-salt, 19

  IHP binding, 16, 19

  polyethylene glycol-grown

    grown with inositol hexaphosphate, 18

    oxygenation, 17–18

cysteine thiol group, S–H vibration, 157–159

deoxygenated, structure

  optical spectroscopy, 59–61

  resonance Raman spectroscopy, 208

deoxygenation for front-face fluorometry, 245

derivatives

  high-salt, 19

  PEG-grown, 17–19

dimeric, 242

dimer–tetramer association and dissociation, 447

  equilibrium, 598–600

    in Adair fitting, 597–606

    thermodynamic model, 608

  kinetic studies, double mixing methods for, 432, 440–442

dromedary

  circular dichroism in near-UV region, 254

  heme–ligand geometry, 280–282

electron transfer reactions with small molecules, 86–94

  conformational gating, 87, 92–94

  rates

    and activation parameters, 91–92

    measurement, 89–91

  reactants, 87–88

extrinsic fluorescent probes

  binding, 240–241, 246

  sensitivity to quaternary structure, 241

Fe–C–O configuration, 280–281

fish, *see* Fish, hemoglobins

fluorescence, intrinsic, effects of organic phosphates and pH, 239–240

fluorescence emission

  ligand dependence, 239

  quaternary structure changes and, 239

fluorescence lifetime, heterogeneity, 43

front-face fluorometry, 231–246

functions, 195–196

guinea pig, ligand infrared band assignments, 144

horse

  deoxy R-state crystals, 20–21

  ligand infrared band assignments, 144

human

  circular dichroism

    in far-UV region, 249–251

    in Soret region, 257–259

    in visible region, 260–261

  protease digest, circular dichroism in far-UV region, 250–251

hydrogen exchange labeling, 26–42

infrared studies, 195–200

intermediates, ligand distribution among, 446

iron

  atom movement on ligation, 272–273

  K-edge X-ray absorption, 284

  XANES spectra, 274–278

isolation for front-face fluorometry, 245

Kansas

  dimers, 598, 600

  dimer–tetramer equilibrium, in Adair fitting, 600–606

  liganded, optical spectroscopy, 62

  linkage equilibrium constants, 601–602

  tetramer Adair binding constants estimation, 600–601

metmyoglobin, acidic conformational
    states, phase diagram, 12–14
Hydrochloric acid
    induction of heme protein folding, 5–6
    solutions for, 7
    induction of heme protein refolding
        apomyoglobin, 7–12
        metmyoglobin, 12–14
Hydrogen–deuterium exchange
    amide I infrared band frequencies and,
        169
    with NMR analysis, 28
    functional labeling method, 29
Hydrogen exchange labeling, 26–42
    chemistry, 31, 39–40
    functional labeling method, 29, 32–35
        problems, 40–41
    gel-filtration passage, 30–31
    kinetic labeling method, 28–29
    local unfolding model, 39–41
    NMR analysis, 28
Hydrogen–tritium exchange, 28
    functional labeling method, 29, 32–35
        tritium loss curve, 38
    kinetic labeling method, 28–29

I

Impulsive Raman scattering, 210
Infrared microspectroscopy, single eryth-
    rocytes, 151–157
Infrared spectroscopy
    in aqueous environments, 169–173, 179
    asymmetric ligand binding at α and β
        subunits of hemoglobin, 150–153
    exogenous ligands bound to heme iron,
        140–151
    Fourier transform, myoglobin conforma-
        tional substrates, 186–187
    hemoglobin, 139–175
        amide I band, 139–140, 167
            secondary structure determination
                and, 167–175
        amide II band, 167
        conditions for, 140
        C–O stretch bands, 145, 148–153
        C–O stretch spectral changes, pH-
            induced, 151, 154–155
        ligand infrared band assignments, 143–
            147

ligand infrared spectra, measurement,
    141–143
mutation-induced environment
    changes at ligand-binding site,
        147–150
N–O stretch bands, 145, 148
O–O stretch bands, 145–147
S–H vibration band, 139
    cysteine thiol group, 157–167
picosecond, 176–204
    hemoglobin, 196–200
    hemoglobin dynamics, rationale for,
        176–178
    in myoglobin, 188–195
    protein dynamics, 200–204
    subtraction of IR absorption due to
        water, 169–173
time-resolved
    ligand dissociation from heme pro-
        teins, 182–187
    myoglobin, 187–195
transient
    proteins, rationale for, 176–178
    pulse repetition rates, 182
    technical problems, 178–182
    up-conversion techniques, 180–182
Inositol hexaphosphate, 98
    allosteric and affinity factors contributed
        by, 511
    as allosteric effector, 485
    binding to hemoglobin crystals, 16, 19
    effect on hemoglobin ligand rebinding
        after photodissociation, 81–84
    effect on optical spectrum of nitrosyl Hb
        A, 62–64
    fluorescence intensity of hemoglobins
        and, 239–240
    PEG-grown hemoglobin crystals grown
        with, 18
Iron
    atoms
        deoxyhemoglobin, distances around,
            271–272
        hemoglobin, movement with ligation,
            272–273
    heme
        in hemoglobin A, 97
        ligand infrared spectra, 140–151
    XANES spectra
        aquomethemoglobin, 274–275

N

ISBN 0-12-182133-1

9 780121 821333

90038